URBAN AIR QUALITY – RECENT ADVANCES

Urban Air Quality – Recent Advances

Proceedings of the Third International Conference on Urban Air Quality – Measurement, Modeling and Management Loutraki, Greece, 19–23 March 2001

Edited by:

RANJEET S. SOKHI

Atmospheric Science Research Group (ASRG), Department of Environmental Sciences, University of Hertfordshire, College Lane, Hatfield AL10 9AB, U.K.

and

JOHN G. BARTZIS

Environemtanl Research Laboratory (EREL), INT-RP, NCSR "Demokritos", Aghia Paraskevi, Attikis 15310, Greece

Organised by
Institute of Physics, U.K., NCSRD, Greece, University of Hertfordshire, U.K.
in collaboration with
SATURN, TRAPOS and COST715 and supported by A&WMA, IUAPPA and EURASAP.

Conference Organising Committee

Dr Ranjeet S Sokhi (Chairperson) University of Hertfordshire, UK
Dr John Bartzis, Demokritos, Greece
Professor Nicolas Moussiopoulos, SATURN
Dr Ruwim Berkowicz, TRAPOS
Professor Bernard Fisher, COST715 and EURASAP
Mr John Sadler, A&WMA
Mr Richard Mills, IUAPPA

International Scientific and Advisory Committee

Professor Ranjeet S Sokhi (Chairperson) University of Hertfordshire, UK
Professor John G Bartzis, NCSR DEMOKRITOS, Greece
Dr Ruwim Berkowicz, NERI, Denmark
Professor Carlos Borrego, University of Aveiro, Portugal
Dr Trond Bohler, NILU, Norway
Professor Eugene Genikhovich, Main Geophysical Observatory, Russia
Dr Norbert Gonzalez-Flesca, INERIS, France
Professor Steven R Hanna, George Mason University, USA
Professor Mark Z Jacobson, University of Stanford, USA
Dr Jaakko Kukkonen, FMI, Finland
Dr Don McKay, Meteorological Service of Canada
Dr Doug R Middleton, Meteorological Office, UK
Professor Nicolas Moussiopoulos, Aristotle University Thessaloniki, Greece
Professor Michael Schatzmann, University of Hamburg, Germany
Professor Sjaak Slanina, ECN, Netherlands
Professor Roberto San Jose, Technical University of Madrid, Spain
Dr Andreas Skouloudis, JRC, Ispra

Conference Secretariat

Jasmina Bolfek-Radovani, Institute of Physics, 76 Portland Place, London, W1N 4AA, UK

SPRINGER-SCIENCE+BUSINESS MEDIA, B.V.

A C.I.P. Catalogue record for this book is available from the Library of Congress.

ISBN 978-94-010-3935-2   ISBN 978-94-010-0312-4 (eBook)
DOI 10.1007/978-94-010-0312-4

TABLE OF CONTENTS

Editorial

The field of urban air quality has continued to develop over the past decade with areas of local and urban scale modelling, urban meteorology and aerosols taking particular prominence. There is greater international cooperation between scientists, through for example, SATURN (Studying Atmospheric Pollution in Urban Areas), a subproject of EUROTRAC II. In addition, networks such as TRAPOS (Optimisation of Modelling Methods for Traffic Pollution in Streets) have been established through the support of the European Commission framework programmes. International cooperation and integration of research activities on urban air quality has also continued through COST initiatives including COST715 (Meteorology Applied to Urban Air Pollution Problems). This series of conferences on urban air quality was initiated, with the support of the Institute if Physics in the UK, in response to this intensive growth of research activity. It aims to continually stimulate discussion of the latest advances in this area and to promote worldwide collaborative effort to address the future research challenges.

The first conference in the series was launched at the University of Hertfordshire in 1996 and the second at the Technical University of Madrid in March 1999 with active participation of SATURN and TRAPOS investigators and the support of IUAPPA (International Union of Air Pollution and Prevention Associations) and A&WMA (Air & Waste Management Association). This, the third conference, was additionally supported by COST715 and EURASAP and attracted 175 scientists from over 30 different countries confirming the series as a major international forum for exchanging, discussing and advancing the latest research findings in this field. More than 130 papers were presented on topics including air quality management and policy, emissions, measurements of air pollution, local and urban scale modelling, aerosols, urban meteorology, remote sensing and indoor air quality and exposure. As usual all papers submitted for publication were subjected to a thorough peer review process in strict accordance with the usual journal review policy. An important feature of the meeting was the high number of presentations on findings from international collaborative programmes including a number of major European Commission funded projects. Additionally, numerous small group meetings were direct evidence of how new research ideas were being transformed into potential proposals for funding.

The success of this conference was clearly built upon the excellent organisation by the main sponsoring and collaborating bodies and especially Professor John Bartzis at NCSRD and his team of researchers. The dedication of the referees, numbering over 30, is fully acknowledged for their diligence during the review process. I am grateful to the IOP, and in particular Jasmina Bolfek-Radovani, for providing efficient administration for the conference. Christine Shepperson, Anna Tod and other members of ASRG are fully acknowledged for their assistance during conference and the collation and preparation of this special issue. Above all, I would like to express my thanks to the researchers in this field whose efforts are leading to major scientific advances and helping not only to increase our understanding of the underlying processes that affect urban air quality but also to the development of new tools necessary for its efficient management.

Ranjeet S. Sokhi
University of Hertfordshire, UK
May 2002

Preface

Funding Science: An Immense Challenge

The real challenge in science should be hypothesis formation, experimental design and conducting the actual research under rigorous control and experimental conditions with the best available equipment/infrastructure. However, in many parts of the world, the real challenge in science is trying to obtain some minimal amount of funding to initiate research. Every country in the world is challenged with an immense number of interconnected environmental and health problems. Examples include biologically and chemically contaminated water and soil, air pollution, waste and sewage treatment, a multitude of infectious diseases in humans, animals and plants, global change, population growth, alternate energy sources, deforestation, floods, crop production and so on.

Solutions to these national and international problems require training and research to ensure the best available people and knowledge are available to manage and/or solve these problems. This can only be accomplished if the research community has the funding to conduct priority research and apply the knowledge globally. The challenge in science should be discovering the unknown and collaborative research/training of new scientists. Too often the challenge is how to conduct research without proper support. Many scientists are likely of the opinion that it is not their jobs/careers that is stressful, but the inability to do their research properly, that is stressful. Good research in a knowledge based economy requires proper funding. A knowledge based economy is not about making as much money as possible, but it is about a peaceful, healthy planet for all people and the proper management of the Earth's resources.

Professor J. T. Trevors (FiBiol: CBiol)
Department of Environmental Biology
Rm 3220 Bovey Building
University of Guelph
Guelph, Ontario
Canada N1G 2W1
Tel: 519-824-4120 ext. 3367
Fax: 519-837-0442
E-mail:jtrevors@uoguelph.ca

INVESTIGATING THE SURFACE ENERGY BALANCE IN URBAN AREAS – RECENT ADVANCES AND FUTURE NEEDS

M. PIRINGER[1]*, C. S. B. GRIMMOND[2], S. M. JOFFRE[3], P. MESTAYER[4], D. R. MIDDLETON[5], M. W. ROTACH[6], A. BAKLANOV[7], K. DE RIDDER[8], J. FERREIRA[9], E. GUILLOTEAU[10], A. KARPPINEN[3], A. MARTILLI[11], V. MASSON[12] and M. TOMBROU[13]

[1] *Central Institute for Meteorology and Geodynamics, A-1190 Vienna, Austria;* [2] *Dept. of Geography, Indiana University, Bloomington, IN;* [3] *Finnish Meteorological Institute, FIN-00101 Helsinki, Finland;* [4] *LMF, CNRS – Ecole Centrale de Nantes, F-44321 Nantes, France;* [5] *Meteorological Office, Bracknell RG12 2SZ, United Kingdom;* [6] *Swiss Federal Institute of Technology, CH-8057 Zuerich, Switzerland;* [7] *Danish Meteorological Institute, DK-2100 Copenhagen, Denmark;* [8] *VITO-TAP, B-2400 Mol, Belgium;* [9] *Instituto de Meteorologia, P-1700 Lisbon, Portugal;* [10] *University Valenciennes, France;* [11] *Swiss Federal Institute of Technology, CH-1015 Lausanne, Switzerland;* [12] *CNRM Meteo France, F-31057 Toulouse, France;* [13] *Dept. of Applied Physics, Univ. of Athens, GR-157 84, Athens, Greece*
(* *author for correspondence, e-mail: martin.piringer@zamg.ac.at, fax: +43 1 36026 74*)

Abstract. Recent advances in understanding of the surface energy balance of urban areas, based on both experimental investigations and numerical models, are reviewed. Particular attention is directed to the outcome of a COST-715 Expert Meeting held in April 2000, as well as experiments initiated by that action. In addition, recent complete parameterisations of urban effects in meso-scale models are reviewed. Given that neither the surface energy balance, nor its components, normally are directly measured at meteorological stations, nor are there guidelines for the set-up of representative meteorological stations in urban areas, this paper also provides recommendations to close these gaps.

Keywords: COST-715, meso-scale models, surface energy balance, surface flux modelling, urban boundary layer

1. Introduction

The requirements of the European Union Framework Directive on air quality assessment and management, adopted by the Council of Ministers of the European Union in September 1996, present new challenges for the meteorological community. Assessments of air quality will be required in urban areas with large populations for up to 13 air pollutants. The Directive requires remedial plans to be drawn for areas assessed to have poor air quality. Reliable air pollution models are needed to supplement measurements, for mapping, and to investigate future emission scenarios. The models are essential management tools. These models will need accurate and representative meteorological input variables consistently applied across EU Member States. Surface fluxes needed in urban air pollution

Water, Air, and Soil Pollution: Focus **2**: 1–16, 2002.
© 2002 *Kluwer Academic Publishers.*

assessments are not routinely measured. Moreover, the number of meteorological stations in urban areas is commonly limited to a few sites, often just at airports.

The surface energy balance is the key component of any model aiming to simulate dynamical and thermodynamical patterns above the surface (e.g., Kallos, 1998; Mestayer, 1998). In its simplest one-dimensional form, it can be written as:

$$Q^* = H + LE + G \tag{1}$$

where H and LE denote turbulent sensible and latent heat fluxes, respectively, and G is the storage heat flux usually not measured but determined as a residual. In many cities, additional sources of energy due to human activities (Q_F, the anthropogenic heat flux) also have to be included. In urban areas, the terms of Equation 1 require special treatment, given the complexity of the materials and morphology of the urban surface. There are marked differences in energy partitioning compared to rural conditions, where most parameterisation schemes and measurements have been performed. There is still considerable uncertainty concerning the partitioning of the components of the surface energy balance in urban areas, and the role of surface cover (e.g. the fractions of built-up areas/greenspace), city surroundings, and prevailing meteorological conditions.

This paper addresses some of these issues. The focus in the experimental section 3.1 is on surface flux measurements conducted in North-American cities during the past 10 years and on ongoing recent European experiments. These latter studies were initiated in part by COST-715 (Co-operation in the Field of Scientific and Technical Research) on 'Meteorology Applied to Urban Air Pollution Problems'. The framework, objectives and work plan of this Action, which will end in September 2003, are discussed in Fisher et al. (2001). Working Group 2 of COST-715 conducted an Expert Meeting on the surface energy budget in urban areas, held in Antwerp, Belgium, on 12 April 2000 (COST-715, 2001). The findings were summarized by Piringer et al. (2001) at the 3rd International Air Quality Conference in March 2001 in Loutraki, Greece. The modelling section 3.2 mainly refers to recent full parameterisations of urban effects in meso-scale models. The paper ends by outlining future needs to further increase our knowledge of the urban surface energy balance. Assessing methods to determine or model the height of the urban boundary layer (UBL), which is itself strongly dependent on the surface heat fluxes, is the second task of WG2 of COST-715.

2. Siting Considerations

In contrast to rural areas, the UBL is more complex. A roughness sub-layer (RS) of much larger vertical extent than found in typical rural areas occupies the first tens of meters above the surface, with the remainder of the surface layer (i.e., the *inertial sublayer*) aloft. The RS includes the urban canopy layer, which is composed of individual street canyons and other roughness elements (Figure 1).

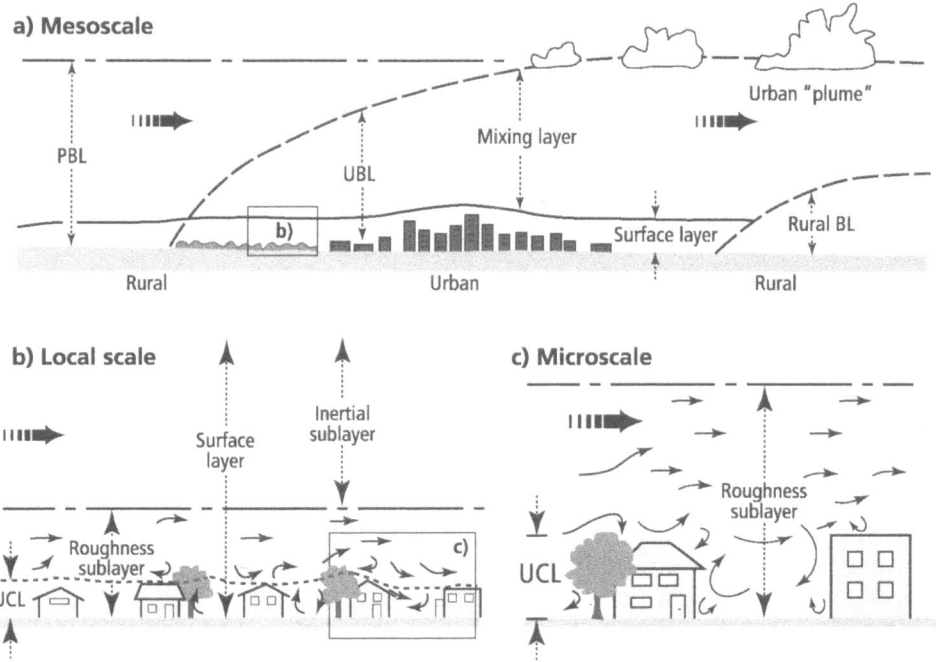

Figure 1. Schematic of the urban boundary layer including its vertical layers and scales. 'UBL' stands for Urban Boundary Layer, and 'UCL' for Urban Canopy Layer (revised by Oke and Rotach after a figure in Oke, 1997).

The Monin-Obukhov similarity theory is not valid within the RS and turbulent fluxes of momentum, energy, moisture and pollutants are height dependent (e.g. Rotach, 2001). Only in the surface layer aloft, contributions from individual surface roughness elements are blended into possibly representative averages (Taha and Bornstein, 2000).

The WMO-guideline for rural stations declares wind measurements as representative if placed 10 m above ground without close obstacles; temperature and humidity measurements have to be conducted at 2 m. For urban areas, no guidelines for proper siting exist, although this issue is presently under review by WMO (Oke, *pers. comm.*, 2000).

Monitoring stations in urban areas should be sited so that their data reflect the characteristic meteorological state of the urban terrain zone (quarter) under consideration, excluding local influences. For surface energy balance observations, instruments must be placed above the RS. This height, z^*, is usually expressed as a function of the roughness element separation and/or the mean building height h_r. When described by the latter, z^*, as a rule-of-thumb, is somewhere between 2–$5\ h_r$ (see Figure 1 b-c). With instruments mounted at such heights, the area for which the measurements are representative (the flux footprint), typically is of the order of 10^2–10^4 m^2. The exact dimensions and area of influence will vary with the exact height of measurements and roughness of the surface, as well as with wind speed, direction and atmospheric stability. The area of influence is much greater at night, a factor that must be considered when interpreting surface controls on urban energy balance fluxes through time. If large observational heights are necessary, the fetch of the observation may become very large and the surface cover inhomogeneous. Achieving uniform upwind fetch may be difficult to meet in parts of typical European cities.

3. The Urban Surface Energy Balance

3.1. EXPERIMENTAL INVESTIGATIONS

Knowledge of surface heat fluxes as well as atmospheric stability and surface roughness is essential, both as input and boundary conditions, in advanced air pollution dispersion models. Normally, however, the surface energy balance or its components are not directly measured at meteorological stations. In the last decade, a series of local-scale energy balance observations have been conducted at a restricted number of sites, largely, though not exclusively, residential areas in North America (see, for example, Grimmond and Oke, 1995, 1999, 2000; Grimmond et al., 1996; Oke et al., 1998, 1999; Spronken-Smith, 1998; Feigenwinter et al., 1999).

A **M**ulti-city **U**rban **H**ydrometeorological **D**atabase (MUHD) has been generated to document local-scale surface heat flux variability in several North-American cities (Grimmond and Oke, in COST-715, 2001). MUHD integrates surface energy balance observations, each 1 to 8 weeks in duration, conducted over a ten-year period. These surface energy balance data were collected primarily in summertime, with the exception of Mexico City (dry winter season). The urban land uses represented include central city, light industrial, and low or medium density residential. Sites were selected to represent different building densities (sometimes in the same city) and climates. Grimmond and Souch (1994) outlined the methods used to develop the surface cover information for the MUHD sites. For all sites, areas of similar surface cover and morphometry in a 2–5 km radius around each measurement site are mapped from aerial photographs, and detailed attributes such as building height and type, density, vegetation amount and type are documented. These databases are then coupled to the Flux Source Area model of Schmid (1994, 1997) to quantify the surface cover influencing local-scale surface energy balance observations taken using instruments mounted on tall towers ($> 2h_r$, the mean height of the surrounding roughness elements).

Some key characteristics of surface energy balance partitioning, derived from MUHD, are summarised in Table I. In general, the radiation fluxes of cities show magnitudes and diurnal behaviour similar to those of rural surfaces (Oke et al., 1998). As expected, areas with little vegetation have extremely small latent heat

TABLE I

Ranges of average daily maximum values of net radiation and fluxes at the MUHD sites (after Grimmond and Oke, in COST-715, 2001)

Parameter	Range (W m^{-2})
Net all-wave radiation Q*	<400–650
Latent heat flux LE	10–235
Sensible heat flux H	120–310
Storage heat flux G	150–280
Average daytime Bowen ratios H/LE:	(Dimensionless)
Residential sites	1.2–2
During irrigation ban Vancouver	~ 2.8
Light industrial site	~ 4.4
Downtown	~ 9.8

flux values. Of the residential sites, the neighbourhood in Vancouver had the lowest measured rates, due to an effective garden irrigation ban that was in force in 1992, an abnormally dry year (Oke *et al.*, 1998). The remaining residential areas have daytime peaks in evaporation ranging from 125 to 235 W m^{-2}; these are significant fluxes, sustained by garden irrigation and/or frequent rainfall (at least 20 to 40% of daytime Q*). At all sites, more energy leaves the surface as sensible rather than latent heat flux; consequently the Bowen ratio values are greater than unity (Table 1). Storage heat flux was determined as a residual (Equation 1; Grimmond and Oke, 1999). This is always a significant term in the surface energy balance in urban areas (Table I), and is considerably larger than for most natural systems, except water. In general, the storage sensible heat flux is most important at central city and light industrial sites (at least 50% of daytime Q*), whereas turbulent sensible heat flux is most important at residential sites (40 to 60% of daytime Q*).

An investigation of urban-rural energy balance differences for three North-American cities with different rural environments revealed contrasting heat and water balance regimes, some even being capable of completely inverting expected urban-rural difference (Oke and Grimmond, in COST-715, 2001). The cities studied were Tucson, Arizona, with surrounding desert, Sacramento, California, with semi-arid grassland and irrigated farmland, and Vancouver, British Columbia, with surrounding moist farmland. Whereas urban-rural differences in net radiation are mostly less than ± 50 W m^{-2}, the Tucson results are distinctly different. Daytime urban Q* in Tucson is as much as 125 W m^{-2} greater than in the surrounding rural desert suggesting the city is a better absorber than the desert by day. This might be interpreted to be due to the lower albedo and/or lower surface temperature of the city, which in turn is related to greater abundance of urban vegetation and irrigation. For all the cities, the storage heat flux is, as expected, greater than the rural values; the urban fabric sequesters more heat by day than the surrounding countryside. At night, the cities release greater amounts of heat from storage. For the turbulent sensible and latent heat fluxes, the station pairs with moist and wet rural sites (Vancouver and Sacramento) confirm conventional expectations, i.e. cities evaporate less and generate a greater sensible heat flux to the air than their rural surroundings. On the other hand, the semi-arid and desert pairs show the reverse. The source of urban moisture here is garden irrigation.

While offering much new insight into surface energy exchanges in urban environments, more data need to be collected to represent cities with different building materials and architectural styles, and for conditions where direct release of energy from human activities (the anthropogenic heat flux, Q_F) is more significant. Moreover, the above results cannot be directly extrapolated to European cities due to land use, climatological and urban metabolism differences. There are several experimental studies of the urban surface energy budget for European cities (e.g. Klysik, 1996; Dupont et al., 1999; Holmer and Eliasson, 1999), but they do not specifically analyse the different components of the surface energy budget in cities. Therefore, three new field campaigns, in Basle, Marseilles and Birmingham, have been initiated, to more explicitly study processes in European cities.

BUBBLE, the Basle Urban Boundary Layer Experiment, is an initiative of six Swiss research institutes (Rotach et al., 2001) financed in the framework of COST 715. It is supported by a large number of groups from all over the world. The objective of BUBBLE is to study the UBL in the city of Basle for a one-year period, monitoring near-surface turbulence characteristics as well as the vertical structure of the entire UBL. Near-surface observations are undertaken at two urban, one suburban and a rural reference sites up to a height greater than twice the mean roughness element height; these comprise conventional meteorological observations, turbulence measurements by sonic anemometers, full radiation balance, and, at the urban sites, detailed measurements of the canyon thermal and radiative properties. A 1290 MHz wind profiler is operated at one urban site. The aerosol distribution within the urban boundary layer and aloft is determined with an aerosol backscatter Lidar. The data set will be used to evaluate and improve the surface exchange parameterisation of Martilli (2001) as described in section 3.2.

Another experiment on the urban boundary layer was undertaken during the large French photochemistry experimental campaign ESCOMPTE in Provence, June-July 2001. The urban meteorological (UBL/CLU-Escompte) campaign took advantage of the project's dense network of meteorological and remote sensing observations both on the ground and from airplanes. Surface energy balance fluxes were measured at five stations within Marseilles, with three above the urban canopy of different parts of the city. A central site, located in the densely built 19th Century city centre, was also equipped with an array of infrared thermoradiometers with a thermal scanner monitoring street sections and individual building elements. These data are to be integrated with those from a small aircraft equipped with a scanning thermal infrared camera, NOAA/AVHRR satellite observations, and a network of 20 T-RH sensors 6 m above ground. The 4D structure of the UBL was documented with two mini-Sodars, one RASS sounder, one UHF radar, and two 3D-scanning O_3 and Doppler Lidars. The data are to be used to test urban energy budget models, improve remote sensing methods to determine the effective urban surface temperature to predict heat fluxes, and to test high resolution meteorological and chemistry-transport models.

A field program to measure surface fluxes at an industrial site at Birmingham, UK, has been set up to better forecast urban air quality and understand past pollution episodes (Ellis and Middleton, 2000). Three data sets (1998, 1999 and 2000) each of about four weeks collected data from ultrasonic anemometers and resistance thermometers on masts at 15, 30 and 45 m as well as radiation, humidity and pressure sensors.

Satellite remote sensing has the potential to provide estimates of urban surface energy balance fluxes (Zhang *et al.*, 1998; De Ridder, in COST-715, 2001). The two most common methods to derive information about surface energy balance fluxes across cities are either the difference between the surface radiation temperature measured by the satellite and the local air temperature, or based on estimating surface parameters from satellite remote sensing and land-use maps and computing the surface energy balance via a SVAT (Soil Vegetation Atmosphere Transfer) module. The use of thermal remote sensing to infer the surface heat flux is potentially prone to large errors, which become even more pronounced in the case of urban environments. This is due to uncertainties in the determination of the surface radiant temperature itself, particularly in the case of large roughness elements, as well as ancillary parameters such as the roughness length and the wind speed, if they are extrapolated from routine meteorological observations (e.g., the discussion in Voogt and Grimmond, 2000). Given these uncertainties accumulate in the determination of the sensible heat flux, the second method may be preferred compared to the first. However, more complete evaluations of the predicted fluxes for urban environments, based on carefully conducted field studies, is urgently needed.

3.2. MODEL RESULTS

Air quality management is mainly concerned to identify pollution 'hot-spots' or areas of exceedence; the modelling of concentration maxima in these is strongly sensitive to the diagnosing of stable conditions. Modelling of the urban energy balance and stability is largely absent from current dispersion models. Across Europe, different cities can be expected to influence local stability to differing extents, but little is known about this. Another possibility to obtain the surface fluxes, needed in urban air pollution assessments, is the use of nested numerical weather prediction and meso-meteorological models. Such models will approach the necessary resolution for the urban scale, but parameterisations of urban effects in most of

the existing operational models are absent or greatly simplified for this purpose
(Baklanov *et al.*, 2001). More research, such as the work in Basle, Birmingham,
and Marseilles, is needed to identify sensible constraints on stability diagnosis,
and to improve parameterisations (as discussed above).

3.2.1. *Detailed Surface Exchange Parameterisations*

The most detailed parameterisations of urban effects within current numerical models are the Town Energy Balance (TEB) scheme of Masson (2000) and the surface
exchange parameterisation in the Finite Volume Model (FVM) of Martilli *et al.*
(in COST- 715, 2001). Both explicitly consider the effects of buildings, roads, and
other anthropogenic building materials on the urban surface energy budget.

The meso-scale atmospheric model MESO-NH in combination with the TEB
scheme has been used to compute urban surface fluxes to investigate the influence
of Paris on the atmospheric boundary layer for an anticyclonic summer day (Lemonsu and Masson, 2001). The TEB model parameterises both the urban surface
layer and the roughness sublayer, so that the atmospheric model only 'sees' a
constant flux layer as its lower boundary. All the turbulent fluxes and the upward
radiation flux are computed for each land cover type (e.g. sea, lake, natural and cultivated terrestrial surfaces, urban) by the appropriate scheme (TEB/ISBA) and then
averaged in the atmospheric model grid mesh, in proportion to the area covered by
each land cover type. The city is represented by generic buildings which have the
same height and width, with the roof level at the surface level of the atmospheric
model. Buildings are located along identical roads; any road orientation is possible,
but all exist with the same probability. A detailed scheme parameterising multiple
reflection of radiation is available in TEB.

Surface cover information comes from the Corine Land Cover database with a
horizontal resolution of 250 m. Atmospheric data come from ECMWF analyses
updated every 6 h. When the simulation output is compared to observations from
30 meteorological stations in and around Paris, and with atmospheric profiles from
radiosondes, the urban heat island is simulated fairly well, though temperature is
over-predicted, especially during the night (Figure 2). Simulated surface energy
balance fluxes are shown in Figure 3. The enhanced storage heat flux documented
in urban areas and the distinct diurnal hysteresis pattern (Grimmond *et al.*, 1991)
both are reproduced well. During the night, storage heat release is enough to maintain a small positive turbulent sensible heat flux (also observed by Grimmond and
Oke in COST-715, 2001).

Urban effects are parameterised in the meso-scale FVM model by considering
roofs, walls, and the canyon floor as active surfaces (Martilli *et al.*, 2002). For the
exchange of momentum, two different roughness lengths are defined for the roof
and canyon floors, respectively, while the contribution of the walls is parameterised
with a drag force approach. The sensible heat fluxes are determined as a function of
the difference between the air temperature and the corresponding surface temperat-

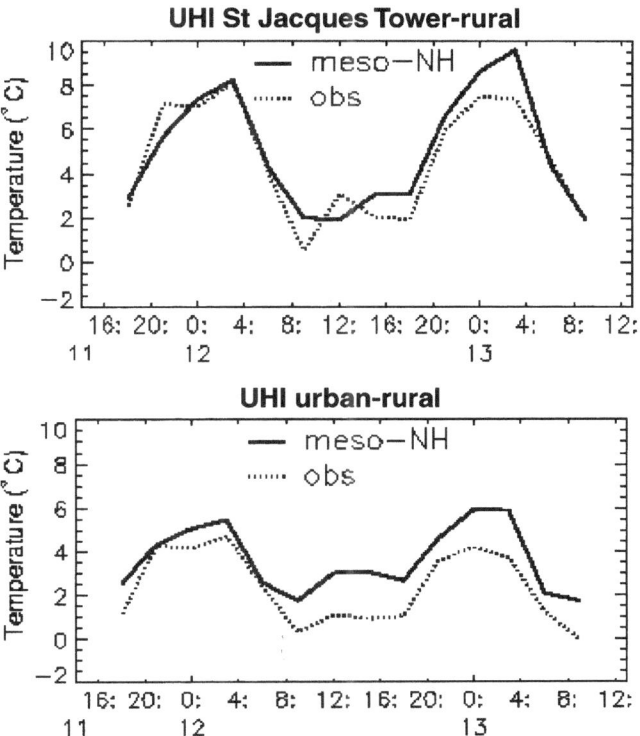

Figure 2. Observed and simulated urban heat island on 11–13 July, 1994. Top: between the center of Paris and the surrounding countryside; bottom: between all dense urban area and countryside (from Masson *et al.*, in COST-715, 2001).

ure. The short and long wave radiative fluxes are computed by taking into account the shadows and multiple reflection effects of the street canyon element.

The model has been run for an idealised 2-D case of an UBL with flat terrain and an idealised city of 10 km, building height 25 m, street width 25 m, surrounded by a rural area. The computations show that walls are the most active surfaces for momentum and turbulent kinetic energy, during day and night. For heat, on the other hand, street and roofs are more active during daytime and walls during night-time.

Determining the impact of vertical and horizontal surfaces on the energy balance is very important in order to estimate the storage heat flux term correctly. To investigate this, Martilli et al (2002) have compared their new surface exchange scheme ('urb' in Figure 4) to the 'traditional' approach ('trad'), in which the roughness length and the soil thermal properties are modified in order to simulate urban influences (Figure 4). Considering only the ground as an active surface with an increase in the roughness length and the soil thermal characteristics of the concrete ('Trad' in Figure 4), there is a tendency to underestimate the magnitude of the storage heat flux during day- and night-time. On the other hand, there is

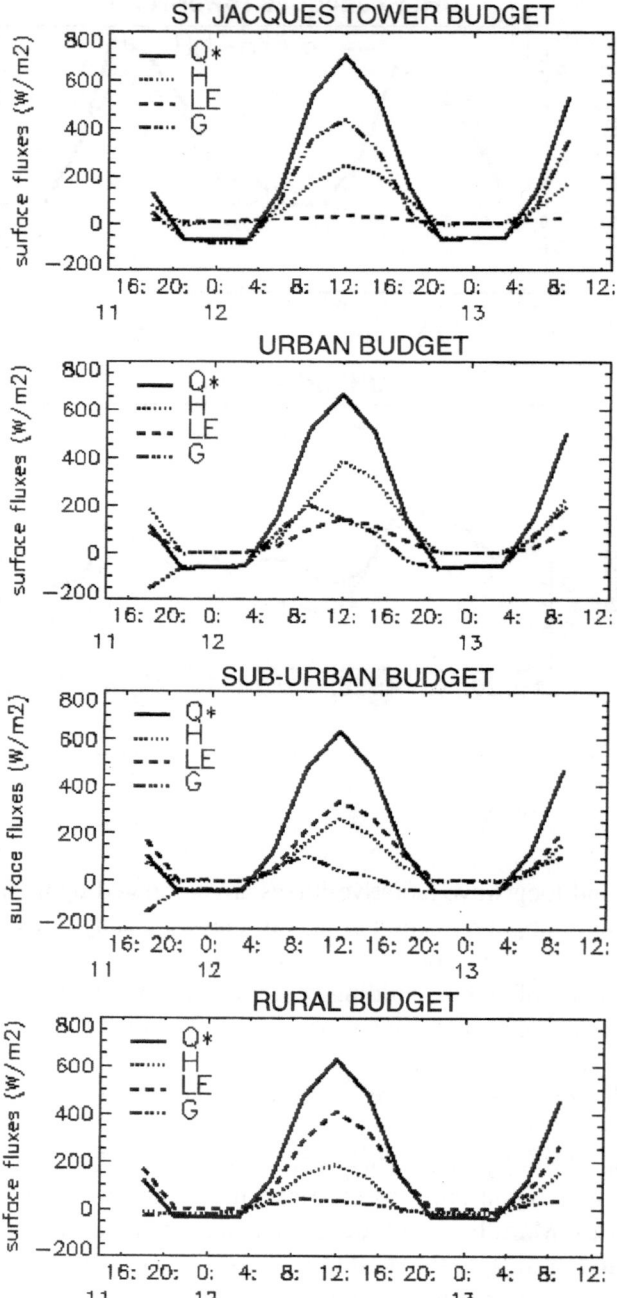

Figure 3. Comparison of surface fluxes at various sites in the City of Paris as calculated by MesoNH using the TEB scheme. The four panels refer to specific sites of areas as indicated (from Masson *et al.*, in COST-715, 2001).

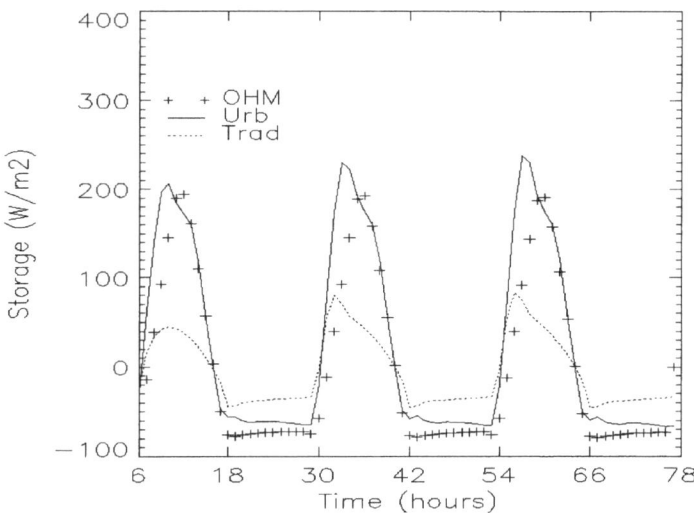

Figure 4. Comparison of different methods to estimate heat storage (from Martilli *et al.*, in COST-715, 2001). OHM refers to the Objective Hysteresis Model (Grimmond *et al.*, 1991), 'trad' and 'urb' are defined in the text.

quite a good agreement in the three-day simulation between the results obtained with the new parameterisation used in FVM and the Objective Hysteresis Model OHM (Grimmond *et al.*, 1991; Arnfield and Grimmond, 1998; Grimmond and Oke, 1999), an empirical formulation for the storage term based on the total radiation of the canyon and its material properties.

Another approach to modelling urban surface energy balance fluxes is to adapt current modules for rural conditions by introducing urban features. The French SUBMESO model, which has a force-restore model for rural soil (derived from Noilhan and Planton, 1989) was used to introduce new parameterisations for urban soil-atmosphere interactions (Guilloteau and Dupont, in COST-715, 2001). Five soil cover types are considered: water, natural soil, vegetation, impervious surface materials (asphalt, concrete), and buildings. The processes considered are evaporation or dewfall, transpiration by the vegetation, interception of water by the vegetation and buildings run-off towards the natural soil and impervious ground materials as well as infiltration of water through the surface to the underlying soil. Return toward equilibrium state is considered for the temperature of the natural and artificial materials and for the humidity of the natural soil. Drainage of the water intercepted by the buildings and the impervious surface to the drainage and sewerage network are also included. Additional heat input from vehicles is considered.

One-dimensional simulations with imposed meteorological data (rural soil of Caumont during the Hapex-Mobilhy experiment) were carried out over typical urban terrain types (Table II) for one year to assess model behaviour. As seen from the example shown in Figure 5, the model simulates the low latent heat flux values expected for a city centre, but transforms most of the net radiation into turbulent

TABLE II

Surface cover characteristics of five typical urban terrain zones (from
Guilloteau and Dupont, in COST-715, 2001)

% grid surface	soil	veg.	water	paved	build.
City centre	0	10	0	40	50
Residential	20	30	0	30	20
Green quarter	10	60	15	10	5
Industrial and commercial	10	0	0	50	40
Tall apartments	5	25	0	40	30

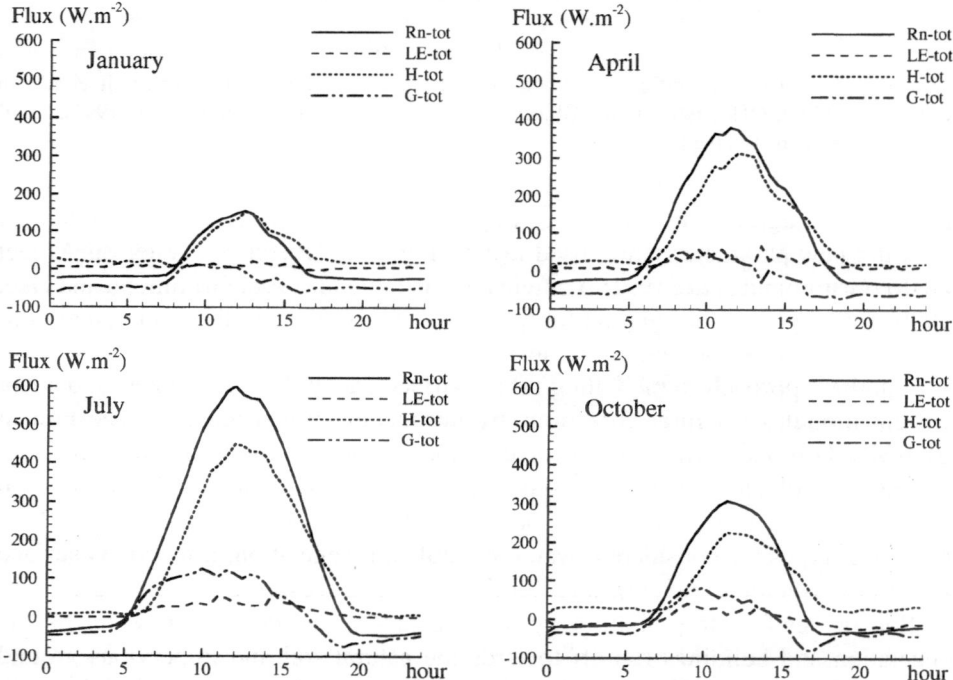

Figure 5. Energy balance over 'City centre' for the average day of four typical months (from Guilloteau and Dupont, in COST-715, 2001). Note: symbols are different to that used in text (Rn is equivalent to Q*).

sensible heat; as a consequence, the heat storage in the fabric is too small. The same model behaviour is found for 'industrial and commercial' and 'tall apartment' districts. For residential areas, the partitioning of fluxes is reasonable. For vegetated zones the calculated latent heat flux is lower than expected. The shortcomings of these simulations were due to the model considering only the horizontal surfaces of the city (grounds, roofs, vegetation cover), demonstrating the importance of the 3-D structure of the urban canopy. The SM2-U version (Dupont *et al.*, 2000) now includes efficient parameterisations for the heat storage by building walls and for an effective albedo depending on building shape and sky view factor, deduced from the work of Masson (2000). The SM2-U model has the advantage of universality for all urban areas, from the rural soils to the dense city centre. Though similar to the TEB scheme, it does not however offer the possibility to explicitly simulate the 3D canopy exchange due to the employed force-restore approach.

3.2.2. *Meteorological Preprocessors*

In recent years, a number of boundary-layer parameterisation schemes have been developed to estimate net radiation, sensible heat flux and other UBL parameters from hourly standard meteorological data. COST-710 (1998) conducted an extensive comparison of these methods and presented various validations against observational data. Most of these models were developed and validated using data from flat, grass-covered environments. These approaches are therefore limited to horizontally homogeneous conditions. Grimmond and Oke (2000, 2002) developed a linked set of equations to calculate heat fluxes, and in turn atmospheric stability, specifically designed for the urban environment. The scheme is basically an urban-specific modification of the approach of Holtslag and van Ulden (1983). This pre-processor scheme (LUMPS -Local-scale Urban Meteorological Pre-processing Scheme) makes use of parameterisations that require standard meteorological observations, supplemented by basic knowledge of the surface character of the target urban area. Ideally LUMPS is forced by observed short- or net radiation data, but these fluxes also can be modelled. Heat storage in the urban fabric, including hysteresis, is parameterised from the radiation and surface cover information using the objective hysteresis model (Grimmond *et al.*, 1991; Grimmond and Oke, 1999). The turbulent sensible and latent heat fluxes are calculated using the available energy and a simplified Penman-Monteith/Priestley-Taylor type of equation using a measure of the surface moisture status, given by the fraction of the surface covered by vegetation, and temperature. LUMPS has been shown to perform well when evaluated using data from North American cities. The scheme is tried to be evaluated with data collected in European cities, notably Basle, Graz and Marseilles.

4. Findings and Recommendations

1. Siting criteria for urban stations are urgently needed. Sites should be characterised with the help of aerial photos, local surveys, maps, building dimensions, GIS, and urban databases.
2. Measurement of surface fluxes at meteorological stations is desirable, but so far such measurements have only been realised in research programs of limited

duration. Urban meteorological masts should extend above the roughness sub-layer into the inertial sub-layer and above. The heights of these layers vary with conditions and fetch (2 to 5 times the building height). For central urban areas with relatively tall buildings, the above requirements may be unrealistic for practical purposes. Therefore, the urban roughness sublayer should be investigated in more detail, specifically with regard to defining appropriate and practical guidelines for the siting of meteorological instruments in urban areas.

3. Available observations of urban heat fluxes demonstrate significant perturbation of surface energy balance partitioning compared to the rural surroundings.

4. The behaviour of turbulent flux profiles in the roughness sub-layer, due to high roughness elements requires more study, both in the field and with models.

5. The state of rural soil moisture, and therefore soil thermal admittance is a very important determinant of city heat island effects; the state of the surrounding countryside therefore must be considered in any such studies.

6. Horizontal inhomogeneity of the canopy means that the diffusivities for heat and water vapour differ, (i.e., $K_E \neq K_H$). This is because while all surfaces are sources of sensible heat, not all urban surfaces are sources of water vapour (Roth and Oke, 1995).

7. A number of European groups run meso-scale models with sub-models of fluxes for urban areas. These models are not operational yet, but advances are encouraging. Preliminary simulations indicate that the influence of the urban canopy, building energy flows and thermal properties, along with effective albedo reduction by radiative trapping between canyon walls is important and need to be explicitly modelled.

8. For applications in connection with dispersion modeling, often no detailed surface exchange parameterisation can (computationally) be afforded. As an alternative, a meteorological pre-processor that has been modified for urban surfaces (LUMPS) is available. Turbulent fluxes (and hence stability) obtained from this scheme apply to heights sufficiently far from the urban fabric. A detailed validation, especially using data from European cities, would complement already existing North American studies and is much encouraged.

9. Satellites can in principle be used to measure the urban surface energy balance, but there are significant issues concerning the ability to estimate the appropriate interface temperature.

References

Arnfield, A. J. and Grimmond, C. S.: 1998, 'An urban canyon energy budget model and its application to urban storage heat flux modelling', *Ener. Build.* **27**, 61–68.

Baklanov, A., Rasmussen, A., Fay, B., Berge, E. and Finardi, S.: 2001, 'Potential and shortcomings of NWP models in providing meteorological data for UAP forecasting', *Ext. abstract to the Third Urban Air Quality Conference*, Loutraki, Greece. CD-ROM, paper PL.4.1. (*Int. J. Water, Air and Soil Poll.*, accepted).

COST-710: 1998, 'Harmonisation of Pre-Processing of Meteorological Data for Atmospheric Dispersion Models, Final Report', European Commission, Report EUR 18195 EN.

COST-715: 2001, 'Surface Energy Balance in Urban Areas', *Extended Abstracts of an Expert Meeting. WG-2 COST Action 715*, Antwerp, Belgium, 12 April 2000. European Commission. Report EUR 19447.

Dupont, S., Guilloteau, E. and Mestayer, P. G.: 2000, 'Energy Balance and Surface Temperatures of Urban Quarters', *Proc. 3rd AMS Symposium on Urban Environment*, Davis, California, 14–18 August 2000, pp. 149–150.

Dupont, E., Menut, L., Carissimo, B., Pelon, J. and Flamant, P.: 1999, 'Comparison between the atmospheric boundary layer in Paris and its rural suburbs during the ECLAP experiment', *Atmos. Env.* **33**, 979–994.

Ellis, N. L. and Middleton, D. R.: 2000, 'Field Measurements and Modelling of Urban Meteorology in Birmingham, UK', *Proc. 3rd AMS Symposium on Urban Environment*, Davis, CA, American Meteorological Society, Boston, MA, 108–109.

Feigenwinter, C., Vogt, R. and Parlow, E.: 1999, 'Vertical structure of selected turbulence characteristics above an urban canopy', *Theoret. Appl. Climatol.* **62**, 51–63.

Fisher, B. E. A., Kukkonen, J., Schatzmann, M.: 2002, 'Meteorology applied to urban air pollution problems COST 715', *Sixth Int. Conf. On Harmonisation within Atmospheric Dispersion Modelling for Regulatory Purposes*, October 11–14, 1999, Rouen, France, 10 p. *Int. J. Environment and Pollution* **16**, 560–569.

Grimmond, C. S. B. and Oke, T.R.: 1995, 'Comparison of heat fluxes from summertime observations in the suburbs of four North American cities', *J. Appl. Meteor.* **34**, 873–889.

Grimmond, C. S. B. and Oke, T. R.: 1999, 'Heat storage in urban areas: Local-scale observations and evaluation of a simple model', *J. Appl. Meteor.* **38**, 922–940.

Grimmond, C. S. B. and Oke, T. R.: 2000, 'A Local-Scale Urban Meteorological Pre-Processing Scheme (LUMPS)', *Proceedings of the COST 715 Workshop 'Preparation of Meteorological Input Data for Urban Site Studies'* (ed.: M. Schatzmann, J. Brechler, B. Fisher), Prague, Czech Republic, June 15, 2000, 73–83.

Grimmond, C. S. B. and Oke, T. R.: 2002, 'Turbulent heat fluxes in urban areas: Observations and local-scale urban meteorological parameterization scheme (LUMPS)', *J. Appl. Meteor.* **41**, 792–810.

Grimmond, C. S. B. and Souch, C.: 1994, 'Surface description for urban climate studies: a GIS based methodology', *Geocarto Int.* **9**, 47–59.

Grimmond, C. S. B., Cleugh, H. A. and Oke, T. R.: 1991, 'An objective urban heat storage model and its comparison with other schemes', *Atmos. Env.* **25B**, 311–326.

Grimmond, C. S. B., Souch, C. and Hubble, M.: 1996, 'The influence of tree cover on summertime energy balance fluxes, San Gabriel Valley, Los Angeles', *Clim. Res.* **6**, 45–57.

Holmer, B. and Eliasson, I.: 1999, 'Urban-rural vapour pressure differences and their role in the development of urban heat islands', *Int. J. Climatol.* **19**: 989–1009.

Holtslag, A. A. M. and Van Ulden, A. P.: 1983, 'A simple scheme for daytime estimates of the surface fluxes from routine weather data', *J. Appl. Meteor.* **22**, 517–529.

Kallos, G.: 1998, 'Regional/Mesoscale Models', *Urban Air Pollution – European Aspects* (Part. IV., Chap. 11.), Kluwer Acad. Publ., Dordrecht, Boston, London, pp. 177–196.

Klysik, K.: 1996, 'Spatial and seasonal distribution of anthropogenic heat emissions in Lodz, Poland', *Atmos. Env.* **30**, 3397–3404.

Lemonsu, A. and Masson, V.: 2001, 'Simulation of a summer urban breeze over Paris', *Boundary-Layer Meteorol.* **104**, 463–490.

Martilli, A.: 2001, 'Development of an Urban Turbulence Parameterisation for Mesoscale Atmospheric Models', *Swiss Federal Institute of Technology Lausanne (EPFL)*, Dissertation 2445, 176 pp.

Martilli, A., Clappier, A. and Rotach, M. W.: 2002, 'An urban surface exchange parameterisation for mesoscale models', *Boundary-Layer Meteorol.* **104**, 261–304.

Masson, V.: 2000, 'A physically-based scheme for the urban energy budget in atmospheric models', *Boundary-Layer Meteorol.* **94**, 357–397.

Mestayer, P. G.: 1998, 'Urban Scale Models', *Urban Air Pollution – European Aspects* (Part. IV., Chap. 11.), Kluwer Acad. Public., Dordrecht, Boston, London, pp. 197–222.

Noilhan, J. and Planton, S.: 1989, 'A simple parameterization of the land surface processes for meteorological models', *Mon. Wea. Rev.* **117**, 536–549.

Oke, T. R.: 1997, 'Urban environments', in W. G. Bailey, T. R. Oke and W. R. Rouse (eds), *The Surface Climates of Canada*, McGill-Queen's University Press, Montréal, 303–327.

Oke, T. R., Grimmond, C. S. B. and Spronken-Smith, R.: 1998, 'On the confounding role of rural wetness in assessing urban effects on climate', *Preprints of the AMS Second Urban Environment Symposium*, 59–62.

Oke, T. R., Spronken-Smith, R., Jauregui, E. and Grimmond, C. S. B.: 1999, 'Recent energy balance observations in Mexico City', *Atmos. Env.* **33**, 3919–3930.

Piringer, M., Baklanov, A., De Ridder, K., Ferreira, J., Joffre, S., Karppinen, A., Mestayer, P., Middleton, D., Tombrou, M., Vogt, R.: 2001, 'The Surface Energy Budget and the Mixing Height in Urban Areas: Status Report of Working Group 2 of COST-Action 715', *Ext. abstract to the Third Urban Air Quality Conference*, Loutraki, Greece. CD-ROM, paper Pl 1.3.

Rotach, M. W.: 2001, 'Simulation of urban-scale dispersion using a Lagrangian stochastic dispersion model', *Boundary-Layer Meteorol.* **99**, 379–410.

Rotach, M. W., Mitev, V., Vogt, R., Clappier, A., Richner, H., Ruffieux, D.: 2001, 'BUBBLE – Current Status of the Experiment and Planned Investigation/Evaluation of the Urban Mixing Height', *Proc. COST-Action 715 Workshop on the Mixing Height and Inversions in Urban Areas*, Toulouse, 3–4 Oct. 2001 (submitted).

Roth, M. and Oke, T. R.: 1995, 'Relative efficiencies of turbulent transfer of heat, mass and momentum over a patchy urban surface', *J. Atmos. Sci.* **52**, 1864–1874

Schmid, H. P.: 1994, 'Source areas for scalars and scalar fluxes', *Boundary-Layer Meteorol.* **67**, 293–318.

Schmid, H. P.: 1997, 'Experimental design for flux measurements: matching scales of observations and fluxes', *Agr. Forest Meteorol.* **87**, 179–200.

Spronken-Smith, R. A.: 1998, 'Comparison of Summer and Wintertime Energy Fluxes over a Suburban Neighborhood', *Preprints of the Second Urban Environment Symposium*, Christchurch, New Zealand, American Meteorological Society, pp. 243–246.

Taha, H. and Bornstein, R.: 2000, 'Urbanization of meteorological models and implications on simulated heat islands and air quality', in R. J. de Dear, J. D. Kalma, T. R. Oke and A. Auliciems (eds), *Biometeorology and Urban Climatology at the Turn of the Millenium*. Selected Papers from the Conference ICB-ICUC '99, Sydney, Australia, 8–12 Nov. 1999, pp. 431–435.

Voogt, J. A. and Grimmond, C. S. B.: 2000, 'Modeling surface sensible heat flux using surface radiative temperatures in a simple urban area', *J. Appl. Meteor.* **39**, 1679–1699.

Zhang, X., Aono, Y., Monji, N.: 1998, 'Spatial variability of urban surface heat fluxes estimated from LANDSAT TM data under summer and winter conditions', *J. Agric. Meteor.* **54**, 1–11.

THE IMPORTANCE OF MIXING HEIGHT IN CHARACTERISING POLLUTION LEVELS FROM AEROSOL OPTICAL THICKNESS DERIVED BY SATELLITE

A. DANDOU[1], E. BOSSIOLI[1], M. TOMBROU[1]*, N. SIFAKIS[2], D. PARONIS[2], N. SOULAKELLIS[3] and D. SARIGIANNIS[4]

[1] *Department of Physics, University of Athens, Greece;* [2] *Institute for Space Applications & Remote Sensing, National Observatory of Athens, Greece;* [3] *Department of Geography, University of the Aegean, Mytilene, Greece;* [4] *Institute for Systems, Informatics and Safety, DG JRC, European Commission, Italy*
(* author for correspondence, e-mail: mtombrou@cc.uoa.gr, fax: 00 3010 7295281)

Abstract. In the present study the horizontal distribution of columnar aerosol optical thickness derived at high spatial resolution from Earth observation satellite data in the Lombardy area (Italy) was converted to the horizontal distribution of optically effective aerosols concentration at the ground level. This was achieved by incorporating information on atmosphere's mixing height, at which pollutants released at ground level are vertically dispersed by convection or mechanical turbulence. The resulted fields compared favourably to pollutant concentration measurements provided by the ground stations. These results show that it is possible to calculate mean concentration fields by using the spatial distribution of aerosol optical thickness (AOT) measured by satellite normalized by the atmospheric mixing height. The advantage of satellites in measuring AOT is that they can capture all actual emissions compared to the models, which are based on inventoried data.

Keywords: aerosol optical thickness, atmospheric boundary layer, mixing height, modeling, normalized concentration, planetary boundary layer, remote sensing, satellite observations

1. Introduction

Many of the latest satellite Earth observation (EO) instruments are devoted to the retrieval of information on a variety of atmospheric parameters and pollutants in order to respond to the recently developed strong multidisciplinary interest in climate change and environmental research (NASA, 1999). The observational and respective study scales depend on each EO instrument's spatial resolution or GSD (ground sampling distance): namely, low resolution sensors – with tens of kilometres GSD – address Global scale observations, moderate resolution sensors – with a hundreds of meters to few kilometres GSD-address continental to regional observations, and high resolution sensors – with less than hundred meter GSD – address regional to local observations. With respect to the retrieval of the spatial distribution and abundance of gaseous atmospheric pollutants there are as yet only low resolution measurements (e.g., by TOMS, GOME, MOPITT and instruments

Water, Air, and Soil Pollution: Focus **2**: 17–28, 2002.
© 2002 *Kluwer Academic Publishers.*

on the forthcoming Envisat and future CHEM). With respect to the particulate pollutants that is, the aerosols distribution and abundance a variety of methods exist to carry out this task over land or over water by using low (e.g., TOMS), moderate (e.g., AVHRR) or high (e.g., Landsat) spatial resolution satellite sensors (King *et al.*, 1999). Independently of the resolution of the sensor, satellite-based information on aerosols is always expressed in terms of aerosol optical thickness (AOT) and related indexes (e.g., aerosol index). Satellite sensors measure columnar AOT defined as the integral of the extinction coefficient, from the ground to the satellite, due to scattering and absorption processes by aerosols. Its magnitude is linked to the total column of optically effective particles, providing information on the available reservoir of these pollutants, above each location.

The information that is of utmost importance, however, for characterising pollution levels required by air quality monitoring and relevant studies, concerns the vertical distribution structure and the height of the extension of AOT in order to translate this non-local pollution field to reliable near-surface mean concentration levels. The assumption that AOT, due to aerosols, is expanded uniformly on the vertical up to the same height in the lower part of the troposphere would rarely be reliable. This is reasonable, since pollution follows the atmospheric movements and thus, in order to translate these non-local pollution fields to reliable mean concentration levels, it is important to know their vertical structure and the height of their extension.

The main source of secondary particles is the atmospheric oxidation of SO_2 to sulphuric acid droplets, and of NO_2 to nitric acid vapour, which react with ammonia gas (produced from the decomposition of animal urine) to make ammonium sulphate and ammonium nitrate salts (Colls, 1997).

As it was mentioned previously, the atmospheric pollutants are strongly affected by the meteorological parameters, which are varied spatially due to the atmospheric conditions and topographical features. More specifically, convection or mechanical turbulence disperses the pollutants released at the surface vertically, within an hour or less, up to the upper level, which is defined by the mixing height (Beyrich *et al.*, 1996). Thus, the pollutants are lifted during the daytime due to the convective activity and remain suspended until the late afternoon when the thermal mixing decreases gradually. During the night, pollution is confined to the lower levels of the atmosphere, where mainly the mechanical turbulence prevails. Moreover, in case of aerosol particles, it is worth to mention that their profiles derived by lidar have been used as tracers to study the atmospheric boundary layer structure and stratification (Devara and Ernest Raj, 1991). Since the mixing height composes the upper limit of pollutant dispersion, it becomes a parameter of great importance in case of converting AOT measurements to surface pollutant concentrations at near ground level based on AOT measurements by satellite.

The current study examines precisely this possibility of integrating information on the mixing height in the derivation of mean concentration fields from satellite-based columnar AOT data. As the study focuses on local and regional pollution transportation over and around urban areas in the lower troposphere, it is necessary to use satellite data of high spatial resolution since only these can provide necessary detail for urban pollution case studies.

2. Methodology

A method was developed that converts AOT values into mean concentration fields by using the spatial distribution of the simulated mixing heights. As a first approach, an intense vertical mixing throughout the whole mixing layer of the pollution was assumed (box model assumption). This assumption is reasonable during atmospheric conditions that favor pollution episodes. In such conditions, potential temperature, humidity and concentrations are considered nearly constant with height. In a further examination, it is planned to involve the process information on the atmospheric vertical structure under certain atmospheric conditions. The method currently followed and proposed by this study can be divided into three stages:

2.1. DERIVATION OF COLUMNAR AOT

HSR satellite data selection – according to meteorological and pollution criteria – followed by ad hoc image processing attain the quantification of AOT. The principle of the processing method is that a differential textural analysis of the radiometric values of satellite images acquired under clear and pollution conditions can allow to quantify the optical atmospheric effects due to particulate pollutants, and derive columnar AOT values by resolving the radiative transfer equation over cloud-free areas (Sifakis et Deschamps, 1992). Sifakis *et al.* (1998) describe the latest image processing code applied in detail and Sifakis *et al.* (2001) describe all the stages of the current application. According to this method the extracted AOT values are relative to the reference satellite data (i.e., acquired under pollution-free atmospheric conditions with AOT quasi zero). This process was carried out by using the Landsat TM 2nd spectral band, and therefore the aerosol optical thickness values were calculated in the visible spectrum (AOTV) and in particular at 550 nm; this measurement is sensitive to aerosols in the accumulation mode that is, with aerodynamic diameters between 0.1 and 2.5 μm.

2.2. CALCULATION OF THE MIXING HEIGHT

Due to the importance of the mixing height to the prediction of pollutant concentrations, large effort is being made by researchers to improve the estimation of mixing heights. Nonetheless, there is still no unique definition and no overall accepted method of mixing height (Beyrich *et al.*, 1996). As a result the determination of mixed layer depth is often ambiguous under realistic atmospheric conditions, even over relatively homogeneous terrain. Apart from the radiosonde ascents that still provide the most common database for mixing height estimation, also sodars, wind profiler/RASS systems and lidars are increasingly used to determine reliable values for the mixing height (Piringer *et al.*, 1988). However, mixing height values derived from measurements are available, if at all, at specific sites and partly also for limited time periods only, so that parameterizations are widely used. The so-called meteorological pre-processors that have been developed to compute the mixing height are based on the 'horizontal homogeneity' approximation. In complex terrain however, the effects of horizontal advection, need also to be included in the calculations. The meteorological dynamic models that provide planetary boundary layer (PBL) height as a part of a predictive quantity as used in the simulation of atmospheric turbulence, seems to give the most reliable calculations especially under conditions of strong horizontal in-homogeneity (Beyrich *et al.*, 1998).

In the present study, the spatial distribution of boundary layer heights were calculated by the fifth-generation Penn State/NCAR Mesoscale Model MM5 (Anthes and Warner, 1978). This is a limited area hydrostatic or non-hydrostatic, terrain following sigma coordinate model designed to simulate or predict mesoscale and regional atmospheric circulation with flexible and multiple nesting capabilities. This model offers a variety of closure schemes. In order to assure a representative estimation of the mixing height, numerical simulations during this study were performed with three different parameterization closure schemes: The traditional local-K approach of Blackadar (Zhang and Anthes, 1982), the non-local diffusion concept based on Troen and Marht representation of counter gradient term and K profile in the well mixed PBL (Hong and Pan, 1996), and the Mellor-Yamada TKE prediction in saturated conditions (Ballard *et al.*, 1991).

2.3. CALCULATION OF POLLUTANT CONCENTRATIONS

In order to derive the spatial horizontal distribution of pollutant concentrations for ambient air quality assessments, it was assumed that the atmosphere is constructed by a set of adjacent vertical box models, each one extending to the corresponding mixing height. Since all the material in each box is assumed to mix almost instantaneously (Beyrich *et al.*, 1996), the concentration of each gas and particle is taken uniform throughout each box. Thereafter, by combining the AOTV fields with the spatial distribution of the calculated mixing heights a quantitative spatial distribution of the optically effective air pollutants is resulted. Due to the fact that AOTV values are clear numbers, it is not possible to provide the real concentration values. Instead the normalized mean concentration values are given which can be easily transformed into real values if at least one reference ground station is available.

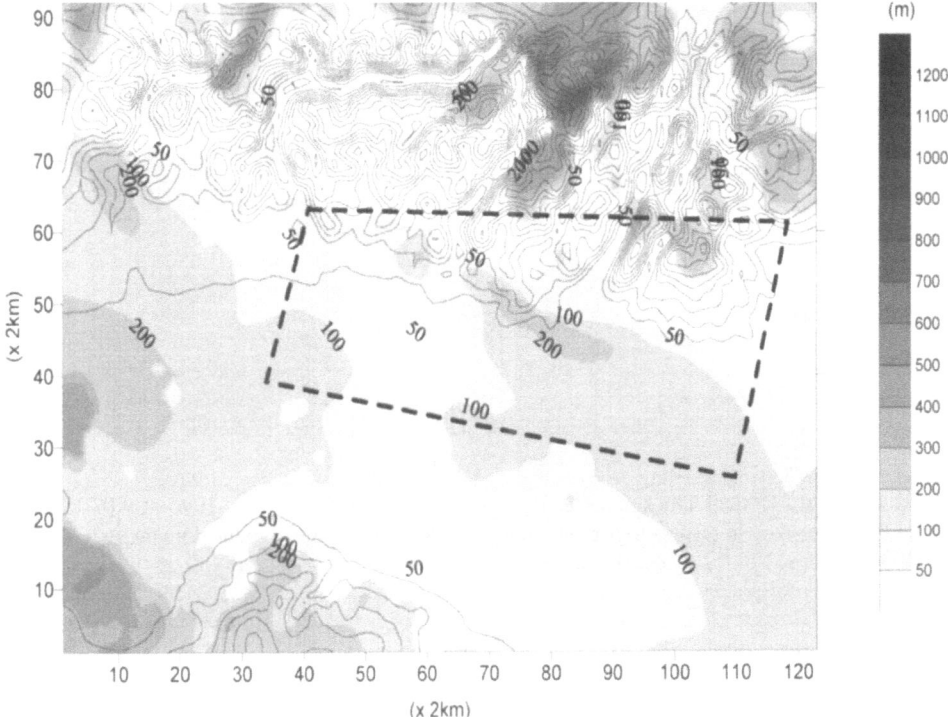

Figure 1. The area of Brescia (Northern Italy). The black dots mark the industrial emission sources. The black solid frame indicates the area where the pollution measurements are collected. The bold and the thin dashed frame refer to the areas covered by the satellite images on 16–1–1996 and on 06–2–1998 respectively. Netting areas indicate inland water bodies.

3. Application and Results

The method previously described used high spatial resolution Landsat satellite data of the Lombardy Area in Northern Italy centred over the city of Brescia (Figure 1) and acquired during two winter days: January 16th, 1996 and February 6th, 1998. The exact acquisition time for the two Landsat images was 9:07 UT and 9:40 UT respectively.

Due to the absence of measurements of ammonium sulphate and ammonium nitrate salts, the main source of secondary particles, the AOTV and the normalized mean concentration fields were compared with the measured concentration fields for various pollutants such as NOx, SO_2, CO, O_3 and TSP (total suspended particles) as these were interpolated from the 142 ground stations. From these measurements, emphasis was given to NO_x concentrations since the SO_2 was at very low concentration levels with insignificant spatial variability. The TSP data are only from a very limited number of stations, therefore unable to provide reliable information on the aerosol spatial distribution.

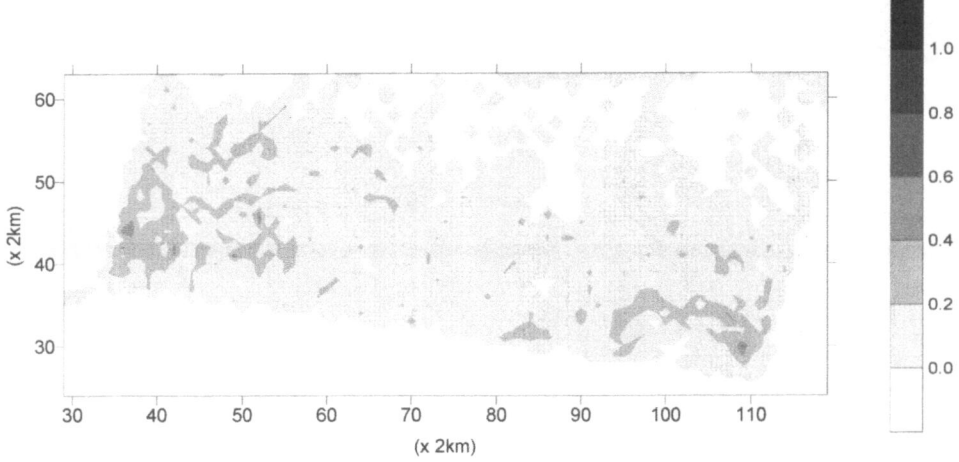

Figure 2. Aerosol Optical Thickness (AOTV), (Landsat TM 2nd) on 16–1–1996 at 9:07 UT. The values are normalized in respect to the maximum AOTV value observed in the domain on 6–2–1998 at 9:40 UT.

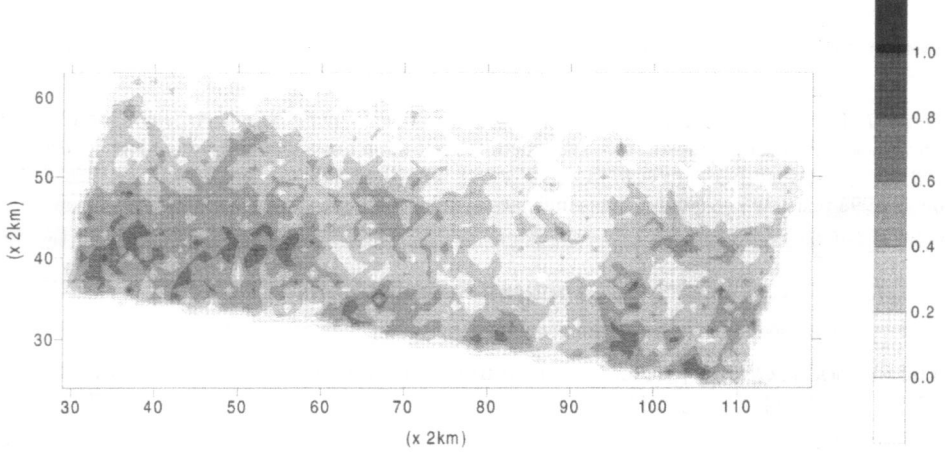

The spatial distributions of AOTV for 16–1–1996 and 6–2–1998 are given in Figures 2 and 3. At a first view, it is apparent the mosaic pattern of pollution as it is anticipated by the capability of satellite images to capture all the pollution sources, which are dispersed in the whole domain (Figure 1). Such a synoptic representation is not possible to be provided by the ground measurements solely since the limited number of stations would lead to severe 'smoothing' of the concentration isopleths. The comparison between the two days shows a larger amount of optically effective particles available in the reservoir, for the second day (6–

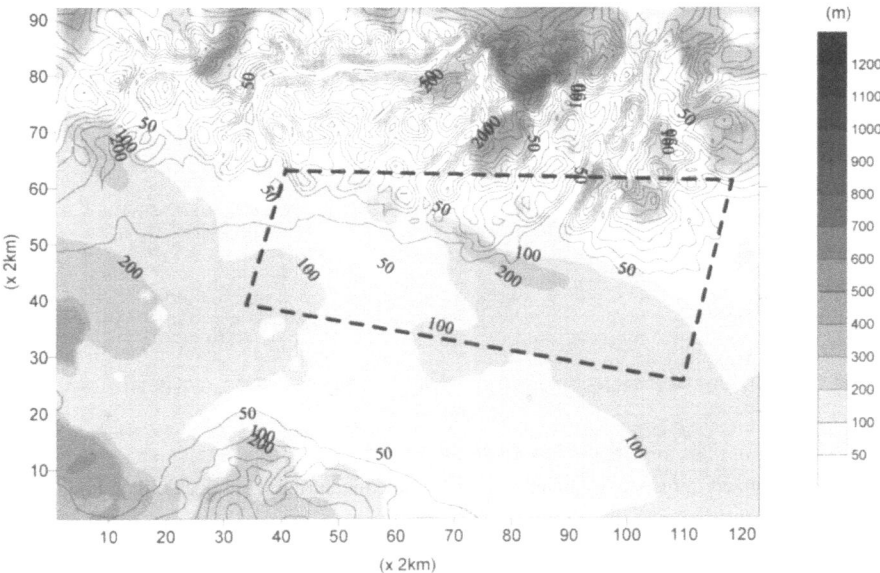

Figure 4. Spatial distribution of the mixing height on 16–1–1996 at 9:00 UT. The dashed frame indicates the area covered by the satellite image.

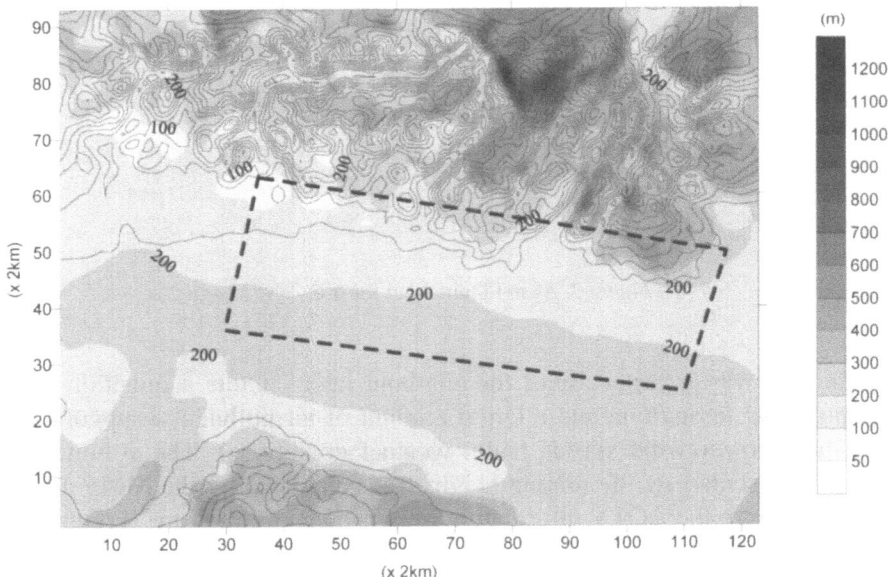

Figure 5. As in Figure 4 but for 6–2–1998 at 10.00 UT.

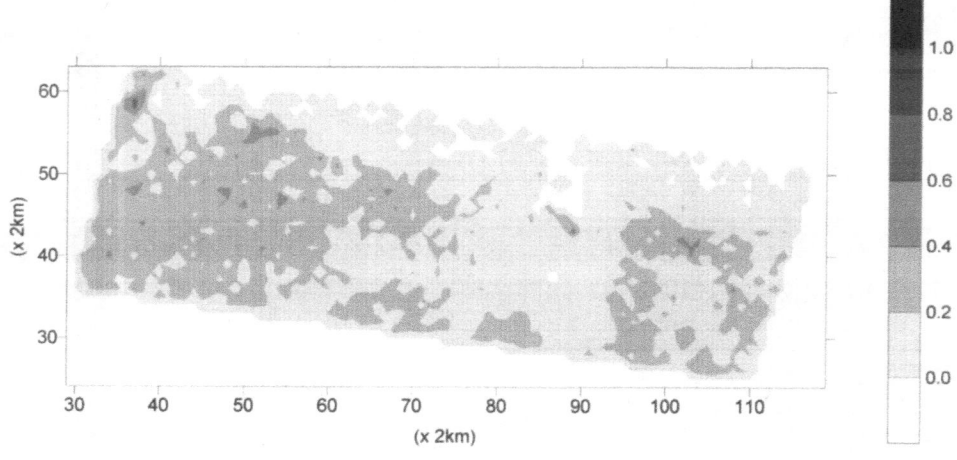

Figure 7. As in Figure 6 but for 6–2–1998.
values are normalized in respect to the maximum mean concentration value calculated on 6–2–1998.

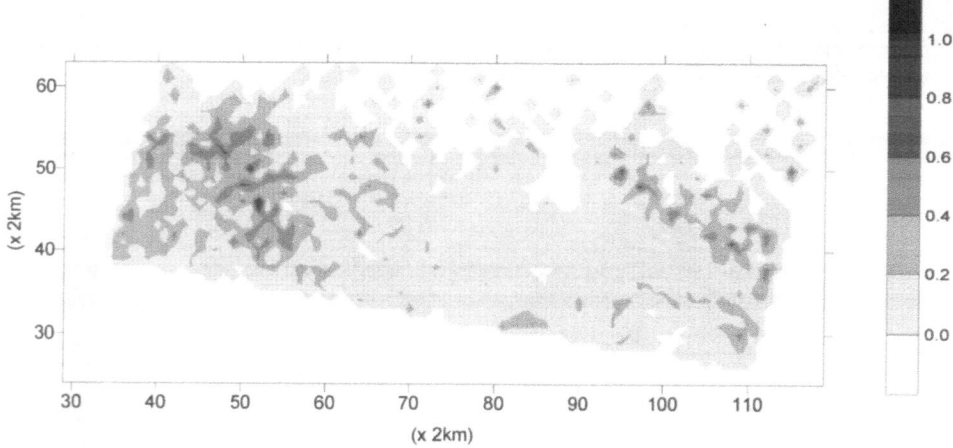

Figure 7. As in Figure 6 but for 6–2–1998.

2–1998). This was expected, since the one-hour later satellite acquisition for that day contributed to agglomerate a larger amount of air pollutants, supported also by the much lower wind speeds (calm weather conditions). The volume of the reservoir represented by the mixing height was calculated by the MM5 model in order to transform the AOTV information into mean concentrations (Figures 4 and 5). Although these two days correspond to different years, present almost the same values for the mixing heights. The larger variation of mixing heights is observed in the mountainous area where valleys and crests are frequently interchanged. In the basin, (the domain which coincide with the satellite images) and especially around

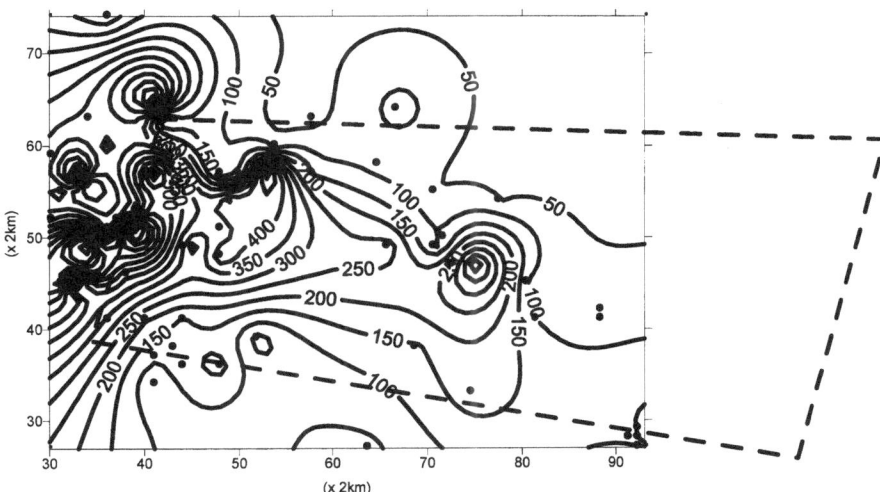

Figure 8. NO$_x$ concentration (ppb) on 16–1–1996 at 9:00 UT. The black points indicate the position of the ground stations that were operating and the dashed frame shows the domain covered by the satellite image.

Milan, smaller values of mixing heights are observed, (up to the height of 250 m), due to the time of the day and season. The slightly larger values for the second day are well explained by the fact that the calculations were for one hour later, in order to coincide with the satellite passage.

The constructed new patterns of normalized mean concentrations are shown in Figures 6 and 7. From these figures, it is apparent that on the 6–2–1998, higher pollution levels prevail in the whole domain. However, the higher isolated values are observed on 16–1–1996, mainly in the greater area of Milan. This behavior is not very far from the measured concentrations of NO$_x$, shown in Figures 8 and 9. For comparison reasons the SO$_2$ values are also given in Figures 10 and 11, although a direct comparison with these two pollutants should not be sought for the reasons described already in the introduction. Moreover it should be noticed that the measured concentrations are collected near the surface while the derived normalized concentration values depict mainly the mean value inside the mixing height. These two patterns come closer when an intense mixing is established in the lower atmosphere. Finally, it is interesting to notice that the higher fluctuations are also apparent in the measured concentration pattern, around the greater Milan area, where a large number of monitoring stations is available.

4. Conclusions

This study produced mean normalized concentration fields by using the spatial distribution of aerosol optical thickness in the visible and concluded that while numer-

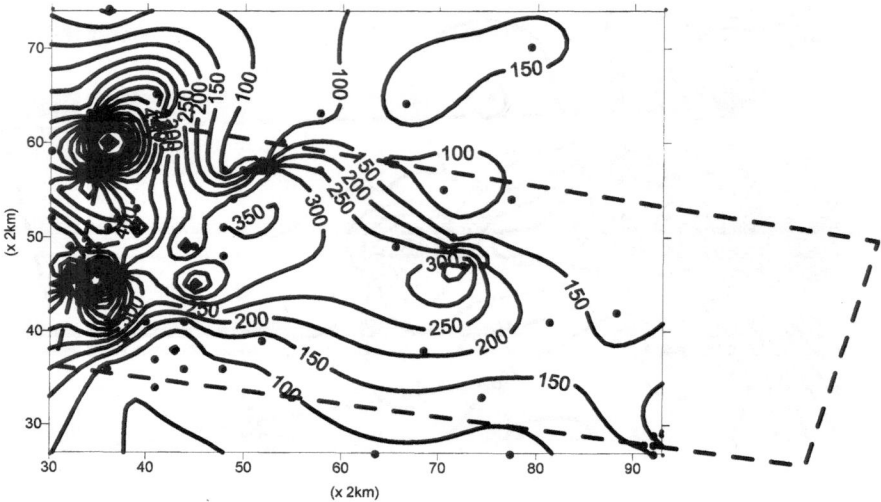

Figure 9. As in Figure 8 but for 6–2–1998 at 10:00 UT.

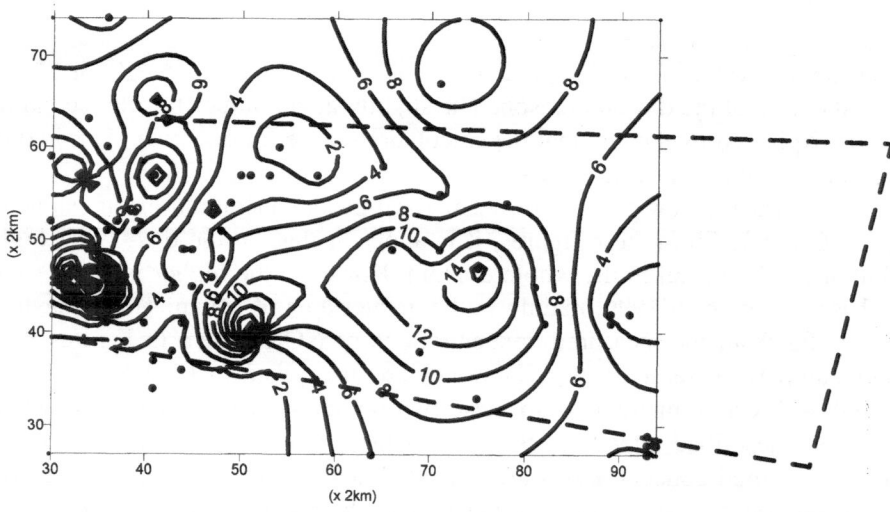

Figure 10. SO$_2$ concentration (ppb) on 16–1–1996 at 9:00 UT. The black points indicate the position of the ground stations that were operating and the dashed frame shows the domain covered by the satellite image.

ical modeling providing information on pollutants spatial distribution is strongly depended on the available emission data, satellite-derived AOTV maps have the advantage of capturing all actual emissions, as far as, these are finally transformed into secondary optically effective particles. Therefore under calm weather conditions, where transport processes are not encouraged, satellite-based AOTV maps may provide an ideal tool for re-examining the emission inventory. However, satellite-

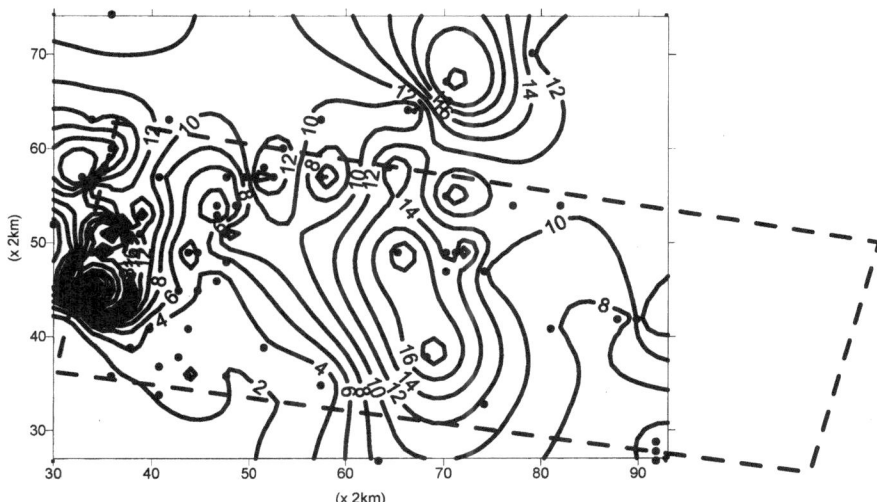

Figure 11. As in Figure 10 but for 06–2–1998 at 10:00 UT.

based AOTV mapping is still a new operational monitoring tool that needs further processing by taking into account basic parameters of the atmospheric boundary layer such as the mixing height. This elaboration was shown to be necessary when information on the spatial distribution of pollutant concentrations is searched.

In order to conclude with more reliable surface concentration patterns, the structure of the concentration profiles should also be accounted. It is planned, in the near future, to collect simultaneous measurements of aerosol profiles and selected meteorological parameters (wind and temperature) to study the structure of aerosol layers in relation to the associated turbulence characteristics.

Acknowledgements

This work was supported by the European Commission funded project ICAROS (Integrated Computational Assessment via Remote Observation System) under contract No. ENV4-CT97-0417.

References

Anthes, R. A. and Warner, T. T.: 1978, 'Development of hydrodynamic models suitable for air pollution and other meso-meteorological studies', *Mon. Wea. Rev.* **106**, 1045–1078.

Ballard, S. P., Golding, B. W. and Smith, R. N.: 1991, 'Mesoscale model experimental Forecasts of the Haar of Northeast Scotland', *Mon. Wea. Rev.* **119**, 2107–2123.

Beyrich, F., Gryning, S. E., Joffre, S., Rasmussen, A., Seibert, P. and Tercier, P.: 1996, 'On the Determination of Mixing Height: A Review', *Proc. of the 4th Workshop on Harmonisation within*

Atmospheric Dispersion Modelling for Regulatory Purposes, 6–9 May 1996, Oostende, Belgium, pp. 155–162.

Beyrich, F., Gryning, S. E., Joffre, S., Rasmussen, A., Seibert, P. and Tercier, P.: 1998, in Sven-Erik Gryning and Nadine Chaumerliac (eds), *Mixing Height Determination for Dispersion Modelling – A Test of Meteorological Pre-Processors. Air Pollution Modelling and its Application XII*, Plenum Press, New York.

Colls J.: 1997, *Air Pollution: An Introduction*, E and FN SPON, London, 341 pp.

Devara, P. C. S. and Ernest, Raj, P., 'Study of atmospheric aerosols in terrain-induced nocturnal boundary layer using bistatic Lidar', *Atmos. Environ.* **25A**, 655–660.

Grell, G. A., Dudhia, J. and Stauffer, D. R.: 1993, 'A Description of the Fifth-Generation Penn State/NCAR mesoscale model (MM5)', *NCAR Technical Note, NCAR/TN-398 + STR*, 117 pp. [Available from NCAR information Services, P.O. Box 3000, Boulder, CO 80307].

Hong, S.-Y. and Pan H.-L.: 1996, 'Nonlocal boundary layer vertical diffusion in a medium-range forecast model', *Mon. Wea. Rev.* **124**, 2322–2339.

King, M. D., Kaufman, Y. J., Tanré, D. and Nakajima, T.: 1999, 'Remote sensing of tropospheric aerosols from space: past, present and future', *Bull. Amer. Meteorol. Soc.* **80**, 2229–2259.

Marsik, F. J., Fischer, K. W., Mcdonald, T. D. and Samson, P. J.: 1995, 'Comparison of methods for estimating height used during the 1992 Atlanta Field Intensive', *J. Appl. Meteorol.* **34**, 1802–1814.

Piringer, M., Baumann K. and Langer, M.: 1998, 'Summertime mixing heights at Vienna, Austria, estimated from vertical soundings and by a numerical model', *Boundary-Layer Meteorol.* **89**, 25–45.

NASA: 1999, in M. King and R. Greenstone (eds), *EOS Reference Handbook: A Guide to NASA's Earth Science Enterprise and the Earth Observation System*, NASA/GSFC, Greenbelt.

Sifakis, N.: 1992, 'Potentiality of the High Spatial Resolution Satellite Imagery for Tracking of Air Pollution in the Lower Troposphere; the Athens Case Study', Ph. D. Thesis, Department of Environmental Studies, University of Paris VII, (in French) 296 pp.

Sifakis, N. and Deschamps, P. Y.: 1992, 'Mapping of Air Pollution Using SPOT Satellite Data', *Photogrammetric Engineering and Remote Sensing*, Vol. 58, no. 10, October 1992, pp. 1433–1437.

Sifakis, N., Soulakellis, N. and Paronis, D.: 1998, 'Quantitative mapping of air pollution density using Earth observations: A new processing method and application on an urban area', *Int. J. Remote Sens.* **19**, no. 17, pp. 3289–3300.

Sifakis, N., Tombrou, M., Sarigiannis, D., Soulakellis, N., Dantou, A., Bossioli, E. and Paronis, D.: 2001, 'Satellite Earth observations: A complementary tool to numerical modelling in urban air quality monitoring', *The 3rd International Conference on Urban Air Quality Measurement, Modelling and Management*, 19–23 March 2001, Loutraki, Greece.

Troen, I. and Mahrt, L.: 1986, 'A simple model of the atmospheric boundary layer: Sensitivity to evaporation', *Bound. Layer Meteorol.* **37**, 129–148.

Zhang, D. and Anthes, R. A.: 1982, 'A high-resolution model of planetary boundary layer – sensitivity tests and comparisons with SESAME-79 data', *J. Appl. Meteor*, 1594–1609.

IMPROVEMENTS IN AIR QUALITY MODELLING BY USING SURFACE BOUNDARY LAYER PARAMETERS DERIVED FROM SATELLITE LAND COVER DATA

N. KITWIROON, R. S. SOKHI*, L. LUHANA and R. M. TEEUW

Atmospheric Science Research Group (ASRG), Department of Environmental Sciences, University of Hertfordshire, College lane, Hatfield, Hertfordshire AL10 9AB, U.K.
(* author for correspondence, e-mail: r.s.sokhi@herts.ac.uk, fax: +44 (0) 1707 285258)

Abstract. Many atmospheric dispersion models include only simple treatment of surface features to estimate the wind profiles and stability parameters. Detailed characterisation of the land cover, particularly in large and complex urban conurbations, is especially important, as the surface features can vary significantly over the area. This paper discusses the use of satellite land cover data to derive spatially resolved surface boundary layer (SBL) parameters. These parameters have been used in an air quality model, PEARL (Prediction Air Quality in Urban and Regional Locations) for estimating monthly and annual CO concentrations. Land cover data, derived from LANDSAT Thematic Mapper Imagery, has been used to estimate SBL parameters (surface roughness length, albeedo, Bowen ratio and anthropogenic heat flux) for a study area of 10000 km^2 encompassing Greater London and the surrounding counties. The SBL parameters have been assigned according to major land cover types for the whole area at a spatial resolution of 1×1 km. Predictions from two versions of the PEARL model (one with land cover data and one without) have been compared with each other and with measured data for annual and monthly CO concentrations from seven London air quality monitoring sites. This comparison shows that differences between predicted and observed values can be reduced by up to a factor of three. The use of SBL parameters derived from land cover data also yields more detailed predicted annual CO spatial patterns especially in and around suburban areas. The performance of both versions of the model for monthly CO concentrations has been compared with a range of statistical measures. This comparison confirms that improved agreement is observed between modelled and measured monthly CO concentrations when use is made of spatially resolved SBL parameters.

Keywords: box model, dispersion modelling, land cover, satellite imagery, surface boundary layer, urban areas

1. Introduction

Atmospheric dispersion models are now widely used for impact assessment and management of air quality in urban areas (Moussiopoulos, 1997 and Sokhi *et al.*, 2000). Various surface boundary layer (SBL) parameters, such as, surface roughness, anthropogenic heat flux, albedo and Bowen ratio (ratio of sensible to latent heat fluxes) feed directly into air pollution models through formulations to describe the stability of lower atmosphere. For example, an important parameter regarding land surface characteristics is the surface roughness length (z_o) (Wieringa, 1992).

Water, Air, and Soil Pollution: Focus **2**: 29–41, 2002.
© 2002 *Kluwer Academic Publishers*.

The surface roughness length is generally defined as the height at which the mean wind speed is zero. In all cases it is smaller than the physical height of the roughness elements. Although it does not change with meteorological conditions, it does change with the height and coverage of obstacles on the earthþs surface and hence higher elements are associated with larger z_0 values (Stull, 1988). The effect of surface roughness is to slow down the near-surface flow and, therefore, retards the dispersion in the horizontal direction. In many air quality assessment models z_0 and other SBL parameters are treated simplistically by taking just one average value for the entire modelled area. In reality urban areas constitute quite complex terrains and these lead to equally complex turbulence and dispersion patterns within the urban canopy (Grimmond et al., 1999).

Remote sensing technology is now increasingly playing a crucial role as a provider of high resolution environmental data (Harding et al., 1994). In particular remote sensing can provide detailed information on land surface characteristics relevant to atmospheric dispersion modelling (Hasager and Thykier-Nielsen, 2001). Traditionally, surface geometry parameters such as z_0 have been derived from meteorological field mast data (see for example Stull, 1988). It can be time consuming and expensive if information is required for a large area. Satellite imagery offers an alternative approach to obtain information on surface characteristics to derive spatially resolved values for SBL parameters, such as, z_0 (Hasager and Jensen, 1999; Jasinski et al., 1999).

This paper reports on the incorporation of detailed land cover data into an air quality model providing improved surface boundary layer description. Land cover data, derived from LANDSAT Thematic Mapper (LANDSAT TM) satellite imagery, has been used with a multibox model, PEARL (Predicting Air Quality in Urban and Regional Locations), to assess the effect of this approach on long-term (monthly and annual) concentrations of CO for an area encompassing Greater London. Air quality models based on the box model formulation have been used extensively for operational uses including forecasting (see for example, Middleton 1998). They offer an efficient option for air quality management purposes especially for calculating long-term concentrations (Luhana and Sokhi 1998a,b). Air quality models in general, however, have not benefited widely from the use of land cover data to improve the treatment of SBL parameters.

In this study surface boundary layer parameters, z_0, anthropogenic heat flux, Bowen ratio and albedo, and in turn mean wind speed and stability, have been determined for each 1×1 km grid cell of the area of study. These spatially resolved parameters are used in the PEARL model and the predictions of monthly and annual CO concentrations are compared to the original version of the model which assumes uniform surface characteristics and hence uniform wind speed and stability for the whole area of interest. Model evaluation has been conducted by comparing the predicted CO concentrations from both versions of the model (with and without land cover data) with measured data from seven air quality monitoring stations located within Greater London for the year 1997.

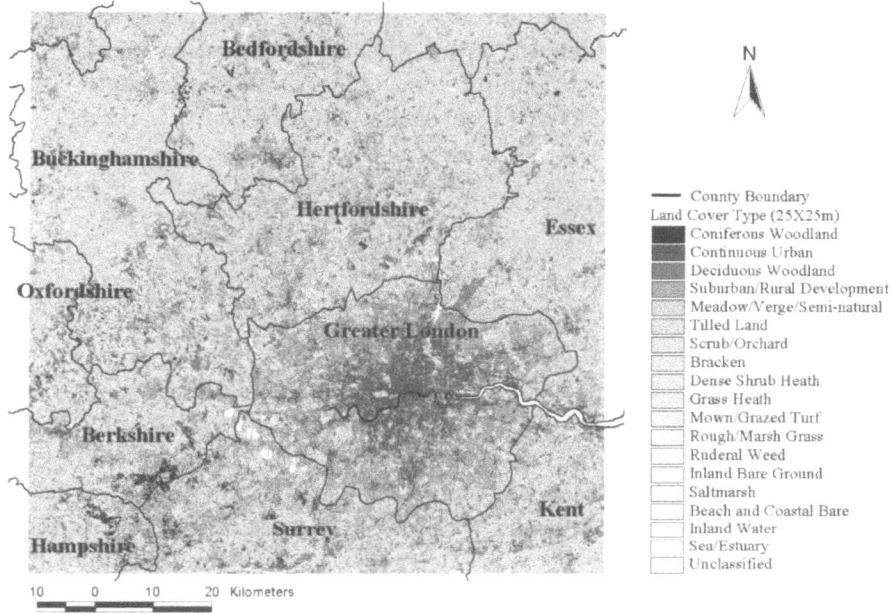

Figure 1. Land cover types at 25 × 25 m spatial resolution within the study area.

2. Methodology

Land cover data derived from LANDSAT TM Imagery was used to derive SBL parameters for the study area at a 1 × 1 km spatial resolution. The study area covers Greater London and surrounding counties, with a total area of 10000 km^2 and 19 different land cover types. Figure 1 shows the land cover types in the study area at a resolution of 25 × 25 m. The data was processed with ArcView GIS based user-interface for PEARL. Details are given below of the modelling formulation as well as how values of SBL parameters were assigned to each grid cell.

2.1. PEARL MODEL

The multibox model PEARL is based on the following formulation, assuming that the surface roughness is uniform (Hanna, 1971 and Luhana and Sokhi, 1998a,b):

$$C = \frac{(2/\pi)^{1/2}(\Delta x/2)^{1-b}}{Ua(1-b)}\left\{Q_0 + \sum_{i=1}^{N} Q_i\left[(2i+1)^{1-b} - (2i-1)^{1-b}\right]\right\} \quad (1)$$

where C is the surface concentration of the pollutant (μg m^{-3}), Δx is the grid spacing or distance across the box (m), set at 1000 m in this case, U is the mean wind speed (m s^{-1}), assumed to be uniform for the whole area, Q_0 is the source emission strength for the box in question (M.tonne yr^{-1} km^{-2}), Q_i represents the source emission strengths in the N upwind boxes, i is 1,2...N, and a and b are empirical parameters which depend upon the Pasquill-Gifford stability class chosen.

Equation 1 can be modified to allow concentrations of a pollutant to be calculated using SBL parameters derived from spatially resolved land cover data:

$$C = \frac{Q_0(2/\pi)^{1/2}(\Delta x/2)^{1-b_0}}{U_0 a_0(1 - b_0)} +$$

$$+ \sum_{i=1}^{N} \frac{Q_i[(2i + 1)^{1-b_i} - (2i - 1)^{1-b_i}](2/\pi)^{1/2}(\Delta x/2)^{1-b_i}}{U_i a_i(1 - b_i)} \quad (2)$$

where U_0 is the mean wind speed (m s^{-1}) of the cell of interest, U_i represents the mean wind speed (m s^{-1}) for cell i in the N upwind boxes, Q_0 is the source emission strength for the cell of interest (M.tonne yr^{-1} km^{-2}), Q_i are the source emission strengths in the N upwind boxes, and i = 1,2...N, a_0 and b_0 are the empirical parameters values which depend upon the land cover type and Pasquill-Gifford stability class for the box in question and a_i and b_i are the equivalent empirical parameters for the N upwind boxes.

In the box model formulation the distribution of source strengths and the resulting surface concentrations in the box are assumed to be homogenous and hence, the horizontal component of the diffusion is neglected compared to the vertical component. The parameters a and b are related to the vertical standard deviation, σ_z, through a power law expression:

$$\sigma_z = ax^b \quad (3)$$

where x is the distance from the source (m). The empirical parameters a and b used in equations (1) and (2) were estimated from procedures detailed in USEPA (1993) and Zannetti (1990). The unmodified model (Equation 1) employs a constant value for z_0 of 0.5 m over the whole study area whereas the modified version (Equation 2) incorporates values of z_0, albedo and Bowen ratio according to the land cover type. Consequently, wind speed and atmospheric stability are also estimated for individual grids as explained in section 2.2. For the unmodified model, therefore, values for wind speed and stability (hence a and b empirical parameters) are assumed to be constant for the whole area of interest. The CO emissions data at a resolution of 1 × 1 km was obtained from NETCEN (2000) and LRC (1998) for the study area.

2.2. ESTIMATION OF SURFACE BOUNDARY LAYER PARAMETERS

In this case, the land cover data, supplied by the Centre for Ecology and Hydrology (CEH), UK, and derived from LANDSAT TM Imagery, was used to assign SBL parameters to individual 1 × 1 km grid cells. The data records include 25 different land cover types, consisting of sea and inland water, beaches and bare ground,

TABLE I

Land cover classes and surface roughness
length $(z_0)^a$

Grouped land cover type	z_0 (m)
(1) Water	0.0002
(2) Grass land	0.001–0.1
(3) Saltmarsh	0.0002–0.1
(4) Cultivated land	0.02–0.2
(5) Suburban	0.55
(6) Continuous urban	1.0
(7) Forest land	0.5–1.2
(8) Shrub land	0.15–0.3

[a] Modified from USEPA (1998), Wieringa
(1993).

developed and arable land and 18 types of semi-natural vegetation (Fuller *et al.*, 1994). For the present study area 19 different land cover types were found to be present in the area of interest. These land cover types were grouped in eight classes as listed in Table I and z_0 values assigned. The classification of Wieringa (1993) and USEPA (1999) for z_0 values according to land cover type has been employed for this work. The original land cover data of 25×25 m pixel resolution has been transformed into z_0 value for each 1x1km grid using the ArcView median filter option. This option allows a particular land type to be assigned to a particular grid cell as shown in Figure 2. Each 1×1 km grid is assigned a particular z_0 value depending upon the predominant type of land cover within that grid (see for example, Hasager and Thykier-Nielsen, 2001). An alternative approach to this method is to take all land cover type into account (Varvayanni *et al.*, 1995). However, for the area of study it was found that in most grid cells one land cover was dominant. Although simple, this method is computationally efficient as not all land types need to be included in the calculations.

To estimate the wind speed and Pasquill-Gifford stability class for each grid, meteorological data was obtained from British Atmospheric Data Centre, BADC (2000), for the London Weather Centre, which is the nearest representative site. The meteorological data included wind speed (m s^{-1}), wind direction, air temperature (°C), total cloud cover (Octas) and global radiation (W m^{-2}). These parameters were processed to provide the average Monin-Obukhov Length, friction velocity and wind speed at the measurement site using Equations 4 and 5:

$$u_* = \frac{kU}{\ln\left(\frac{z_{ref}}{z_o}\right) - \psi + \psi_o} \tag{4}$$

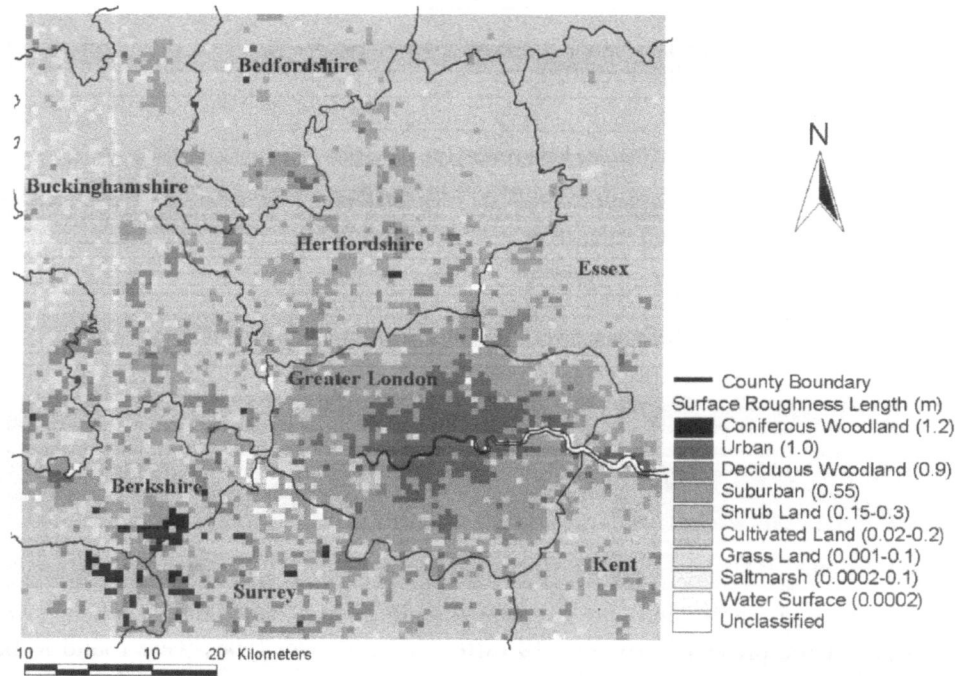

Figure 2. Land cover types and z_0 at 1 x 1km spatial resolution within the study area.

where u_* is friction velocity, k is von Karman's constant (0.4), U is wind speed (m s^{-1}), z_{ref} is the height of anemometer (taken as 10 m). ψ and ψ_0 are the universal dimensionless functions (Stull, 1988). Using u_* the values for the Monin-Obukhov length (L) can be calculated from:

$$L = \frac{-\rho c_p T u_*^3}{kgH} \qquad (5)$$

where ρ is air density, c_p is specific heat capacity of air, T is ambient temperature (K), k is von Karman's constant (0.4), g is acceleration due to gravity, and H is sensible heat flux. H was calculated using the following equation:

$$H = \frac{(1.0 - c_g)(R_N + Q_f)}{1 + \frac{1}{B_o}} \qquad (6)$$

where c_g is the fraction of the net radiation absorbed at the ground, R_N is net radiation, Q_f is anthropogenic heat flux, and B_o is Bowen ratio. These parameters were varied according to land cover types for each 1×1 km grid cell using the classification scheme given in USEPA (1999).

Using the values from equation 4 and 5 the following expressions can be used to estimate the average wind speed and stability class for each grid cell:

$$U_1 \mu_{*1} \approx U_2 \mu_{*2} \qquad (7)$$

$$L_2 = L_1 \, (\mu_{*2}/\mu_{*1})^3 \tag{8}$$

where U_1 and U_2 are the wind speeds at the measurement site and the grid in question respectively, u_{*1} and u_{*2} are the friction velocities and L_1 and L_2 are the Monin-Obukhov lengths. Monin-Obukhov length (L) is then converted to the equivalent Pasquill-Gifford stability class using the plots contained in Golder (1972) which then allow parameters a and b to be estimated (USEPA, 1993 and Zannetti, 1990).

3. Results and Discussion

3.1. LAND COVER DATA AND SURFACE ROUGHNESS

Figure 1 shows the land cover data for the study area at a 25 × 25 m resolution. The highest z_0, 1.2 m, is assigned to coniferous forest that makes $\sim 0.8\%$ of the total area and can be seen as the darkest patch in the bottom left corner of the study area. A z_0 value of 1.0m is given to continuous urban areas and constitutes about 4% of the study area, mainly in the centre of London. Deciduous forest type land cover ($z_0 = 0.9$ m) account for nearly 5% of the study area but this land cover type is distributed as small patches around the suburbs of Greater London. Suburban type land cover comprises 17% of the study area. The major land cover type prevalent in the study area is cultivated land ($z_0 = 0.2$ m) and makes up nearly 70% of the study area. Grassland and water surface account for about 4% and 1% of the study area respectively and have z_0 value < 0.1 m. The 25 × 25 m data has been grouped into 1 × 1 km resolution so as to be compatible with the PEARL model inputs. Land cover data and z_0 for the study area at 1 × 1 km spatial resolution is shown in Figure 2.

3.2. MODEL PERFORMANCE

The predicted annual CO concentrations for the entire area from the modified and unmodified versions of PEARL are shown in Figure 3. The unmodified version of PEARL assumes uniform land surface across the study area and hence the meteorological parameters are considered to be constant over the region. As a result, the spatial variations in pollutant concentrations will depend only on variations in emissions across the study area. In contrast the modified model predicts concentrations which will vary according to emissions as well as SBL characteristics. When comparing the outputs of the two versions (Figures 3a and 3b), the land cover version of PEARL gives lower CO concentration values in the centre of London but predicts higher concentrations in the surrounding areas, essentially following the major road networks and urban conurbations. The areas immediately surrounding London are either suburban or small urban centres and are predominantly of suburban type land cover with a mean z_0 greater than the value of 0.5 m used in the

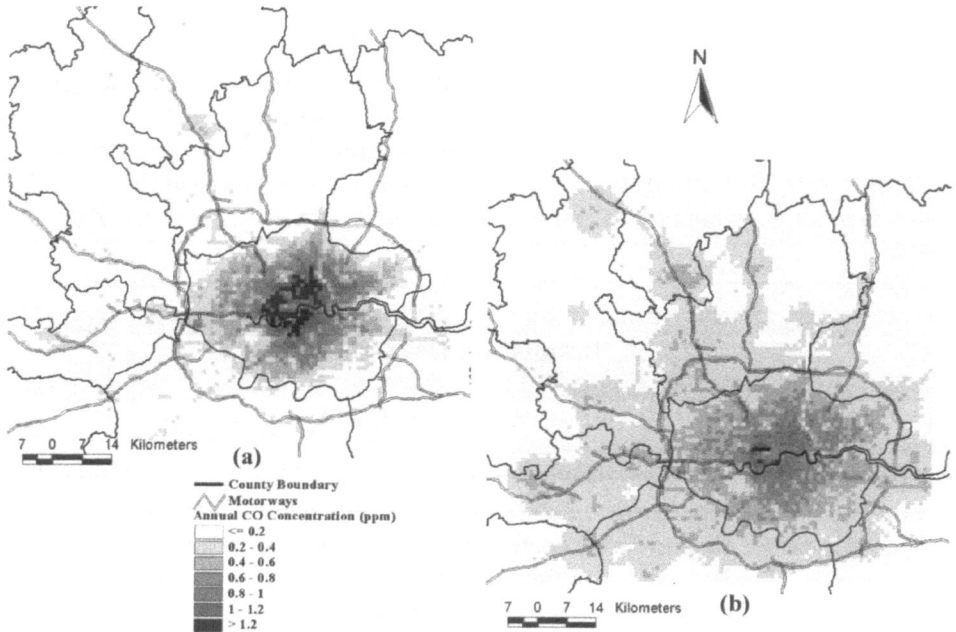

Figure 3. Annual (1997) CO concentration (ppm) predictions from (a) unmodified model assuming uniform surface and (b) modified model using land cover data.

unmodified model. Hence the modified model predicts higher CO concentrations in these grid cells. The use of land cover data in the model, therefore, clearly reveals a more detailed spatial pattern for the CO concentrations across the region.

Beyond the urban and suburban areas of Greater London, the rural and semirural areas consist of small sized urban centres and routes with high traffic levels such as motorways and main trunk roads. In the smaller urban centres emissions can also be high in a given area and land cover patterns can also be quite similar to the suburban areas of greater London resulting in similarly high concentrations in certain grids. In the grids through which motorways pass, the terrain is generally quite flat resulting in low z_0 values but relatively high CO values are predicted in these cells because of the high traffic emissions.

To evaluate the model performance the predicted annual CO concentrations were compared with the measurements from seven London stations, Hillingdon, Brent, Bexley, North Kensington, Bridge Place, Bloomsbury and West London for the year 1997. The location of these staions are shown in Figure 4. Out of these seven sites Hillingdon and Bexley are classified as suburban while the remaining five are categorised as urban background sites. Comparison between measured and predicted annual data at these stations is shown in Figure 5. For Hillingdon the modified model gives slightly higher concentration values than the unmodified model. However, the predicted values agree to the measured values to within 10%.

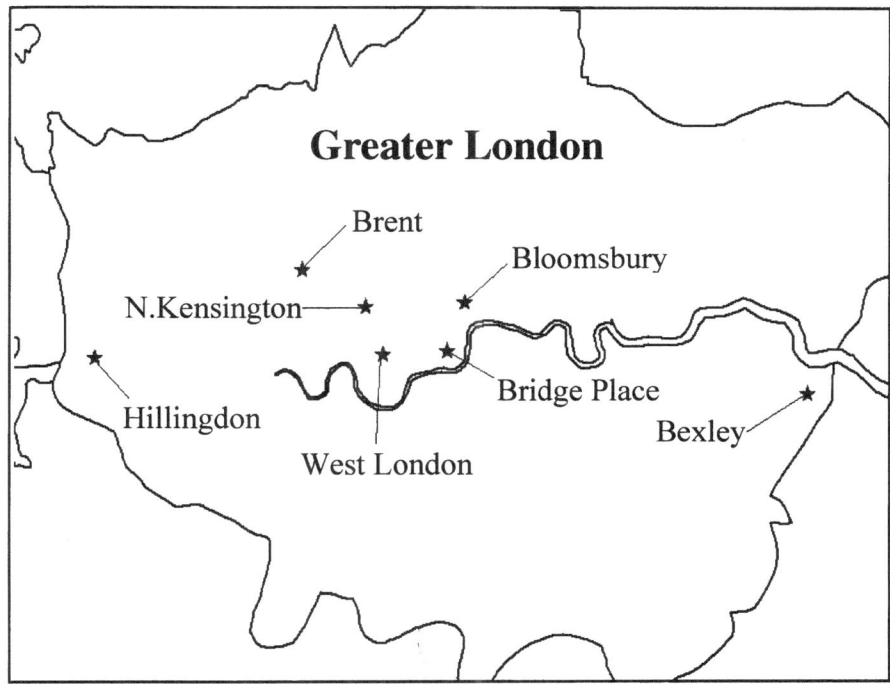

Figure 4. Locations of air quality monitoring stations within London used for model performance evaluation.

Very little difference is observed between the predictions for Brent as the use of land cover data produced values of z_o similar to 0.5 m used in the unmodified model. For all other stations, however, the predicted values are significantly improved by using land cover data. For example, the difference in the annual mean CO concentrations between measured and modelled is reduced to 20% from nearly 64% for Bexley, 15% from 50% for N. Kensington, 5% from nearly 27% for Bridge Place, 25% from nearly 71% for Bloomsbury, and 77% from nearly 145% for West London. As the same emissions data has been used for both versions of the model these improvements may be attributed to the use of more representative land cover data.

The two versions of the model have been further evaluated for monthly averaged concentrations for the same year using the data from these seven stations. The scatter plot of the comparison of measured and modelled CO concentrations is shown in Figure 6a (unmodified model) and 6b (modified model). The figures show the scatter of the values around the y = x line as well as the factor of two lines. Several statistical measures have also been calculated to quantify the comparison and these are listed in Table II. More details on these measures can be found in Henmi (2000) and Kukkonen *et al.* (2001). Although the correlation coefficient (R) value is lower for the modified model the spread of the values is more even as indicated by a lower

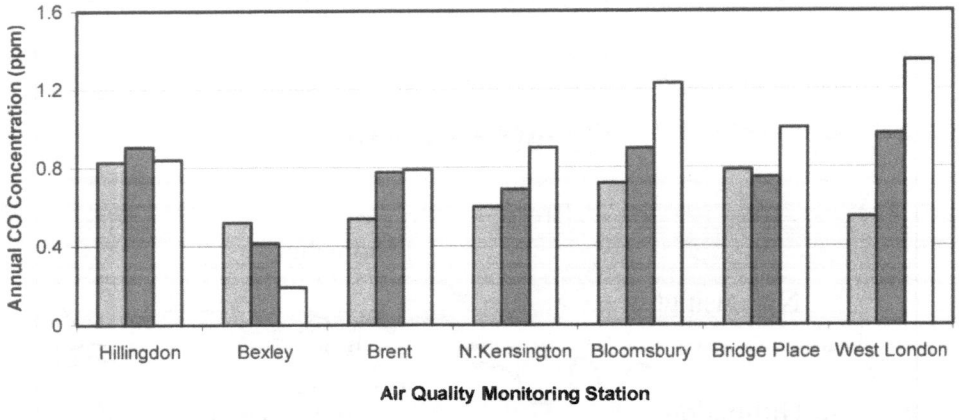

Figure 5. Comparison of predicted annual (1997) CO concentrations (ppm) with measured values from seven air quality monitoring stations in London.

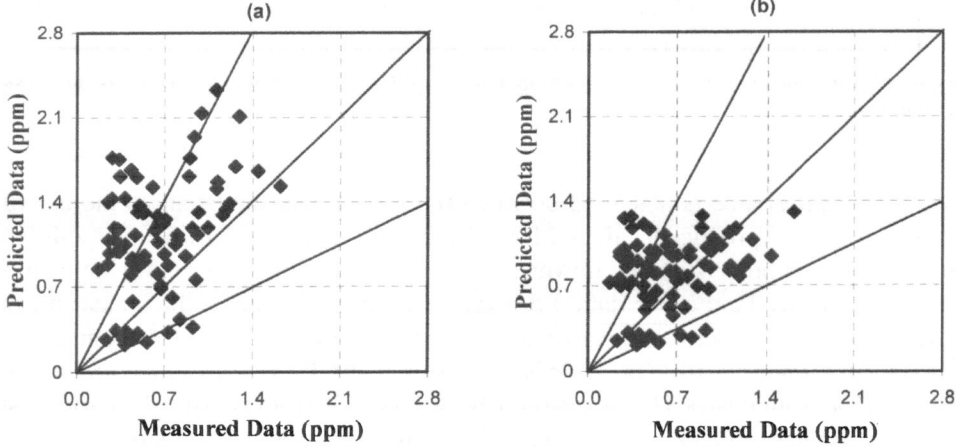

Figure 6. Scatter plot of predicted and measured monthly CO concentrations (ppm) for 1997 using data from seven air quality monitoring stations in London. (a) Unmodified model assuming uniform surface and (b) modified model using land cover data. The y = x and factor of two lines are also shown.

fraction of exceedance (FOEX). The mean fractional bias (MFB) and the normalised mean square error (NMSE) both indicate that the modified model exhibits lower deviations from measured values. Similarly, the mean absolute difference (AD) and the root mean square error (RMS) also show significant improvements.

TABLE II

Statistical parameter comparisons using monthly CO concentrations from seven air quality monitoring stations in London (80 data points)

Predicted CO Concentrations	R[a]	MFB[b]	NMSE[c]	FOEX[d] (%)	AD[e]	RMSE[f]
Uniformal and Surface scheme	0.39	1.64	0.25	35	0.53	0.65
Land cover Data scheme	0.25	0.48	0.13	16	0.33	0.40

[a] = Correlation Coefficient, [b] = Mean Factional Bias, [c] = Normalised Mean Square Error, [d] = Faction of exceedance, [e] = Mean Absolute Difference, [f] = Root Mean Square Error.

4. Conclusions

This study demonstrates the use of satellite land cover data for improving the surface boundary layer parameters for use in an air pollution model for an area of 10000 km^2 of Southeast England including Greater London. Values for z_o, albedo, Bowen ratio and anthropogenic heat flux are attributed according to land cover type at a spatial resolution of 1×1 km. An air quality model PEARL has been used to assess the improvements that may result from the use of more representative SBL parameters. The model was used to predict annual and monthly CO concentrations for the study area. Two versions of the model (one using land cover data and the other assuming uniform surface features) have been compared with each other and with measured data from seven air quality monitoring stations within London. Significant improvements were observed when land cover assigned SBL parameters were used to calculate wind speed and stability and hence the CO concentrations. For example, for most stations differences between modelled and measured annual values were reduced by up to a factor of three. The version of the model using land cover data gave lower CO values in the centre of London but in the surrounding areas higher predictions were obtained. Overall, the use of land cover data provided more detailed spatial patterns for CO annual concentrations across the study area. Several statistical measures were used to assess the performance of both versions of the model for monthly CO predictions. This evaluation confirmed that agreement between predicted and measured data is significantly improved with the use of more representative spatially resolved land cover data.

Acknowledgements

The Royal Thai Government has provided the financial support for this project. The Centre for Ecology and Hydrology (CEH) is acknowledged for supplying the land cover data. Dr Simon Kingham, University of Canterbury, is also acknowledged for providing guidance during the initial stages of the project. The work reported in this paper contributed towards the SATURN EUROTRAC II subproject as well as to COST715 Working Group 3 (Meteorology during peak pollution episodes) activities.

References

British Atmospheric Data Centre (BADC): 2000, http://www.badc.rl.ac.uk/.

Fuller, R. M., Groom, G. B. and Jones, A. R.: 1994, 'The land cover map of Great Britain: an automated classification of Landsat thematic mapper data', *Photogram. Eng. Rem. Sen.* **60**, 553–562.

Golder, D.: 1972, 'Relations among stability parameters in the surface layer', *Bound. Lay. Meteo.* **3**, 47–58.

Grimmond, C. S. B. and Oke, T. R.: 1999, 'Aerodynamic properties of urban areas derived from analysis of surface form', *J. Appl. Meteo.* **38**, 1262–1292.

Hanna, S. R.: 1971, 'A simple method of calculating dispersion from urban area source', *J. Air Poll. Cont. Assoc.* **21**, 774–777.

Harding, D. J., Bufton, J. L. and Frawley, J. J.: 1994, 'Satellite laser altimetry of terrestrial topography: vertical accuracy as a function of surface slope, roughness and cloud cover', *IEEE Trans. Geosci. Rem. Sen.* **32**, 329–339.

Hasager, C. B. and Jensen, N. O.: 1999, 'Surface-flux aggregation in heterogeneous terrain', *Quart. J. Roy. Meteo. Soc.* **125**, 2075–2102.

Hasager, C. B. and Thykier-Nielsen, S.: 2001, 'IRS-IC LISS III land cover maps at different spatial resolutions used in real-time accidental air pollution deposition modelling', *Rem. Sen. Environ.* **76**, 326–336.

Henmi, T.: 2000, 'Comparison and Evaluation of Operational Mesoscale Models MM5 and BFM over White Sands Missile Range (WSMR)', *The Battlespace Atmospheric and Cloud Impacts on Military Operations (BACIMO)*, Fort Collins, Colorado, USA, 24–27 April 2000.

Jasinski, F. J. and Crago, R. D.: 1999, 'Estimation of vegetation aerodynamic roughness of natural regions using frontal area density determined from satellite imagery', *Agric. Forest Meteo.* **94**, 65–77.

Kukkonen, J., Harkonen, J., Kappinen, A., Pohjola, M., Pietarila, H. and Kostentalo, T.: 2001, 'A semi-empirical model for urban PM10 concentrations, and its evaluation against data from an urban measurement network', *Atmos. Environ.* **35**, 4433–4442.

London Research Centre: 1998, 'London Atmospheric Emission Inventory', Release 2 Dated April 3 1998, *On CD-ROM Supplied by London Research Centre, UK*.

Luhana, L. and Sokhi, R. S.: 1998a, 'Application of an Urban and Regional Air Quality Model (PEARL) to Estimate the Impact of Traffic Management Policies', *Fifth International Conference on Harmonisation within Atmospheric Dispersion Modelling for Regulatory Purposes*, Rhodes, Greece, 18–21 May 1998.

Luhana, L. and Sokhi, R.-S.: 1998b, 'Application of the Box Model (PEARL) to Simulate the Seasonal Air Quality in Urban Centres', *The Proceedings of the World Clean Air and Environmental Congress*, Durban, South Africa, 13–18 September 1998.

Middleton, D. R.: 1998, 'A new box model to forecast urban air quality: boxurb', *Environ. Mon. Assess.* **52**, 315–335.

Moussiopoulos, N.: 1997, 'State of the art of air pollution models needs and trends', *Int. J. Environ. Pollut.* **8**, 250–259.

National Environmental Technology Centre (NETCEN): 2000, 'The UK National Air Quality Information Archive', http://www.aeat.co.uk/netcen/airqual/.

Sokhi, R. S., San Jose, R., Moussiopoulos, N. and Berkowicz, R.: 2000, *Urban Air Quality: Measurement, Modelling and Management*, Kluwer Academic Publishers, Dordrecht, The Netherlands, 484 pp.

Stull, R. B.: 1988, *An Introduction to Boundary Layer Meteorology*, Kluwer Academic Publishers, Dordrecht, The Netherlands, 666 pp.

USEPA: 1993, *Selection Criteria for Mathematical Models Used in Exposure Assessment: Atmospheric Dispersion Models*, U.S. Environmental Protection Agency, Office of Research and Development, Washington DC, 96 pp. Report No: EPA/600/8-91/038.

USEPA: 1999, *PCRAMMET User's Guide*, U.S. Environmental Protection Agency, Emissions, Monitoring, and Analysis Division, Office of Air Quality Planning and Standards, Research Triangle Park, NC, 94 pp. Report No: EPA-454/B-96-001.

Varvayanni, M., Catsaros, N., Bartzis, J. G., Konte, K. and Horsch, G. M.: 1995, 'Wind simulation over greater Athens area with highly resolved topography', *Atmos. Environ.* **29**, 3593–3604.

Wieringa, J.: 1992, 'Updating the Davenport roughness classification', *J. Wind Eng. Indust. Aerody.* **41**, 357–368.

Wieringa, J.: 1993, 'Representative roughness parameters for homogeneous terrain', *Bound. Lay. Meteo.* **63**, 323–363.

Zannetti, P.: 1990, *Air Pollution Modelling: Theories, Computational Methods and Available Software*, Van Nostrand Reinhold, New York, 444 pp.

POTENTIAL AND SHORTCOMINGS OF NUMERICAL WEATHER PREDICTION MODELS IN PROVIDING METEOROLOGICAL DATA FOR URBAN AIR POLLUTION FORECASTING

ALEXANDER BAKLANOV[1]*, ALIX RASMUSSEN[1], BARBARA FAY[2], ERIK BERGE[3] and SANDRO FINARDI[4]

[1] *Danish Meteorological Institute (DMI), Copenhagen Ø, Denmark;* [2] *German Weather Service (DWD), Offenbach, Germany;* [3] *Norwegian Meteorological Institute (DNMI), Blindern, Oslo, Norway;* [4] *Arianet, viale Elvezia 42, Monza (MI), Italy*
(* *author for correspondence, e-mail: alb@dmi.dk, fax: +4539157460)*

Abstract. The last decade progress in numerical weather prediction (NWP) modelling and studies of urban atmospheric processes for providing meteorological data for urban air pollution forecasting is analysed on examples of several European meteorological centres. Modern nested NWP models are utilising land-use databases down to 1 km resolution or finer, and are approaching the necessary horizontal and vertical resolution suitable for city scale. The recent scientific developments in the field of urban atmospheric physics and the growing availability of high-resolution urban surface characteristics data promise further improvements of the capability of NWP models for this aim. A strategy to improve NWP data for the urban air pollution forecasting is suggested.

Keywords: numerical weather prediction models, operational urban air quality information and forecasting systems, urban air pollution episodes, urban boundary layer

1. Introduction

The main problem in forecasting **U**rban **A**ir **P**ollution **(UAP)** is the prediction of episodes with high pollutant concentration in urban areas causing health problems for the population and possibly requiring counter-measures from authorities. The demand from authorities and public for precise forecasts of urban air quality, in particular during episodes of pollution levels above the threshold values of acute health effects, has increased during the last years, especially after the implement-ation of higher air quality standards (EC/92/72, EC/96/62, EC/99/33). To handle air pollution episodes in European cities most authorities have established emer-gence preparedness systems, and in several cities operational **U**rban **A**ir **Q**uality **I**nformation and **F**orecasting **S**ystems **(UAQIFSs)** are implemented. Such systems are very likely to be more widespread in the future. The use of reliable UAQIFS can help to move urban air quality management from the mitigation to the prevention of pollution episodes.

The quality of UAP forecasts depends mainly on three factors: the mapping of emissions, the UAP model and the quality of meteorological forecast data. While

Water, Air, and Soil Pollution: Focus **2**: 43–60, 2002.
© 2002 *Kluwer Academic Publishers.*

many studies and projects are aimed at developing UAP dispersion models and chemical transformation mechanisms and at improving knowledge about pollutants and emissions, no significant efforts are put to improving the forecasts of operational Numerical Weather Prediction (NWP) models that provide the meteorological parameters in UAP models for characteristic air pollution episodes (low winds, stable stratification, local air circulations, topographic effects, breeze conditions, internal boundary layers). These meteorological fields constitute a main source of uncertainty in UAP models which may be larger than the uncertainty in the chemical part (Sistla *et al.*, 1996; Hanna *et al.*, 1998).

Historically, UAP forecasting and NWP models were developed separately and there is no tradition for co-operation between the modelling groups. This was plausible in the previous decades when the resolution of NWP models was too poor for city-scale air pollution forecasting, but the situation has now changed and it is obvious that a revision of the conventional conception of urban air pollution forecasting is required.

2. Potential of NWP Models

During the last decade substantial progress in NWP modelling and in the description of urban atmospheric processes was achieved. The following paragraphs are intended to highlight some of the recent developments.

2.1. INCREASED GRID RESOLUTION AND NESTING OF NWP MODELS

The growth of computer power and the implementation of grid nesting techniques allowed modern NWP models to approach the resolution of the city-scale (e.g. Sattler, 1999; Rasmussen *et al.*, 1999; Saito, 1998; LM, 1999). For example, the Danish operational system consists of several nested models named DMI-HIRLAM 'G', 'N', 'E' and 'D', where the high resolution model 'D', covering an area around Denmark, has 5 km horizontal grid and uses boundary values from the large scale model 'E'. Last year, DMI started to run an experimental version of DMI-HIRLAM over the Zealand Island, including the Copenhagen area, with the horizontal resolution of 1.4 km. The vertical resolution of an experimental version of DMI-HIRLAM was increased up to 52 vertical levels. Examples of forecasted wind fields at 10 m height and of 2 m air temperature for the Copenhagen metropolitan area are presented in Figure 1.

The DWD's Local Model LM is currently operated as a nest within the Global Model (GME). It has 7 km resolution for Central and Western Europe and will be made operational for that area with 2.8 km resolution in 2003. In the framework of the European non-hydrostatic modelling consortium COSMO, triple nesting of LM will be realised to produce weather forecasts for the Olympic games in Athens (2006) on a 1.1 km grid and for possible other purposes (airports, cities).

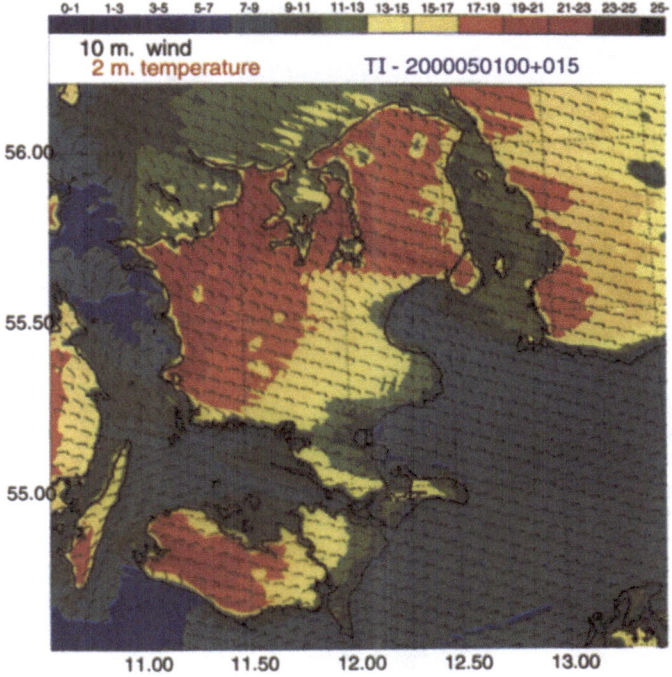

Figure 1. One example of forecasted wind fields at 10 m height and of 2 m air temperature for the Copenhagen metropolitan area by the experimental version of DMI-HIRLAM with the horizontal resolution of 1.4 km.

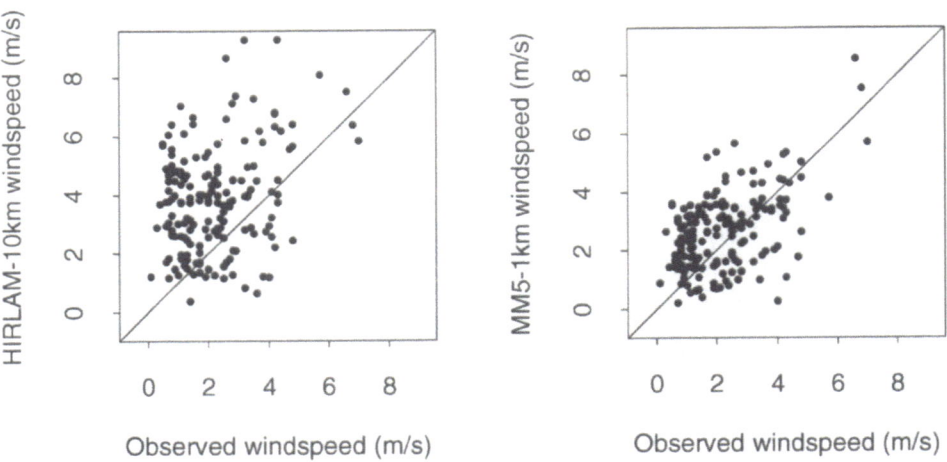

Figure 2. Observed and modeled wind speeds at Valle Hovin, a meteorological station in Oslo, for 22 days during the winter 1999/2000.

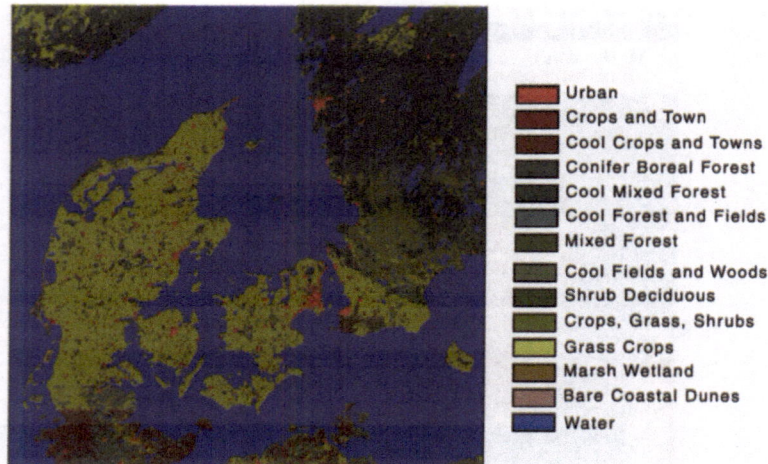

Figure 3. Land-use classification over Denmark and south Sweden from GLCC- and KMS-data for the DMI-HIRLAM-D model with 1 km resolution (Sattler, 1999).

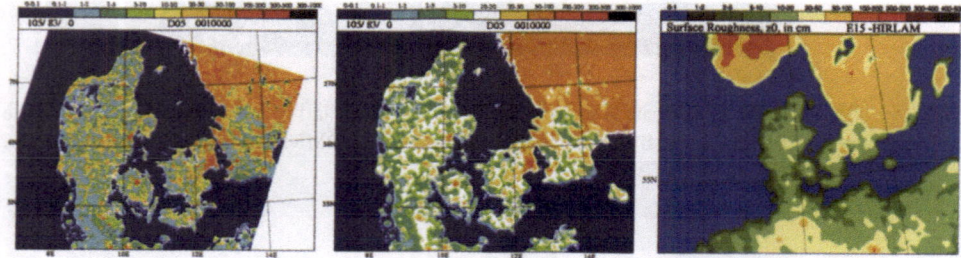

Figure 4. The simulated roughness length (z_0) for Denmark from the high resolution physiographic data for DMI-HIRLAM: (left) experimental version in 1.4 km resolution, (centre) D-version in 5 km resolution, and (right) E-version in 15 km resolution.

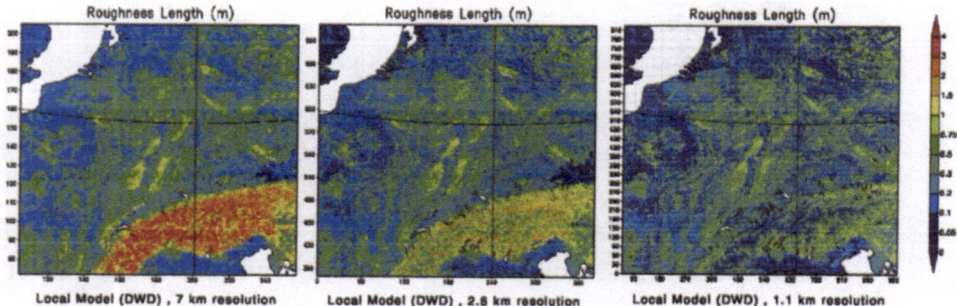

Figure 5. DWD Local Model simulation of the roughness length in 7, 2.8 and 1.1 km resolution (all based on a 1 km × 1 km database).

At DNMI the non-hydrostatic MM5 model is nested with HIRLAM. The HIR-LAM model is operated on a 10 km horizontal resolution for north west Europe. A domain with 3 km resolution has been set up for the Oslo region in which MM5 is one-way nested with HIRLAM. A two-way nesting takes place between the 3 km domain and a domain with 1 km resolution covering the city of Oslo. The output data from MM5 are employed to an Air Quality Model developed by the Norwegian Institute for Air Research. The model was run daily for the period 1 November 2000 until 1 May 2001, which is the season of the largest air quality problems in Oslo. For the winter 1999/2000 the model was run for 22 cases (see Clean City Air, 2000; Berge *et al.*, 2002). An evaluation of the meteorological predictions based on this dataset has been performed. In Figure 2 the wind speed prediction at 25 m from HIRLAM with 10 km resolution and MM5 with 1 km resolution are compared with measurements from a wind mast. We clearly see that the wind speed predictions are improved employing MM5-1 km instead of HIRLAM-10 km. A clear improvement is also found for the predictions of wind direction (not shown). Since the complex topography in the Oslo region has a great impact on the local winds, we would ascribe a large part of this improvement to the increased horizontal resolution. However, some of the improvements may also be due to differences in model formulations. The latter has not been investigated and would require a set-up of the HIRLAM model on a 1 km horizontal grid for the Oslo region. However, for the temperature predictions (not shown), the strength of the near surface inversion is somewhat overestimated by MM5 (and also HIR-LAM). Studies have been undertaken to improve this, and a first order PBL-closure scheme has been modified in order to give a better description of the near surface temperature profiles during stable conditions (Sorteberg, 2001).

2.2. EXTERNAL PARAMETERS

The results of existing NWP models for urban areas will only be reliable when refined databases for external parameters with 1 km resolution or finer are also provided and used.

Utilising land-use databases down to 1 km resolution or finer, models can reach a detailed description of surface features, including the identification of urban areas. Some land-use data of modern NWP models already include urban classes as well. For example, the DMI-HIRLAM model uses a land-use classification with 1 km resolution from GLCC- and KMS-data (Figure 3). The database describes 21 types of landcover/sea including a separate class of urban areas. Each grid cell includes a percentage of different classes in the cell, thus urban areas are presented differently in a city centre and suburb territories. Correspondingly, the surface parameters in NWP models, like the albedo, surface fluxes and roughness length (z_0), are simulated from the high resolution physiographic data. Figure 4 shows the roughness length, simulated in different DMI-HIRLAM versions: for Denmark in

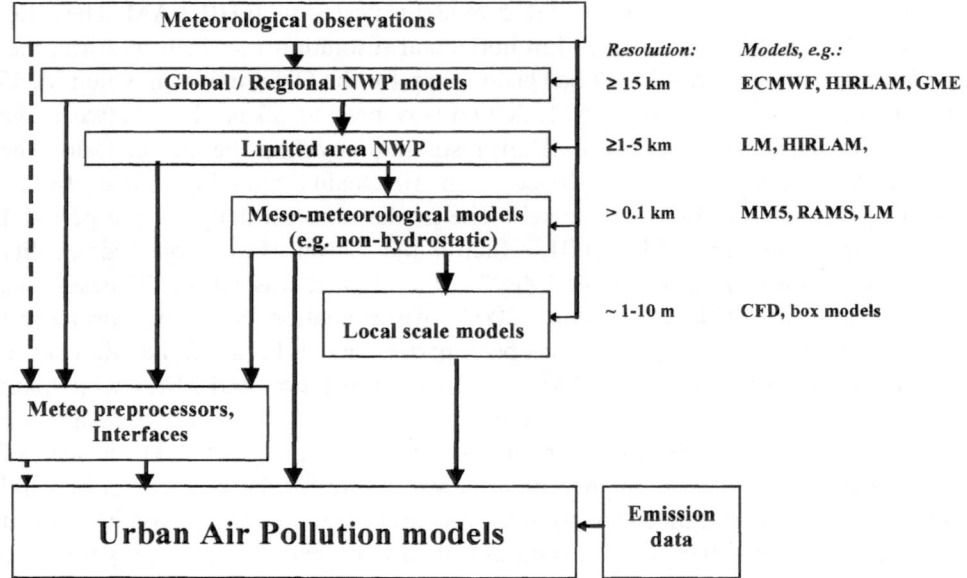

Figure 6. Current regulatory (dash line) and suggested (solid and dash lines) ways for systems of forecasting of urban meteorology for UAQIFS.

1.4 km (left) and 5 km (centre) resolutions, and for the European territory in 15 km resolution (right).

Figure 5 presents the roughness length for the central European territory, simulated in the DWD Local Model LM with 7, 2.8 and 1.1 km resolution (all based on a 1 km × 1 km database). z_0 not only shows much more detail (valleys etc.), but also decreases with increasing resolution, because the subscale orographic influence decreases with increased NWP model resolution, whereas the portion of the plant cover stays approximately the same. However, with the 1.1 km resolution map in Figure 5 the limitations of using the common 1 km × 1 km data base become apparent. z_0 appears too low especially in mountainous areas due to the absent roughness in the sub-grid scale of the database which has a resolution close to the accuracy of the calculated field.

2.3. METEOROLOGICAL PRE-PROCESSING

Existing operational UAQIFS and the corresponding UAP models often employ simple local measurements and meteorological pre-processors with a poor description of the temporal and spatial evolution of meteorological variables on the urban scale (Fisher *et al.*, 1999). Modern UAP models demand a lot more of additional meteorological input data (e.g. Herrmann *et al.*, 2000), such as humidity distribution, cloud characteristics, intensity and type of precipitation, radiation characteristics etc. Existing meteorological pre-processors, based on simple met-

LM (dx=7km)   15sep2000 12 UTC   + 3 h

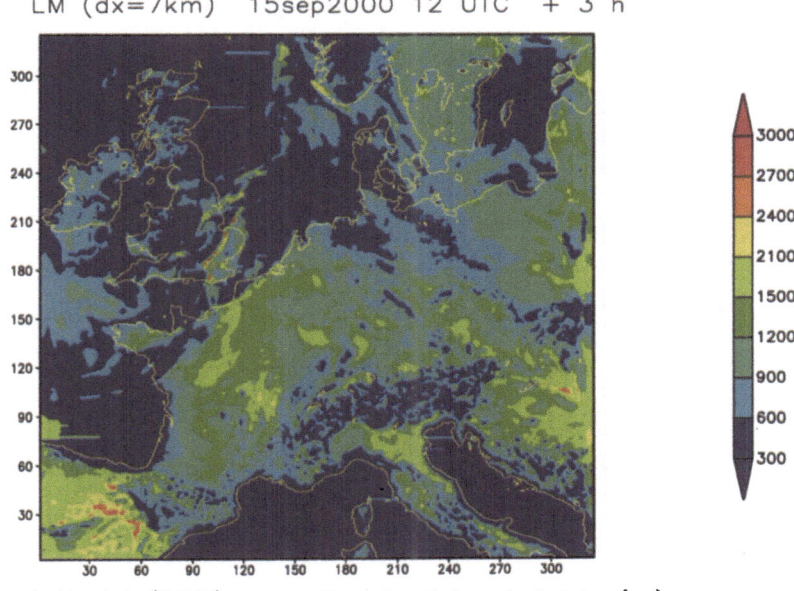

Local Model (DWD) operational mixing heights (m)

Figure 7. DWD Local Model LM daily operational mixing height (m) calculations for central/western Europe.

eorological measurements and parameterisations, cannot provide this information. Clearly, present UAP models could greatly benefit from utilising meteorological data from NWP models to give a physically consistent basis for urban air quality forecasts (see Figure 6).

Some European countries (e.g. Norway and France) perform research in this direction and have shown this to be a very promising approach (Clean City Air, 1999; Masson, 2000). The U.S. EPA still uses simple UAP regulatory methods, however U.S. scientists (Pielke and Uliasz, 1998; Byun and Ching, 1999) have a good understanding of the perspectives of this approach.

Increasingly, NWP modellers start considering the potential of forecasting meteorological fields for atmospheric pollution forecasting purposes. The COST Action 710 of 1994 to 1997 was devoted to the harmonisation of modelling and input data albeit on a coarser scale (Fisher *et al.*, 1998), whereas urban scale data and modelling are the concern of the current COST Action 715 (e.g. Piringer *et al.*, 2001).

The PBL height, for example, is one of the important characteristics for UAP models and is already included in several operational NWP models, e.g., in DMI-HIRLAM and DWD Local Model, as an output field (Figure 7).

Methods for estimating the **M**ixing **H**eight (MH) based on different methods, including a bulk Richardson number method, the Vogelezang and Holtslag method, and parcel methods are studied (Rasmussen *et al.*, 1997; Zilitinkevich and Bak-

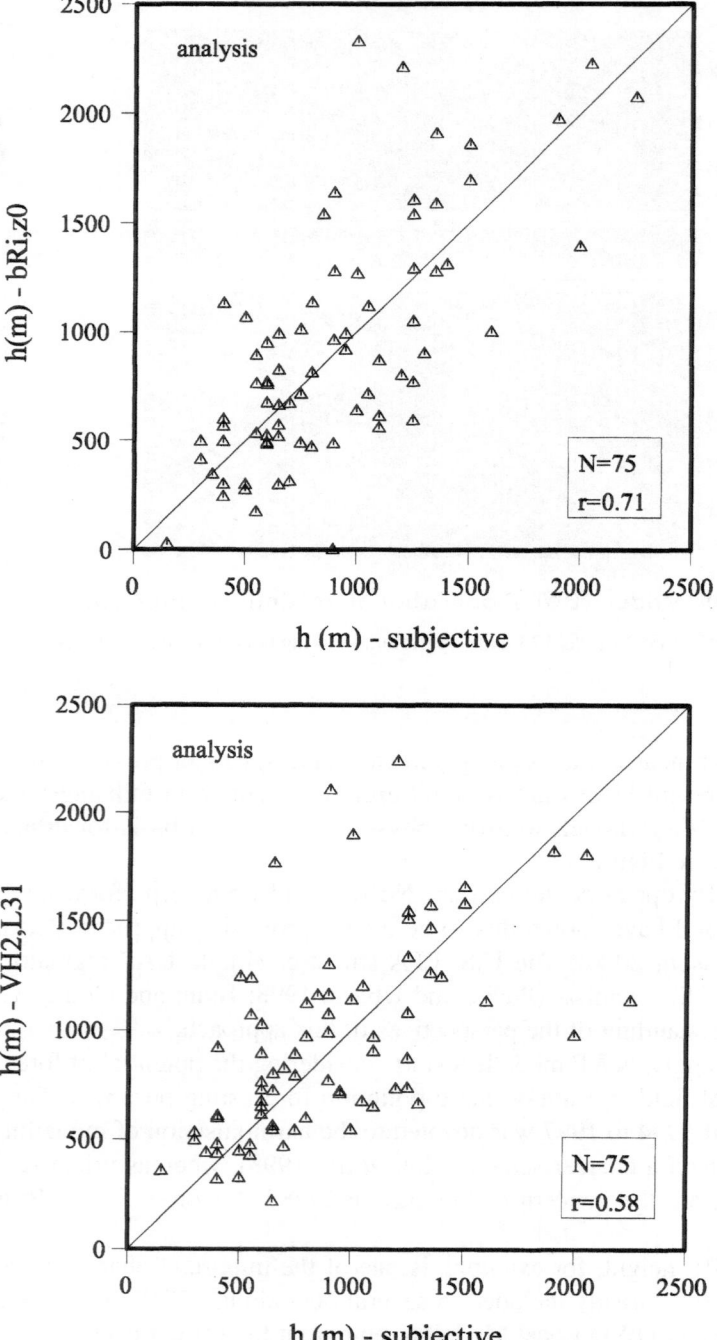

Figure 8. Scatter plots of subjective MH from Jægersborg, and the MH calculated from DMI-HIRLAM data by standard bulk Richardson method (top) and by Vogelezang-Holtslag method (bottom).

lanov, 2001). The methods are intercompared, and MHs based on data from DMI-HIRLAM are compared with the corresponding MHs valuated from radiosonde data. Figure 8 presents the results, based on data from the Jægersborg station in Copenhagen. The results based on data from 1994–1995 show that the resemblance between measured profiles and the DMI-HIRLAM profiles is fairly good in general. Also the estimates of the MH based on DMI-HIRLAM data is in general of nearly the same quality as estimations based on observed data. However, especially in unstable and convective conditions there can be large errors in the temperature profiles generally caused by a too weak development of the boundary layer in these situations. Under these conditions, also the MHs from the operational NWP model of the DWD for 1997 show the largest systematic underestimations (20 to 30%) compared to estimated radiosounding MHs and wind profiler measurements (Fay et al., 1997). MHs were also calculated for different seasons and weather situations and compared to radiosoundings and MHs (calculated from various other methods commended in COST 715) showing the generally high quality of the MHs derived from the bulk Richardson number in the DWD model.

2.4. SENSITIVITY OF UAP MODELS TO GENERATED PROGNOSTIC AND DIAGNOSTIC METEO-FIELDS

The meteorological fields needed by air quality models can be generated by either diagnostic or prognostic modelling approaches. Diagnostic models are simple and fast to use and can exploit local data, but their effectiveness is limited by spatial and temporal measurements density. Prognostic NWP models have the advantage to guarantee the physical consistency of meteorological variables and to be able to model different scale simultaneous phenomena (Seaman, 2000).

Urban air pollution episodes are influenced by both local and mesoscale meteorological conditions. The characteristics of mesoscale circulation are particularly relevant for photochemical pollution episodes, when the background concentrations advected over the city area can tune the chemical reactions of pollutants. Moreover, with ozone maximum concentrations generally observed just outside the urban area, a proper description of transport pathways is important to be able to identify the area affected by the maximum impact. A test case over the Lombardia Region (northern Italy) was set up to verify the sensitivity of the photochemical models to the technique employed to reconstruct the meteorological input field. A summer ozone episode (5–7 June, 1996) was simulated with the STEM-FCM model (Silibello et al., 2001) driven by the diagnostic model CALMET (Scire et al., 1995; Pirovano et al., 1999) and by the prognostic model RAMS (Pielke et al., 1992). The air quality model used the same emissions and boundary conditions for both simulations, the meteorological models had a similar computational domain (about 200 km × 200 km) and the same horizontal resolution (4 km). The diagnostic models used both ECMWF meteorological analyses and local surface and upper air observations to reconstruct 3D fields. The prognostic model was

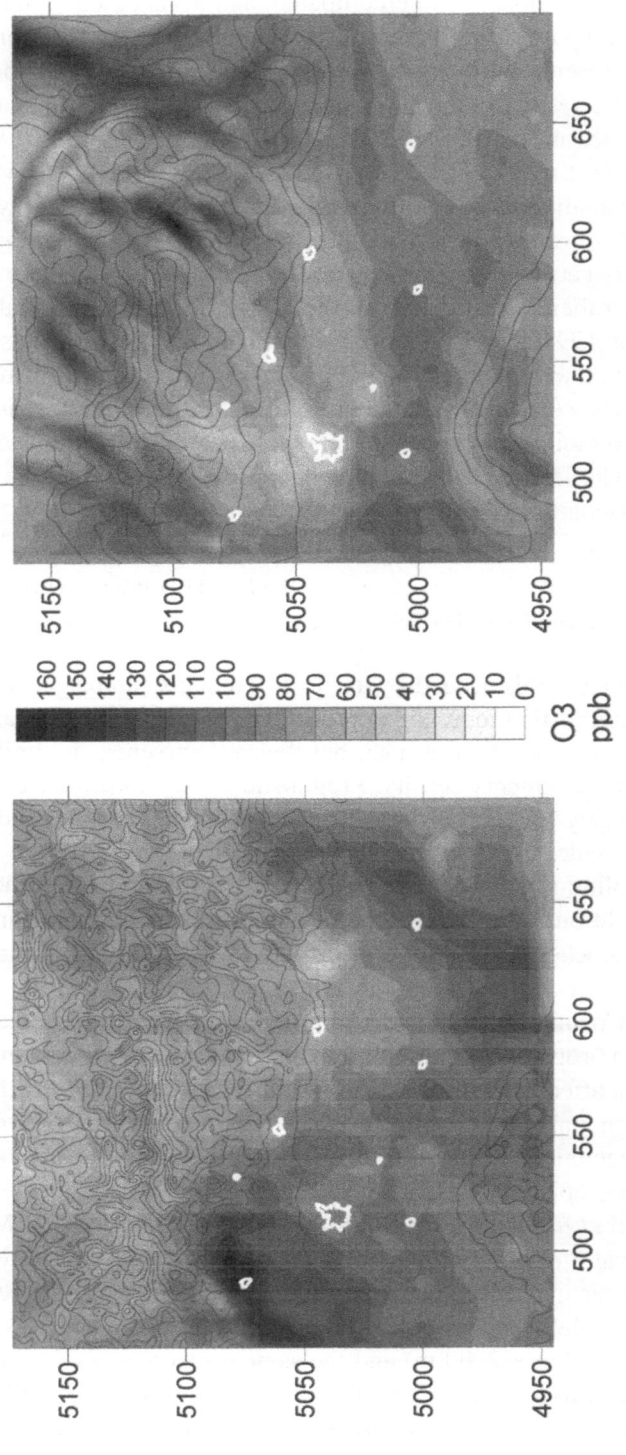

Figure 9. Surface O₃ concentrations over the Lombardia Region obtained from CALMET+STEM-FCM (left) and RAMS+STEM-FCM (right) on 7 June 1996 at 16:00.

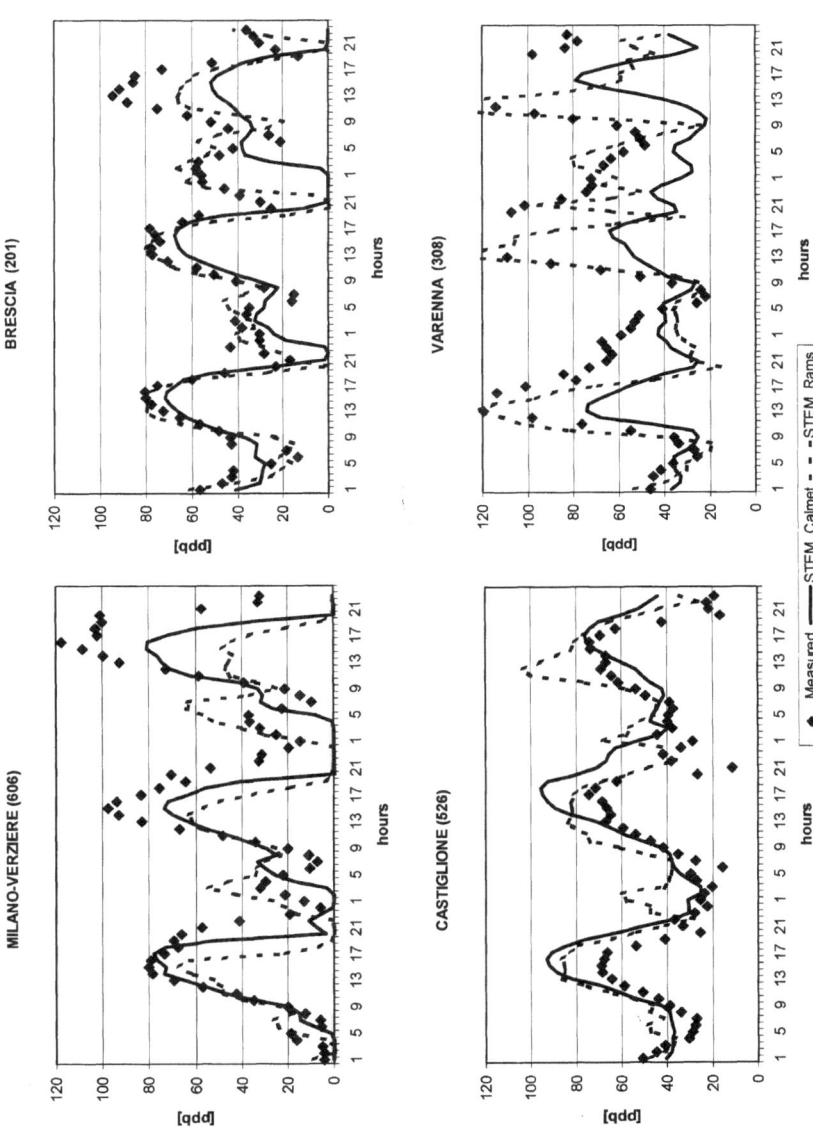

Figure 10. Comparison among ozone concentrations computed by CALMET+STEM-FCM (solid line), RAMS+STEM-FCM (dashed line) and observations (dotted line), for the period 5–7 June 1996. The plotted stations refer to Milan city (upper left), Brescia city (upper right), central Po Valley countryside (lower left) and Como Lake shore (lower right).

initialised by ECMWF analyses and surface observations and driven at the boundaries by ECMWF fields. A two-way nesting technique was employed to downscale
synoptic scale information to the target area. The air pollutant space distributions
obtained with the two meteorological drivers showed relevant differences. Figure 9
shows the ground level ozone concentrations on 7 June 1996, at 16:00. Comparing
model results with available measurements it was observed that the use of prognostic meteorological fields improved the air quality results at mountain sites
and in cities located at the foot of mountains. Figure 10 shows the comparison of
O_3 measured and predicted concentrations in different geographic locations. The
RAMS+STEM-FCM system clearly gives better results than CALMET+STEM-
FCM for the site of Varenna (located on the Como lake shore) and for the town of
Brescia. Even if the prognostic model results overestimate surface winds and do
not fit all the available observations, the overall air quality model results suggest
that RAMS is superior to CALMET in describing the structure and the evolution
of the circulation, that determine pollutant transport, dispersion and reaction over
the studied region.

Thus, the recent scientific developments in the field of urban atmospheric physics and the growing availability of high-resolution earth surface characteristics
data sets, e.g. from satellite remote sensing, promise further improvements of the
capability of NWP models to provide reliable urban meteorological data.

3. Shortcomings of NWP Models

In general, NWP models are not designed for atmospheric pollution modelling
and, therefore, have limitations and strengths unadapted to dispersion problems
(Pielke and Uliasz, 1998; Seaman, 2000). Therefore, new parameterisations need
to be introduced and/or their output needs to be processed in order to be used as
input to urban- and meso-scale air quality models. Thus, the development of met
pre-processors (from NWP) and interfaces to UAP models is a very important item.

Despite the exemplified recent developments in NWP modelling, urban and
non-urban areas are also still treated similarly in most NWP models, e.g., through
similar sub-surface, surface, and boundary layer formulations (Tana and Bornstein,
1999). No dedicated or additional methods or parameterisations exist to calculate the needed meteorological input fields for specifically urban dynamics and
energetics and for their impact on the atmospheric boundary layer.

Recent efforts suggest various modelling approaches to link the different physics of the canopy and urban boundary layers (Masson, 2000; Martilli et al., 2001).
Therefore, one important objective for further studies is the improvement of the
resolution and the parameterisations of urban effects in NWP models.

For example, let us give a summary of results from a sensitivity study of DMI-
HIRLAM profiles of temperature and wind for the radiosonde station Jægersborg
in Copenhagen (Rasmussen et al., 1999). The Jægersborg station is situated in

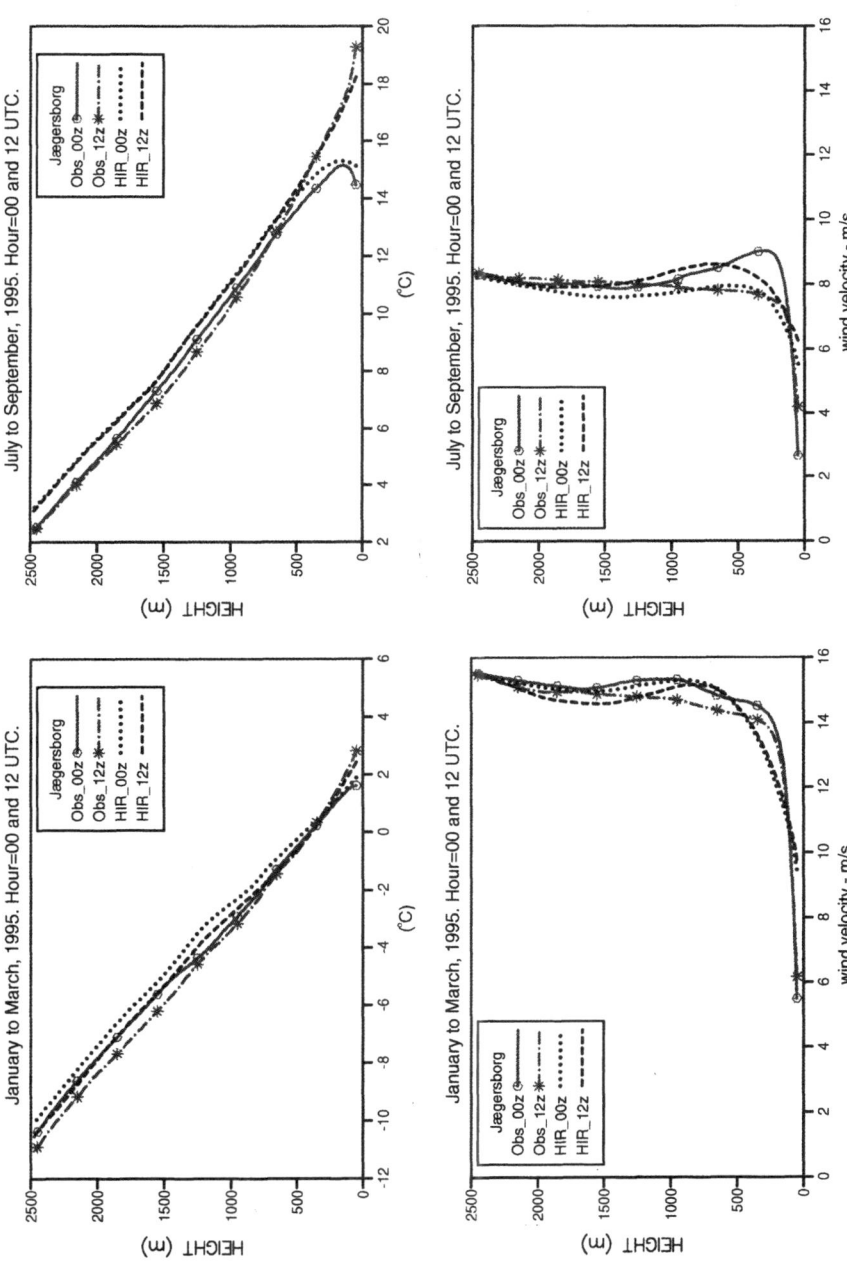

Figure 11. The mean profiles of the temperature (upper) and wind-velocity (down) at 00 and 12 UTC for the periods January to March (left) and July to September (right), 1995.

a northern part of the Copenhagen city and, according to its position and surrounding, it can be called a semi-urban station. Figure 11 shows two examples of vertical profiles of temperature and wind observed at the radiosonde station Jægersborg and calculated by DMI-HIRLAM (analysis). The mean profiles for the temperature at 00 and 12 UTC, for the periods January to March (left) and July to September (right), 1995, are shown in the upper graphs of Figure 11. In general, the correspondence between the NWP and measured profiles are quite suitable. However, near the ground the HIRLAM temperature was too low during the day, and too high during the night, and above 400 m the HIRLAM temperature was slightly higher, both during day and night. The mean profile of wind velocity at 00 UTC and 12 for the periods January to March (left) and July to September (right), 1995, are shown in the lowest graphs of Figure 11. Generally, the HIRLAM wind velocity was too high near ground, and the diurnal variation was not correctly described, possibly causing an erroneous gradient of the model wind velocity. This is especially important if one uses local gradients of, e.g., the Richardson number. It is necessary to stress that the absolute errors are relatively small, but they can play an important role in typical low-wind speed pollution episodes in the urban areas.

Besides the improved prediction of urban temperature and wind fields, the suitable parameterisation of urban *turbulence and turbulence-dependent parameters and fluxes* is essential and unsolved. Many high-level pollution episodes happen during periods with *stably-stratified* boundary layer (SBL). The Monin-Obukhov similarity theory, used in most of the models, does not work in case of the stable-static atmospheric boundary layer and needs to be improved (Zilitinkevich *et al.*, 1998). Existing atmospheric dispersion models for the stably-stratified boundary layer have a poor description of pollution processes due to, first of all, poor estimation of the *SBL height*, vertical profile parameterisation, non-adequate dispersion parameterisation and topographical effects (Fisher *et al.*, 1998). For example, for strongly stably-stratified ABL, the conventional methods for the SBL height estimation (like the bulk Ri-method) can give a unrealistic mixing height (Zilitinkevich and Baklanov, 2001). New approaches and parameterisations, based on an improved similarity theory, non-local characteristics of turbulence, and specifics of urban areas should be realised for NWP/UAP models.

Adapted external parameters for urban modelling are also important. A highly refined database is required, the effects of insufficient resolution on roughness length having been pointed out above. Urban roughness parameterisation as a critical point should be improved e.g. by implementing a geometrical model for computing the roughness parameters of urban areas (Grimmond and Oke, 1999; Guilloteau and Mestayer, 1999). In this context, a new and very important problem is the provision and assimilation of surface characteristics into the urban scale NWP – based on high resolution (up to 10 m) satellite data, e.g. remotely sensed heat fluxes of urban areas (Parlow, 1999). Algorithms for assimilation of surface temperature, albedo and snow cover have been developed and preliminarily tested (Clean City,

1999; De Ridder, 2000). This information is certainly crucial for the modelling quality.

The improved prediction of *urban surface processes* is also necessary, especially for water evaporation, and the implementation and verification of the full force-restore soil model for urban areas. Urban effects on *precipitation patterns* can be considerable, but nowadays these studies are based solely on investigator intuition rather than on a description of the contribution of the physical processes in numerical models (Lin and Bornstein, 1999). Thus, improvement of the urban effect parameterisations in NWP models will also increase understanding of this important problem.

An important task in the evaluation of UAP-model input and results is the investigation of *spatial effects and scale interaction* between non-obstacle resolving NWP models and obstacle resolving micro-scale UAPs (e.g. Baklanov, 2000). The new European regulations (EC/99/30) require prediction of local peak values (very high percentiles) at specific points (hot spots), which can only be determined by the obstacle resolving UAP-models. Therefore, the delivery of input parameters for those obstacle resolving UAPs, as well as the sensitivity analysis of improved parameterisations and physiographic fields on meteorological input fields for UAP models should be a focus point of further studies.

4. Suggested Improvement Strategy

In order to resolve the above discussed issues, new innovative studies are needed in the UAQIFSs, that are specifically tailored to the following objectives (Figure 12):

- An improved urban meteorology and air pollution modelling system suitable to be applied to any European urban area on a basis of available operational weather forecast.
- A model interface capable to connect mesoscale meteorological model results to updated UAP and atmospheric chemistry models.
- Improvement of the boundary layer parameterisation, e.g. surface/turbulent sensible and latent heat fluxes, revised roughness and land use parameters and models.
- Assimilation of surface characteristics based on satellite data and additional urban meteorological measurements for urban scale NWP models; development of assimilation algorithms for surface temperature, albedo and snow for urban areas.
- Evaluation and sensitivity studies of these improvements on the meteorological input fields for UAP models and the resulting air pollution simulations.

For many practical and historical reasons NWP and UAQ models were developed separately and are used by different work groups. These models often implement different coordinate projections, grid systems and advection schemes. This usage

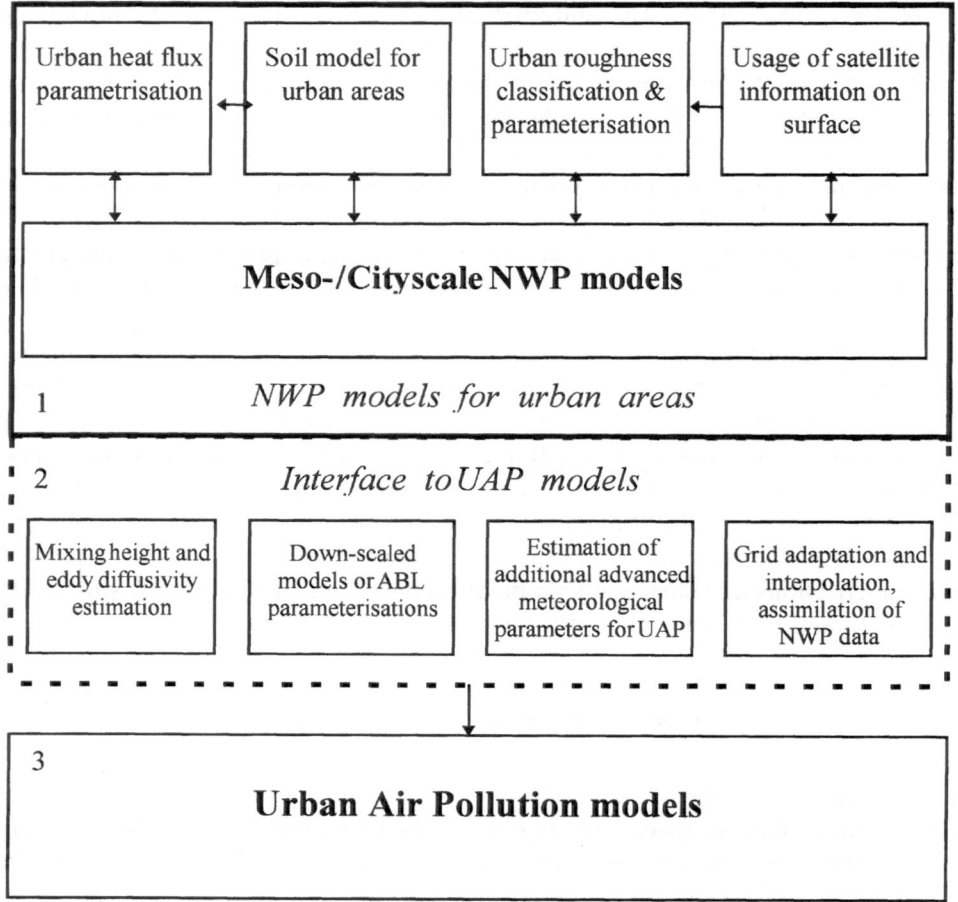

Figure 12. Scheme of the suggested improvements of meteorological forecasts (NWP) in urban areas and interfaces to UAP models for the Urban Air Quality Information Forecasting Systems.

is not justified theoretically and its effect on pollutant dispersion modelling results should be investigated, e.g. comparing concentration fields with those obtained applying consistent numerical methods to solve horizontal and vertical transport equations in both NWP and UAQ models.

As a practical step towards realising an improvement strategy for NWP/UAP modelling, eighteen research groups, including meteorological institutes, UAP modelling organisations and end-users responsible for urban air quality, from ten European countries develop a proposal '*Integrated Systems for Forecasting Urban MEteorology for Urban Air Quality Information Systems*' to the Fifth Framework Programme of the European Commission. The results of the suggested studies should improve forecasting capabilities of meteorological parameters for urban areas, and could be used by NWP centres and in met pre-processors.

Acknowledgements

This work was done partly during our work within the COST-715 action and we are grateful to many COST-715 colleagues from several EU countries for valuable discussions and comments. AB and AR thank Jess U. Jorgensen, Kai Sattler and Bjarne Amstrup for providing DMI-HIRLAM simulations.

References

Baklanov, A.: 2000, 'Application of CFD methods for modelling in air pollution problems: Possibilities and gaps', *J. Environ. Monitor. Assess.* **65**, 181–189.

Berge, E., Walker, S.-E., Sorteberg, A., Lenkopane, M., Eastwood, S., Jablonska, J. and Ødegaard, M.: 2002, 'A real-time operational forecast model for meteorology and air quality during peak air pollution episodes in Oslo, Norway', *Water, Air and Soil Pollut.* this issue.

Byun, D. W. and Ching, J. K. S.: 1999, *Science Algorithms of the EPA Models-3 Community Multiscale Air Quality (CMAQ) Modelling System*, EPA/600/R-99/030, Washington, DC.

Clean City Air Status Report: 1999, 'Development and Testing of a Pilot Model for Air Quality in Oslo and Bergen', Results from Oslo. DNMI, Norway.

De Ridder, K.: 2000, 'Modelling the Impact of Urban/Regional Scale Land-use Patterns on Atmospheric Flow and Pollutant Dispersion', in *SIMPAQ-2000, VIT 2000/TAP/55*, Antwerp, pp. 32–37.

Fay, B., Schrodin, R., Jacobsen, I. and Engelbart, D.: 1997, 'Validation of Mixing Heights Derived from the Operational NWP Models at the German Weather Service', in *EURASAP Workshop Proc. on the Determin. of the Mixing Height – Current Progress and Problems*, Risø-R-997(EN), pp. 55–58.

Fenger, J., Hertel, O. and Palmgren, F.: 1998, *Urban Air Pollution – European Aspects*, Kluwer Academic Publishers, The Netherlands.

Fisher, B. E. A., Erbrink, J. J, Finardi, S., Jeannet, P., Joffre, S., Morselli, M. G., Pechinger, U., Seibert, P. and Thomson, D. J.: 1998, 'Harmonisation of the Pre-processing of Meteorological Data for Atmospheric Dispersion Models. COST Action 710', *Fin. Rep., EUR 18195*.

EC/92/72: *Directive 92/72/EC of 21 Sept on Air Pollution by Ozone*, J. Europ. Comm. L297/1.

EC/96/62: *Directive 96/62/EC of 27 Sept on Ambient Air Quality Assess. a. Managem.*, J. EC L296/55.

EC/99/30: *Directive 99/30/EC of 22 April Relating to Limit Values for Sulphur Dioxide, Nitrogen Dioxide and Oxides of Nitrogen, Particulate Matter and Lead in Ambient Air*, J. EC L163/41.

Grimmond, C. S. B. and Oke, T. R.: 1999, 'Aerodynamic properties of urban areas derived from analysis of surface form', *J. Appl. Met.* **38**, 1262–1292.

Guilloteau, E. and Mestayer, P. G.: 2000, 'Numerical simulations of the urban roughness sublayer: A first attempt', *J. Environ. Monitor. Assess.* **65**(1/2), 211–219.

Hanna, S., Yang, R. and Yin, X.: 1998, 'Evaluations of Numerical Weather Prediction (NWP) models from the point of view of inputs required by atmospheric dispersion models', *Int. J. of Environ. and Pollut.*

Herrmann, H., Ervens, B., Jakobi, H.-W., Wolke, R., Nowacki, P. and Zellner, R.: 2000, 'CAPRAM.3: A chemical aqueous phase radical mechanism for tropospheric chemistry', *J. Atmosph. Chem.* **36**(3), 231–284.

Lin, Q. and Bornstein, R.: 1999, 'Summary of Urban Effects on Precipitation', in *Proc. of Int. Conf. on Biomet. a. Urban Climat. at the Turn of the Millennium*, Sydney, 8–12 November 1999, ICUC16.1.

LM: 1999, *Quarterly Report of the Operational NWP-models of the Deutscher Wetterdienst*, No. 20, 1–63, Department of Research and Development, DWD.

Martilli, A., Clappier, A. and Rotach, M. W.: 2001, 'An urban surface exchange parameterisation for mesoscale models', submitted to *Bound.-Layer Meteor.*

Masson, V. : 2000, 'A physically-based scheme for the urban energy budget in the atmospheric models', *Bound.-Layer Meteor.* **94**, 357–397.

Parlow, E.: 1999, 'Remotely Sensed Heat Fluxes of Urban Areas', *Int. Conf. on Biomet. and Urban Climatol. at the Turn of the Millennium*, Sydney, Australia, 8–12 November 1999, ICUC13.3.

Pielke, R. A. and Uliasz, M.: 1998, 'Use of meteorological models as input to regional and mesoscale air quality models – Limitations and strengths', *Atmosph. Environ.* **32**(8), 1455–1466.

Pielke, R. A., Cotton, W. R. *et al.*: 1992, 'A comprehensive meteorological modeling system – RAMS', *Meteorol. Atmos. Physics* **49**, 69–91.

Piringer, M., Baklanov, A., De Ridder, K., Ferreira, J., Joffre, S., Karppinen, K., Mestayer, P., Middleton, D., Tombrou, M. and Vogt, R.: 2001, 'The surface energy budget and the mixing height in urban areas: Status report of Working Group 2 of Cost-Action 715', in *Proc. of the Third Urban Air Quality Conference*, Loutraki, Greece. CD-ROM, paper Pl 1.3. 8 pp.

Pirovano, G., Brusasca, G., Calori, G., Desiato, F., Finardi, S., Lena, F., Longhetto, A. and Morselli, M. G.: 1999, 'Meteorological input for photochemical modelling in the Milan Region', in P. M. Borrel and P. Borrel (eds), *Proc. EUROTRAC Symp. '98*, WIT Press, Southampton, pp. 767–771.

Rasmussen, A., Sørensen, J. H. and Nielsen, N. W.: 1997, 'Validation of Mixing Height Determined from Vertical Profiles of Wind and Temperature from the DMI-HIRLAM NWP Model in Comparison with Radiosoundings' in *EURASAP Workshop Proc. on the Determination of the Mixing Height – Current Progress and Problems*, Risø-R-997(EN), pp. 101–105.

Rasmussen, A., Sørensen, J. H., Nielsen, N. W. and Amstrup, B.: 1999, *Uncertainty of Meteorological Parameters from DMI-HIRLAM*, RODOS(WG2)TN(99)12.

Saito, K., Doms, G., Schaettler, U. and Steppeler, J.: 1998, '3-D Mountain Waves by the Lokal-Modell of DWD and the MRI Mesoscale Nonhydrostatic Model', *Met. a. Geophys.* **49**(1), 7–19.

Sattler, K.: 1999, 'New high resolution physiographic data and climate generation in the HIRLAM forecasting system at DMI, an overview', *HIRLAM Newsletter* **33**, 96–100.

Scire, J. S., Insley, E. M., Yamartino, R. J. and Fernau, M. E.: 1995, *A User's Guide for the CALMET Meteorological Model*.

Seaman, N. L.: 2000, 'Meteorol. Model. for Air-quality Assessments', *Atm. Env.* **34**, 2231–2259.

Silibello, C., Calori, G., Pirovano, G. and Carmichael, G. R.: 2001, 'Development of STEM-FCM Modeling System: Chemical Mechanisms Sensitivity Evaluated on a Photochemical Episode', in *Proc. of 2nd Int. Conf. on Air Pollution Modell. a. Simul.*, Champs-sur-Marne (F), 9–12 April 2001.

Sistla, G., Zhou, N., Hao, W., Ku, J. Y., Rao, S. T., Bornstein, R. and Freedman, F.: 1996, 'Effects of uncertainties in meteorological inputs on urban airshed model predictions and ozone control strategies', *Atmosph. Environ.* **30**, 2011–2025.

Sorteberg, A.: 2001, 'The Sensitivity of Inversion Strength to the Formulation of the Non-dimensional Momentum and Heat Profiles', *Research Rep. No. 117*, Norwegian Met. Inst., Oslo.

Tana, H. and Bornstein, R.: 1999, 'Urbanisation of Meteorological Models and Implications on Simulated Heat Islands and Air Quality', in *Proc. of Int. Conf. on Biometeorology and Urban Climatology at the Turn of the Millennium*, Sydney, Australia, 8–12 November 1999, ICUC1.1.

Zilitinkevich, S. and Baklanov, A.: 2001, 'Calculation of the height of stable boundary layers in operational models', *Bound.-Layer Meteorol.* (accepted).

Zilitinkevich, S., Johansson, P.-E., Mironov, D. V. and Baklanov, A.: 1998, 'A similarity-theory model for wind profile and resistance law in stably stratified planetary boundary layers', *J. Wind Engineer. Industr. Aerodyn.* **74–76**, 209–218.

TRIP-BASED EXPLANATORY VARIABLES FOR ESTIMATING VEHICLE FUEL CONSUMPTION AND EMISSION RATES

YONGLIAN DING[1] and HESHAM RAKHA[2]*

[1] *Ricondon & Associates, Inc., 8610 N. New Braunfels, San Antonio, TX 78217, U.S.A.;* [2] *Virginia Tech Charles Via Department of Civil and Environmental Engineering, Virginia Tech Transportation Institute, Blacksburg, VA 24061, U.S.A.*
(* *author for correspondence, e-mail: hrakha@vt.edu, fax: (540) 231-1555*)

Abstract. The current state-of-practice in the US for estimating vehicle emissions is based on a single traffic-related explanatory variable, namely average speed. Research, however, has demonstrated that the use of average speed as a single traffic-related variable is insufficient for the estimation of vehicle emissions. For example, although the Environmental Protection Agency (EPA) MOBILE5 model would indicate that a slowing of traffic typically increases emissions, empirical research indicates the opposite in many cases. The objective of this paper is to identify critical aggregate trip variables as potential explanatory variables for the estimation of a vehicle's fuel consumption and emissions. Subsequently, statistical models for estimating fuel consumption and emissions of hydrocarbon (HC), carbon monoxide (CO), and oxides of nitrogen (NO_x) are developed using these critical variables that include the average speed, speed variability, the level of deceleration, and the level of acceleration. The proposed models are demonstrated to be consistent with microscopic energy and emission model estimates that are based on the vehicle's instantaneous speed and acceleration levels (coefficient of determination ranges from 0.88 to 0.96).

Keywords: environmental modeling, transportation modeling, vehicle emissions

1. Introduction

The current state-of-practice in the US for estimating vehicle emissions is based on the vehicle's average speed, however, research has demonstrated that the use of average speed alone is insufficient in estimating vehicle emissions. The microscopic models that were developed by Ahn *et al.* (1999) and Rakha *et al.* (2000a) demonstrate that while both the instantaneous speed and acceleration significantly impact vehicle emissions, vehicle acceleration becomes a more dominant factor on HC and CO emissions, especially at high speeds (Ahn *et al.*, 1999 and Rakha *et al.*, 2000b). However, NO_x emissions are less sensitive to high engine loads. This paper identifies aggregate trip variables for the estimation of a vehicle's fuel consumption and emission rates. Subsequently, statistical models for estimating fuel consumption and emissions of hydrocarbon (HC), carbon monoxide (CO), and oxides of nitrogen (NO_x) are developed using these critical trip variables that include the average speed, speed variability, acceleration noise, kinetic energy, and power exerted. These variables are described in detail later in the paper. While

Water, Air, and Soil Pollution: Focus **2**: 61–77, 2002.
© 2002 *Kluwer Academic Publishers.*

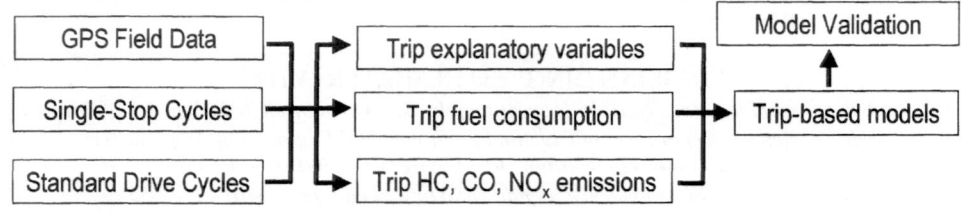

Figure 1. Overview of research approach.

some of these trip-based variables require second-by-second speed profiles typical explanatory variables could be estimated depending on the facility type, type of traffic control, and level of congestion.

2. Methodology

In conducting this analysis, initially typical driver acceleration and deceleration levels were characterized using field data that were collected along a signalized arterial in Phoenix, Arizona (Rakha *et al.*, 2000a). Using these typical acceleration and deceleration levels, a number of artificial trips were constructed in order to systematically isolate the impact of various driver and traffic-related factors on vehicle fuel consumption and emission rates, as summarized in Figure 1. These artificial data were augmented with field data that were collected using vehicles equipped with a Global Positioning System (GPS). In addition, a number of standard drive cycles (FTP City cycle and the New York City cycle) were included in the analysis. Instantaneous fuel consumption and emission models (Ahn *et al.*, 1999 and Rakha *et al.*, 2000b) were applied to the three sets of data to compute vehicle fuel consumption and emission rates per unit distance. In addition, a number of variables were identified as potential explanatory variables for the development of regression models. The models were then validated against laboratory data that were gathered by the Environmental Protection Agency (EPA) on a chassis dynamometer.

In summary three data sets were utilized to conduct the analysis. These data sets included a set of 301 trips that was collected using GPS technology along a signalized arterial in Phoenix (field data); a set of 1080 single-stop synthetic trips that was derived from a base single-stop cycle using typical acceleration and deceleration levels; and a set of 120 modified drive cycles that were derived from the Federal Test Procedure (FTP) city cycle and the New York city cycle.

2.1. CONSTRUCTION OF DATA SETS

2.1.1. *Field Data*
The field data were collected along a signalized arterial corridor in Phoenix using GPS-equipped vehicles (Rakha *et al.*, 2000a). These vehicles were driven along the study corridor for three days (Tuesday through Thursday) using several test

subjects before and after changes were made to the signal timings during the AM peak (7am–8am), the off-peak (11am–1pm), and the PM peak (4–6pm). A total of 301 trips were recorded over the 9.6 km study section. The GPS unit measured the vehicle's location, its heading, and its speed every second to an accuracy of 1 m s^{-1}. Subsequently, acceleration was computed based on the second-by-second speed measurements. Because of occasional losses in the GPS signal significant errors in speed measurements were observed, the acceleration estimates resulted in occasional unrealistic observations, which exceeded the maximum feasible acceleration that a vehicle can attain at a specific speed. Consequently, a robust form of acceleration smoothing was applied to the acceleration estimates, which in turn resulted in more realistic speed estimates (Rakha *et al.*, 2001a).

The maximum attainable vehicle acceleration can be derived from the residual tractive force assuming a constant vehicle power (force = power/speed). This relationship results in a non-linear function between the maximum acceleration and the vehicle speed with the maximum acceleration decreasing as a function of the vehicle speed. It should be noted that a linear relationship does provide a reasonable approximation to the maximum acceleration/speed relationship and is consistent with the range of data coverage within the Oak Ridge National Lab (ORNL) data (Rakha *et al.*, 2000b) (West *et al.*, 1997).

The distribution of speed and acceleration estimates for the entire 301 trips resulted in a mean acceleration rate of 19% of the maximum feasible acceleration and a mean deceleration rate of –0.52 m s^{-2}. Consequently, in constructing the single-stop drive cycles a default acceleration rate of 20% the maximum feasible acceleration rate and a deceleration rate of –0.5 m s^{-2} were utilized.

2.1.2. *Synthetic Single-Stop Data*

Single-stop trips were derived from base single-stop cycles by varying the vehicle's cruise speed, level of deceleration, and level of acceleration. In constructing the base single-stop cycles, a complete stop was introduced to the 4.5 km trip using typical acceleration and deceleration levels that were characterized from the field data above to ensure consistency between these hypothetical drive cycles and field observations.

Four factors were varied for the single-stop trip. The first of these factors was the cruise speed, which was varied from 20 to 120 km h^{-1} at increments of 20 km h^{-1}. The second factor was the level of deceleration, which was varied from –1.50 m s^{-2} to –0.25 m s^{-2} at increments of 0.25 m s^{-2}. These deceleration levels are consistent with the field observed data that were described earlier. The third factor was the level of acceleration, which was varied from 20 to 100% the maximum acceleration level at increments of 20%. The final factor that was varied was the level of speed variability using a scale factor (k_1), as summarized in Equations 1 and 2. This scale factor varied the profile around the mean speed to generate partial stop cycles.

The combination of 6 speed levels, 6 deceleration levels, 5 acceleration levels, and 6 speed variability levels resulted in a total of 1080 trip combinations ($6 \times 6 \times 5 \times 6 = 1080$).

$$\bar{u} = \frac{\sum_{i=1}^{n} u(t_i)}{n} \tag{1}$$

$$\tilde{u}(t_i = k_1(u(t_1) - \bar{u}) + \bar{u} \tag{2}$$

Where:

t_i	=	The time instant for observation i, $t_0 + i\Delta t$
$u(t)_i)$	=	The instantaneous speed at instant t_i,
$\bar{u}$	=	The average speed for original speed profile,
$\tilde{u}(t_i$	=	The modified instantaneous speed at instant t_i,
n	=	The number of observations along the trip, and
k_1	=	Speed variability factor.

2.1.3. *Modified Drive Cycles*

Two standard drive cycles, namely, the FTP City cycle and the New York City cycle, were modified to construct a number of additional hypothetical drive cycles. The first modification was to apply the speed variability factor (k_1) that was presented earlier in Equations 1 and 2 to each drive cycle. By altering k_1, different speed profiles were generated for each cycle while maintaining the same average speed. The second modification involved an increase in the average speed by adding a constant (k_2) to instantaneous speeds, as expressed in Equation 3. The addition of a constant to each speed ensures that the acceleration levels remain constant, however, the speed acceleration combination is altered.

The third modification involved altering both the mean speed and speed variability. The modification was conducted by adding a constant (k_3) to speeds that exceeded the average speed. In total 120 cycles were generated from the standard FTP City cycle and the New York City cycle.

$$\tilde{u}(t_i) = u(t_i) + k_2 \forall i \tag{3}$$

2.2. ESTIMATION OF VEHICLE FUEL CONSUMPTION AND EMISSIONS

The Virginia Tech Microscopic energy and emission model (VT-Micro) version 1.0 was developed from experimentation with numerous polynomial combinations of speed and acceleration levels (Ahn *et al.*, 1999 and Rakha *et al.*, 2000b). Specifically, linear, quadratic, cubic, and quartic terms of speed and acceleration were tested using chassis dynamometer data collected at the Oak Ridge National Laboratory (ORNL). The final regression model included a combination of linear, quadratic, and cubic speed and acceleration terms because it provided the least number of terms with a relatively good fit to the original data (R^2 in excess of 0.72 for all measures of effectiveness). While a more detailed description of the derivation of the model is provided in the literature, it is sufficient to note at this point that the final structure of version 1.0 of the model, which is summarized in Equation 4, involved a logarithmic transformation of a single-regime third order polynomial. It should be noted that the VT-Micro model Version 2.0 was enhanced by developing a dual-regime relationship that provided a better fit to the ORNL data (R^2 in excess of 0.72 for all measures of effectiveness), as described in Ahn *et al.* (2002).

$$\ln(MOE_e) = \sum_{i=0}^{3}\sum_{j=0}^{3}(K_{i,j}^e \times u^i \times a^j) \qquad (4)$$

Where:

MOE_e = Instantaneous fuel consumption or emission rate (l s^{-1} or mg s^{-1}),

$K_{i,j}^e$ = Model regression coefficient for MOE 'e' at speed power 'i' and acceleration power 'j',

u = Instantaneous vehicle speed (km h^{-1}), and

a = Instantaneous vehicle acceleration (m s^{-2}).

2.3. IDENTIFICATION OF POTENTIAL EXPLANATORY VARIABLES

Potential explanatory variables were identified including average speed, speed variability, number of vehicle stops, acceleration noise, deceleration noise, total acceleration noise, kinetic energy consumed, and the power exerted by the vehicle.

Commonly used trip variables like average speed, number of vehicle stops, and speed variability are computed using Equations 1, 5, and 6, respectively. The computation of vehicle stops that is presented in Equation 5 was proposed by Rakha *et al.* (2001b) to capture both partial stops and multiple stops at over-saturated signalized intersections. This formulation computes the partial stop at instant t_i as the ratio of the speed reduction that occurs between instants t_i and t_{i-1} to the roadway free-speed (maximum speed).

Alternatively, the acceleration noise is a less common variable that was developed in the late 1950's in order to quantify the smoothness of flow in a traffic stream. Specifically, Drew (1968) mentions, 'the term *noise* is used to indicate the disturbance of the flow, comparable to the coined phrase *video noise*, which is used to describe the fluttering of the video signal on a television set'. Drew mentions that acceleration noise received considerable attention as a possible measurement of traffic flow quality for two basic reasons. First, it is dependent on the three basic elements of the traffic stream, namely, (1) the driver, (2) the road, and (3) the traffic condition. Second, it is in effect, a measurement of the smoothness of flow in a traffic stream. Specifically, the acceleration noise (standard deviation of accelerations) can be considered as the disturbance of the vehicle's speed from a uniform speed.

The acceleration noise that is present on a road in the absence of traffic is termed the *natural noise* of the driver on the road (Drew, 1968). Several factors affect acceleration noise, such as the roadway geometry, the type of control on the roadway, and the level of congestion on the roadway. Specifically, a field study in the mid 1950's indicated that the acceleration noise increased with an increase in congestion (Jones and Potts, 1955).

Jones and Potts (1955) developed a mathematical equation for approximating the acceleration noise. Specifically, using an acceleration profile Jones and Potts computed the average acceleration and the acceleration noise as the standard deviation of the acceleration. The details of the derivation are beyond the scope of this paper, however, it is worthwhile mentioning that the formulation only computes the acceleration noise when the vehicle is in motion (speed is greater than zero). A modified acceleration noise estimate is proposed for purposes of this analysis, as demonstrated in Equation 7. The first modification is that for long trips (*T* large) the average acceleration that is computed typically tends to zero. Consequently, in Equation 6 it is assumed that the average acceleration is zero. The second modification to the Jones and Potts formulation is that Equation 7 weights each acceleration observation by the vehicle speed because acceleration levels at higher speeds result in higher fuel consumption and emission estimates than equivalent acceleration levels at lower speeds. Equations 8 and 9 further separate the positive acceleration noise (accelerations) from the negative acceleration noise (decelerations) because fuel consumption and emission estimates are more sensitive to acceleration levels than they are to deceleration levels.

It should be noted that Drew (1968) demonstrated that the kinetic energy of a traffic stream can be computed using Equation 10 where β is a unitless constant. Furthermore, Drew demonstrated that there is an internal energy or lost energy associated with the traffic stream, which manifests itself in erratic motion and is nothing but the acceleration noise. Consequently, Capelle (1966) hypothesized that the internal energy or acceleration noise measured over a segment of roadway is equal to the total fuel consumed. The model was tested against freeway data and demonstrated a good fit, however that was not the case for arterial streets (Rowan, 1967). It is felt that the proposed modification to the computation of acceleration noise, together with the separation of positive and negative acceleration noise could provide better explanatory variables. Furthermore, it is proposed that the total kinetic energy also be computed for the entire trip for a single vehicle using Equation 10, which computes the kinetic energy for a single vehicle in the traffic stream.

A final explanatory variable that is considered is related to the power exerted along a trip. The power is equal to the force multiplied by the speed, which means that the instantaneous residual power is a function of the product of the instantaneous speed and acceleration levels [$u(t_i)a(t_i)$]. Consequently, the total power exerted is computed using Equation 11 where α is the mass of the vehicle that is a constant and $a(t_i)$ is greater than zero.

$$S = \sum_i \frac{u(t_i) - u(t_{i-1})}{u_f} \forall i \ni u(t_i) < u(t_{i-1}) \tag{5}$$

$$\sigma_u^2 = \frac{\sum_{i=1}^{n}(u(t_i) - \bar{u})^2}{n} \tag{6}$$

$$A = \sqrt{\frac{\sum_{i=1}^{n} a(t_i)^2 u(t_i)}{\sum_{i=1}^{n} u(t_i)}} \tag{7}$$

$$A^+ = \sqrt{\frac{\sum_{i=1}^{n} a(t_i)^2 u(t_i)}{\sum_{i=1}^{n} u(t_i)}} \forall i \ni a(t_i > 0) \tag{8}$$

$$A^- = \sqrt{\frac{\sum_{i=1}^{n} a(t_i)^2 u(t_i)}{\sum_{i=1}^{n} u(t_i)}} \forall i \ni a(t_i > 0) \tag{9}$$

$$E_k = \beta \sum_{i=1}^{n} u(t_i)^2 \tag{10}$$

$$P = \alpha \sum_{i=1}^{n} u(t_i) a(t_i) \tag{11}$$

Where:
$u(t_i)$ Instantaneous speed at instant 'i'

$\bar{u}$ Average speed for original speed profile

n Number of observations along a trip

u_f Facility free-speed (km h^{-1})

a_i Instantaneous acceleration (km h^{-1} s^{-1})

σ_u^2 Speed variability

S   Number of vehicle stops

A   Total acceleration noise

A$^+$ Acceleration noise

A$^-$ Deceleration noise

m   Vehicle mass (kg)

k   Traffic density (veh km^{-1})

E_k Kinetic Energy

P   Power

α, β Constants

3. Results and Discussion

Utilizing the estimates of the various independent variables for each trip together with the corresponding Measures of Effectiveness (MOEs), statistical regression models were developed for a typical average vehicle. Prior to developing the models, each dependent variable (fuel consumption, HC, CO, and NO$_x$ emissions) was plotted against each independent variable in order to identify potential independent variables and model structures. Initially, simple statistical models including single explanatory variables were considered. Next, multiple explanatory variables were systematically added to the regression models. The criteria for selecting the optimum model focused on the model with the highest adjusted coefficient of determination and the fewest number of independent variables. In addition, an attempt was made to ensure that multi-colinearity did not exist or at least minimum multi-colinearity existed among the independent variables, as measured by the Variance Inflation Factor (VIF). The general format of these regression models included a logarithmic transformed model, as demonstrated in Equation 12 and a basic model, as summarized in Equation 13. The logarithmic transformation ensured that the MOE estimates were non-negative.

$$\ln(MOE_e) = a_e + \sum_{i=1}^{n} b_e^i v_i \tag{12}$$

$$MOE_e = a_e + \sum_{i=1}^{n} b_e^i v_i \qquad (13)$$

Where:

MOE_e = Dependent variable for fuel consumption, HC, CO, and NO_x estimates

n = Number of independent variables included in model ($n \leq 10$)

a_e = Constant MOE 'e'

b_e^i = Coefficient of independent variable i for MOE e

v_i = Independent variable i

3.1. CONTRIBUTION OF VARIABLES TO MODEL ERROR

Each of the eight variables identified above was used as a single independent variable for each of the dependent variables (fuel consumption, HC, CO, and NO_x emissions). Since the relationship between each dependent variable (fuel, HC, CO, and NO_x) and the average speed was found to be non-linear, two additional potential explanatory variables were introduced. These variables include the inverse of the average speed and the average speed squared. To avoid any negative MOE estimates, a log-transformation was applied to each model. The resulting coefficient of determination (R^2) indicated the portion of the sum of squared error that was explained by the independent variable, as summarized in Table I. For example, speed variability was found to explain most of the sum of squared error associated with fuel consumption, HC, and NO_x emissions, while the power consumed explained most of the variability in CO emissions. Table I demonstrates that speed variability provided a good explanation of the sum of squared error for each of the four MOEs.

3.2. ONE AND TWO INDEPENDENT VARIABLE MODELS

The initial objective was to develop simple models of one or two variables for the estimation of vehicle fuel consumption and emissions. For example, Dion *et al.* (1999) developed mesoscopic fuel consumption and emission models as a function of average speed, number of vehicle stops, and stopped delay. In addition, several models were developed for estimating fuel consumption estimates using the reciprocal of average speed (Evans and Herman, 1978 and Lam 1985). Consequently, the initial regression models that were developed used the combination of average speed and vehicle stops as independent variables. The models were developed using the same data set with and without a log-transformation, as summarized in Table II. While the use of a log-transformation reduced the accuracy of the model, it ensured that only positive MOE estimates were computed.

The optimal model was selected using two criteria. The first criterion was to ensure that positive regression coefficients were utilized; otherwise a log-transformation was applied to the data in order to ensure positive MOE estimates. The second

Y. DING AND H. RAKHA

TABLE I

Fuel consumption and emission models for single independent variables

Model sequence	Model	R^2
$Ln(Fuel) = f(x)$		
1	$Ln(Fuel) = f(\sigma_u^2)$	0.6026
2	$Ln(Fuel) = f(A)$	0.5664
3	$Ln(Fuel) = f(E_k)$	0.5581
4	$Ln(Fuel) = f(S)$	0.3918
5	$Ln(Fuel) = f(A^-)$	0.3568
6	$Ln(Fuel) = f(\bar{u})$	0.3305
7	$Ln(Fuel) = f(\bar{u}^2)$	0.2821
8	$Ln(Fuel) = f(1/\bar{u})$	0.2708
9	$Ln(Fuel) = f(A^+)$	0.1145
10	$Ln(Fuel) = f(P)$	0.0025
$Ln(HC) = f(x)$		
1	$Ln(HC) = f(\sigma_u^2)$	0.5811
2	$Ln(HC) = f(A)$	0.4082
3	$Ln(HC) = f(P)$	0.3157
4	$Ln(HC) = f(A^-)$	0.2918
5	$Ln(HC) = f(S)$	0.2046
6	$Ln(HC) = f(A^+)$	0.0961
7	$Ln(HC) = f(\bar{u}^2)$	0.0359
8	$Ln(HC) = f(\bar{u})$	0.0142
9	$Ln(HC) = f(E_k)$	0.0133
10	$Ln(HC) = f(1/\bar{u})$	0.0001
$Ln(CO) = f(x)$		
1	$Ln(CO) = f(P)$	0.6259
2	$Ln(CO) = f(\bar{u}^2)$	0.4137
3	$Ln(CO) = f(\bar{u})$	0.3776
4	$Ln(CO) = f(\sigma_u^2)$	0.3338
5	$Ln(CO) = f(1/\bar{u})$	0.2307
6	$Ln(CO) = f(A)$	0.1625
7	$Ln(CO) = f(E_k)$	0.1379
8	$Ln(CO) = f(A^-)$	0.1194
9	$Ln(CO) = f(S)$	0.0364
10	$Ln(CO) = f(A^+)$	0.0186

TABLE I

continued

Model sequence	Model	R^2
$Ln(NO_x) = f(x)$		
1	$Ln(NO_x) = f(\sigma_u^2)$	0.7185
2	$Ln(NO_x) = f(A)$	0.4622
3	$Ln(NO_x) = f(P)$	0.3502
4	$Ln(NO_x) = f(A^-)$	0.2595
5	$Ln(NO_x) = f(S)$	0.1983
6	$Ln(NO_x) = f(1/\bar{u})$	0.1333
7	$Ln(NO_x) = f(\bar{u})$	0.1151
8	$Ln(NO_x) = f(\bar{u}^2)$	0.1039
9	$Ln(NO_x) = f(A^+)$	0.0195
10	$Ln(NO_x) = f(E_k)$	0.0001

criterion was to select the model that provided the highest coefficient of determination (R^2).

As demonstrated in Table II, the use of average speed and speed variability as independent variables provided a best fit to the microscopic fuel consumption and emission estimates. Specifically, the coefficient of determination ranged from 62% for HC emissions to 90% for NO_x emissions and fuel consumption.

3.3. SELECTION OF OPTIMUM STATISTICAL MODELS

Having developed initial simple regression models, the next step was to include additional variables to improve the model estimates. Specifically, all possible 8 independent variables in addition to the reciprocal of average speed and the square of average speed were considered, to give a total of 10 possible independent variables. The selection of the optimum model was achieved by utilizing the function of variable selection in SAS to each dependent variable (both without a log-transformation and with a log-transformation), and then several candidate models are selected based on the two criteria that were described earlier, namely the highest coefficient of determination with the lowest number of independent variables.

The best model was selected as being the model that produced the highest adjusted coefficient of determination (Adjusted R^2), included the fewest independent variables as possible, and had minimum multi-colinearity between independent variables as measured by the Variance Inflation Factor (VIF).

Table III lists the models that were selected as candidate models for further comparison for each dependent variable using log-transformed data. It should be noted that the first candidate model for each dependent variable included the full model

TABLE II

Fuel consumption and emission models for two independent variables

Model sequence	Model	R^2
$Ln(Fuel) = f(x)$		
1	$Ln(Fuel) = f(\bar{u})$	0.3305
2	$Ln(Fuel) = f(\bar{u}, S)$	0.5256
3	$Ln(Fuel) = f(\bar{u}, \sigma_u^2)$	0.8472
4	$Ln(Fuel) = f(1/\bar{u}, \sigma_u^2)$	0.9080
$HC = f(x)$ or $Ln(HC) = f(x)$		
1	$HC = f(\bar{u})$	0.0162
2	$Ln(HC) = f(\bar{u}, S)$	0.3011
3	$Ln(HC) = f(\bar{u}, \sigma_u^2)$	0.6214
$Ln(CO) = f(x)$		
1	$Ln(CO) = f(\bar{u})$	0.3776
2	$Ln(CO) = f(\bar{u}, S)$	0.5851
3	$Ln(CO) = f(\bar{u}, \sigma_u^2)$	0.7960
$Ln(NO_x) = f(x)$		
1	$Ln(NO_x) = f(\bar{u})$	0.1151
2	$Ln(NO_x) = f(\bar{u}, S)$	0.5963
3	$Ln(NO_x) = f(\bar{u}, \sigma_u^2)$	0.9052

with all 10 independent variables. The optimal model was selected progressively based on the selection criteria that were described earlier. These optimal models are indicated in the table and presented in Equations 14, 15, 16 and 17.

The proposed statistical models were found to compute fuel consumption and emission rates to within 88–90% of those computed using the microscopic models that were described earlier. A comparison of the proposed model estimates to the microscopic model estimates demonstrated a fairly good fit with most observations symmetric around the line of perfect correlation, as illustrated in Figure 2. It should be noted, however, that multi-colinearities of independent variables existed for the HC and CO models. If these variables were removed from the regression model, the R^2 would have been reduced significantly. Consequently, these variables were kept in the models.

$$\ln(\mathit{fuel}) = a_0 + \frac{a_1}{\bar{u}} + a_2\bar{u}^2 + a_3\sigma_u^2 + a_4 S + a_5 A + a_6 E_k \quad (R^2 = 0.9564) \quad (14)$$

TABLE III

Candidate models for estimating vehicle fuel consumption and emission rates

#	Model	Number of independent variables	Adjusted R^2	Multi-colinearity
Ln(Fuel) = f(x)				
1	Ln(Fuel) = f($\bar{u}$, $1/\bar{u}$, $\bar{u}^2$, σ_u^2, S, A, E_k, P)	7	0.9632	
2 (Selected)	Ln(Fuel) = f($1/\bar{u}$, $\bar{u}^2$, σ_u^2, S, A, E_k)	6	0.9564	
3	Ln(Fuel) = f($\bar{u}$, $1/\bar{u}$, σ_u^2, S, A, E_k)	6	0.9528	
4	Ln(Fuel) = f($1/\bar{u}$, σ_u^2, S, A, E_k)	5	0.9328	
Ln(HC) = f(x)				
1	Ln(HC) = f($1/\bar{u}$, $\bar{u}^2$, σ_u^2, S, A^+, E_k)	6	0.8831	$\bar{u}^2$, E_k
2 (Selected)	Ln(HC) = f($\bar{u}$, $\bar{u}^2$, σ_u^2, A^+, E_k)	5	0.8820	$\bar{u}$, $\bar{u}^2$, E_k
3	Ln(HC) = f($1/\bar{u}$, $\bar{u}^2$, σ_u^2, A^+, E_k)	5	0.8690	$\bar{u}^2$, E_k
4	Ln(HC) = f($\bar{u}$, $\bar{u}^2$, σ_u^2, A^+)	5	0.8642	$\bar{u}$, $\bar{u}^2$
Ln(CO) = f(x)				
1	Ln(CO) = f($\bar{u}^2$, σ_u^2, A^+, E_k, P)	5	0.9113	E_k
2	Ln(CO) = f($1/\bar{u}$, $\bar{u}^2$, σ_u^2, A^+, E_k)	5	0.9090	E_k
3	Ln(CO) = f($\bar{u}^2$, σ_u^2, A^+, E_k)	5	0.9044	E_k
4 (Selected)	Ln(CO) = f($\bar{u}$, $\bar{u}^2$, σ_u^2, A^+)	5	0.9043	$\bar{u}$, $\bar{u}^2$
Ln(NO$_x$) = f(x)				
1	Ln(NO$_x$) = f($1/\bar{u}$, $\bar{u}^2$, σ_u^2, A, E_k)	5	0.9650	
2 (Selected)	Ln(NO$_x$) = f($\bar{u}^2$, σ_u^2, A, E_k)	5	0.9582	
3	Ln(NO$_x$) = f($\bar{u}$, σ_u^2, A, E_k)	5	0.9557	
4	Ln(NO$_x$) = f($\bar{u}^2$, σ_u^2, E_k)	3	0.9446	

$$\ln(HC) = b_0 + b_1\bar{u} + b_2\bar{u}^2 + b_3\sigma_u^2 + b_4 A^+ + b_5 E_k \quad (R^2 = 0.8820) \qquad (15)$$

$$\ln(CO) = c_0 + x_1\bar{u} + c_2\bar{u}^2 + c_3\sigma_u^2 + c_4 P \qquad (R^2 = 0.9043) \qquad (16)$$

$$\ln(NO_x) = d_0 + d_1\bar{u}^2 + + d_2\sigma_u^2 + d_3 A + d_4 E_k \qquad (R^2 = 0.9582) \qquad (17)$$

3.4. MODEL VALIDATION

The next step in the model development was to validate the model by comparing the model estimates with independent field measurements. Specifically, data collected

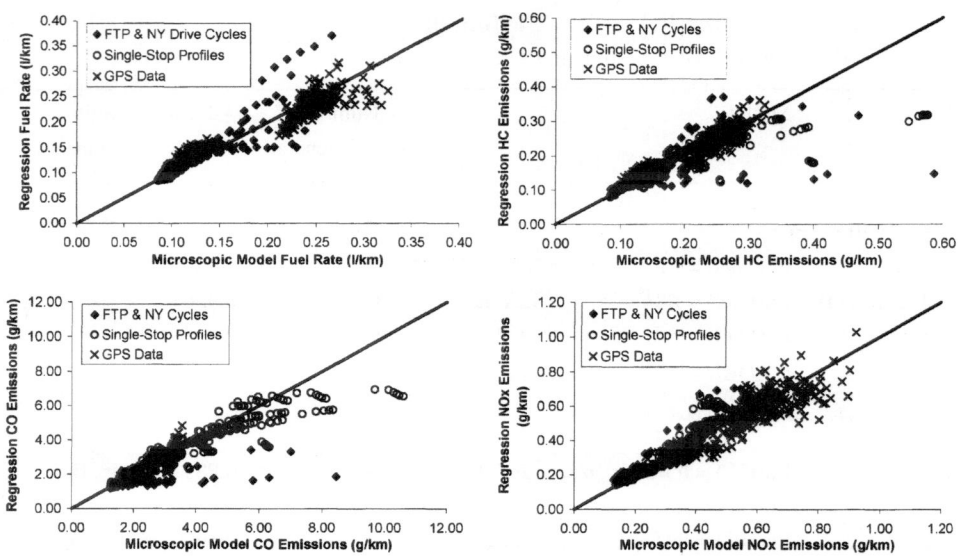

Figure 2. Comparison of trip-based and microscopic models for estimating vehicle fuel consumption and emission rates.

on a dynamometer by the US Environmental Protection Agency (EPA, 1997) were utilized for the validation of the emission models. Unfortunately, because similar fuel consumption field data were not available, it was not possible to validate the fuel consumption model.

The EPA database included second-by-second measurements of speed and emissions of HC, CO, CO_2, and NO_x for a total of 101 vehicles over a minimum of 14 drive cycles. All test vehicles in the EPA database were classified as being clean, normal, high or very high emitters for each of the emissions. The classification was based on the vehicle's emission rate for the FTP city cycle. Because the proposed models are developed for normal vehicles, the validation exercise only included clean and normal vehicles for each of HC, CO and NO_x emissions in the EPA database. The mean drive cycle emission rate per unit distance was computed as the average emission rate across all clean and normal test vehicles for the cycle (63 vehicles for HC emissions, 69 vehicles for CO emissions, and 75 vehicles for NO_x emissions). The 95th-percentile and 5th-percentile field emission rates per unit distance were also calculated for each cycle, as illustrated in Figure 3 as the whiskers of the plot. It was assumed that the speed-acceleration profile for each EPA drive cycle was the same for all test vehicles, thus the regression models were applied once per drive cycle. Figure 3 illustrates that the HC, CO, and NO_x model estimates fell within the confidence limits of the EPA laboratory measurements. It should also be noted, however, that the statistical models tended to under-estimate the HC and NO_x emission rates when compared to the mean rate while they tended to over-estimate the CO emission rates compared to the mean. These differences in

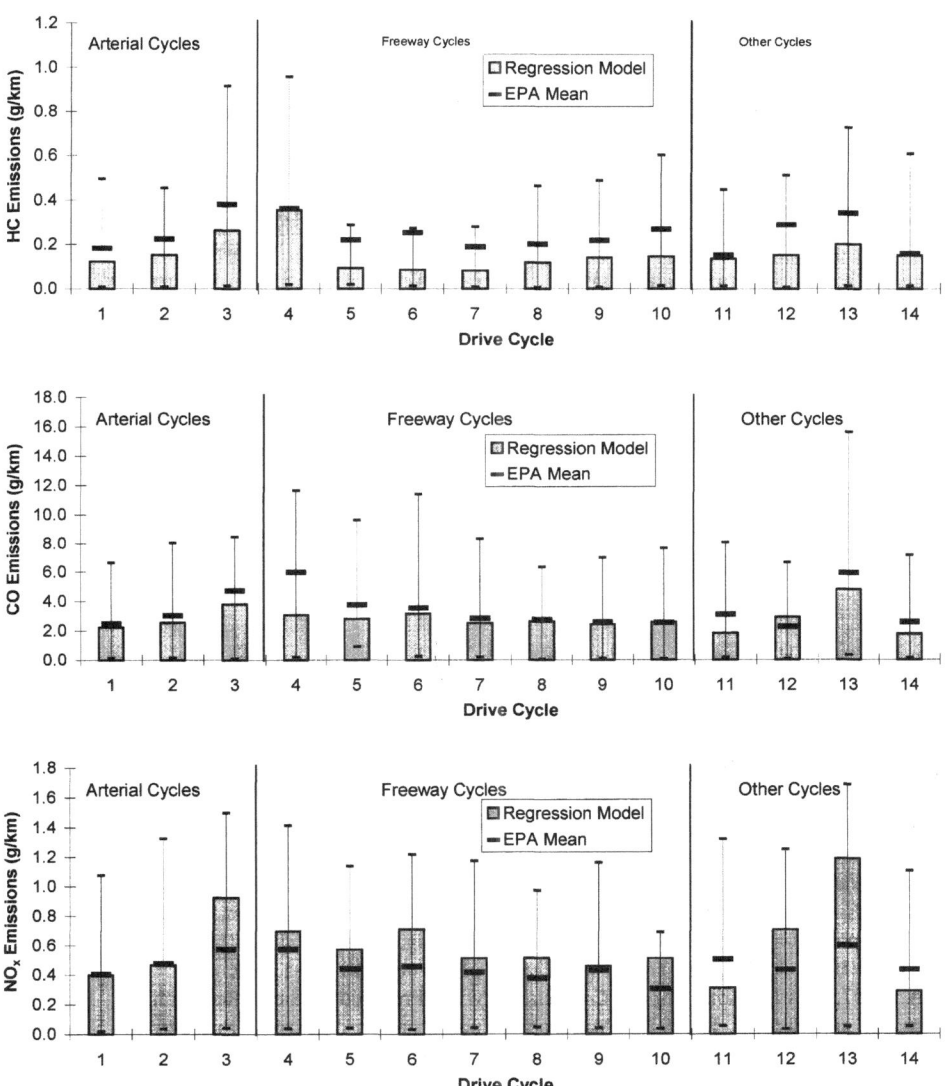

Figure 3. Model validation using EPA dynamometer measurements.

the mean rates could be attributed to differences in the vehicle population utilized for the development of the models and validation of the models. Finally, the figure indicates that, in general, the variation in regression emission estimates across the various cycles appears to be consistent with the variation in the mean EPA measurements.

4. Conclusions

This research effort demonstrated that average speed and speed variability were two critical variables for estimating vehicle fuel consumption and emission rates. Specifically, these two variables explain between 62 and 90 percent of the squared error for all four MOEs. Furthermore, the study demonstrated that the reciprocal of average speed, average speed squared, speed variability, number of vehicle stops, total acceleration noise, positive acceleration noise, and kinetic energy are critical explanatory variables for predicting fuel consumption and emission rates. The regression models that were developed were found to predict fuel consumption and emission rates of HC, CO, and NO_x to within 0.88 to 0.96 of instantaneous microscopic emission estimates. Furthermore, the emission model estimates were found to fall within the margin of variability of typical normal vehicles in the US.

References

Ahn K., Rakha H., Trani A. and Van Aerde, M.: 2002, 'Estimating vehicle fuel consumption and emissions based on instantaneous speed and acceleration levels', Accepted for publication in the ASCE Journal of Transportation Engineering, March Issue.

Ahn, K., Trani, A. A., Rakha, H. and Van Aerde, M.: 1999, *Microscopic Fuel Consumption and Energy Emission Models*', Presented at the Transportation Research Board 78th Annual Meeting, January.

Capelle, D. G.: 1966, '*An Investigation of Acceleration Noise as a Measure of Freeway Level of Service*', Doctoral dissertation, Texas A&M University, College Station.

Dion F., Van Aerde, M. and Rakha H.: 1999, *Mesoscopic Fuel Consumption and Vehicle Emission Rate Estimation as a Function of Average Speed and Number of Stops*, Paper presented at the Transportation Research Board 79th Annual Meeting, Washington, D.C.

Drew, D.: 1968, *Traffic Flow Theory and Control*, McGraw Hill, 68–13626.

EPA.: 1997, *Development of Speed Correction Cycles*, Prepared by Sierra Research Inc., June.

Evans, L. and Herman, R.: 1978, 'Automobile Fuel Economy on Fixed Urban Driving Schedules', Transportation Science, May.

Herman, R., Montroll, E., Potts, R. and Rothery, R.: 1959, 'Traffic dynamics: analysis of stability in car following', *Operations Res.* **7**, 86–106.

Jones, T. and Potts, R.: 1955, 'The measurement of acceleration noise: a traffic parameter', *Operations Res.* **10**, 755–763.

Lam, T.: 1985, 'Estimating Fuel Consumption from Engine Size', ASCE paper No. 19881.

Rakha, H., Medina, A., Sin, H., Dion, F., Van Aerde M. and Jenq, J.: 2000a, 'Coordination of Traffic Signals across Jurisdictional Boundaries: Field and Modeling Results', Presented at the Transportation Research Board 79th Annual Meeting, January, [Paper # 00–1133].

Rakha, H., Van Aerde, M., Ahn, K. and Trani, A.: 2000b, 'Requirements for evaluation of environmental impacts of intelligent transportation systems using speed and acceleration data', *Transportation Research Record* **1738**, 56–67.

Rakha, H., Dion, F. and Sin, H.: 2001a, 'Field Evaluation of Energy and Emission Impacts of Traffic Flow Improvement Projects using GPS Data: Issues and Proposed Solutions', Presented at the Transportation Research Board 80th Annual Meeting, January, [Paper # 01–2427].

Rakha, H., Kang, Y. and Dion, F.: 2001b, 'Estimating Vehicle Stops at Under-Saturated and Over-Saturated Fixed-Time Signalized Intersections', Presented at the Transportation Research Board 80th Annual Meeting, January, [Paper # 01–2353].

Rowan N.: 1967, 'An Investigation of Acceleration Noise as a Measure of the Quality of Traffic Service on Major Streets', Doctoral dissertation, Texas A&M University, College Station.

Transportation Research Board (TRB): 1995, 'Expanding Metropolitan Highways Implications for Air Quality and Energy Use', Special Report 255, National Academy Press.

West, B., McGill, R., Hodgson, J., Sluder, S. and Smith, D.: 1997, 'Development of Data-Based Light-Duty Modal Emissions and Fuel consumption Models', Society of Automotive Engineers Paper No. 972910.

RESULTS FROM THREE FIELD TRACER EXPERIMENTS ON THE NEIGHBOURHOOD SCALE IN THE CITY OF BIRMINGHAM UK

R. E. BRITTER[1], S. DI SABATINO[1*], F. CATON[3], K. M. COOKE[4],
P. G. SIMMONDS[5] and G. NICKLESS[5]

[1] *University of Cambridge, Department of Engineering, Trumpington Street, Cambridge, CB2 1PZ, U.K.;* [2] *Cambridge Environmental Research Consultants, 3 Kings Parade, Cambridge, CB2 1SJ, U.K.;* [3] *Laboratorie d'Energétique et de Mécanique Théorique et Appliquée CNRS UMR 7563, 160 54404 Vandoeuvre Cedex, France;* [4] *SIRA Ltd, South Hill, Chislehurst, Kent, BR7 5EH, U.K.;* [5] *University of Bristol, School of Chemistry, Cantock's Close, Bristol, BS8 1TS, U.K.*
(* *author for correspondence, e-mail: silvana@cerc.co.uk, fax: +44 (0) 1223 357 492*)

Abstract. The physical processes governing flow and pollutant dispersion at the neighbourhood scale, a spatial scale intermediate between the street scale and the city scale, is not well understood. Furthermore, it is not clear whether a traditional approach using averaged characteristics such as the aerodynamic roughness length is sufficient to predict the concentration field at this scale. To investigate pollutant dispersion in a real urban area, three field experiments were designed within the UK-URGENT programme sponsored by NERC. The experiments were performed in the City of Birmingham using a finite duration release of inert, non-toxic and non-depositing tracers, vis. perfluoromethylcyclohexane (PMCH) and perfluoromethylcyclopentane (PMCP). Measurements were taken using air bag samplers placed in an arc at 3.5 km (first experiment) and 1 km (second and third experiments) from the source; some trap samplers were placed outside the main arc in the outskirts of the city. Measurements were analysed in the laboratory using a novel gas-chromatography technique. Data so obtained were compared with predictions from a simple steady-state model and a time-dependent model. The concentration-time series were very asymmetrical with a sharp rise, a plateau followed by a relatively slow decrease and finally a long-lived plateau above (or possibly very slow decrease to) the background level.

Keywords: atmospheric dispersion, neighbourhood scale, tracer experiments, urban air quality

1. Introduction

In recent years much research has been devoted to the study of dispersion mechanisms in the urban environment. However, the physical processes governing flow and dispersion at the scale intermediate between the street scale and the city scale have not been specifically addressed. This spatial scale is referred to as the neighbourhood scale. Most urban dispersion models are based on either a description of street scale, as in the OSPM model (Berkowicz, 2000), or on the city scale, through the aerodynamic roughness length concept, or using both scales together. Only very recently have urban dispersion models attempted to incorporate an explicit description of the intermediate scale.

At the level of operational modelling, questions arise as to whether a traditional approach using averaged gross characteristics such as the aerodynamic roughness length is sufficient to predict the pollutant concentration field at the neighbourhood scale. The analysis of full-scale field experiments could improve our understanding of the dispersion processes over and within areas characterised by a spatially inhomogeneous distribution of buildings of different height and horizontal dimensions.

Field experiments at the neighbourhood scale are very rare in the literature and data are often not openly available as the experiments are performed in a confidential context. Experiments with releases from elevated sources were performed in the late 1960's in St Louis (McElroy and Pooler 1968). An experiment in Tel Aviv is known to have been performed but data are not available. An experiment in the framework of the US large-scale urban diffusion programme (URBAN) was performed in October 2000 in the Salt Lake City Basin (Shinn *et al.*, 2000) but data are not yet openly available. Further similar experiments are planned for in the US.

To test out an experimental technique in an urban area and to provide field data, three experiments were designed within the UK Urban Regeneration and the Environment (URGENT) programme, sponsored by the Natural Environment Research Council (NERC) as a result of a collaborative effort between the University of Bristol, the University of Cambridge and Cambridge Environment Research Consultants (CERC).

The experiments were performed in the City of Birmingham using a finite duration release of perfluoromethylcyclohexane (PMCH) and perfluoromethylcyclopentane (PMCP). Both PMCH and PMCP have a very low background concentration ($\sim 10^{-5}$ μg m^{-3}) in the atmosphere. They are also inert, non-toxic and non- depositing. These specific characteristics of the tracers, together with our ability to measure very low concentrations with a recently developed gas chromatography/negative ion chemical-ionisation mass spectrometry (GC/NICI/MS) technique (Cooke *et al.*, 2000a; Cooke *et al.*, 2000b), allowed us to perform tracer experiments in a real urban environment. These experiments were the first to provide data for a near-ground level release within a city.

2. Experimental Arrangement

The experiments were designed based on the following requirements and constraints. Firstly we wanted a simple experiment in order to test the technique and to facilitate the interpretation of results. Secondly, we wanted to perform the experiments in a neutrally stratified atmosphere; this being the easiest meteorological conditions to study. Finally, the experimental arrangement had to take into account the intrinsic limitations of the technique in terms of the maximum and minimum measurable concentrations, and the financial and personnel resources available.

The three experiments took place in Birmingham (UK) in July 1999, February 2000 and August 2000 respectively. They were performed during the PUMA (Pollution of the Urban Midlands Atmosphere) campaigns which were also part of the URGENT programme and which provided the necessary meteorological data. The PUMA campaigns collected extensive data from various meteorological sites within the Birmingham urban area. Standard meteorological data such as wind speed, wind direction, temperature and relative humidity were available as 15 min averages for the three tracer experiments. A summary of the meteorological data available for the tracer experiments is reported in Cooke *et al.* (2000c).

Near-neutral stability conditions with a wind speed of about 4–5 ms^{-1} were chosen to satisfy our requirements for the simplest meteorological conditions for the experiment.

Each tracer experiment was set-up with a 'top hat' temporal evolution of the tracer release rate. The spatial and temporal experimental design was determined using simple steady-state models (as described in Hanna *et al.*, 1982) and an advanced operational dispersion model ADMS3 (CERC, 2000).

Most receptors were placed in an arc within the city centre, while some other receptors were placed well outside the city centre. The source-receptor distance for the inner arc was initially chosen to be 3–4 km for the July 1999 experiment (see Britter *et al.*, 2000) and later reduced to 1 km for both the February and August 2000 experiments.

Two types of sampling systems were employed. The first (air bag samplers) consisted of a 10-port multiple base fitted with small solenoid valves onto which a series of either 5-L or 10-L Tedlar bags (SKC Ltd, UK) were attached. The air sample was pumped into each bag in turn by electrical actuation of the appropriate solenoid (see Cooke *et al.*, 2000 for further details). The air bag samplers were used mainly in the inner arc. The second sampling system (trap samplers) consisted of a 16-position multiport valve (Valco Instruments Co. Inc., Houston), fitted with 15 stainless steel adsorption traps, containing 150 mg Carboxen 564 (Sigma-Aldrich, Bellefonte, PA). The multiport valve system was incorporated into a robust stand-alone metal sampling box system. Air was sampled sequentially via the valve through each trap at a predetermined flow rate controlled by a mass flow controller (Unit Instruments Ltd, Ireland). The trap samplers were used outside the inner arc in the outskirts as checkpoints. Further information on the samplers and the concentration measurement technique are available in Cooke *et al.*. (2000a).

Due to a limited number of available air bag samplers (constrained by the financial resources and personnel available), the receptors were distributed within an arc of about 60°, suitable for south-westerly wind conditions; these being, for the city of Birmingham, the most frequent winds.

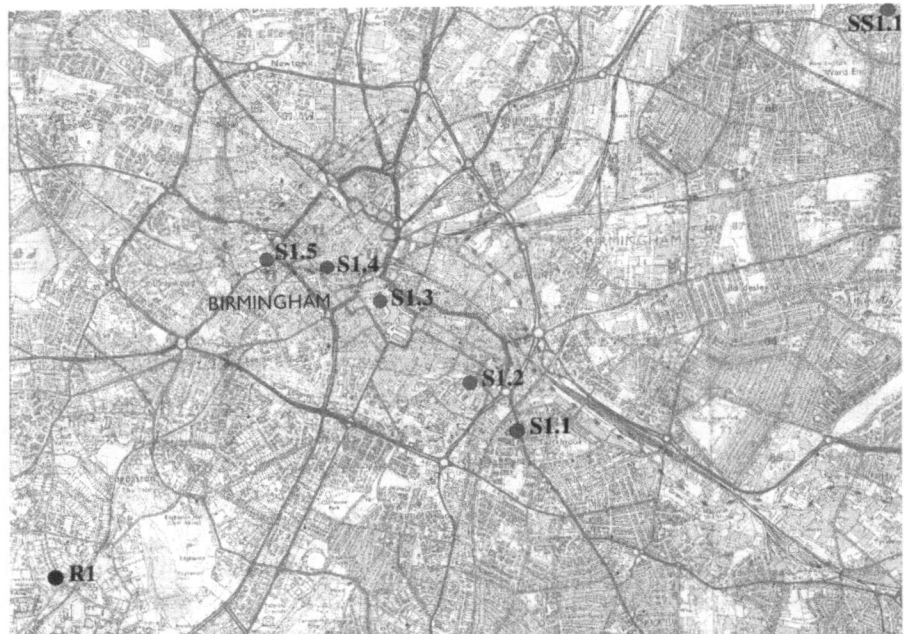

Figure 1. First experiment: tracer release (R1) and sampling sites (S1.n). Birmingham, UK, 1st July 1999. The distance between the release and the sampling sites is about 3.5 km.

2.1. FIRST EXPERIMENT

A release of PMCH took place on 1st July 1999, from a heated chimney of 4.5 m height at 13:00, local time. The release rate was selected in order to make the most of the dynamic range of the analytical instrumentation. The release time was selected in order to record the total temporal evolution of the puff. The averaging time of the receptors was set at 15 min; an averaging time that, with hindsight, was too long. The release apparatus was situated at the University of Birmingham, Pritchatts Road (R1 in Figure 1) and the effective release rate was 4 gs^{-1} over a 40 min period.

Five samplers were placed in an arc at approximately 3.5 km from the release point with a cross-wind spacing of about 500 m. The sampling height was 2 m above ground for four of the samplers while one sampler was positioned on a roof top at 40 m. The optimum spacing was deduced on the basis of the estimate of the plume width by means of simple steady-state models (see Hanna *et al.*, 1982). The samplers are referred to as Sites S1.1, S1.2, S1.3 (on the roof top), S1.4 and S1.5 (see Figure 1). A secondary receptor Site SS1.1 (trap sampler) was placed on the outskirts of Birmingham at about 9 km from the release source. It was expected that measurements at a distance comparable with the city scale would be easily interpreted through standard plume dispersion models. A detailed description of the sites can be found in Cooke *et al.* (2000c).

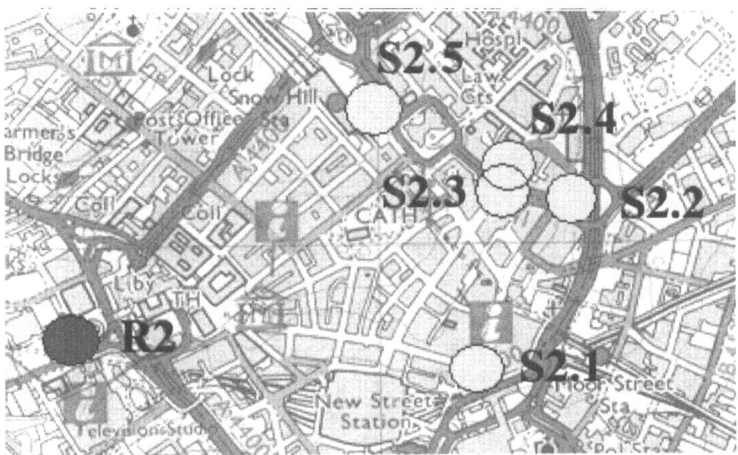

Figure 2. Second experiment: tracer release (R2) and sampling sites (S2.n) in the main arc. Birmingham, UK, 1st February 2000. The distance between the release and the sampling sites is about 1 km.

2.2. SECOND EXPERIMENT

The second experiment took place on the 1st February 2000 at 13:00, local time. It was decided, on the basis of the results obtained in the previous experiment, to reduce the distance from the source to receptor and to perform this experiment in an area where the plume depth would be more comparable with the building heights.

Five samplers, of which one was on a 30 m high roof (S2.4), were placed on an arc at about 1 km from the release site (R2; see Figure 2). The optimum cross-wind spacing between receptors was determined to be around 300 m. The averaging time for the samplers in the main arc was set at 3 min. This shorter time resolution was expected to give more data during the rise and the decay of the concentration levels, which, at this distance from the source, was expected to be more rapid. Three secondary receptors were placed at about 6 km from the release site. Due to a technical failure only one receptor (SS2.3) worked. The release rate was 0.23 gs^{-1} over a period of 35 min. The reduction in the release rate from the first experiment was made because the source to receptor distance was reduced and also because the first experiment showed that the technique had more than adequate resolution.

2.3. THIRD EXPERIMENT

The third experiment performed on the 2nd August 2000, was designed on the basis of the results from the first two.

Eight samplers (see Figure 3), of which three were at elevation on roofs (11 m and 20 m high), were placed at 1 km from the release site with a cross-wind spacing between the receptors of about 250 m. A second tracer PMCP was used in this experiment in conjunction with the first tracer PMCH. The secondary tracer

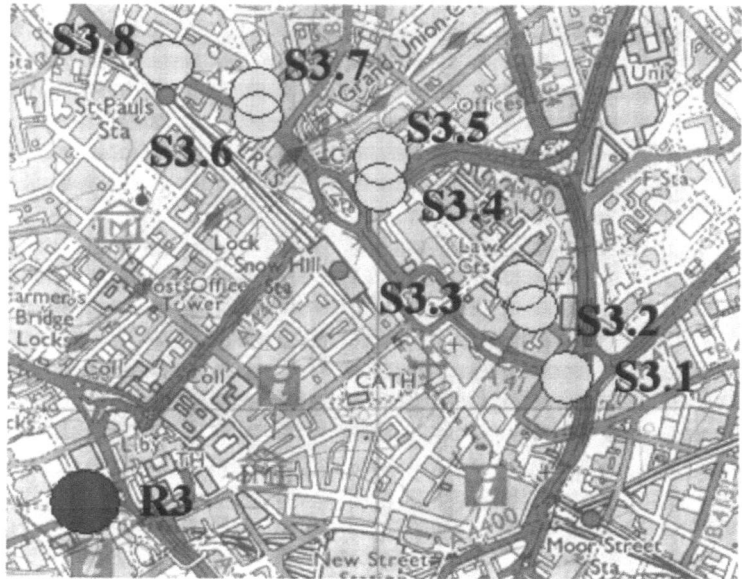

Figure 3. Third experiment: tracer release (R3) and sampling sites (S3.n) in the main arc. Birmingham, UK, 2nd August 2000. The distance between the release and the sampling sites is about 1 km.

release was delayed by half the sampling period (averaging time). The averaging time was set at 6 min. The primary tracer, PMCH, was released on 2nd August at 13:00, local time, while the secondary tracer, PMCP, was released at 13:03. The motivation for this was to obtain the same time resolution as in the February experiment over a longer measuring period without using more samplers and to test if a multiple tracer release could be performed. The effective release rate was 0.22 gs^{-1} for PMCH and 0.18 gs^{-1} for PMCP over a period of 20 min.

Atmospheric thermal stratification was estimated from the available meteorological data provided by the PUMA campaigns, to be near neutral for all the experiments.

3. Results and Discussion

Concentration measurements taken from the main arc for the three experiments are shown as a function of time in Figures 4, 5, and 6. Only the PMCH results are shown as the PMCP results from the August 2000 experiment were very similar to those with PMCH.

The July 1999 and August 2000 experiments were successful while the plume was partially missed by the receptors for the February 2000 experiment and, due to a technical failure, the most important receptor (S2.5) failed. The partial missing of the plume was the result of an unexpected change from the forecasted wind

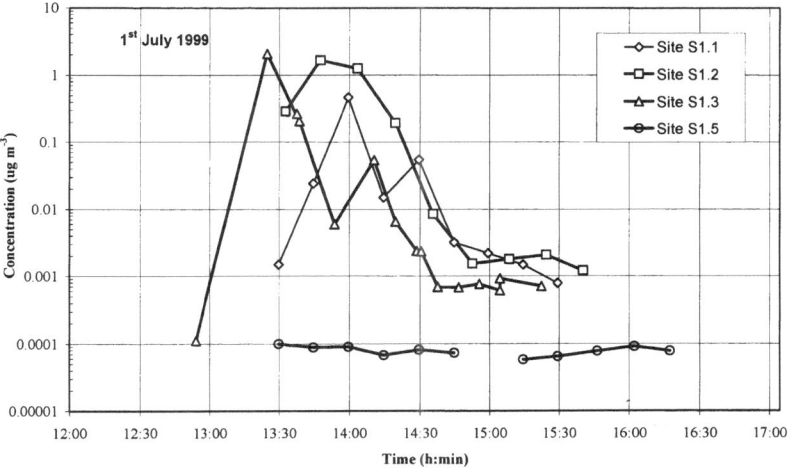

Figure 4. PMCH concentration measurements for the July 1999 experiment.

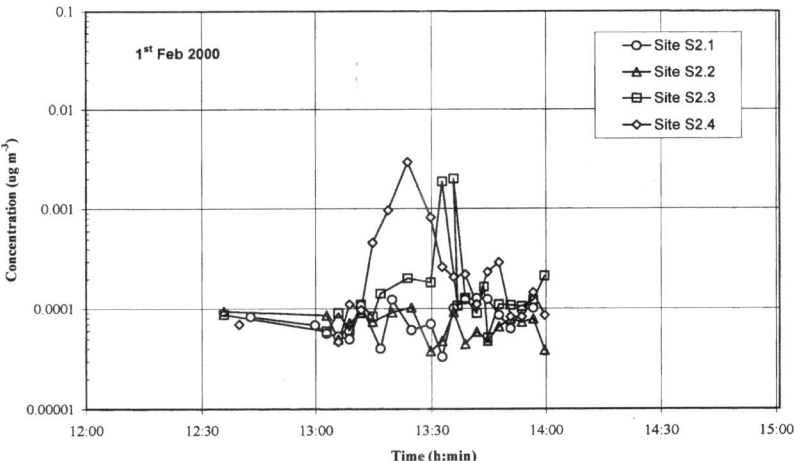

Figure 5. PMCH concentration measurements for the February 2000 experiment.

direction. Obviously a good forecast of the meteorological conditions is crucial for the success of this type of experiment.

The August 2000 experiment clearly shows that the receptors that see the maximum concentrations (and are presumably close to the centre line of the plume) experience a short time delay (roughly equal to the source-receptor distance divided by the wind speed). The concentration then rapidly rises to maximum values that are maintained for a time comparable to the release time. This is followed by a slow decrease towards the background level. A similar, though less clear cut, interpretation can be made for the July 1999 experiment.

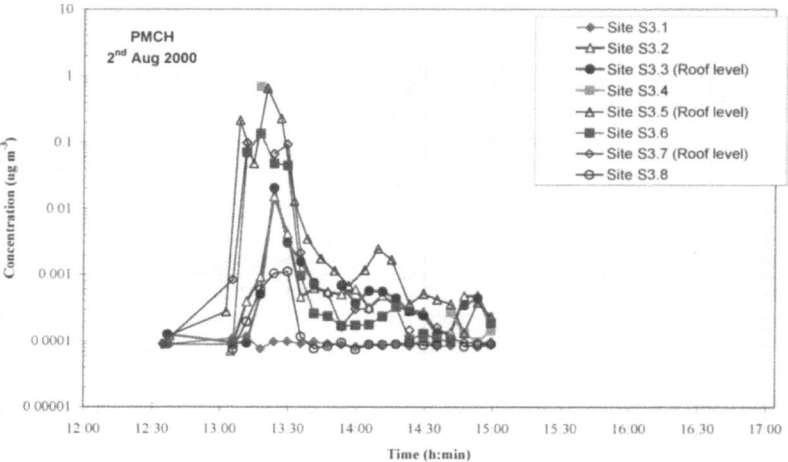

Figure 6. PMCH concentration measurements for the August 2000 experiment.

All the results show considerable variability; a variability consistent with dispersion experiments in the field but probably made worse by the urban complexity, the quite short sampling times being used and that the results are from one realisation of an unsteady problem.

Figures 7 and 8 also show selected experimental concentration histories for both the July 1999 and the August 2000 experiments with the origin moved to the release time. The rapid rise, elevated plateau, and slow decrease in concentration towards the background level can be discerned. The slow decrease is most probably due to material being retained in the wakes of the buildings and slowly released. The concentration data can be fitted with an exponential decay curve with a well defined time constant. This time constant for the concentration decrease of the July experiment was estimated to be about 7 min and about 4 min for the August experiments. These estimates are based on data from 50–100 min for the July 1999 experiment and 20–50 min for the August 2000 experiments. However in both experiments the concentration tends to remain above the background levels for very long periods. There is, apparently, a very slow decay that occurs over a long period, probably some hours beyond the total duration of the experiments. We believe that this long-lived plateau might be due to material being taken inside the complex configuration of buildings through natural and forced ventilation and then released with a time constant comparable to building air-exchange rates.

A comparison of the maximum measured concentration values with results from simple Gaussian plume models was performed to verify that those values were broadly consistent with expectations. The simplest way to do that is by calculating the concentration C using a Gaussian plume model for a continuous source of constant tracer release rate Q and uniform wind speed U (Pasquill 1961, 1974). The constant release rate Q has been taken to be the average release rate over the

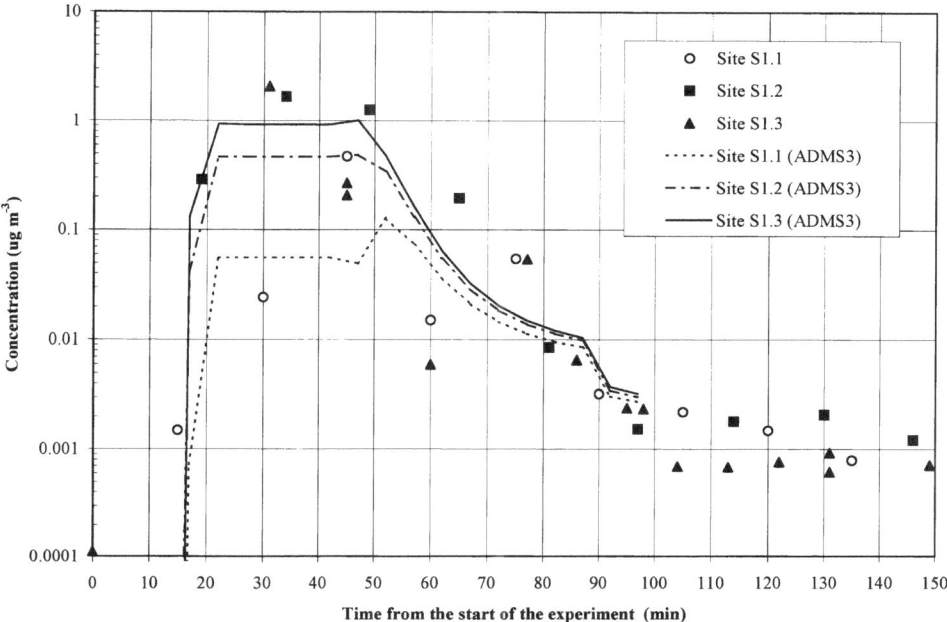

Figure 7. Comparison between measured concentration and ADMS3 model predictions for the July 1999 experiment.

release duration i.e. $Q = 4$ gs^{-1} for the first experiment, $Q = 0.23$ gs^{-1} for the second experiment and $Q = 0.22$ gs^{-1} for the third experiment.

Many parameterisations for the lateral and vertical standard deviation of concentration σ and σ are available in the literature. The concentration estimated through the formulas for σ and σ proposed by Briggs for urban conditions (Hanna *et al.*, 1982) give the values reported in Table I. The comparison was donemade for both Pasquill stability classes C and D.

From the table it can be observed that the measured maximum concentration for both July 1999 and August 2000 experiments are comparable with values obtained by a simple steady-state model. Looking again at measurements from the July 1999 experiment (Figure 4) it is seen that the maximum concentration occurred for receptor S1.3, which is consistent with a wind direction of about 230°. This is about 30° less than the actual wind direction. The reason for this could be an initial offset of the plume produced by the buildings situated in front of the release point (Theurer, 1993) but further investigation is still required.

A wind direction of 230° was used in subsequent model calculations for the July 1999 experiment shown in Figure 7. The use, for model calculations, of a wind direction deriving from the alignment between the source and that receptor which recorded the maximum concentration was done to distinguish between uncertainties due to the wind direction and other uncertainties. The measurements showed

TABLE I

Measured and modelled maximum concentrations

Distance from release km	Measured concentration μgm^{-3}	Briggs curves class C μgm^{-3}	Briggs curves class D μgm^{-3}
3.5	2.16 (Site S1.3)	1.0	3
9	0.08 (Site SS1.1)	0.09	0.15
1	0.003 (Site S2.4)	0.4	0.9
6.6	0.017 (Site SS2.3)	0.015	0.05
1	0.69 (Site S3.4)	0.5	1
6.6	0.008 (Site SS2.3)	0.02	0.07

that the plume did not travel in the direction expected from the meteorological data available and this is noted separately from the diffusion of the tracer cloud.

The maximum concentration for the August 2000 experiment was recorded at Site 3.4 (Figure 6) that is consistent with a wind direction of 219°. This wind direction is about 20° less than that measured at the PUMA meteorological stations. As for the July 1999 experiment we used the bearing of the site that recorded the maximum concentration as wind direction in the model calculations for the August 2000 experiment.

For the February 2000 experiment (Site S2.4) the maximum concentration is about one order of magnitude less than analytical model predictions. The failure to obtain data from the most relevant receptor restricts any significant interpretation of theise data.

We analysed the measurements further by modelling the experiments with the time-dependent finite-duration release module of ADMS3 and compared the measurements with those results. The comparison between measurements and model outputs for the July 1999 is shown in Figure 7. The comparison between measurements for one site only (site S3.5) and model outputs for the August 2000 experiment is shown in Figure 8. The maximum value of the concentration is predicted within a factor of two by ADMS3. Figure 8 also shows results for the simultaneous PMCH and PMCP releases but with the PMCP release commencing 3 min after the PMCH release. Thus the results for the PMCP release should be shifted 3 min earlier (that is, to the left) in Figure 8 bringing the two sets of data into very good alignment. It is also clear from Figure 8 that the advection velocity is smaller in the experiment than in the model suggesting that at the 1 km scale a more refined 'neighbourhood scale' may be required for modelling transient problems.

The general consistency between the measurements and the model predictions support the experimental techniques and procedures. However, the long-lived very

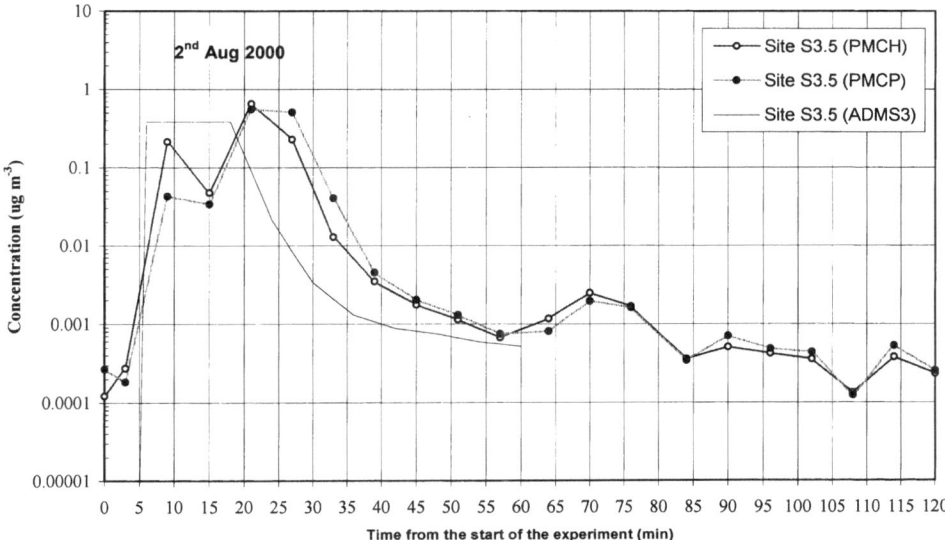

Figure 8. Comparison between measured concentration and ADMS3 model predictions for the August 2000 experiment for Site S3.5.

slow decrease in concentration that is observed for both the July 1999 and August 2000 experiments is, unsurprisingly, not well predicted by the model.

4. Conclusions

Three field experiments based on tracer releases were designed and performed in the City of Birmingham. The requirements for the experiment were a finite duration tracer release, fast sampling, neutral thermal stratification and a source to receptor distance less than the city scale.

The analysis of measurements from the July 1999 experiment performed at a scale of 3.5 km and the August 2000 experiment at a scale of 1 km showed that the concentration-time history is asymmetrical; the rise occurring much more rapidly than the decrease. The decrease occurred with a time constant of about 7 min for the July 1999 experiment and about 4 minutes for the August 2000 experiment.

The good agreement between the measurements and a dispersion model predictions (ADMS 3) for the July 1999 experiment suggested that models based on gross flow characteristics such as the roughness length are likely to be valid for the 3.5 km scale and larger. This is probably due to the fact that the concentration plume at the scale of the July 1999 experiment is very deep compared with the building heights. In this case, a very detailed description of the flow and dispersion through and above the buildings may not be necessary; the concentration plume is not likely to be affected when it is very deep compared with the height of the buildings.

The second and third experiments were performed with a source to receptor distance of about 1 km. The results from the second experiment were not conclusive. The third experiment also showed a strongly asymmetrical concentration-time history. Comparison of measurements with ADMS3 model results showed a good qualitative agreement. However, the observed advection speed was less than predicted. It would appear that at a source to receptor distance of 1 km or less a more refined 'neighbourhood scale' model may be required for modelling transient problems.

An important observation from both the July 1999 and August 2000 experiments was that the maximum plume concentrations were at a bearing some 20–30 degrees different to the wind direction given by the meteorological stations.

An archived data set is available (Cooke *et al.*, 2000c) for these field experiments.

Acknowlegments

We are indebted to all the people who were involved in the measuring campaign and made the sites available. One of the authors, SDS, acknowledges the financial support of the European TMR project TRAPOS.

References

Berkowicz, R.: 2000 'OSPM – A parameterised street pollution model', *J. Environ. Monit. Assess.* **65**, 323–331.

Britter, R. E., Caton, F., Di Sabatino, S., Cooke, K. M., Simmonds, P. G. and Nickless, G.: 2000, 'Dispersion of a Passive Tracer Within and Above an Urban Canopy', *Proceeding of the Third Symposium on the Urban Environment*, American Meteorological Society, 30–31.

CERC.: 2000, *ADMS3, USER Guide*. Cambridge Environmental Research Consultants, 3 King's Parade, Cambridge, CB2 1SJ, UK.

Cooke, K. M., Simmonds, P. G. and Nickless, G.: 2000a, 'The Development of a Highly Sensitive and Selective Technique to Monitor Tracer Dispersion within an Urban Environment', *Proceeding of the Third Symposium on the Urban Environment*, American Meteorological Society, 32–33.

Cooke, K. M., Simmonds, P. G., Nickless, G. and Makepeace, A. P. W.: 2000b 'The use of capillary gas chromatography with negative ion-chemical ionisation mass spectrometry for the determination of perfluorocarbon tracers in the atmosphere', Accepted for publication in Analytical Chemistry.

Cooke, K. M., Di Sabatino, S., Simmonds, P., Nickless, G., Britter, R. E. and Caton, F.: 2000c, 'Tracer and Dispersion of Gaseous Pollutants in an Urban Area'. Birmingham Tracer Experiments. Technical Paper CUED/A-AERO/TR.27. Department of Engineering, of the University of Cambridge University.

Hanna, S., Briggs, G. and Hosker, R.: 1982, *Handbook on Atmospheric Diffusion*, Technical Information Center, U.S. Department of Energy.

McElroy, J. and Pooler, F.: 1968, 'St. Louis Dispersion Study', Report AP-53. U.S. Public Health Service, National Air Pollution Control Administration.

Theurer, W.: 1995, 'Point sources in urban areas: Modelling of neutral gas clouds with semi-empirical models', in Cermak *et al.* (eds), *Wind Climates in Cities*, Kluwer Academic Publishers, 485–502,

THREE-DIMENSIONAL GROUND-BASED MEASUREMENTS OF URBAN AIR QUALITY TO EVALUATE SATELLITE DERIVED INTERPRETATIONS FOR URBAN AIR POLLUTION

K. SCHÄFER[1]*, G. FÖMMEL[1], H. HOFFMANN[1], S. BRIZ[1], W. JUNKERMANN[1], S. EMEIS[1], C. JAHN[1], S. LEIPOLD[1], A. SEDLMAIER[1], S. DINEV[2], G. REISHOFER[2], L. WINDHOLZ[2], N. SOULAKELLIS[3], N. SIFAKIS[4] and D. SARIGIANNIS[5]

[1] *Institut für Meteorologie und Klimaforschung, Bereich Atmosphärische Umweltforschung, Forschungszentrum Karlsruhe GmbH (former Fraunhofer-Institut für Atmosphärische Umweltforschung IFU), Garmisch-Partenkirchen, Germany;* [2] *Institut für Experimentalphysik, Technische Universität Graz, Graz, Austria;* [3] *Department of Geography, University of the Aegean, Mytilini, Greece;* [4] *Institute for Space Applications and Remote Sensing, National Observatory of Athens, Greece;* [5] *Institute for Health and Consumer Protection, EC-DG Joint Research Centre, Ispra (VA), Italy*
(* *author for correspondence, e-mail: klaus.schaefer@imk.fzk.de, Fax: +49 8821 73573*)

Abstract. Urban air quality and meteorological measurements were carried out in the region of Brescia (Italy) simultaneously to the acquisition of satellite data during winter and summer smog conditions in 1999. The main objectives of the campaigns were: delivery of data for the validation of air pollution interpretations based on satellite imagery, and determination of the aerosol optical thickness in spectral ranges similar to those used by satellites. During the winter campaign the ground-based network was complemented by local stations and by SODAR, DOAS, and FTIR remote sensing measurements. Size distributions of aerosol particles up to 4,000 m a.s.l. were measured by means of an ultra-light aircraft, which was also equipped with meteorological sensors and an ozone sensor. During the summer campaign an interference filter actinometer, an integrating nephelometer and an ozone LIDAR were operated additionally. The satellite images acquired and processed were taken from SPOT. Optical thickness retrieved from interference filter actinometer measurements were compared with the retrievals from the satellite imagery in the same spectral intervals. It is concluded that remaining aerosols in the reference image yield an off-set in the satellite retrieval data and that information about the vertical structure of the boundary layer is very important.

Keywords: aircraft measurements, DOAS, FTIR spectrometry, *in situ* measurements, interference filter actinometer, LIDAR, remote sensing, satellite, smog, SODAR

1. Introduction

Satellite images of high spatial resolution (i.e., horizontal resolution of tens of metres), such as taken by Landsat or SPOT, contain useful environmental inform-ation not only on land cover and vegetation but also on tropospheric pollution (Sifakis, 1992). This information, in terms of aerosol optical thickness (AOT), can be used as an indicator of air pollution's density and spatial distribution over ex-tended areas (e.g., Sifakis *et al.*, 1998): The AOT is correlated to the particle mass

Water, Air, and Soil Pollution: Focus **2**: 91–102, 2002.
© 2002 *Kluwer Academic Publishers.*

loading (Cachorro and Tanre, 1997) which, in the case of small-sized particles in urban and industrial areas, is due to ammonium sulphate and/or nitrate salts. These, in turn, are coupled with sulphuric acid and nitric acid droplets and consequently with their gaseous precursors NO_2 and SO_2 (Colls, 1997). Furthermore, aerosols are necessary reaction surfaces in the air for ozone production processes.

The work presented here is part of the investigations aiming at defining the exact correlation between the atmospheric information retrieved by satellite (in terms of optical thickness) and air quality (in terms of pollutant concentrations). The algorithms and results are presented in Sifakis *et al.* (1999) and Sarigiannis *et al.* (2001). More specifically, this research is intended to find out if reliable information on air pollution distribution can be extracted on the basis of columnar atmospheric AOT values retrieved from high spatial resolution satellite data during cloud-free days. To examine this, two ground-based measurement campaigns were performed simultaneously with the satellite overpasses. The objectives of these campaigns were:

1. Delivery of necessary input data for modelling air pollution transport;
2. Delivery of data for the evaluation of air pollution interpretations based on satellite;
3. Determination of the AOT in spectral ranges similar to those used by the satellites;
4. Elaboration of a proposal for the optimisation of the existing air-pollution monitoring network in the city of Brescia on the basis of satellite imagery.

The present article will concentrate on items 2 and 3. The background for these investigations is the EU Council Directive on ambient air quality assessment and management (96/62/EC), which recommends the use of operational instruments such as air pollution monitoring sensors, networks and modelling. The greater area of the city of Brescia in the region of Lombardy (Northern Italy) was selected as investigation area because it is a city in a high industrialised region with an air pollution monitoring network. The air quality in Brescia is influenced by advection of air from industrial areas north of Brescia, and sometimes by polluted air from Milano and Bergamo during west wind episodes an area with frequent smog epis-odes. Air pollution episodes in Alpine valleys in Northern Italy had already been investigated (Dosio *et al.*, 2001). Brescia has approximately 200,000 inhabitants and a complex topography dominated by three main features: (1) in the south of the city there is flat terrain (the Po plain with heights around 120 m a.s.l.), (2) the city itself is situated at the end of the 'Val Trompia', a valley, which runs from north to south, (3) five kilometres north from Brescia the Nave valley branches off from the Val Trompia to the east. East and west of Brescia there are mountains (first mountains are Monte Maddalena (874 m) in the east and Monte Picastello (383 m) in the west). The altitude of the mountains in the northern part of Val Trompia is about 1,000 m a.s.l.

2. Methodology

Air pollution investigations were performed simultaneously with the acquisition of the satellite images in the greater Brescia area during winter (18 January–19 February, 1999) and summer (30 August–23 September, 1999). During winter several days with high air pollution due to stable atmospheric stratification occurred and during summer photochemical smog conditions happened. In addition to the standard measurements carried out systematically by the existing air pollution monitoring network of the city of Brescia operated by ASM Brescia, the following devices were used:

- A Fourier Transform Infrared Spectrometer system (FTIR) addressing path-integrated concentrations measurements of CO, CH_4, N_2O, CO_2, NO and O_3 (Haus et al., 1994).
- A system for Differential Optical Absorption Spectroscopy (DOAS) addressing path-integrated concentrations measurements of NO, NO_2, SO_2 and O_3 (Grant et al., 1992).
- A Sound Detection and Ranging system (SODAR) measuring the vertical profiles of wind vector (Reitebuch and Emeis, 1998).
- In situ CO (IFU: Horiba, ASM: Thermo Environmental Instruments), NO (IFU: winter Horiba, summer Monitor Labs, ASM: Thermo Environmental Instruments), NO_2 (IFU: winter Horiba, summer Monitor Labs, ASM: Thermo Environmental Instruments), SO_2 (IFU: Horiba, ASM: Thermo Environmental Instruments), O_3, (IFU: winter Horiba, summer Thermo Environmental Instruments, ASM: Thermo Environmental Instruments), CH_4 (Horiba), dust (FAG Kugelfischer), PM10 (Sierra Andersen) and meteorological (wind, temperature, humidity, pressure, solar radiation) measurement instruments (Friedrichs, Thies, Ammonit, Zeno).
- An interference filter actinometer measuring the AOT from the ground (Wiegner and Rabus, personal communication).
- An integrating nephelometer measuring the aerosol scattering coefficient in sampled air (SCI).
- A scanning Light Detection and Ranging system (LIDAR) measuring the vertical profiles of ozone concentration (Löscher and Windholz, 2001).
- Finally, an ultra-light aircraft was operated to measure in situ ozone, meteorological parameters (temperature, solar radiation, dew point, pressure) and total extinction in 15 channels (Junkermann, 2001).

The positions of all previous instruments at the ground were selected according to the criterion that as little as possible, mobile and fixed emission sources are in the surroundings of the measurement sites (see Figure 1). Further the instruments were installed at sites which accomplished the existing monitoring network.

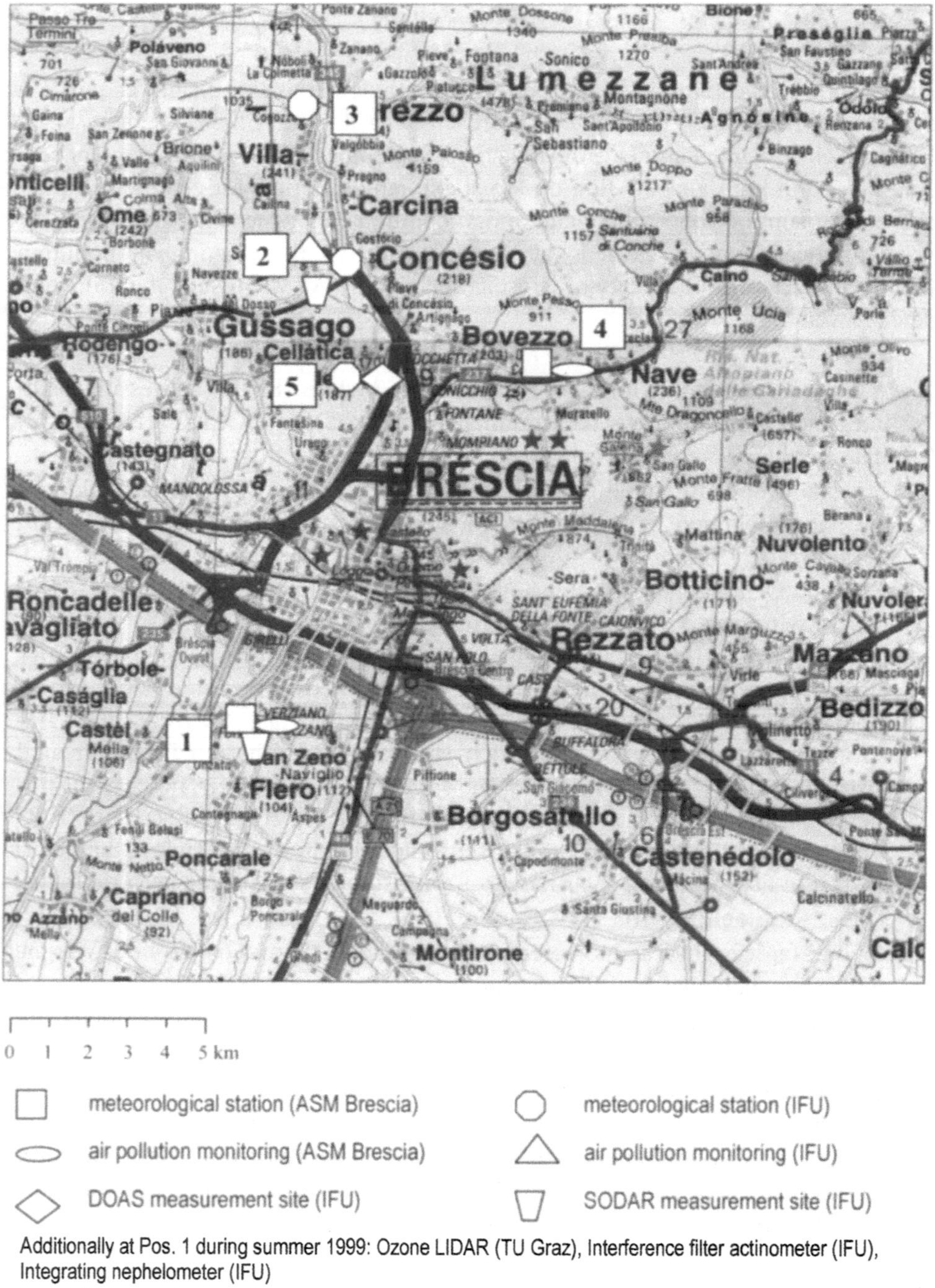

0   1   2   3   4   5 km

☐   meteorological station (ASM Brescia)   ◯   meteorological station (IFU)

⬭   air pollution monitoring (ASM Brescia)   △   air pollution monitoring (IFU)

◇   DOAS measurement site (IFU)   ▽   SODAR measurement site (IFU)

Additionally at Pos. 1 during summer 1999: Ozone LIDAR (TU Graz), Interference filter actinometer (IFU),
Integrating nephelometer (IFU)

Figure 1. Survey of measurement sites during the campaigns in the region of Brescia in January/February 1999 and August/September 1999. **1**: Waste water treatment Verziano; **2**: Scuola Media Statale di Concesio; **3**: water plant Fonte di Cogozzo; **4**: ENEL substation between Bovezzo and Nave; **5**: fountain 'Pozzo Roma', Collebeato.

E.g. in the Nave valley there is no measurement station, therefore the devices measuring meteorological and air pollution data were installed in the Val Trompia (Pos. 2, 3 and 5 in Figure 1) as well as in the valley near Nave (Pos. 4 in Figure 1). Extensive information of the wind system over Brescia was obtained by a long-range SODAR (up to 800 m, vertical resolution 30 m), which was situated 5 km south of Brescia in the imaginary prolongation of the Val Trompia axis (Pos. 1 in Figure 1). During the summer campaign at the same site as the long-range SODAR in Verziano the dominant air pollutant ozone was characterised by a LIDAR giving vertical profiles of ozone concentrations and aerosol optical parameters from 100 m up to 2,000 m altitude (vertical resolution 100 m). Additionally, the AOT in 10 spectral ranges was measured from solar radiation by an interference filter actino-meter at that site which was available during this campaign only to determine the total atmospheric aerosol content. The determination of the optical thickness of the atmosphere from the ground is done in similar spectral ranges as from satellite im-ages. This kind of ground-truthing of space-based sensors was discussed by King *et al.* (1999) and is one of the goals of the AERONET (AErosol RObotic NETwork) program (internet http://aeronet.gsfc.nasa.gov:8080/) as well as the Multi-Filter Rotating Shadow-band Radiometer (MFRSR) network (Alexandrov *et al.*, 2001). The FTIR, LIDAR and interference filter actinometer measurements were per-formed during appropriate weather conditions from the morning until the evening. All the other instruments were running.

During each campaign an inter-comparison of all instruments at a place of well-mixed air was performed. If necessary, mean values among the measurement results of the different devices (*in situ* analyses of trace gas concentrations from air sampling and open-path spectroscopic measurements by FTIR and DOAS) were calculated to correct each data set. High spatial resolution SPOT satellite images, which were taken during the two campaigns at different days at about 10:00 a.m. local time were acquired and processed (see Sarigiannis *et al.*, 2001). These satellite images covered an area of 60 km × 60 km.

3. Results and Discussion

3.1. SYNOPTIC SITUATION AND BOUNDARY-LAYER STRUCTURE

The first days of the winter campaign were influenced by high pressure over Italy and surface inversions. Low temperatures in the evening and night (lower than 5 °C) resulted in formation of strong fog. In the south of Brescia the air was highly polluted, resulting in bad visibility. A periodic wind system dominated in the Val Trompia and also in the valley of Nave: winds coming down-valley during the evening, night and morning during the coldest hours of the day (mountain breeze) and winds coming from southern directions during the warm hours of the day (valley breeze). So the northern parts of Brescia in the Val Trompia and valley

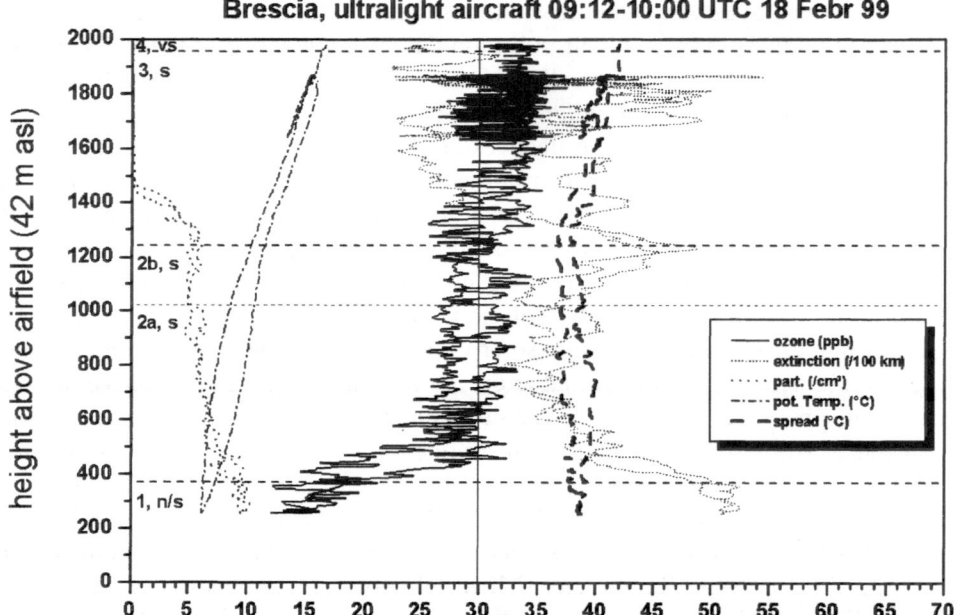

Figure 2. Vertical profiles measured by the ultra-light aircraft at 18 February, 1999, 09:12–10:00 UTC: ozone concentration (ppb), extinction (1/100 km), potential temperature (°C), spread (i.e. temperature minus dewpoint temperature) (°C). Please note: The origin for spread has been shifted to 30 (vertical line). 'n/s' denotes neutral/stable, 's' stable, and 'vs' very stable thermal stratification.

of Nave were influenced by air pollution coming from Brescia and the Po plain during the day. In the night, the air quality in Brescia was influenced by cold air flux coming from the Val Trompia. The Val Trompia and particularly the ancillary valleys (Łumezzane) are characterised by strong traffic and industry (metal fusion and treatment), so the cold air flux during the night was 'injected' with polluted air.

The complex orographic structure results in a difficult vertical structure of the boundary layer with up to 5 sub-layers characterised by different thermal stability and different degrees of pollution as shown in Figures 2 and 3 containing the measured data of the ultra-light aircraft. The structure of the boundary layer above Brescia was more complex and extended up to greater heights during the summer campaign than during the winter campaign. The top of the boundary layer – indicated by less than 1 particle per cm^3 (thin leftmost curve in Figures 2 and 3) – was around 1,500 m in February and 3,500 m in September. The long-range SODAR gave the valley/mountain wind regime structure in the boundary layer up to 800 m above ground as well as information about the layering of the atmosphere in this altitude range. During the aircraft flights the boundary layer height could be found on 18 February at about 1,250 m (above layer 2 in Figure 2) and on 19 February at 1,300 m. There was a frequent peak in the dust concentration between 08:00 and 10:00 a.m. At 4 February peak values in near-ground air pollution (3.5 mg m^{-3}

Brescia, ultralight aircraft 08:38-09:13 UTC 09 Sept 99

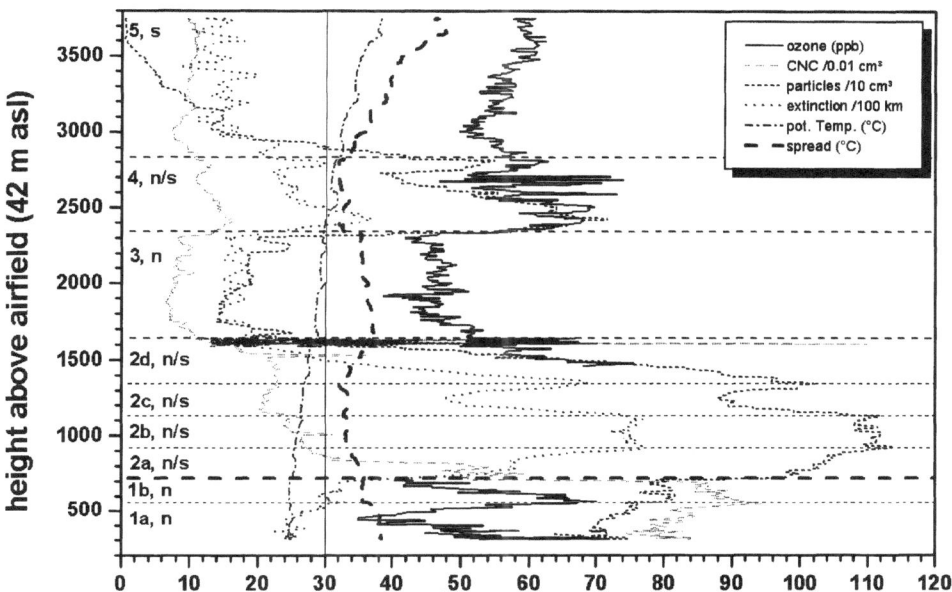

Figure 3. Vertical profiles measured by the ultra-light aircraft at 9 September 1999, 08:38–09:13 UTC: ozone concentration (ppb), extinction (1/100 km), potential temperature (°C), spread (°C). Please note: The origin for spread has been shifted to 30 (vertical line). 'n/s' denotes neutral/stable, 's' stable, and 'n' neutral thermal stratification.

CO, 140 μg m^{-3} NO$_2$ and 45 μg m^{-3} SO$_2$) were reached nearby the city. Peak values of near-ground ozone concentration of about 100 ppb were found from 7 September until 15 September. During the aircraft flights from 5 until 9 September the boundary layer height could be found at about 3,200 m (5 September), higher than 3,200 m (6 and 7 September) and at 3,500 m (9 September) near the top of layer 5 shown in Figure 3.

The scanning LIDAR measurements were processed to give vertical profiles of ozone concentration up to 1,700 m height with a vertical resolution of better than 100 m from the morning until the evening in the period 2 September until 21 September, i.e. delivering a temporal extension of the data from the aircraft flights and showing the spatial development of ozone concentration during the day. A comparison of ozone height profiles from measurements of the LIDAR and the ultra-light aircraft on 5 September is showing a typical deviation of ±10% (see Figure 4).

3.2. AEROSOL OPTICAL THICKNESS

The AOT was measured together with the solar radiation by the ground-based interference filter actinometer. As the AOT is an integral quantity it is not effected

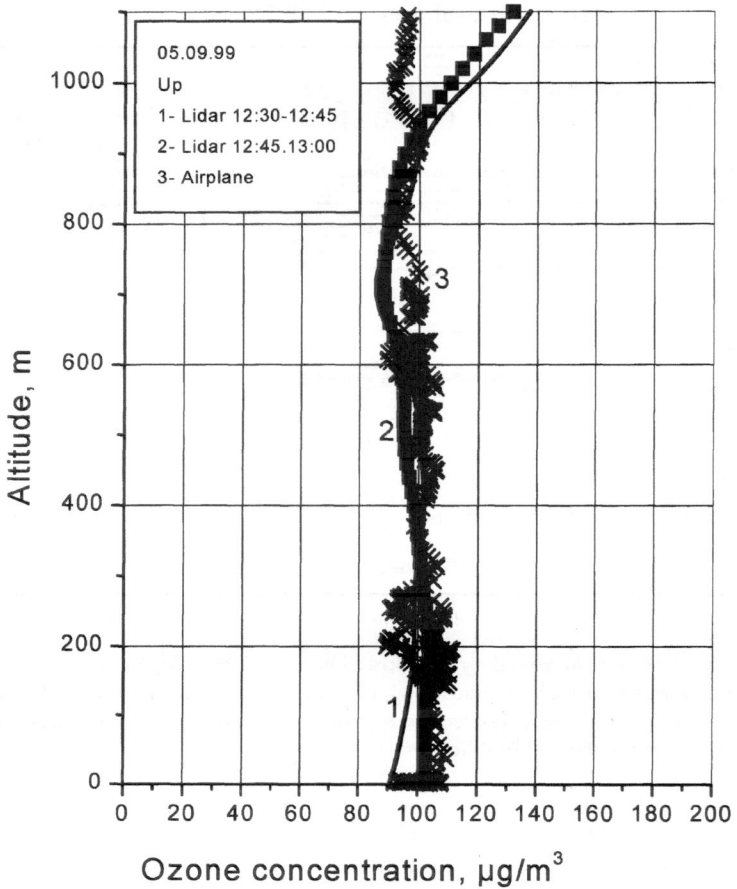

Figure 4. Comparison of ozone profiles measured by LIDAR and ultra-light aircraft at 5 September, 1999.

by the complex structure of the boundary-layer. But during comparisons of AOT with near-surface aerosol concentrations the boundary-layer structure is decisive.

The comparison with the AOT retrieved from satellite images was performed for the days 9, 12 and 13 September. High-quality satellite images were available for these days only. A channel, corresponding best to the 2nd spectral band of Landsat and to the 1st spectral band of SPOT, was the wavelength range 545.5–565.2 nm. Measurements in the time frame from 09:30 until 10:30 a.m. local time, i.e., the time of satellite passes, were used for comparison. The analysis of the direct-beam extinction measurements was performed after calibration by the Langley regression which is used for the retrieval of the MFRSR instruments as well (Harrison and Michalsky, 1994). The AOT measured from the ground at those days was 1.03, 0.500 and 0.516. These values were compared with the following AOT values extracted by SPOT satellite at the average wavelength of 550 nm at the location

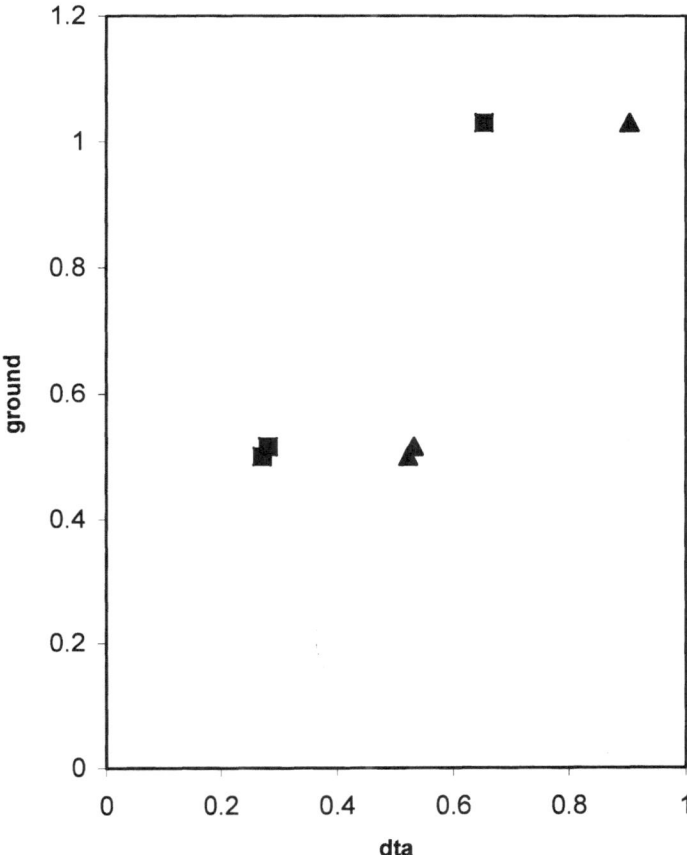

Figure 5. Quadratic dots show the actual aerosol optical thickness from satellite retrieval (dta) versus ground-based interference filter actinometer measurements (ground) at 9, 12 and 13 September, 1999. The triangular dots are with dta values shifted by 0.25 because the reference image was not totally free of aerosols. The deviation between these values and the ground values is less than ±10%.

of the ground-based instrument: 0.653, 0.271, 0.281. The algorithms are described in detail in Sarigiannis *et al.* (2001) as well as retrieval results are given in the last one. It seems that there is a shift of 0.25 in these values in comparison to the ground-based values giving values of 0.903, 0.521, 0.531 after summation. The deviation between these values and the interference filter actinometer data is less than ±10%, which is a satisfactory evaluation of the values retrieved from satellite images by these ground-based measurements of optical thickness (see Figure 5).

The main reason for the AOT shift is that the reference image which was used for the retrieval of the AOT from the actual satellite images was not totally free of aerosols.

4. Conclusions

The most important input parameter for the extraction of information about air pollutants from the AOT retrieval of satellite images is the vertical structure of the boundary layer (PBL).

The pollution level in the PBL above Brescia is influenced by local as well as regional emission sources. The dynamic structure of the PBL is dominated by the near-by Alps and the mountain breezes. Elevated heating and cooling surfaces in the nearby mountains modify the atmosphere. By advection, layers from the mountains are brought over Brescia resulting in a very complicated PBL structure and quite large PBL heights.

Future ground-truthing of this kind of satellite data retrievals for determination of air pollution is necessary and should include:

- Information about structure and height of boundary layer during satellite passes by remote sensing (backscatter LIDAR and/or DIAL ozone LIDAR up to 4,000 m altitude and/or long-range SODAR in homogeneous terrain);
- Evaluation of aerosol column density retrievals from satellite images with data from the regional or global atmospheric aerosol ground-based measurement networks (AERONET, MFRSR) which can be directly compared with high spatial resolution satellite data as e.g. from the satellite IKONOS.

The methods developed within this work will be applied at additional test sites and developed further in the framework of a new research and development project.

Due to the high horizontal resolution of satellite imagery and the information content about aerosol loading and air pollution satellite imagery will be a useful tool to study the effectiveness of ground-based air pollution networks. These satellite data will be used in the future to optimise the spatial and technical configuration of air pollution networks to deliver data with most information for the objectives for their operation.

The data described in this article were used as input to calculate secondary aerosol by an appropriate model. The modelled AOT were correlated with the AOT retrieved from satellite images during the same time (Sarigiannis *et al.*, 2001). This procedure was an important step towards the use of satellite images to characterise air pollution.

Acknowledgements

The financial support by the Commission of the European Community under Grant ENV4-CT97-0417 (project Integrated Computational Assessment via Remote Observation System (ICAROS): internet http://mara.jrc.it/icaros.html, including reports and publications) is gratefully acknowledged.

The authors like to thank Mr. Percesepe, Mr. Bissolati and Mr. Bonetti from ASM Brescia for the operation and data management of the ground-based air pollution monitoring network in Brescia and the *in situ* monitoring station in the valley near Nave (including PM10) during the campaigns as well as Dr. Wiegner and Mr. Rabus from the Meteorologisches Institut der Universität München for the possibility to work with their interference filter actinometer and for their support during data analysis.

References

Alexandrov, M. D., Lacis, A. A., Carlson, B. E. and Cairns, B.: 2001, 'MFRSR-based climatologies of atmospheric aerosols, trace gases and water vapor', in J. E. Russell, K. Schäfer and O. Lado-Bordowsky (eds), *Remote Sensing of Clouds and the Atmosphere V*, Proc. SPIE 4168, pp. 256–264.

Cachorro, V. and Tanre, D.: 1997, 'The correlation between particle mass loading and extinction: Application to desert dust aerosol content estimation', *Remote Sensing Environ.* **60**, 187–194.

Colls, J.: 1997, *Air Pollution: An Introduction*, E&FN SPON, London, 341 pp.

Dosio, A., Emeis, S., Graziani, G., Junkermann, W. and Levy, A.: 2001, 'Assessing the meteorological conditions of a deep Italian valley system by means of a measuring campaign and simulations with two models during a summer smog episode', *J. Atmosph. Environ.* **35**, 5441–5454.

Grant, W. B., Kagann, R. H. and McClenny, W. A.: 1992, 'Optical remote measurement of toxic gases', *J. Air Waste Manage. Assoc.* **42**, 18–30.

Harrison, L. and Michalsky, J.: 1994, 'Objective algorithms for the retrieval of optical depths from ground-based measurements', *Appl. Optics* **33**, 5126–5131.

Haus, R., Schäfer, K., Bautzer, W., Heland, J., Mosebach, H., Bittner, H. and Eisenmann, T.: 1994, 'Mobile FTIS-monitoring of air pollution', *Appl. Optics* **33**, 5682–5689.

Junkermann, W.: 2001, 'An ultra-light aircraft as platform for research in the lower troposphere: system performance and first results from radiation transfer studies in stratiform aerosol layers and broken cloud conditions', *J. Atmosph. Ocean Technol.* **18**, 934–946.

King, M. D., Kaufman, Y. J., Tanré, D. and Nakajima, T.: 1999, 'Remote sensing of tropospheric aerosols from space: Past, present and future', *Bull. Amer. Meteorol. Soc.* **80**, 2229–2259.

Löscher, A. and Windholz, L.: 2001, *Analyse von LIDAR-Ozon-Messungen der OLAK Messkampagne im August 1999 in Berlin auf die Möglichkeit einer Optimierung der Auswerteparameter, um die Schwankungsbreite der Messwerte zu reduzieren und die Ergebnisse zu verbessern, Interne Berichte 37*, Institut für Experimentalphysik der Technischen Universität Graz, Petersgasse 16, A-8010 Graz, 85 pp.

Reitebuch, O. and Emeis, S.: 1998, 'SODAR-measurements for atmospheric research and environmental monitoring', *Meteorologische Zeitschrift, N. F.* **7**, 11–14.

Sarigiannis, D., Soulakellis, N., Schäfer, K., Tombrou, M., Sifakis, N., Assimakopoulos, D., Lointier, M., Bossioli, E., Dantou, A. and Saisana, M.: 2001, 'ICAROS: Integrated Computational Assessment via Remote Observation System', *Final Report for the European Community funded Project ENV4-CT97-0417*, Institute for Health and Consumer Protection, EC–DG Joint Research Centre, Ispra (VA), I-21020, Italy.

Sifakis, N.: 1992, 'Potentiality of the High Spatial Resolution Satellite Imagery for Tracking of Air Pollution in the Lower Troposphere; The Athens Case Study', Ph.D. Thesis, University Paris-7, 296 pp.

Sifakis, N., Soulakellis, N. and Paronis, D.: 1998, 'Quantitative mapping of air pollution density using Earth observations: A new processing method and application on an urban area', *Internat. J. Remote Sens.* **19**, 3289–3300.

Sifakis, N. I., Sarigiannis, D., Assimakopoulos, D., Bonetti, A., Nicoloyanni, E., Lointier, M., Schäfer, K., Soulakellis, N. and Tombrou, M.: 1999, 'ICAROS: Integrated Computational Assessment via Remote Observation System using Satellite Earth Observations', in *Proceedings of the HELECO'99 International Exhibition and Conference*, Thessaloniki, 3–6 June 1999, pp. 486–495.

AIR QUALITY IN MAJOR PORTUGUESE URBAN AGGLOMERATIONS

F. FERREIRA*, H. TENTE and P. TORRES

Faculdade de Ciências e Tecnologia, UNL, DCEA, Quinta da Torre, P-2825-114 Caparica, Portugal
(* author for correspondence, e-mail: ff@fct.unl.pt, fax: (+351) 21 294 8374)

Abstract. This paper presents some recent research work that has been developed for the major Portuguese agglomerations. Three main topics are developed: the methodology used in Portugal to limit agglomerations (as they are defined by the European Air Quality Framework Directive 96/62/EC), the preliminary assessment of the air quality levels in the most densely populated agglomerations over the last five years and their influence on the air quality levels across the country (a requirement by Directive 96/62/EC), and the use of an air quality index to raise public awareness about air quality levels. It is concluded that particulate matter is the critical pollutant in Portuguese populated urban areas. In Lisbon and Oporto, based in 1999 data, in all monitoring stations, the daily average limit value of 50 μg m^{-3} for particulate matter (PM$_{10}$) is exceeded more times during a year period then allowed by Directive 99/30/EC. In the same areas, nitrogen dioxide concentrations are above the annual limit value of 40 μg m^{-3} for the protection of human health set by Directive 99/30/EC, and influence pollution concentrations within a few tens of kilometres surrounding the urban areas.

Keywords: air quality index, diffusive sampling, urban air quality monitoring, Lisbon, Oporto

1. Introduction

The main goal of this paper is to present four major topics related with the implementation of the European Air Quality Framework Directive 96/62/EC and the first Daughter Directive 99/30/EC in Portugal: the definition of most populated areas (agglomerations), the air quality levels of the most populated agglomerations (Lisbon and Oporto), and the use of an air pollution index for public information purposes.

The impacts of urban areas on the national air quality levels were also evaluated as a result of a diffusive sampling national campaign performed for nitrogen dioxide, sulphur dioxide, and ozone. Diffusive sampling technique is an ideal tool for large scale air pollution surveys. Pollutant concentrations can be mapped over the sampling area by interpolation of the measurements. The technique has been applied for pollutants such as sulphur dioxide, nitrogen dioxide, ozone and benzene, both as a complement of monitoring networks data, as part of assessment studies or as a support to optimise the location of monitoring stations (Aalst *et al.*, 1998). Air quality has been successfully assessed in many European cities such as Madrid and Brussels through the referred technique (De Seager *et al.*, 1995). The distribution density of samplers varied according to the representativeness of each

Water, Air, and Soil Pollution: Focus **2**: 103–114, 2002.
© 2002 *Kluwer Academic Publishers.*

measurement site. In the United Kingdom, a national monitoring network using diffusive samplers was set up to evaluate nitrogen dioxide concentrations in traffic hot-spots and urban background sites (Bush *et al.*, 2000). In the United Kingdom, this technique was selected to assess air quality levels based in the quality of results, cost, and easiness of use.

Most of the applications of diffusive sampling cover large urban areas with a high spatial resolution. However, examples of coverage of background levels in an entire country at the same time were not found in the literature. Due to the relatively small size of Portugal (approximately 91,000 square kilometres), this experiment was feasible with an acceptable spatial resolution, with low cost and easy operation. The uncertainty associated with the method was not a restriction since the major objective was to evaluate the range of concentrations of the different air pollutants and the general spatial trends throughout the country.

2. Methodology

2.1. THE IDENTIFICATION OF AGGLOMERATIONS

As a result of the transposition of the European Framework Directive to Portuguese law (Decreto-Lei n° 276/99), an agglomeration is defined as 'a zone where the number of inhabitants is higher than 250,000 or when the total population is smaller or equal to that number, but higher than 50,000 inhabitants, and the population density is higher than 500 inhab. km^{-1}. The methodology used to identify such zones used the lowest political boundary defined in Portugal (freguesias). A municipality is a group of one or more of these spatial units. Population from the 1991 census was used and temporary summer population for certain areas was also considered. A total of thirteen agglomerations were defined according to the described criteria of both number of inhabitants and population density.

2.1.1. *The Evaluation of Existing Air Quality Levels of Monitoring Stations in Lisbon and Oporto*

As part of the preliminary assessment required by Directive 99/30/EC, an evaluation of the concentration levels for three pollutants was performed at the country scale:
– Particulate matter as $PM_{10}(PM_{10})$;
– Nitrogen dioxide (NO_2);
– Sulphur dioxide (SO_2).

A more detailed analysis was developed for the most populated agglomerations of Lisbon and Oporto. In this task, trends for the last five years (1995–1999) were identified. Table I presents the stations and pollutants measured in each station used for this purpose.

TABLE I

Stations and pollutants measured in Lisbon and Oporto agglomerations

Stations	PM_{10}	NO_2	SO_2
Oporto			
Faculdade de Engenharia		x	x
Formosa		x	x
Vila Nova de Gaia		x	x
Paranhos	x	x	x
Custóias	x	x	x
Baguim		x	
Ermesinde	x		x
Vila Nova da Telha	x	x	x
Lisbon			
Rua da Prata		x	
Chelas		x	x
Beato		x	x
Olivais	x	x	x
Entrecampos	x	x	x
Avenida Casal Ribeiro		x	
Benfica		x	x
Avenida da Liberdade	x	x	

Hourly and daily data for the year 1999 were used to check compliance with the Directive limits, both with the 1999 year limits with the considered margin of tolerance and the limits fixed for 2005 and 2010, depending on the specific pollutant legislation. For example, a daily average limit value of 50 μg m^{-3} for particulate matter (PM10) should not be exceeded more then thirty-five times during a year period by 2005.

2.1.2. *The Development of an Air Quality Index for the Major Urban Areas*
Public information is considered to be an important part of an air quality strategy. The development of an index facilitates the way people understand air quality levels in each agglomeration. A review of several air quality indexes such as the Air Quality Index from USEPA, USA and ATMO from ADEME, France was performed. A proposal for a specific index to be applied consistently across all the Portuguese territory, particularly in the most relevant areas such as Lisbon and Oporto, was developed. The calculation basis for the Air Quality Index is the

TABLE II

Classification of the Portuguese air quality index (example for year 1999)

Pollutant concentrations $(\mu g\ m^{-3})/$ Classification	8 h – CO		1 h – NO$_2$		1 h – O$_3$		Daily – PM$_{10}$		1 h – SO$_2$	
	From	To	From	To	From	To	From	To	From	To
Bad	10000	–	300	–	240	–	75	–	500	–
Weak	5000	10000	200	300	180	240	50	75	300	500
Medium	3000	5000	100	200	100	180	30	50	125	300
Good	1000	3000	50	100	50	100	10	30	50	125
Very good	0	1000	0	50	0	50	0	10	0	50

CO – carbon monoxide; NO$_2$ – nitrogen dioxide; O$_3$ – ozone; PM$_{10}$ – particulate matter < 10 μm; SO$_2$ – sulphur dioxide.

following: the worst daily (for particulate matter in terms of PM$_{10}$, only), 8 h (for carbon monoxide, only) or hourly concentration (for all other pollutants) is selected from the different monitoring stations that belong to the same agglomeration. Then, for each pollutant, the worst average is classified in five categories according to Table II.

These categories were selected both in terms of the limit and alert values (including the decrease in the limit values that is a consequence of the decrease of tolerance margins). There are minimum requirements to define a certain index for any agglomeration, such as a minimum of two monitoring stations (all automatic monitoring stations are included, independently of their type) for each pollutant with, at least, 75% of data capture. This index is calculated on a daily basis. An estimate is published daily at 17.00 h based on data from midnight to 15.00 h.

2.2. THE DIFFUSIVE SAMPLING TECHNIQUE APPLIED TO PORTUGAL: THE EVALUATION OF THE URBAN INFLUENCE

This measurement campaign took place during two weeks in the summer period, between July, 17th and July, 31st 2000. The samples were exposed during one week in background locations. A systematic sampling was used such that the country's territory was divided in squares of 20 by 20 km. The centre of each square was always used to locate a sample, except where this was impossible due to accessibility problems or proximity of major air pollution sources, since the objective was to access background concentration levels. Radial diffusive samplers from Radiello® were used to measure nitrogen dioxide, sulphur dioxide and ozone. Diffusive sampling is one of the recommended air quality assessment methods by the European Environmental Agency (Aalst et al., 1998). The option of using diffusive sampling to perform this research was based: a) by the possibility of working in the same time scale for all the country, allowing a comparison of the air quality in different areas for the same period, and b) by the spatial distribution that you can cover with this kind of equipment (Stevenson *et al.*, 1999).

3. Results and Discussion

3.1. THE IDENTIFICATION OF AGGLOMERATIONS

The agglomeration of Lisbon has a total population of 1,740,288 inhabitants with an area of 482 km². Oporto's agglomeration has 1,248,952 inhabitants and an area of 697 km². Figure 1 shows the relative location of the two major agglomerations of Lisbon and Oporto as well as the number and type of monitoring stations that where used in this study. Some other stations exist but they are not working yet.

Currently, all the stations for the Lisbon agglomeration are concentrated only in the Lisbon municipality while for Oporto their geographic cover is better.

3.1.1. *The Evaluation of Existing Air Quality Levels of Monitoring Stations in Lisbon and Oporto*

Tables III and IV present the status and trends of particulate matter and nitrogen dioxide. Particulate levels are above the 99/30/EC European Directive limit values in most monitoring stations, while for nitrogen dioxide problems are also arising. Between 1995 and 1999, and through a graphical analysis of the data, a general decreasing trend was found for most stations. For nitrogen dioxide, there is no relationship between the station type and the levels, even though an extended analysis of the data shows that these levels are mostly due to traffic (Ferreira *et al.*, 1998a; Ferreira *et al.*, 1999b).

3.1.2. *The Development of an Air Quality Index for the Major Urban Areas*

Figure 2 shows a significant number of days with weak and bad quality, predominantly due to the particulate levels, with a worse situation for Oporto. The intensity of traffic plays a definite role in this result.

3.2. THE DIFFUSIVE SAMPLING TECHNIQUE APPLIED TO PORTUGAL: THE EVALUATION OF THE URBAN INFLUENCE

Figure 3 shows that background nitrogen dioxide levels are relatively low throughout the country. However, concentrations are clearly higher close to Lisbon, Oporto and Braga urban areas within the direction of predominant winds measured during the sampling period. Traffic and some industrial influence contribute to this result (Hargreaves *et al.*, 2000; Lebret *et al.*, 2000).

Sulphur dioxide higher concentrations are excellent mirrors of the locations of the Portuguese larger industrial areas, all located near the Atlantic coast (Figure 4). The influence of a power plant and a refinery and petrochemical industrial

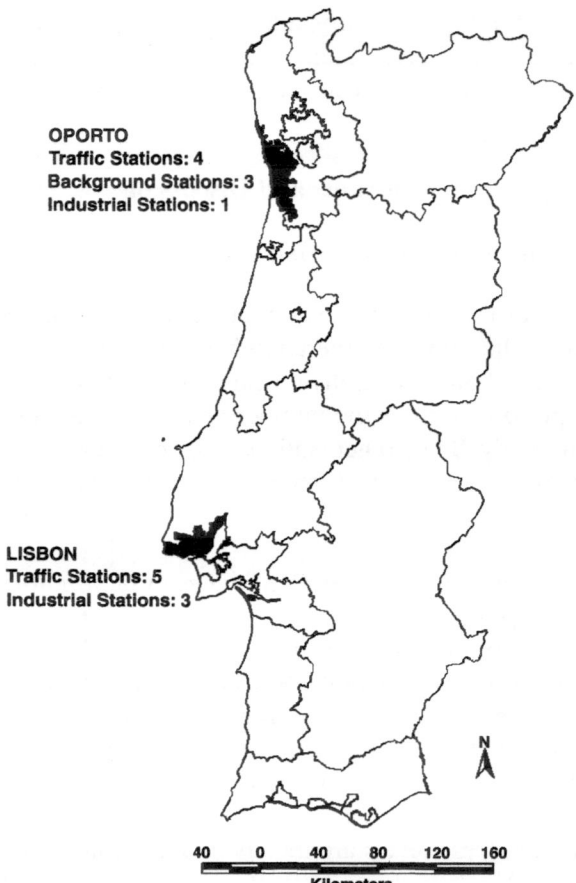

OPORTO
Traffic Stations: 4
Background Stations: 3
Industrial Stations: 1

LISBON
Traffic Stations: 5
Industrial Stations: 3

```
40    0    40    80    120    160
                Kilometers
```

Figure 1. Number of operating air quality monitoring stations in Lisbon and Oporto agglomerations used in this study.

complex located in Sines, in the Southwest of Portugal is clearly visible. Also, other industrial areas located Southwest of Lisbon (power plant and chemical industry), in Estarreja (chemical industry), and Oporto (refinery), are easily identified.

High ozone levels are located inland from the major urban areas (Figure 5). Observations of the wind patterns during the sampling period suggested that those ozone levels do reflect the precursors emissions, mainly located near the two major cities (Lisbon and Oporto) in the coastal area, since wind was blowing almost exclusively from West. Peak concentrations coincide with higher altitudes. This may be related with higher solar radiation levels in these areas.

TABLE III

Trends and evaluation of particulates levels (PM$_{10}$) measured continuously by β-attenuation in the agglomerations of Lisbon and Oporto

Station	Station type	Trend	Health protection				
			Daily values				Annual values
			LV2000	LV2005	UAT	LAT	
Oporto							
Paranhos	Traffic	?	☺	☺	☺	☺	> LV2000
Custóias	Industrial	?	☺	☺	☺	☺	> LV2000
Ermesinde	Background	?	☺	☺	☺	☺	> LV 2000
Vila Nova da Telha	Background	?	☺	☺	☺	☺	[LV2005, LV2000]
Lisbon							
Olivais	Background	?	✓	☺	☺	☺	[UAT, LV2005]
Entrecampos	Traffic	↘	✓	☺	☺	☺	[LV2005, LV2000]
Avenida da Liberdade	Traffic	↘	☺	☺	☺	☺	> LV2000

Trends based on data from the years 1995 to 1999; Limits and thresholds evaluation for the year 1999.
? – No detectable trend; only 1999 year available.
Different intensity of concentration trends from upward (↑) to downward (↓).
✓ – below the limit or threshold.
☺ – above the limit or threshold.
LV – limit value; UAT – upper assessment threshold; LAT – lower assessment threshold.

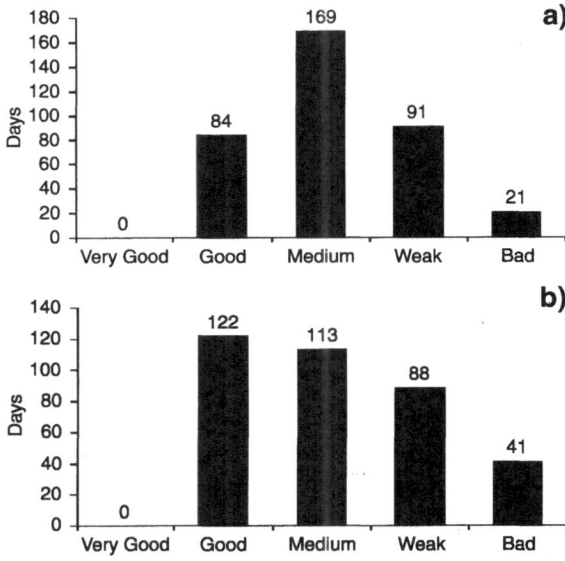

Figure 2. Air quality index results for days in 1999 for Lisbon (a) and Oporto (b).

TABLE IV

Trends and evaluation of nitrogen dioxide (NO_2) measured continuously by chemiluminescence in the agglomerations of Lisbon and Oporto

Station	Station type	Trend	Alert threshold	Health protection				Annual values
				Hourly values				
				LV2000	LV2005	UAT	LAT	
Oporto								
Faculdade Engenharia	Traffic	↘	✓	✓	✓	✓	✓	[UAT, LV2010]
Formosa	Traffic	↑	✓	✓	✓	✓	✓	[LAT, UAT]
Vila Nova de Gaia	Traffic	?	✓	✓	☺	☺	☺	> LV2000
Paranhos	Traffic	?	✓	✓	✓	☺	☺	[LV2010, LV2000]
Custóias	Industrial	?	✓	✓	✓	✓	☺	[UAT, LV2010]
Baguim	Background	?	✓	✓	✓	✓	✓	✓
Vila Nova da Telha	Background	?	✓	✓	✓	✓	✓	✓
Lisbon								
Rua da Prata	Traffic	↘	✓	✓	✓	☺	☺	> LV2000
Chelas	Background	↘	✓	✓	✓	☺	☺	[UAT, LV2010]
Beato	Background	→	✓	✓	✓	☺	☺	✓
Olivais	Background	↗	✓	☺	☺	☺	☺	[LV2010, LV2000]
Entrecampos	Traffic	↘	✓	✓	✓	☺	☺	[UAT, LV2010]
Avenida Casal Ribeiro	Traffic	↘	✓	✓	☺	☺	☺	[LV2010, LV2000]
Benfica	Traffic	→	✓	✓	✓	☺	☺	[LV2010, LV2000]
Avenida da Liberdade	Traffic	↘	✓	✓	☺	☺	☺	[LV2010, LV2000]

Trends based on data from the years 1995 to 1999; Limits and thresholds evaluation for the year 1999.
? – No detectable trend; only 1999 year available.
Different intensity of concentration trends from upward (↑) to downward (↓).
✓ – below the limit or threshold.
☺ – above the limit or threshold.
LV – limit value; UAT – upper assessment threshold; LAT – lower assessment threshold.

4. Conclusions

Particulate matter is the critical pollutant in the two major Portuguese agglomerations of Lisbon and Oporto. Nitrogen dioxide generated in urban areas extends its influence within a few tens of kilometres but with relatively low concentrations,

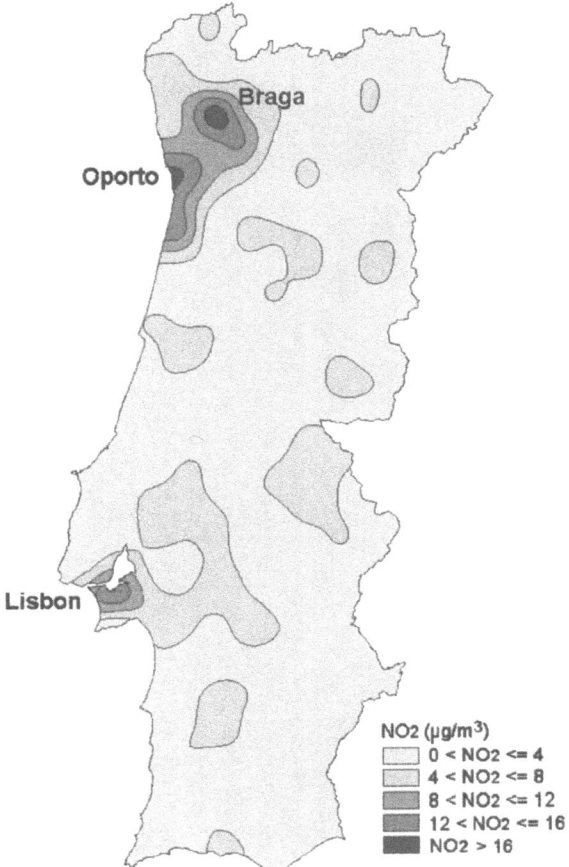

Figure 3. Distribution of nitrogen dioxide in Portugal, one week averaging period, July 2000 (in μg m^{-3}).

particularly in the North. Traffic emissions play a major role. Ozone presents high concentrations, especially in the inland area. This result seems to be related with the precursors emissions located in the major urban areas in the coastal area and their wind transport from those areas. It is also clear that ozone concentrations are lower where nitrogen dioxide is higher. This situation reflects the primary reactions involving ozone and ozone precursors, such as nitrogen oxides, in urban areas. This conclusion highlights, with a high degree of probability, another important consequence of the urban activity in terms of air quality because ozone emerges as an inter-regional problem displacing sources and effects and making harder to implement control measures. However, future studies should evaluate in detail the causes of the high concentration levels detected. As a tool to provide better public information and awareness, a daily air quality index was successfully developed since it reflects the high pollutant levels found according to European legislation.

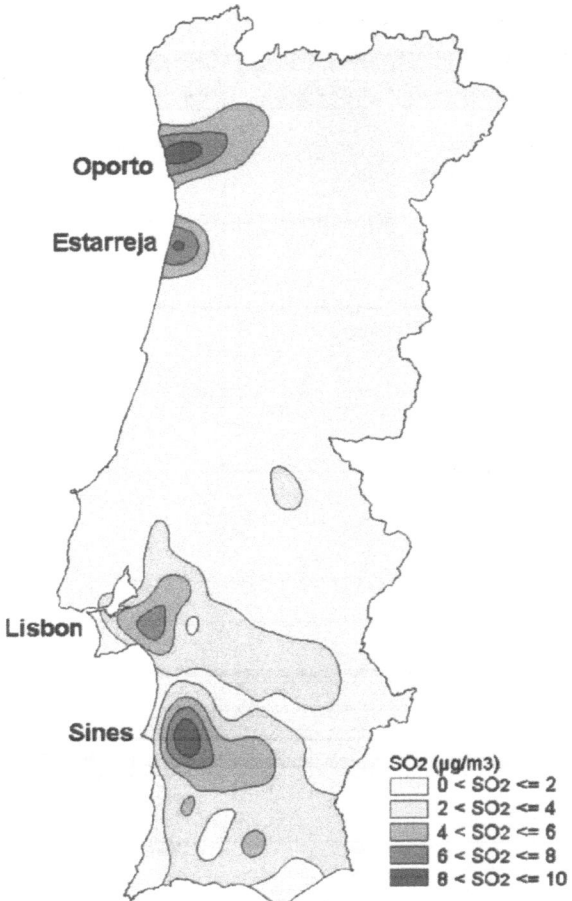

Figure 4. Distribution of sulphur dioxide in Portugal, one week averaging period, July 2000 (in μg m^{-3}).

Acknowledgements

The work presented is the result of a research and advising project developed by the Department of Sciences and Environmental Engineering of the Faculdade de Ciências e Tecnologia/Universidade Nova de Lisbon to the Ministério do Ambiente e do Ordenamento do Territćrio through Direcção Geral do Ambiente (DGA). We would like to acknowledge all the efforts from staff at DGA and at the State environmental regional offices (Direcções Regionais do Ambiente e do Ordenamento do Território).

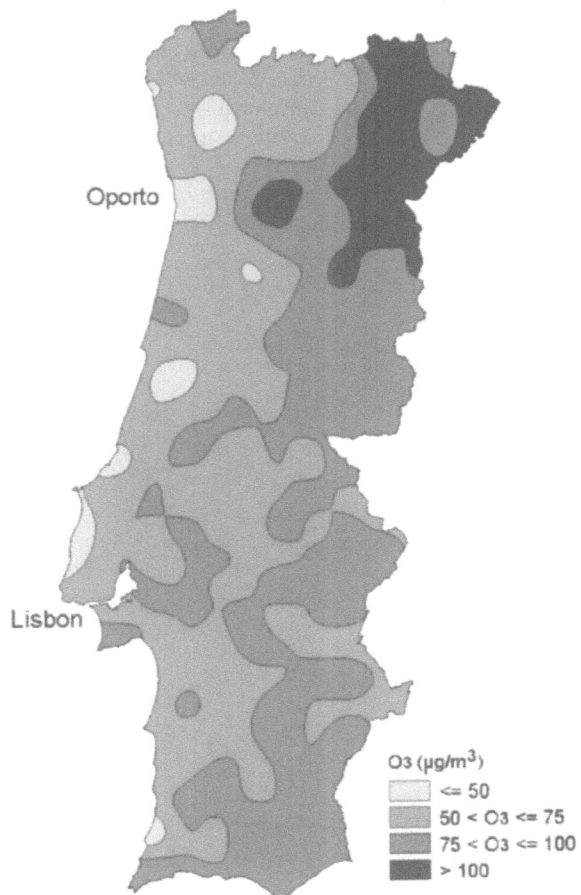

Figure 5. Distribution of ozone in Portugal, one week averaging period, July 2000 (in μg m^{-3}).

References

Aalst, R., Edwards, L., Pulles, T., De Saeger, E., Tombrou, M. and Tønnesen, D.: 1998, 'Guidance Report on Preliminary Assessment under EC Air Quality Directives', European Environment Agency.

Bush, T., Mooney, D. and Loader, A.: 2000, 'UK Nitrogen Dioxide Diffusion Tube Network Instruction Manual', AEA Technology.

De Seager, E., Gerbolès, M., Perez Ballesta, P., Amantini, L. and Payrissat, M.: 1995, 'Air Quality Measurements in Brussels (1993–1994) – NO$_2$ and BTX Monitoring Campaigns by Diffusive Samplers', EUR Report 16310 EN.

Ferreira, F., Seixas, J., Nunes, C. and Silva, J. P.: 1998, 'Air pollution space-time analysis', in I. Linkov and R. Wilson (eds), *Air Pollution in the Ural Mountains – Environmental, Health and Policy Aspects*, Kluwer Academic Publishers, Dordrecht, 357–360.

Ferreira, F., Tente, H., Torres, P., Cardoso, S. and Palma-Oliveira, J. M.: 1999, 'Air Quality Monitoring and Management in Lisbon, 2nd Urban Air Quality Conference', Institute of Physics, Madrid, 3–5 March 1999.

Hargreaves, P. R., Leidi, A., Grubb, H. J., Howe, M. T. and Mugglestone, M. A.: 2000, 'Local and seasonal variations in atmospheric nitrogen dioxide levels at Rothamsted, UK, and relationships with meteorological conditions', *Atmos. Environ.* **34**, 843–853.

Lebret, E., Briggs, D., van Reeuwijk, H., Fischer, P., Smallbone, K., Harssema, H., Kriz, B., Gorynski, P. and Elliott. P.: 2000, 'Small area variations in ambient NO_2 concentrations in four European areas', *Atmos. Environ.* **34**, 177–185.

Stevenson, K., Bush, T. and Mooney D.: 1999, 'Five years of nitrogen dioxide measurement with diffusion tube samplers at over 1000 sites in the UK', *Atmos. Environ.* **35**, 281–287.

EMISSION MODELS AND INVENTORIES FOR LOCAL GREENHOUSE GAS EMISSIONS ASSESSMENT: EXPERIENCE IN THE EAST MIDLANDS REGION OF THE UK

P. H. WEBBER[1]* and P. D. FLEMING[2]

[1] *Leicester Energy Agency, Leicester, UK;* [2] *De Montfort University, Leicester, U.K. (* author for correspondence, e-mail: webbp900@leicester.gov.uk, fax: +44 (0)116 299 5137; Present address: Leicester Energy Agency, Leicester Energy Efficiency Centre, 2-4 Market Place South, Leicester. LE1 5HB, U.K.)*

Abstract. This paper considers how regional greenhouse gas emission inventories can be determined. It presents a greenhouse gas emissions inventory, by source, for the East Midlands that has been compiled as part of a regional study into climate change impacts in the United Kingdom. This has used available local data, and national emissions data with appropriate scaling factors. Total greenhouse gas emissions for the region are estimated to be 59 million tonnes of carbon dioxide equivalent (CO_2 equivalent) for 1997. Of these emissions, approximately 86% were carbon dioxide emissions, 7% methane emissions, and 5% were nitrous oxide, with emissions of hydrofluorocarbons (HFCs), perfluorocarbons (PFCs) and sulphur hexafluoride (SF_6) contributing less than 2% of total emissions.

Keywords: East Midlands of UK, emission inventories, emission models, emissions assessment, local greenhouse gas emissions

1. Introduction

Developed countries have set greenhouse gas emissions reduction targets following the United Nations Conference of the Parties to the Framework Convention on Climate Change at Kyoto in December 1997. In order to meet these targets greenhouse gas emissions are being assessed and action is taking place not only at the national level (e.g. DETR, 2000) but also at the local and regional administrative levels. Local energy planning and greenhouse gas emissions reduction activities have been taking place worldwide, with examples in Europe and the United Kingdom (UK).

This paper describes how an inventory of greenhouse gas emissions has been produced for the East Midlands Sustainable Development Round Table as part of their work looking at potential climate change impacts in the region (Kersey *et al.*, 2000; Shackley *et al.*, 2001).

The UK's East Midlands has a population of about 4.2 million, and contains Derbyshire, Leicestershire, Lincolnshire, Northamptonshire and Nottinghamshire County Councils, Derby, Leicester, Nottingham, and Rutland Unitary Councils, and thirty six other local authorities at district level. A map of the East Midlands,

Water, Air, and Soil Pollution: Focus **2:** 115–126, 2002.
© 2002 *Kluwer Academic Publishers.*

Figure 1. A map of the East Midlands region, naming the main cities and towns and showing major power station emissions sources.

showing the main cities and towns and major power stations in the region is given in Figure 1.

Previous studies in the East Midlands include the use of the Dynamic Regional Energy Analysis Model (DREAM), for the city of Leicester (Titheridge *et al.*, 1996), and the county of Leicestershire (Leicestershire County Council and De Montfort University, 1995). Different energy end-use sectors and related carbon dioxide emissions were considered.

2. Methodology

Many local greenhouse gas emissions inventories have been prepared worldwide, for example as part of the International Council for Local Environmental Initiatives' Cities for Climate Protection campaign. In Europe, several local authorities have taken an active role in addressing climate change, and have undertaken energy/carbon dioxide emission studies and greenhouse gas emissions reduction activities (e.g. Collier, 1997).

Different approaches have been used to compile local inventories. Studies have used either a spreadsheet approach or energy-related greenhouse gas emission models. Examples from the East Midlands include the use of an energy model ('DREAM') to simulate energy flows through European Commission supported projects in the city of Leicester (Titheridge *et al.*, 1996) and in the county of Leicestershire (Leicestershire County Council and De Montfort University, 1995), where the domestic, industry, services and transport energy end use sectors and related carbon dioxide emissions were considered.

2.1. REVIEW OF AVAILABLE ENERGY-RELATED EMISSIONS MODELS

Several computer models have been used at the local level in different parts of the world to determine energy related greenhouse gas emissions. Examples of the main local energy and emission models suitable for use at the local level in the UK are:

i) TEMIS (Total Emission Model for Integrated Systems)

The software has been developed by the Oko-Institut and the University of Kassel, Germany, and has been used in a number of areas including Heidelberg and Hanover (Germany), Rovigo (Italy), and Newcastle upon Tyne (UK) (Newcastle City Council, 1992). A particular feature of TEMIS is that it considers lifecycle emissions, while a potential disadvantage for its use in the UK is that a range of UK specific data is required.

ii) EEP (Energy and Environmental Prediction) Model

The model has been developed in the UK by the University of Wales, Cardiff, (University of Wales Cardiff, 2001) and has been used in Cardiff and for part of Leicester, for example. An advantage of the EEP model is that it can estimate local energy use and carbon dioxide emissions, while a potential disadvantage is that the software has been under development.

iii) DREAM (Dynamic Regional Energy Analysis Model)

This was developed by the Open University's Energy and Environment Research Unit (UK) and has been used for Barcelona (Spain), and Leicester (Titheridge *et al.*, 1996), Leicestershire, and Milton Keynes (UK). An advantage of the DREAM model is that it can determine energy consumption and carbon dioxide emissions, while a disadvantage is that a range of input data is required.

iV) CCP (Councils for Climate Protection) Greenhouse Gas Emission Software

The software has been developed by Torrie Smith Associates (Canada) for the International Council for Local Environmental Initiatives (ICLEI) Climate Protection campaign, and has been used by several local governments in different parts of the world, including the United States, Australia, and the United Kingdom. The software (TSA, 2000) is relatively easy to use and, while concentrating on carbon dioxide (CO_2) emissions related to energy use and methane (CH_4) emissions from landfill sites an advantage of the software is that it can be used to consider the six greenhouse gas emissions in the Kyoto climate change agreement.

Comparative information about the models is summarised in Table I.

In preparing an inventory for the East Midlands, the greenhouse gases and their sources were determined from available regional data, national data and from previous regional energy and greenhouse gas emissions studies.

TABLE I

Example energy and emission models for assessing local greenhouse gas emissions

TEMIS	EEP	DREAM	CCP software
Description			
Determines the life cycle emissions related to material, transport or energy processes	Model may be used to calculate energy consumption and related emissions in energy end-use sectors of an area both for the present and the future	Simulates energy flows for certain area and able to be used to develop future energy use scenarios	Aids in preparing a greenhouse gas emission inventory and in monitoring progress towards an emission reduction target
Sectors considered			
Domestic, services/ commercial, industry, transport	Domestic, non-domestic buildings, industrial processes, transport	Domestic, services/ commercial, industry, transport	Residential, commercial, industrial, transport, waste, other
Data input			
Fuels and process data for UK, energy consumption, passenger-km, tonne-km	Includes data on housing built form, and output of industrial product	Includes temperature, population, rate of domestic heat loss, building floor area, fuel shares, for example	Includes emission factors for UK, energy consumption, vehicle- km, waste quantities
Data outputs			
Include SO_2, NO_x, particulates, CO_2, CH_4, and N_2O	Include energy consumption and carbon dioxide emissions	Include energy demand, CO_2 and other emissions	Include CO_2, N_2O, CH_4, NO_x, SO_x, CO

Other local greenhouse gas emissions inventories are currently being prepared in different areas in the UK, as part of the Councils for Climate Protection pilot project of 24 local authorities in England and Wales. Six local administrative areas in the East Midlands region are participating in this pilot project which uses the CCP greenhouse gas emissions software originally developed as part of the International Council for Local Environmental Initiatives (ICLEI) Cities for Climate Protection campaign (Torrie Smith Associates, 2000). This software has been designed specifically for use in the preparation of greenhouse gas emission inventories, at the local level rather than the national level. The software requires data on energy use in the domestic, commercial and industrial sectors, and transport and waste information, which is used to prepare an inventory of the six greenhouse gases (CO_2, CH_4, N_2O, HFCs, PFCs, and SF_6) included in the Kyoto agreement.

The software is used to estimate local greenhouse gas emissions for a base year (e.g. 1990) for both a local authority area (community) and for a local authority's own operations (corporate). This information is then used to specify an emissions reduction target (e.g. for 2010). The local authority can then implement measures to reduce greenhouse gas emissions and monitor its progress. An advantage of this software for local authorities is that it is quick and easy to use, it considers a number of sectors, and may be used for listing emission reduction measures and monitoring progress towards a target. Because the software requires less data than some other models it could be argued that it is less accurate.

2.2. DEVELOPMENT OF A GREENHOUSE GAS EMISSIONS INVENTORY FOR THE EAST MIDLANDS

Nationally, it is possible to calculate greenhouse gas emissions by multiplying a figure for the quantity of a certain activity, such as fuel use, by the appropriate emissions factor (e.g. greenhouse gas emissions per unit of fuel use) (e.g. NETCEN, 1999). Determining greenhouse gas emissions by this approach was not possible in the East Midlands, because there was insufficient relevant activity data at the regional level. Instead a spreadsheet approach was used and, for most sources, emissions were derived pro-rata from national data sets using greenhouse gas emissions data compiled by the National Environmental Technology Centre (NETCEN).

In the North West region's inventory (Mander *et al.*, 2000) regional emissions (E_R) from most sources were estimated by multiplying a range of pro-rata factors (using regional and national statistics, S_R and S_N respectively) by national emissions data (E_N) (Lindley *et al.*, 1996) as in:

$$E_R = (S_R/S_N) E_N$$

For the East Midlands inventory, available data on actual local greenhouse gas emissions has been used directly in the inventory. Data on local greenhouse gas emissions from large industrial sites (including power stations) was provided by the Environment Agency. However, as much relevant regional data was not available, in order to compile the initial inventory shares of national emissions data by source category have been assigned to the region, using scaling factors for the East Midlands for the different emission sources. Regional and national data (from

TABLE II

East Midlands greenhouse gas emissions, by gas (1997) (Shackley *et al.*, 2001)

Gas	East Midlands emissions (tonnes of gas)	East Midlands emissions (million tonnes CO_2 equivalent)
Carbon dioxide	13.9 Mt carbon	50.8
Methane	0.2 Mt CH_4	4.4
Nitrous oxide	9.3 kt N_2O	2.9
Hydrofluorocarbons	0.1 kt HFC	0.7
Perfluorocarbons	4 t PFC	0.03
Sulphur hexafluoride	6 t SF_6	0.1
Total emissions		59.1

Government published statistics) were used to obtain the pro-rata factors to scale national greenhouse gas emissions data for different sources to the East Midlands, for example for population, household numbers, gross domestic product, number of cars licensed, livestock numbers, and aircraft movements. Also regional and national information on waste disposal was obtained from the Environment Agency and on coal production from The Coal Authority.

Estimates of the East Midlands emissions for the six greenhouse gases included in the Kyoto climate change agreement (carbon dioxide (CO_2), nitrous oxide (N_2O), methane (CH_4), hydrofluorocarbons (HFCs), perfluorocarbons (PFCs), and sulphur hexafluoride (SF_6)) have been obtained. The inventory was compiled for 1997, which was the most recent year that necessary data was available. The greenhouse gas emissions estimates for the East Midlands were expressed both in terms of tonnes of each gas and also, by considering the relative global warming potential of the gases, as carbon dioxide equivalent.

3. Results and Discussion

3.1. EAST MIDLANDS RESULTS, BY GAS

The East Midlands assessment has estimated total greenhouse gas emissions for the region to be 59 million tonnes CO_2 equivalent. Emissions of carbon dioxide provided the largest quantity of greenhouse gas emissions (about 51 Mt CO_2 equivalent), while methane and nitrous oxide emissions contributed about 4 and 3 Mt CO_2 equivalent respectively (Table II). The accuracy of these results is discussed below.

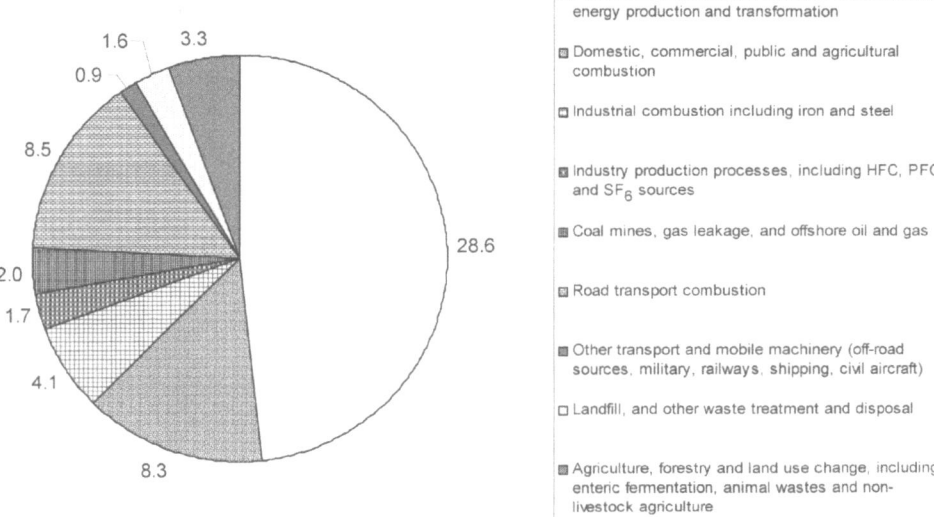

Figure 2. East Midlands greenhouse gas emissions by source (1997) (Mt CO_2 equivalent) (Shackley *et al.*, 2001).

3.2. EAST MIDLANDS RESULTS, BY SOURCE

Details of East Midlands greenhouse gas emissions by source (1997) (kt CO_2 equivalent) are given in Figure 2 and Tables III and IV.

The largest carbon dioxide emission source was from fossil fuel combustion in power generation (53% of CO_2 emissions), while other significant sources were combustion in road transport (16%), domestic sector combustion (12%), industry sector combustion (8%), and combustion in the commercial, public and agricultural sectors (4%).

The main methane emissions source was estimated to be from coal mines (31% of methane emissions) while landfill sites and enteric fermentation in livestock contributed 26% and 21% of methane emissions respectively. The majority of nitrous oxide emissions (73%) were from non-livestock agriculture (e.g. agricultural soils) with combustion in power stations and road transport contributing 12% and 8% respectively.

In the East Midlands it is estimated that the main HFC emission sources are refrigeration/air conditioning and aerosols, while the main PFC emission source is electronics (and training shoes), and the main SF_6 source is magnesium manufacture.

3.3. DISCUSSION

It is difficult to prepare accurate regional inventories owing to limited data availability. As scaled national data has been used with locally available data the data is

TABLE III

East Midlands greenhouse gas emissions, by source (1997) (Mt CO$_2$, equivalent) - CO$_2$, CH$_4$, N$_2$O (Shackley et al., 2001)

Emission source	Carbon dioxide			Methane			Nitrous oxide		
	UK	East Mids	% of Total EM GHG	UK	East Mids	% of Total EM GHG	UK	East Mids	% of Total EM GHG
Power stations, refineries, and other combustion in energy production and transformation	185.9	28.2	47.7	0.4	0.07	0.1	2.0	0.3	0.6
Domestic, commercial, public and agricultural combustion	116.0	8.2	13.9	0.8	0.06	0.1	0.2	0.02	0.0
Industrial combustion including iron and steel	87.7	4.1	7.0	0.2	0.05	0.0	0.5	0.02	0.0
Industry production processes	12.1	0.8	1.4	0.2	0.01	0.0	21.5	0	0.0
Coal mines, gas leakage, and offshore oil and gas	0.6	0.04	0.1	15.5	2.0	3.3			
Road transport combustion	116.3	8.2	13.9	0.4	0.03	0.1	3.3	0.2	0.4
Other transport and mobile machinery	15.9	0.8	1.3	0.05	0.003	0.0	0.9	0.07	0.1
Landfill, and other waste treatment and disposal	5.7	0.4	0.7	18.5	1.2	2.1	0.2	0.01	0.0
Agriculture, forestry and land use change	1.3	0.1	0.2	21.1	1.1	1.8	30.6	2.2	3.7
Totals	**541.7**	**50.8**	**86.1**	**57.3**	**4.4**	**7.5**	**59.3**	**2.9**	**4.9**

TABLE IV

East Midlands greenhouse gas emissions, by source (1997) (Mt CO_2 equivalent) – HFCs, PFCs, SF_6 (Shackley *et al.*, 2001)

Emission source	HFCs			PFCs			SF_6		
	UK	East Mids	% of Total EM GHG	UK	East Mids	% of Total EM GHG	UK	East Mids	% of Total EM GHG
HFC sources (halocarbon production, fire-fighting, electronics, foams, refrigeration, aerosols)	18.4	0.7	1.2						
PFC sources (fire-fighting, aluminium production, halocarbon production, refrigeration, electronics and training shoes)				0.7	0.03	0.05			
SF_6 sources (magnesium manufacture, electrical insulation, electronics and training shoes)							1.3	0.1	0.3
TOTALS	**18.4**	**0.7**	**1.2**	**0.7**	**0.03**	**0.05**	**1.3**	**0.1**	**0.3**
E. MIDS TOTAL GREENHOUSE GAS EMISSIONS	**59.1 Mt CO_2 equivalent**								

not accurate but has provided a first estimate of greenhouse gas emissions for the East Midlands region.

The accuracy of the results depends particularly on the accuracy of the UK emissions data used and how closely the pro-rata factors used for different emission sources relate to the ratio of actual local emissions and national emissions. The inaccuracies associated with the pro-rata factors are uncertain. However, for the national emissions data used from the 1997 National Atmospheric Emissions Inventory Report NETCEN considered the uncertainties to be: carbon dioxide ± 4%, methane ± 17%, nitrous oxide –55% to + 234%, HFCs ± 25%, PFCs ± 20%, SF_6 ± 13%.

The region's emissions represented about 8.7% of UK emissions, which is a larger proportion than would be expected by considering the share of the UK's population in the East Midlands (7%). This variation is largely due to the above average concentration of power generation in the region.

A large share of total greenhouse gas emissions (48%) was estimated to be from combustion in energy production and transformation (predominantly from power station emissions). Since emissions by source rather than by end user have been considered, and in the East Midlands more power is generated than is used a larger emissions estimate has resulted than would otherwise be the case.

While the use of pro-rata factors with national emissions data has enabled baseline figures to be estimated, the use of a similar approach with updated data to monitor trends over time could give a limited indication of regional emissions trends although this would not be particularly sensitive to local variations.

The lack of easily available energy data at the local level in recent years has been found to be a difficulty in local greenhouse gas emissions assessment in the UK. Although some local energy data has been made available recently to local authorities participating in the ongoing CCP pilot project the lack of data will cause difficulties when trying to monitor the Region's progress in reducing emissions. A recent regional energy study (Land Use Consultants and IT Power, 2001) includes energy-related targets for the Region, including for energy efficiency, and considers the effect on carbon dioxide emissions. In order to monitor progress to such targets availability of relevant data will be a crucial issue.

4. Conclusions

There are several methodologies for compiling local greenhouse gas emissions inventories in the UK. Examples include using a spreadsheet approach, models such as the TEMIS, DREAM and EEP models and the CCP software, using actual local data where available with estimates of local emissions by scaling from national figures, or a combination of approaches.

The comprehensive greenhouse gas emissions inventory for the East Midlands was produced through a spreadsheet apportionment approach and the use of available local data. On a source basis, approximately 59 Mt CO_2 equivalent of the six greenhouse gases included in the Kyoto agreement were estimated for the East Midlands (1997). The majority of greenhouse gas emissions were of carbon dioxide with 53% of CO_2 emissions associated with power generation. 7% of total greenhouse gas emissions were methane emissions, with the largest sources being coal mines, landfill sites and livestock.

The lack of easily available energy data at the local level is a common difficulty in local greenhouse gas emissions assessment in the UK. In preparing the East Midlands inventory, difficulties have been encountered obtaining sufficient local data and the need for better data has been recognised by the East Midlands Sustainable Development Round Table, the Councils for Climate Protection pilot authorities, and others. This will be a particular problem when the Region tries to monitor its emissions over time to measure its progress in reducing greenhouse gas emissions.

Acknowledgements

The compilation of the greenhouse gas emissions inventory for the East Midlands was carried out at the Institute of Energy and Sustainable Development, De Montfort University. It was part of the work funded by the East Midlands Sustainable Development Round Table on the Impacts of Climate Change on the East Midlands (Kersey *et al.*, 2000) undertaken by ENTEC UK Ltd, University of Manchester Institute of Science and Technology (UMIST), the University of Derby and De Montfort University.

References

Collier, U.: 1997, 'Local authorities and climate protection in the European Union: putting subsidiarity into practice?', *Local Environ.* **2**, 39–57.
DETR: 2000, *Climate Change: the UK Programme*, DETR, London.
Kersey, J., Shackley, S., Wilby, R. and Fleming, P.: 2000, *The Potential Impacts of Climate Change in the East Midlands. Summary report. July 2000*, East Midlands Sustainable Development Round Table, East Midlands.
Land Use Consultants and IT Power: 2001, 'Viewpoints on Sustainable Energy in the East Midlands: A Study of Current Energy Projects and Future Prospects', Final Report, Land Use Consultants, London.
Leicestershire County Council and De Montfort University: 1995, 'Identification and Utilisation of Energy Resources to Minimise the Effect on the Environment in Leicestershire (UK)', Final Report, Leicestershire County Council and IESD, De Montfort University, Leicester.
Lindley, S. J., Longhurst, J. W. S., Watson, A. F. R. and Conlan, D. E.: 1996, 'Procedures for the estimation of regional scale atmospheric emissions – an example from the North West of England', *Atmos. Environ.* **30**, 3079–3091.
Mander, S., Buchdahl, J., Shackley, S. and Connor, S.: 2000, 'Carbon Counting: North West England's First Inventory of Greenhouse Gas Emissions', Summary and Full Technical Report. UMIST, ARIC and North West Climate Group, Sustainability North West, Manchester.

NETCEN: 1999, 'National Atmospheric Emissions Inventory 1997 report', http://www.aeat.co.uk/netcen/airqual/naei/annreport/naei97.html.

Newcastle City Council: 1992, *Energy and the Urban Environment Strategy for a Major Urban Centre Newcastle upon Tyne, UK*, Newcastle City Council, Newcastle.

Shackley, S., Kersey, J., Wilby, R. and Fleming, P.: 2001, *Changing by Degrees. The Potential Impacts of Climate Change in the East Midlands*, Ashgate, Aldershot.

Titheridge, H., Boyle, G. and Fleming, P.: 1996, 'Development and validation of a computer model for assessing energy demand and supply patterns in the urban environment', *Energy & Environ.* **7**, 29–40.

Torrie Smith Associates: 2000, *Greenhouse Gas Emissions Software-U.K. Edition, User's Guide.* Release 4.6, December 2000, Torrie Smith Associates, Ottawa.

University of Wales Cardiff: 2001, 'Energy and Environmental Prediction', http://www.cf.ac.uk//archi/research/eep/index.html.

MEASUREMENTS OF TRAFFIC-INDUCED TURBULENCE WITHIN A STREET CANYON DURING THE NANTES'99 EXPERIMENT

G. VACHON[1], P. LOUKA, J-M. ROSANT*, P.G. MESTAYER and J-F. SINI

Laboratoire de Mécanique des Fluides UMR 6598 CNRS – Ecole Centrale de Nantes, 1 rue de la Noë, BP 92101, F-44321 Nantes Cedex 3, France; [1] Present address: Centre Scientifique et Technique du Bâtiment – 11, rue Henri Picherit, BP 82341, 44323 Nantes Cedex 3, France (author for correspondence, e-mail: Jean-Michel.Rosant@ec-nantes.fr, fax: +33 (0)2 40 37 25 56)*

Abstract. A full scale experiment has been performed in a street canyon, the Rue de Strasbourg in Nantes (France). This campaign, the Nantes'99 experiment, provided a detailed data base documenting, amongst others, the production of turbulent kinetic energy (TKE) by vehicles within the street. Airflow and CO concentration measurements have been analysed during days with low wind perpendicular to the street axis, i.e. for conditions expected to greatly favour the enhancement of the turbulence produced by traffic on flow and dispersion within the street canyon. It is shown that traffic is associated with an increase in turbulent kinetic energy at the lowest levels of the street, especially at the leeward side of the street. It is suggested that turbulent kinetic energy increases with the number of vehicles up to a threshold value and then decreases when vehicles form a 'block' shape that limits the additional production of turbulence. Moreover, it is suggested that traffic-produced turbulence affects pollutant dispersion reducing CO concentration at the lower levels of the leeward side of the street from a threshold value of TKE equal to about $0.15 \ \mathrm{m}^2 \cdot \mathrm{s}^{-2}$. On the other hand, high traffic density generates less turbulence which in turn leads to a lower pollutant dispersion.

Keywords: full-scale experiment, low wind conditions, pollutant dispersion, street canyon, traffic-produced turbulence

1. Introduction

The pollutant dispersion in a street canyon is influenced by several parameters such as street geometry, wind direction and speed, temperature and traffic conditions. Many studies on dispersion of vehicular emission in an urban street canyon have been conducted (Väkevä *et al.*, 1999; Bauman *et al.*, 1982; Georgii *et al.*, 1967) but only a few full-scale experimental studies have been focused on investigating the influence of the vehicles motion on the airflow and turbulence in a street canyon. Qin and Kot (1993) observed a large influence of the movement of the vehicle fleet on the airflow and turbulence near the bottom of the canyon and considered that the vehicle wake and the hot exhaust gases generate mechanical and thermal turbulence. They could measure the influence of vehicle movements on the airflow in the canyon up to 12 m above the road surface, i.e. $z/H = 0.8$. DePaul and Sheih's (1986) observations indicate that additional turbulence generated by traffic has a marked

Water, Air, and Soil Pollution: Focus **2:** 127–140, 2002.

influence on the turbulent velocity distribution up to a height of approximately 7 m, i.e. $z/H = 0.2$. Recently, Kastner-Klein et al. (1998) studied the traffic-induced turbulence effect on concentration fields within a canyon by means of wind-tunnel experiments. In one-way configuration, the moving traffic leads to a pronounced transport of pollutants along the canyon axis. It also appears that the turbulence in the street has an evident diurnal variation which follows that of the traffic quite well, and that in-street local concentrations may decrease when the traffic increases due to turbulence created by the vehicle motion.

This paper deals with the effect of traffic on turbulence within a street canyon based on the experimental results of the Nantes'99 campaign.

2. Experiment

The Nantes'99 experiment was a full-scale experiment performed in the centre of the city of Nantes (France) during June–July 1999. The Rue de Strasbourg is a three-lane street canyon with nearly North-to-South orientation (332°). It is a high-traffic one-way street, with a great homogeneity in building construction; its width W is approximately 15 m. The experimental section was located midway between two crossroads which are 60 m apart. In this section, the building height at the west side is $H_w = 19.4$ m, and $H_e = 22.8$ m at the east side, i.e. the mean height of the buildings is $H = 21$ m leading to an aspect ratio, $H/W = 1.4$. It is noted here that due to the orientation of the street axis, the east side of the street corresponds to its leeward side for easterly ambient winds.

The velocity components and turbulence were measured with sonic and 3-D propeller anemometers at three levels on each side of the street : $z/H = 0.07$, 0.19 and 0.69 at the west side and $z/H = 0.07$, 0.19 and 0.5 at the east side of the street (z is the height above road level). Carbon monoxide, CO, was chosen as a car pollutant emission tracer; this gas is a good indicator of the dispersion and dilution of pollutants within the street as its chemical transformation is very slow. CO concentration was measured at the same levels as the wind within the street. Traffic (number and velocity of vehicles) was measured at eight different places within the street and within all lateral streets. Reference wind and background pollution were measured on the west building roof, at $z/H = 1.3$ (Figure 1).

The air temperature and the temperature of the wall surfaces were also measured within the street and their analysis is presented and discussed in Louka et al., 2001. A detailed description of the experimental set-up can be found in Vachon et al. (1999).

In order to analyse the traffic-induced turbulence at its greatest intensity, twelve hourly periods have been selected (Table I) with nearly the same meteorological conditions : a reference wind direction D_{ref} perpendicular to the street axis and a reference wind speed V_{ref} less than 1.1 m s^{-1} (Figure 2). In such conditions the shear wind production of turbulence is assumed to be locally weaker than the traffic

(a)

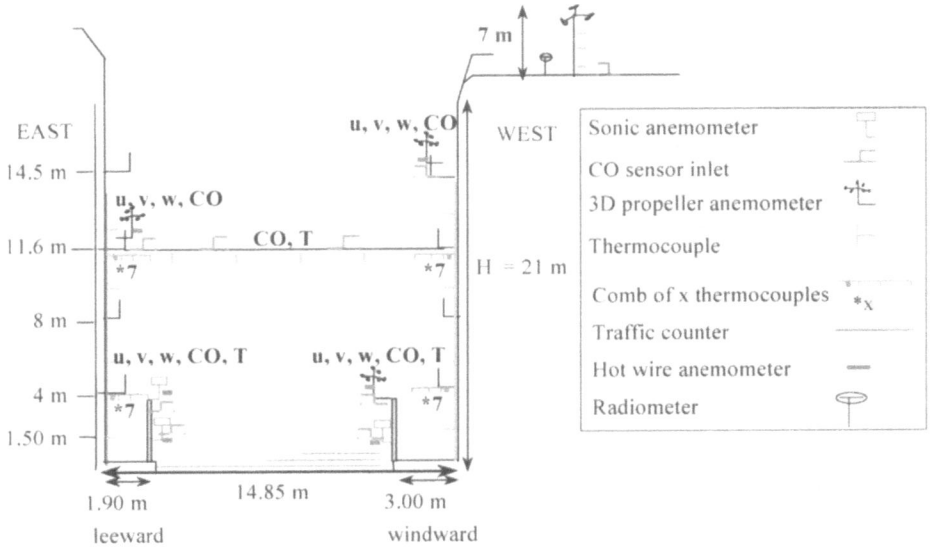

(b)

Figure 1. (a) The study section and (b) two-dimensional schematic representation of the experimental set-up.

TABLE I

Determination of the hourly periods selected for the analysis of the
traffic induced turbulence

No.	Hourly periods
1	22/06/99, 8h45–9h45
2	22/06/99, 17h00–18h00
3	23/06/99, 9h15–10h15
4	23/06/99, 12h00–13h00
5	23/06/99, 13h30–14h30
6	23/06/99, 18h00–19h00
7	24/06/99, 9h30–10h30
8	25/06/99, 9h00–10h00
9	25/06/99, 10h00–11h00
10	25/06/99, 11h00–12h00
11	11/06/99, 12h00–13h00
12	15/06/99, 17h00–18h00

production. The recirculation wind speed is lower in the street when the reference wind is normal than when oblique or parallel to the street axis. Measurements performed in Copenhagen, Denmark (Berkowicz *et al.*, 1996) and in Guangzhou, South China (Qin and Kot, 1993) showed that in the case of a low wind speed, when the thermal stratification is important, the airflow created by the vehicle motion is evident and the traffic-induced turbulence may be the dominant mixing mechanism at the bottom of the street canyon.

The wind data acquisition rate is 4 Hz for all measurement levels. Wind components fluctuations are obtained by applying the moving average technique with an averaging period of 1 min to the original signal and then subtracting the resulted values from the initial signal. The TKE calculation is discussed in detail in Vachon (2001).

The number of vehicles per lane has been selected as the parameter to study the traffic-produced turbulence instead of the total number of vehicles. This choice was based on the fact that, as shown by Figure 3, the vehicles density at the leeward side of the street (i.e. at the east side) is different from that at the windward side; a maximum difference of 300 vehicles per hour can be observed for the hourly period n°12. Therefore, the contribution of traffic to turbulence production may differ in the three lanes. Although the aim of the experimental campaign was to investigate the canyon airflow characteristics for different wind directions, mostly easterly winds were observed, while westerly winds were rare. For this reason, this

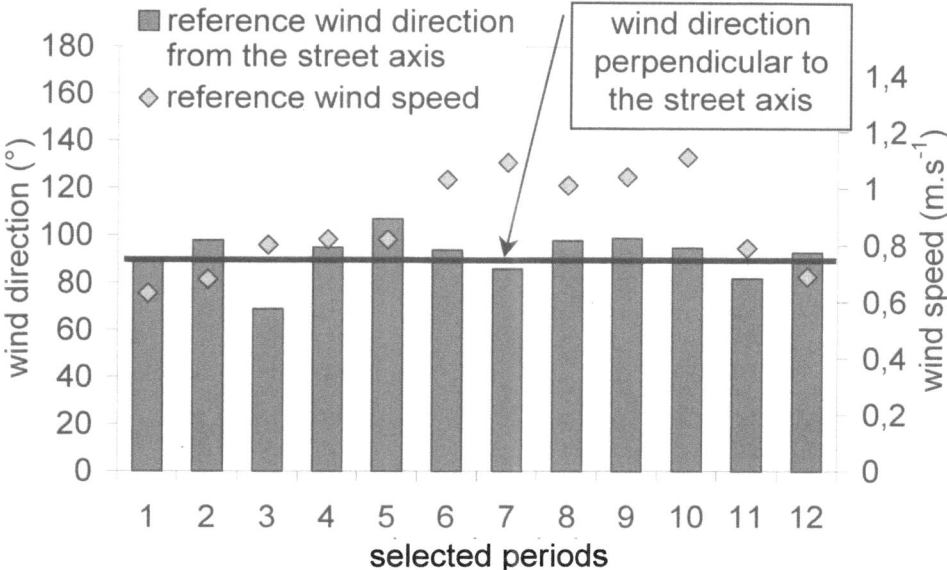

Figure 2. Reference wind speed and direction of the hourly periods selected for the traffic induced turbulence analysis. The 0° corresponds to the street direction, −28° from the North. The numbers of the selected periods are defined in Table I.

paper is concerned with the turbulent kinetic energy produced in Rue de Strasbourg for easterly ambient winds.

3. Results and Discussion

Mostly previous studies aimed at characterising the airflow and its influence on pollution dispersion within 'vehicle-free' street canyons. In a situation without traffic a decrease in turbulent kinetic energy (TKE) appears as the ambient air enters the canyon. Production of TKE is greatest at the roof level due to the presence of the shear layer at this level (Louka *et al.*, 2000), therefore, at the bottom of the canyon, where shear is weaker production of turbulence is less significant. When the ambient wind is perpendicular to the street axis a vortex develops within the street (Dabberdt *et al.*, 1973; Sini and Mestayer, 1996) transporting turbulence. Then due to its dissipation, TKE is expected to be weaker at the leeward side than close to the windward side of the street.

During the Nantes'99 experiment, a re-circulation develops within the street for easterly wind perpendicular to the street axis, even for low reference winds, i.e. $V_{ref} < 1.1$ m s^{-1} as shown in Figure 4.

Figure 5 displays the turbulent kinetic energy within the street in the presence of traffic; the profile numbers refer to the selected periods presented in Table I. It is shown that TKE increases in the lower part of the street, close to the vehicles.

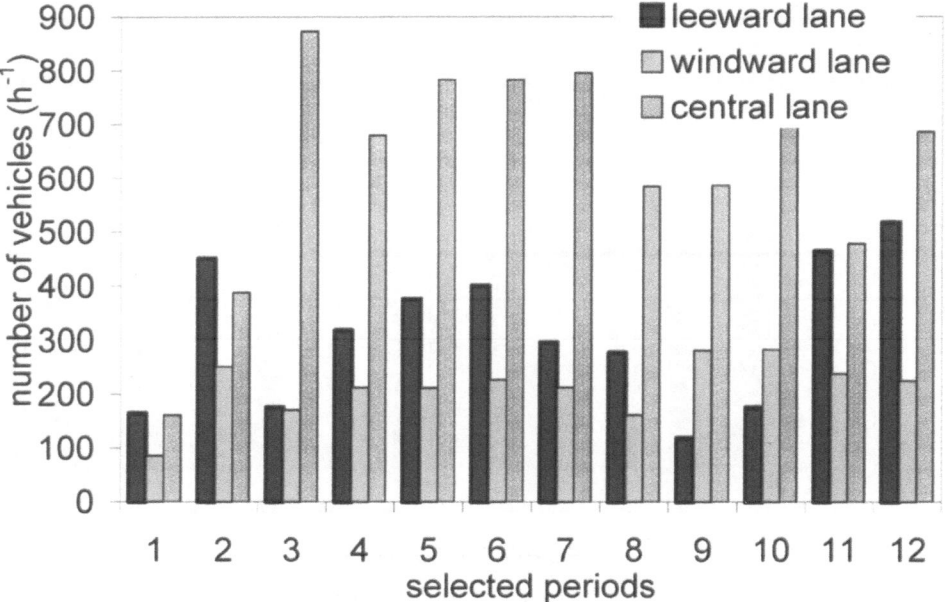

Figure 3. Number of vehicles per hour at each lane of the street for the selected periods.

In particular, at the leeward side, TKE becomes equal or even greater than at the windward side. As in the absence of vehicles TKE is expected to be weaker at the leeward than at the windward side, the behaviour of the TKE profile in this case may be attributed to the transfer of the additional turbulence produced by the vehicles to the leeward side by the vortex.

The TKE dependence on the number of vehicles is illustrated by Figure 6. On the leeward side (Figure 6a) the influence of traffic on TKE is more effective in the lower part of the street. The TKE increases up to a threshold value of approximately 400 vehicles per hour and per lane and then decreases. Above this value the vehicles are close to each other and their velocity is lower resulting in a reduction of further turbulence production. In the study section, vehicles velocity is not constant because of the presence of traffic lights at the end of the section causing traffic irregularities. Due to these irregularities vehicles velocity was not measured in the study section. Nevertheless, it was measured 250 m south from the study section, where traffic is less intermittent. The analysis of these measurements may be used for qualitatively associating the number and the velocity of vehicles. Figure 7a associates the vehicle number on the leeward lane, in the study section, with the vehicles velocity measured 250 m away from the study section at the same side of the street. A decrease of the velocity with the number of vehicles is observed especially for a number greater than 300 vehicles per hour.

On the windward side (Figure 6b) TKE increases with traffic up to 300 vehicles per hour and per lane. This behaviour is mostly observed at the lowest level of the

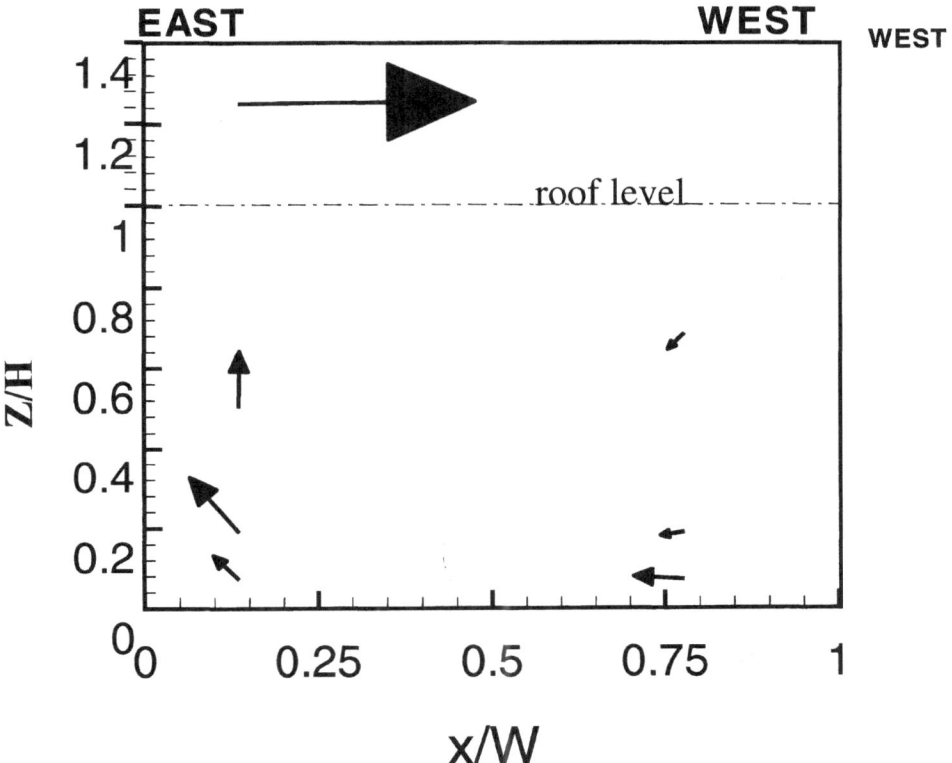

Figure 4. Flow structure within the street in case of a reference wind perpendicular to the street axis and a reference wind speed less than 1 m s^{-1}.

street, z/H = 0.07. Unfortunately, for the selected periods, the number of vehicles does not exceed the 300 vehicles on the windward lane. Therefore the reduction of TKE observed on the leeward side for the highest vehicle density can't be notice on the windward side. Likewise, the decrease of velocity with vehicles number above 300 vehicles per hour noticed on the leeward side can not be observed on the windward side (Figure 7b) since the maximum of vehicles number measured per hour is 300 on the windward lane. Since the number of vehicles on the central lane is high compared to the vehicle density on the others lanes (Figure 3), the TKE independence on the number of vehicles on the central lane has been verified. Indeed, TKE measured on the leeward lane (Figure 8a) and on the windward lane (Figure 8b) is relatively constant with vehicle density on the central lane at each level of the street. It should be noted, however, that at the centre lane, TKE is probably influenced by the traffic present in the same lane. Unfortunately, no measurements could be realised in the centre of the street during Nantes '99 experiment.

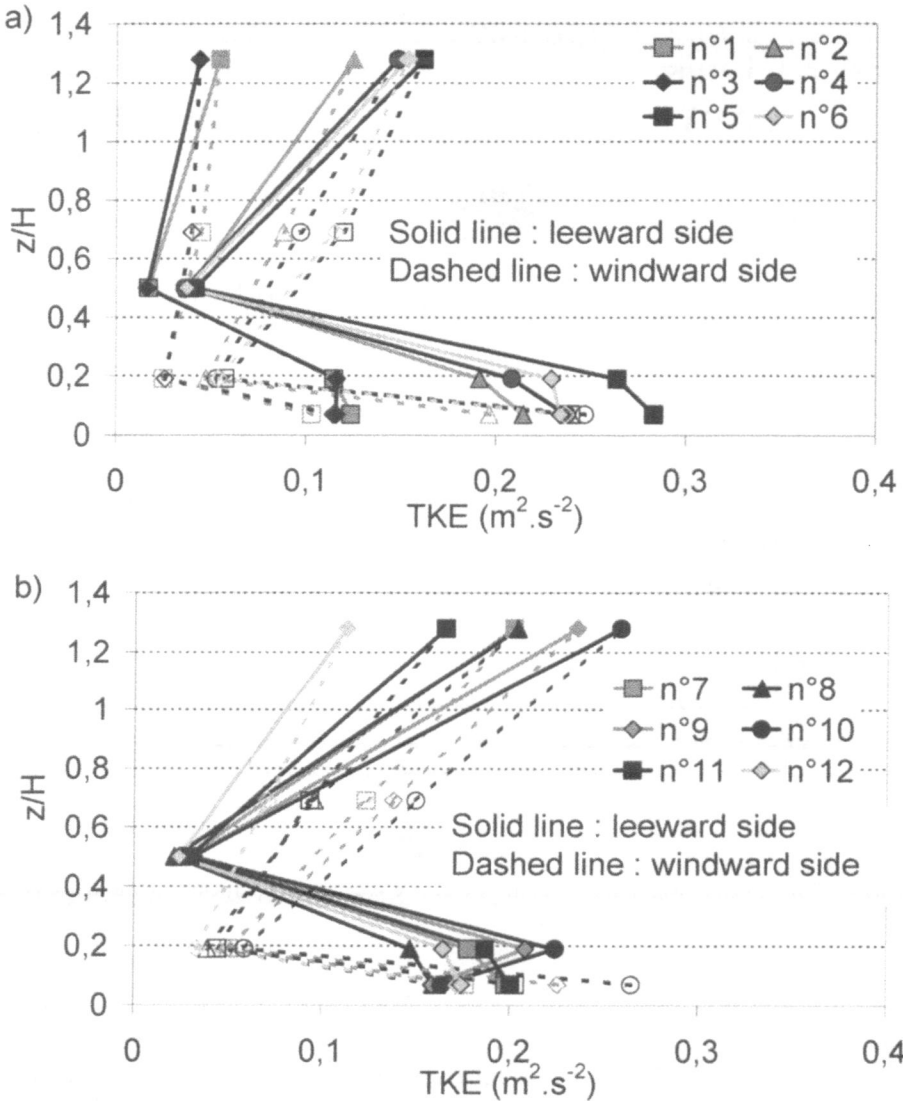

Figure 5. Vertical profiles of turbulent kinetic energy (TKE) in the street for the selected periods n° 1 to 6 (a) and n° 7 to 12 (b).

The dependence of CO concentration on the number of vehicles per lane and per hour is shown in Figure 9. CO_{ref} refers to the background CO concentration measured on the roof and CO is the measured CO concentration within the street. Therefore, the CO-CO_{ref} represents the CO emitted by traffic in the street. At the leeward side CO concentration remains relatively low up to the threshold value of 400 vehicles per hour and per lane, then a sharp increase is mainly observed at the

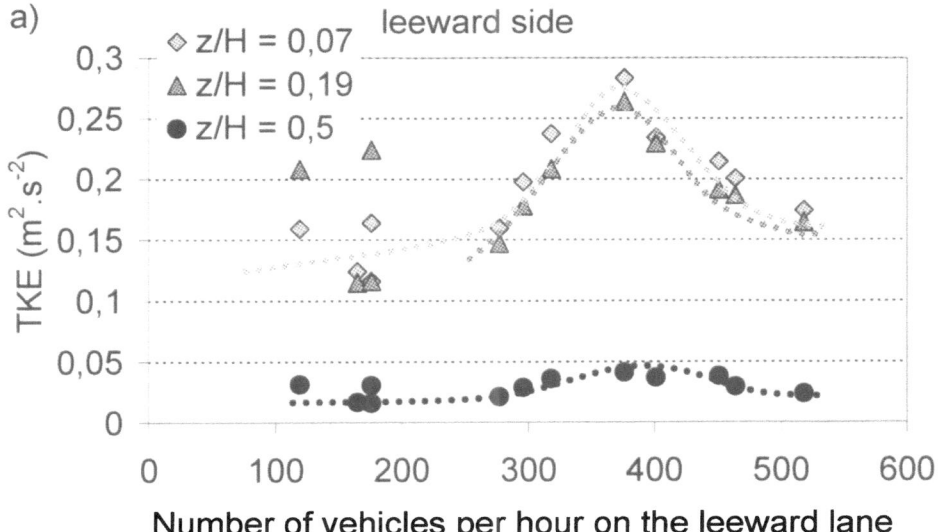

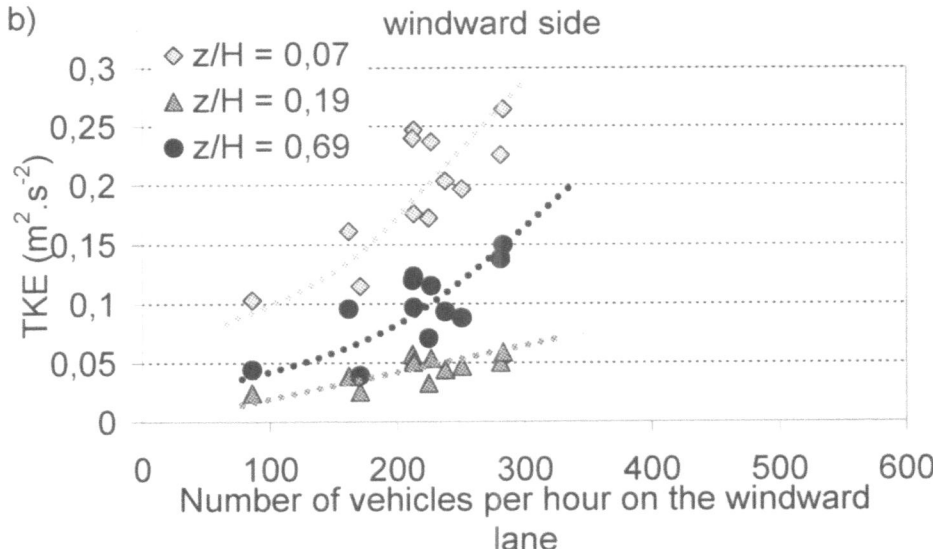

Figure 6. Influence of the traffic intensity per lane on turbulent kinetic energy (TKE) within the street, on the leeward side (a) and on the windward side (b).

lowest level. This figure suggests that up to 400 vehicles per hour and per lane, the vehicles motion creates an additional turbulence strong enough to disperse the vehicular exhausts. On the other hand, at the windward side a weak increase of CO concentration may be observed at $z/H = 0.07$, at least up to 300 vehicles per

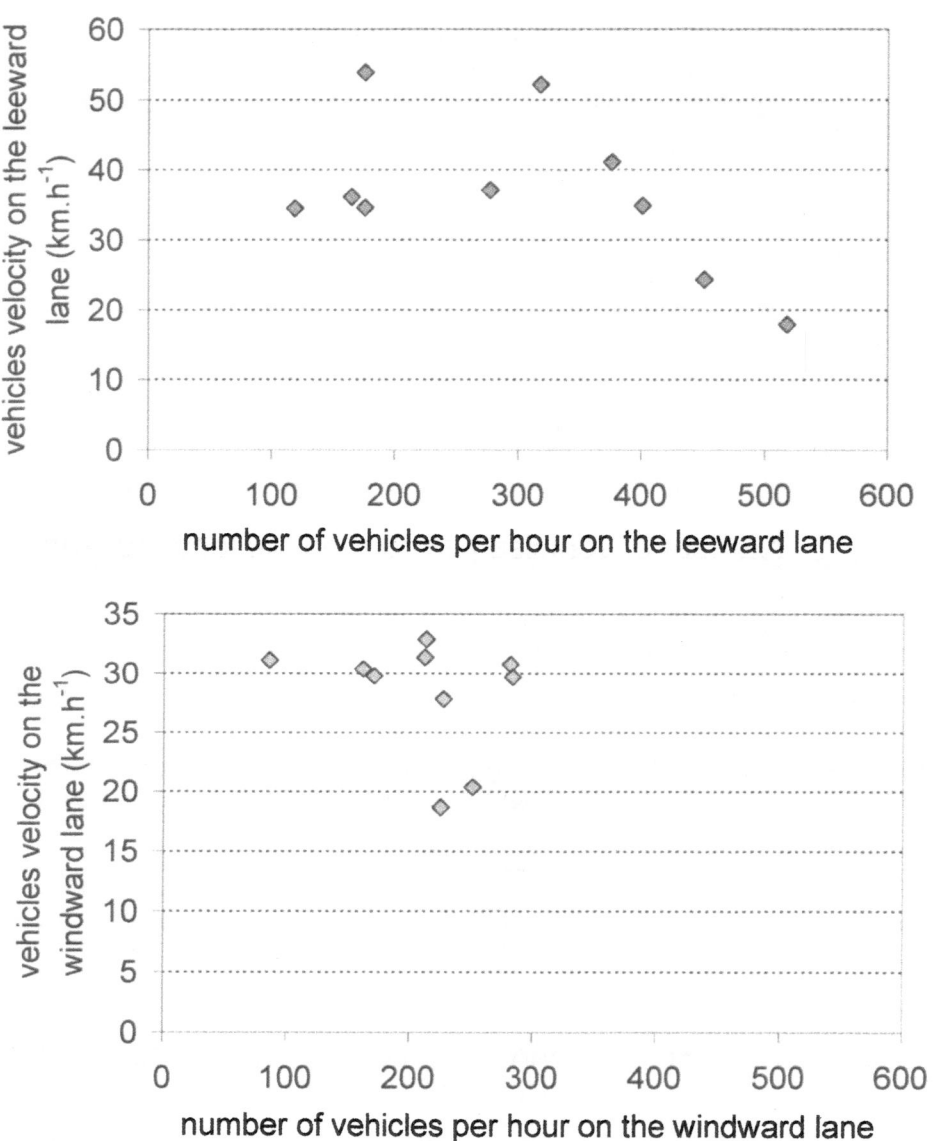

Figure 7. Evolution of vehicles velocity with the number of vehicles, on the leeward lane (a) and on the windward lane (b).

hour and per lane. For other measuring locations, it is suggested that TKE has no significant influence on CO concentration.

As CO emission rate is closely related to vehicles velocity the direct effect of TKE on CO concentration was also considered. Figure 10 shows that for TKE greater than 0.15 m^2 s^{-2} and at the lowest levels of the street ($z/H = 0.07$ and

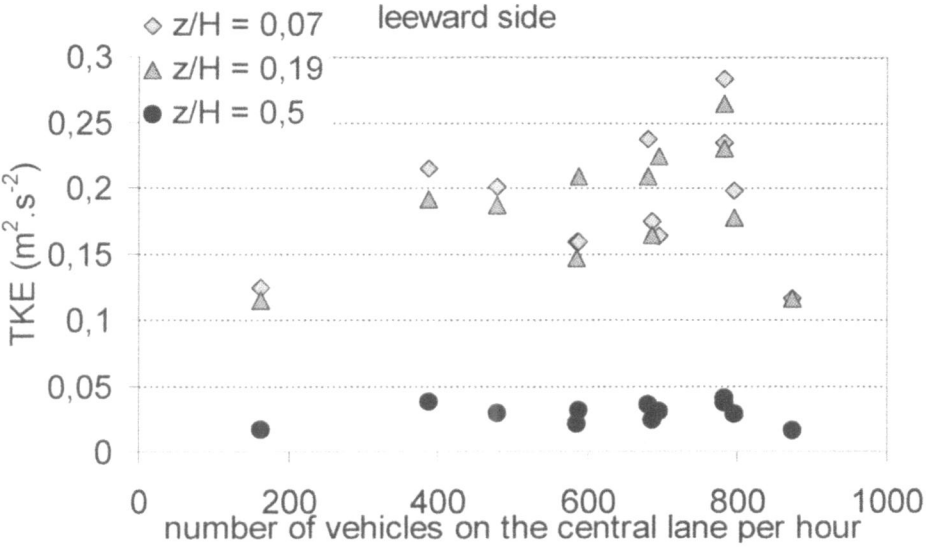

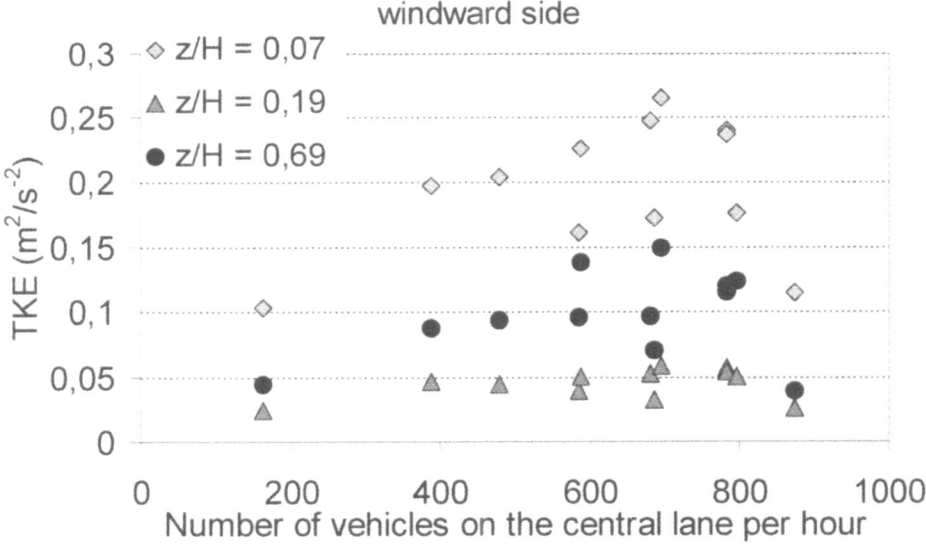

Figure 8. Measured turbulent kinetic energy (TKE) on the leeward side (a) and on the windward side (b) versus traffic intensity on the central lane.

z/H = 0.19) at the leeward side, an increase in TKE corresponds to a decrease in concentration.

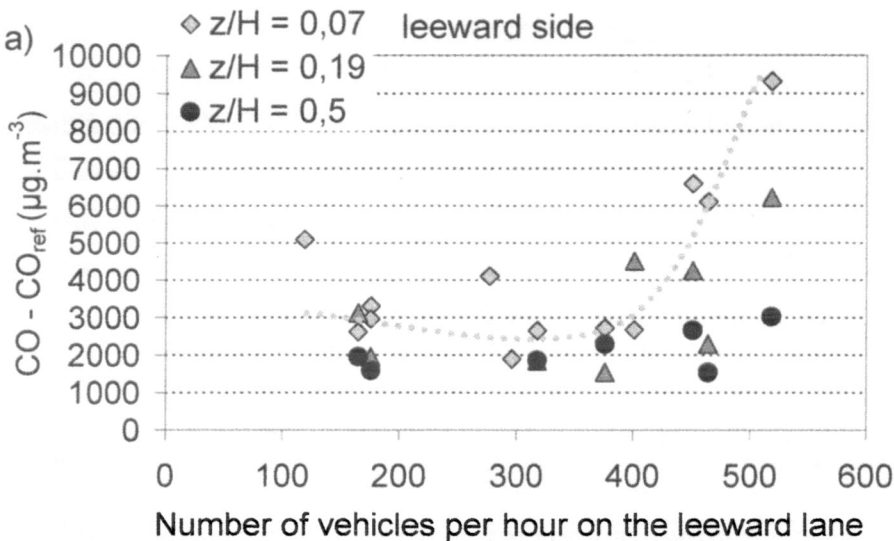

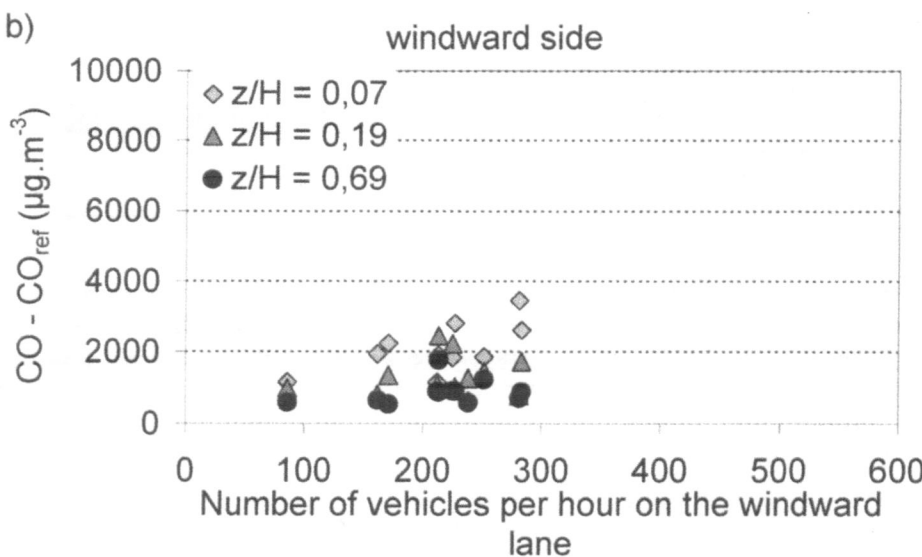

Figure 9. Influence of the traffic intensity on CO concentration (CO-CO$_{ref}$) within the street, on the leeward side (a) and on the windward side (b). The dashed curve corresponds to a trend of the CO concentration at the lower street levels, z/H = 0.07 and z/H = 0.19 on the leeward side.

4. Conclusions

In low wind speed conditions (reference wind speed less than 1.1 m s^{-1}) with a wind direction perpendicular to the street axis, the measurements of the Nantes'99 experiment showed that additional turbulence is created by traffic at the lower

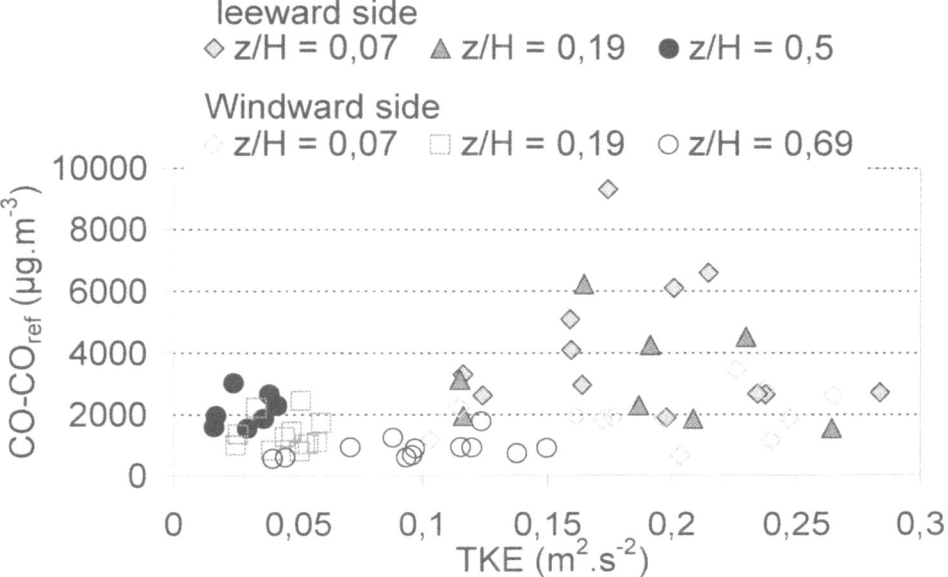

Figure 10. Relationship between the turbulent kinetic energy (TKE) and the CO concentration ($CO-CO_{ref}$) within the street.

levels of the street, and is observed at least up to 4 m high (z/H = 0.19) at the leeward side of the street. The turbulent kinetic energy is strongly correlated with the traffic intensity. It was shown that the highest traffic conditions (larger than 400 vehicles per hour and per lane), leading to a closer distance between vehicles and smaller vehicles velocity, generate relatively lower TKE. The measurements further suggested that the traffic-produced turbulence influences pollution dispersion by reducing CO concentration especially at the lower levels of the street on the leeward side. Consequently, with regard to the applications of our experimental results to the numerical modelling, a parameterisation associating turbulence production with traffic density, distance between vehicles and vehicles velocity is necessary.

Acknowledgements

The authors gratefully acknowledge support from the PRIMEQUAL-PREDIT (contrat INERIS 1998), the Agence de l'Environnement et de la Maîtrise de l'Energie (ADEME), PSA Peugeot Citroën and the European Commission (TMR contract ERNFMRXCT97-0105). This work was performed within the framework of the European Research Network on 'Optimisation of Modelling Methods for Traffic Pollution in Streets' (TRAPOS).

References

Bauman, S. E., Williams, E. T. and Finiston, H. L.: 1982, 'Street level versus rooftop sampling: carbon monoxide and aerosol in New York city', *Atmos. Environ.* **16**, 2489–2496.

Berkowicz, R., Palmgren, F., Hertel, O. and Vignati, E.: 1996, 'Using measurements of air pollution in streets for evaluation of urban air quality – meteorological analysis and model calculations', *Sci. Tot. Environ.* **189/190**, 259–265.

Dabberdt, W. F., Ludwig, F. L. and Johnson, W. B.: 1973, 'Validation and applications of an urban diffusion model for vehicular pollutants', *Atmos. Environ.* **7**, 603–618.

DePaul, F. T. and Sheih, C. M.: 1986, 'Measurements of wind velocities in a street canyon', *Atmos. Environ.* **20**, 455–459.

Georgii, H. W., Busc, E. et Weber, E.: 1967, 'Untersuchung über die zeitliche und raumliche Verteilung des Imissions-Konzentration des Kohlenmonoxid in Frankfurt am Main', IMG Report, Univ. Frankfurt am Main, no. 11, 1–60.

Kastner-Klein, P., Berkowicz, R., Rastetter, A. and Plate, E. J.: 1998, 'Modelling of Vehicles Induced Turbulence in Air Pollution Studies for Streets', 5th International Conference on Harmonisation within Atmospheric Dispersion Modelling for Regulatory Purposes, Rhodes, Greece, 18–21 May 1998.

Louka, P., Belcher, S. E. and Harrison, R. G.: 2000, 'Coupling between air flow in streets and the well-developed boundary layer aloft', *Atmos. Environ.* **34**, 2613–2621.

Louka, P., Vachon, G., Sini, J-F., Mestayer, P. G. and Rosant, J-M.: 2001, 'Thermal effects on the airflow in a street canyon – Nantes'99 experimental results and model simulations', *Water, Air, and Soil Pollut.*, this issue.

Qin, Y. and Kot, S. C.: 1993, 'Dispersion of vehicular emission in street canyons, Guangzhou city, South china (P.R.C.)', *Atmos. Environ.* **27B**, 283–291.

Sini, J.-F. and Mestayer, P. G. : 1996, 'Pollutant dispersion and thermal effects in urban street canyon', *Atmos. Environ.* **15**, 2659–2677.

Vachon, G.: 2001, 'Transferts des Polluants des Sources Fixes et Mobiles dans la Canopée Urbaine: Évaluation Expérimentale', Doctoral Thesis, University of Nantes, France.

Vachon, G., Rosant, J.-M., Mestayer, P. G. and Sini, J.-F.: 1999, 'Measurements of Dynamic and Thermal Field in a Street Canyon, URBCAP Nantes '99', *Proceedings of the 6th International Conference on Harmonisation within Atmospheric Dispersion Modelling for Regulatory Purposes*, Rouen, France, 11–14 October 1999, CD-ROM proceed, paper 124.

Väkevä, M., Hämeri, K., Kulmala, M., Lahdes, R., Ruuskanen, J. and Laitinen, T.: 1999, 'Street level versus rooftop concentrations of submicron aerosol particles and gaseous pollutants in an urban street canyon', *Atmos. Environ.* **33**, 1385–1397.

ADAPTING THE SPECIATION OF THE VOCS EMISSION INVENTORY IN THE GREATER ATHENS AREA

E. BOSSIOLI[1]*, M. TOMBROU[1] and C. PILINIS[2]

[1] Department of Physics, University of Athens, Greece; [2] Department of Environmental Studies, University of the Aegean, Mytilene, Greece

(* author for correspondence, e-mail: ebossiol@phys.uoa.gr, fax: 00 3010 7295281)

Abstract. An adaptation procedure of a new emission inventory of the Greater Athens Area is attempted, based on a sensitivity analysis on the treatment of the VOC emissions. Through this procedure the impact that a more detailed treatment of the VOCs emissions might have on the atmospheric chemistry simulations, is examined. For this analysis three different chemical mechanisms were applied for two different locations (urban and city plume) with different VOC and NO_x mixture characteristics. Finally, this study recommends new carbon fractions, reflecting the local conditions in Athens basin.

Keywords: chemical mechanisms, emission inventory, photochemical box model, speciation, street canyon, urban and rural air pollution

1. Introduction

A great portion of uncertainties in atmospheric chemistry modeling is associated with emission inventory inaccuracies and especially those of Volatile Organic Compounds (VOC) emissions. In European countries for example, the emission inventories include only a few VOC species while hundreds of them are detected in the ambient air. Furthermore, the current chemical mechanisms involve only a limited number of chemical species and consequently they require a further lumping of the emissions. Thus, the methodology of properly characterizing the VOC emissions becomes very crucial considering that some of the VOC categories produce substantial different amount of O_3, influenced by local conditions (mixture composition, NO_x levels). In this study it is examined the effectiveness of the existing VOC profiles in Athens in relation to ozone productivity. Then, this profile is upgraded by considering VOC compounds, which are included in more detailed emission inventories (e.g. Los Angeles). In order to examine the impact of these VOC compounds on the atmospheric chemistry simulations, a sensitivity analysis is carried out with three different chemical mechanisms. In this analysis emphasis is given mainly on ozone production for two different locations (street canyon, city plume) with different VOC and NO_x mixture characteristics.

Water, Air, and Soil Pollution: Focus **2**: 141–153, 2002.
© 2002 *Kluwer Academic Publishers.*

2. Methodology

2.1. COMPARISON OF AMBIENT MIXTURES

At the beginning of this study, the available hydrocarbons (HC) measurements in Athens' ambient air (Moschonas et al., 1996; Rappengluck et al., 1998) have been compared with those measured in urban mixtures of other cities. In particular, the measurements from a London street (Blake et al., 1993), a tunnel in Sidney (Duffy et al., 1996) and from various urban areas around the United States (Jeffries et al., 1989) have been used. From this comparison the common and different features between the urban mixtures have been derived and been related to possible emission sources.

Another comparison is also attempted between the detailed emission profile of the Central Los Angeles area, two other sets of individual HC emission rates, from a heavy traffic street canyon in the center of Athens and a data set with city plume characteristics (Kourtidis et al., 1999; Klemm and Ziomas, 1998). However, it should be mentioned that the emission rates for the Athens two data sets were based on concentration measurements from the MEDCAPHOT-TRACE experiment (20 August to 20 September 1994) (Rappengluck et al., 1998; Kourtidis et al., 1999; Klemm and Ziomas, 1998). In particular, these traffic emission rates, were estimated based on the measured enhancement ratios of individual VOCs versus NO_x/CO. Especially in the study of Kourtidis et al., (1999), the acquired data were only during sea breeze days at wind speeds < 2 m s^{-1}, during rush hours. Under such meteorological conditions, the emissions source strength, in a street canyon with heavy traffic load, is high enough compared to the magnitude of other processes such as transport, dry deposition and chemical losses. Thus, the assumption of Kourtidis et al., (1999), relating the concentration measurements to emissions rates, is considered quite reasonable.

2.2. CHEMICAL MECHANISMS AND LUMPING PROCEDURES

As it is impossible to include chemical reactions for the hundreds of organic compounds present in the atmosphere, the limitation of the number of reactions and species included in the mechanisms is achieved by the lumping procedures. The sensitivity analysis on the treatment of the VOC emissions was carried out with three different chemical mechanisms: the widely used CB-IV (Gery et al., 1988; 1989) and two approaches of the detailed SAPRC99 mechanism (one with variable parameters and the other one with fixed parameters, Carter, 2000).

The CB-IV mechanism uses the 'lumped structure' approach, in which organics are grouped according to the number and type of carbon bonds, treating the different parts of the molecule as if they react independently (Gery et al., 1988). This permits representation of a large number of compounds with a relatively small number of model organic species.

In contrast to the CB-IV mechanism, SAPRC99 considers a totally different lumping approach, the lumped molecule and the lumped parameter approach. In the lumped molecule approach, classes of compounds that are either believed to react very similarly, or they are not sufficiently important to justify more complex approaches are being represented by lumped molecule groups. In the lumped parameter approach a group of VOCs that react with similar rate constants is represented by the model species, of which the kinetic and product yield parameters are weighted averages of the VOCs mixture. The main difference between the two modes of the SAPRC99 mechanism concerns the estimation of the parameters for the lumped species. In the variable parameter approach, all the parameters have been derived from the available information on the mixture of VOCs, in the Athens basin, reflecting its composition. In the fixed parameter approach, the parameters have been derived from the ambient VOC mixture obtained from urban atmospheres in the United States (Jeffries *et al.*, 1998) and from oxygenate measurements in the California South Coast Air Basin (Carter, 1994). In both modes of the SAPRC99 mechanism the lumped molecule approach is considered.

2.3. SCENARIOS

Two basic emission profiles were considered for the case of Athens; the one reflecting the characteristics of a street canyon in the center of the city, 'AB street canyon' and the other one the characteristics of a city plume, 'AB city plume'. The species that have been used to construct the two basic emission profiles (around 50) are commonly used in the emission inventories and were selected under the condition that have been measured in Athens (alkanes, cycloalkanes, alkenes, acetylene, aromatics, limonene). In particular, the daily VOCs emission rates were estimated based on the available measured enhancement ratios of individual VOCs, either versus NO_x (weight basis) as in the case of a street canyon (Kourtidis *et al.*, 1999) or versus CO (volume basis) for both street canyon and city plumes (Klemm and Ziomas, 1998). Thereafter, the absolute VOCs values were estimated from the latest yearly total emission rates of NO_x and CO (Ministry of Environment). For this calculation, the enhancement ratios at a heavy traffic street were assumed to be representative for the whole of Athens (Kourtidis *et al.*, 1999). The total VOCs emission rates for the street canyon case, being estimated according to the above methodology, exhibit minor differences (3.3%) in comparison with the emission rates from traffic provided by the Ministry of Environment. But for the plume case the corresponding emission rates were found 30% higher. According to Klemm and Ziomas, this discrepancy is attributed to emissions from other sources that are about as high as those of traffic in areas away from the street canyons.

Starting from the AB emission profile the extra VOC species contained in the detailed Los Angeles' (LA) profile were added, for both street canyon and city plume cases. In this procedure all the available information from the Athens' ambient air has been taken into account. In particular, the ratios of the measured emission rates to the isopentane's were used. For those compounds that there was no information, the emission rates were estimated by using the LA profile, although it is recognized that there are differences between Athens and LA profiles.

Finally, 150 species were added to the AB profile in order to approach the one from LA. The ozone-forming capability of the added individual organics, in terms of incremental reactivity, was calculated with the SAPRC99 mechanism. From the added compounds, nearly half of them had relatively low impact on the production of O_3. The most reactive compounds, in terms of the amount of O_3 formed per unit of VOC, that thereafter are being added to the 'AB street canyon' mixture, are presented in Table I. So, instead of adding one by one the species, subsets of VOC categories with similar reactivity, were gradually added to the AB profile, in order to approach the LA's profile. From now on, the final modified emission profile is called as Athens modified according to Los Angeles profile (ALA scenario).

All the simulations were carried out with the SAPRC Atmospheric Photochemical Mechanism program (Carter, 1988). During the simulations a very simple atmospheric physical model was considered. In particular, the daily total amount of emissions was initially injected and the system was left to react, without supplying it with any fresh ones. The temperature was set at 300°K, a representative value for the Athens basin and it was held constant throughout the simulation. The rate constants for the photolytic reactions were calculated by using the intensity and the spectral distribution of the natural sunlight.

The initial concentrations of VOCs, concerning all scenarios for the street canyon mixture and only the initial and the final one for the city plume, are presented in Table II. In the same table, the initial concentrations of NO_x are also given.

3. Results and Discussion

3.1. COMPARISON OF AMBIENT MIXTURES

In Figure 1, distribution of HC mixtures at various sites, together with those measured in the Athens' ambient air, are presented. To facilitate the comparison of distributions, that may correspond to different absolute mixing ratio scales, the hydrocarbons' concentrations have been normalized to isopentane's, a major component of gasoline and one of the most abundant species in urban atmosphere. The normalized HC concentrations from 39 US cities (ambient measurements from the EPA data base) are declared as 'All city average' mixture.

From this comparison it is evident that acetylene, ethene, aromatic compounds and 1,3-butadiene are less abundant in the US ambient air than in other countries. The much lower levels of all these species and especially of acetylene are associated with the extensive use of catalytic converters in cars, in US cities (Andersson et al., 1991; McCabe et al., 1992). In European cities larger concentrations of aromatic compounds have been measured (Moschonas et al., 1996). For example,

TABLE I

Calculated incremental reactivity (mole O_3/moleC) of selected added VOCs for the ALA street canyon scenario

Compound	Mole O_3/ moleC	Compound	Mole O_3/ moleC
3-Methyl-1-butene	5.73	Trimethylbenzene C9-BEN3	15.98
Trans-2-Hexene	8.14	C10 dialkyl benzenes	9.11
Cis-2-hexene	7.91	Dimethyl ethylbenzenes C10-BEN3	15.86
Cis-3-hexene	8.64	Tetramethylbenzene C10-BEN4	15.80
Trans-3-hexene	8.67	Methyl butylbenzene	9.21
4-Methyl-1-pentene	5.46	C12-BEN2	8.98
3,3-Dimethyl-1-butene	5.52	C12-BEN3	15.79
2-Methyl-2-pentene	10.85	Formaldehyde	4.41
4-Methyl-trans-2-pentene	7.89	Acetaldehyde	3.87
Trans-3-heptene	8.14	Propionaldehyde	6.05
C7 internal alkenes	8.01	Butyraldehyde	6.29
C8 internal alkenes	7.49	Isovaleraldehyde	6.13
C9 internal alkenes	6.93	Acrolein (2-propenal)	5.13
C10 internal alkenes	6.54	Crotonaldehyde	9.33
Cyclopentene	5.60	4-Methyl-2-pentanone	4.47
Cyclohexene	4.50	2-Ethyl-1-hexanol	2.28
Methylcyclopentene	7.81	Ethyl isopropyl ether	3.51
1,2,4-Trimethylcyclopentene	7.59	Tetrahydrofuran	3.00
2 Butyne	11.83	2-Ethoxyethanol	3.40
Nonadiene	6.80	Ethyl acrylate	10.15
124-Trimethylbenzene	11.55	Vinyl acetate	3.55
123-Trimethylbenzene	17.82	Ethanolamine	4.22
1-Methyl-2-ethylbenzene	9.33	Chloroform	10.34

both the benzene and toluene concentrations were much different between Athens and 'All city average' mixtures (Jeffries *et al.*, 1989). The current Athens mixture is however, more relevant to the one in the United States, before catalyst-converters, pre-1970 period (Singh *et al.*, 1985). On the other hand US mixture seems to be enriched in ethane and propane species, which are not much abundant in motor vehicle exhausts. High concentrations of both species were also measured in London downwind plume. Both of these species could be associated with the leakage of natural gas (Blake *et al.*, 1993; Goldan *et al.*, 1995). Finally, in Sidney tunnel the high concentrations of propane were associated with the use of LPG in taxis (Duffy *et al.*, 1996).

TABLE II

The initial concentrations of VOCs, for the CB-IV and SAPRC99 mechanisms (the fixed and variable parameter approaches use the same lumping procedure) for the different scenarios and for both street canyon and city plume cases. VOC categories are added gradually

	Street canyon VOC (ppbC) $NO_x =120$	City plume VOC (ppbC) $NO_x = 24$ ppb
AB = Athens Basic	421	121
S1 = AB + aldehydes	431	
S2 = S1 + other alkanes	459	
S3 = S2 + other cycloalkanes	464	
S4 = S3 + other alkenes	470	
S5 = S4 + alkadienes, alkines	470	
S6 = S5 + other aromatics	483	
S7 = S6 + other ketones	488	
S8 = S7 + alcohols, glycols	506	
S9 = S8 + ethers, glycol ethers	526	
S10 = S9 + esters	531	
S11 = S10 + chlorides, bromides, amines	537	
S12 = S11 + cresol, pinene	537	
ALA (Athens modified) = S12 + naphthal., phenol	538	196

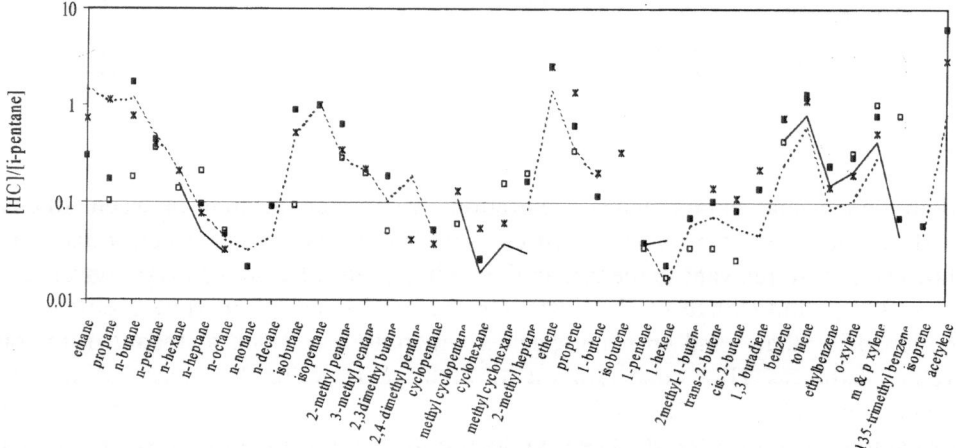

Figure 1. HCs' concentrations from different cities. All concentrations are normalized to isopentane's concentration.

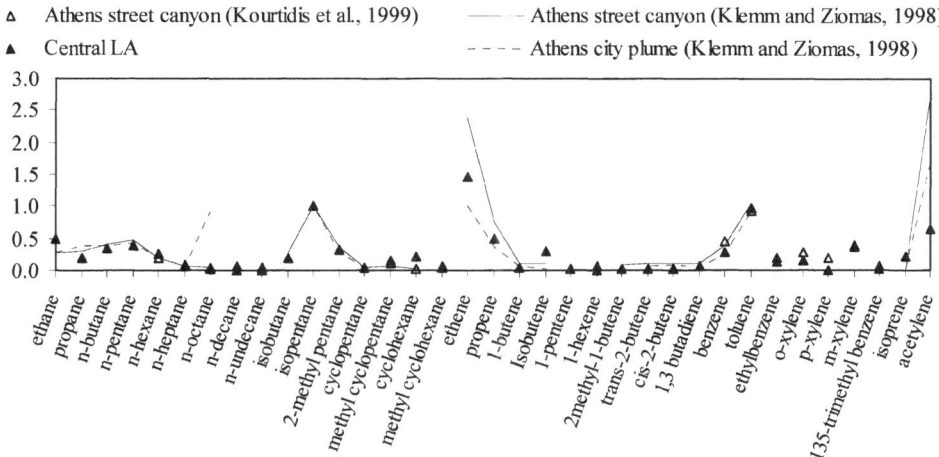

Figure 2. Comparison of HCs' emission rates of Athens and Central LA. All emission rates are normalized to isopentane's emission rate.

In Figure 2, the detailed emission profile of the Central Los Angeles area, the two data sets from a heavy traffic street canyon in the center of Athens as well as a data set with city plume characteristics (normalized to isopentane's), are also compared. It should be mentioned that this comparison was performed only for the hydrocarbons for which there were available measurements in Athens. Only for few species, related to traffic emissions, the normalized rates given by Kourtidis are somehow higher, than those of Klemm. This is well explained since Kourtidis *et al.*, (1999) used only data under specific meteorological conditions.

Concerning the common oxygenated organic compounds, only Amanatidis *et al.*, (1997), measured the ratio of formaldehyde versus acetaldehyde concentrations in the center of Athens as well as in a suburb and he concluded that anthropogenic sources of carbonyls are predominant in both sites. Unfortunately, due to the different atmospheric conditions (on January 1993, under very low wind speeds and during April 1993 with high wind speed, rainfall and cloudiness), these measurements could not be embodied to the previous data sets in Athens. Thus, although the oxygenated organic compounds are significant species of the inventories, they have not been included in the basic scenarios but they have been extracted from the one from Los Angeles.

In general, the comparison between the two data sets from the center of Athens and the Central LA, exhibits similarities for most of the species, a fact that encourages the idea to modify the existing Athens emission profile according to the detailed LA's one.

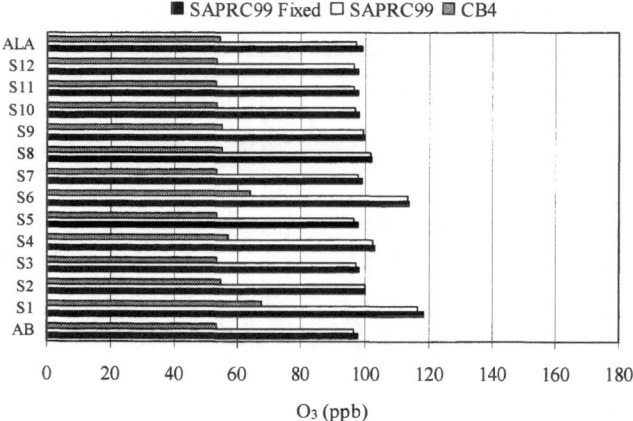

Figure 3. Calculated maximum hourly O₃ concentrations of the street canyon scenarios, for the three chemical mechanisms. VOC categories are added separately.

3.2. SCENARIOS

In the beginning, the effects from each of the VOC categories, to the basic scenarios, were examined separately. The street canyon case is presented in Figure 3. All the examined scenarios, as well as the initial concentrations of the added VOC categories are presented in Table II. Both mechanisms reflect the same fluctuations on ozone, although the amount predicted by the CB-IV mechanism is almost half from that predicted by the SAPRC. Apart from the aldehydes and aromatics (S1 and S6 scenarios) most of the other VOC categories contribute almost equally to the increase of O₃ concentrations. It is also interesting to notice that the addition of the alkenes had a great impact even though the added concentration of alkenes is relatively low (S4 scenario). Among the alkenes, those, that contribute more, are the internal alkenes with carbon atoms up to 7, while from the aromatics those that contribute more are the trimethylbenzenes, dimethyl ethylbenzenes and triethylbenzenes. Concerning alcohols (S8), although they were found significant for the ozone production, they should be considered with caution since their emission rates in LA's profile may not be representative for Athens (these compounds are abundant in US fuels), until further specific measurements in the Athens' mixture take place. The differences in ozone amount produced by the two versions of the SAPRC are almost negligible, with higher values when the parameters are taken fixed.

The calculated maximum hourly concentrations of O₃, for all the street canyon scenarios (successively from AB to ALA), for the three chemical mechanisms, are presented in Figure 4. The transition from the basic AB to the ALA scenario resulted in a 28% increase of the initial concentration of VOC (Table II) while the maximum concentration of O₃ increased by 63–77% depending on the mechanism. In the case of CB-IV the ozone from 53 ppb (AB scenario) reached 94 ppb when

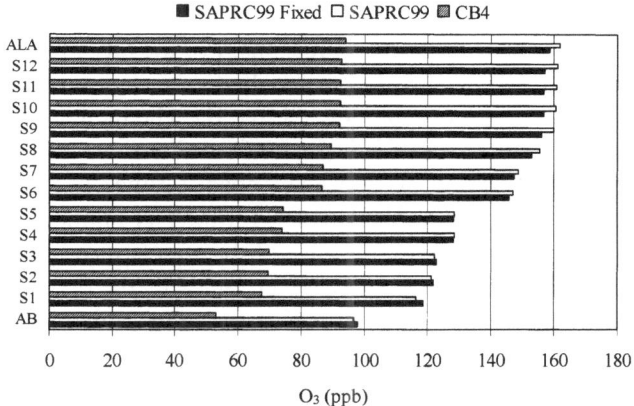

Figure 4. As in Figure 3 but VOC categories are added gradually.

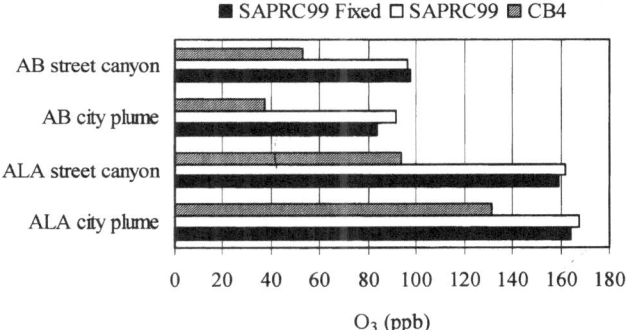

Figure 5. Calculated maximum hourly O_3 concentrations for the AB and ALA city plume scenarios, for the three chemical mechanisms. Street canyon scenarios are also given for comparison reasons.

the full ALA scenario was applied. Thus, even with the detailed VOC profile this mechanism could not predict O_3 concentrations higher than 100 ppb, which is frequently being measured (up to 160 ppb), during ozone episodes in the Athens basin. In particular, the ozone limit of $180\mu g\ m^{-3}$ was exceeded, 45 days during a typical year in Athens. In contrast, SAPRC99 (both versions) predicts higher concentrations of O_3 compared with the CB-IV for all the examined scenarios. From nearly 100 ppb in the basic scenario the calculated ozone reached the concentration of 165 ppb for the ALA scenario. It is quite interesting to notice that although the fixed parameter approach predicted higher concentrations for the initial 4 scenarios, as had already been foreseen, this changed after the addition of aromatics.

The addition of chlorides, bromides, amines, cresols (S11 and S12 scenarios) do not have a great impact on ozone production (Figure 4). Thus these compounds could be reasonably ignored from a modified Athens inventory.

Thereafter, the mixture with plume characteristics was examined. In Figure 5 the calculated maximum hourly O_3 concentrations for the AB and ALA city plume scenarios, are presented, for the three chemical mechanisms. The street canyon scenarios are also given for comparison reasons. For the city plume case the transition from the AB to the ALA scenario resulted in a larger percentage increase (60%) of the initial concentration of VOC (Table II). The increase of the maximum concentration of O_3 varied from 83% (SAPRC99 Fixed) to 250% (CB-IV). The addition of aromatics (S6) had the greatest impact on the city plume mixture. In particular, the ozone produced by the SAPRC for the city plume mixture was comparable with the ozone produced in the case of the street canyon, though the initial concentration of VOCs was significantly lower. Between the two mechanisms, CB-IV turned to be more sensitive on the VOC/NO_x ratio. This became much more evident for the transition from the AB city plume to the ALA city plume profile. The differences between the two mechanisms are mainly attributed to the different number of photolytic reactions as well as to the different photolytic rate constants considered by the two mechanisms. Moreover, the detailed SAPRC mechanism, has the capacity to incorporate the chemistry related to a much larger number of species. In particular, SAPRC mechanism encounters 18 categories, while CB-IV only 8.

It is also interesting to notice that both mechanisms predicted higher O_3 concentrations for the AB street canyon profile than for the AB city plume profile while this behavior was reversed for the corresponding ALA profiles. From the above results, it is demonstrated that the proper characterization of the VOC emission profile becomes very important for the models predictions.

Finally, in Table III the carbon fractions for the AB and ALA scenarios for both the street canyon and city plume cases for the CB-IV chemical mechanism are presented. For comparison reasons the carbon fractions (according to CB-IV) proposed by Smylie et al., (1991), for the Athens case, as well as those derived by Jeffries et al., (1989) for urban simulations in US, are also presented. The carbon fractions for the apportionment of VOCs, derived from the SAPRC mechanism are not included, since the categories in which VOCs are lumped differ from those in the CB-IV mechanism and this would lead to a misleading.

For the city plume case the carbon fractions are higher for isoprene while for the street canyon the higher ones are alkenes and the aromatics. Concerning the alkanes it is evident that their participation in the city plume mixture increased. Ethene's participation in the city plumes mixture is lower than that in the street canyon since, its enhancement ratio versus CO was estimated higher in the street canyon (Klemm and Ziomas, 1998). Moreover, Amanatidis et al., (1997), measured higher than expected carbonyl levels in an Athens suburb, which he actually related to photochemical produced carbonyls transported there or emitted by the biogenic activity. Since no aldehydes' measurements were used in this work, the carbon fractions for HCHO and ALD2 (for AB scenarios) have been extracted from the measurements of alkenes and alkines (according to the 'lumped structure' approach of the CB-IV). Klemm and Ziomas (1998) measured most of these compounds higher in street canyon than in the city plume.

TABLE III

Carbon fractions (according CB-IV) for the AB and ALA scenarios for both street canyon and city plume cases. For comparison reasons the carbon fractions proposed by Smylie *et al.* (1991), Jeffries *et al.* (1989) are also given

	AB street canyon	ALA street canyon	Athens Smylie *et al.* (1991)	US Jeffries *et al.* (1989)	AB City plume	ALA City plume
PAR *single-carbon bond*	0.45	0.492	0.56	0.564	0.618	0.581
ETH *ethene*	0.075	0.059	0.095	0.037	0.048	0.030
OLE *double-carbon bond*	0.042	0.034	0.067	0.035	0.025	0.0176
TOL *toluene*	0.193	0.154	0.124	0.089	0.156	0.115
XYL *xylene*	0.102	0.1003	0.119	0.117		0.098
HCHO *formaldehyde*	0.003	0.0095	0.0108	0.021	0.002	0.0097
ALD2 *acetaldehyde, higher ald.*	0.04	0.041	0.0207	0.052	0.0276	0.029
ISOP *isoprene*	0.0012	0.0009			0.002	0.001
MEOH *methanol*		0.004				0.0054
ETOH *ethanol*		0.017				0.0201

4. Conclusions

An adaptation procedure of a new emission inventory of the Greater Athens Area is attempted based on a sensitivity analysis on the treatment of the VOC emissions. This study indicates the most important species in terms of ozone productivity for the Athens case. From the comparison between air samples and emission profiles from Athens and other cities, including the US, the main differences were observed in acetylene, ethene, aromatic compounds and 1,3-butadiene concentrations. These are well explained from the different car fleet (catalytic, taxis-LPG) in the cities.

Thus, it is considered reasonable to use a detailed emission profile, specifically LA's, in order to expand Athens' inventory with those significant, to ozone production, VOC groups. With the available information for Athens, two basic emission profiles were constructed, reflecting the characteristics either of the city center or of the city plume. Starting from these profiles, subsets of additional VOC species, contained in the detailed LA's profile, were added in order gradually to approach the LA's profile. For the significant to ozone production VOC compounds, (such as aldeydes and alcohols) for which there is not sufficient information for Athens, it is considered necessary to carry out a measurement campaign. Until then, their emission rates (according the LA's profile) should be considered with caution.

The transition from the basic to the modified profiles resulted in 28% (street canyon) and 60% (city plume) increase of the initial concentration of VOCs while the maximum concentration of O_3 increased by 63–77% (street canyon) and 83–250% (city plume) depending on the chemical mechanism. Most of the VOC categories contribute almost equally to the increase of O_3 concentrations. Only the aldehydes, the aromatics and the alkenes, exhibit a significant difference. In the city plume case, the modified emission profile predicted a maximum O_3 concentration comparable to the one predicted for the street canyon case, even though the initial concentration of VOCs was significantly lower. From the two chemical mechanisms always the SAPRC99 (either with fixed or variable parameters) predicts higher O_3 concentrations in comparison to CB-IV.

The speciation of the VOCs found to be very crucial for different air mixtures (VOC/NO_x). CB-IV turned to be the more sensitive mechanism on the VOC/NO_x ratio while this sensitivity became much more evident for the city plume mixture. Therefore, the possibility to consider different profiles during the Eulerian photochemical simulations, could be very important in cases where the air masses travel above different land uses.

Acknowledgements

This work was financially supported by the Commission of the European Community, through the General Secretariat for Research and Technology (GSRT) of the Ministry of Development of Greece, under the contract 97YP84. We would like to acknowledge Dr. Carter and Dr. Allen for their help and the useful information they provided us with.

References

Amanatidis, G. T., Viras, L. G., Kotzias, D. and Bartzis, J. G.: 1997, 'Carbonyl levels in Athens during a winter air pollution episode', *Fresenius Envir. Bull.* **6**, 372–377.
Andersson, S., Frestad, A., Dempster, N. M. and Shore, P. R.: 1991, 'The Effect of Catalyst Ageing on the Composition of Gasoline Engine Hydrocarbon Emissions', SAE Technical Paper, 910174.
Blake, N. J., Penkett, S. A. and Clemitshaw, K. C.: 1993, 'Estimates of atmospheric hydroxyl radical concentrations from the observed decay of many reactive hydrocarbons in a well-defined urban plume', *J. Geophys. Res.* **98**, NO. D2, 2851–2864.
Carter, W. P. L.: 1988, 'Documentation for the SAPRC Atmospheric Photochemical Mechanism. Preparation and Emissions Processing Programs for Implementation in Airshed Models', California Air Resources Board Contracts No. A5-122-32.
Carter, W. P. L.: 1994, 'Development of ozone reactivity scales for volatile organic compounds', *J. Air Waste Manage. Assoc.* **44**, 881–899.
Carter, W. P. L.: 1994, 'Calculation of Reactivity Scales Using an Updated Carbon Bond IV Mechanism', Systems Applications International Under Funding from the Auto/Oil Air Quality Improvement Research Program.
Carter, W. P. L.: 2000, 'The SAPRC-99 Chemical Mechanism and Updated VOC Reactivity Scales', Final Report, California Air Resources Board Contracts No. 92–329 and 95–308.

Duffy, B. L. and Nelson, P. F.: 1996, 'Non-methane exhaust composition in the Sidney harbor tunnel: A focus on benzene and 1,3 butadiene', *Atmos. Environ.* **30**, 2759–2768.

Gery, M. W., Whitten, G. Z. and Killus, J. P.: 1988, 'Development and Testing of the CBM-IV for Urban and Regional Modeling', U.S. Environmental Protection Agency, Research Triangle Park, North Carolina by Systems Applications International, San Rafael, California (SYSAPP-88/002).

Gery, M. W., Whitten, G. Z., Killus, J. P. and Dodge, M. C.: 1989, 'A photochemical kinetics mechanism for urban and regional scale computer modeling', *J. Geophys. Res.* **94**, 12925–12956.

Goldan, P. D., Trainer, M., Kuster, W. C., Parrish, D. D., Carpenter, J., Roberts, J. M., Yee, J. E. and Fehsenfeld, F. C.: 1995, 'Measurements of hydrocarbons, oxygenated hydrocarbons, carbon monoxide, and nitrogen oxides in an urban basin in Colorado: Implications for emission inventories', *J. Geophys. Res.* **100**, NO. D11, 22771–22783.

Jeffries, H. E., Sexton, K. G., Arnold, J. R. and Kale, T. L.: 1989, 'Validation Testing of New Mechanisms with Outdoor Chamber Data. Volume 2: Analysis of VOC data for the CB4 and CAL Photochemical Mechanisms', Final Report, EPA-600/3-89-010b.

Klemm, O. and Ziomas, I. C.: 1998, 'Urban emissions measured with aircraft', *J. Air and Waste Manage. Assoc.* **48**, 16–25.

Kourtidis, K. A., Ziomas, I. C., Rappengluck, B., Proyou, A. and Balis, D.: 1999, 'Evaporative traffic hydrocarbon emissions, traffic CO and speciated HC traffic emissions from the city of Athens', *Atmos. Environ.* **33**, 3831–3842.

McCabe, R. W., Seigle, W. O., Chun, W. and Perry, J. J.: 1992, 'Speciated hydrocarbon emissions from the combustion of single component fuels. II. Catalyst effects', *J. Air and Waste Manage. Assoc.* **42**, 1071–1077.

Moschonas, N. and Glavas, S.: 1996, 'C3-C10 Hydrocarbons in the atmosphere of Athens, Greece', *Atmos. Environ.* **30**, 2769–2772.

Rappengluck, B., Fabian, P., Kalabokas, P., Viras, L. G. and Ziomas, I.: 1998, 'Quasi-continuous measurements of Non-Methane Hydrocarbons (NMHC) in the Greater Athens Area during MEDCAPHOT-TRACE', *Atmos. Environ.* **32**, 2103–2121.

Singh, H. B., Salas, L. J., Cantrell, B. K. and Redmond, R. M.: 1985, 'Distribution of aromatic hydrocarbons in the ambient air', *Atmos. Environ.* **19**, 1911–1919.

Smylie, M., Fieber, J. L., Myers, T. C. and Burton, S.: 1991, 'A Preliminary Examination of the Impact on Emissions and Ozone Air Quality in Athens, Greece from Various Hypothetical Motor Vehicle Control Strategies', European Conference on New Fuels and Clean Air, 19 June 1991, Antwerp, Belgium.

Ziomas, I. C., Suppan, P., Rappengluck, B., Balis, D., Tzoumaka, P., Melas, D., Papayannis, A., Fabian, P. and Zerefos, C. S.: 1995, 'A contribution to the study of photochemical smog in the Greater Athens Area', *Beitr. Phys. Atmosph.* **68**, 191–203.

RISK ANALYSIS OF VOLATILE ORGANIC COMPOUNDS THROUGH DAILY LIFE CYCLE IN THE INDUSTRIAL CITY IN KOREA

BYEONG-KYU LEE* and JUNGBUM CHO

Department of Civil and Environmental Engineering, University of Ulsan, Ulsan, Korea
(* author for correspondence, e-mail: bklee@mail.ulsan.ac.kr, fax: 82 52 259 2629)

Abstract. This study analyzed the risk of exposure to volatile organic compounds (VOCs) through a study of activity patterns in the Korean industrial city, Ulsan. The daily life cycle patterns (LCPs) of 331 people in Ulsan were surveyed and the average LCPs in Ulsan were obtained by statistical analysis. Nine to twelve personal air samples of VOC exposure at the breathing zones were collected at each LCP. This included hours for sleeping, cooking and eating, going to and from work, working, participating in field or outdoor activities, reading, watching TV, and shopping. The components and concentrations of the collected VOCs were identified by a Gas Chromatography-Mass Detector (GC-MS). The overall reproducibility of all GC analytical procedures of the simultaneously collected duplicate sample pairs represented a mean of percent differences ranging from about 9 to 13%. For the general population of Ulsan, the carcinogenic and non-carcinogenic risk of exposure to the VOCs during the LCPs was evaluated. The carcinogenic risk was analyzed using both the chronic daily exposure or lifetime average daily exposure (CDI) and the cancer potency factor. The non-carcinogenic risk was analyzed using both the CDI and the chronic reference dose. The major chemical forms of the identified VOCs were oxidized forms (43%), aliphatic alkanes (29%) and aromatics (15%). Even though the highest total exposure strength per unit time of each activity was observed during shopping, the highest total amount of exposure to VOCs was identified as the exposure during work. The total carcinogenic risk of exposure to the carcinogenic VOCs through daily life cycle in Ulsan was 2.0×10^{-4} which is substantially exceeding the permissible carcinogenic risk level, $10^{-5} \sim 10^{-6}$. The carcinogenic risk during most of the life cycle activities, except for reading, mainly performed indoors, was higher than that of the activities performed outdoors. The carcinogenic risk by benzene exposure was about 56% (time weighted average) of the total carcinogenic risk by the exposure to the carcinogenic VOCs. During cooking and eating, shopping and out door activities, however, the carcinogenic risk by the exposure to chlorinated compounds like chloroform exceeded the exposure to benzene. The overall hazard index (non-carcinogenic risk) by a chronic exposure to carcinogenic and non-carcinogenic VOCs through daily life cycle in Ulsan was evaluated as 3.91×10^{-1}, which is much less than 1.0 considered as a hazard level to human health, and thus it seems likely not to produce a severe health hazard.

Keywords: acetaldehyde, benzene, carcinogenic, chloroform and dichloromethane, exposure, hazard index, life cycle, risk, VOCs

1. Introduction

Recently, many people living in urbanized or industrialized cities are worrying about risk or potential to be exposed to hazardous or toxic air pollutants, such as volatile organic compounds (VOCs) and heavy metals (Thorsson and Eliasson,

Water, Air, and Soil Pollution: Focus **2:** 155–171, 2002.
© 2002 *Kluwer Academic Publishers.*

2001; Kornartit *et al.*, 2001; Jurvelin *et al.*, 2001; Lee and Na, 2000). VOCs are precursors for formation of tropospheric ozone and thus they could be important causes of photochemical smog formation (Wark *et al.*, 1988). Also, some of VOCs like benzene and chloroform are regulated as carcinogenic (suspecting) compounds (Keith and Walker, 1995; Harte *et al.*, 1991). In urbanized or industrialized cities major VOC emission sources include painting, printing, solvent uses, storage losses, and traffic related processes such as exhaust pipe emissions and evaporation losses (Air and Waste Management Association (A), 2000; Han, 1996). Even though VOCs produced from industrial activities or sources are excluded, there are lots of VOC emission sources in the middle of our modern life. In fact, VOCs are produced everywhere and thus many people have been exposed to various VOCs through their daily life cycle (DLC). This is because we have encountered considerable amounts of VOCs in our modern lifestyle – for example, during cooking, commuting and traveling, working, shopping and even outdoor activities. According to the recent reports (Kornartit *et al.*, 2001; Baek *et al.*, 1998), most of our time is spent indoors where the VOC concentrations indoors are much greater than those outdoors. Thus many people are exposed to a lot of VOCs through their indoor activities like cooking, watching TV, sleeping, reading, and so on. Even though some VOCs found indoors may have outdoor origins, indoor VOCs mainly come from various indoor stuffs such as clothes, curtains, carpets, furniture, reading materials, and cooking processes (Lee and Ellenbecker, 1998; Vainiotalo and Matveinen, 1993; Samet *et al.*, 1987; Hollowell and Miksch, 1981).

Ulsan is the largest industrial city in Korea, and has various industrial complexes, such as huge petrochemical and chemical, non-ferrous metallic, metallic and shipbuilding industries. Also, the city of Ulsan is the 7th biggest metropolitan city in Korea having a population of more than a million. Thus there have been a lot of environmental pollution problems. Especially, because of the numerous various emission sources of air pollutants, the city of Ulsan has been troubled about its serious air pollution. In Ulsan a large amount of VOCs as emitted from sources such as industries and traffic – and thus the metropolitan Ulsan area has been designated a special area where VOC emissions at sources should be reduced within a designated period (Lee, 2000). Thus many VOC control technologies, such as adsorption, biofilteration and combustion, to reduce VOC emissions have been developed and applied to industry sources in Ulsan (Air and Waste Management Association (B), 2000; Lee and Cho, 1999; Lawson and Adams, 1999; Ergas and McGrath, 1997; Ruddy and Carroll, 1993). Nevertheless, the general public in Ulsan is very worried about exposure potential and risk of carcinogenic VOCs such as benzene and chloroform. From this point of view, this study analyzed carcinogenic and non-carcinogenic risks of exposure to VOCs through various life activities in Ulsan, a representative industrial city in Korea.

2. Methods

The daily life cycle patterns (LCPs) of the general public in Ulsan were analyzed through a survey of 331 citizens over the age of 20. The survey considered gender, location, residence, age, occupation, eating locations, subscriptions, transportation, shopping areas, and field or outdoor activities. The survey also included a time breakdown of a general daily activity, which is the information about how much of time was spent in a day for the following activities: sleeping, cooking and eating, going to and from work, outdoor activities, reading, watching TV, and shopping. To obtain average LCPs, the statistical analysis of the collected survey data was performed using the SPSS for Windows release 10.0.5 (27 November, 1999) standard version.

We collected personal air samples of VOCs at the breathing zone of the people to be exposed to VOCs through daily life cycles. The air samples were taken during the same time as the time spent for the analyzed average LCPs. The personal air samples of 9–12 were obtained at each LCP by using calibrated personal air sampling pumps (Gillian Co., U.S.A.) with flow rates ranging from 20 to 40 mL min^{-1} and tedlar air bags (OMI, Japan) of 10 L according to the analyzed LCPs (Ness, 1991). The VOC air samples collected in tedlar bags were adsorbed into the clean and conditioned Tenax TA adsorption tube (Teckma, U.S.A.) and then desorbed and concentrated by using a cryo-focusing module (DS 5000, Donam Systems Inc., Korea). The desorbed VOCs in the cold trap, after being purged by helium gas, were injected into the analytical column (HP-PONA, 50 × 0.2 mm × 0.5 μm) of the Gas Chromatography (HP 5890 series II, Hewlett Packard) by using a temperature program. The components and concentrations of the eluted VOCs were identified by a mass detector (HP 5971 series, Hewlett Packard) and a data system (HP59944C MS ChemSystem, Hewlett Packard). Each VOC standard (Aldrich, U.S.A.) was used to identify GC retention time for qualitative analysis of the VOC standard mixture solution (Revised PVOC/GRO Mix, Supleco, U.S.A.) and the VOC standard mixture gas (Matheson Gas Products Inc., U.S.A.). The reproducibility of the VOC analytical method was identified in terms of retention time and peak area within a day of the VOC standards. The relative standard deviation of the reproducibility of the retention time and the peak area was less than 1% and 10%, respectively, except for a few standards. The method detection limit (MDL), identified by analyzing VOCs after spiking 5 ng of each VOC standard into 10 Tenax tubes (Baek *et al.*, 1999; U.S. EPA, 1990), ranged from 0.9 to 1.8 ng. This result means that the converted MDL of air sample of 10 L ranged from 0.012 to 0.035 ppb. The overall reproducibility of all GC analytical procedures of the simultaneously collected duplicate sample pairs represented a mean of percent differences ranging from about 9 to 13% (U.S. EPA, 1997a, b).

The identified VOCs were classified into two categories: carcinogenic and non-carcinogenic compounds, according to the carcinogenic classification by the U.S. Environmental Protection Agency (U.S. EPA) and the International Agency for

Research on Cancer (IARC) (U.S. EPA, 1986; IARC, 1982). The carcinogenic risk by exposure to the carcinogenic VOCs was analyzed using both the chronic daily exposure or lifetime average daily exposure (CDI) calculated from both the geometric mean of the carcinogenic VOCs analyzed in this study and the cancer potency factor (CPF) of the VOCs obtained from the U.S. EPA and the IARC (Lee, 1998; U.S. EPA, 1996). The overall non-carcinogenic risk (hazard index) by exposure to the VOCs was also analyzed using both the CDI calculated from geometric mean concentrations of the carcinogenic and non-carcinogenic VOCs and the chronic reference dose (RfD) obtained from the U.S. EPA and the IARC (Jarabek and Segal, 1994). This risk analysis was performed based on the daily activity patterns and time of activity, of people who live in Ulsan. Thus each activity fraction or factor was included in the evaluation processes of the carcinogenic risk and the non-carcinogenic risk. Then the total lifetime risk of the general public who spend their whole life (based on 70 yr) in the industrial city was evaluated by an addition of each risk obtained from the life pattern.

3. Results and Discussion

3.1. DAILY LIFE CYCLE PATTERNS

Table I shows the analysis of life cycle patterns of the general public in an industrial city in Korea. In an analysis of residential types or sleeping habits, most of the citizens in Ulsan live in apartment complexes. The major eating areas for people in Ulsan are cafeterias and homes. The most popular shopping places were big-sales markets and open markets. The most popular reading material was newspapers and the place of major outdoor activities were those near their residential area. Generally citizens in Ulsan spend 85 to 90% of their time indoors and are outdoors about 10 to 15% of their time during the weekdays. The major areas for indoor activities were at home (46%) and in the workplace (38%). The major reasons for outdoor activities were the commute to and from work and the outdoor activities such as walking or jogging or outdoor sports. Thus, almost the exposure to VOCs for the general population of Ulsan was due to the exposure to VOCs experienced during indoor activities. Table I also shows the approximate or high probable or average time breakdown of daily life cycle patterns of the general public in Ulsan. The carcinogenic risk and the non-carcinogenic risk by exposure to VOCs discussed later on were calculated based on the average time breakdown.

3.2. EXPOSURE TO VOCs

Table II shows the analysis of average concentrations of VOCs experienced during different daily activities in Ulsan. The concentration range of benzene identified during daily life in Ulsan was 3.5–32.2 μ m^{-3}. The benzene concentrations during sleeping and watching TV were higher than those during other activities. This

TABLE I

Analysis of life cycle patterns of the general public in an industrial city, Ulsan

Activity	Time (hr)	Relative %	Most probable or average time (hr)	Comments
Sleeping	6.0–6.5	19.5	7.0	APT: 55.9%
	6.5–7.0	24.8		House: 44.1%
	7.0–7.5	25.7		
	More than 7.5	14.3		
Cooking and	0.5–1.0	26.9	1.5	Home: 41%
having a meal	1.0–1.5	27.5		Cafeteria: 29.8%
	1.5–2.0	15.8		Work cafe: 25.5%
	2.0–2.5	15.2		Other: 3.3%
Going to work	1/3–0.5	35.3	1.0	Bus: 37.3%
and leaving	0.5–1.0	22.1		Gas car: 19.4%
from work	1.0–1.5	21.5		Diesel car: 7.6%
Work	6.0–7.0	14.5	8.5–8.7	Self: 15.7%
	7.0–8.0	15.8		Bureau: 10.3%
	8.0–8.5	10.3		Technical: 16.0%
	8.5–9.0	19.6		Student: 21.8%
	9.0–10.0	18.4		Housewife: 23.6%
Outdoor	0.5–1.0	22.4	2.0	Near residence: 57.8%
activities	1.0–1.5	17.5		School: 14.9%
	1.5–2.5	22.4		Near downtown: 18.5%
	More than 2.5	20.2		
Shopping	1/3–0.5	41.7	0.8–1.0	Market: 34.3%
	0.5–1.0	24.8		Dept. store: 13.1%
	1.0–1.5	17.2		Mall: 37.4%
Reading	less than 1/3	11.2	1.0	Newspaper: 62.7%
	1/3–2/3	32.3		Magazine: 14.2%
	2/3–1.0	27.5		Books: 23.0%
	1.0–2.0	22.1		
Watching TV	0.5–1.0	17.2	2.0	APT: 55.9%
	1.0–1.5	18.4		House: 44.1%
	1.5–2.0	17.8		
	2.0–3.0	20.9		
	More than 3.0	13.6		

means that benzene concentrations indoors are much greater than outdoors. The concentration range of dichloromethane identified in the middle of daily activities in Ulsan was 6.1–70.8 μg m^{-3}. The average concentrations of chloroform and dichloromethane during cooking and meals were higher than those during other activities. This means that some compounds containing chlorine like chloroform and dichloromethane could be produced in the cooking process. This might be due to the cooking of food containing chlorine compounds. The origin of these chlorine compounds could be food containing agricultural chemicals and/or some salty Korean food. Having a meal in Korea is also one source of high exposure to VOCs. This is because many Koreans like to eat boiled, heated or fried food prepared at high temperature. Therefore, many Koreans might be exposed to VOCs produced from the cooked food during meals. The highest average concentration of dichloromethane was observed during shopping. This high concentration is possibly due to the emissions from clothes and/or stuffs like electric appliances or packaging materials. Thus the general public in Ulsan has a high potential for exposure to chlorinated compounds like chloroform and dichloromethane during cooking and shopping. This means that housemen or cooks and workers in shopping places like department stores or big sales markets might have high exposure potential to chloroform and/or dichloromethane.

In this study the highest average concentrations of acetaldehyde, toluene, xylenes and n-hexane were also observed during shopping.

The VOCs identified in this study were categorized into groups based on their chemical structure. Table III shows the relative concentrations of the chemical groups, obtained by simply adding exposure to the VOCs identified in each daily activity. Oxidized forms, such as aldehydes, ketones and alcohols, were the most prominent chemical groups for exposure during daily activities except for sleeping. Relatively high exposure to aromatics, such as benzene and toluene, were observed during sleeping, shopping and reading as compared to other activities. A relatively high exposure to chlorinated compounds was identified during cooking and meals. Relative high exposure to aliphatic alkanes was obtained during sleeping, outdoor activity and shopping. The time weighted average (TWA) of the relative exposure values to the chemical groups exposed during daily life in Ulsan were as follows: oxidized forms 43%, aliphatic alkanes 29%, aromatics 15%, aliphatic alkenes 9%, and chlorine compounds 4% (see Figure 1). The TWA values of VOC exposure differ from those of the arithmetic mean (AM). Especially, the relative proportions of the aliphatic alkanes and the chlorinated compounds which are very different from each other. This is probably due to the high AM value for high concentration of the chlorinated compounds measured during cooking and meals. However, the TWA values were relatively low during cooking and meals that have relatively short time occupancy. The difference between TWA and AM values is also due to the very broad concentration range of the aliphatic alkanes and the difference in the time breakdown of the daily activities.

TABLE II

Analysis of average concentration of VOCs as a function of the daily life cycle patterns in Ulsan

VOCs	Average concentration during activity ($\mu g\ m^{-3}$)								Range	A.M.[a]	S.D.[b]	G.M.[c]
	Sleeping	Cooking/having a meal	Work	Reading	Going to/leaving from office	Watching TV	Shopping	Outdoor activity				
Carcinogenic VOCs												
Benzene	26.14	7.66	3.97	8.81	5.04	32.21	15.21	3.51	3.5–32.2	12.82	10.87	9.40
Acetaldehyde	11.44	11.20	10.62	15.41	5.45	15.16	60.99	31.36	5.5–60.9	20.20	18.14	15.49
Chloroform	1.57	33.41	14.83	6.14	2.31		6.69	6.89	1.5–33.4	10.17	13.22	5.62
Dichloromethane	19.85	65.53			12.75	9.11	70.80	8.80	6.1–70.8	25.98	26.41	17.44
Non-Carcinogenic VOCs												
Acetone	12.87	21.10	113.02	19.07	10.85	11.42	62.73	25.26	11.4–113	34.54	35.91	24.13
Acetonitrile	18.49	40.53	161.21	15.17	5.47		27.43	4.54	4.5–161	38.98	55.34	19.93
n-Butanol					2.83					2.83		2.83
Ethyl benzene	9.14	4.73	6.31	19.60	1.09	8.16	9.66	8.14	1.0–19.6	8.35	5.33	6.63
n-Heptane	11.26		1.88				4.78		1.8–11.2	5.97	4.80	4.66
M.E.K.	7.79	19.96	9.31	35.00		7.69	30.24	4.44	4.4–35.0	16.35	12.20	12.67
Toluene	18.19	32.59	9.19	15.90	17.24	7.83	83.18	5.60	5.6–83.1	23.71	25.47	16.48
Xylenes	14.81	13.63	20.23	22.62	2.75	15.09	25.48	11.55	2.7–25.4	15.77	7.11	13.54
Benzaldehyde		10.63			1.35		5.66		5.6–10.6	8.14	3.51	7.76
Carbon disulfide	2.70	51.33	5.47						1.3–51.3	15.21	24.14	5.65
n-Hexane	54.66	22.66	19.33	11.02	9.09	4.70	115.15	7.69	4.7–115	30.54	37.72	17.78
d-Limonene	16.54	23.78	8.80	20.56		24.10	15.29	6.09	6.0–24.1	16.45	7.03	14.85
Styrene	11.39		24.16	8.05	2.98	36.27	17.38	3.77	2.9–36.2	14.86	12.07	10.67

[a] A.M.: Arithmetic mean. [b] S.D.: Standard deviation. [c] G.M.: Geometric mean.

TABLE III

Relative exposure concentration to VOCs depending upon chemical structure during daily life cycle

VOC category (chemical group)	Relative exposure concentration to VOCs during daily life cycle activity (%)								Arithmetic mean	Time weighted average
	Sleeping	Cooking/ having a meal	Work	Reading	Going to/ leaving from work	Watching TV	Shopping	Outdoor activity		
Oxidized	25	47	60	53	52	46	38	40	45	41
Aliphatic alkenes	12	5	4	7	14	28	5	4	10	11
Aliphatic alkanes	31	14	24	11	16	7	25	32	20	28
Chlorinated	7	23	2	2	8	5	9	6	8	4
Aromatics	25	11	10	27	10	14	23	18	17	16

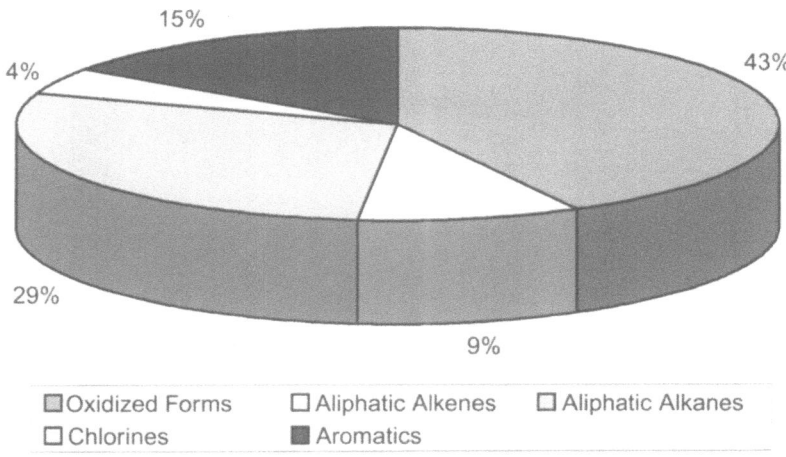

Figure 1. Relative exposure concentration versus chemical structure of VOCs.

In analysis of the total personal exposure strength per unit time of each activity to the VOCs identified in the classified life activities, shopping represented the highest total exposure strength among the activities (see Figure 2). The total exposure strength during shopping was followed by that during work or cooking and eating. As the windows of the vehicles used for VOC air samples going to work and leaving work investigated in the study were opened to the atmosphere, these could be considered as outdoor air samples. The total exposure strength to the VOCs during the outdoor activities and going to work and leaving work was relatively low as compared to that during shopping, working, and cooking and meals, which were typically performed indoors. This means that the total exposure strength to VOCs during the indoor activities is much greater than that during the outdoor activities.

Figure 3 shows the total amounts exposed to VOCs during the daily activities. The highest total amount exposed to VOCs was observed as the exposure during the work hours. This was followed by sleeping hours. Even though the highest total exposure strength to VOCs during shopping was observed, the total amount of exposure to VOCs during shopping was relatively small. This is due to the short time spent shopping as compared to the long time spent working.

3.3. CARCINOGENIC RISK BY EXPOSURE TO CARCINOGENIC VOCs

The carcinogenic VOCs identified in this study were benzene, acetaldehyde, chloroform, and dichloromethane. Table IV represents the carcinogenic risk by exposure to the carcinogenic VOCs depending on the classified life patterns. The total carcinogenic risk by exposure to the carcinogenic VOCs through daily life in Ulsan was 2.0×10^{-4} which is exceeding substantially the permissible carcinogenic risk level, $10^{-5} \sim 10^{-6}$. Most of the life activities, except for reading, mainly performed

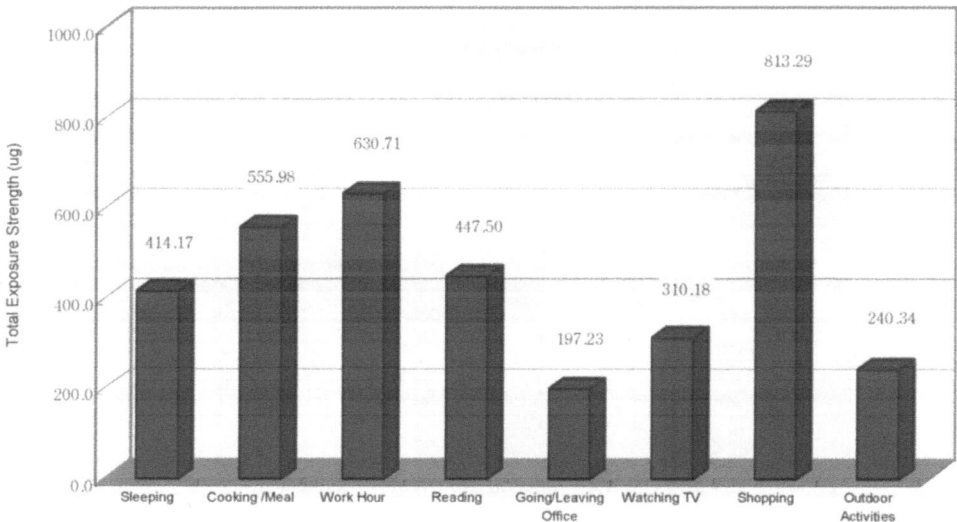

Figure 2. Total exposure strength to VOCs per unit time of each activity through daily life cycle in Ulsan.

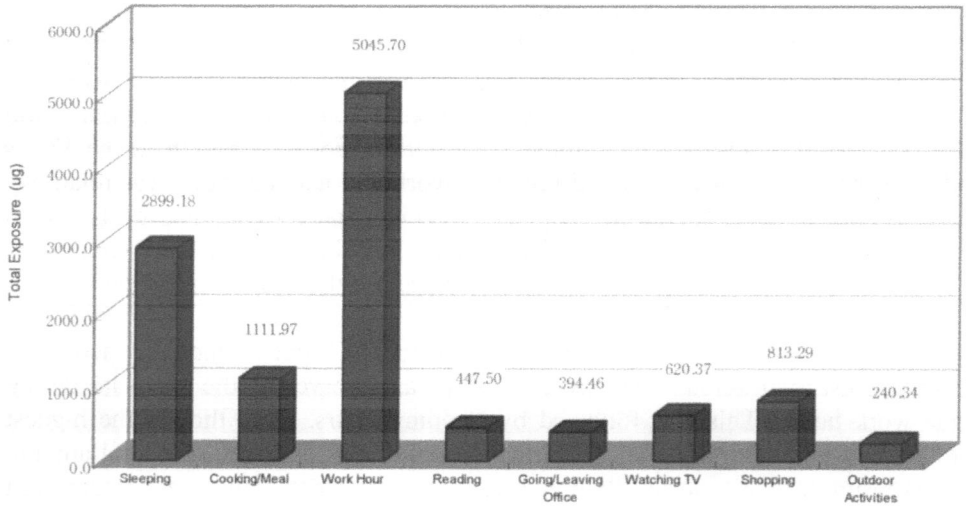

Figure 3. Total amounts exposed to VOCs through daily life cycle in Ulsan.

indoors, the carcinogenic risk also greatly exceeded the permissible carcinogenic risk level. In addition, the carcinogenic risk of the activities, e.g. going to work and leaving work and the outdoor activities such as walking and jogging outside the general residential area, mainly performed outdoors was ranged from 4.1×10^{-6} to 5.5×10^{-6}. However, the carcinogenic risk during the outdoor activities near the industrial complex area was about 1.5×10^{-5} and it was about 3 times that during the outdoor activities near the general residential area. The relative high

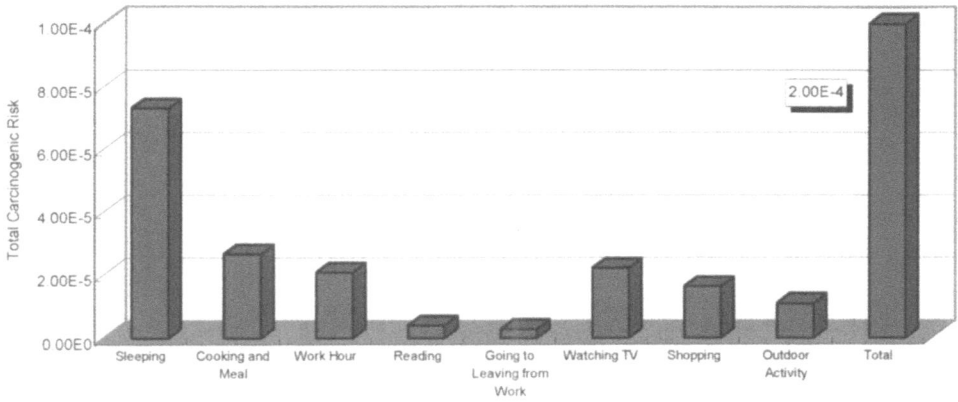

Figure 4. Carcinogenic risk by exposure to carcinogenic VOCs through daily life cycle in Ulsan.

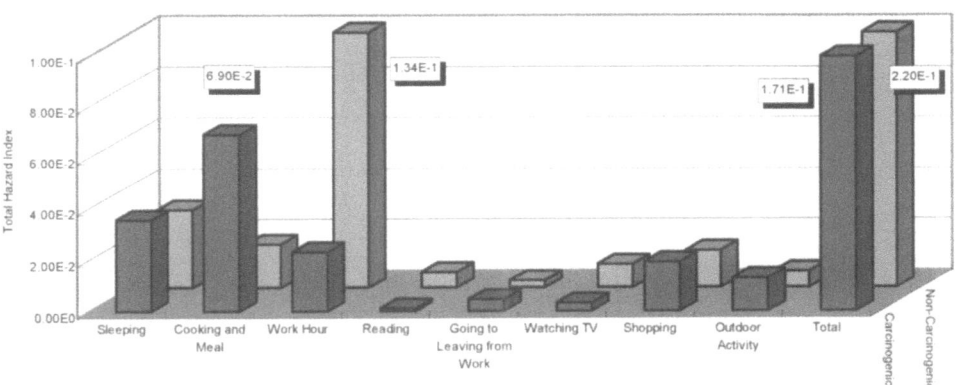

Figure 5. Non-carcinogenic risk (hazard index) by exposure to VOCs through daily life cycle in Ulsan.

carcinogenic risk near the industrial complex area is probably due to the exposure to the VOCs transferred from the industrial complex area.

The highest carcinogenic risk of the life activities classified in this study was the exposure to carcinogenic VOCs during sleeping. Especially, the carcinogenic risk of sleeping at the newly built apartments was about 1.0×10^{-4} that is a very serious value. Also, in comparison the carcinogenic risk of sleeping in the newly built and the in theold apartments, the value at the new ones was 2 times as high as the old ones. This means that the total amount of carcinogenic VOCs emitted from new furniture, new painting and new walls is much greater than that from the old ones. Therefore, at the new apartments it is especially required to keep a good ventilation system, such as wide open windows and forced ventilation by exhaust fans, for reduction of the exposure to carcinogenic VOCs.

The second highest carcinogenic risk (2.65×10^{-5}) in this study was identified as exposure to carcinogenic VOCs during cooking and eating. The carcino-

TABLE IV

Carcinogenic risk by exposure to carcinogenic VOCs through daily life cycle activities in Ulsan

VOCs (Carcinogenic group)	Average carcinogenic risk of daily life cycle								Total
	Sleeping	Cooking/ having a meal	Work	Reading	Going to/ leaving from work	Watching TV	Shopping	Outdoor activity	
Benzene (A)	5.52×10^{-5}	3.46×10^{-6}	1.08×10^{-5}	2.66×10^{-6}	1.52×10^{-6}	1.94×10^{-5}	4.58×10^{-6}	1.59×10^{-6}	9.92×10^{-5}
Acetaldehyde (B2)	6.38×10^{-6}	1.34×10^{-6}	7.63×10^{-6}	1.23×10^{-6}	4.37×10^{-7}	2.42×10^{-6}	4.88×10^{-6}	3.75×10^{-6}	2.81×10^{-5}
Chloroform (B2)	9.16×10^{-6}	2.09×10^{-5}	–	–	1.93×10^{-6}	–	5.57×10^{-6}	8.61×10^{-6}	6.71×10^{-5}
Dichloromethane (B2)	2.38×10^{-6}	8.40×10^{-7}	2.28×10^{-6}	1.05×10^{-7}	2.18×10^{-7}	3.12×10^{-7}	1.21×10^{-6}	2.27×10^{-7}	8.41×10^{-6}
Total carcinogenic risk in each activity (TCR)	7.31×10^{-5}	2.65×10^{-5}	2.07×10^{-5}	$4.00 \times 1\text{-}^{-6}$	2.82×10^{-6}	2.21×10^{-5}	1.62×10^{-5}	1.42×10^{-5}	2.03×10^{-4}
Contribution to TCR by benzene exposure (%)	75.5	13.0	52.1	66.5	54.3	87.8	28.3	14.6	49 (A.M.)[a] 56 (T.W.A.)[b]
Other prominent contributing chemicals to TCR (%)	CHCl$_3$ (78.9)						CHCl$_3$ (30.1) CH$_2$Cl$_2$ (34.3)	CHCl$_3$ (39.5) CH$_3$CHO 34.4	

[a] A.M. stands for an arithmetic mean.
[b] T.W.A. stands for a time weighted average.

TABLE V

Non-carcinogenic risk (hazard index) by eposure to VOCs through daily life cycle in Ulsan

VOCs	Chronic average non-carcinogenic hazard index of daily life cycle								
	Sleeping	Cooking/ having a meal	Work	Reading	Going to/ leaving from work	Watching TV	Shopping	Outdoor activity	Total
Carcinogenic VOCs									
Benzene									
Acetaldehyde									
Chloroform	1.14×10^{-2}	5.20×10^{-2}	5.20×10^{-2}		2.39×10^{-3}	2.39×10^{-3}	6.92×10^{-3}	1.07×10^{-2}	8.62×10^{-2}
Dichloromethane	2.40×10^{-2}	1.70×10^{-2}	2.31×10^{-2}	1.06×10^{-3}	2.20×10^{-3}	3.15×10^{-3}	1.22×10^{-2}	2.28×10^{-3}	8.50×10^{-2}
Total hazard index	3.54×10^{-2}	6.90×10^{-2}	2.31×10^{-2}	1.06×10^{-2}	4.59×10^{-3}	3.15×10^{-3}	1.91×10^{-2}	1.30×10^{-2}	1.71×10^{-1}
Non-carcinogenic VOCs									
Acetone	9.33×10^{-3}	3.29×10^{-3}	1.05×10^{-1}	1.98×10^{-3}	1.13×10^{-3}	2.37×10^{-3}	6.50×10^{-3}	3.93×10^{-3}	1.34×10^{-1}
n-Butanol					2.94×10^{-4}				2.94×10^{-4}
Ethyl benzene	6.62×10^{-3}	7.37×10^{-4}	5.88×10^{-3}	2.03×10^{-3}	1.13×10^{-4}	1.69×10^{-3}	1.00×10^{-3}	1.26×10^{-3}	1.93×10^{-2}
M.E.K.	9.43×10^{-4}	5.17×10^{-4}	1.44×10^{-3}	6.04×10^{-4}		2.65×10^{-4}	5.22×10^{-4}	1.15×10^{-4}	4.41×10^{-3}
Toluene	6.59×10^{-3}	2.54×10^{-3}	4.30×10^{-3}	8.23×10^{-4}	8.95×10^{-4}	8.13×10^{-4}	4.31×10^{-3}	4.35×10^{-4}	2.07×10^{-2}
Xylenes	5.37×10^{-4}	1.06×10^{-4}	9.43×10^{-4}	1.17×10^{-4}	1.43×10^{-5}	8.63×10^{-5}	1.32×10^{-4}	8.97×10^{-5}	2.03×10^{-3}
Benzaldehyde		1.65×10^{-3}					5.88×10^{-4}		2.24×10^{-3}
Carbon disulfide	1.96×10^{-3}	8.03×10^{-3}	5.10×10^{-3}		1.40×10^{-4}				1.52×10^{-2}
Ethyl acetate	2.35×10^{-4}	1.14×10^{-5}			2.55×10^{-5}		3.37×10^{-4}	2.13×10^{-4}	8.22×10^{-4}
Styrene	4.13×10^{-3}		1.13×10^{-2}	4.17×10^{-4}	1.55×10^{-4}	3.76×10^{-3}	9.00×10^{-4}	2.93×10^{-4}	2.10×10^{-2}
Total hazard index	3.03×10^{-2}	1.69×10^{-2}	1.34×10^{-1}	5.97×10^{-3}	2.76×10^{-3}	8.98×10^{-3}	1.43×10^{-2}	6.34×10^{-3}	2.20×10^{-1}
Hazard index ratio (carcinogenic: non-carcinogenic)	65 : 35	80 : 20	45 : 55	15 : 85	29 : 71	21 : 79	57 : 43	67 : 33	44 : 56

genic risk identified during cooking and having meals at the houses was about 9.39×10^{-7}. However, the carcinogenic risk (5.27×10^{-5}) identified during cooking and eating at the restaurants was much greater (50 to 60 times as high as) than that at the houses. The high exposure to chloroform contributed greatly to the high carcinogenic risk obtained during cooking and eating at the restaurants. Especially, about 80% of the carcinogenic risk during the cooking and having meals at restaurants was due to the exposure to chloroform. However, chloroform was not detected during inside houses. This means that a considerable amount of the chlorinated VOCs like chloroform and dichloromethane were produced during the cooking in restaurants where the cooking fumes include benzene, toluene and chlorinated compounds and were not effectively ventilated. In fact, because of this poor ventilation, a considerable amount of the cooking fumes were deposited in the chairs, tables and on floors at the restaurants and thus they became very sticky or slippery.

In an analysis of the carcinogenic risk by the exposure to carcinogenic VOCs during shopping, the risk was highly dependent upon the shopping place. The average carcinogenic risk of the department stores, the big sales markets and the open markets was 3.88×10^{-5}, 1.91×10^{-5} and 7.89×10^{-6}, respectively. The carcinogenic risk at the open markets located outdoors opened to the atmosphere was identified as the lowest value as compared to the shopping places located indoors. The carcinogenic risk at the department stores was 2 times as high as that at the big sales markets. This is possibly due to the difference in operating ventilation systems and dealing goods. Especially, the ventilation conditions at one of the department stores was as bad one could smell the odorous compounds and feel nose-irritating compounds during the air sampling periods. Therefore, it is necessary to operate better effective ventilation systems for reduction of the workers exposure to carcinogenic VOCs.

The carcinogenic risk (3.57×10^{-5}) by the exposure to carcinogenic VOCs identified during the work hour in the factory offices located in industrial complex was 3 times as high as that in the general city areas. This is due to higher emissions of carcinogenic VOCs from industrial processes or facilities. These results imply that the total carcinogenic risk in almost all activities in Ulsan, is exceeding the permissible carcinogenic risk level.

Benzene is highly regulated as a very strong carcinogenic compound. Table IV shows the relative contributions of only benzene exposure on the total carcinogenic risk. The carcinogenic risk by benzene exposure was about 56% (time weighted average) of the total carcinogenic risk by the exposure to the carcinogenic VOCs through daily life cycle in Ulsan. Especially, the relative carcinogenic risk by benzene exposure during watching TV and sleeping was 87.8 and 75.5%, respectively. This result represents that it is necessary to maintain a good ventilation system at home in order to reduce the carcinogenic risk and reduce benzene emissions indoors or home. However, the relative carcinogenic risk by the exposure to only benzene during the outdoor activities or cooking or shopping was relatively low. During these activities the carcinogenic risk by the exposure to chlorinated compounds like chloroform and dichloromethane exceeded the carcinogenic risk by exposure to benzene. During the cooking and eating, especially, the relative contribution of the total carcinogenic risk by chloroform exposure was very high (78.9%).

3.4. HAZARD INDEX BY EXPOSURE TO VOCs

Table V shows non-carcinogenic risk (hazard index) by the chronic exposure to the VOCs through daily life cycle in Ulsan. The total or overall hazard index was evaluated as 3.91×10^{-1} and that is much less than the 1.0 considered as a hazard level to human health. This implies the overall hazard by the chronic exposure to the VOCs in Ulsan not serious to produce a severe hazard. The non-carcinogenic risk by the exposure to non-carcinogenic VOCs contributed to 56% of the total non-carcinogenic risk by exposure to all VOCs. However, the relative contributions of carcinogenic and non-carcinogenic VOCs to the total non-carcinogenic risk differ from each activity. For example, the carcinogenic VOCs contributed greatly to the total non-carcinogenic risk during cooking and eating – having the non-carcinogenic VOCs contributed greatly to the total non-carcinogenic risk during reading.

4. Conclusions

The total carcinogenic risk, 2.0×10^{-4}, by exposure to the carcinogenic VOCs through daily life in a representative industrial city, Ulsan, was exceeding substantially the permissible carcinogenic risk level of $10^{-5} \times 10^{-6}$. The carcinogenic risk of most of the life activities, except for reading, performed indoors was much greater than that of activities performed outdoors. The carcinogenic risk by benzene exposure was about 56% (TWA) of the total carcinogenic risk by exposure to the carcinogenic VOCs in Ulsan. During cooking and meal, shopping and outdoor activities such as walking and jogging near the residential areas, however, the relative carcinogenic risk by exposure to chlorinated compounds was higher than that by exposure to benzene. The overall non-carcinogenic risk (hazard index), 3.91×10^{-1}, by chronic exposure to VOCs in Ulsan was not serious to produce severe hazard. It is necessary to reduce VOC emissions, indoor and outdoor sources, and prepare proper measures such as better ventilation systems for reduction of personal exposure to VOCs.

Acknowledgements

The authors wish to acknowledge the partial financial support of the Ulsan Regional Environmental Technology Development Center.

References

Air and Waste Management Association (A): 2000, 'Evaporative Emissions and Surface Coatings', Wayne T. Davis (ed.), *Air Pollution Engineering Manual*, 2nd ed., John Wiley & Sons, Inc., pp. 301–332.

Air and Waste Management Association (B): 2000, 'Control of gaseous pollutants', in Wayne T. Davis (ed.), *Air Pollution Engineering Manual*, 2nd ed., John Wiley & Sons, Inc., pp. 22–65.

Baek, S. O., Song, H. B., Shin, D. C., Hong, S. H. and Chang, H. S.: 1998, 'Seasonal and locational concentrations of particulate air pollutants in indoor air of public facilities in Taegu area', *J. Korea Air Pollut. Assoc.* **14**, 163–175.

Baek, S. O., Hwang, S. M., Park, S. K., Jeon, S. J., Kim, B. J. and Heo, G. S.: 1999, 'Evaluation of methodology for the measurement of VOCs in the air by adsorbent sampling and thermal desorption with GC analysis', *J. Korean Soc. Atmosph. Environ.* **15**, 121–138.

Ergas, S. J. and McGrath, M. S.: 1997, 'Membrane bioreactor for control of volatile organic compound emissions', *J. Environ. Engineer.* **June**, 593–598.

Han, H. J.: 1996, *A Study on VOC Emission Estimation on Emission Sources*, Korea Petroleum Association, pp. 1–186.

Harte, J., Holdren, C., Schneider, R. and Shirley, C.: 1991, *Toxics A to Z*, Univ. of California Press, pp. 233–235.

Hollowell, C. D. and Miksch, R. R.: 1981, 'Sources and concentrations of organic compounds in indoor environment', *Bull. New York Academy of Medicine* **57**(10), 962–977.

IARC Working Group: 1982, *IARC Monographs on the Evaluation of Carcinogenic Risk of Chemicals to Humans: Supplement 4*, IARC, Lyon, pp. 270.

Jarabek, A. M. and Segal, S. A.: 1994, 'Noncancer toxicity of inhaled toxic air pollutants: available approaches for risk assessment and risk management', in D. R. Patrick (ed.), *Toxic Air Pollution Handbook*, Van Nostrand Reinhold, New York, pp. 10–130.

Jurvelin, J., Vartiainen, M., Jantunen, M. and Pasanen, P.: 2001, 'Personal exposure levels and microenvironmental concentrations of formaldehyde and acetaldehyde in the Helsinki Metropolitan area, Finland', *J. Air and Waste Manag. Assoc.* **51**, 17–34.

Keith, L. H. and Walker, M. M.: 1995, *Handbook of Air Toxics*, Lewis Publishers, pp. 183–186.

Kornartit, C., Sokhi, R. S., Burton, M. A. S., Gonzalez-Flesca, N., Tod, A. M. and Kingham, S. P.: 2001, 'Personal Exposure to Benzene and Nitrogen Dioxide resulting from Indoor and Outdoor Sources', *Proceedings of the 3rd International Conference on Urban Air Quality and the 5th Saturn Workshop*, PE.3.5.

Lawson, P. L. and Adams, C. D.: 1999, 'Enhanced VOC absorption using the ozone/hydrogene peroxide advanced oxidation process, *J. Air and Waste Manag. Assoc.* **49**, 1315–1323.

Lee, B. K. and Cho, S. W.: 1999, 'Emission reduction of air pollutants produced from chemical plants', *J. Korean Soc. Atmosph. Environ.* **15**, 29–38.

Lee, B. K. and Ellenbecker, M. J.: 1998, 'Effective local exhaust ventilation on cooking fumes of seasoned meats', *Environ. Sci.* **2**, 49–56.

Lee, B. K. and Na, D.: 2000, 'A study on the characteristics of PM-10 and airborne metallic elements produced in the industrial city', *J. Korean Soc. Atmosph. Environ.* **16**, 23–36.

Lee, J. H.: 1998, 'Risk assessment of volatile organic compounds' in Korea Air Pollution Association (eds), *Air Environment and Volatile Organic Compounds*, Junghaeang Publishers, Inc., pp. 189–219.

Lee, K. Y.: 2000, 'Policy strategies for air environment management, *J. Environ. Hi-Technol.* **8**, 2–8.

Ness, S. A.: 1991, 'Sampling methods', in S. A. Ness (ed.), *Air Monitoring for Toxic Exposures: An Integrated Approach*, Van Nostrand Reinhold, New York, pp. 29–40.

Rudy, E. N. and Carroll, L. A.: 1993, 'Select the best VOC control strategy', *Chemical Engineering Progress* **July**, 28–35.

Sammet, J. M., Marbury, M. C. and Spengler, J. D.: 1987, 'Health effects and sources of indoor air pollution', *American Review of Respiratory Disease* **136**(6), 1486–1508.

Thorsson, S. and Eliasson, I.: 2001, 'Personal Exposure of Different Urban Environments – A Case Study', *Proceedings of the 3rd International Conference on Urban Air Quality and the 5th Saturn Workshop*, PE.3.4.

U.S. Environmental Protection Agency: 1990, 'Definition and Procedure for the Determination of the Method Detection Limit', Code of Federal Regulations, Part 136, Appendix B, pp. 537.

U.S. Environmental Protection Agency: 1997a, *Compendium Method TO-17: Determination of VOCs in Ambient Air Using Active Sampling onto Sorbent Tubes, Compendium of Method for the Determination of Toxic Organic Compounds in Ambient Air*, 2nd ed., EPA/625/R-96/010b.

U.S. Environmental Protection Agency: 1997b, *Compendium Method TO-15: Determination of VOCs in Air Collected in Specially-Prepared Canister and Analyzed by Gas Chromatography/Mass Spectrometry*, EPA/625/R-96/010b, pp. 1–15–66.

U.S. Environmental Protection Agency: 2000, Integrated Risk Information System (IRIS).

U.S. Environmental Protection Agency: 1986, Guidelines for Cancer Risk Assessment, *Federal Register* **51**(185), 33992–34003.

Vainiotalo, S. and Matveinen, K.: 1993, 'Cooking fumes as a hygienic problem in the food and catering industries', *American Industrial Hygiene Assoc. J.* **July**, 376–382.

Wark, K., Warner, C. F. and Davis, W. T.: 1998, *Air Pollution: Its Origin and Control*, Addison Wesley, pp. 471–488.

THE RELATIVE INFLUENCES OF PRIMARY AND SECONDARY PARTICULATES TO URBAN AIR QUALITY IN THE UNITED KINGDOM

TIM CHATTERTON[1]*, STEPHEN DORLING[1], ANDREW LOVETT[1] and MICHAEL STEPHENSON[2]

[1] *School of Environmental Sciences, University of East Anglia, Norwich, U.K.;* [2] *Environmental Health Department, Norwich City Council, Norwich, U.K.*

(* author for correspondence, fax: 01603 507719)

Abstract. This study uses a combination of data from U.K. monitoring stations and from modelling undertaken with the U.K. Meteorological Office's NAME Model to investigate the relative influences of primary and secondary particulates on total PM_{10} levels at sites in the United Kingdom. Co-located PM_{10} and sulphate aerosol measurements indicate that sulphate has a disproportionately large influence on the variation of PM_{10} levels in comparison to its contribution to their total mass. Comparisons of measured PM_{10} at urban centre, roadside and rural sites suggest that local primary sources have very little influence on daily mean levels. NAME has been used to model both primary particles and sulphate aerosol from sources across the whole of Europe. The discrepancies between modelled and observed PM_{10} suggest that coarse particles, such as windblown dust and resuspended roaddust, may comprise a very large, if not dominant, proportion of observed PM_{10} levels. The apparently minor role of primary particles (especially locally-sourced ones) raises a number of issues regarding the suitability of current U.K. and European legislation to addressing the particle problem.

Keywords: aerosol, air pollution, modelling, NAME, particles, particulate, PM_{10}, rural, sulphate, urban

1. Introduction

When the U.K. National Air Quality Strategy (DoE, 1997) was initially launched in 1997 the air quality objective for particulates was set, on the basis of the air quality standard, at $50 \ \mu g \ m^{-3}$, measured as a running 24 hr mean, to be achieved by 2005. To avoid the cost implications of enforcing rigid, 100% compliance to the objective an allowance was made for PM_{10} levels to exceed the objective for four days per year to allow for uncontrollable natural sources, adverse weather conditions and national festivals such as Bonfire night.

In 1999 the U.K. government judged that 'in many cases it will not be feasible for local authorities to meet the PM_{10} objective through action at a local level' (DETR, 1999). This situation was true for most of the country – but it was likely to be particularly relevant to areas in eastern England, such as Norfolk, where influxes of large quantities of particulates from both the European mainland and

Water, Air, and Soil Pollution: Focus **2:** 173–187, 2002.
© 2002 *Kluwer Academic Publishers.*

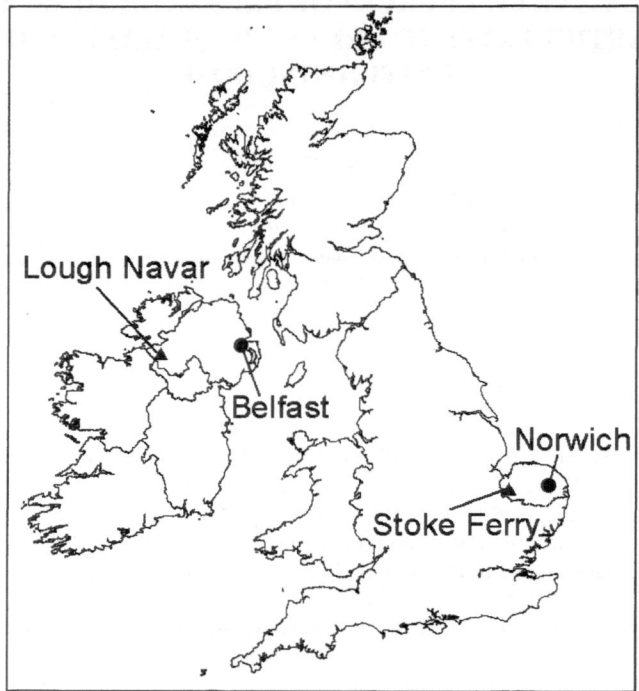

Figure 1. Locations of monitoring sites.

the rest of the United Kingdom are balanced against comparatively low-levels of local emissions. In response to this, the objective has been temporarily relaxed with the allowance of 35 days of exceedences in any year (although the 50 μg m^{-3} level was now specified as a gravimetric, rather than a TEOM measurement, making the limit level somewhat stricter). Levels of PM$_{10}$ in the U.K. are, however, still likely to exceed the health-based air quality standard and the objective is likely to be made more stringent following a further review commencing in 2001. At the latest it will take place when policy takes into account the second stage of the European Union Air Quality Daughter Directive (Directive 96/62/EC).

2. Methodology

This study examines monitoring data for both particulates and sulphate aerosol at a number of sites across the U.K. and has related it to results of modelling work undertaken using the U.K. Meteorological Office's NAME model (Ryall and Maryon, 1996). This paper focuses primarily on data from monitoring sites in Norwich (urban) and Stoke Ferry (rural), in Norfolk (see Figure 1). As a contrast to Norfolk, data has also been presented from Lough Navar (rural) and Belfast (urban) in Northern Ireland. These additional sites have been chosen for a number of reasons:

- Lough Navar is the only location in the national monitoring network where both PM_{10} and sulphate particles are routinely measured.
- Northern Ireland on the western edge of the United Kingdom has a very different pattern of exposure to sulphate aerosol compared to Norfolk.
- Belfast has very different emissions characteristics to Norwich, having very high levels of PM_{10} and SO_2 emissions from domestic heating sources (CE-HOGNI, 1999).

2.1. MONITORING

Monitoring data has been used from four sources:

- the Automatic Urban and Rural Network and the Rural Sulphur Network operated by the U.K. DETR/NETCEN (http://www.aeat.co.uk/netcen/airqual/welcome.html);
- particulate monitoring work carried out by IMC Ltd for the Department of Trade and Industry (King *et al.*, 1999) at Stoke Ferry in Norfolk, between 24/3/98 and 27/9/98;
- measurements carried out by Norwich City Council using their mobile monitoring unit.

The IMC data has been particularly valuable as it provided both rural background measurements of particulates (PM_{10} and $PM_{2.5}$) and co-located particulate and sulphate data (only available elsewhere in the U.K. at Lough Navar, in Northern Ireland). It should be noted that all PM_{10} monitoring data have been collected using TEOMs rather than Gravimetric devices (for a recent investigation as to the implications of this distinction see Green *et al.* (2001)). PM_{10} levels have been analysed on the basis of daily means in order to compare directly with the sulphate aerosol data and the air quality Objective.

2.2. MODELLING

The NAME model is a Lagrangian, multiple-particle model incorporating a sulphate chemistry scheme. It uses 3-dimensional meteorological fields covering an area slightly larger than Europe to trace the movement of particles. These particles are released from pollution sources, set by emission grids. Within each emission grid and within each time-step, the model incorporates a random element to create an even spread of emissions. At the time of release each particle has a certain quantity of pollutants attributed to it. Over time, processes of dispersion, wet and dry deposition and chemical reactions change the quantity of pollutant on each particle. At each timestep the model records the concentration of pollution present

Figure 2. Domains of model runs (emissions cells shown 2, 10 and 50 km, respectively).

at a number of pre-defined receptors. Comprehensive details of the NAME model have been given in Ryall and Maryon (1996), and in Ryall and Maryon (1998) while the sulphate chemistry scheme is well described in Malcolm *et al.* (2000).

The model was set up for 4 different runs (see Table I) in order to attribute pollution to one of three different emission regions (see Figure 2). Time-series of pollution concentration were created for a number of sites across the U.K. corresponding with the eight sulphate monitoring locations on the national network and a number of PM_{10} monitoring stations distributed across the country. Time series were also generated for a number of sites in Norfolk without monitoring equipment in order to examine variations in model output across the county.

Setup 1:   A high resolution (both spatially and temporally) run to deal with emissions in the East Anglian region. Data from the U.K. National Atmospheric Emissions Inventory (http://www.aeat.co.uk/netcen/airqual/naei/index.html) for 1997 was used and aggregated from a 1 km resolution to a 2 km resolution grid. Emissions sources were divided into traffic and industrial sources with traffic sources being given a diurnal emission cycle. Release heights for traffic and industrial sources were set respectively at 0–50 and 75–175 m.

Setup 2:   A run with a lower resolution emissions grid, output grid and timestep to allow for the much greater area covering Great Britain and Northern Ireland (the United Kingdom). Again U.K. NAEI data was used and aggregated to create a 10 km resolution grid. In addition to the 10 km resolution grid, 43 emission sources (the top 30 emitters of NO_x, SO_2 and PM_{10} in the U.K.) were included as 1 km resolution 'point' sources, N.B. none of these were within the area covered by Setup 1. Release heights for these point sources were set at 175–225 m for the major power stations and 100–175 m for other emitters. All 10 km resolution release heights were set to 0–100 m with no differentiation between traffic and industry.

Setup 3:   This was a run using the 50 km EMEP emissions grid for Europe (EMEP, 1997) between 35° East and 35° West, and 30 and 70° North. Emissions of SO_2 and

TABLE I

Model configurations for each modelling domain

Setup	Area	Emission data	Input grid resolution (km)	Output grid resolution (km)	Timestep (sec)	Sulphur chemistry	Emissions used
1	East Anglia (EA)	U.K. NAEI 1996	2	2	150	N	PM_{10}
2	United Kingdom (excl. EA)	U.K. NAEI 1996	10	10	900	N	PM_{10}
3	Europe	EMEP 1997	50	10	900	Y	SO_2, NH_3
4	Europe (excl. U.K.)	EMEP 1997, TNO 1997	50	10	900	N	PM_{10}

NH_3 were used and the chemistry scheme calculated sulphate aerosol (H_2SO_4 and $(NH_4)_2SO_4$), a major component of secondary particulate.

Setup 4:   A proxy emissions inventory for primary particulates was created using emissions totals from each European country from the Dutch TNO emissions inventory for 1993 (TNO, 1997 – quoted in APEG, 1999). These were then disaggregated spatially using NO_x emissions from the EMEP inventory. Emissions from the U.K. were excluded from this run.

Within this paper where reference is made to coarse particles this refers to those particles other than 'primary' (directly emitted by combustion processes) and 'secondary' (formed by chemical reactions in the atmosphere) as opposed to the fraction of particles greater than 2.5 and less than 10 μm in diameter. These coarse particles are presumed to consist of material such as sea-salt, wind-blown dust, resuspended road dust, particles from mining and quarrying activities.

3.   Monitored Concentrations

3.1. PM_{10}

Figures 3 and 4 show monitoring data comparing PM_{10} levels between the rural monitoring site at Stoke Ferry, the Urban Centre/Background site in Norwich and roadside data from Norwich City Council's mobile monitoring unit. The data from the mobile unit has been taken from two locations:

- Pitt Street – next to a busy, and frequently congested roundabout on the City's Inner Ring Road (Average Annual Daily Flow $\approx$20 000) – from March 1999 to 8 July 1999; and
- St. Stephens Street – a narrow shopping street with large numbers of buses, taxis and private cars (AADF $\approx$9000) – from 15 July 1999 to March 2000.

Over the whole of the periods considered, daily mean levels of particulates were comprable with the rural and urban background measurements differing by an average of 2 μg m^{-3} and urban background and roadside sites differing by an average of just over 1 μg m^{-3}. Whilst individual days could see substantial differences, the surprising fact that urban background levels were frequently higher than roadside levels or lower than rural levels meant that on average the differences were minimal. Although further work is being carried out analysing the data at a finer time resolution (Doyle *et al.*, 2001) it is the daily mean that is relevant to the U.K. and EU objectives. King and Dorling (1997) and King *et al.* (1999) also demonstrated minimal differences and similar behaviour between urban and rural PM_{10} climates.

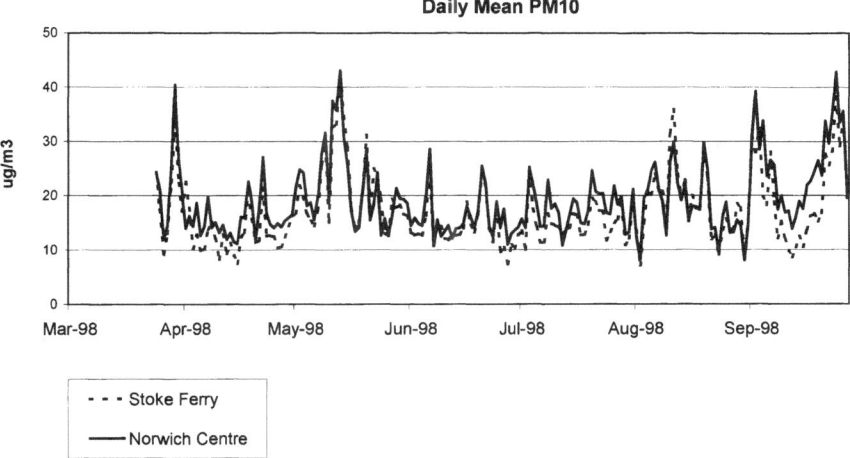

Figure 3. Daily mean PM$_{10}$ levels at rural and uban centre locations in Norfolk.

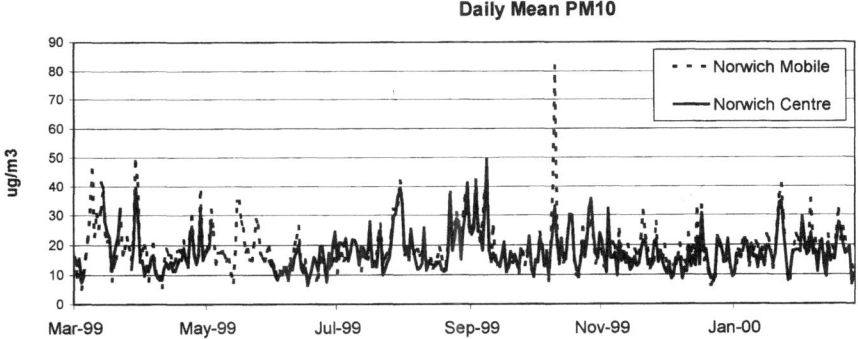

Figure 4. Daily mean PM$_{10}$ levels at roadside (mobile) and urban centre locations in Norwich, Norfolk.

3.2. SULPHATE

Monitoring data for sulphate has been taken from the DETR's Rural Sulphur Network. This records sulphate particles by volume of sulphur. Sulphate aerosol is largely comprised of ammonium sulphate ((NH_4)$_2$$SO_4$) and sulphuric acid ($H_2SO_4$) although the very reactive H_2SO_4 can also take up other elements such as sodium, magnesium and calcium depending on the history of the airmass. The sulphate component generally contributes the majority of the mass of these particles and so sulphate aerosol measurements reported here are calculated simply by multiplying the measured sulphur component by 3 to give the mass of sulphate alone (Atomic Masses: Sulphur = 32, Oxygen = 16). This method leads to an underestimation of the total mass of the sulphate aerosol.

Observed PM10 and SO4 - Stoke Ferry 24/3/98-28/9/98

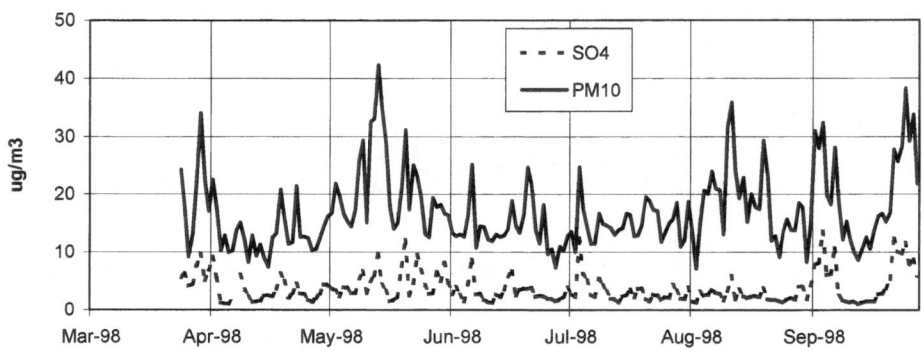

Figure 5. Observed PM$_{10}$ and SO$_4$ at Stoke Ferry.

Figure 5 shows the contribution of sulphate aerosol to measured PM$_{10}$ at Stoke Ferry during the time PM$_{10}$ monitoring was undertaken. During this period SO$_4$ comprised between 6 and 38% of PM$_{10}$ (mean = 17%, r^2 = 0.58). As the typical formation of SO$_4$ occurs over a timescale of 30–50 hr, during which an air mass may travel 500–750 km (APEG, 1999) it can be assumed that sulphate measurements at Stoke Ferry are likely to be largely representative of levels in Norwich itself (except in certain, extreme conditions). Modelling results suggest a mean difference in sulphate levels of <0.3% for 1998 between the two sites. On the basis of this assumption, during the same period sulphate aerosol accounted for between 4 and 48% of daily mean PM$_{10}$ (mean = 15%, r^2 = 0.50) at Norwich Centre. Similarly, for the period from March 1999 to March 2000 where comparisons have been undertaken for Norwich Mobile and Norwich Centre sites, the same pattern emerges:

- % SO4 Norwich Centre: max. = 60, min. = 1, mean = 14, r^2 = 0.36;
- % SO4 Norwich Mobile: max. = 52, min. = 1, mean = 13, r^2 = 0.35.

This suggests that although sulphate, on average, accounts for quite a low proportion of PM$_{10}$, the variation of sulphate levels has a disproportionately large effect on fluctuations of overall PM$_{10}$.

Figure 6 shows levels of sulphate observed at Stoke Ferry since 1981. Over this period there is a noticable downward trend (mean 1981–1990 = 5.7 μg m^{-3}, 1991–1999 = 3.8 μg m^{-3}). The last period where there were any widescale sulphate related PM$_{10}$ episodes was in 1996 and these have been well studied (APEG (Airborne Particles Expert Group), 1999; King and Dorling, 1997; Stedman, 1997; Malcolm *et al.*, 2000). Recent work by Dixon and Middleton (2001) suggests that although the difference in PM$_{10}$ levels between 1996 and 1998 (chosen as the 'worst' and 'cleanest' years for PM$_{10}$ in recent years) was considerable in terms of

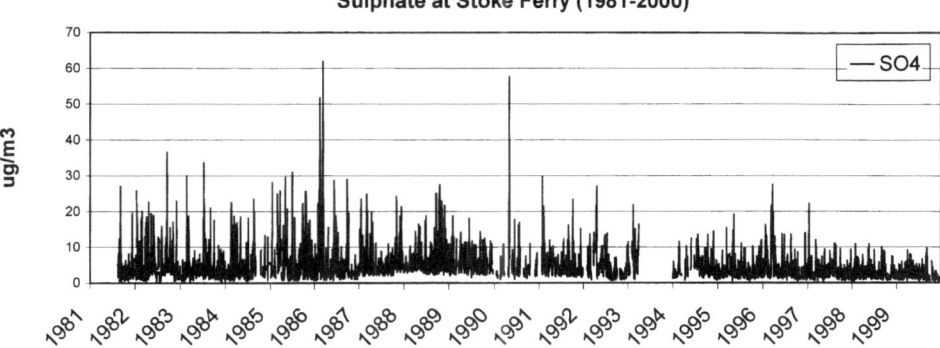

Figure 6. Daily mean sulphate at Stoke Ferry 1981–2000.

TABLE II

Analyses of proportion of PM_{10} comprised of SO_4, March–September 1998

	Mean PM_{10} ($\mu g\ m^{-3}$)	Range SO_4 (% of PM_{10})	Mean SO_4 (% of PM_{10})	r^2
Norwich Centre	19.24	4–48%	15%	0.50
Belfast Centre	19.74	1–32%	7%	0.54
Stoke Ferry	17.11	6–38%	17%	0.58
Lough Navar	9.24	0–52%	13%	0.81

the number of PM_{10} exceedences, the meteorology was 'not as different as might be supposed'. The increased number of episodes of PM_{10} in 1996 appear to have been caused by a slightly greater easterly flow in this year. Analysis of the pattern of PM_{10} in the 1996 episodes is hampered by the fact that routine monitoring of PM_{10} in the U.K. only commenced in 1992 (Stedman, 1997) and in 1996 there were only 12 PM_{10} monitors on the U.K. government's Automatic Urban and Rural Network (AURN) (1995 = 10, 2000 = 55). The time-period covered by this study was governed by the fact that 1998 was the first complete year for which PM_{10} monitoring data was available for Norwich.

3.3. COMPARISON WITH NORTHERN IRELAND

Tables II and III present figures describing the relationship between sulphate aerosol and total PM_{10} at the rural Lough Navar and Stoke Ferry sites and the urban centre sites in Belfast and Norwich for both the 6 month monitoring period at Stoke Ferry and for the whole of 1998.

TABLE III

Analyses of proportion of PM_{10} comprised of SO_4, January–December
1998

	Mean PM_{10} ($\mu g\ m^{-3}$)	Range SO_4 (% of PM_{10})	Mean SO_4 (% of PM_{10})	r^2
Norwich Centre	19.66	2–48%	14%	0.65
Belfast Centre	20.95	0–34%	6%	0.63
Lough Navar	9.61	0–52%	13%	0.78

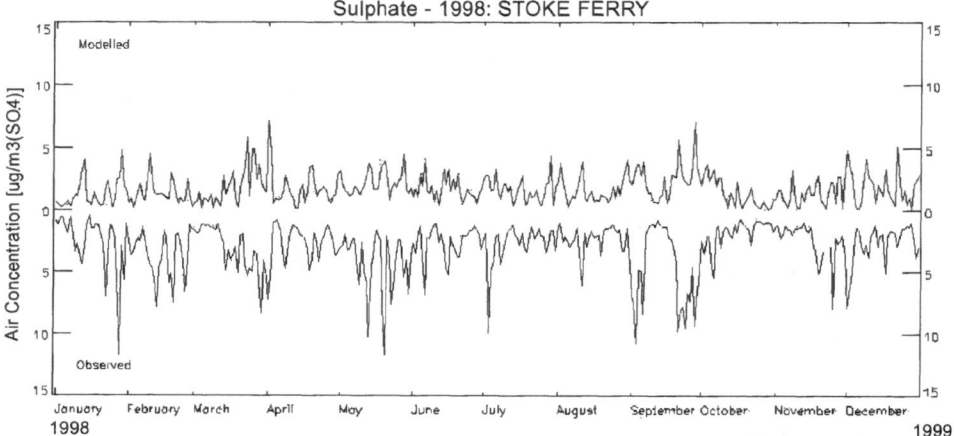

Figure 7. Comparison of modelled and observed sulphate for 1998 at Stoke Ferry MBE = −1.08 μg
m^{-3}, $r = 0.56$.

4. Modelling Results

4.1. SULPHATE MODELLING

Good results were produced for sulphate at all sites modelled. With mean observed sulphate levels between 1.1 and 3.0 $\mu g\ m^{-3}$, Mean Bias Error (MBE) varied
between −0.47 and −1.2 $\mu g\ m^{-3}$ for daily mean values across the eight sites that
measure sulphate particles in the U.K. The greatest modelling error was found at
the sites in the South East of the U.K. where the levels of sulphate were highest.
Values of r ranged between 0.43 and 0.73 for the whole year.

Figure 7 shows a comparison of modelled and observed data for Stoke Ferry,
1998. From the graph, it is clear that although the model has done well at determining the occurance of peaks in sulphate levels it performs less well in predicting
the scale of the peaks, frequently underestimating the higher concentrations.

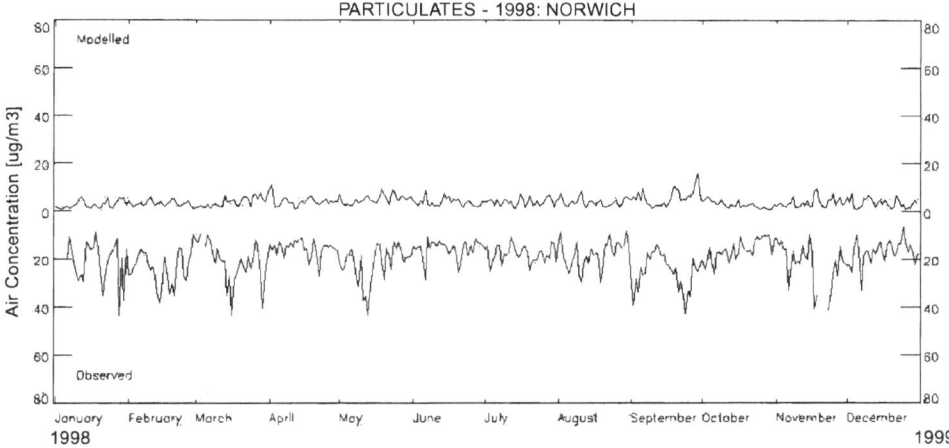

Figure 8. Comparison of modelled and observed PM_{10} (1998) at Norwich Centre MBE = -15.69 μg m^{-3}, $r = 0.34$.

4.2. PM_{10} MODELLING

Modelling of the total PM_{10} fraction was less accurate than for sulphate (across the 9 receptors mean observed levels ranged from 9.7 to 21.2 μg m^{-3}, MBE from -8.4 to -17.7 μg m^{-3}, r from 0.55 to 0.23). Figure 8 illustrates the lack of detail replicated by the model for the receptor point in Norwich. Some of the error is due to factors such as poor emissions data, the effect of disaggregating emissions from concentrated point or line sources to area sources or to the use of sub-optimal parameters such as release heights although work by the Met Office using different parameters has produced slightly better results (Malcolm and Manning, 2001). However, as only primary and some secondary particles were modelled it was never anticipated that the NAME would produce extremely accurate results for PM_{10}. By looking closer at the discrepancies between observed and modelled data it should be possible to infer information about those elements not modelled.

4.3. COMPARISON OF MODELLING RESULTS FOR NORFOLK AND NORTHERN IRELAND

Table IV shows some basic summary statistics for the model results at the receptor points in Norfolk and Northern Ireland. The figures highlight a number of points:

- Both observed and modelled values are in the same 'intuitive' order.
- SO_4 is modelled more accurately than PM_{10} (in terms of both error and fit).
- Variations in PM_{10} are replicated better in Belfast than other sites.
- Model results appear better for Northern Ireland than for Norfolk.
- The underestimation of PM_{10} is greater at the urban sites (as mass but not as %).

TABLE IV

Comparison of model results for PM_{10} and SO_4 for 1998

		Mean O (μg m^{-3})	Mean P (μg m^{-3})	Mean P/ mean O	MBE (μg m^{-3})	MAE (μg m^{-3})	r^2
Norwich	(PM_{10})	19.66	3.91	0.20	−15.69	15.69	0.12
Belfast	(PM_{10})	20.93	4.16	0.20	−16.80	16.80	0.30
Stoke Ferry	(PM_{10})[a]	17.11	3.28	0.19	−13.83	13.83	0.09
Lough Navar	(PM_{10})	9.69	1.32	0.14	−8.38	8.38	0.19
Stoke Ferry	(SO_4)	2.72	1.64	0.60	−1.08	1.32	0.31
Lough Navar	(SO_4)	1.64	0.87	0.53	−0.60	0.91	0.33

P = Predicted, O = Observed, MBE = Mean Bias Error, MAE = Mean Actual Error).
[a] 6 months (March–September) only.

TABLE V

Mean percentage component of modelled PM_{10} (1998)

		% Primary (U.K.)	% Primary (EU)	% Sulphate
Norwich	(PM_{10})	35	14	51
Belfast	(PM_{10})	60	5	35
Stoke Ferry	(PM_{10})[a]	19	16	65
Lough Navar	(PM_{10})	3	14	84

[a] 6 months (March–September) only.

The first two points give us some reassurance that the model processes are working reasonably well and that we can probably trace most of the error back to factors other than the physics of the model (this is also confirmed by statistics for the structured/unstructured RMSE). With respect to the composition of the modelled PM_{10} (Table V) the fact that Belfast performs much better suggests that the diffusion of particles from the dispersed, low-level domestic sources is probably being reasonably represented. The final two points may be explained by coarse particles. If coarse particles form a significant proportion of the missing PM_{10}, then the generally wetter climate in Northern Ireland would help to prevent resuspension of settled dust thereby reducing the effect that these particles can have on overall variability of PM_{10} levels. The issue of coarse particles is discussed further in the next section.

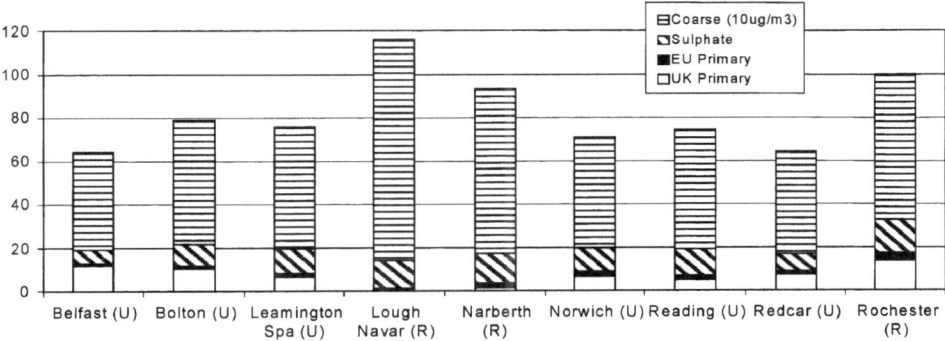

Figure 9. % Contribution of modelled PM$_{10}$ to observed levels at (U)rban and (R)ural locations.

5. Discussion

Comparing the modelling results for different categories of monitoring/receptor location allows deductions to be made regarding the composition and origin of those particles that have not been included in the model. The major portion of the model 'error' can be accounted for by coarse particulates which, in preliminary calculations, have been taken as an average of 10 μg m^{-3} in line with guidance provided for U.K. local authorities by NETCEN (http://www.aeat.co.uk/netcen/air-qual/laqm/pm10_sec/pm2top.html).

Figure 9 shows the contribution of modelled PM$_{10}$ (with coarse particulates as a constant 10 μg m^{-3}) as a percentage of observed levels. Where the modelled contribution exceeds 100% at Lough Navar this suggests an over-prediction of the coarse fraction which may be due to the location of the monitor, in a very remote location surrounded by woodland, leading to very low levels of resuspended particles in the area. The shortfall at a number of sites suggests that even with coarse particles taken in to consideration there is a significant underprediction occurring, especially in urban areas. The likeliest causes for this may be the weakening of the effect of disaggregating emissions to a 2 or 10 km grid and the absence of re-suspended road dust in the model both of which would explain the greater disparity between modelled and observed levels in urban locations.

6. Conclusions

The modelling work shows that primary particles appear to be a minor constituent of PM$_{10}$, especially in rural locations in the West of the U.K. The monitoring carried out in Norfolk suggests that the increased pollution sources in an urban environment have a minimal effect on overall PM$_{10}$ levels (around 2 μg m^{-3} or 10%) whilst the immediate proximity of a busy road with lots of diesel vehicles may only increase the urban background level by half this amount over the length of a day. Early results from modelling work being undertaken with ADMS-Urban reinforce this.

Increasingly, it is appearing that legislation based on controlling particulate levels using a 'simple' metric based on mass of PM_{10} may be entirely unsuited to the problem at hand. This is largely due to the apparent lack of influence that may be achievable over this figure because of the high component of secondary and coarse particles. Debate has been continuing for sometime over whether $PM_{2.5}$ would form a better focus for legislation, however the recent report of the Expert Panel on Air Quality Standards has recently advised that there is no overall advantage with this approach in the U.K. (EPAQS, 2001). This study has found that it may well be impossible for certain local authorities in the U.K. to have any significant effect on levels of particulates in their areas, even (or especially) during pollution episodes. Whilst there is a great deal of uncertainty as to the exact source of some particles in the U.K., the current legislation forms a very blunt instrument in terms of approaching the problem. The problems highlighted by this paper, primarily the large variation in PM_{10} composition within a single nation are also applicable on a much larger scale, as illustrated in a recent paper looking at the effect of Saharan dust in Spain (Rodríguez et al., 2001). Here the authors find that influxes of coarse particles during Saharan dust events are responsible for a large number of the PM_{10} exceedences recorded in parts of Spain. They conclude that 'the anthropogenic contribution to atmospheric particulate matter in Southern and Northern European regions cannot be monitored with one criterion given the higher natural input in Mediterranean countries'. If the reduction of secondary particulate can only be achieved through work at a national or European level and the prospect for the reduction of coarse particles is limited due to their nature, perhaps new legislation on controlling particles should focus on something more specific.

Acknowledgements

Funding for the modelling work described in this report has been provided by: Breckland District, Council, Great Yarmouth Borough Council, North Norfolk District Council, Norwich City Council, South Norfolk District Council, The Borough Council of Kings Lynn and West Norfolk and Norfolk County Council.

The work was carried out as part of a Ph.D. in the School of Environmental Sciences at the University of East Anglia under the supervision of Dr. Steve Dorling. The Ph.D. has been funded by the Natural Environment Research Council and Norwich City Council.

Thanks are also due to the U.K. Meteorological Office who provided the NAME model and in particular Karl Kitchen, Ben Miners Andrew Mirza and Dr. Alison Redington for their time and assistance in learning how to use the model.

References

APEG (Airborne Particles Expert Group): 1999, *Source Apportionment of Airborne Particulate Matter in the United Kingdom*, Department of the Environment, Transport and the Regions, London.

CEHOGNI: 1999, *Air Quality in Northern Ireland*, Chief Environmental Health Officers' Group for Northern Ireland, Belfast, Northern Ireland.

DETR: 1999, *Review of the United Kingdom National Air Quality Strategy*, Department of the Environment, Transport and the Regions, London, U.K.

DoE: 1997, *The United Kingdom National Air Quality Strategy*, Department of the Environment and the Scottish Office, London.

Dixon, J. and Middleton, D. R.: 2001, 'An Analysis of the Meteorology Associated with Air Pollution Episodes in the U.K.', *3rd International Conference on Urban Air Quality*, Extended Abstracts, Institute of Physics, Loutraki, Greece.

Doyle, M., Chatterton, T. and Dorling, S. R.: 2001, 'An Examination of PM_{10} and $PM_{2.5}$ in a Busy Shopping Area in Norwich, U.K.', *3rd International Conference on Urban Air Quality*, Extended Abstracts, Institute of Physics, Loutraki, Greece.

EMEP: 1997, *Transboundary Air Pollution in Europe*, MSC-W, Norwegian Meteorological Institute, Status Report 1/97, Oslo, Norway.

EPAQS: 2001, *Airborne Particles: What is the Appropriate Measurement on Which to Base a Standard?*, Expert Panel on Air Quality Standards, London, U.K.

Green, D., Fuller, G. and Barratt, B.: 2001, 'Evaluation of TEOM 'correction factors' for assessing the EU Stage 1 limit values for PM_{10}', *Atmosph. Environ.* **35**(14), 2589–2593.

King, A. and Dorling, S., 1997, 'PM_{10} particulate matter – The significance of ambient levels', *Atmosph. Environ.* **31**(15), 2379–2381.

King, A., Jones, A., Dorling, S., Merefield, J., Stone, I., Hall, K., Garner, G., Hall, P. and Stokes, B.: 1999, 'Study for Particulate Sampling, Sizing and Analysis for Composition', *Report Commissioned by the Department of Trade and Industry* (Energy Technology Support Unit), ETSU N/01/00049/REP, IMC Technical Services Ltd, available at http://www.etsu.com/en_env/html/rep49.html.

Malcolm, A. L., Derwent, R. G. and Maryon, R. H.: 2000, 'Modelling the long-range transport of secondary PM_{10} to the U.K.', *Atmosph. Environ.* **34**, 881–894.

Malcolm, A. L. and Manning, A. J.: 2001, 'Testing the skill of a lagrangian dispersion model at estimating primary and secondary particles', *Atmosph. Environ.* **35**(9), 1677–1685.

Rodríguez, S., Querol, X., Alastuey, A., Kallos, G. and Kakaliagou, O.: 2001, 'Saharan dust contributions to PM_{10} and TSP levels in Southern and Eastern Spain', *Atmosph. Environ.* **35**(14), 2433–2447.

Ryall, D. B. and Maryon, R. H.: 1996, *The NAME 2 Dispersion Model: A Scientific Overview*, U.K. Meteorological Office, Turbulence and Diffusion Note 217b, U.K.

Ryall, D. B. and Maryon, R. H.: 1998, 'Validation of the U.K. Met. Office's name model against the ETEX dataset', *Atmosph. Environ.* **32**(24), 4265–4276.

Stedman, J.: 1997, 'A U.K.-wide episode of elevated particle (PM_{10}) concentration in March 1996', *Atmosph. Environ.* **31**(15), 2381–2383.

TNO: 1997, *Particulate Matter Emissions in Europe in 1990 and 1993*, TNO, TNO-MEP-R96/472, The Netherlands.

THE SPATIAL AND TEMPORAL VARIATION OF MEASURED URBAN PM$_{10}$ AND PM$_{2.5}$ IN THE HELSINKI METROPOLITAN AREA

M. A. POHJOLA[1]*, A. KOUSA[2], J. KUKKONEN[1], J. HÄRKÖNEN[1],
A. KARPPINEN[1], P. AARNIO[2] and T. KOSKENTALO[2]

[1] *Finnish Meteorological Institute, Air Quality Research, Helsinki, Finland;* [2] *Helsinki Metropolitan Area Council, Helsinki, Finland*

(* *author for correspondence, e-mail: Mia.Pohjola@fmi.fi, fax: +358 9 1929 5403*)

Abstract. We have studied particulate matter (PM) concentrations, PM$_{10}$ and PM$_{2.5}$, measured in an urban air quality monitoring network in the Helsinki Metropolitan Area during 1997–1999. The data includes PM$_{10}$ concentrations measured at five locations (two urban traffic, one suburban traffic, one urban background and one regional background site) and PM$_{2.5}$ concentrations measured at two locations (urban traffic and urban background sites). The concentrations of PM$_{10}$ show a clear diurnal variation, as well as a spatial variation within the area. By contrast, both the spatial and temporal variation of the PM$_{2.5}$ concentrations was moderate. We have analysed the evolution of urban PM concentrations in terms of the relevant meteorological parameters in the course of one selected peak pollution episode during 21–31 March, 1998. The meteorological variables considered included wind speed and direction, ambient temperature, precipitation, relative humidity, atmospheric pressure at the ground level, atmospheric stability and mixing height. The elevated PM concentrations during the 1998 March episode were clearly related to conditions of high atmospheric pressure, relatively low ambient temperatures and low wind speeds in predominantly stable atmospheric conditions. The results provide indirect evidence indicating that the PM$_{10}$ concentrations originate mainly from local vehicular traffic (direct emissions and resuspension), while the PM$_{2.5}$ concentrations are mostly of regionally and long-range transported origin.

Keywords: air pollution, episode, meteorology, particulate matter, PM$_{2.5}$, PM$_{10}$, resuspension, urban

1. Introduction

Particulate matter concentrations, measured as PM$_{10}$ and particularly as PM$_{2.5}$, have been associated with hospital admissions and mortality in several studies conducted both in Europe and the U.S.A. (Katsouyanni *et al.*, 1997; Schwartz *et al.*, 1996; Dockery and Pope, 1994; Pope *et al.*, 1995). Airborne particulate matter originates not only directly from combustion processes, but also from resuspension from street surfaces. In spring, after the snow has melted and streets dry out, particles from street surfaces are resuspended by traffic induced turbulence and wind. The influence of resuspension depends on the mechanical wear of the street surfaces, on street maintenance and cleaning – particularly the winter time sanding of streets –, on traffic-induced turbulence and on meteorological conditions.

 Water, Air, and Soil Pollution: Focus **2:** 189–201, 2002.
© 2002 *Kluwer Academic Publishers.*

In Finland, studded tyres are in widespread use in vehicular traffic in winter, increasing street surface wear. Resuspension also commonly takes place in late autumn, probably due to the start of street salting and sanding (e.g., Kukkonen *et al.*, 1999, 2000). Snow can melt several times before a permanent snow cover is obtained. In Northern and Central European cities, resuspended PM can have a substantial influence particularly on total suspended particles (TSP), but also on PM_{10} concentrations (for instance, Hämekoski and Salonen (1996) and Johansson *et al.* (1999)).

Previously, measured urban PM_{10} and $PM_{2.5}$ concentrations and their spatial and temporal variations in Europe have been discussed by, e.g. Monn *et al.* (1995), Harrison *et al.* (1997), Kingham *et al.* (2000), Johansson *et al.* (1999) and Ojanen *et al.* (1998). Buzorius *et al.* (1999) have analysed PM number concentrations (from 10 to 500 nm in particle diameter) in the city of Helsinki. However, the amount of data and the time periods considered in the above-mentioned studies have been fairly limited. In some studies carried out in North America, more extensive time periods have been addressed (Brook *et al.*, 1999; Darlington *et al.*, 1997). Most previous studies have not addressed systematically the influence of the relevant meteorological parameters on urban PM concentrations; Monn *et al.* (1995), however, did consider briefly the dependencies of PM_{10} concentrations on some selected meteorological parameters.

The objective of this article is to analyse and interpret $PM_{2.5}$ and PM_{10} measurements carried out in the Helsinki Metropolitan Area. We have previously analysed the PM concentrations in these years on a seasonal basis (Pohjola *et al.*, 2000). We have investigated the diurnal variation of $PM_{2.5}$ and PM_{10} during 1997–1999. We also analysed a spring episode of exceptionally high PM concentrations in 1998.

2. Methodology

2.1. MONITORING SITES

Figure 1 shows the location of the air quality monitoring stations in the Helsinki metropolitan area and the pollutants monitored in 1999. Many of these stations have been located for the purpose of monitoring 'hot spots' near busy traffic environments, or major local energy production sources. The network contains six permanent multicomponent stations; these are located in Helsinki city districts (Töölö, Vallila and Kallio 2), in suburban in the cities of Espoo and Vantaa centres (Leppävaara and Tikkurila), and in a rural area in Espoo (Luukki).

The stations used in this study represent urban (Töölö and Vallila) and suburban traffic environments (Leppävaara), together with the urban background (Kallio 2). Regional background concentrations were monitored in a rural environment in Luukki, approximately 20 km to the north-west of downtown Helsinki.

Two urban monitoring stations, Töölö and Vallila, are located in the Helsinki downtown area. The station of Töölö is situated in a small square in a busy cross-

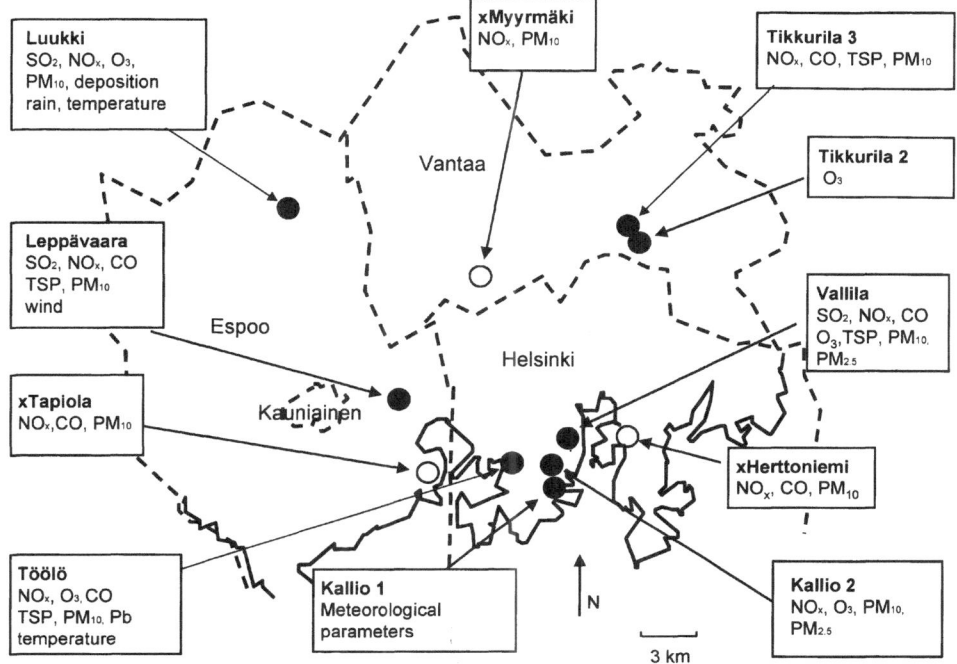

Figure 1. Urban air quality monitoring network in the Helsinki Metropolitan Area in 1999. The legends show the name of the station and the pollutants measured. X indicates a mobile station.

ing area, surrounded by several buildings. This station is situated at a distance of less than 10 m from streets with traffic volumes of approximately 50 000 vehicles per day during weekdays. The station of Vallila is situated in a small park, at the distance of about 14 m from a street with 13 000 vehicles per day during weekdays.

The suburban station of Leppävaara in Espoo is situated in a shopping and residential area. The distances of this station to three nearest roads are approximately 25, 50 and 100 m; the traffic volumes of these roads are 11 000, 29 000 and 63 000 vehicles per day during weekdays, respectively.

The monitoring height is 4.0 m at all stations, except for the regional background station at Luukki, where the probe is at a height of 7.0 m. The $PM_{2.5}$ measurements were started at the Vallila site in March 1997, and at the Kallio 2 and Luukki sites in the beginning of 1999.

2.2. MONITORING METHODS

At the stations of Töölö and Leppävaara, the concentration of PM_{10} was measured with TEOM (Tapered Element Oscillating Microbalance). At the stations of Vallila, Kallio 2 and Luukki, the concentration of PM_{10} was measured with Eberline FH 62 I-R that is based on β-attenuation method. At the stations of Vallila and Kallio 2, the concentration of $PM_{2.5}$ was also measured with Eberline FH 62 I-R. The flow rate of the PM analysers was calibrated twice a year and the mass measurement once a year.

The accuracy of continuous PM monitoring devices is dependent on the chemical composition of the PM and local meteorological conditions, due to possible evaporation of semi-volatile material. The daily average concentrations of PM_{10} and $PM_{2.5}$, determined by the Eberline FH 62 I-R analysers, were therefore compared with the corresponding results obtained by virtual impactors. These comparisons were performed at the station of Vallila during one year, from June 1999 to May 2000.

Plotting the concentrations obtained with the Eberline analyser against those obtained with the virtual impactor yields the following results. For the concentrations of PM_{10}, the slope of the regression line (k) was 1.02, the constant term (d) was $- 0.31$ μg m^{-3}, and the correlation coefficient squared (R^2) was 0.95. For the concentrations of $PM_{2.5}$, k = 0.98, d = +0.03 μg m^{-3}, and R^2 = 0.91. Field intercomparisons have also been conducted, in which the concentrations of PM_{10} obtained by TEOM monitors have been compared with those obtained by other corresponding methods; these have also indicated a good agreement of results (Sillanpää et al., 2001).

2.3. METEOROLOGICAL METHODS

We used the meteorological database of the Finnish Meteorological Institute, which contains weather and sounding observations. We used a combination of the data from the stations at Helsinki-Vantaa airport (about 15 km north of Helsinki town centre) and Helsinki-Isosaari (an island about 20 km south of Helsinki). The mixing height of the atmospheric boundary layer was evaluated using the meteorological pre-processor, based on the sounding observations made at Jokioinen (90 km northwest of Helsinki) and on routine meteorological observations.

The relevant meteorological parameters were evaluated using a meteorological pre-processing model, adapted specifically for urban environment (Karppinen et al., 1998, 2000). The model is based mainly on the energy budget method of Van Ulden and Holtslag (1985). The model utilises meteorological synoptic and sounding observations, and its output consists of hourly time series of relevant atmospheric turbulence parameters (the Monin-Obukhov length scale, the friction velocity and the convective velocity scale) and the boundary layer height. Atmospheric stability was evaluated using a dimensionless stability parameter (Karppinen et al., 1998).

3. Results and Discussion

3.1. DIURNAL VARIATION OF $PM_{2.5}$ AND PM_{10} CONCENTRATIONS

Data has been compiled on the diurnal variation of the PM_{10} and $PM_{2.5}$ concentrations at the various monitoring stations during 1997–1999. As an example, we have presented the diurnal variation of PM_{10} concentrations at the various stations in Figures 2a–c, averaged over each of the years 1997–1999, and $PM_{2.5}$ concentrations at Vallila averaged over the same years and additionally at Kallio 2 in 1999 in Figures 3a–b. Clearly, the diurnal variation of traffic-originated pollutant concentrations depends on the day of the week; for these figures we have selected data from working days (Monday–Friday) only.

The concentrations of PM_{10} show a clear diurnal variation. The concentrations increase continuously during the morning rush hours, from approximately 6 to 8 a.m.; as expected, this increase took place irrespective of the year (during the period 1997–1999) and the season of the year. Subsequently the concentrations decrease slowly during the rest of the daytime hours, also showing in some cases peak values during the afternoon rush hours, from approximately 3 to 6 p.m. The more moderate diurnal variation of the PM_{10} concentrations, compared with the traffic flows, could be caused by resuspension of PM from street surfaces. Resuspended PM increases with increasing traffic flow; however, it can reach a saturated state, in which a further increase of traffic cannot cause any more resuspension.

During working days, there is a very clear diurnal variation of local vehicular traffic. Despite this, the $PM_{2.5}$ concentrations are temporally fairly uniform during working days, except for a moderate increase during the morning rush hours. The diurnal variation of local vehicular traffic flows seems to have no substantial correlation with the $PM_{2.5}$ concentrations. In 1999, the temporal variation of $PM_{2.5}$ concentrations at both monitoring stations was also very similar. This indicates that the PM resuspended from street surfaces and other sources has only a minor effect on the $PM_{2.5}$ concentrations, and that a large fraction of the $PM_{2.5}$ concentrations most likely originates from regional or long-range scale sources. This result is qualitatively in agreement with the results by Ojanen *et al.* (1998); for the station of Vallila in 1996–1997, they evaluated that approximately 40% of the $PM_{2.5}$ concentration originated from local sources, and the rest of the $PM_{2.5}$ mass from regional or long-range transported pollution.

Both for PM_{10} and $PM_{2.5}$, the concentrations at night are higher than would be expected based on the diurnal variation of local vehicular traffic emissions. This is partly caused by commonly occurring unfavourable atmospheric diffusion conditions at night, e.g., low wind speed, stable stratification and ground-level or low-altitude inversions (Karppinen *et al.*, 2000).

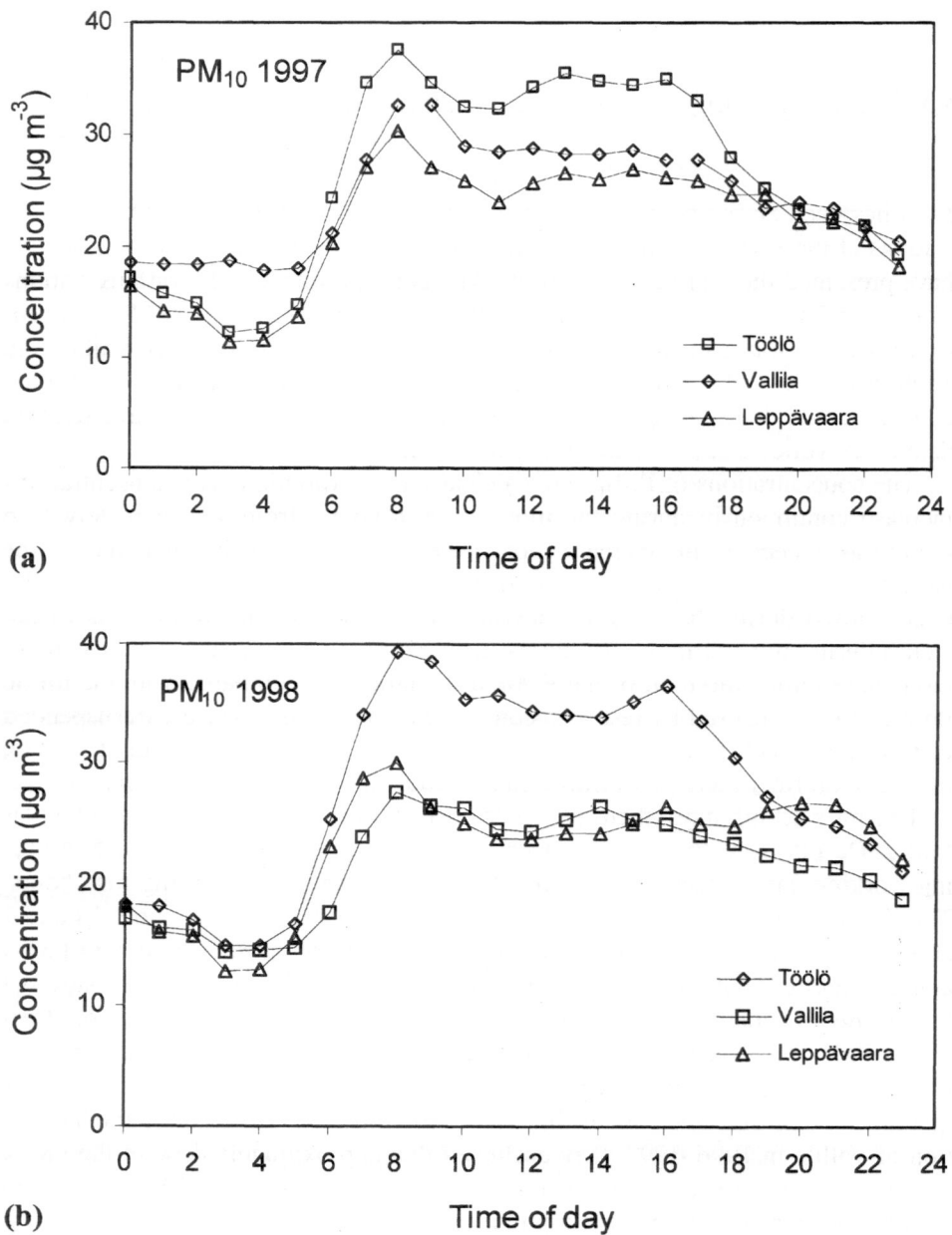

(a)

(b)

Figure 2. The diurnal variation of PM_{10} concentrations at the stations of Töölö, Vallila and Leppävaara during 1997–1999, and at the station of Luukki, during 1999. Figures include data from working days only.

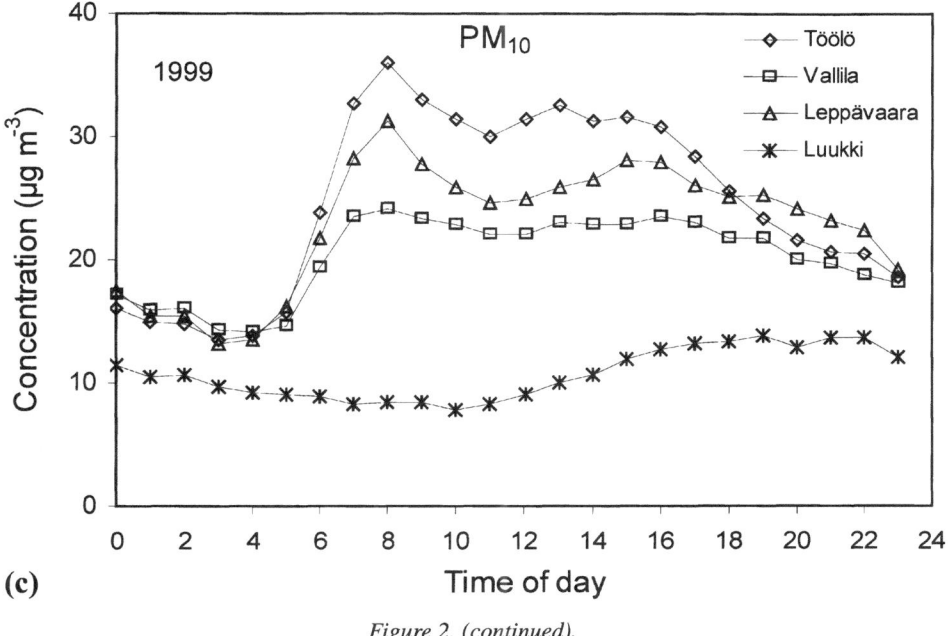

(c)

Figure 2. (continued).

3.2. THE PM EPISODE DURING 21–31 MARCH 1998

In the Helsinki Metropolitan Area as in other major Finnish cities, episodes of high PM concentrations take place particularly in spring. An example has been presented in Figure 4 of the evolution of PM concentrations in the course of such an episode. The PM_{10} concentrations were approximately at their highest level during 21–31 March 1998. The weekends are on 21–22 March and 28–29 March. During this episode, the highest hourly average PM_{10} concentration was measured at Töölö (306 $\mu g \, m^{-3}$) on 24 March. The second highest hourly PM_{10} concentration occurred at Töölö on 27 March (180 $\mu g \, m^{-3}$). The PM_{10} concentrations were at an elevated level (compared with the longer term average values) during two periods, 22–24 and 27–31 March.

We have analysed the hourly evolution of the relevant meteorological parameters during the episode. The atmospheric pressure, shown in Figure 5a, rapidly increased during 21–23 March, reaching a maximum value (approximately 1045 mbar) on 23 March. The pressure subsequently decreased during 24–27 March, and increased again to reach another local maximum value (approximately 1023 mbar) on 29 March. The ambient temperature (measured at a height of 2 m), shown in Figure 5a, had a substantial diurnal variation, during 21–25 March it varied from a minimum value of –10 °C (at night) to a maximum value of +4 °C (in the afternoon), while during the period 26–31 March the corresponding variation was from –4 to +6 °C.

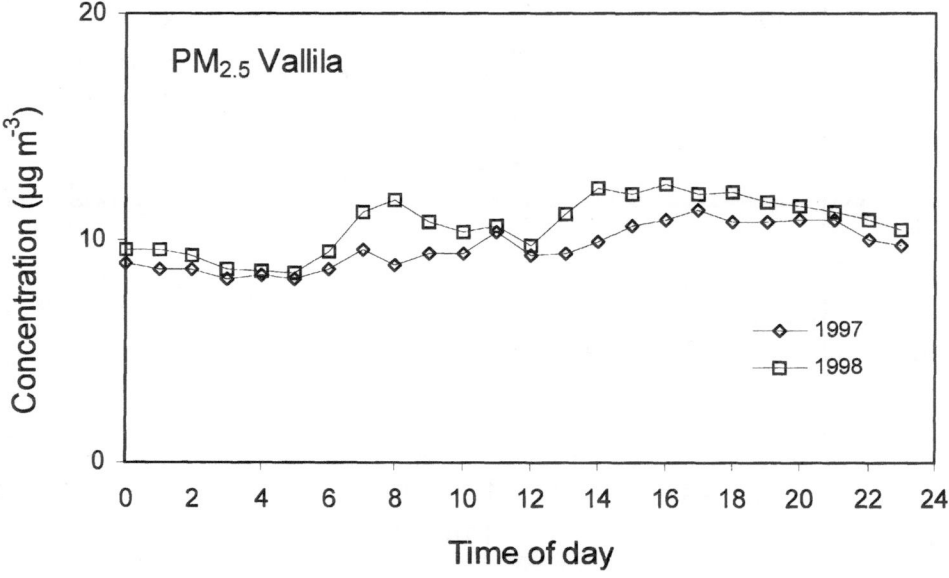

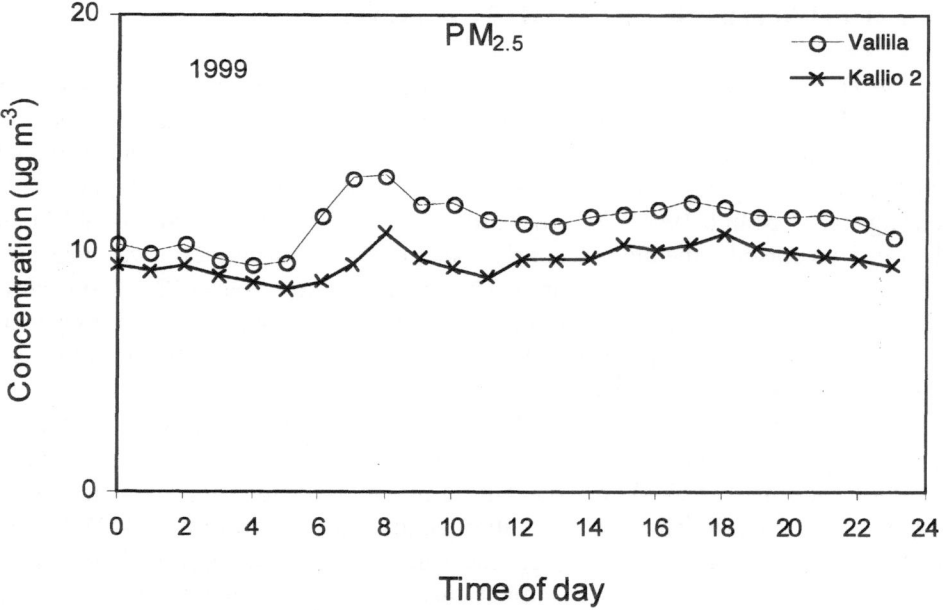

Figure 3. The diurnal variation of PM$_{2.5}$ concentrations at the station of Vallila during 1997–1998, and at the stations of Vallila and Kallio 2 during 1999. Figures include data from working days only.

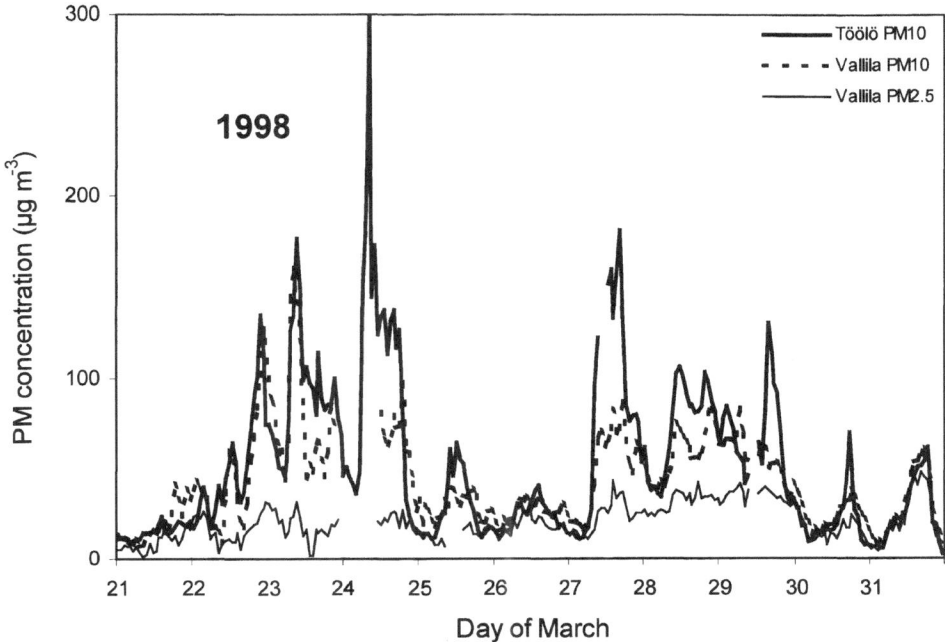

Figure 4. The evolution of the PM_{10} and $PM_{2.5}$ concentrations in the course of the air pollution episode during 21–31 March 1998, at the stations at Töölö and Vallila.

In the early stages of the episode during the period 22–24 March, the wind speed (measured at a height of 10 m), shown in Figure 5b, was very low, ranging from 1.0 to 2.5 m s^{-1}; during the latter period of 25–31 March it was occasionally higher, ranging from 1.0 to 5.0 m s^{-1}. The wind direction was initially predominantly southerly and south-westerly. The atmospheric stability was evaluated based on the dimensionless stability parameter p_q (as defined by Karppinen *et al.*, 1998). Atmospheric stratification showed a substantial diurnal variation (the parameter p_q ranged from a minimum value of –2.5 to a maximum of +2.2). Moderately or extremely stable atmospheric conditions prevailed during most of the period.

The period considered contains two shorter periods with substantially elevated PM concentrations, during 22–24 and 27–31 March (Figure 4). Both of these periods of high concentrations were clearly related to conditions of high atmospheric pressure and relatively low ambient temperatures. The elevated PM concentrations during both 22–24 and 27–31 March were caused by low wind speeds in predominantly stable atmospheric conditions. It is interesting to compare the atmospheric conditions during the two above-mentioned time periods. During the former period, the maximum value of atmospheric pressure was higher, the ambient temperatures were slightly lower and the wind speeds were on the average lower, compared with the latter period. All of these meteorological parameter values would imply a more severe episode during the former period; the maximum concentrations were indeed

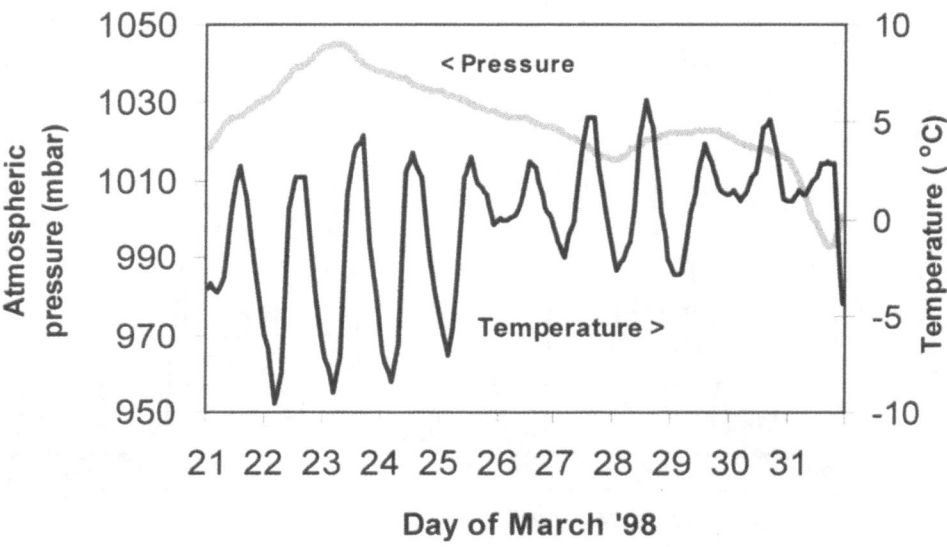

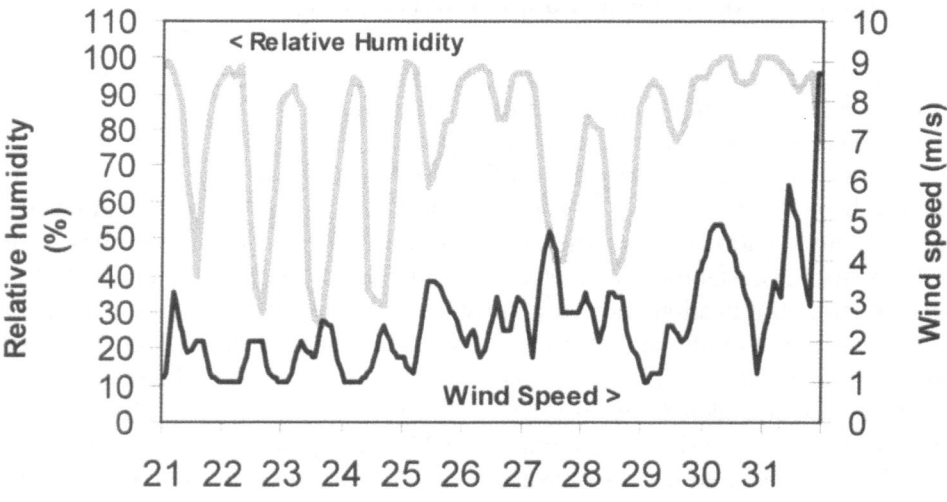

Figure 5. The evolution of atmospheric pressure and relative humidity (left-hand-side vertical axis) and temperature and wind speed (right-hand-side vertical axis) in the course of the air pollution episode during 21–31 March 1998.

higher on 24 March, compared with the corresponding maximum values during the latter period on 27 March.

The latter period of elevated PM concentrations occurred partly during the weekend on 28 and 29 March. The relatively lower concentrations during those days, compared with those on 27 March, were probably mainly caused by lower emissions of PM originated from local traffic. For instance, the annual averages of the total traffic volumes during weekends were approximately 62% of those during weekdays from Monday to Friday in central Helsinki in 2000 (Lilleberg *et al.*, 2001).

At the Vallila monitoring site, the highest hourly average $PM_{2.5}$ concentration (49 μg m^{-3}) was measured on 31 March (unfortunately, part of the $PM_{2.5}$ measurements are not available on 24 March). The $PM_{2.5}$ concentrations on this particular day were almost identical to the corresponding PM_{10} concentrations at both monitoring stations considered. This indicates that the influence of resuspension was negligible, and that the polluted air masses could have been of long-range transported origin on that particular day.

4. Conclusions

The diurnal variation of the PM_{10} concentrations was clear, irrespective of the year and the season of the year. This variation partly follows the corresponding variation of local vehicular traffic flows. On the other hand, both the spatial and temporal variation of the fine particle ($PM_{2.5}$) concentrations was moderate. The results provide indirect evidence indicating that the PM_{10} concentrations are originated mainly from local vehicular traffic (direct emissions and resuspension), while the $PM_{2.5}$ concentrations are mostly of regionally and long-range transported origin. This result is qualitatively in agreement with source apportionment studies in the same area (Ojanen *et al.*, 1998).

We have also analysed the evolution of urban PM concentrations in terms of the relevant meteorological parameters in the course of a selected peak pollution episode during 21–31 March, 1998. The episode was related to an anticyclonic high-pressure system, and the highest concentrations were connected with conditions of exceptionally high atmospheric pressure and relatively low ambient temperatures. The elevated PM concentrations were caused by low wind speeds and predominantly stable atmospheric stratification. The polluted air masses could have partly been of long-range transported origin; however, a detailed analysis would require air parcel trajectory analyses.

Acknowledgements

This study has been part of the National Research Programme on Environmental Health (SYTTY), the project 'Development of a population exposure model, using atmospheric dispersion modelling together with measured concentrations and personal exposures – EXPAND', 1998–2001, and the international research programme EUROTRAC-2, SATURN, 'Studying Atmospheric Pollution in Urban Areas', 1998–2001. The funding from the Academy of Finland for both studies is gratefully acknowledged.

References

Brook, J. R., Dann, T. F. and Bonvalot, Y.: 1999, 'Observations and interpretations from the Canadian Fine Particle Monitoring Program', *J. Air Waste Manage. Assoc.* **49**, PM-35–44.

Buzorius, G., Hämeri, K., Pekkanen, J. and Kulmala, M.: 1999, 'Spatial variation of aerosol number concentration in Helsinki city', *Atmosph. Environ.* **33**, 553–565.

Darlington, T. L., Kahlbaum, D. F., Heuss, J. M. and Wolff, G. T.: 1997, 'Analysis of PM_{10} trends in the United States from 1988 through 1995', *J. Air Waste Manage. Assoc.* **47**, 1070–1078

Dockery, D. W. and Pope, C. A.: 1994, 'Acute respiratory effects of particulate air pollution', *Ann. Rev. Publ. Health* **15**, 107–132.

Harrison, R. M., Deacon, A. R., Jones, M. R. and Appleby, R. S.: 1997, 'Sources and processes affecting concentrations of PM_{10} and $PM_{2.5}$ particulate matter in Birmingham (U.K.)', *Atmosph. Environ.* **31**, 4103–4117.

Hämekoski, K. and Salonen, R. O.: 1996, 'Particulate matter in northern climate of Helsinki Metropolitan Area, Finland', in J. Lee, and R. Phalen (eds), *Proceedings of the Second Colloquium on Particulate Air Pollution and Human Health*, 1–3 May, 1996, Park City, Utah, U.S.A.

Johansson, C., Hadenius, A., Johansson, P.-Å. and Jonson, T.: 1999, 'SHAPE the Stockholm Study of Health Effects of Air Pollution and their Economic Consequences. Part I: NO_2 and Particulate Matter in Stockholm – Concentrations and Population Exposure', *AQMA Report*, Vol. 6(98), Swedish National Road Administration, Stockholm.

Karppinen, A., Kukkonen, J., Nordlund, G., Rantakrans, E. and Valkama, I.: 1998, 'A Dispersion Modelling System for Urban Air Pollution', Finnish Meteorological Institute', *Publications on Air Quality*, Vol. 28, Helsinki, 58 p.

Karppinen, A., Joffre, S. M. and Kukkonen, J.: 2000, 'The refinement of a meteorological preprocessor for the urban environment', *Internat. J. Environ. Pollut.* **14**, 565–572.

Katsouyanni, K., Toulumi, G., Spix, C., Schwartz, J., Balducci, F., Medina, S., Rossi, G., Wojtyniak, B., Sunyer, J., Bacharova, L., Schouten, P., Pönkä, A. and Andersen, H. R.: 1997, 'Short term effects of ambient sulphur dioxide and particulate matter on mortality in 12 European cities: results from time series data from the APHEA project', *British Medical J.* **314**, 1658–1663.

Kingham, S., Briggs, D., Elliott, P Fischer, P. and Lebret, E.: 2000, 'Spatial variation in the concentrations of traffic-related pollutants in indoor and outdoor air in Huddersfield, England', *Atmosph. Environ.* **34**, 905–916.

Kukkonen, J., Salmi, T., Saari, H., Konttinen, M. and Kartastenpää, R.: 1999, 'Review of urban air quality in Finland', *Boreal Environ. Resear.* **4**, 55–65.

Kukkonen, J., Konttinen, M., Bremer, P., Salmi, T. and Saari, H.: 2000, 'The seasonal variation of urban air quality in northern European conditions', *Internat. J. Environ. Pollut.* **14**(1–6), 480–487.

Lilleberg, I. and Hellman, T.: 2001, *Variation of Traffic Flows in Helsinki in 2000* (in Finnish), Publications of the Helsinki City Planning Office, 2001, **4**, Helsinki, 33 pp. + App.

Monn, Ch., Braenli, O., Schaeppi, G., Schindler, Ch., Ackermann-Liebrich, U., Leuenberger, Ph. and SAPALDIA team: 1995, 'Particulate matter <10 μm (PM_{10}) and total suspended particulates (TSP) in urban, rural and alpine air in Switzerland', *Atmosph. Environ.* **29**, 2565–2573.

Ojanen, C., Pakkanen, T., Aurela, M., Mäkelä, T., Meriläinen, J., Hillamo, R., Aarnio, P., Koskentalo, T., Hämekoski, K., Rantanen, L. and Lappi, M.: 1998, 'Hengitettävien hiukkasten kokojaukauma, koostumus ja lähteet pääkaupunkiseudulla (Distribution, composition and sources of the respiratory particles in the Helsinki Metropolitan Area)', *Pääkaupunkiseudun julkaisusarja C 7*, YTV, Helsinki.

Pohjola, M., Kousa, A., Aarnio, P., Koskentalo, T., Kukkonen, J., Härkönen, J. and Karppinen, A.: 2000, 'Meteorological interpretation of measured urban $PM_{2.5}$ and PM_{10} concentrations in the Helsinki Metropolitan Area', in J. W. S. Longhurst, C. A. Brebbia and H. Power (eds), *Air Pollution VIII*, Wessex Institute of Technology Press, Southampton, U.K., pp. 689–698.

Pope, C. A., Thun, M. J., Namboodri, M. M., Dockery, D. W, Evans, J. S., Speizer, F. E. and, Heath Jr., C. W.: 1995, 'Particulate air pollution as a predictor of mortality in a prospective study of U.S. adults', *Amer. J. Respirat. Crit. Care Medic.* **151**, 669–674.

Schwartz, J., Dockery, D. W. and Neas, L. M.: 1996, 'Is daily mortality associated specifically with fine particles?', *J. Air Waste Manage. Assoc.* **46**, 927–939.

Sillanpää, M., Hillamo, R., Kerminen, V.-M., Koskentalo, T. and Aarnio, P: 2001, 'PM_{10} in downtown Helsinki – Comparison between different devices', *J. Aerosol Sci.* **32**, S777–S778.

Van Ulden, A. and Holtslag, A.: 1985, 'Estimation of atmospheric boundary layer parameters for diffusion applications', *J. Clim. Appl. Meteor.* **24**, 1196–1207.

DIURNAL PROFILE OF PARTICLE-BOUND POLYCYCLIC AROMATIC HYDROCARBON (PPAH) CONCENTRATION IN URBAN ENVIRONMENT IN TOKYO METROPOLITAN AREA

TASSANEE CHETWITTAYACHAN[1]*, RYOSUKE KIDO[1], DAI SHIMAZAKI[2] and KAZUO YAMAMOTO[3]

[1] *Department of Urban Engineering, Graduate School of Engineering, The University of Tokyo, Hongo, Bunkyo-ku, Tokyo, Japan;* [2] *Engineering Research Institute, Graduate School of Engineering, The University of Tokyo, Yayoi, Bunkyo-ku, Tokyo, Japan;* [3] *Environmental Science Center, The University of Tokyo, Hongo, Bunkyo-ku, Tokyo, Japan*
(* author for correspondence, e-mail: tassanee@env.t.u-tokyo.ac.jp, fax. +81 3 5841 8533)

Abstract. Traffic emits particles under 1 μm. The particles are the most responsible to particle-bound polycyclic aromatic hydrocarbon (pPAH) which can impact human health. To assess them as health hazards, we monitored diurnal changes in the concentration and distribution of pPAH near roads in Tokyo. The total pPAH concentration was determined using a photoelectric aerosol sensor (PAS) which ionized PAH-adsorbing particles. The total pPAH concentration was compared with chemical analyses by gas chromatography/mass spectrometry (GC/MS). Two sampling sessions, one in August and one in September 2000, were done at three sampling sites at the Hongo Campus of the University of Tokyo. Monitoring was every two minutes for six consecutive days for the first session and for seven consecutive days for the second session. Correlation of the pPAH concentration with traffic flow and with meteorological conditions were also assessed. The pPAH concentration varied in the same manner on all days: it sharply increased in the early morning by a sudden burden of traffic, and it rapidly decreased during the daytime, probably owing to photodegradation and/or dilution by rising in the mixing zone. The local wind field, and consequently the transportation of pPAH from the road, were strongly influenced by the configuration and location of the surrounding buildings. The pPAH clearly changed in 1- and 0.5 day cycles, particularly at the roadside.

Keywords: diurnal profile, particle-bound polycyclic aromatic hydrocarbon (pPAH), photoelectric aerosol sensor (PAS), road traffic, Tokyo Metropolitan, urban environment

1. Introduction

Polycyclic aromatic hydrocarbons (PAHs) have drawn attention as air pollutants because they are toxic, mutagenic, and carcinogenic. The United States Environmental Protection Agency (U.S. EPA) and the International Agency for Research and Cancer (IARC) have identified PAHs are possible human carcinogens. PAHs are generally emitted by incomplete combustion of organic materials including, coal, crude oil, wood and gasoline fuel. Atmospheric PAH concentrations can be high in towns and cities because they are mainly emitted by vehicles. It has been indicated that, although stationary sources can account for up to 90% of the total

Water, Air, and Soil Pollution: Focus **2:** 203–221, 2002.
© 2002 *Kluwer Academic Publishers.*

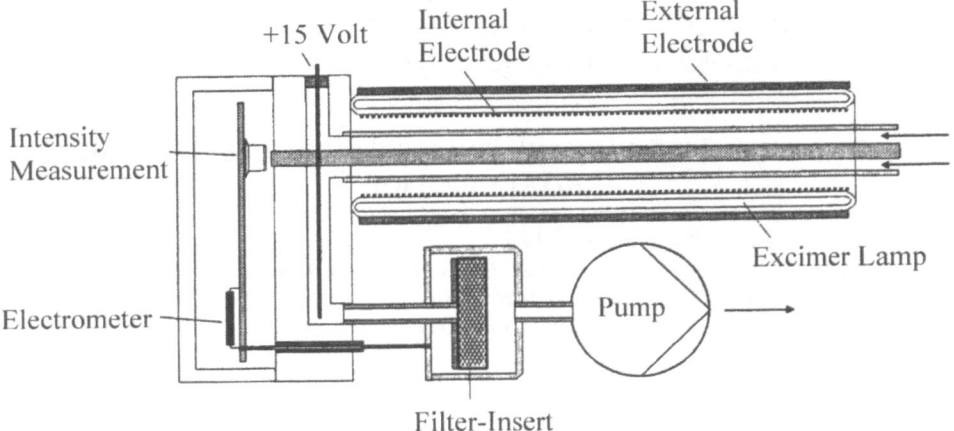

Figure 1. Process schematic for the photoelectric aerosol sensor.

PAH emission, this is not true in city centers and suburbs where mobile sources are prevalent (Caricchia *et al.*, 1999).

Atmospheric PAHs are principally partitioned between particulate matter and gas phase, but they exhibit a strong adsorption affinity to aerosol particles because of their relatively low vapour pressure and aromaticity. A large mass fraction of PAHs especially recognized as carcinogenic is mostly associated with particulate matter, in particular ultrafine particles: 90 to 95% of particulate PAHs are associated with particle diameters less than 3.3 μm, and the peak distributions are localized between 0.4 and 1.1 μm (Beak *et al.*, 1991). Similarly, it has been indicated that they are predominantly associated with particles with an aerodynamic diameter (d_{ae}) of less than 1–2 μm (Kaupp and McLachlan, 1999).

Furthermore, motorized-vehicle can also produce respiratory particles smaller than 1 μm, and they are the most responsible to particle-bound PAH (pPAH) (Agnesod *et al.*, 1996).

The impact of airborne particles on human health is strongly dependent on the size of the particles. Particles with a diameter above 1 μm are typically deposited in the nose, whereas those below 1 μm are deposited in the bronchial and pulmonary part of the respiratory system (Siegmann and Siegmann, 1998). As mentioned above, particles under 1 μm are respirable and the most responsible to pPAH. Hence pPAH is particularly hazardous because it is respirable and the hazard is further increased because people are actually exposed to a mixture of PAHs. This warrants monitoring of pPAH levels in cities. Therefore, the aim of this study was to obtain a short-time scale profile of pPAH near city roads in order to assess potential exposure risk to the human body. Time series of daily pPAH concentration were also evaluated by the corresponding power spectra.

2. Methodology

2.1. MATERIALS AND METHOD

The concentration of total PAHs concentration absorbed on airborne particles under 1 μm were determined using a photoelectric aerosol sensor (PAS 2000CE manufactured by EcoChem Messtechnik GmbH, Germany). The PAS (Figure 1) works on the principle that particles under 1 μm are usually very good photo-emitters. The PAS has a KrBr-excimer lamp which emits high intensity ultraviolet (UV) light that can ionize PAH absorbed particles. The emitted UV light was at a wavelength about 207 nm. At this wavelength, PAH-absorbing particles are ionized, but gas molecules and non-carbon aerosols are not. The PAH molecules absorbed on particles surface emit electrons, which are subsequently removed when an electric field is applied. The positively charged particles are collected onto a filter inside an electrometer, where the charge is measured. The resulting electric current establishes a signal, which is proportional to the concentration of total pPAH.

Before using the PAS2000CE, its function was first checked against total concentration of some PAHs in airborne particulate obtained by a high volume air sampler, HV-500-5S (SIBATA Scientific Technology Ltd.) and analyzed by a GC/MS (HP 6890 Series). Eleven PAHs, with three or more rings, which have been identified by the U.S. EPA as priority pollutants or potential carcinogens, were selectively analyzed and compared to the total pPAH concentration determined by the PAS2000CE. A HV-500-5S was set at an airflow rate of 400 L min^{-1} and operated together with the on-line monitor. Twelve 4-hr samplings were performed between 12–15 June, 2000. After sampling, each filter was stored at 4 °C in the dark. Each filter was also referred to a blank one that handled an equivalent treatment. Samples were extracted in 10 mL dichloromethane in an ultrasonic bath for 30 min and concentrated in a rotating evaporator and further concentrated by a stream of nitrogen. Instrumental analysis was carried out by a GC/MS (HP 6890 Series) using the single ion monitoring (SIM) technique.

The output obtained by PAS2000CE was relative higher than the total concentration of the 11 PAHs as shown in Figure 2. This indicates that the PAS2000CE can measure a wider range of PAHs. By linear regression, there was good agreement ($R^2 = 0.827$) between the two techniques. This relationship might give a bit simpler way to assess human exposure risk. Therefore, the PAS2000CE is a practical and useful technique for continuous real-time monitoring of total pPAH concentration in urban environments.

2.2. ON-LINE MEASUREMENT AND SITE LOCATION

Monitoring was done in two sessions, six days in August from 23 to 28 and seven days in September from 19 to 26, 2000, at the Hongo Campus of the University of Tokyo (Figure 3). In August, three monitoring sites were chosen near Hongo

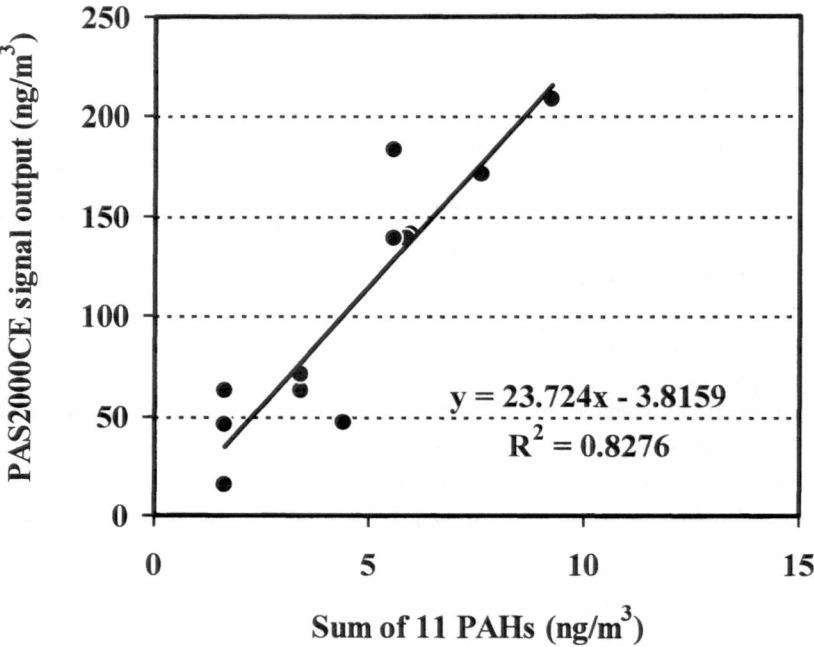

Figure 2. Relationship between PAS2000CE signal and total concentrations of the eleven PAHs selectively determined.

Street: the Department of Urban Engineering building (bldg. #14); the Department of Literature building (DOL); and Yasuda Hall (YSH). In September, two different sites were chosen: the Department of Architecture building (bldg. #11); and the Environmental Science Center building (ESC). The location and configuration of each sampling site is shown in Figure 3. The pPAH concentrations were continuously measured by the PAS2000CE and the data were stored at 2 min intervals.

2.3. METEOROLOGICAL DATA AND TRAFFIC SURVEY

Throughout both sessions, the direction and speed of the wind were observed at 10 min intervals at each site using a wind vane and a cup anemometer. The volume of traffic was measured manually on Wednesday, 23 August in the first session and on Monday 25 September in the second session. In the first session, it was measured on Hongo Road and in the second session it was measured on four main roads surrounding the sampling sites (Figure 3). Traffic survey condition of each sampling session is summarized in Table I. The vehicles were categorized into three types: light-duty vehicles (car, van, minivan), heavy-duty vehicles (truck, mini-truck and bus), and motorcycles. Meteorological data from Japan Meteorological Agency was used to investigate the relationship bewteen total pPAH concentration and

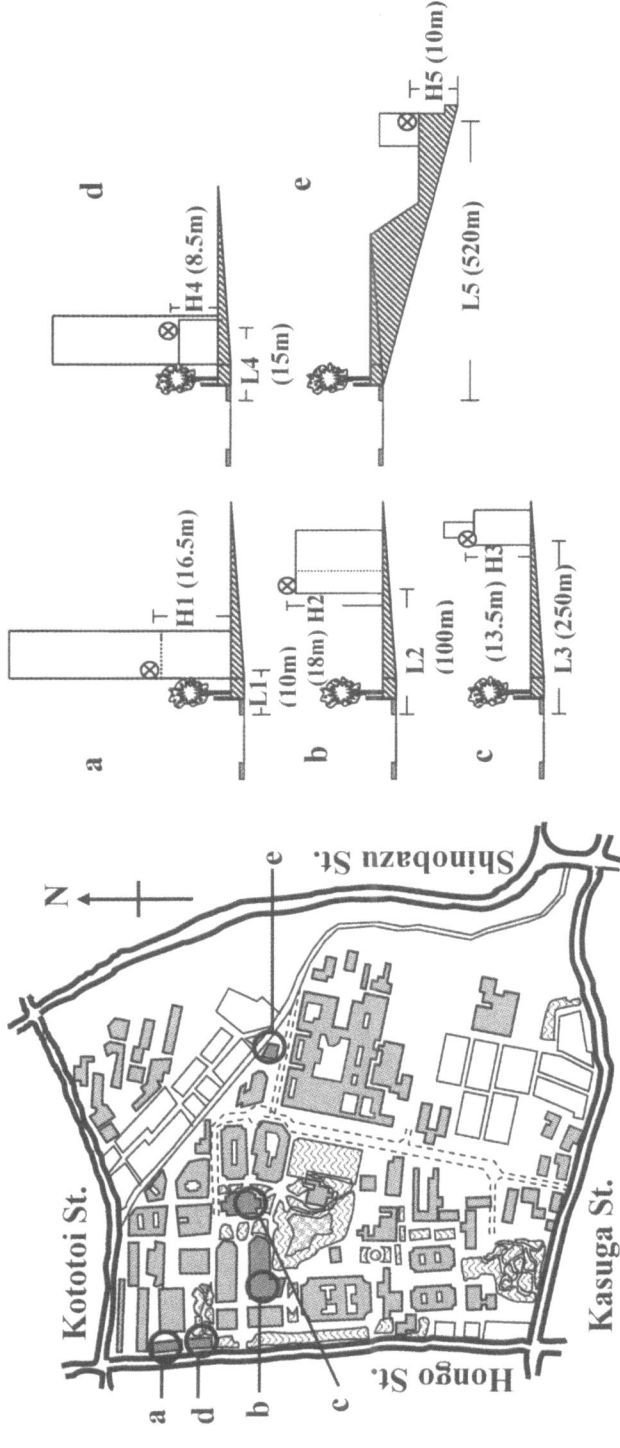

Figure 3. Location and configuration of sampling sites. (a) In front of an emergency exit of 4th floor of bldg. #14, (b) on the roof of Department of Literature building, (c) on the roof of Yasada Hall, (d) on the roof of 2nd floor of bldg. #11, and (e) on the ground beside Environmental Science Center Building.

TABLE I

Traffic survey conditions

Sampling period	Date	Location	Survey period
1st period	23 August (Wed)	Hongo St.	0:00–24:00
2nd period	25 September (Mon)	Hongo St.	0:00–8:00, 8:00–24:00[a]
		Kasuga St.	0:00–8:00, 8:00–24:00[a]
		Shinobazu St.	0:00–8:00, 8:00–24:00[a]
		Kototoi St.	0:00–8:00, 8:00–24:00[a]

[a] 8:00–24:00: 20 min observation in every two hours.

meteorological conditions. To investigate a cyclic pattern of pPAH concentration, the spectral analysis was performed.

3. Results and Discussion

3.1. DIURNAL PROFILE OF pPAH CONCENTRATION AND ITS RELATIONSHIP TO METEOROLOGICAL CONDITION

The diurnal changes in the total pPAH concentration were similar every day. The pPAH concentrations rose sharply in the early morning, with the peak between 7:00 to 8:00 a.m., followed by a significant and rapid reduction during the daytime (Figure 4). The total pPAH concentration was much higher on weekdays than on weekends, in accordance with the traffic flow. The pPAH concentration distribution profiles at the DOL and YSH sites (first session) were almost the same – there was no significant difference ($p = 0.05$) between their average pPAH concentrations. This implies that the dispersion of pPAH on this scale was relatively uniform. On the other hand, the pPAH concentration at the roadside was about 1.5 fold that at the DOL and YSH sites. As for the second monitoring session, the concentration of pPAH at the ESC was 1.25 fold that at the DOL site ($p = 0.05$). Since the ESC site was further from the main road, Hongo Street is not the only source of pPAH. The pPAH concentration distribution profiles in the second sampling period differed from those in the first period. The average pPAH concentration at the roadside was 1.8 and 1.3 times higher than that at the DOL and ESC sites, respectively.

The pPAH concentration on most days markedly decreased around midday. The possible explanation is an enhancement of photolytic degradation due to a rise in sunlight intensity, as well as dilution due to rising in the mixing zone caused by temperature increase in an afternoon (Figure 5). Kamens *et al.* (1988) reported that an impact of sunlight on PAH during the first several hours of sunlight exposure was found, while the PAH was stable during the followings hours of darkness.

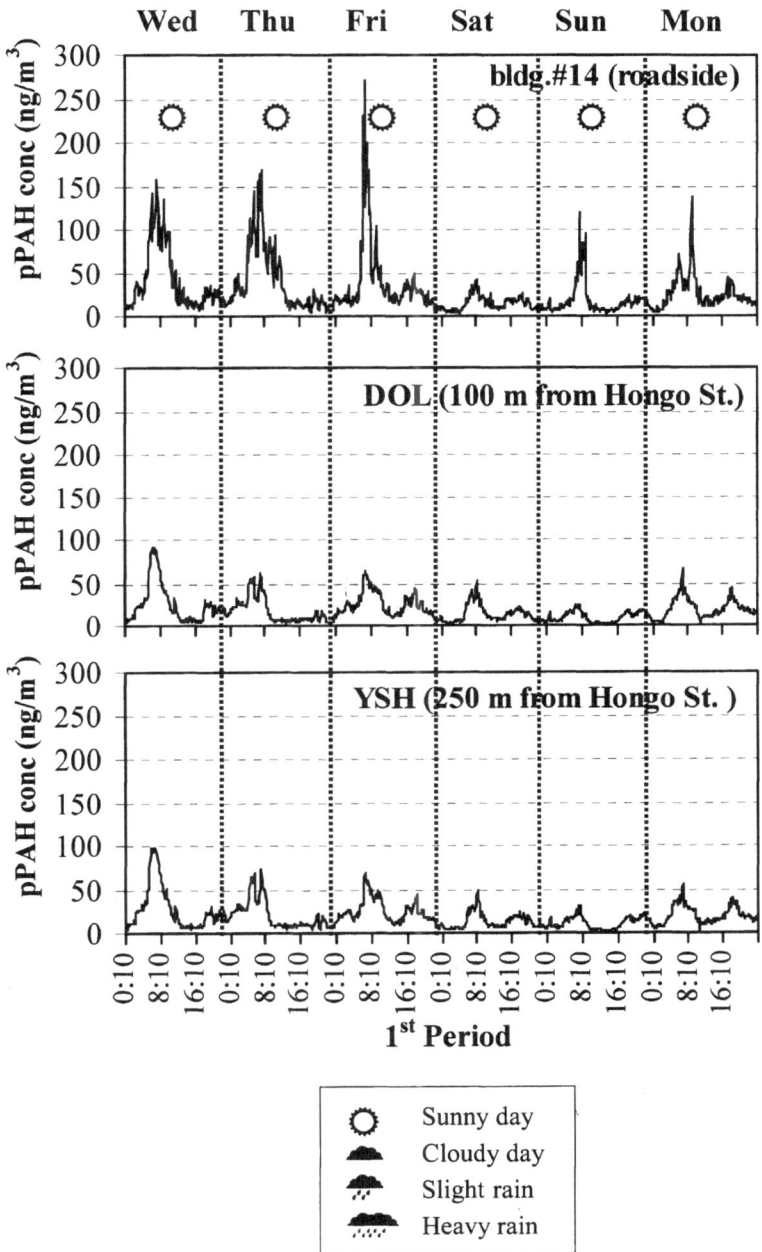

Figure 4. Diurnal profiles of pPAH concentration at all sampling sites during the whole sampling periods. Note that the concentration is the average pPAH concentration of 10 min interval.

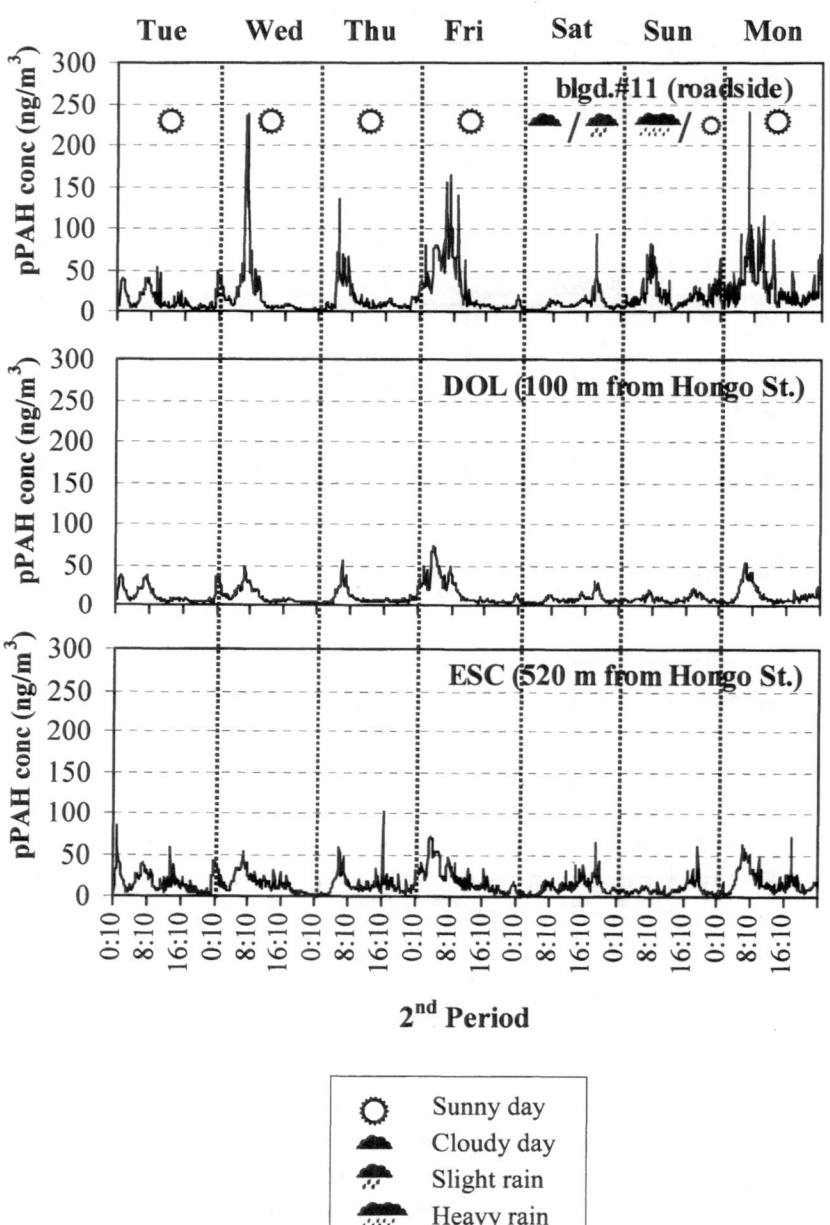

Figure 4. (Continued).

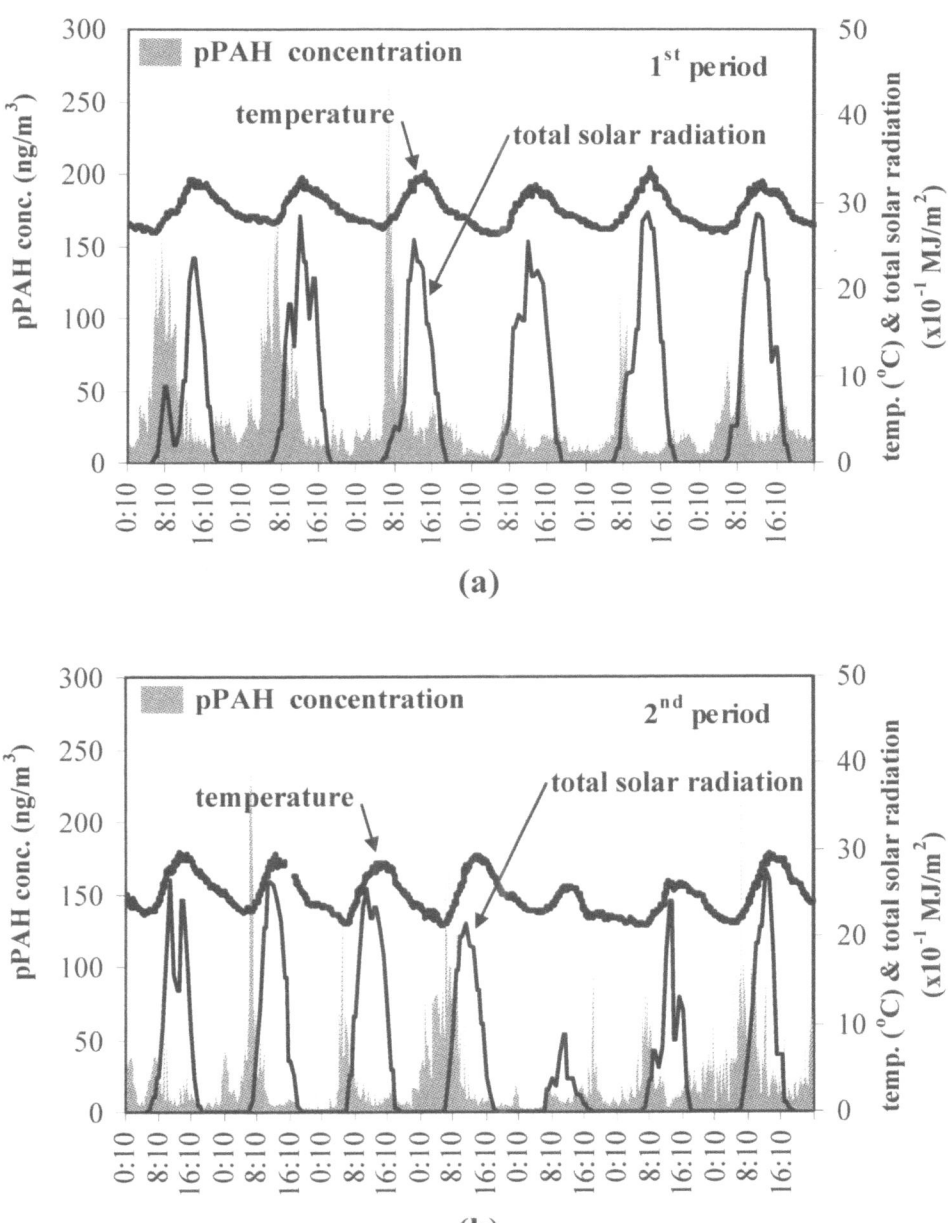

Figure 5. Profiles of the pPAH concentration at sampling sites near the road, and its relationship to temperature and solar radiation during the whole sampling periods. (a) The profile at bldg. #14 during 23–28 August, 2000. (b) The profile at bldg. #11 during 19–26 September, 2000.

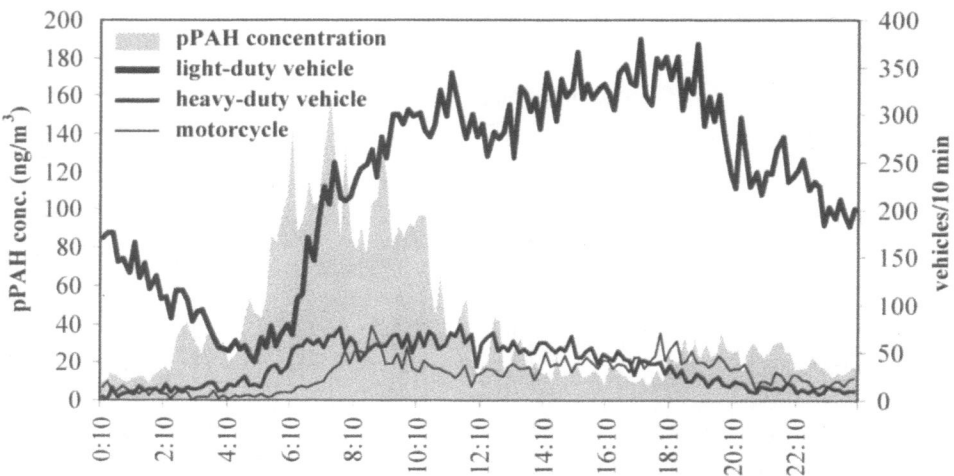

Figure 6. Profile of the pPAH concentration at bldg. #14 and traffic flow on the Hongo Street on 23 August, 2000.

Fan *et al.* (1995) also investigated photodegradation of PAHs such as fluoranthene, pyrene, benzo(a)anthracene, and benzo(b)fluoranthene. The result showed that the PAHs were rapidly decayed at the beginning of experiment (1 hr) and then appeared to be stable. These events can, therefore, explain the diurnal changes of pPAH in this study especially on a sunny day. During the second sampling period, it rained from Saturday afternoon until Sunday morning, and the pPAH concentration peaks at the DOL and ESC sites were not clearly observed, particular Sunday morning. In this study, though a relationship between the rain and the pPAH concentration change is unclear because the traffic volume on that day also changed. Nevertheless, Kaupp and McLachlan (1999) have reported rain resulting in a 50% reduction of the pPAH concentration, mainly by the removal of atmospheric small particles with $d_{ae} < 1 \ \mu$m.

3.2. pPAH CONCENTRATION VERSUS TRAFFIC AMOUNT

The changes in the traffic flow on 23 August and 25 September are shown in Figures 6 and 7. On both days, the number of all vehicles types rapidly increased early in the morning, became relatively constant during the daytime and gradually decreased from the evening till midnight. To obtain a more explicit relationship between pPAH concentration and each vehicle type, the traffic profiles were divided into three periods, 0:00–8:00, 8:00–16:00, and 16:00–24:00. Between 0:00–8:00 a.m. on 23 August and on 25 September, there were strong correlations (R^2 = 0.836 and R^2 = 0.715, Table II) between the number of heavy-duty vehicles and pPAH concentration. Junker *et al.* (2000) has also found a significant correlation between pPAH concentration and the number of heavy-duty vehicles. In this study, however, the strong correlation was found only in the early morning hours.

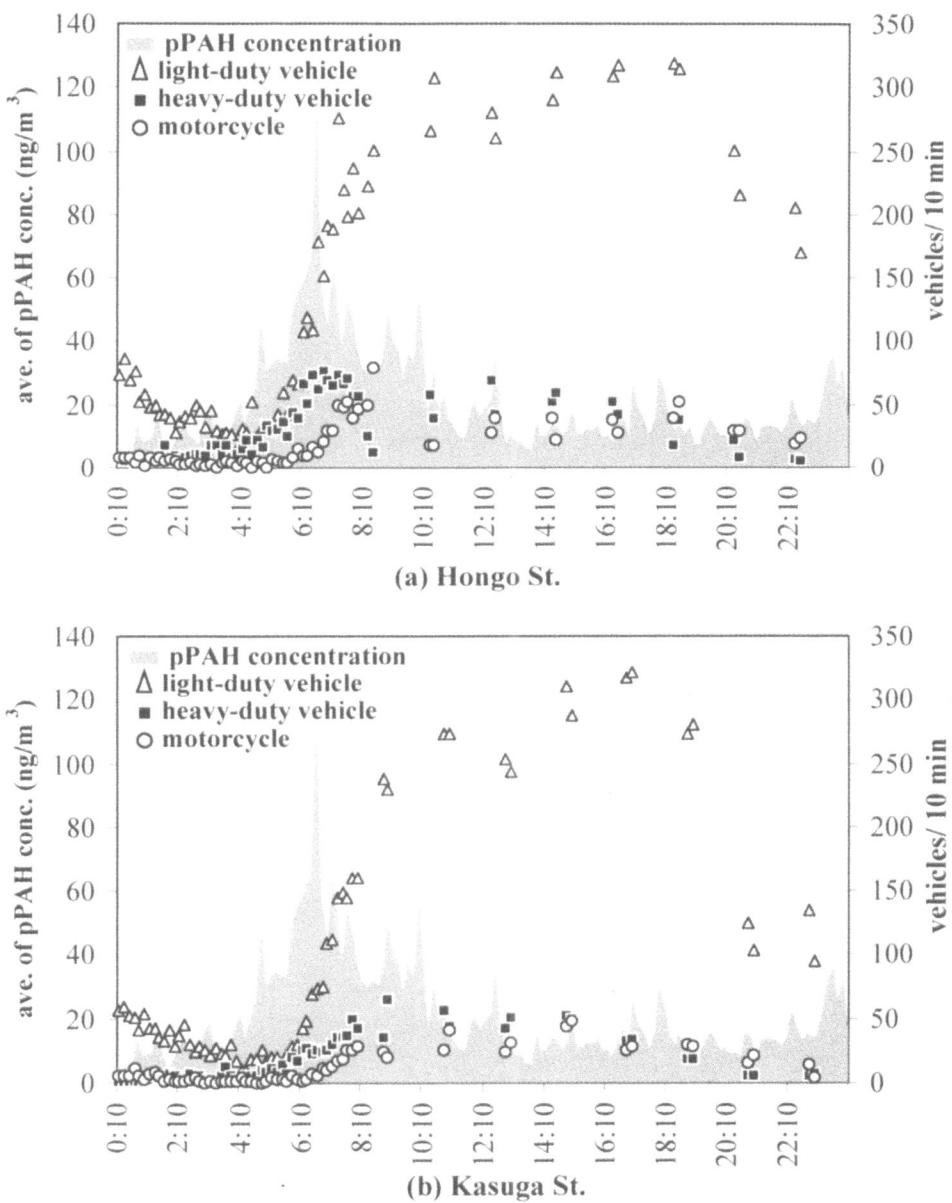

Figure 7. Profile of the average of pPAH concentration at all sampling sites and traffic flows on four main street surrounding the sampling sites on 25 September, 2000.

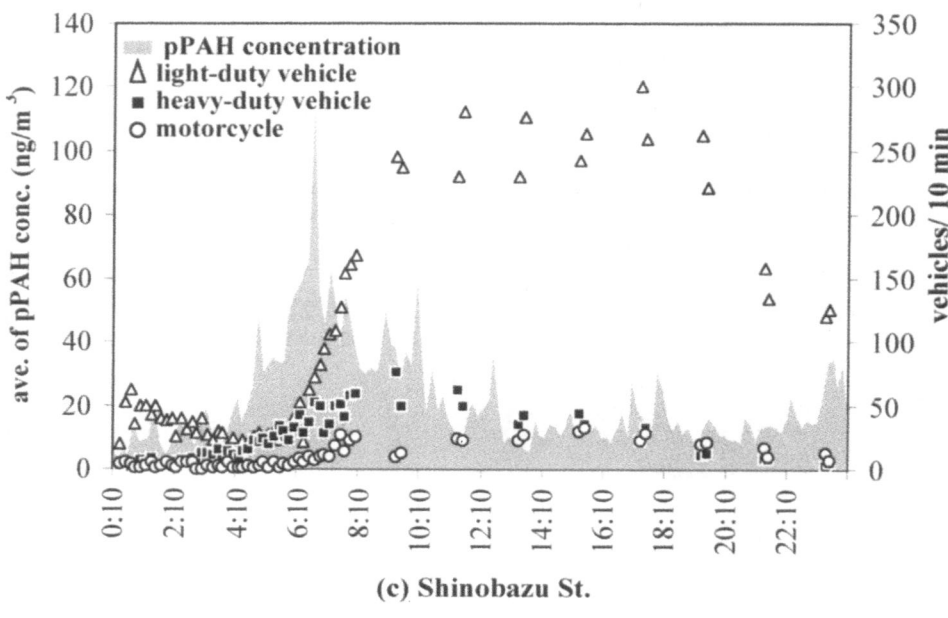

(c) Shinobazu St.

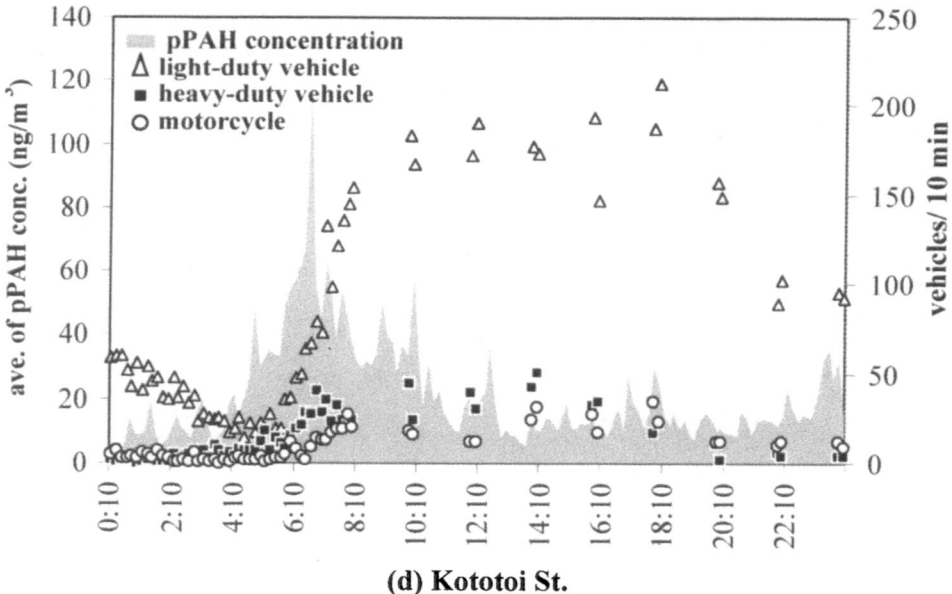

(d) Kototoi St.

Figure 7. (Continued).

TABLE II

Coefficient of determination (R^2) of correlation between pPAH concentration and vehicle types

Observed period	23 August			25 September		
	Light-duty	Heavy-duty	Motorcycle	Light-duty	Heavy-duty	Motorcycle
0:00–8:00	0.076	0.836	0.240	0.274	0.715	0.227
8:00–16:00	0.384	0.105	0.265	0.065	0.037	0.049
16:00–24:00	0.004	0.127	0.002	0.026	0.006	0.004

TABLE III

Local wind characteristics at all sampling sites during the whole sampling periods

Parameter	23 to 28 August			19 to 25 September		
	Bldg. #14	DOL	YSH	Bldg. #11	DOL	ESC
Predominant wind direction	SE, WNW	SSE	SE	NE	NNE	NE, ESE
Average wind speed (m s^{-1})	1.2	0.9	0.9	1.5	0.8	0.8
Wind direction causing high pPAH concentration	W	SSE	ESE	WSW	NNE	NE

Therefore, it is especially important to know the pPAH concentration in the early morning in order to evaluate the human exposure risk due to traffic.

3.3. LOCAL WIND CHARACTERISTICS AND pPAH FLUX

To investigate the dispersion of the pPAH by the wind from Hongo road, the local wind characteristics at all sampling points during the whole period were observed. Figure 8 shows the relationship of wind speed and pPAH concentration with wind direction for all sampling sites during both sessions. The main wind direction and average wind speed are summarized in Table III.

The winds with pPAH concentrations higher than 60 ng m^{-3} at bldg. #14 and bldg. #11 blew at less than 2 m s^{-1}, mainly from the west and west-south-west, respectively. At these roadside sites, the winds, carrying high pPAH concentrations, came from the main road, whose direction much differed from those of the other sites. As for diurnal changes of pPAH concentration, the high concentrations were mostly found in the early morning or morning rush hour. It has been reported that at low wind speeds, air pollutants are predominantly dispersed by traffic-created turbulence, especially during morning rush hour and evening rush hour (Berkowicz

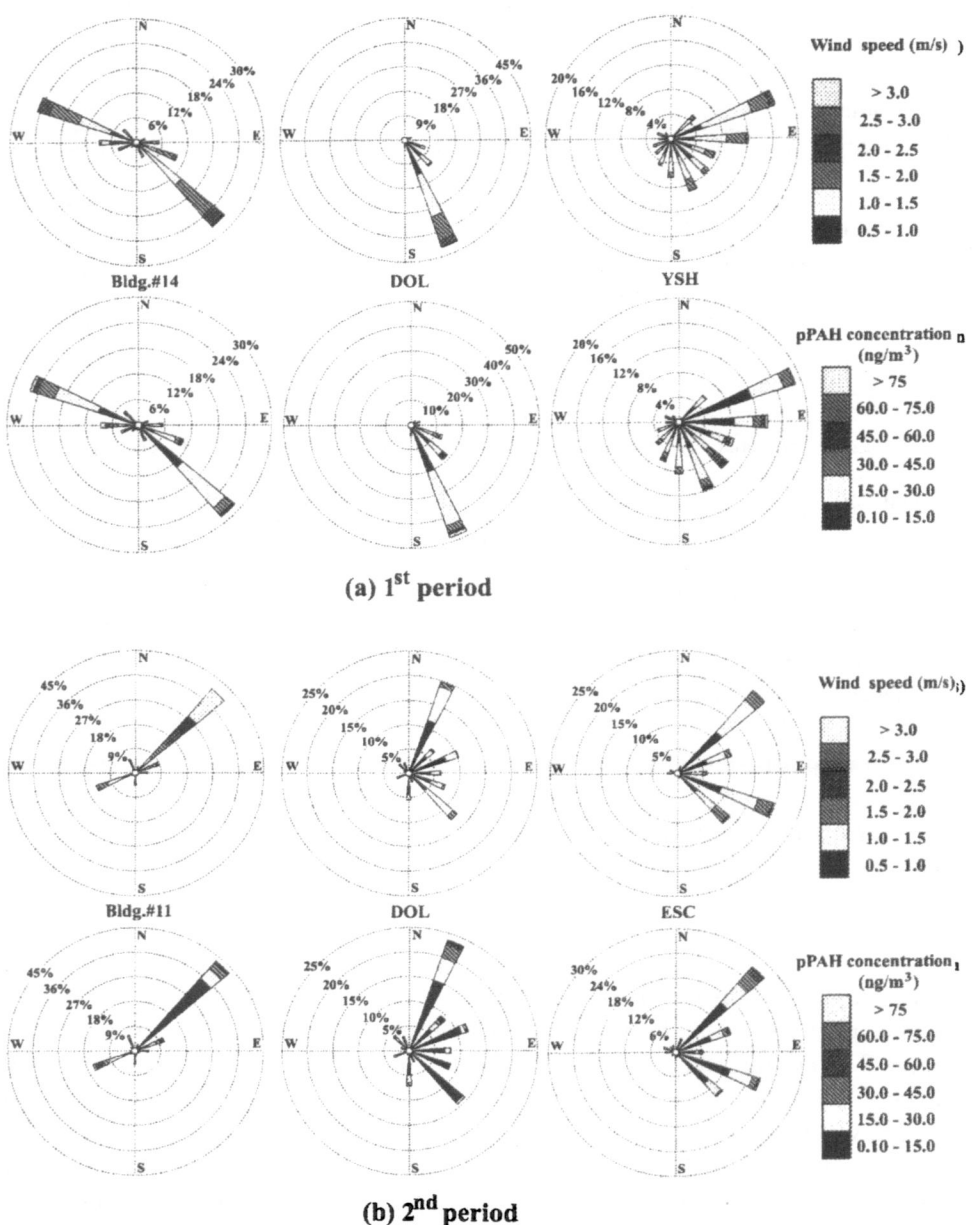

Figure 8. Local wind characteristics at all sampling sites during the whole sampling periods. (a) The relationships of wind speed (upper row) and of pPAH concentration (lower row) with wind direction during 1st monitoring period. (b) The relationships of wind speed (upper row) and of pPAH concentration (lower row) with wind direction during 2nd monitoring period.

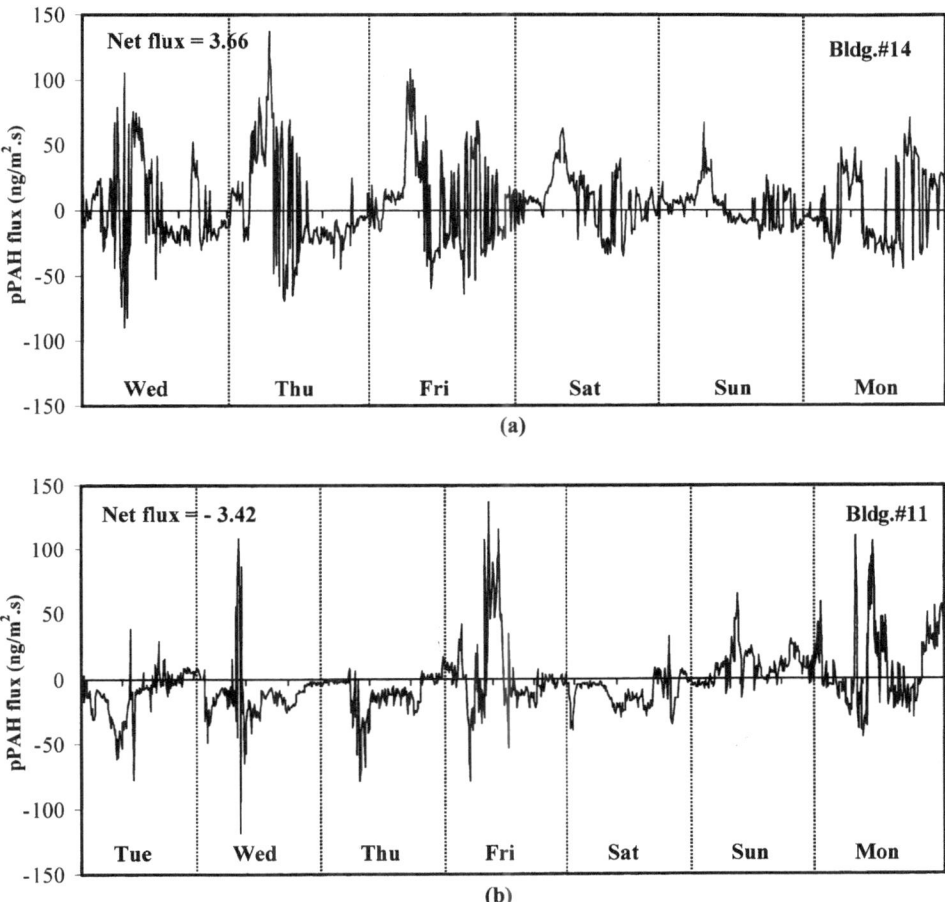

Figure 9. pPAH flux profiles at sampling sites near the Hongo Street during both campaigns. (a) The profile at bldg. #14 during 1st period. (b) The profile at bldg. #11 during 2nd period.

et al., 1996). In this study, therefore, the high pPAH concentration dispersed from the road into the sites might be strongly affected by traffic turbulence.

The pPAH flux at bldg. #14 and bldg. #11 was calculated in order to clearly evaluate pPAH transportation from the Hongo road into the campus. The north and south winds along the road were assumed not to transport the pPAH and the air drawn west was assumed to entirely transport the pPAH. The pPAH flux as follows:

$$\text{pPAH flux}_t (\text{ng m}^{-2}\text{ s}) = C_t * V_t \sin(\theta), \tag{1}$$

where C_t is the pPAH concentration at time t, V_t is the observed wind velocity at time t, and θ is the counterclockwise angle between the wind direction at time t and N direction.

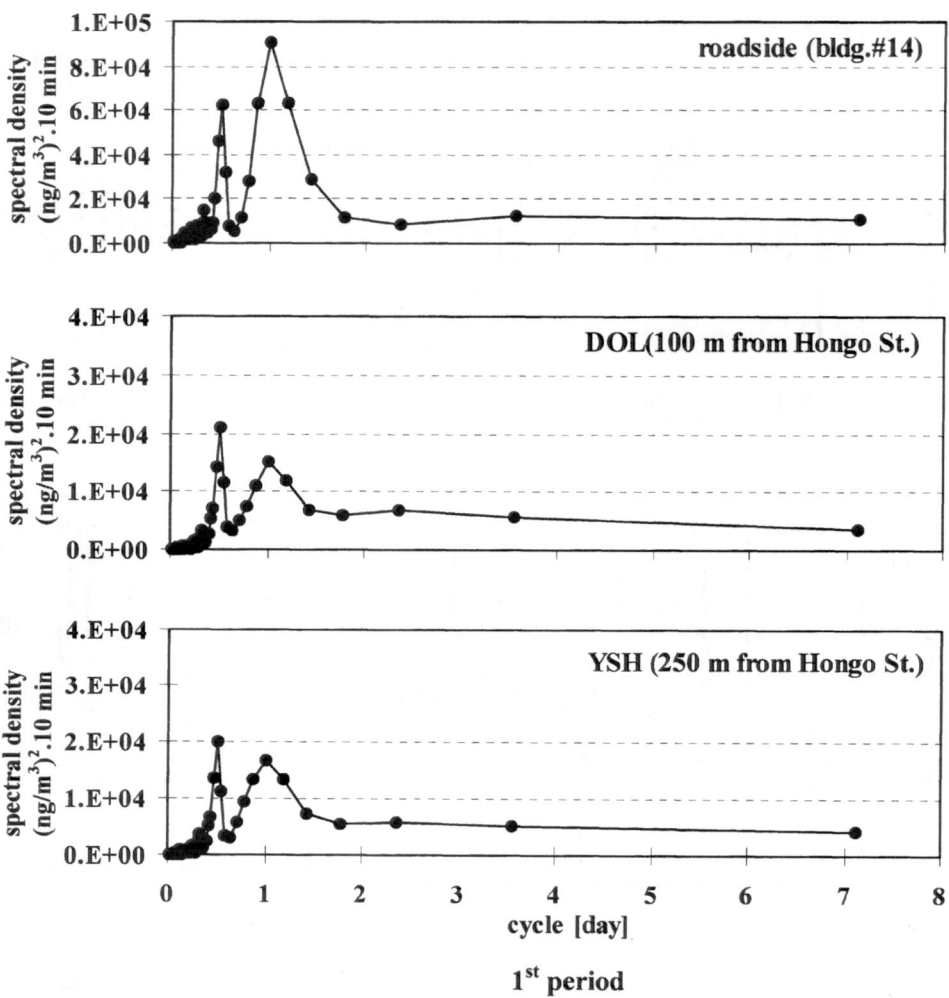

Figure 10. Spectrum density of pPAH time series.

The resulting pPAH flux profile is given in Figure 9. At roadside, the net flux during the first sampling period was positive at bldg. #14 but negative at bldg. #11 during the second session. The sampling site at bldg. #14 was like a street canyon, where a generated wind vortex markedly alters the wind direction, influencing the net flux. Thus, even though the predominant wind flow was toward the road, the net pPAH flux was towards the Hongo Campus because the local wind direction in a street canyon is opposite to the predominant wind field. The local wind field and pPAH transportation, therefore, strongly depend on the configurations and locations of the surrounding buildings.

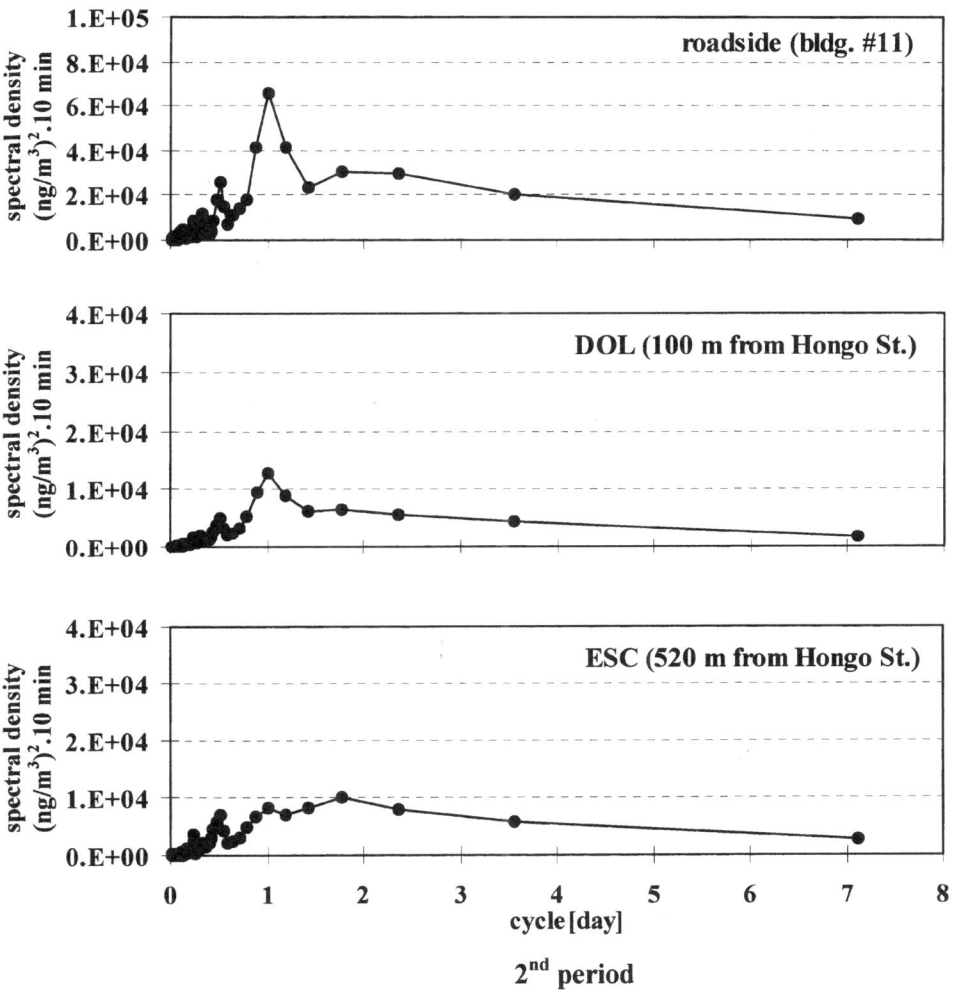

Figure 10. (Continued).

3.4. SPECTRAL ANALYSIS OF pPAH CONCENTRATION

Two significant peaks with cycles of 1- and 0.5 day were found for all sites during both sampling periods as shown in Figure 10. Additionally, the peaks at the roadside gave relatively higher spectral density than those at other sites because the pPAH concentrations were distinctly higher at the roadside. The profiles of the spectral peaks were similar to each other in various areas within the campus, with the exception of at the ESC site. The ESC site might get pPAH not only from road traffic but also from other sources, most likely a nearby incinerator and boiler.

4. Conclusions

Similar diurnal changes in pPAH concentration were observed during the whole sampling periods: pPAH concentration increased early in the morning in accordance with a sudden burden of road traffic and rapidly decreased in the daytime probably due to photodegradation and/or dilution by rising in the mixing zone. There was a strong correlation between pPAH concentration and the number of heavy-duty vehicles (trucks, mini-trucks, and buses), particularly in the early morning between 0:00–8:00. Therefore, to assess the risk of exposure to pPAHs created by road traffic in urban areas, it is particular important to know the pPAH concentration early in the morning. The local wind characteristics, especially at the roadside, had quite different patterns from those at other sites inside the campus. On this study scale, the local wind field and the transportation of pPAH from the road can be strongly influenced by the location and geometry of the surrounding buildings. The pPAH clearly changed in 1- and 0.5 day cycles, particularly at the roadside.

Acknowledgements

The authors would like to thank the Department of Urban Engineering, the Department of Literature, the Department of Architecture, the Yasuda Hall office, and the Environmental Science Center for providing us with the sites for sampling. We greatly express our thanks to Prof. Mikami Takehiko and Assist. Prof. Kimura Keiji, the Tokyo Metropolitan University, for lending us the wind measurement instruments. This study has been partially supported by the grant from Thailand-Japan Technology Transfer Project (TJTTP), Chulalongkorn University, Bangkok, Thailand.

References

Agnesod, G., De Maria, R., Fontana, M. and Zublena, M.: 1996, 'Determination of PAH in airborne particulate: Comparison between off-line sampling techniques and an automatic analyzer based on a photoelectric aerosol sensor', *Sci. Total Environ.* **189/190**, 443–449.

Beak, S. O., Field, R. A. and Goldstone, M. E.: 1991, 'A review of atmospheric polycyclic aromatic hydrocarbons: Source, fate, and behavior', *Water, Air, and Soil Pollut.* **60**(3–4), 279–300.

Berkowicz, R., Palmgren, F., Hertel, O. and Vignati, E.: 1996, 'Using measurements of air pollution in streets for evaluation urban air quality-meteorological analysis and model calculations', *Sci. Total Environ.* **189/190**, 259–265.

Caricchia, A. M., Chiavarini, S. and Pezza, M.: 1999, 'Polycyclic aromatic hydrocarbons in the urban atmospheric particulate matter in the City of Naples (Italy)', *Atmosph. Environ.* **33**, 3731–3738.

Fan, Z., Chen, D., Birla, P. and Kamens, R. M.: 1995, 'Modeling of nitro-polycyclic aromatic hydrocarbon formation and decay in the atmospheric', *Atmosph. Environ.* **29**, 1171–1181.

Junker, M., Kasper, M., Roosli, M., Camenzind, M., Kunzli, N., Monn, Ch., Theis, G. and Braun-Fahrlander, Ch.: 2000, 'Airborne particle number profiles, particle mass distributions and particle-bound PAH concentrations within the city environment of Basel', *Atmosph. Environ.* **34**, 3171–3181.

Kamens, R. M., Guo, Z., Fulcher, J. N. and Bell, D. A.: 1988, 'Influence of humidity, sunlight, temperature on the daytime decay of polyaromatic hydrocarbons on atmospheric soot particles', *Environ. Sci. Technol.* **22**, 103–108.

Kaupp, H. and McLachlan, M. S.: 1999, 'Atmospheric particle size distribution of polychlorinated dibenzo-p-dioxins and dibenzofurans (PCDD/Fs) and polycyclic aromatic hydrocarbons (PAHs) and their implications for wet and dry deposition', *Atmosph. Environ.* **33**, 85–95.

Siegmann, K. and Siegmann, H. C.: 1998, 'Molecular precursor of soot and quantification of the associated health risk', in *Current Problems in Condensed Matter*, Plenum Press, New York, pp. 143–160.

FORMATION AND TRANSPORT OF ATMOSPHERIC AEROSOL OVER ATHENS, GREECE

I. COLBECK[1]*, MENG-CHEN CHUNG[1] and K. ELEFTHERIADIS[2]

[1] *Department of Biological Sciences, University of Essex, Colchester, U.K.;* [2] *Environmental Radioactivity Laboratory, Institute of Nuclear Technology, N.C.S.R. 'Demokritos', 15310 Ag. Paraskevi, Attiki, Greece*

(* *author for correspondence, e-mail: colbi@essex.ac.uk, fax: ++44 1206 872592*)

Abstract. The effects of meteorology on ambient aerosol concentrations and aerosol transport, within the Greater Athens Area during the summer period, was investigated. Measurements of size fractionated anions and cations were made at two sites (inland at Ag. Stefanos and on the coast at Pireas) within the Greater Athens Area. The wind regime exhibited a distinct influence such that the sea-breeze circulation strongly enhanced the formation of secondary aerosols. For sulphate the difference in concentration between the two sites was, on average, 8 times greater on sea-breeze days compared with Etesian days (warm days with NE winds). During 'normal' days, any differences in concentrations were possibly due to local emissions. Elevated concentrations in the fine mode were detected at both sites during the sea-breeze days. The sea-breeze circulation enhances the development of secondary aerosols which was clearly shown at the inland site. Nitrous acid, hydrochloric acid and particulate nitrate, sulphate and ammonium increase during sea-breeze days. Elevated levels of nitrate, 4 μm diameter, were particularly observed on the days with a strong sea-breeze circulation. Sulphate was well correlated with both sulphur dioxide and ammonium suggesting the production of $NH_4HSO_4/(NH_4)_2SO_4$ aerosols, formed through the neutralisation of NH_3 with sulphuric acids. Ammonium sulphate was found to be the major ammonium component in Athens.

Keywords: atmospheric aerosol, composition, size distribution

1. Introduction

Air pollution in Mediterranean urban centres is characterised by intense photochemical activity, high temperatures and low relative humidities during the summer. The effect of these climatic conditions on atmospheric chemistry coupled with complex topography and specific emissions can be the generation of secondary pollutants different in composition and concentration to those measured in other parts of the world.

Athens is one of the largest coastal urban agglomerations in the world with a dense population (about four million). The Athens basin, which is about 450 km^2, is open towards the sea (Saronikos Bay) to the south-west and is surrounded by mountains to the north-west, north-east and south-east. The mountains standing as physical barriers with small gaps between them. These are the routes for air masses to enter or exit the basin by near-surface convection over the land. During

Water, Air, and Soil Pollution: Focus **2**: 223–235, 2002.
© 2002 *Kluwer Academic Publishers.*

the warm period of the year, the existence of persistent northeasterly winds known as Etesians efficiently ventilates the basin (Melas *et al.*, 1995). There is very heavy automobile traffic ($\sim$1.5 million vehicles) and industrial activity within the Greater Athens Area. Two main industrial areas lie in the Greater Athens Area, one is in the southwest near the harbour of Pireas and the other is located outside the Athens basin, to the west of the city.

Numerous scientific papers have been published in the literature on air pollution in Athens (see Ziomas, 1998, for details) and, in 1994, Athens was the focus of an extensive experimental and modelling investigation of the chemical and meteorological evolution of ozone and related trace gases (Ziomas, 1998, and that issue of Atmospheric Environment). The role of the sea-breeze circulation on photochemical pollution episodes has been investigated in many studies (e.g. Lalas *et al.*, 1983, 1987; Cvitas *et al.*, 1985; Prezerakos, 1986; Mantis *et al.*, 1992; Pilinis *et al.*, 1993). Worse air pollution episodes occur during the days with a critical balance of different wind regimes or sea-breeze circulation. This has been investigated by model simulations (Kallos *et al.*, 1993) and by experimental identification (Kambezidis *et al.*, 1995; Klemm *et al.*, 1998; Suppan *et al.*, 1998).

In the Greater Area of Athens, the predominant winds blow from SSW/SW for the early summer and from N and NE in late summer (Etesians). The SSW wind directions are associated with a sea-breeze circulation in the basin. For the sea breeze days, a weak northerly flow dominates during the late night and early morning hours. The sea-breeze cell develops over the Athens basin during daytime then a strengthening sea-breeze circulation occurs in the early afternoon due to maximisation of the sea/land temperature difference. During this time, SW winds prevail in the area.

Despite their importance in the atmospheric environment, ambient aerosols have been poorly studied in Greece. Most work has concentrated on the chemical speciation of urban aerosol (Koliadima *et al.*, 1998; Eleftheriadis *et al.*, 1998; Ziomas *et al.*, 1995; Kirkitsos and Sikiotis, 1993; Tsitouridou and Samara, 1993). Previously we have investigated the concentration of common ionic species in the atmospheric aerosol at the centre of Athens, as well as that of secondary gaseous pollutants such as nitric and hydrochloric acids (Eleftheriadis *et al.*, 1998). It was shown how the build-up of atmospheric aerosol and acidic species in the centre of the city was dependant upon the wind flow patterns. In this article we present further investigations of the impact of wind regime on aerosol concentrations together with size fractionated concentrations of atmospheric aerosols at two sites: one upwind and one downwind of Athens.

2. Methodology

Two sampling sites were selected, located at the coastal and inland boundaries of the Athens Metropolitan area along the NE-SW axis. The distance between the two

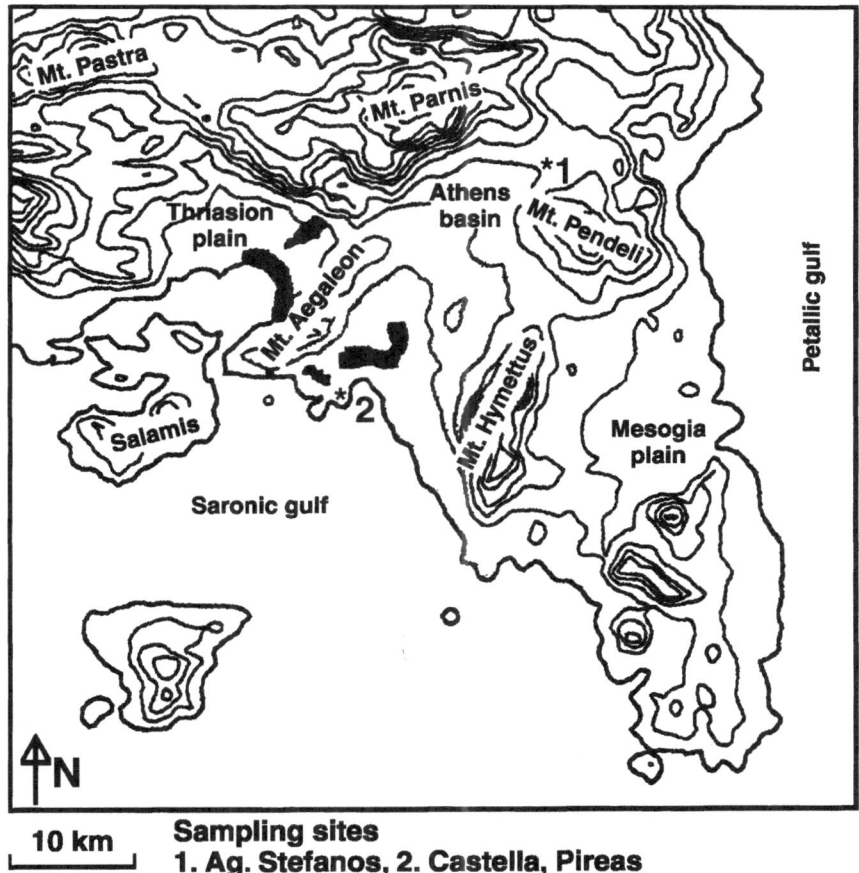

Figure 1. Map of the Greater Athens Area displaying the sampling locations.

sites was 22 km. The southwesterly wind coincides with the main surface wind flow pattern over the summer period. The coastal site was located on the top of Castella hill 90 m a.s.l. in Pireas and the inland site on the edge of the Ag. Stefanos residential area (Figure 1).

Sampling was performed over 24 hr periods with an annular denuder system and a May cascade impactor. Three annular denuder tubes were used for SO_2, HNO_3, HCl, HONO and NH_3 measurements. Air was drawn through the annular denuder by means of a vacuum pump operating at 10 L min^{-1}. The first and the second denuder tubes were coated with 1% Na_2CO_3 and glycerol in a methanol:H_2O (1:1) solution. Determination of ammonia was made by the third denuder coated with 1% citric acid in a methanol solution. A PTFE filter pack with three filters in series followed the annular denuder tube: Teflon (0.45 μm Whatman), Nylon (Nylon 66, 0.45 μm Sartorius) and Whatman 41 filters, in 47 mm diameter, were used in tandem. Nylon filters were washed with de-ionised water then dried in a vacuum desiccator prior to use in order to reduce the background levels of nitrate and chloride. The Whatman 41 filters were impregnated with 5% phosphoric acid solution for ammonia collection and then dried in a vacuum desiccator prior to use. The cascade impactor sampled at 5 L min^{-1} resulting in size cuts of 0.5, 1.0, 2.0, 4.0, 8.0, 16.0 and 32.0 μm aerodynamic diameter. Standard microscopy slides (76 $\times$ 26 mm, thickness = 1.0 mm) were coated with 1% silicone oil in hexane solution. One slide was used for each stage. A Teflon filter (47 mm diameter, pore size 0.45 μm) was used as the back up filter.

All samples were analysed for Cl^-, NO_3^-, SO_4^{2-}, NH_4^+, Na^+, K^+, Mg^{2+} and Ca^{2+} by ion chromatography (Dionex 2000i/SP). Over the three-week period from 17 July to 6 August 1997, seventeen samples were obtained for each site (no samples were obtained on four days during this period). Meteorological parameters, such as wind speed and direction and relative humidity, were provided by routine measurements of the Greek Meteorological Service. These data were taken at Elliniko and Dekeleia, 6 and 7 km from the inland and coastal sites, respectively. They were selected as the closest representative measurements for the ambient air at the sampling sites.

Over the sampling period, the temperature was in the range of 20–37 °C and relative humility ranged from 20–80% with the highest values occurring for only a few hours of the night. Under these conditions, nitric and hydrochloric acid in the gaseous phase may be responsible for artifact formation on filter packs or impactor substrates. The methodology used here did not provide artifact prevention for the impactor by means of an upstream denuder (Wall *et al.*, 1988). However, mass concentrations of ionic species obtained by the integrated impactor samples can be examined against those obtained by the denuder filter pack, which are free from artifact formation. From the sampling characteristics of the denuder system a 50% aspiration efficiency cut off for its inlet at 8 μm was obtained. In Figure 2, the results for denuder versus impactor concentrations are plotted for both sites. A large discrepancy is observed for Cl^- and Na^+ whilst NO_3^-, SO_4^{2-} and NH_4^+ show relatively good agreement (correlation coefficients = 0.23, 0.31, 0.66, 0.53 and 0.69, respectively). For the first two species it is more likely that they are part of NaCl which was inefficiently sampled by both devices. It is concluded that artifact formation does not make a significant contribution to the mass concentration of the examined species sampled by the impactor.

3. Results and Discussion

The 24 hr average concentrations for atmospheric aerosol and gaseous species at Ag. Stefanos and Pireas are shown, respectively, in Figures 3 and 4. It is evident that much higher concentrations of all the aerosol species were found at Ag. Stefanos during the first half of the sampling period, while similar or slightly lower concentrations were observed during the remainder of the period. This is clearly shown

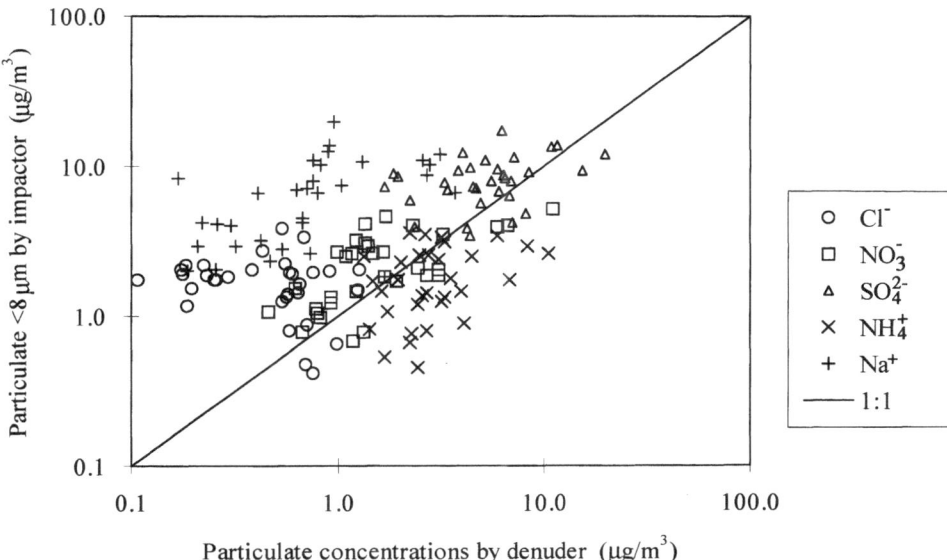

Figure 2. Mass concentrations measurend in parallel by a May Cascade impactor (<8 μm) and denuder/filter pack.

in Figure 5 where the difference in concentration between the two sites is plotted. The concentration patterns at the two sites can be explained by the dominant wind flow patterns present in the area during the measurements.

Following the work of Suppan *et al.* (1998) each day was classified according to the wind pattern:

- Etesian days: strong northerly winds, which suppress the land/sea-breeze and lead to low pollution levels inside the basin.
- Sea-breeze days: weak northerly winds and a fully developed sea-breeze cell affecting the whole basin of Athens.
- Normal days: moderate northerly winds and development of a weak sea-breeze cell near the coast line.

Table I shows the classification of the days into the three categories. Two typical situations dominated during this period. During the first half of the campaign, 17–26 July 1997, southwesterly winds, due to the sea-breeze circulation, were dominant. From 29 July 1997 onward, the Etesian winds prevailed in the Athens basin. For typical sea-breeze days higher pollution levels are observed at Ag. Stefanos whilst lower concentrations coincide with cleaner northeasterly winds (see Figure 5). During 30–31 July 1997, the concentrations detected at both sites were similar. This was due to the northerly winds with relatively higher speeds ($\sim$6 m s^{-1}).

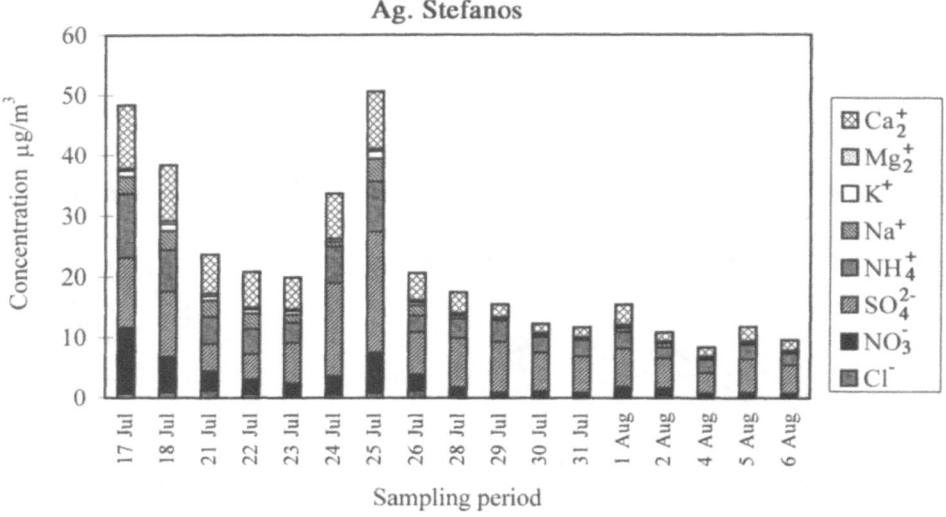

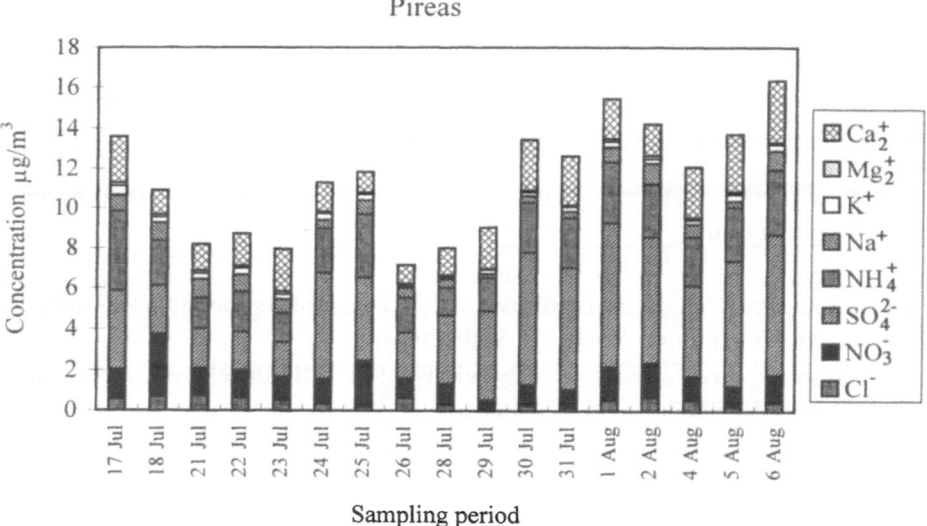

Figure 3. Ambient aerosol concentrations during the campaign.

When a strong sea-breeze circulation occurred, high levels of secondary pollutants, such as HNO_3, NO_3^- and SO_4^{2-} were detected. For instance, during the period of 24–25 July, there was very weak northerly winds or calm states that enhanced the formation of secondary aerosols. The pollutants were then transported by the southwesterly sea-breeze and reached the NE of the Athens basin, e.g. Ag. Stefanos. This is confirmed by the higher levels of nitrous acid that were observed

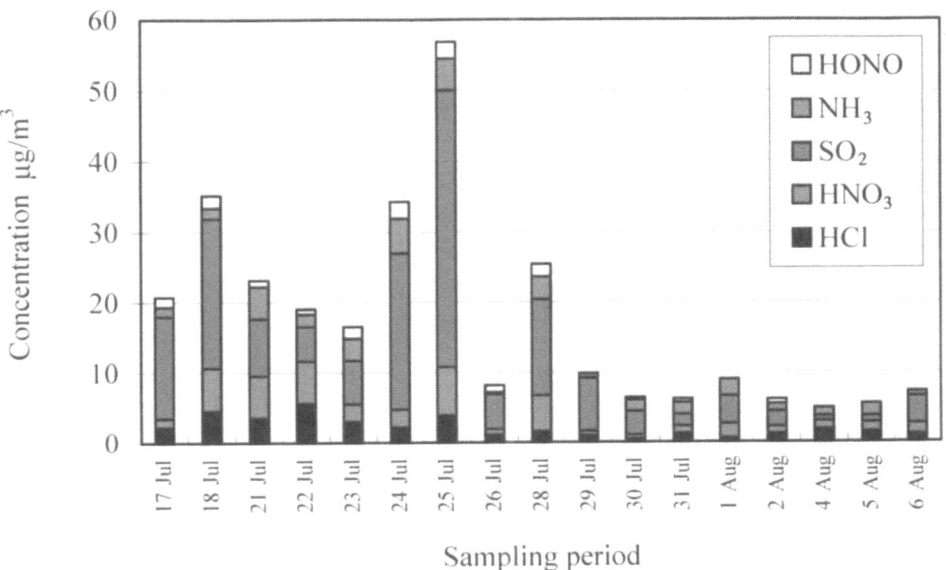

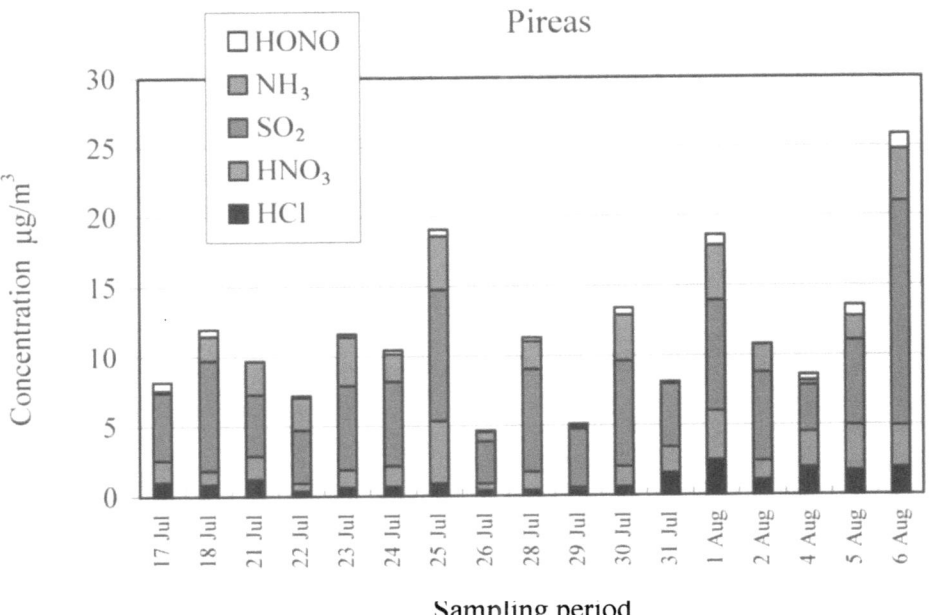

Figure 4. Ambient gaseous concentrations during the campaign.

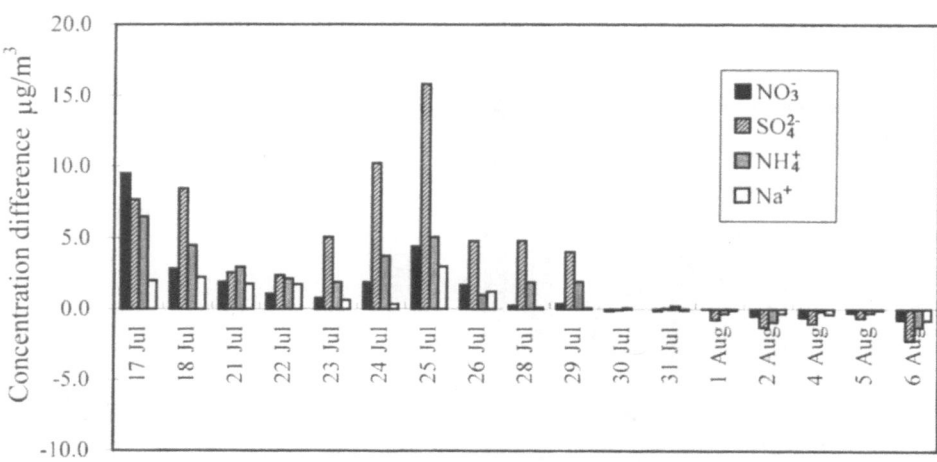

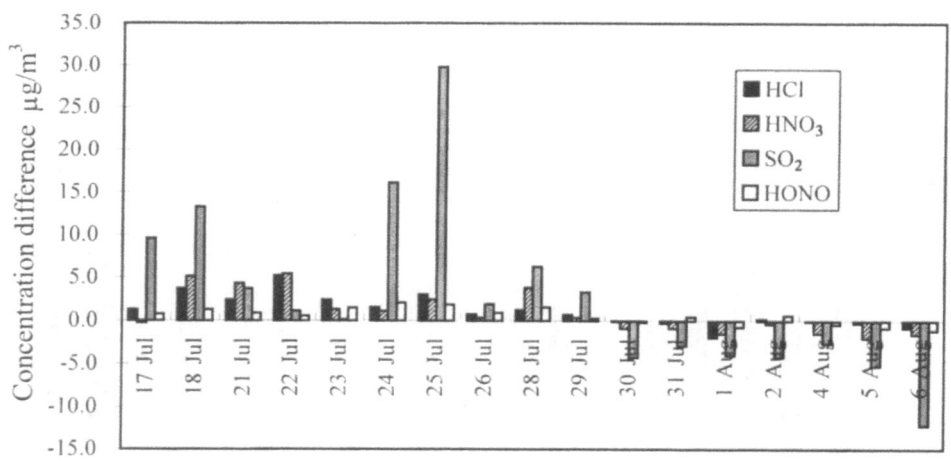

Figure 5. Difference in the concentrations of aerosol and gaseous pollutants between Ag. Stefanos and Pireas.

TABLE I

Classifications of the days into meteorological conditions

Sea-breeze days	Normal days	Etesian days
17–26 July		
|	27–28 July	
|	|	29 July–1 August
2 August	3 August	4–7 August
(10 days)	(3 days)	(7 days)

TABLE II

Increased pollutant levels during sea-breeze days over the basin (displayed by the ratios of differences between the two sites on sea-breeze days to those on Etesian days)

ΔSea/ΔEtesian	HONO	HNO_3	HCl	SO_2	Cl^-	NO_3^-	SO_4^{2-}	NH_4^+
Average	3.3	2.0	7.4	2.0	1.8	8.8	8.2	10.5
Max.	5.9	4.9	17.0	7.5	4.8	32.0	20.8	22.9

during these sea-breeze days (Figure 5). Nitrous acid, an important secondary air pollutant, promotes photochemical reactions due to its photolytic generation of hydroxyl radicals (OH·). This promotes the production of nitric acid:

$$HONO + h\nu \; (\lambda < 400 \text{ nm}) \rightarrow OH\cdot + NO$$

$$2HONO \rightarrow NO + NO_2 + H_2O$$

$$NO_2 + OH\cdot + M \rightarrow HNO_3 + M$$

Danalatos and Glavas (1999) indicated that this gas phase formation is predominant, with a maximum diurnal value in the afternoon during the summer.

The degree of enhanced pollutant levels by the sea-breeze circulation is examined in Table II by using the differences of the measured concentrations between the two sites. The results clearly show that secondary pollutants are dramatically increased during the sea-breeze days together with the efficient clearance of pollution by the Etesian winds over the basin. On average, primary SO_2 on sea-breeze days is double that on Etesian days. HCl exhibits a high degree of intensification and as well the secondary particulate NO_3^-, SO_4^{2-} and NH_4^+.

The illustrations of particulate SO_4^{2-}, NO_3^- and Cl^-, in measured size ranges under three different wind regimes, are presented in Figure 6. During the Etesian days, the well-mixed atmosphere leads to a relatively small difference between the two sites. Slightly higher concentrations of nitrate and sulphate were measured at Pireas. The increment was exhibited predominantly within the size range of 1–2 μm diameter. This suggests that the northerly wind might have transported the secondary particles, formed in the centre of Athens, to the Saronic Gulf. During normal days, northerly winds prevail at Ag. Stefanos and southerly winds dominant over the coastal area. The difference in size distribution between the two sampling stations on the normal days is due to the local emissions at each site. Fine sulphate was higher at Pireas than at Ag. Stefanos. This may possibly be associated with industrial emissions, however a firm conclusion can not be made from the few samples classified as normal days in this survey. During the sea-breeze days, elevated concentrations of the fine mode particles were detected at both sites. A significant increase in the condensation mode for sulphate ($d < 0.5$ μm) was evident as the sea-breeze cell developed. Over this period, particulate nitrate in both the fine and coarse modes also increased. Elevated levels of nitrate, 4 μm diameter, were particularly observed on the days with a strong sea-breeze circulation. The formation of nitric acid with marine aerosols is assumed as the main reaction.

The ion balance between measured ions at both sites is shown in Figure 7. The large deficiency of anions may be due to carbonate/bicarbonate ions ($CaCO_3$ is a major component of loose soil and dust in Greece) and also due to the possible existence of other earth materials such as CaO, MgO, Na_2O and K_2O. It was found that correlations amongst the measured species were generally significantly higher at Ag. Stefanos than that at Pireas. Sulphate showed good correlations between its precursor SO_2 as well with ammonium (r = 0.94 and 0.76, respectively). This suggests the production of $NH_4HSO_4/(NH_4)_2SO_4$ aerosols, formed through the neutralisation of NH_3 with sulphuric acids. The molar ratio of ammonium to sulphate obtained was (NH_4^+/SO_4^{2-}) 1.3±0.4 at Ag. Stefanos and 1.5±0.4 at Pireas. This indicates that the components of ammonium sulphate aerosol were mainly present as NH_4HSO_4 and $(NH_4)_3H(SO_4)_2$ during this period.

4. Conclusions

The wind regimes exhibit a distinct effect on ambient aerosols in Athens. Analysis of the size distributions, for different wind regimes indicated that on Etesian days concentrations of all species are marginally higher at Pireas. During 'normal' days, the size distributions were similar and any difference in the distribution between the two sampling sites on such days maybe due to the local emissions. Elevated concentrations in the fine mode were detected at both sites during the sea-breeze days. The sea breeze circulation enhances the development of secondary aerosols which was clearly shown at the inland site. Nitrous acid, hydrochloric acid and particulate nitrate, sulphate and ammonium increase during sea-breeze days. Elevated levels of nitrate, 4 μm diameter, were particularly observed on the days with

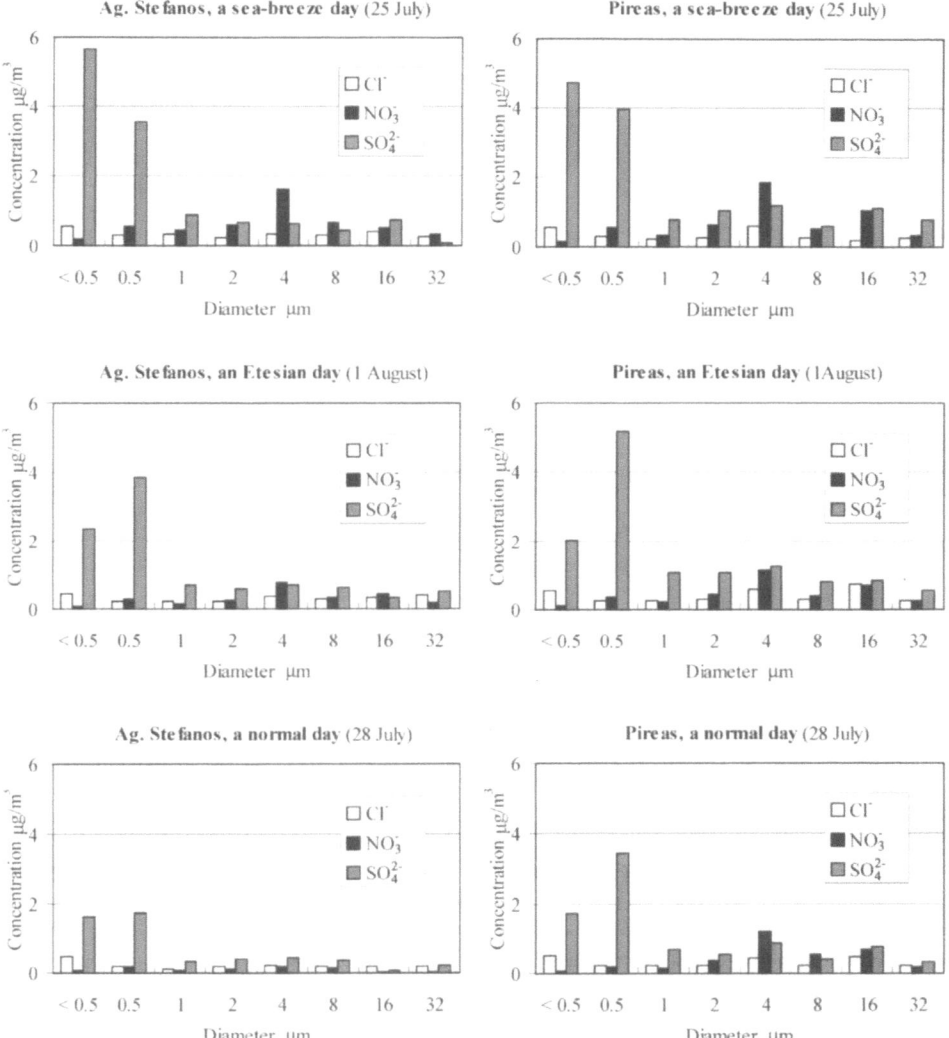

Figure 6. Concentrations of sulphate, nitrate and chloride, as a function of particle size, during sea-breeze, Etesian and normal days.

a strong sea-breeze circulation. The formation of nitric acid with marine aerosols was assumed as the main reaction. Ammonium sulphate was found to be the major ammonium component in Athens. As only seventeen days data were collected in the present study, further evaluation involving more samples will be required for a better understanding and confirmation of the results.

Ion Balance

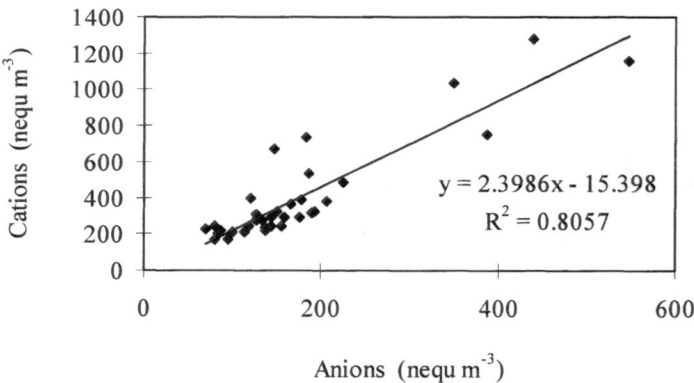

$$y = 2.3986x - 15.398$$
$$R^2 = 0.8057$$

Figure 7. Ion balance of ambient aerosols.

References

Cvitas, T., Güsten, H., Heinrich, G., Klasinc, L., Lalas, D. P. and Petrakis, M.: 1985, 'Characteristics of air pollution during the summer in Athens, Greece', *Staub-Reinhalt Luft.* **45**, 297–301.

Danalatos, D. and Glavas, S.: 1999, 'Gas phase nitric acid, ammonia and related particulate matter at a Mediterranean coastal site, Patras, Greece', *Atmos. Environ.* **33**, 3417–3425.

Eleftheriadis, K., Galis, D., Ziomas, I., Colbeck, I. and Manalis, N.: 1998, 'Atmospheric aerosol and gaseous species in Athens, Greece', *Atmos. Environ.* **32**, 2183–2191.

Kallos, G., Kassomenos, P. and Pielke, R. A.: 1993, 'Synoptic and mesoscale weather conditions during air pollution episodes in Athens, Greece', *Bound.-Layer Meteor.* **62**, 163–184.

Kambezidis, H. D., Peppes, A. A. and Melas, D.: 1995, 'An environmental experiment over Athens urban area under sea breeze conditions', *Atmosph. Resear.* **36**, 139–156.

Kirkitsos, P. D. and Sikiotis, D.: 1993, 'Chemical composition of the $NH_3 - HNO_3 - H_2SO_4 - NaCl$ system in the atmosphere of Athens, Greece', *Environ. Monit. Assess.* **28**, 61–81.

Klemm, O., Ziomas, I. C., Balis, D., Suppan, P., Slemr, J., Romero, R. and Vyras, L. G.: 1998, 'A summer air-pollution study in Athens, Greece', *Atmos. Environ.* **32**, 2071–2087.

Koliadima, A., Athanasopoulou, A. and Karaiskakis, G.: 1998, 'Particulate matter in air of the cities of Athens and Patras (Greece): Particle-size distributions and elemental concentrations', *Aerosol Sci. Technol.* **28**, 292–300.

Lalas, D. P., Asimakopoulos, D. N, Deligiorgi, D. G. and Helmis, C.: 1983, 'Sea-breeze circulation and photochemical pollution in Athens, Greece', *Atmos. Environ.* **17**, 1621–1632.

Lalas, D. P., Tombrou-Tsella, M., Petrakis, M., Asimakopoulos, D. N. and Helmis, C.: 1987, 'An experimental study of the horizontal and vertical distribution of ozone over Athens', *Atmos. Environ.* **17**, 1621–1632.

Mantis, H. T., Repapis, C. C., Zerefos, C. S. and Ziomas, I. C.: 1992, 'Assessment of the potential for photochemical air pollution in Athens: A comparison of emissions and air-pollutant levels in Athens with those in Los Angeles', *J. Appl. Meteor.* **31**, 1467–1476.

Melas, D., Ziomas, I. C. and Zerefos, C. S.: 1995, 'Boundary layer dynamics in an urban coastal environment under sea breeze conditions', *Atmos. Environ.* **29**, 3605–3617.

Pilinis, C., Kassomenos, P. and Kallos, G.: 1993, 'Modeling of photochemical pollution in Athens, Greece. Application of the RAMS-CALGRID modeling system', *Atmos. Environ.* **27B**, 353–370.

Prezerakos, N. G.: 1986, 'Characteristics of the sea breeze in Attica, Greece', *Bound.-Layer Meteor.* **36**, 245–266.

Suppan, P., Fabian, P., Vyras, L. and Gryning, S. E.: 1998, 'The behaviour of ozone and peroxyacetyl nitrate concentrations for different wind regimes during the Medcaphot-Trace Campaign in the Greater Area of Athens, Greece', *Atmos. Environ.* **32**, 2089–2102.

Tsitouridou, R. and Samara, C.: 1993, 'First results of acidic and alkaline constituents determination in air particulates of Thessaloniki, Greece', *Atmos. Environ.* **27B**, 313–319

Wall, S. M., John, W. and Ondo, J.: 1988, 'Measurement of aerosol size distributions for nitrate and major ionic species', *Atmos. Environ.* **22**, 1649–1656.

Ziomas, I. C.: 1998, 'The Mediterranean Campaign of Photochemical Tracers – Transport and Chemical Evolution (MEDCAPHOT-TRACE): An Outline', *Atmos. Environ.* **32**, 2045–2053.

Ziomas, I. C., Paliatsos, A. G., Viras, L. G. and Zerefos, C. S.: 1995, 'A note on smoke concentrations in urban Athens, using data measured by two different methods', *Meteorol. Atmos. Phys.* **55**, 215–221.

THE CHARACTERISATION OF SETTLED DUST BY SCANNING ELECTRON MICROSCOPY AND ENERGY DISPERSIVE X-RAY ANALYSIS

VAUGHAN SHILTON*, PAUL GIESS, DAVID MITCHELL and CRAIG WILLIAMS

University of Wolverhampton, School of Applied Sciences, Wolverhampton, West Midlands, U.K.
(* author for correspondence, e-mail: V.F.Shilton@wlv.ac.uk)

Abstract. Settled dust has been collected inside the main foyers of three University buildings in Wolverhampton City Centre, U.K. Two of the three buildings are located in a street canyon used almost exclusively by heavy duty diesel vehicles. The dust was collected on adhesive carbon spectro-tabs to be in a form suitable for analysis by scanning electron microscope and energy dispersive X-ray analysis. Using these analytical techniques, individual particle analysis was undertaken for morphology and chemistry. Seasonal variations and variations due to location were observed in both the morphological measurements and chemical analysis. Many of the differences appear attributable to the influence of road traffic, in particular, the heavy duty diesel vehicles, travelling along the street canyon.

Keywords: airborne particulate matter, diesel emissions, dust soiling, indoor dust, scanning electron microscope, settled dust

1. Introduction

Air pollution caused by road traffic is an important factor when considering the appraisal of road schemes. This appraisal currently focuses on the road side concentrations of traffic related pollutants, which are potentially harmful to human health or ecological systems. However, particulate vehicle emissions, especially from heavy duty diesel engines, can cause nuisance problems, including the soiling of buildings through the accumulation of dust (QUARG, 1996) and, may also, be partly responsible for sick building syndrome (Gyntelburg *et al.*, 1994). The soiling of buildings on the outside due to traffic emissions is a well known occurrence (Smith and Warke, 1995). However, inside a building, in close proximity to a busy road, the rate of dust soiling and the composition of the settled dust has not received as much attention, although papers have been published in related areas (Brooks and Schwar, 1987; Ford and Adams, 1999; Raza *et al.*, 1990; Williams and McCrae, 1995). In terms of the nuisance that dust soiling causes indoors, factors such as the colour of the dust and the surface upon which the dust settles are likely to be important. As dust produced by traffic emissions tends to be darkly coloured and surfaces in commercial buildings are often light in colour, it may be assumed that

Water, Air, and Soil Pollution: Focus **2**: 237–246, 2002.
© 2002 *Kluwer Academic Publishers.*

relatively low levels of traffic derived dust can produce a serious soiling problem inside commercial buildings.

This article introduces a method which can be used for the collection of settled dust in a form suitable for physical and chemical analysis by Scanning Electron Microscope (SEM) and Energy Dispersive X-ray Analysis (EDX). Using this method the settled dust has been characterised chemically and morphologically. Three buildings have been used to investigate the variations in the chemistry and morphology of settled dust dependent upon location. In addition, seasonal variations have also been examined.

2. Materials and Methods

Settled dust was collected inside the main foyers of three University buildings located in the centre of Wolverhampton, U.K., and analysed by SEM (Camscan SV2) and EDX (Link Analytical). Dust soiling rates were also determined for each location, using a gravimetric method. All monitoring was carried out on a monthly basis for one year between November 1999 and October 2000. Exposure periods were for one calendar month beginning on the 15th of each month, for example, the exposure for the month of November would be between 15 November and 14 December.

The foyers are located at ground level and are similar in terms of size and the degree of usage by students and staff. Buildings 1 and 2 are located on either side of the main approach road to the city's primary bus station. This road is, therefore, used by several thousand heavy duty diesel buses each day. The road is closed to other vehicles, so the buses contribute almost 100% of the traffic using the road. Traffic is held up at traffic lights at one end of the road regularly causing queues of vehicles idling outside the two buildings. Four and five storey buildings on both sides of the road produce a small street canyon effect which may impede the dispersion of traffic related emissions. Previous work has demonstrated that high concentrations of airborne particulate matter exist both inside and immediately outside these two buildings (Giess, 1998). Building 3 is located approximately 300 m away from the other two buildings in a different road and is not within a street canyon. This road is only occasionally used by heavy duty diesel vehicles and does not carry high traffic flows.

The measurement of dust soiling rates was achieved by using 5 pre-weighed glass microscope slides coated in a thin, even layer of grease (petroleum jelly) to prevent the removal of dust from the slide during collection, transport and subsequent weighing. To ensure an even layer of grease the slide was gently heated until the grease had melted evenly over the slide surface. Post-weighing of the slides enabled the determination of dust soiling rates in mg m^{-2} day^{-1}. Control slides left in dust free enclosures open to the air demonstrated that there was no significant weight loss due to evaporation or drying of the grease, with this loss being less than 0.1 mg for a period of one month. Slides spiked with a known quantity of dust showed that there was no measurable loss of dust during transport of the slides. Therefore, any error in the gravimetric measurements can be assumed to be less then ±0.1 mg, being less than 5% of a typical dust loading of 2 to 3.5 mg. All weighing of glass slides was performed on a Mettler Toledo micro-balance with a readability of 0.1 mg. The locations of the slides inside the building foyers were away from radiators and air conditioning outlets/inlets. Cleaning staff were also instructed not to interfere with the slides. The cleaning regime is the same for each of the buildings and does not alter throughout the year. Therefore, the exposure environments of the slides from building to building was almost identical.

Settled dust for SEM and EDX analysis was collected on carbon SEM stubs with carbon low contaminant spectro-tabs attached (Supplied by Agar Scientific, Stansted, U.K.). The spectro-tabs are weakly adhesive on both sides hence they easily attach to the stubs and once dust particles settle, they adhere to the surface of the tab. Each stub was left in the appropriate location at approximately 1.5 m above ground for a month to enable the collection of sufficient dust for analysis. The stubs and slides were placed in locations where they were not interfered with by students and staff using the buildings.

Using SEM measurements of particle size, aspect ratio and chemistry (determined semi-quantitatively by EDX) were made of particles on each stub.

For individual particle analysis, all particles were randomly selected. Particle size and aspect ratio was measured by obtaining an image of 202 individual particles from each stub. Image analysis software (Scion Image) was then used to measure the particles and to determine aspect ratio. For 100 of these particles elemental content was determined semi-quantitatively by EDX. The Link Analytical EDX used for this analysis is capable of analysing for all elements with an atomic mass greater than neon.

3. Results and Discussion

It is not possible to directly relate the dust soiling rates published in other papers with the values measured here due to the wide range of methods used. However, other researchers have obtained comparable results of between 22 and 37 mg m^{-2} day^{-1} (Rufus *et al.*, 1998). For the month of April in Building 3 there is a high peak in the dust soiling rate. Although there was a small amount of construction work being carried out in the vicinity of this building, it is unlikely that this was the cause of this peak. A possible explanation is contamination of the slides during the exposure period leading to an artificial increase in the mass of dust collected. However, no obvious contamination mechanism was observed during the study. Dust soiling rates for the three buildings show no noticeable seasonal variation, with soiling rates for Building 3 being fairly constant after the removal of the

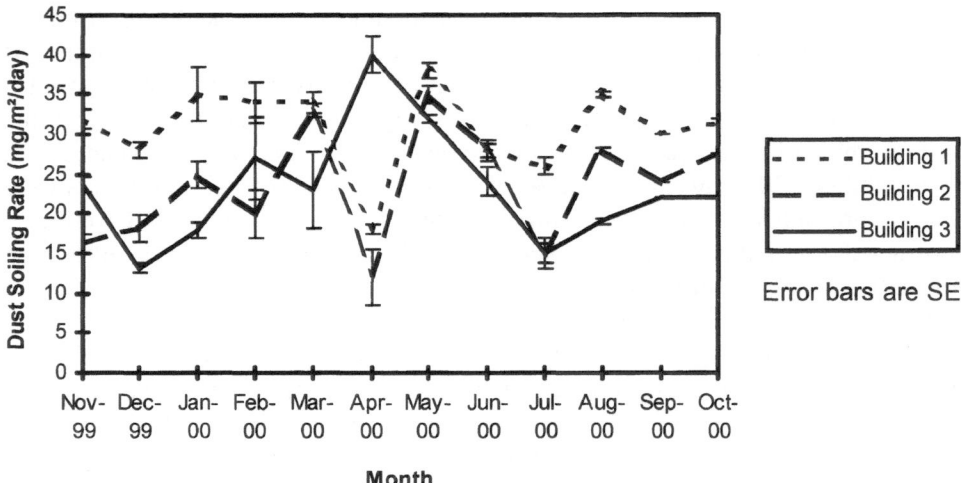

Figure 1. Variation in dust soiling rate for the three sampled buildings in the centre of Wolverhampton.

anomalous result for April (Figure 1). The soiling rates for Buildings 1 and 2 are more variable, but also follow no obvious seasonal trend.

Bate and Coppin (1990) have suggested a threshold limit of 200 mg m^{-2} day^{-1} for dust soiling to be considered a severe nuisance. While other literature propose threshold limits varying from 133 to 350 mg m^{-2} day^{-1}. When compared to these suggested thresholds the soiling rates measured in these buildings can be considered as relatively low. Results show that the foyer in Building 1 has a higher average dust soiling rate than the other two buildings. ANOVA analysis of the three sets of data indicate that this difference is statistically significant ($p < 0.05$, n = 12). After removal of the anomalous results for April this difference is still statistically significant ($p < 0.05$, n = 12). A likely explanation for the elevated dust soiling rates in Building 1 is due to the main door to this building being almost in constant use and also being poorly fitted leaving visible gaps around the door when closed. This allows air from outside to easily penetrate the building shell. When the building is not in use particulate matter can then continue to settle on surfaces inside the building. In contrast Building 2, which is located on the opposite side of the road, has a double door system which appears to offer a more efficient barrier to ambient particulates, therefore, reducing the rate of dust soiling inside the foyer. A paired *t*-test comparing the dust soiling rates inside Building 2 to the dust soiling rate inside Building 3, indicates no significant difference between the two sets of data, further suggesting the effectiveness of the double door system of Building 2 filtering out ambient particles generated within the street canyon.

However, although soiling rates are consistently greater in Building 1 than in Building 2, the soiling rates inside Buildings 1 and 2 are well correlated with a similar trend over time. This may indicate that the dust settling inside these two

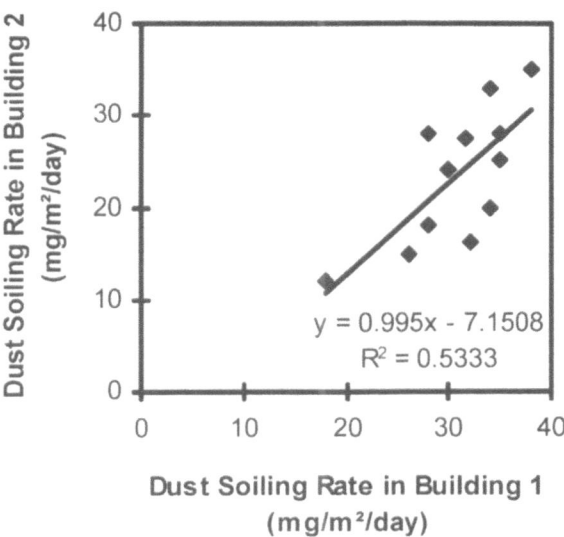

Figure 2. Regression analysis of dust soiling rate (mg m^{-2} day^{-1}) in Building 1 compared with dust soiling rate in Building 2.

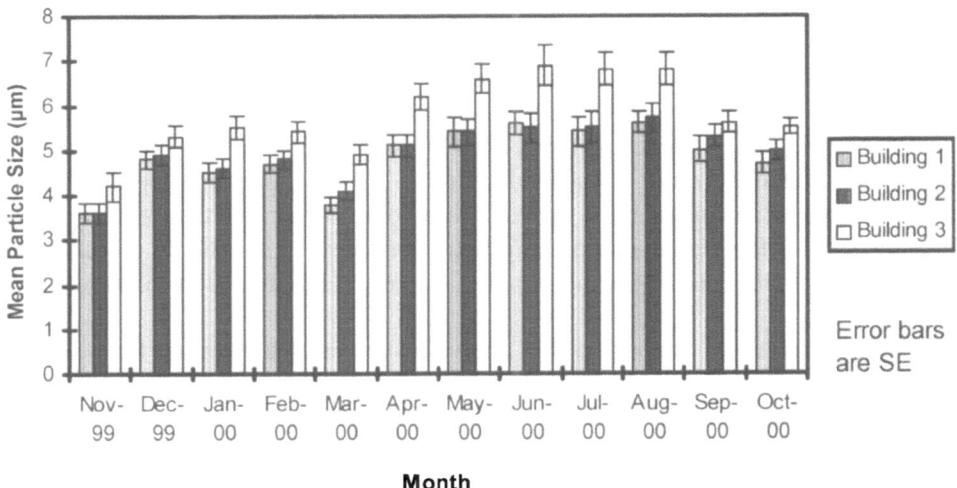

Figure 3. Monthly variation in mean particle size of settled dust in all three buildings.

buildings is from similar sources, likely to be re-suspended road dust and diesel exhaust particulate. A regression analysis of the dust soiling rates inside Building 1 against dust soiling rates inside Building 2 further illustrates this relationship (Figure 2), giving an R^2 value of 0.533 and with the removal of the anomalous results for April this correlation increases to 0.623 ($p < 0.02$, n = 12), showing a significant correlation.

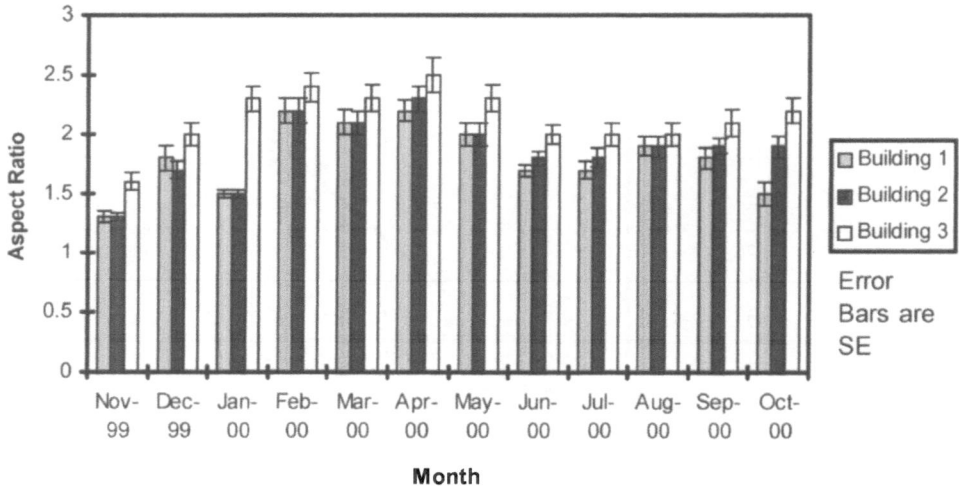

Figure 4. Monthly variation in aspect ratio for all three sampled buildings.

Settled dust particle size shows a seasonal variation with the average particle size being larger during the summer months than the winter months (Figure 3). Rufus *et al.* (1998) in a study of deposited dusts in houses found that larger particles were deposited during the summer. A *t*-test between winter samples (November, December and January) and summer samples (June, July and August) shows that the summer/winter difference is statistically significant for all three buildings ($p < 0.05$, n = 12). These larger particles possibly originate from wind blown soil or street dust, which were not re-suspended during the damper winter period. In the two buildings located adjacent to the bus route, for every sample the particle size is smaller than for the corresponding Building 3 samples. This is likely to be due to the presence of smaller diesel exhaust particles present in samples from Buildings 1 and 2.

The aspect ratio of the settled dust shows no discernible seasonal trend in any of the three buildings and is highly variable from month to month (Figure 4). However, the aspect ratio for settled dust inside Buildings 1 and 2 is lower than Building 3 in every sample. Visual examination of the electron micro-graphs of particles collected inside these two buildings show the presence of sub-micron, spherical particles (Plates 1 and 2). Particles of this type tend to be produced by combustion processes, with the only form of combustion in the local vicinity being the traffic on the adjacent road. Occasional agglomerations of very small particles were also observed inside Buildings 1 and 2 (Plate 3). Diesel particles often occur in this form (Berube *et al.*, 1999), hence this may indicate the presence of diesel particle contamination. Both the small spherical particles and the agglomerations were absent in the Building 3 samples. Regression analysis between the aspect ratio of particles inside Building 1 with the aspect ratio of particles inside Building 2 shows a very strong correlation between the two samples (Figure 5), giving an R^2

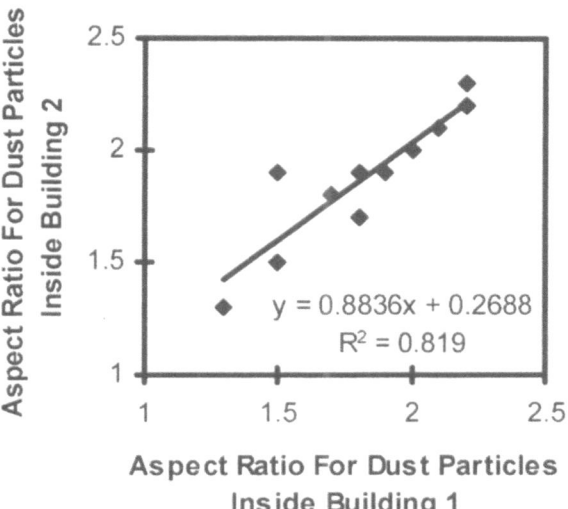

Figure 5. Regression analysis between the aspect ratio of Building 1 and the aspect ratio of Building 2.

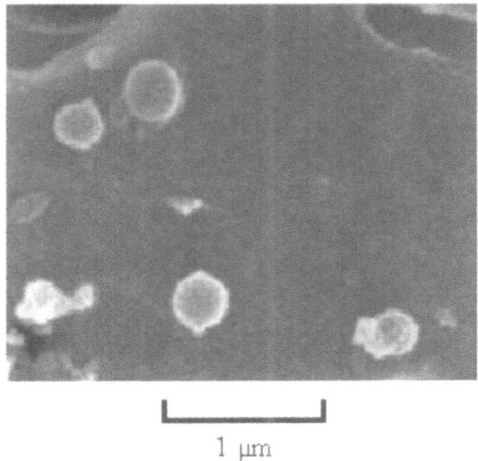

1 µm

Plate 1. Electron micro-graph of settled dust particles collected in Building 1.

value of 0.819 ($p < 0.001$, n = 12). This may also be indicating that the particles inside the two buildings originate from the same sources.

The influence of traffic related particulate matter on the indoor settled dust is clearly shown by the EDX analysis. The percentage of Na and Cl rich particles (defined as particles containing greater than 70% Na and Cl) present in the settled dust is highly dependent upon season (Figure 6). The percentage of Na and Cl rich particles present in the samples is far greater during the winter than the summer, with these particles likely to be originating from re-suspended rock salt applied

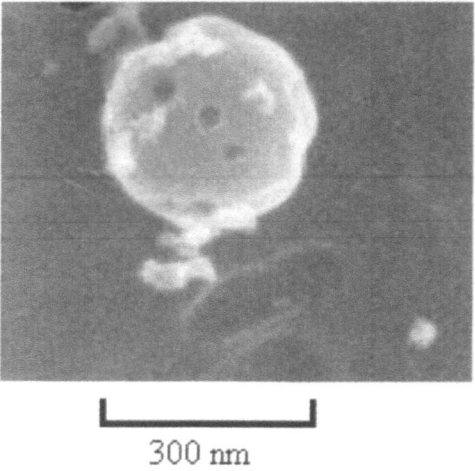

300 nm

Plate 2. Electron micro-graph of individual settled dust particle collected in Building 1.

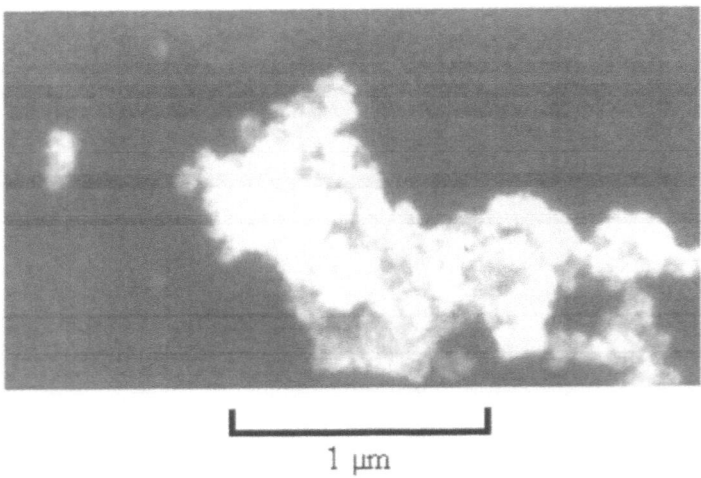

1 μm

Plate 3. Electron micro-graph of particle agglomeration inside Building 2.

to roads as a de-icing agent. The results also show that Na and Cl rich settled dust particles are present in equal amounts for all particle sizes, suggesting that re-suspended road dust can contribute a range of particle sizes to airborne particulate matter. It is worth noting that marine contributions to the presence of sodium and chlorine rich particles in these buildings is likely to be lower than in other U.K. cities as Wolverhampton is located approximately 125 km away from the nearest coastline.

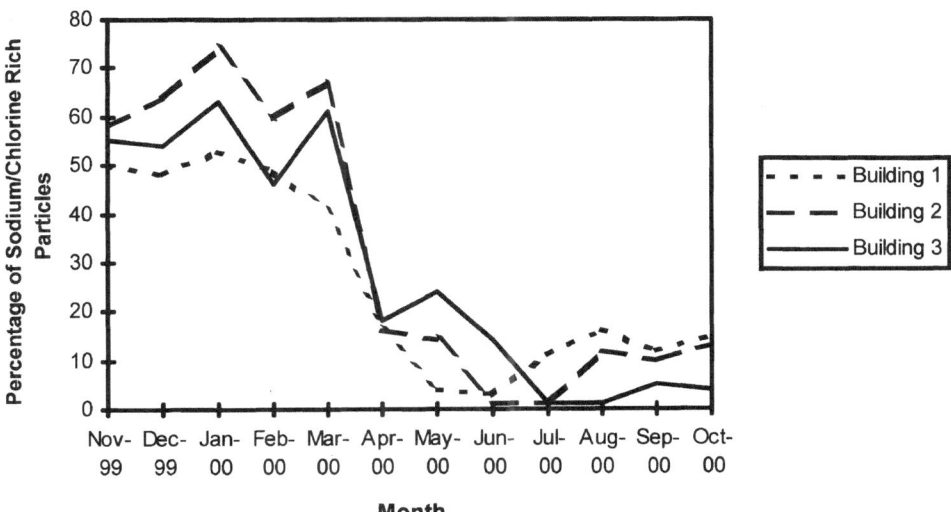

Figure 6. Monthly variation in the percentage of sodium and chlorine rich particles in each building.

4. Conclusion

The results presented here indicate that the proximity of a building to a busy road can affect both the chemical and physical characteristics of indoor settled dust, but may not affect the rate of dust soiling in terms of mg m^{-2} day^{-1}. Other factors may be of more importance, such as air exchange rate and building use.

Even though an increase in particle size was found during the summer months, there was not an associated increase in the dust soiling rate, as would be expected. This is likely to be because the increase of particle size did not cause a sufficient increase in mass to be detectable by the balance.

Acknowledgements

David Crane of the University of Wolverhampton for his patient assistance in using the scanning electron microscope. The Aerosol Society for kindly awarding the C.N. Davies Scholarship.

References

Bate, K. J. and Coppin, N. J.: 1995, 'Impact of Dust from Mineral Workings', *Paper presented to County Planning Officers Society Committee No. 3. Conference*, Loughborough, September 1990.

Berube, K. A., Jones, T. P., Williamson, C., Winters, A. J., Morgon, A. J. and Richards, R. J.: 1999, 'Physiochemical characterisation of diesel exhaust particles: Factors for assessing biological activity', *Atmosph. Environ.* **33**, 1599–1614.

Brooks, K. and Schwar, M. J. R.: 1987, 'Dust deposition and the soiling of glossy surfaces', *Environ. Pollut.* **43**, 129–141.

Ford, D. and Adams, S.: 1999, 'Deposition rates of particulate matter in the internal environment of two London museums', *Atmosp. Environ.* **33**, 4901–4907.

Giess, P.: 1998, 'Respirable particulates and oxides of nitrogen measured inside buildings alongside a busy road', *Indoor and Built Environ.* **7**, 308–314.

Gyntelberg, F., Suadicani, P., Nielson, J. W., Skov, P., Valbjorn, O., Nielson, P. A., Schneider, T., Jorgensen, O., Wolkoff, P., Wilkins, C. K., Gravesen, S. and Norn, S.: 1994, 'Dust and the sick building syndrome', *Indoor Air.* **4**, 223–238.

Quality of Urban Air Review Group: 1996, *Airborne Particulate Matter in the United Kingdom*, Department of the Environment, United Kingdom.

Raza, S. H., Nirmala, B. and Murthy, M. S. R.: 1990, 'Indoor dust fall and its composition in two public areas of a city in India', *Environ. Internat.* **16**, 53–56.

Rufus, D. E., Edward, J. Y. and Paul, J. L.: 1998, 'Seasonal deposition of housedusts onto household surfaces', *Sci. Total Environ.* **224**, 69–80.

Schwar, M. J. R.: 1998, 'Nuisance dust deposition and soiling rate measurements', *Environ. Technol.* **19**, 223–229.

Smith, B. J. and Warke, P. A.: 1995, 'Processes of Urban Stone Decay', *Proceedings of Swapnet '95 Stone Weathering and Atmospheric Pollution Network Conference*, The Queens University, Belfast, 1995.

Williams, I. D. and McCrae, I. S.: 1995, 'Road traffic nuisance in residential and commercial areas', *Sci. Total Environ.* **169**, 75–82.

SOURCE APPORTIONMENT IN THE TOWN OF LA SPEZIA (ITALY) BY CONTINUOUS AEROSOL SAMPLING AND PIXE ANALYSIS

SILVIA NAVA[1], PAOLO PRATI[1]*, FRANCO LUCARELLI[2],
PIER ANDREA MANDÒ[2] and ALESSANDRO ZUCCHIATTI[1]

[1] *Dipartimento di Fisica and INFN, Genova, Italy;* [2] *Dipartimento di Fisica and INFN, Firenze, Italy*

(author for correspondence, e-mail: prati@ge.infn.it, fax: +39 01031 4218)*

Abstract. We describe the results of an aerosol sampling campaign performed in 1999 in the medium-size industrial town of La Spezia, in the Northwest of Italy. We used two-stage continuous streaker samplers in three different sites and periods of the year. This kind of samplers allows the separation of the PM_{10} and $PM_{2.5}$ fractions of the particulate matter. Moreover, the hourly resolution in the aerosol collection is particularly useful in an urban environment where, typically, many pollution sources with fast variations are present. Up to 1700 samples have been analysed by Particle Induced X-ray Emission (PIXE) at the INFN accelerator facility in Florence, obtaining hourly concentration for about 20 elements from Na to Pb, with a sensitivity ranging from below 1 to about 10 ng m^{-3}. The total hourly aerosol mass has been estimated with an optical analysis of the same samples performed (before the PIXE analysis) by an equipment designed and mounted in Genoa. An extensive statistical analysis of the data included standard and Absolute Principal Component Factor Analysis (PCFA and APCFA) to deduce the composition and the weight of the major aerosol sources in both fractions. Thorough different statistical approaches, we generally resolved contributions from vehicle emission, fossil fuel combustion, soil-road dust and sea salt aerosol.

Keywords: APFCA, atmospheric pollution, optical methods, particulate matter (aerosol), PIXE

1. Introduction

The town of La Spezia (Italy) has about 100 000 inhabitants. It lies along the sea, at the centre of a narrow gulf. It faces the Ligurian sea (to the south, south-east) and is closely surrounded by the hills of the Appennino ligure (average height: 400 m) in the other directions. Possible pollution sources are vehicle emissions, electromechanical industries, a thermoelectric power plant and the large commercial and military harbour. The town is provided with some stations for air quality monitoring, but none of them is equipped with PM_{10} particulate samplers. Our campaign was the first study of the aerosol composition in this area: very preliminary results have been published elsewhere (Bongiovanni *et al.*, 2000); we present here a complete and deeper discussion, including the statistical analysis of the data. In the following we put on evidence the methodological aspects of our approach and we deduce information on the aerosol composition, within the limit of the relatively short sampling time.

Water, Air, and Soil Pollution: Focus **2:** 247–260, 2002.
© 2002 *Kluwer Academic Publishers.*

2. Methodology

We used two-stage streaker samplers provided by PIXE International Corporation. The samplers and the control units are described in detail elsewhere (Formenti *et al.*, 1996). Briefly, the streaker consists of a pre-impactor that stops particles with aerodynamic diameter $(D_{ae}) > 10$ μm, a thin Kapton film that collects by impaction particles with 2.5 μm $< D_{ae} < 10$ μm (coarse stage) and of a Nuclepore filter that stops all smaller particles (fine stage) with an efficiency very close to 1. The Kapton impaction plate and the Nuclepore filter are paired on a cartridge, which slowly rotates continuously, at an angular speed of about 45° day^{-1}. At the end of a week of sampling therefore, a circular continuous deposit ('streak') is produced on each of the two stages and can be analysed 'point by point' in order to reconstruct the hourly behaviours (total mass, composition, see below, of the collected particulate).

The aerosol sampling was performed during two periods. The first sampling campaign started on Wednesday 10 March 1999, and lasted for 21 days at two sites, both inside the town. The first site (Site A) was along a street immediately outside the precinct of the commercial harbour while the second one (Site B) was along the main street in the centre of La Spezia. The distance between the two sites was about 800 m. From Friday 17 to 30 September we sampled at a third site (Site C): a very narrow street in the downtown area, about 1 km west from site B. We exposed a streaker sampler in each site collecting a total of 8 pairs of frames (Nuclepore filter + Kapton impactor).

In order to measure the total deposited mass, streaker frames have been first analysed with an optical method, based on the hypothesis that, for thin samples, the visible light is exponentially attenuated with thickness. We used a simple device (Filippi *et al.*, 1999) which allows the determination of the thickness (μg cm^{-2}) of the deposition along the streaks with a sensitivity of 10 μg cm^{-2} and an accuracy of about 25%. The instrument has been calibrated by measuring several TSP, PM$_{10}$ and PM$_{2.5}$ samples, whose thickness was known by previous weighing. The method is not very accurate because the attenuation depends not only on the target thickness, but also on its composition; nevertheless this device makes possible to measure the hourly total mass concentration of aerosol with streaker samplers, which can not be obtained by gravimetric measurement.

PIXE analysis was performed with 3 MeV protons at the Van de Graaff accelerator of INFN, in the Physics Department of the University of Florence. Details of this external beam facility, routinely used for aerosol analysis, are given elsewhere (Del Carmine *et al.*, 1990). Only part of the exposed frames were analysed by PIXE: those collected in the first 14 days for Sites A and B and in the first week for Site C. The streak was moved in front of the beam line in steps of 1.25 mm, corresponding to 1 hour of aerosol sampling. For each spot we collected the emitted X-rays for 5 min: the scanning of one streaker stage required therefore about 15 hours of beam time. The beam current (about 8 nA) was integrated on a graphite Faraday cup behind the streaker. The sensitivity curve as a function of the detected element (X-ray counts $\mu C (\mu g\ cm^{-2})$) was obtained in the same geometry with a set of thin standards of certified areal densities (Micromatter Inc.). Spectra were fitted for 22 elements (Na, Mg, Al, Si, S, P, Cl, K, Ca, Ti, V, Cr, Mn, Fe, Co, Ni, Cu, Zn, Br, Sr, Zr, Pb) using the GUPIX software package (Maxwell et al., 1995). Minimum detection limits (MDL) were $\sim$10 ng m^{-3} for light elements and $\sim$1 ng m^{-3} (or lower) for medium and heavy elements. The uncertainty on the detected concentrations comes out to be usually around 5%, mainly coming from the calibration uncertainties. The errors are obviously higher when concentrations approach the MDL. Vanadium, Cr, Co, Sr, Zr were often below their MDL and were discarded in statistical analysis. No significant contamination was detected in the analysis of blank filters.

3. Results and Discussion

Figure 1 shows the aerosol total mass concentration determined by the optical analysis of the streaks. The time series show a quasi-periodic pattern peaked twice a day in all the sites, around hours 9 and 18, probably correlated with traffic intensity. The fine stage is about 80–90% of the total (fine + coarse stage) with average concentrations of 53 μg m^{-3} (Site A), 34 μg m^{-3} (Site B) and 82 μg m^{-3} (Site C), while the coarse stage average concentrations are: 11 μg m^{-3} (Site A), 5 μg m^{-3} (Site B) and 9 μg m^{-3} (Site C).

PIXE analysis produced elemental concentration time series as those shown in Figure 2. As can be seen, the quasi-periodic pattern peaked twice a day is enhanced in the Br and Pb trends. The concentration of Pb and Br in Italian towns is generally decreased in the last years, but the high sensitivity of PIXE analysis makes it possible to still detect these elements in short measurement times. In 1999, in Italy, more than 20% of cars still used leaded fuel; therefore Pb and Br can be considered good tracers for the traffic source (their concentration time series, indeed, are usually well correlated with the CO's (Lucarelli et al., 2000)). In Site C, on 22 September, a partial car ban was imposed in a limited number of roads around the site: in this occasion the daily average concentrations of Pb and Br resulted to be lower of about a factor two. Fine S pattern shows a very high slowly varying background. This element is present in the fine particulate mainly as sulphate (Cahill, 1995); the sulphate particles can be emitted directly from fossil fuel combustion process, but mainly they are produced by oxidation (in the atmosphere) of SO$_2$, which is also emitted from fossil fuel combustion process. Both fine stage sulphate particles and SO$_2$ have a long persistency time in atmosphere; therefore it is reasonable that local sources (diesel engines, domestic heating) and slowly varying regional background both contribute to the overall S concentration.

The aerosol composition, shown in Figure 3, was deduced by optical and PIXE analysis and is substantially the same in the three sites. The concentration of light

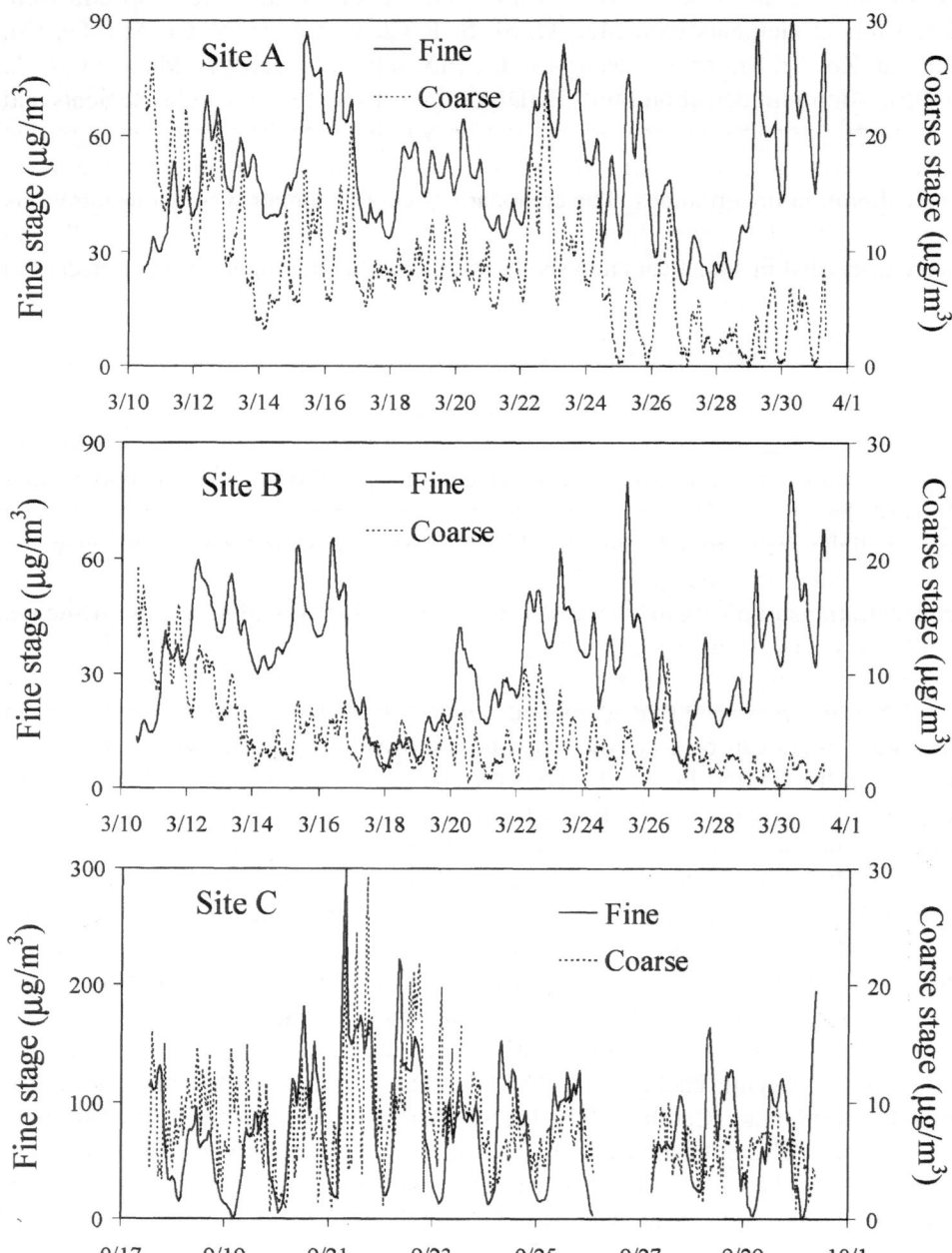

Figure 1. Total aerosol mass measured by the optical method in the fine (D_{ae} < 2.5 μm) and coarse (2.5 μm < D_{ae} < 10 μm) stages.

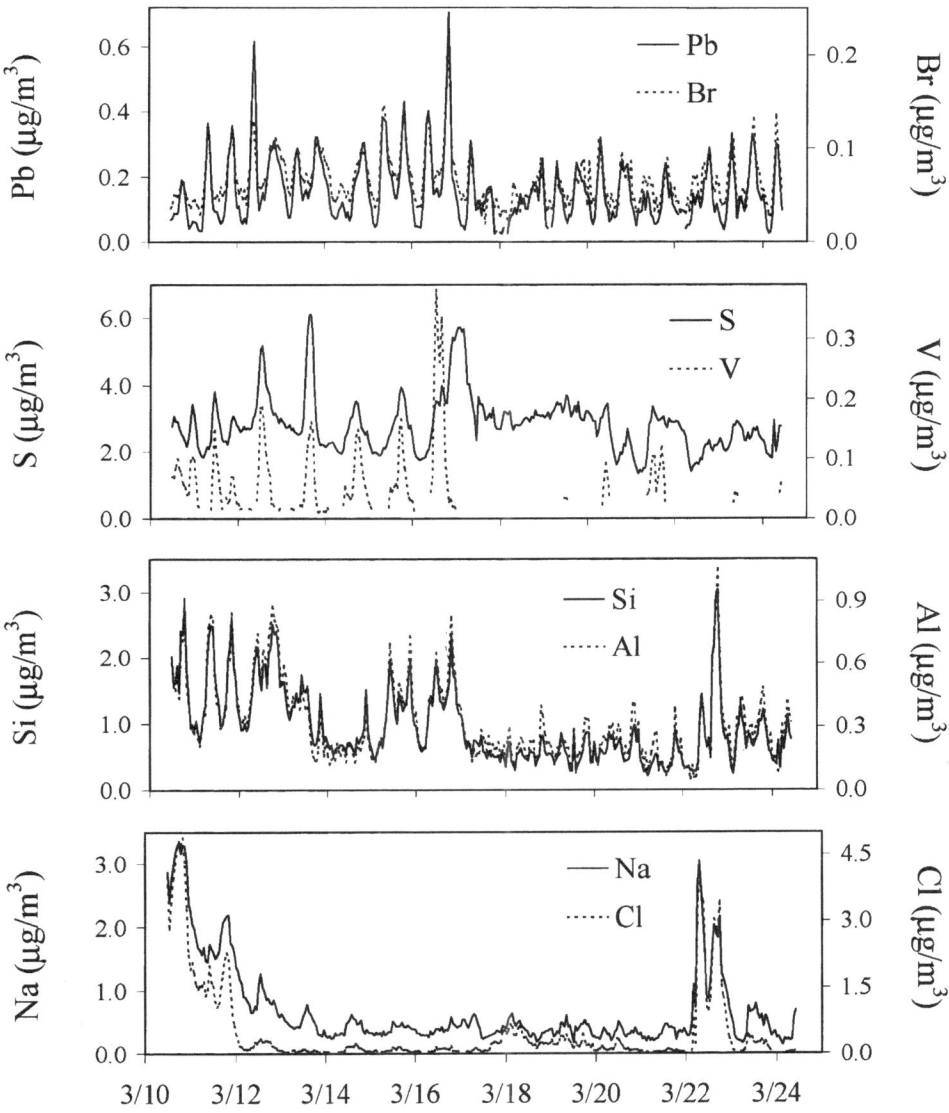

Figure 2. Concentrations in air of some elements measured in Site A, in the fine stage (Br, Pb, S, V) and in the coarse stage (Si, Al, Na, Cl).

elements (below Na), not detectable by PIXE, was evaluated by subtracting the sum of the elemental concentrations measured by PIXE to the total mass measured optically. It turned out that light elements account for about 85% and 45% of the total mass in the fine and coarse fraction, respectively.

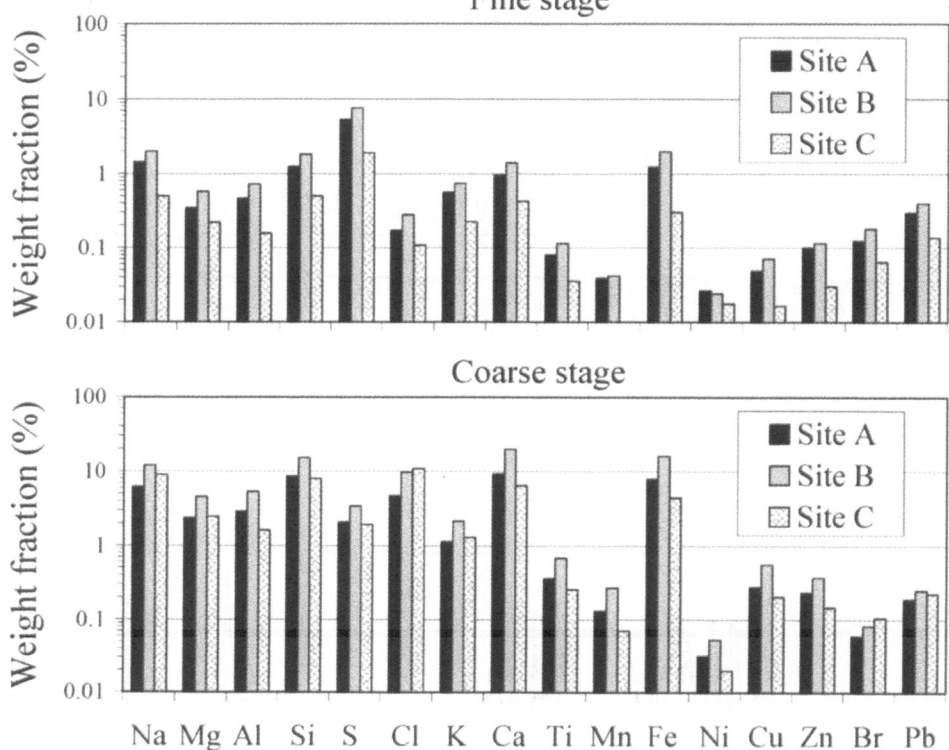

Figure 3. Average aerosol composition in the fine (top) and coarse (bottom) stages.

3.1. STATISTICAL ANALYSIS

We identified aerosol sources using PCFA with VARIMAX rotation on the elemental concentrations in the fine and coarse stages (Heidam, 1982). Briefly, in this analysis the correlation matrix of the elemental concentrations is diagonalized, in order to determine uncorrelated linear combinations of them. These linear combinations are called factors and can be associated to the aerosol sources by the analysis of the factor loadings, which are the correlation coefficients among the factors and the elemental concentrations. If z_{ij} is the standardized concentration of the element i in the hour j at the receptor site, g_{kj} is the value of the standardized factor k relative to hour j and b_{ik} is the factor loading of factor k relative to element i, it is assumed that:

$$z_{ij} \cong \sum_{k=1}^{p} b_{ik} \cdot g_{kj} , \tag{1}$$

$$b_{ik} = \sqrt{\lambda_k} \cdot \lambda_{ik} , \tag{2}$$

$$g_{kj} = \frac{\sum_{i=1}^{n} \lambda_{ki} \cdot z_{ij}}{\sqrt{\lambda_k}} \, , \tag{3}$$

where λ_k is the k-th eigenvalue of the correlation matrix of the elemental concentrations and λ_{ki} is the component i of the k-th eigenvector (normalised to one). The sum is extended on the p factors with greatest eigenvalues, with p lower than the number n of elements included in the analysis: this corresponds to consider only the factors explaining a large fraction of the variance in the data set (time series of elemental concentrations). The VARIMAX rotation is an orthonormal transformation of the selected factors that keeps the factors uncorrelated and makes their interpretation as aerosol sources easier. According with Heidam (Heidam, 1982), good estimates for the factor loading standard deviations are:

$$\sigma_i(b_{ik}) = \sqrt{\frac{1 - h_i^2}{n - p}} \, , \tag{4}$$

where h_i^2 is the 'communality' of the element i, i.e. the variance of the element i explained by the selected factors ($h_i^2 = \Sigma_k b_{ik}^2$); the standard deviations do not depend on the factors and do not change after VARIMAX rotation since this transformation maintains the communalities.

The results of this analysis are reported in Tables I (fine stage) and II (coarse stage). In the fine stage three factors explain about 80% of the total data variance in all three sites. We identified them as: soil dust (soil), vehicle emission (traffic) and fossil fuel combustion (oil). In the coarse stage the factor identified as 'oil' is not detectable and we resolved a new different source: the sea salt aerosol (sea). In Site B 'soil' and 'traffic' got mixed; this is probably related to the site position in the town centre, far from any green area: most of the soil dust, in this case, could be road dust re-suspended by vehicles. Factor 'traffic' is mainly characterised by Pb and Br in both stages, but also by Cu, which is emitted from diesel engines (Swietlicki et al., 1996). Factor 'soil' is characterised by crustal and soil-related elements, like Al, Si, and Ca (in both stages), Mg (in the fine stage), Ti and Fe (in the coarse stage). Zincum, which is known to be abundant in tyre wear particles, comes out to be well correlated with factor 'soil' in the coarse stage. Factor 'oil' is characterised by S, whose gaseous precursor, SO_2, is mainly emitted from fossil fuel combustion process; Ni, strongly correlated with this factor in Site A, is a tracer element for oil combustion.

The PCFA gives only qualitative indications about the sources: to obtain quantitative results we performed also an Absolute PCFA (APCFA). The main purpose of this statistical model is to determine the absolute weights of the aerosol sources at the receptor site, just by the knowledge of the concentrations measured at the same site, that is to say without the use of source inventory and source composition data sets. The model itself generates a hypothesis regarding the number of sources and their composition (source *profile*). It should be noted that the calculated profiles

TABLE I

VARIMAX-rotated loadings obtained by a PCFA of the fine stage concentration data set

Source	Site A				Site B				Site C			
	Soil	Traffic	Oil	St. dev.	Traffic	Soil	Oil	St. dev.	Soil	Traffic	Oil	St. dev.
Mg	0.73	0.23	-0.23	0.19	-0.01	0.82	-0.14	0.17	0.89	0.10	-0.11	0.14
Al	0.77	0.47	0.06	0.13	0.51	0.69	0.23	0.15	0.94	0.24	0.12	0.08
Si	0.86	0.39	0.08	0.10	0.50	0.68	0.22	0.16	0.94	0.10	0.27	0.07
S	-0.09	0.1	0.83	0.17	-0.05	-0.10	0.91	0.12	0.13	-0.23	0.80	0.18
K	0.18	0.71	0.25	0.20	0.71	0.20	0.48	0.15	-0.04	0.19	0.89	0.14
Ca	0.82	0.44	0.03	0.11	0.32	0.87	0.02	0.12	0.89	0.20	-0.10	0.13
Mn	0.87	0.15	0.26	0.13	0.47	0.61	0.33	0.18	–	–	–	–
Fe	0.59	0.76	0.03	0.08	0.90	0.10	-0.04	0.14	0.79	0.41	-0.05	0.15
Ni	0.39	-0.08	0.72	0.18	0.50	0.32	0.40	0.22	-0.38	0.66	0.24	0.20
Cu	0.24	0.9	0	0.11	0.89	0.29	0.06	0.11	0.43	0.77	0.13	0.15
Zn	0.78	0.1	0.42	0.15	0.42	0.47	0.53	0.18	0.19	0.87	0.33	0.11
Br	0.22	0.93	-0.07	0.09	0.77	0.46	0.09	0.14	0.26	0.82	-0.28	0.14
Pb	0.24	0.9	-0.09	0.11	0.71	0.49	0.11	0.15	0.31	0.76	-0.18	0.18
Variance	35%	32%	12%	–	33%	29%	13%	–	36%	30%	15%	–

TABLE II

VARIMAX-rotated loadings obtained by a PCFA of the coarse stage concentration data set

Source	Site A				Site B			Site C			
	Soil	Sea	Traffic	St. dev.	Soil-tr.	Sea	St. dev.	Soil	Sea	Traffic	St. dev.
Na	0.20	0.96	0.00	0.05	−0.04	0.97	0.07	0.06	0.95	0.14	0.08
Mg	0.54	0.72	0.25	0.10	0.33	0.80	0.14	0.51	0.83	0.02	0.06
Al	0.90	0.17	0.33	0.07	0.84	0.42	0.10	0.86	0.44	−0.01	0.08
Si	0.88	0.27	0.31	0.07	0.82	0.51	0.07	0.92	0.29	0.07	0.07
S	0.39	0.88	0.09	0.07	0.25	0.94	0.06	0.23	0.89	0.35	0.06
Cl	0.15	0.96	0.03	0.07	−0.07	0.94	0.09	−0.10	0.92	0.32	0.06
K	0.75	0.53	0.24	0.09	0.63	0.72	0.08	0.50	0.78	0.01	0.12
Ca	0.85	0.43	0.21	0.06	0.74	0.59	0.09	0.90	0.34	−0.12	0.08
Ti	0.85	0.28	0.25	0.10	0.75	0.46	0.13	–	–	–	–
Mn	0.84	0.30	0.33	0.09	0.74	0.37	0.16	–	–	–	–
Fe	0.72	0.08	0.64	0.07	0.83	−0.12	0.15	0.85	0.00	0.45	0.09
Ni	0.47	0.51	0.24	0.19	0.55	0.05	0.23	0.63	0.28	0.32	0.20
Cu	0.49	−0.11	0.77	0.11	0.85	−0.07	0.14	0.74	−0.12	0.56	0.10
Zn	0.67	0.35	0.15	0.18	0.79	0.40	0.13	0.87	0.04	0.37	0.10
Br	0.13	0.31	0.88	0.09	0.40	0.67	0.17	0.01	0.50	0.83	0.07
Pb	0.32	0.06	0.87	0.10	0.74	0.42	0.15	0.17	0.24	0.91	0.09
Variance	40%	28%	19%	–	41%	38%	–	39%	31%	17%	–

do not necessarily reproduce the source profiles at the emission points, because they include transformations taking place during long range transport (like oxidation of gaseous SO_2 to sulphates). If x_{ij} is the measured concentration of element i in the j-th hour, a_{ik} the fraction of element i in the source k composition at the receptor site (source *profile*) and f_{kj} the concentration of the aerosol (at the same site) produced by source k during hour j (*weight* of the source k in the j-th hour), it is assumed that:

$$x_{ij} \cong \sum_{k=1}^{p} a_{ik} \cdot f_{kj} , \tag{5}$$

Using Equation (1) and recalling that z_i and g_k are the standardized versions of x_i and f_k, it can be proved (Swietlicki et al., 1996) that:

$$a_{ik} = \sigma_k^{-1} \cdot \sigma_i \cdot b_{ik}^R \tag{6}$$

$$f_{kj} = \sigma_k \cdot \sum_{s=1}^{p} \sum_{i=1}^{n} u'_{ks} \cdot \lambda_s^{-1} \cdot b'_{si} \cdot \sigma_i^{-1} \cdot x_{ij} , \tag{7}$$

TABLE III

Average source weights, obtained by an APCFA (Methods I and II) of the concentration data set; the values are expressed as percentages (%) of the measured total mass

	Fine stage						Coarse stage					
	Site A		Site B		Site C		Site A		Site B		Site C	
	I	II	I	II	I	II	I	II	I	II	I	II
Traffic	34	28	19	21	73	63	17	17	–	–	–	–
Soil	9	8	44	33	31	34	42	39	–	–	71	60
Traffic + soil	–	–	–	–	–	–	–	–	53	55	–	–
Sea	12	9	14	16	–	–	17	15	43	40	1	4

where σ_k and σ_i are the standard deviations of factor k and element i concentration, b_{ik}^R are the VARIMAX rotated loadings, b_{is} are the loadings before rotation, u_{sk} are the rotation matrix elements and the apex indicates the trasposed matrix. The standard deviations σ_k, however, are not known explicitly and cannot be determined unless the total aerosol mass is also measured. Without this further data item, it is possible to calculate just the $a_{ik}^* \equiv a_{ik}\sigma_k$ and the $f_{kj}^* \equiv f_{kj}\sigma_k^{-1}$.

In another method to determine the f_{kj}^*, a virtual sample (labelled with $j = 0$) in which all concentrations are zero ($x_{io} = 0$) is added to the data set. It can be shown (Thurston and Spengler, 1985) that the f_{kj}^* can be expressed in terms of the rotated standardized factors (obtained by a standard PCFA) by the following equation:

$$f_{kj}^* = g_{kj}^R - g_{k0}^R . \tag{8}$$

To include the aerosol total mass concentration in the analysis it is necessary to consider that it is not an independent variable like the measured elemental concentrations (because it is partially due to their sum). However, in many cases it can be considered to be independent with good approximation, because the sum of the measured elemental concentrations is typically less than 20–30% of the total. In order to calculate the source profiles and weights, we followed two methods, in which the total aerosol mass is considered as an independent and as a dependent variable, respectively. In the following we refer to the two methods as Method I and II.

In Method I, we included the total mass concentration in the PCFA; we found that it does not influence significantly the factor loading values, which came out to be equal to those reported in Tables I and II within one standard deviation. Using the rotated factor loading relative to the total mass ($b_{M,k}^R$), we calculated the standard deviations of the factors:

$$1 = a_{M,K} = \sigma_M b_{M,k}^R / \sigma_k \ \Leftarrow \ \sigma_k = \left(\sigma_M b_{M,k}^R \right) . \tag{9}$$

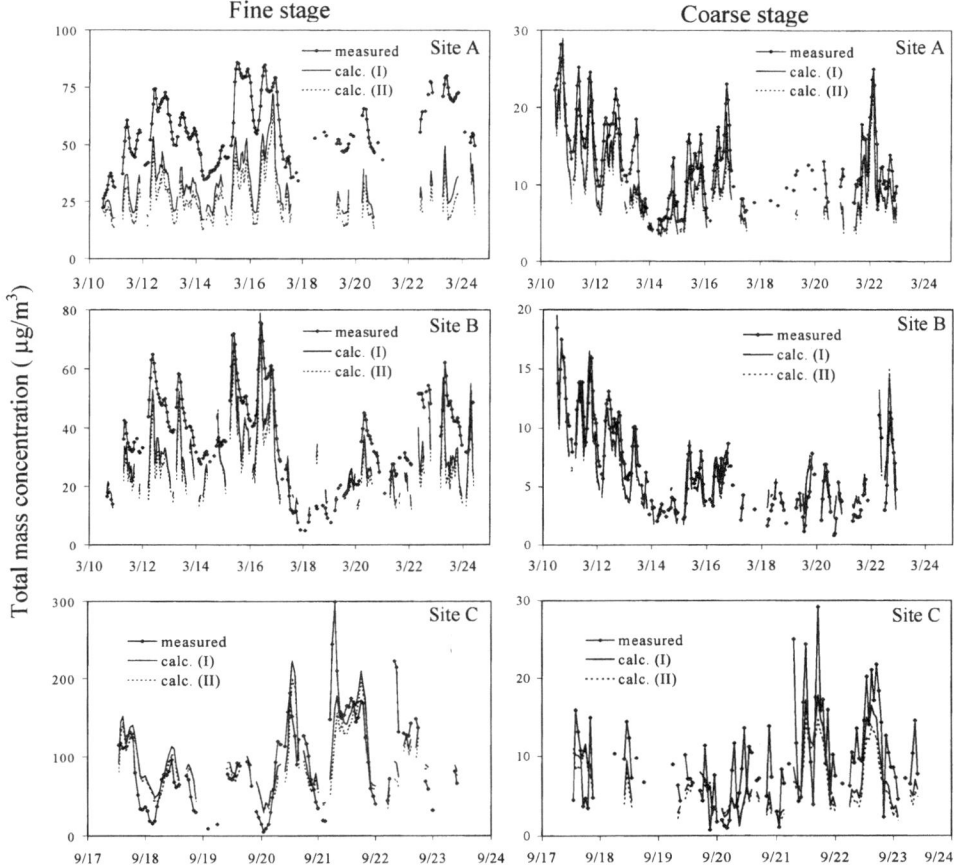

Figure 4. Comparison between measured (optically) and calculated total mass concentrations: Calc. (I) and Calc. (II) are the mass concentrations calculated by Methods I and II, respectively.

We calculated the source profiles and weights using these standard deviations and the Equations (6) and (7).

In Method II, we calculated the source weights by a multiple linear regression with the total aerosol mass used as a dependent variable and the f_k^* obtained by the Equation (8) as independent variables:

$$x_{PM,j} = c + \Sigma_k c_k f_{kj}^* \tag{10}$$

$$f_{kj} = c_k f_{kj}^* . \tag{11}$$

Then we determined the source profiles by n multiple linear regressions (one regression for each detected element) with the elemental concentrations as dependent variables and the source weights as independent variables:

$$x_{ij} = a_{io} + \Sigma_k a_{ik} f_{kj} . \tag{12}$$

TABLE IV

Source profiles obtained by an APCFA (Method I) of the fine stage concentration data set

Source	Site A			Site B			Site C	
	Soil	Traffic	Oil	Traffic	Soil	Oil	Soil	Traffic
Mg	0.01	0.003	–	–	0.006	–	0.005	–
Al	0.02	0.01	–	0.01	0.01	0.02	0.005	0.001
Si	0.05	0.02	0.01	0.03	0.02	0.04	0.01	–
S	–	0.01	0.26	–	–	0.34	0.007	–
K	–	0.01	0.01	0.01	–	0.03	–	0.0004
Ca	0.05	0.03	–	0.02	0.03	–	0.02	–
Mn	0.002	–	0.001	0.001	–	0.001	–	–
Fe	0.03	0.04	–	0.06	–	–	0.01	–
Ni	0.001	–	0.004	–	–	0.001	–	0.001
Cu	–	0.002	–	0.002	–	–	0.0002	0.0002
Zn	0.006	0.001	0.008	0.002	0.001	0.007	0.0001	0.0003
Br	0.001	0.004	–	0.003	0.001	0.001	0.0003	0.0005
Pb	0.004	0.013	–	0.011	0.005	0.006	0.001	0.002
Sum	0.17	0.14	0.3	0.14	0.08	0.44	0.06	0
Si/Al	2.3	–	–	–	2.4	–	2.8	–
Br/Pb	–	0.27	–	0.29	–	–	–	0.32

The average source weights, expressed as percentages of the measured total mass, are reported in Table III. The two methods give similar results for all the sampling sites. For the fine stage, in Site C the source apportionment shows high weights for traffic and soil dust, as we expected since the site was along a very narrow, heavy traffic urban road. In Sites A and B a significant contribution from oil combustion is present, possibly linked to the activities of the harbour nearby. Moreover, Site A is the nearest to the thermoelectric power plant located in the east area of the town. In the coarse stage the marine aerosol source turns out to be much more feeble in Site C, which is in fact the farthest from seaside and also the least airy. In Figure 4 the comparison is reported between the optically measured aerosol mass concentration and the mass concentrations calculated by the two methods; the agreement between the two methods is good, but they both give a lower value for the mass with respect to the optically measured one, especially in the fine stage. We remember that the optical method gives simply an estimate of the total mass with an accuracy of about the 25%. Furthermore secondary nitrates, elemental carbon and organic carbon compounds, which constitute a rather large part of ambient aerosol, are not detected by PIXE.

TABLE V

Source profiles obtained by an APCFA (Method I) of the coarse stage concentration data set

Source	Site A			Site B		Site C	
	Soil	Sea	Traffic	Soil-tr.	Sea	Soil	Sea
Na	0.04	0.22	–	–	0.22	0.01	0.24
Mg	0.03	0.04	0.03	0.03	0.04	0.02	0.05
Al	0.05	0.01	0.04	0.07	0.02	0.02	0.02
Si	0.13	0.05	0.12	0.23	0.09	0.1	0.05
S	0.03	0.07	0.02	0.02	0.05	0.01	0.05
Cl	0.04	0.32	0.02	–	0.32	–	0.47
K	0.02	0.01	0.01	0.03	0.02	0.01	0.02
Ca	0.2	0.12	0.13	0.35	0.17	0.16	0.09
Ti	0.007	0.003	0.005	0.01	–	–	–
Mn	0.003	0.001	0.003	0.005	0.001	–	–
Fe	0.1	0.01	0.22	0.27	–	0.06	–
Ni	–	–	–	0.001	–	–	–
Cu	0.003	–	0.01	0.009	–	0.002	–
Zn	0.004	0.003	0.002	0.007	0.002	0.002	–
Br	–	–	0.002	0.001	0.001	–	0.002
Pb	0.001	–	0.008	0.004	0.001	0.001	0.002
Sum	0.65	0.87	0.61	1.04	0.94	0.39	0.98
Na/Cl	–	0.69	–	–	0.68	–	0.5
Si/Al	2.9	–	–	3.1	–	4.4	–
Br/Pb	–	–	0.28	–	–	–	–

In Tables IV and V the source profiles calculated by Method I are reported (the source profiles calculated by Method II came out to be equal within 20%). The source profiles turn out to be similar for Sites A and B, while quite different values are found for Site C. It should be said that sampling in this site took place in quite peculiar situations: on Monday 20 September, there was a very strong rain, so hard that it even prevented the opening of the schools (after the summer holidays); on Tuesday 21 September, schools opened and on Wednesday 22 September, the mentioned car ban was imposed. The Br/Pb concentration ratio in the 'traffic' profiles resulted quite similar in all the sampling sites, with values in the range 0.27–0.32, in agreement with the values reported in literature (Thurston and Spengler, 1985). The Na/Cl concentration ratio in the 'sea' profile in both Sites A and B (0.69 and 0.68, respectively) results similar to what expected for NaCl. The Si/Al concentration ratio in the 'soil' profiles got values in the range 2.3–4.4, compatible with what expected in typical rock composition.

4. Conclusions

The first $PM_{2.5}$ and PM_{10} characterisation in the town of La Spezia was performed in 1999 by continuous streaker samplers. The use of streakers and subsequent PIXE analysis was a good method to obtain the hourly concentrations of several elements in the aerosol. Moreover, the optical analysis of the streaks gave a useful estimate of the hourly concentration of the total mass. Receptor model statistical analysis of this kind of data produced interesting information, as the weights and the profiles of the identified aerosol sources. The main purpose in this work was to test different statistical approaches, which actually gave compatible results. Future methodological improvements and wider sampling campaigns are however necessary for a more accurate and statistically significant aerosol characterisation.

References

Bongiovanni, S. F., Prati., P., Zucchiatti, A., Lucarelli, F., Mandò, P. A., Ariola, V. and Bertone, C.: 2000, 'Study of the aerosol composition in the town of La Spezia with continuous sampling and PIXE analysis', *Nucl. Instr. and Meth.* **B161–163**, 786–791.

Cahill, T. A.: 1995, 'Aerosol Collection and Compositional Analysis for Improve', *NPS Annual Report* (July 1994–June 1995), Department of Physics, University of California, 46 pp.

Del Carmine, P., Lucarelli, F., Mandò, P. A., Moscheni, G., Pecchioli, A. and MacArthur, J. D.: 1990, 'PIXE measurements of air particulate elemental composition in the urban area of Florence, Italy', *Nucl. Instr. and Meth.* **B45**, 341–346.

Filippi, E., Prati, P., Zucchiatti, A., Lucarelli, F., Mandò, P. A., Ariola, V. and Corvisiero, P.: 1999, 'Hourly measurement of particulate concentrations with streaker samplers and optical methods', *Nucl. Instr. and Meth.* **B150**, 370–374.

Formenti, P., Prati, P., Zucchiatti, A., Lucarelli, F. and Mandò, P. A.: 1996, 'Aerosol study in the town of Genova with a PIXE analysis', *Nucl. Instr. and Meth.* **B113**, 359–362.

Heidam, N. Z.: 1982, 'Atmospheric aerosol factor models, mass and missing data', *Atmosph. Environ.* **16**(8), 1923–1931.

Lucarelli, F., Mandò, P. A., Nava, S., Prati, P. and Zucchiatti, A.: 2000, 'Elemental composition of urban aerosol collected in Florence, Italy', *Nucl. Instr. and Meth.* **B161–163**, 819–824.

Maxwell, J. A., Teesdale, W. J. and Campbell, J. L.: 1995, 'The Guelph PIXE package II', *Nucl. Instr. and Meth.* **B95**, 407–421.

Swietlicki, E., Puri, S. and Hansson, H. C.: 1996, 'Urban air pollution source apportionment using a combination of aerosol and gas monitoring techniques', *Atmosph. Environ.* **30**(15), 2795–2809.

Thurston, G. D. and Spengler, J. D.: 1985, 'A quantitative assessment of source contributions to inhalable particulate matter pollution in metropolitan Boston', *Atmosph. Environ.* **19**(1), 9–25.

ROADWAY-2: A MODEL FOR POLLUTANT DISPERSION NEAR HIGHWAYS

K. SHANKAR RAO

Atmospheric Turbulence and Diffusion Division, Air Resources Laboratory, NOAA, Oak Ridge, TN, U.S.A.
(E-mail: rao@atdd.noaa.gov, fax: 865 576 1327)

Abstract. The formulations and evaluation of ROADWAY-2, a near-highway pollutant dispersion model, are described. This model incorporates vehicle wake parameterizations derived from canopy flow theory and wind tunnel measurements. The atmospheric velocity and turbulence fields are adjusted to account for velocity-deficit and turbulence production in vehicle wakes. A turbulent kinetic energy closure model of the atmospheric boundary layer is used to derive the mean velocity, temperature, and turbulence profiles from input meteorological data. ROADWAY-2 has been evaluated using SF_6 tracer data from General Motors Sulfate Dispersion Experiment. The model evaluation results are presented and discussed.

Keywords: air pollution, dispersion model, wake parameterizations, SF_6 tracer data, model evaluation

1. Introduction

Pollutants such as carbon monoxide, nitrogen oxides, and particles emitted from motor vehicles contribute to the degradation of air quality near major highways. Concern for the resulting health hazards guided regulations in the U.S., leading to the development of mathematical dispersion models for estimating air quality near the existing or proposed roads. These models are of varying complexity and sophistication, ranging from simple line-source Gaussian plume models to elaborate numerical models in two and three dimensions. The open-highway Gaussian models are easy to use, even with modifications for non-ideal situations, and require only simple meteorological input data. However, these models have well-known limitations such as the inability to deal with complex wind fields or near-calm conditions. The grid-based numerical models are capable of handling light winds and accounting for the variability of terrain, flow and diffusion conditions near highways, but these models are complex and often require elaborate inputs.

Among the difficulties in highway dispersion modeling are the specification of emissions and meteorological input data. The emission rates are highly variable depending on the vehicle type and its operating mode, but are often approximated by a constant aggregate value. In addition, many models do not account for vehicle-induced turbulence and buoyancy, though their importance for the prediction of dispersion near highways is well-documented in the literature. For example, Rao *et al.*(1979) showed that moving traffic produces an augmentation of energy in the high frequency end of the spectra and cospectra of the velocity components. Only a few research-grade field studies, such as the General Motors (GM) experiment (Cadle *et al.*, 1976), were performed near highways where detailed emissions, concentrations, and meteorological data were collected for model evaluation.

This article describes the formulations of the mean variables and turbulence fields, and the evaluation of a near-highway dispersion model, ROADWAY-2, at the Atmospheric Turbulence and Diffusion Division under a collaborative research agreement with the Federal Highway Administration (FHWA). This work is based on U.S. EPA's ROADWAY (Eskridge and Catalano, 1987) model, which used mean velocity and temperature profiles and eddy diffusivities based on the equilibrium surface layer similarity theory. It also incorporated vehicle wake formulations derived by Eskridge and Hunt (1979) from a perturbation solution to the equations of motion. The wake effects were linearly superposed on the atmospheric velocity and eddy diffusivity fields. ROADWAY-2 model incorporates an atmospheric boundary layer (ABL) model with turbulent kinetic energy (TKE) closure and up-to-date surface parameterizations to derive the mean and turbulence profiles from input meteorological data. The atmospheric mean velocities and TKE are adjusted to account for the velocity deficit and turbulence production in vehicle wakes. The wake parameterizations in ROADWAY-2 are derived from the vegetation canopy flow theory (e.g., Kaimal and Finnigan, 1994) and wind tunnel measurements in vehicle wakes by Eskridge and Thompson (1982). The model is evaluated using data from the GM experiment (Cadle *et al.*, 1976).

2. Model Formulations

ROADWAY-2 model numerically solves a two-dimensional equation for pollutant species concentration C:

$$\partial C/\partial t + \partial(uC)/\partial x + \partial(wC)/\partial z =$$
$$\partial(K_x\partial C/\partial x)/\partial x + \partial(K_z\partial C/\partial z)/\partial z + S_e + R, \tag{1}$$

with the boundary conditions:

$$\partial C/\partial z = 0 \text{ at } z = 0, \quad C = C_B \text{ at inflow and top,}$$
$$\partial C/\partial x = 0 \quad \text{at outflow boundary.} \tag{2}$$

In the above, S_e and R are the emissions and chemical reactions terms, respectively, and C_B is the background concentration. The y axis is oriented along the roadway, and the derivative parallel to the road (i.e., variation in the direction of travel) is assumed to be zero. This is justified since sufficient information to perform a three-dimensional analysis of the wind field along a roadway is seldom available. The numerical grid varies depending on the wind speed and direction. The logarithmic vertical grid extends from $z = 1$ m at the surface to a height of 70 m at the top. The horizontal grid is internally generated to cover all traffic lanes, with a higher resolution near the road. Numerical solution of Equations (1) and (2) is obtained by a fractional step finite-difference method (Marchuk, 1975). The advection steps are based on a flux-corrected transport algorithm (Zalesak, 1979) with an upstream difference scheme that ensures non-negative concentration values.

2.1. ABL MODEL

The mean velocity and temperature profiles used in Equation (1) are obtained from a 1-D numerical ABL model:

$$\partial u/\partial t = \partial(K_m \partial u/\partial z)/\partial z + f(v - v_g), \tag{3}$$

$$\partial v/\partial t = \partial(K_m \partial v/\partial z)/\partial z - f(u - u_g), \tag{4}$$

$$\partial \theta/\partial t = \partial(K_h \partial \theta/\partial z)/\partial z, \tag{5}$$

where u and v are horizontal mean velocity components along x (east-west) and y (north-south) directions, u_g and v_g are the corresponding geostrophic wind components, f is the Coriolis parameter, and θ is the mean potential temperature.

The eddy diffusivities for momentum and heat (K_m and K_h) are calculated from a turbulent kinetic energy closure model described by Rao and Snodgrass (1981):

$$K_{m,h} = \ell_{m,h}(aE)^{1/2}, \; \ell_{m,h} = kz/\phi_{m,h}, \tag{6}$$

where ℓ_m and ℓ_h are turbulent mixing lengths for momentum and heat, $a = 0.2$ is a proportionality constant and $k = 0.4$ is the von Kármán constant. The TKE (E) is calculated from the prognostic equation:

$$\partial E/\partial t = \partial(K_m \partial E/\partial z)/\partial z + K_m \left[(\partial u/\partial z)^2 + (\partial v/\partial z)^2\right]$$
$$-(g/\theta_o)K_h \partial \theta/\partial z - (aE)^2/K_m. \tag{7}$$

The nondimensional mean velocity and temperature gradients (ϕ_m and ϕ_h) are specified from the boundary layer similarity relations:

For $L < 0$,

$$\phi_m(z/L) = (1 - \gamma_m z/L)^{-1/4}, \; \phi_h(z/L) = Pr_t(1 - \gamma_h z/L)^{-1/2}, \tag{8}$$

For $L > 0$,

$$\phi_m(z/L) = (1 + \beta_m z/L), \qquad \phi_h(z/L) = (Pr_t + \beta_h z/L).$$

Here $L = -u_*^3/[k(g/\theta_o)H_o/(\rho c_p)]$ is Monin-Obukhov length, u_* is surface friction velocity, $H_o/(\rho c_p)$ is surface temperature flux, θ_o is surface potential temperature, and Pr_t is the turbulent Prandtl number. The empirical parameter values used in Equation (8) are $\gamma_m = 19.0$, $\gamma_h = 11.6$, $\beta_m = 5.3$, $\beta_h = 8.0$, and $Pr_t = 0.95$, which were recommended by Högström (1996) after a critical review of similarity formulations based on various experimental data sets.

At $z = z_b$, the lower boundary conditions are specified as

$$S = (u^2 + v^2)^{1/2} = u_* F(z_b/L, z_o/L), u = S\cos\beta, v = S\sin\beta,$$

$$\theta = \theta_o + T_* G(z_b/L, z_o/L), E = 5u_*^2, \tag{9}$$

where z_o is surface roughness height, $T_* = -H_o/(\rho c_p u_*)$ is the temperature scale, $\beta = \tan^{-1}(v/u)$ is the angle of the mean wind vector relative to the x-axis. The functions F and G are given by

$$F(z/L, z_o/L) = \int_{z_o}^{z} (kz')^{-1}\phi_m(z'/L)\, dz',$$

$$G(z/L, z_o/L) = \int_{z_o}^{z} (kz')^{-1}\phi_h(z'/L)\, dz'. \tag{10}$$

Nickerson and Smiley (1975) and Benoit (1977) gave the expressions for F and G obtained from Equation (10) for the stable and unstable cases after substituting for ϕ_m and ϕ_h from Equation (8). At $z = z_t$, the upper boundary conditions are set as

$$S = u_* F(z_t/L, z_o/L), \qquad u = S\cos\beta, \quad v = S\sin\beta,$$

$$\theta = \theta_o + T_* G(z_t/L, z_o/L), \quad \partial E/\partial z = 0. \tag{11}$$

Numerical integration of Equations (3) to (11) is performed using a Dufort-Frankel explicit finite difference scheme. The input data consist of hourly average mean wind velocities and temperatures measured at two heights on a 10 m meteorological tower upwind of the highway. Approximate initial profiles for the variables and surface parameters used in the ABL model are derived from these input data and the surface-layer similarity relations; this procedure is described in Hosker et al. (2001). The TKE calculated from Equation (7) is partitioned into its components by using the diagnostic equations for $\overline{u^2}$ and $\overline{v^2}$ (see Luhar and Rao, 1993); $\overline{w^2}$ is obtained by subtracting these two components from $2E$.

3. Vehicle Wake Parameterizations

Automobile wakes are characterized as momentum wakes and contain trailing organized vortices which rapidly mix the pollutants released in the turbulent wake. This mixing is expected to minimize the buoyant rise of hot exhaust gas. Eskridge and Hunt (1979) presented a theory for the 3-D wake behind a vehicle moving in still air. From a perturbation analysis of the equations of motion, assuming that the strength of the wake is determined by the overturning moment acting on the vehicle, they derived analytical expressions for mean velocity deficit and turbulent velocity variances in the wake. Eskridge and Thompson (1982) investigated the wake behind a block-shaped vehicle fixed in position over a moving surface in the wind tunnel, which is equivalent to simulating the wake behind a vehicle traveling along a highway under calm conditions. This momentum wake did not have a simple self-preserving form. However, they showed it was possible to collapse the velocity deficit data with one length scale and one velocity scale, and the turbulence data with different length and velocity scales.

In the present study, we utilized the vegetation canopy flow theory to derive the parameterizations for vehicle wakes. The canopy flows have been extensively studied (e.g., Dubov and Bykova, 1974; Shaw and Seginer, 1985; Wilson, 1988; Kaimal and Finnigan, 1994) in both models and measurements. Large-eddy simulations of turbulent flow above and within a forest canopy (e.g., Shaw and Schumann, 1992) have also been performed in the past decade. All these studies led to an improved understanding of the canopy flows. The key difference between the two flows is that the canopy elements are stationary, while the vehicles are moving. The latter requires use of the wind speed relative to the moving vehicle. In a canopy flow, complex energy transformations are caused by viscous and form (pressure) drag on vegetation. The rate of work done by the mean flow against the form drag results in a positive contribution to the TKE, i.e., wake production. Experimental evidence suggests that the rate of wake production of TKE is comparable in magnitude to the shear production rate. However, the wake turbulence is of small scale and dissipates very rapidly.

The time-dependent aerodynamic drag force exerted on the canopy per unit volume is equal to the product of the foliage area density (foliage area per unit volume), a constant drag coefficient, and the square of the local velocity. In an analogous way, for the vehicle wakes, the aerodynamic drag force can be written as

$$\partial u_D / \partial t \propto C_d A_v Q^2, \tag{12}$$

where u_D is the wake velocity deficit, and C_d is the drag coefficient; Q is the relative wind speed and A_v is the vehicle area density (vehicle area normal to the wind per unit volume), which are given by

$$Q = [u_\infty^2 + (v_\infty - V_h)^2]^{1/2}, \, A_v = (\ell_v \sin \alpha + w_v \cos \alpha)h \,/\,(\ell_v w_v h). \tag{13}$$

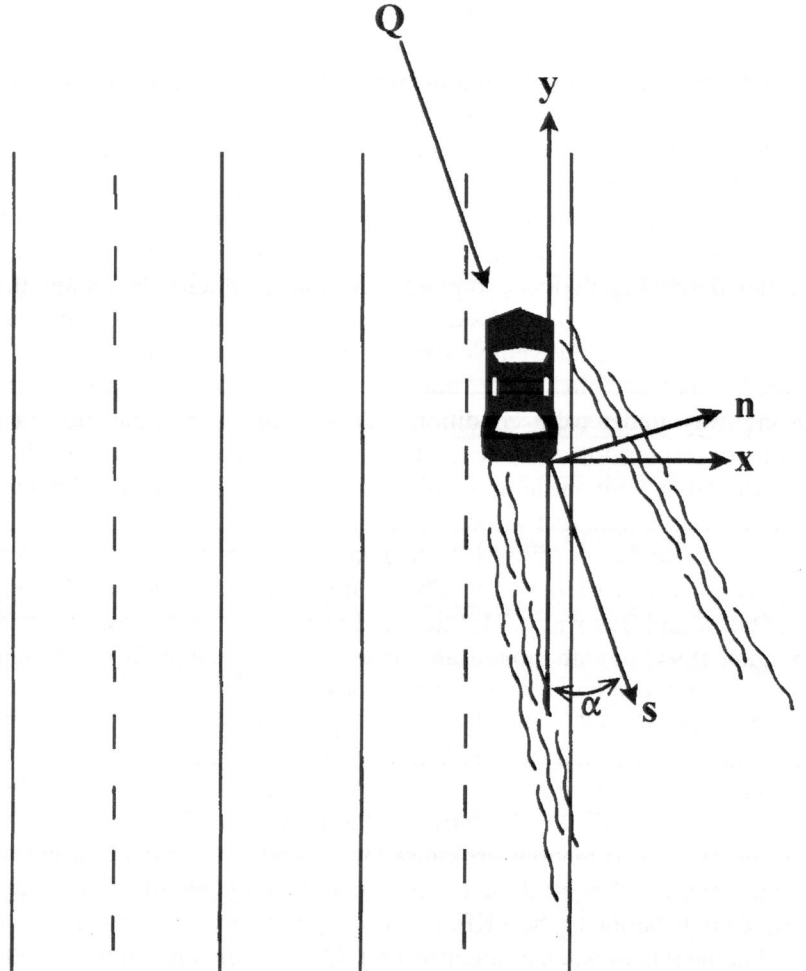

Figure 1. Schematic diagram of the wake-aligned (s, n, z) coordinate system and its relation to the fixed (x, y, z) coordinate system.

Here, u_∞, v_∞ are the ambient velocity components, and V_h is vehicle velocity; ℓ_v, w_v, h are length, width, and height, respectively, of the vehicle; α is angle between the mean wind direction and the road (y axis), as shown in Figure 1.

Assuming an integral time scale $\tau \propto h/Q$ exists, an expression for u_D can be derived from Equation (12) as

$$u_D \propto C_d A_v Q^2 \tau \propto C_d A_v Q h. \tag{14}$$

Incorporating the u_D variations in terms of the wake-aligned coordinates s, n, z (see Figure 1) determined from wind tunnel measurements by Eskridge and Thompson (1982), this can be written as

$$u_D = C_u C_d A_v Q h (s/h)^{-3/4} Y(\eta) T(\zeta), \tag{15}$$

where C_u is a constant, s is the downwind distance measured along the wake axis from its origin, and η and ζ are nondimensional lateral and vertical coordinates in the wake, defined as

$$\eta = n \big/ \big[\lambda (w_v/h) \ell_D \big], \zeta = z/\ell_D, \ell_D = \gamma A h (s/h)^{1/4}. \tag{16}$$

Here, ℓ_D is the length scale for the velocity deficit, $\gamma = 0.095$ and $\lambda = 1.14$ are empirical constants, and $A = \big[C_d/(32\pi e^{1/2} \lambda \gamma^3) \big]^{1/4}$. The functions Y and T, determined from theory and wind tunnel measurements, were given by Eskridge and Catalano (1987).

Similarly, following the canopy flow parameterizations (e.g., Dubov and Bykova 1974), the rate of vehicle wake production of TKE (E_v) can be written as

$$\partial E_v / \partial t \propto C_d A_v Q^3, \tag{17}$$

which leads to

$$E_v \propto C_d A_v Q^3 \tau \propto C_d A_v Q^2 h. \tag{18}$$

Incorporating variations of E_v with respect to (s, n, z) measured in the wind tunnel, this can be written as

$$E_v = C_e C_d A_v Q^2 h (s/h)^{-1.2} F_c(\chi, \omega), \tag{19}$$

Here C_e is a constant, and $F_c(\chi, \omega)$ is a least-squares orthogonal polynomial fit to the wind tunnel data given by Eskridge and Catalano (1987).

These expressions for vehicle wake velocity deficit and turbulence production are analogous to those given by Eskridge and Hunt (1979). In principle, the constants C_u and C_e can be independently determined if suitable data are available. In the present study, their values were set by equating the expressions for u_D and E_v to the corresponding equations given by Eskridge and Catalano (1987). The wake parameterizations given above for a single vehicle can be extended to multiple vehicles passing an observation point as shown in Eskridge and Catalano (1987). The effective pollutant advection velocity components in x and y directions were obtained by adding the ambient velocity components from the ABL model to the corresponding components of the total wake velocity deficit. Similarly, the x, y, z components of the integrated wake turbulence variances from multiple vehicles

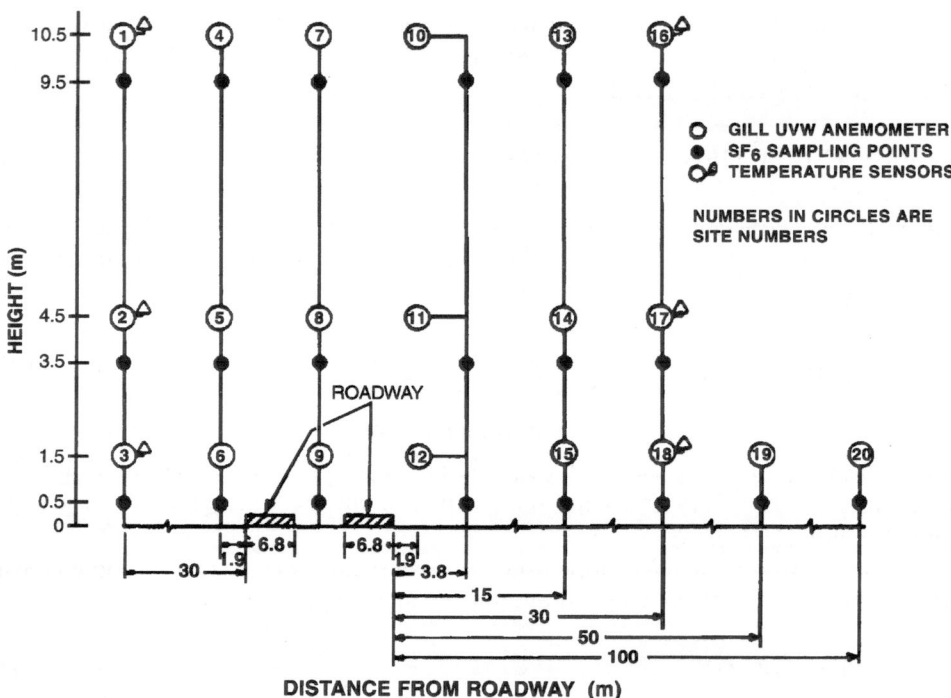

Figure 2. Locations of meteorological towers and instruments in the 1975 GM experiment.

were added to the corresponding ambient values to obtain the total TKE components. The latter were used in the calculation of the eddy diffusivity profiles in Equation (1). These details are given in Hosker *et al.* (2001).

4. Model Evaluation

The ROADWAY-2 model was evaluated with SF_6 dispersion data from the 1975 General Motors experiment at Milford, MI. A fleet of 352 automobiles were driven at 80 km hr^{-1} on a 10 km, 4 lane, north-south test track for 17 days during the morning hours. Cadle *et al.* (1976) described the experiment and measurements, and gave tables of meteorological and tracer data. Meteorological instruments and SF_6 samplers were located on six towers and two short stands 2.4 km north of the south end of the test track. Figure 2 shows the locations of the towers relative to the track and the heights of various instruments. A total of 58 sampling periods (30 min each) for SF_6 concentrations were available. These were classified depending on the mean wind direction and speed at 4.5 m height, and the observed temperature difference ΔT between 4.5 and 1.5 m heights, on the upwind tower, as follows:

TABLE I

ROADWAY-2 evaluation statistics[a]

	Perpend. wind		Parallel wind		Oblique wind		All wind dir.	
	Obs.	Pred.	Obs.	Pred.	Obs.	Pred.	Obs.	Pred.
Mean	7.96	7.13	8.99	10.18	12.01	6.84	9.64	8.06
SD	6.27	6.93	5.87	5.28	7.97	5.53	6.93	6.11
O_i range	0.01–27.7		1.39–22.8		0.09–32.7		0.01–32.7	
N_s	56		56		55		167	
R	0.81		0.88		0.88		0.77	
FB	0.11		−0.12		0.55		0.18	
NMSE	0.30		0.10		0.52		0.29	

[a] Concentration units for mean, SD, and O_i range are ppb × 10. N_s = sample size, SD = standard deviation, R = correlation coefficient, $FB = 2(\overline{O} - \overline{P})/(\overline{O} + \overline{P})$ is fractional bias, $NMSE = \overline{(P_i - O_i)^2}/(\overline{O}\ \overline{P})$ is normalized mean square error, where $\overline{O}$ and $\overline{P}$ are means of observed and predicted concentrations over the sample size N_s.

(1) perpendicular wind (direction was within 30° of the east-west or x axis); parallel wind (direction was within 30° of the north-south or y axis); oblique wind (direction was within the remaining 30° sectors).

(2) high wind (speed ≥ 2.5 m s^{-1}); moderate wind (speed = 1 to 2.5 m s^{-1}); low wind (speed ≤ 1 m s^{-1}).

(3) unstable atmosphere ($\Delta T < 0$); neutral ($\Delta T = 0$); stable ($\Delta T > 0$).

Four sampling periods for each of the perpendicular, parallel, and oblique wind cases were selected for ROADWAY-2 evaluation. For each wind direction, half of these test periods had unstable conditions and the other half had stable conditions. Several periods had low winds; the rest had moderate winds. Thus, the model performance has been evaluated under different meteorological conditions (characterized by the input data on wind direction, atmospheric stability, and wind speed) that are likely to be encountered in operation. The designation of the sampling periods follows Cadle et al. (1976). For a sampling period numbered as 279080959, for example, 279 was the Julian day, 08 was the hour, and 09 and 59 were minutes and seconds, respectively, in local time at the end of the sampling period.

4.1. EVALUATION RESULTS

ROADWAY-2 model predictions are compared to the corresponding 30-min observed tracer concentrations. Figure 3 shows the comparison for the perpendicular wind. The corresponding results for the parallel and the oblique winds are presented in Figures 4 and 5. The model evaluation statistics for the three cases are listed

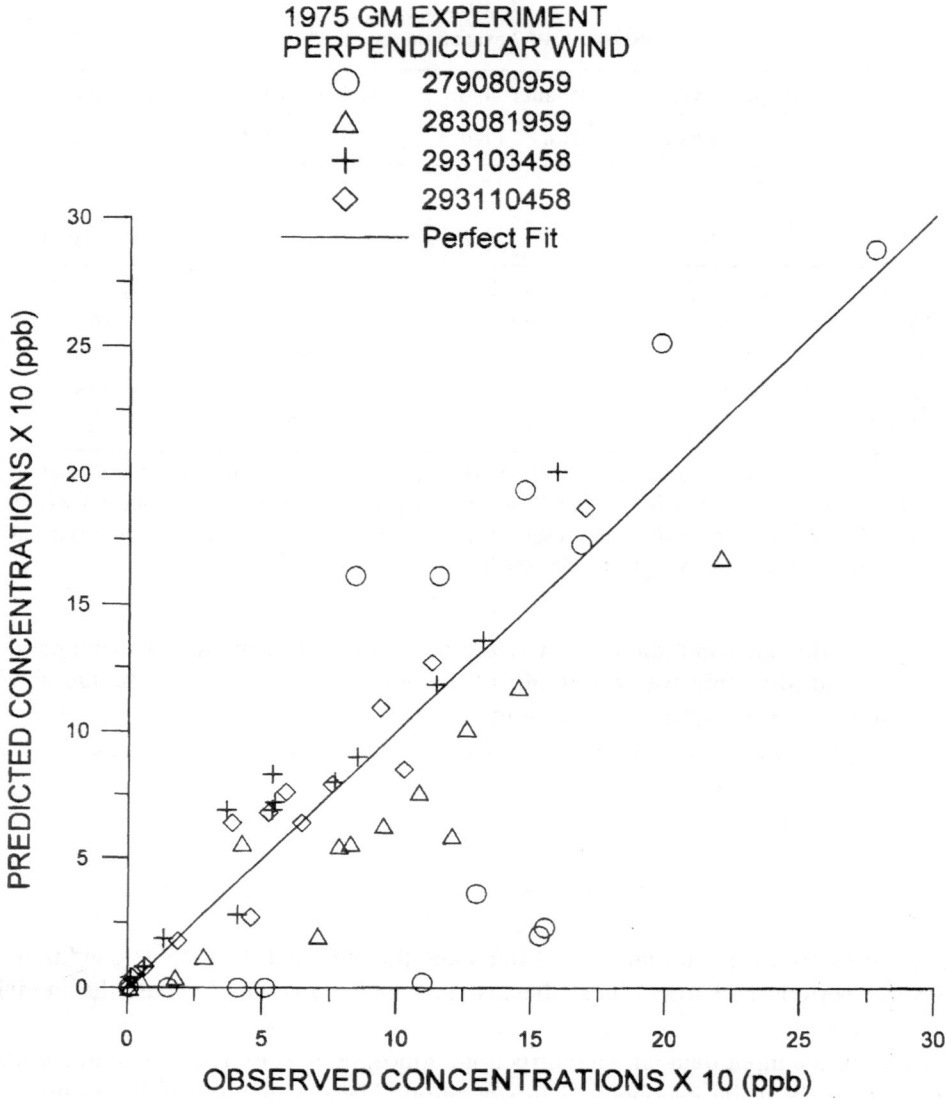

Figure 3. Comparison of predicted and observed concentrations for perpendicular winds.

in Table I. The observed and predicted mean concentrations (ppb × 10) for the perpendicular wind are 7.96 and 7.13, respectively. The fractional bias (FB) is 0.11 (slight underprediction), the normalized mean square error ($NMSE$) is 0.30, and the correlation coefficient (R) is 0.81. FB and $NMSE$ values close to 0 and R value close to 1 indicate good model performance. These statistical measures, defined under Table I, are widely used for air quality model evaluation (e.g., Tangirala *et al.*, 1992). The parallel wind case, which is generally considered to be more

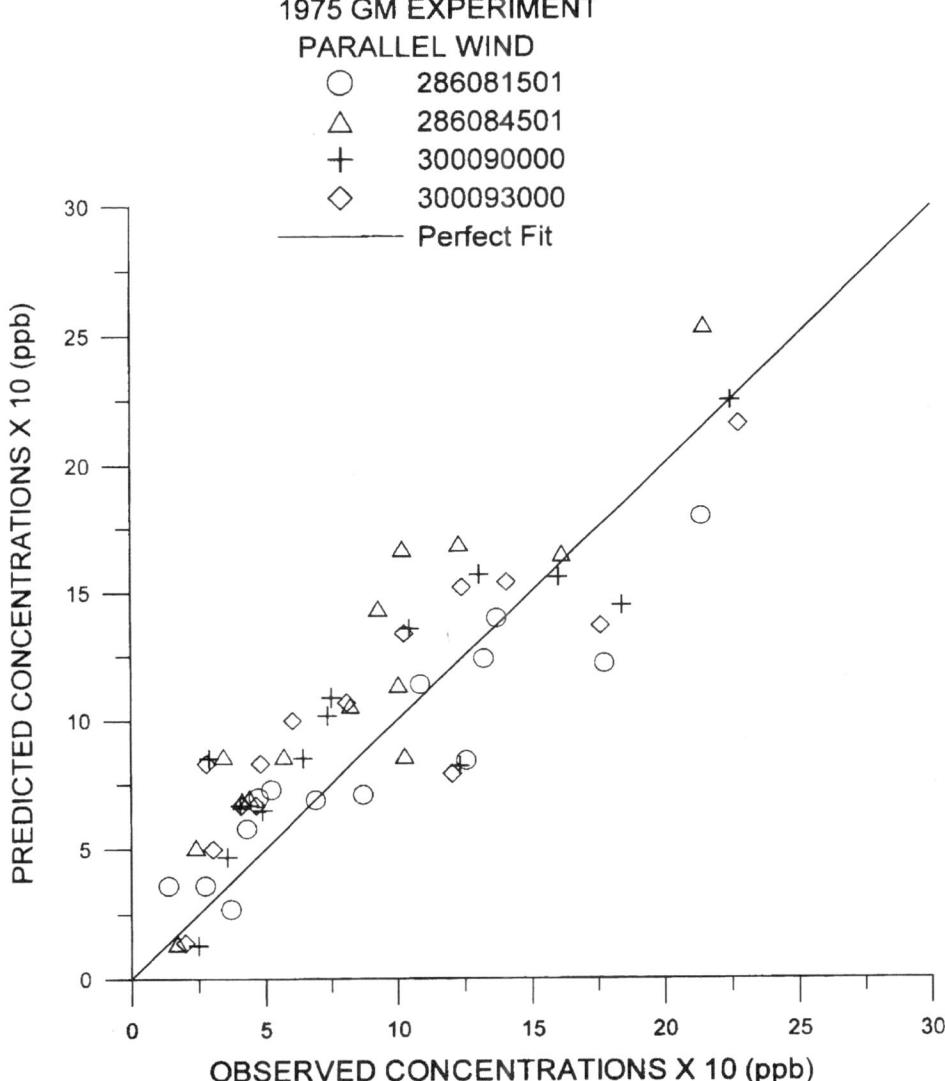

Figure 4. Comparison of predicted and observed concentrations for parallel winds.

difficult to model, yielded even better results: $FB = -0.12$ (slight overprediction) and $NMSE = 0.10$ (small scatter), and $R = 0.88$. These results confirm the good performance of ROADWAY-2 for the perpendicular and parallel wind cases.

The oblique wind case (Figure 5) is systematically underpredicted, with a mean observed concentration $(\overline{O})$ of 12.01 and a mean predicted concentration $(\overline{P})$ of 6.84; however, the R value is quite high (0.88). Further analysis showed that 84%

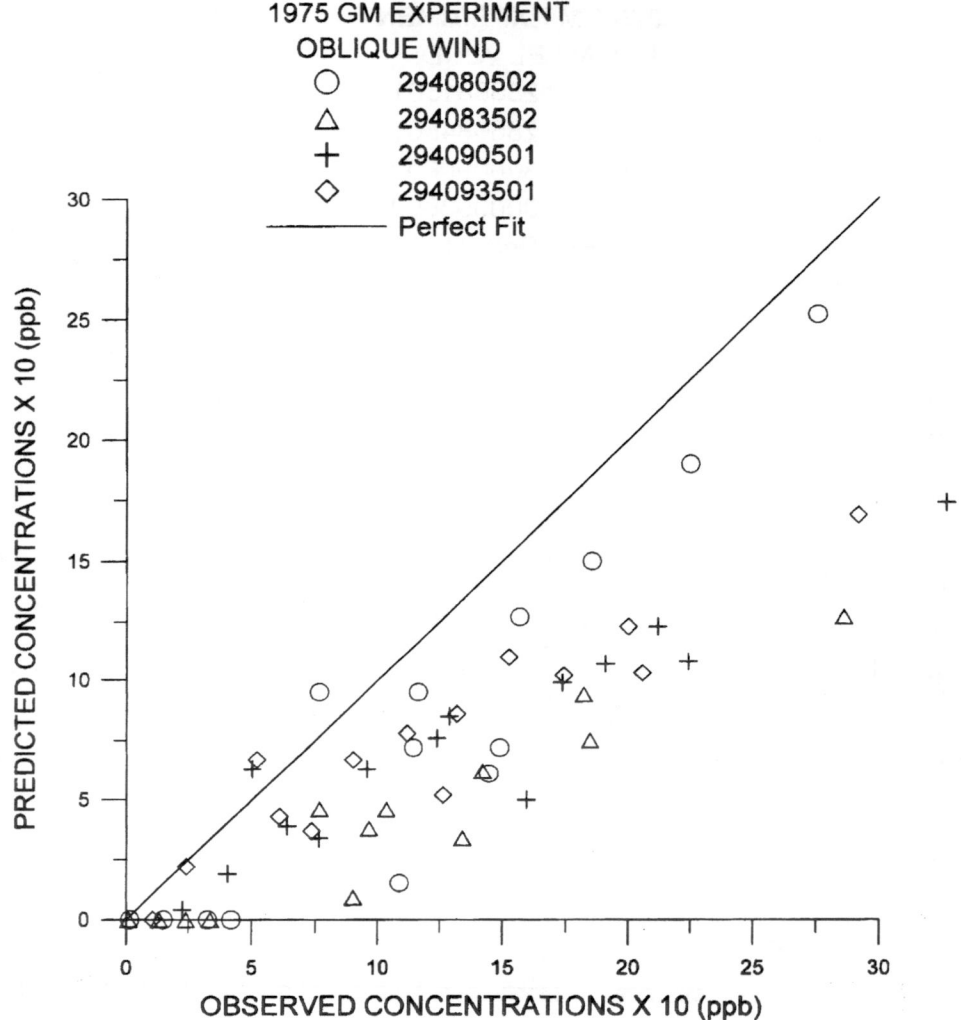

Figure 5. Comparison of predicted and observed concentrations for oblique winds.

of the mean squared error is systematic, suggesting that the model performance can be improved by modifying the model parameterizations for this case.

This was not attempted however, since all four oblique wind cases were sequential half-hour periods on the same day (Julian day 294) and had wind speeds only in the range 1.1 to 1.6 m s^{-1}. Further testing of the model using additional data, preferably with higher wind speeds, is necessary to evaluate the model performance for this case.

A comparison of observed and predicted concentrations for all wind directions is presented in Figure 6. The model evaluation statistics for all wind data are also shown in Table I. The $\overline{O}$ and $\overline{P}$ for all data are 9.64 and 8.06, respectively.

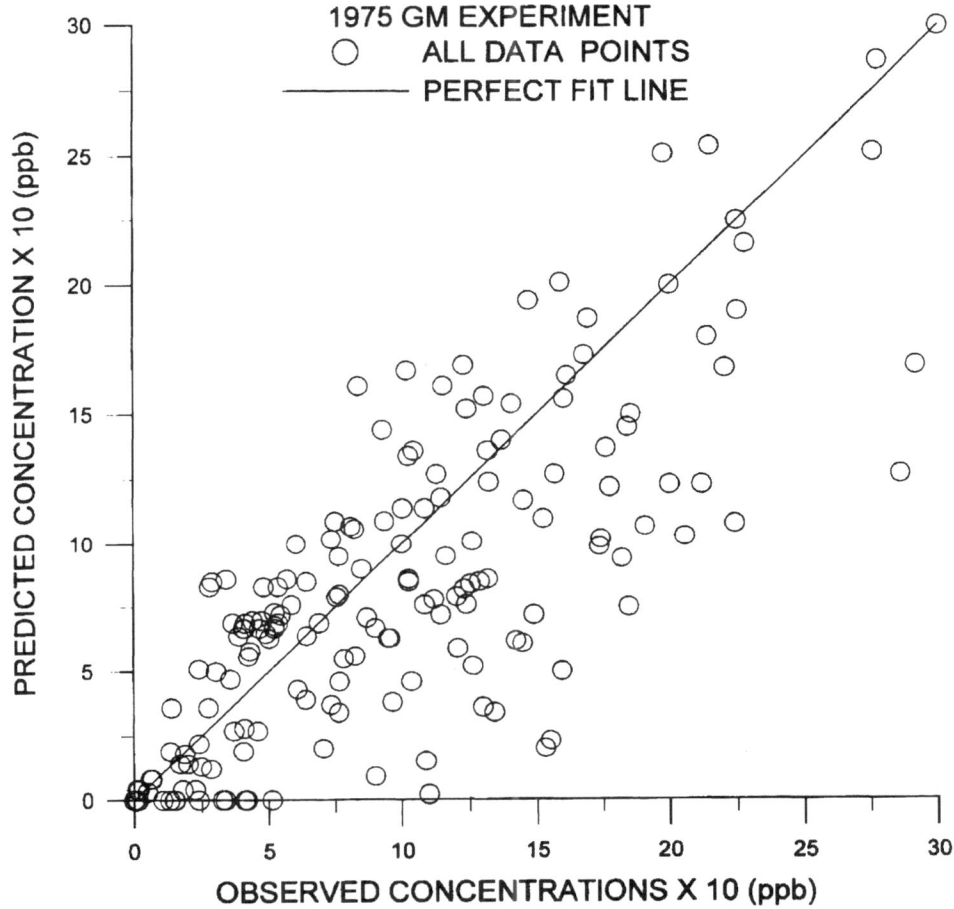

Figure 6. Comparison of predicted and observed concentrations for all wind directions.

The small FB and $NMSE$, and high R values indicate the good performance of ROADWAY-2 over all wind and stability conditions.

A cumulative frequency distribution of the ratio $F = P/O$ (or $F = O/P$ such that $F > 1$), shown in Figure 7, indicates that over 80% of the model predictions are within a factor of 2 of the corresponding observations, and 95% are within a factor of 5.

5. Conclusions

The mean velocities and turbulence formulations and the evaluation of a highway dispersion model, ROADWAY-2, are described in this paper. This model is applicable in the near field (within 200 m of the highway), where the effects of interac-

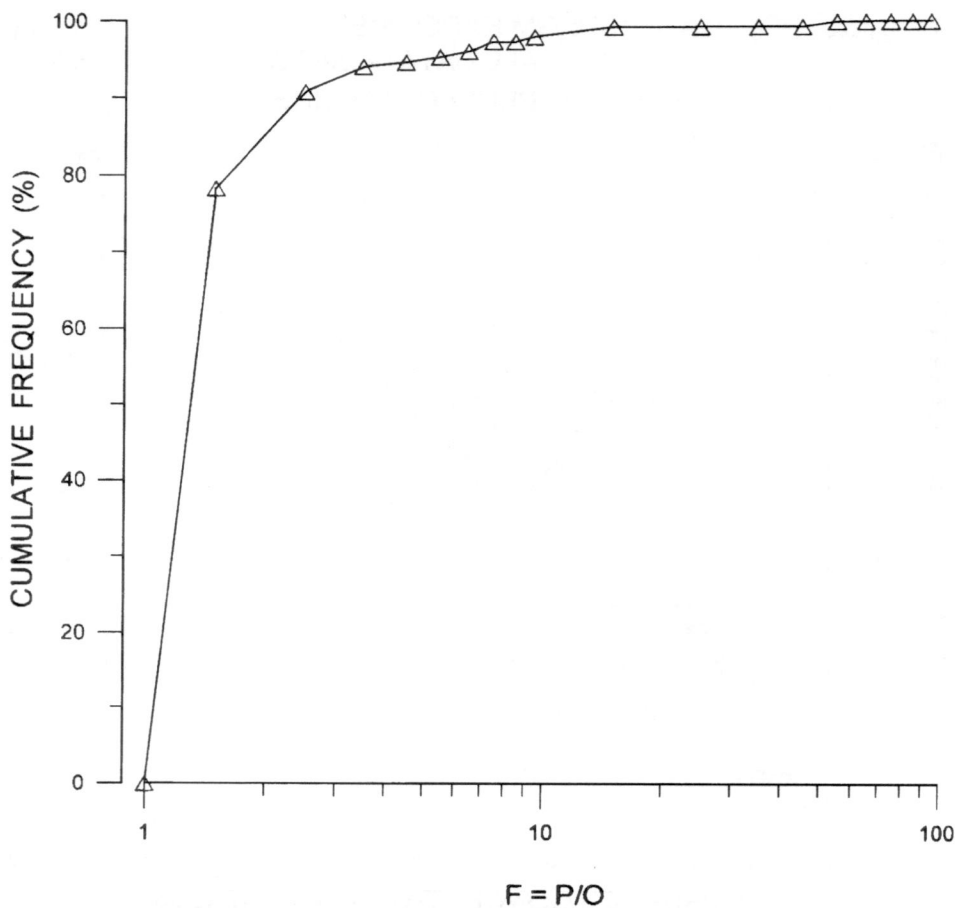

Figure 7. Cumulative frequency distribution of the ratio of predicted and observed concentrations.

tions between the vehicle traffic and the atmospheric surface flow are important. At longer distances, dispersion is dominated by the atmospheric turbulence, and other modeling tools, such as HIWAY-2 (Petersen, 1980) and CALINE (Benson, 1984), which are based on Gaussian plume dispersion algorithms, would be better suited for calculating the impacts of pollution from highway traffic. ROADWAY-2 includes an option for simulating simplified chemical reactions involving nitric oxide (NO), nitrogen dioxide (NO_2), and ozone (O_3), as described by Eskridge and Catalano (1987); this option requires specification of the emission rates, conversion factors, and background concentrations of the species. The model can include up to ten traffic lanes and internally generate a receptor grid which covers all the lanes. ROADWAY-2 is capable of treating light wind conditions, though it does not explicitly account for the wind meander which becomes important when the mean wind is nearly parallel to the highway.

The time-dependent ABL model incorporated in ROADWAY-2 avoids the need for assuming that the equilibrium surface layer similarity relations are applicable near the highways, as done in the ROADWAY model. The linear addition of eddy diffusivities of the atmosphere and the vehicle wake assumed in the latter model is also not valid, since the turbulence length scales of the atmosphere and the wake differ widely. However, eddy energies of the atmosphere and the wake are additive, and the TKE equation facilitates this addition to account for the wake turbulence. The TKE closure model also offers the flexibility for incorporating on-site turbulence measurements in the input data, with only slight modifications to the model. Such measurements are becoming feasible and increasingly attractive due to recent advances and affordability of instrumentation.

The performance of ROADWAY-2, based on the results presented above, is considered quite satisfactory. Direct comparison with the results of EPA's ROADWAY model could not be performed, because of an error found in its code that was determined to affect its predictions significantly. Additional testing of ROADWAY-2 with other roadside data sets is recommended to check if the model parameterizations need adjustment. This is particularly important for the oblique winds, since the GM Experiment data had mostly low wind speeds for this case. It is well-known that lower wind speeds result in greater initial dispersion, and the parameterization of the so-called 'mixing zone' of uniform emissions over the highway becomes critical. In general, low to moderate wind speeds and neutral to slightly stable ambient conditions lead to observations of significant concentrations adjacent to roadways. Under low wind speeds and stable conditions, the plume meandering caused by mesoscale eddies and other mechanisms (e.g., Rao and Nappo, 1998) may become important, especially for the parallel wind direction case. In general, the pollutant concentrations for this case decrease more rapidly with increasing distance from the road than for the perpendicular wind. For the latter case, however, the traffic pollutant plume travels the farthest distance and affects the widest area downwind. Direct measurements of mean velocity profiles and turbulence parameters are preferred, whenever possible, as inputs to the model. Suitable data from field or laboratory experiments should be obtained to independently determine the constants C_u and C_e in the wake parameterizations. There were no known field measurements in the wakes behind vehicles until Hosker *et al.* (2001) used a towed array of 3-D sonic anemometers to measure the mean velocities and turbulence within the wake behind a full-sized van traveling on an airport runway in a rural area.

Acknowledgements

This work was performed under an interagency agreement between the National Oceanic and Atmospheric Administration (NOAA) and the FHWA. The author

276 K. SHANKAR RAO

thanks Dr. H. Jongedyk of FHWA and Dr. R. Hosker of ATDD for their interest and support.

References

Benoit, R.: 1977, 'On the integral of the surface layer profile-gradient functions', *J. Appl. Meteor.*, **16**, 859–860.

Benson, P. E.: 1992, 'A review of the development and application of the CALINE3 and 4 models', *Atmos. Environ.*, **26B**, 379–390.

Cadle, S. H., Chock, D. P., Heuss, J. M. and Monson, P. R.: 1976, 'Results of General Motors Sulfate Dispersion Experiments', *Report GMR-2107*, General Motors Research Labs., Warren, MI, 179 pp.

Dubov, A. S. and Bykova, L. P.: 1974, 'Turbulence in forest canopies', *Izv. Atmos. Ocean. Phys.*, **10**, 650–652.

Eskridge, R. E. and Hunt, J. C. R.: 1979, 'Highway modeling – I. Prediction of velocity and turbulence fields in the wakes of vehicles', *J. Appl. Meteor.*, **18**, 387–400.

Eskridge, R. E. and Thompson, R. S.: 1982, 'Experimental and theoretical study of the wake of a block-shaped vehicle in a shear-free boundary flow', *Atmos. Environ.*, **16**, 2821–2836.

Eskridge, R. E. and Catalano, J. A.: 1987, *ROADWAY – A Numerical Model for Predicting Air Pollutants Near Highways. User's Guide*, EPA-600/S8-87-010, U.S. EPA, Research Triangle Park, NC. Available as PB 87-171 906, NTIS, Springfield, VA, 134 pp.

Högström, U.: 1996, 'Review of some basic characteristics of the atmospheric surface layer', *Bound.-Layer Meteor.* **78**, 215–246.

Hosker, R. P., Rao, K. S., Gunter, R. L., Nappo, C. J., Meyers, T. P., Birdwell, K. R. and White, J. R.:2001, 'Issues Affecting Dispersion Near Highways: Light Winds, Intra-urban Dispersion, Vehicle Wakes, and the Roadway-2 Dispersion Model', *Draft Report*, NOAA/ATDD, Oak Ridge, TN. To be published as NOAA Technical Memorandum.

Kaimal, J. C. and Finnigan, J. J.: 1994, *Atmospheric Boundary Layer Flows*, Oxford University Press, New York, 289 pp.

Luhar, A. K. and Rao, K. S.: 1993, 'Random-walk model studies of the transport and diffusion of pollutants in katabatic flows', *Bound.-Layer Meteor.* **66**, 395–412.

Marchuk, G. I.: 1975, *Methods of Numerical Mathematics*, Springer-Verlag, New York, 316 pp.

Nickerson, E. C. and Smiley, V. E.: 1975, 'Surface layer and energy budget parameterization for mesoscale models', *J. Appl. Meteor.*, **14**, 297–300.

Petersen, W. B.: 1980, *User's Guide for HIWAY-2: A Hiway Air Pollution Model*, EPA-600/8-80-018, U.S. EPA, Research Triangle Park, NC., 69 pp.

Rao, K. S. and Snodgrass, H. F.:1981, 'A nonstationary nocturnal drainage flow model', *Bound.-Layer Meteor.* **20**, 309–320.

Rao, K. S. and Nappo, C. J.: 1998, 'Turbulence and dispersion in the stable atmospheric boundary layer', in M. P. Singh and S. Raman (eds), *Dynamics of Atmospheric Flows: Atmospheric Transport and Diffusion Processes*, Computational Mechanics Publications, Southampton, U.K., pp. 39–92.

Rao, S. T., Sedafian, L. and Czapski, U. H.: 1979, 'Characteristics of turbulence and dispersion of pollutants near major highways', *J. Appl. Meteor.*, **18**, 283–293.

Shaw, R. H. and Schumann, U.: 1992, 'Large-eddy simulation of turbulent flow above and within a forest', *Bound.-Layer Meteor.* **61**, 47–64.

Shaw, R. H. and Seginer, I.: 1985, 'The Dissipation of Turbulence in Plant Canopies'. *Seventh Symp. on Turb. and Diff., Preprints*, Boulder, CO, Amer. Meteor. Soc., Boston, pp. 200–203.

Tangirala, R. S., Rao, K. S. and Hosker, R. P.: 1992, 'A puff model simulation of tracer concentrations in the nocturnal drainage flow in a deep valley', *Atmos. Environ.*, **26A**, 299–309.

Wilson, J. D.: 1988, 'A second-order closure model for flow through vegetation', *Bound.-Layer Meteor.* **42**, 371–392.

Zalesak, S. T.: 1979, 'Fully multidimensional flux-corrected transport algorithm for fluids', *J. Computational Phys.* **31**, 335–362.

THREE-DIMENSIONAL NUMERICAL SIMULATION OF AIR POLLUTANT DISPERSION IN STREET CANYONS

JOHN M. CROWTHER[1]* and ABDEL GALEIL A. A. HASSAN[2]

[1] *School of the Built and Natural Environment, Glasgow Caledonian University, Glasgow, U.K.;*
[2] *Faculty of Science, South Valley University, Qena, Egypt*
(* *author for correspondence, e-mail: j.crowther@gcal.ac.uk, fax: +44 141 331 3696*)

Abstract. The PHOENICS Computational Fluid Dynamics (CFD) software package has been used with a standard k-ε turbulence model to simulate the three-dimensional dispersion of air pollutants in an urban street canyon. In all cases, a vortex was formed within the street canyon, characterized by updrafts near the upwind buildings and down-drafts near the downwind buildings. Contours of pollutant concentrations over a transverse vertical plane at mid-canyon show pollutants circulating within the vortex, with higher concentrations at the leeward face than at the windward faces, and higher concentrations above downwind buildings than above upwind buildings. Longitudinal distributions of pollutant concentrations at leeward and windward faces are characterized by higher concentrations at mid-block and lower concentrations at the ends. These results agree qualitatively with previous wind tunnel findings such as those of Hoydysh and Dabberdt (1988) and Wedding *et al.* (1977). The results also suggest that the k-ε turbulence model is satisfactory for simulating the effect of turbulence on dispersion of pollutants in street canyons

Keywords: air pollution, carbon monoxide, Computational Fluid Dynamics, modelling, PHOENICS, street canyon

1. Introduction

Air flow characteristics in street canyons have been studied by field observation (e.g. DePaul and Sheih, 1986; Yamartino and Goz, 1986; Nakamura and Oke, 1988; Arnfield and Mills, 1994) and by wind tunnel simulation (e.g. Dabberdt *et al.*, 1981; Dabberdt and Hoydysh, 1991; Hoydysh and Dabberdt, 1988). Their observations all showed that a vortex can be maintained within a canyon when the ambient winds are perpendicular to the street, and that this affects pollutant transport from street canyons. Advances in computer software and hardware technology have enabled realistic numerical simulations of the air flow in urban street canyons (e.g. Hunter *et al.*, 1992; Johnson *et al.*, 1990).

Models for predicting pollutant concentrations within an urban street canyon can take several forms: some using field observations to develop an empirical model (e.g. Johnson *et al.*, 1973; Nicholson, 1975; Qin and Kot, 1993; Hassan and Crowther, 1998a) and others using a full simulation of the flow field and pollutant

Water, Air, and Soil Pollution: Focus **2:** 279–295, 2002.
© 2002 *Kluwer Academic Publishers.*

transport by CFD (e.g. Lee and Park, 1994; Moriguchi and Uehara, 1993; Hassan and Crowther, 1998b).

Hassan and Crowther (1998b) used the PHOENICS CFD package with a standard k-epsilon turbulence model to simulate the two-dimensional case of winds incident normally to a street canyon of infinite length. Simulations for the street canyon geometries studied by Hoydysh and Dabberdt (1988) produced reasonably consistent results for the vortex velocity (i.e. differences typically around 10%). The field measurements of wind speeds in a street canyon by DePaul and Sheih (1986) were simulated with similar success. In addition Hassan and Crowther (*op. cit.*) compared simulated pollutant concentrations with field measurements of carbon monoxide in Glasgow's Hope Street. Taking cases where the winds were normal to the street, good agreement was obtained between predicted and measured concentrations. Encouraged by the results of the two-dimensional simulations, the purpose of the study reported here was to extend the use of PHOENICS to describe the three-dimensional transport and dispersion of pollutants from an urban street canyon and the effects of finite street canyon length and of non-normal wind direction.

2. Equations and Boundary Conditions

The PHOENICS CFD package has been configured to solve the Navier Stokes equations for the mean flow field in a street canyon for various street configurations and for various different wind directions. The package includes the standard k-ε turbulence model presented by Launder and Spalding (1972) and by Mohammadi and Pironneau (1994). The technique has been applied successfully to the air flow characteristics within street canyons by Johnson et al. (1990), Hunter et al. (1992) and by Hassan and Crowther (1998b) as described above. The k-ε turbulence model represents the effects of turbulence by including two more variables: namely, the turbulent kinetic energy, k, and its dissipation rate, ε. Those two quantities are treated as variables in transport equations, which have to be solved together with the usual Reynolds-averaged Navier Stokes equations, involving continuity and momentum conservation. The conservation equation for species concentration of pollutants must also be solved together with the above-mentioned equations which describe the flow characteristics.

The solution domain is illustrated in Figure 1a, and is a simple canyon consisting of a street with identical rectangular buildings on each side. The pollutant sources are motor vehicles passing between the two buildings (assuming 1000 vehicles per hour). In this study the carbon monoxide, CO, dispersion was predicted, since the main source of CO is from motor vehicles and it is a relatively non-reactive gas in the atmosphere. The carbon monoxide emission factors (Bardeschi et al., 1991; Jounard, 1987) for petrol and diesel engines are assumed to be 35 and 3.5 g km^{-1}, respectively, for vehicles running at 20 km hr^{-1} average speed.

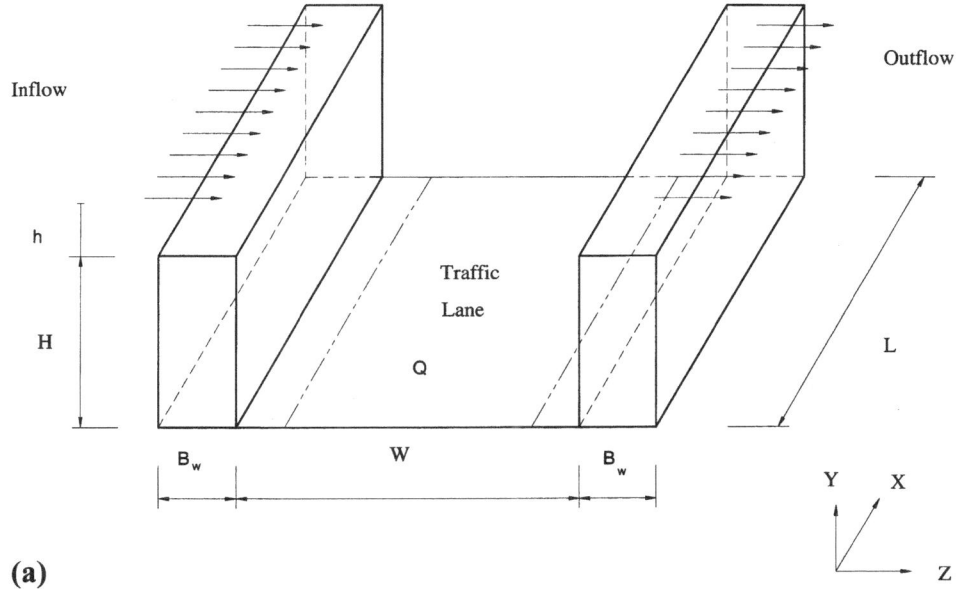

Figure 1. (a) Street canyon solution domain.

The chosen solution grid is shown in Figures 1b and c and is designed to allow more detail of the flow patterns to be seen close to the building walls. At the inflow boundaries, the wind field is specified with both the inlet velocity, w_{in} (horizontal velocity) and mass flow rate (expressed as a pressure boundary condition) assumed to be constant. At the outflow, where the fluid leaves the domain, the (relative) pressure is set to zero. For the free stream boundary condition, we assume that the flow is free at a distance 10 m above the canyon and 10 m beyond the building ends, as suggested by the data analysis of DePaul and Sheih (1984). Turbulent kinetic energy k and its rate of dissipation ε are specified at the inlet, assuming a turbulence intensity of 10% and a realistic value of turbulent kinematic viscosity, v_t. The solid wall friction is described by the PHOENICS wall function, assuming the velocity components parallel to the wall vary logarithmically with the distance from the wall, as found experimentally. The vehicular exhaust emission source term Q is then added as a fixed-flux boundary condition at ground level over the road but not over the pavements, which are assumed to be 1.8 m wide on both sides of the road. The eddy diffusivities are assumed to be isotropic and equal to the effective kinematic viscosity.

3. Results and Discussion

The first part of this study simulates dispersion of pollutants within street canyons of finite length when the wind direction is perpendicular to the street. The second

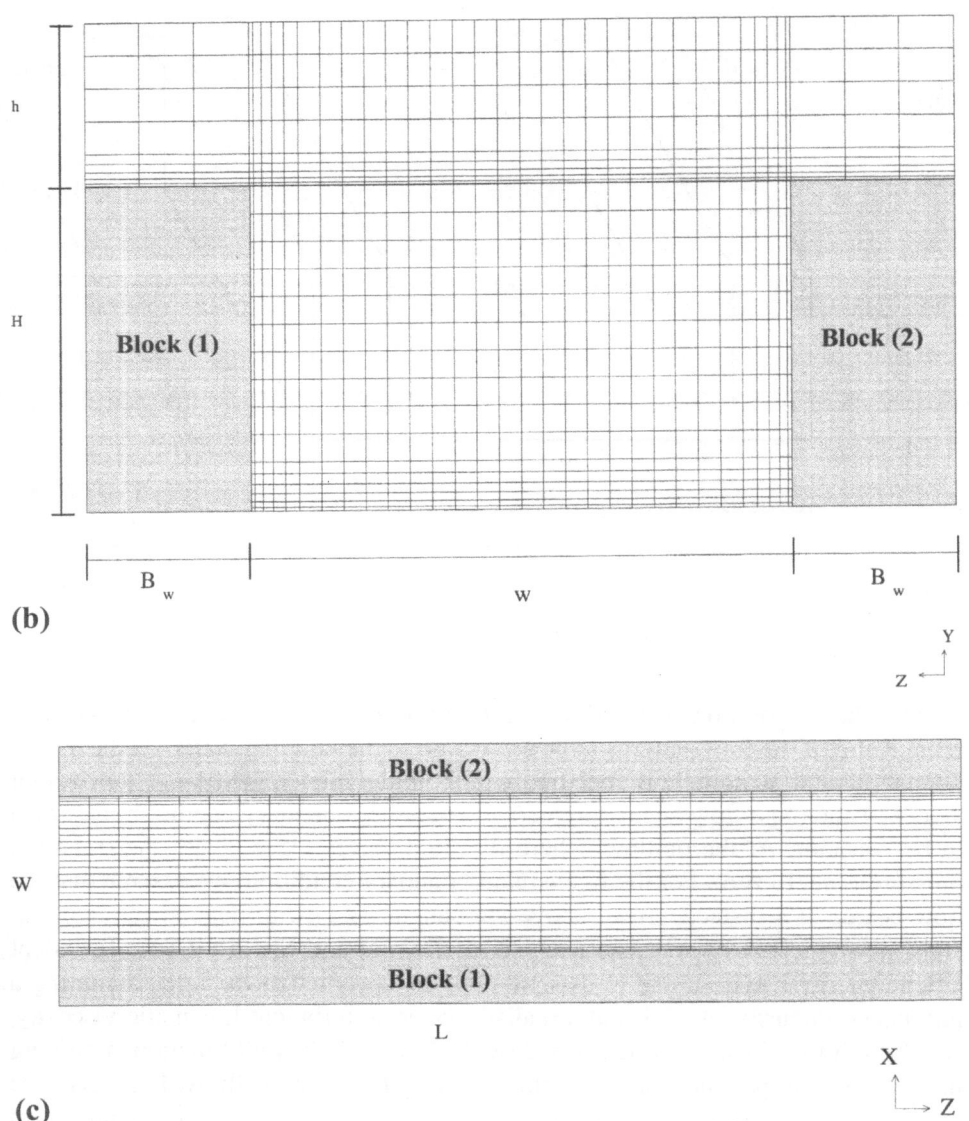

Figure 1. (b) Transverse solution grid for PHOENICS. (c) Longitudinal/plan solution grid for PHOENICS.

part simulates the effect of non-normal wind directions on distribution of pollutants within the canyons.

3.1. WIND DIRECTION NORMAL TO THE STREET

For all cases, a vortex was formed within the street canyon, characterized by up-drafts near the upwind building (leeward face) and down-drafts near the downwind

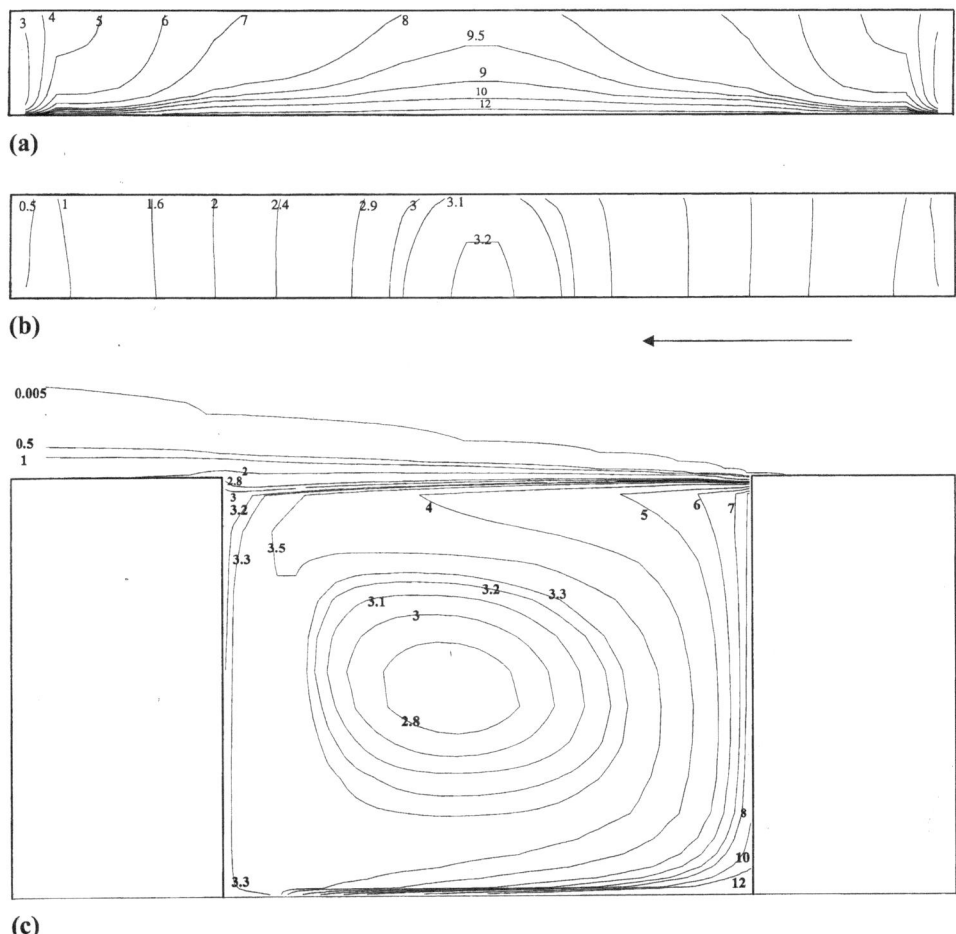

Figure 2. (a) Leeward face CO concentration contours, ppm, (L = 180, H = 20, W = 20 m). (b) Windward face CO concentration contours, ppm, (L = 180, H = 20, W = 20 m). (c) CO contours, ppm, (L = 180, H = 20, W = 20 m) for a transverse vertical plane at mid-canyon.

buildings (windward face). The distribution of pollutants within the street canyon is governed by the air flow characteristics within the canyon, with the pollutants re-circulating within the vortex formed within the canyon. Figures 2a and b show that the pollutant concentrations at leeward and windward faces along the street are characterized by higher concentrations at mid-block, with the street-level gradient directed from the end of the street to the mid-block. Also, the concentrations at the leeward face are greater than at the windward face. At the ends of the street the pollutants are mixed with cleaner air and expelled, resulting in lower concentrations in these regions. These results confirmed, qualitatively, the previous wind tunnel findings, such as those of Hoydysh and Dabberdt (1988) and Wedding *et al.* (1977).

Vertical plane contours of pollutant concentrations at mid-canyon (e.g. Figure 2c) show re-circulating pollutants within the vortex and concentrations of pollutant above downwind building that are higher than above the upwind buildings, as expected.

3.1.1. *Effect of Canyon Geometry on Pollutants Distribution*

Street width, W, street length, L, and building height, H, determine the geometry of any street. Therefore, in the present study, the effect on pollutant concentration of aspect ratio (H/W) of the street canyon and street canyon length, L, have been considered.

3.1.1.1. *Effect of Aspect Ratios (H/W)*

The pollutant concentrations were calculated at mid-canyon (0.13 m away from the upwind and downwind wall) for different aspect ratios. In general, the results for all studied cases (when the street length is constant) indicated that the pollutant concentrations increase when the aspect ratio increases, confirming the previous research results obtained using wind tunnels (e.g. Hoydysh *et al.*, 1974; Kennedy and Kent, 1977). The predicted relationship between the aspect ratios and pollutant concentration (CO) is shown in Figures 3a and b for canyons with L = 180, H = 20 m but differing widths for an incoming velocity = 5 m s^{-1} and an assumed traffic flow of 1000 vehicles per hour. The figures are designed to show how the width influences pollution at various fixed heights, y. For aspect ratios less than or equal 1.25, the pollutant concentrations at the leeward face increased with increasing aspect ratios. For aspect ratios greater than 1.25, there was little variation in concentrations at heights of 0.82 and 2.64 m from the ground. For aspect ratios greater than 1.25, it was found that the concentration of pollutant near the roof of the buildings decreases slightly when aspect ratio increases. For wider canyons (low aspect ratio), the exchange of air between street-level, where pollutants are emitted, and above roof-level is sufficient to dilute the pollutant concentrations within street. If the street canyons become deeper (high aspect ratio), air exchange is restricted, resulting a high street-level concentrations. Further increasing this ratio, the street-level concentrations become constant.

The roof-level concentrations above downwind building were examined for different aspect ratios, showing a decrease with increasing aspect ratio as the pollutants undergo greater dilution (Figure 4).

3.1.1.2. *Effect of Street Length*

The relationship between the calculated mid-canyon concentrations for different street lengths is illustrated by Figures 5a and b for canyons with aspect ratio = 0.8, traffic flow = 1000 vehicles per hour and the ventilated wind = 5 m s^{-1}. The figures show that, for L less than 120 m, the

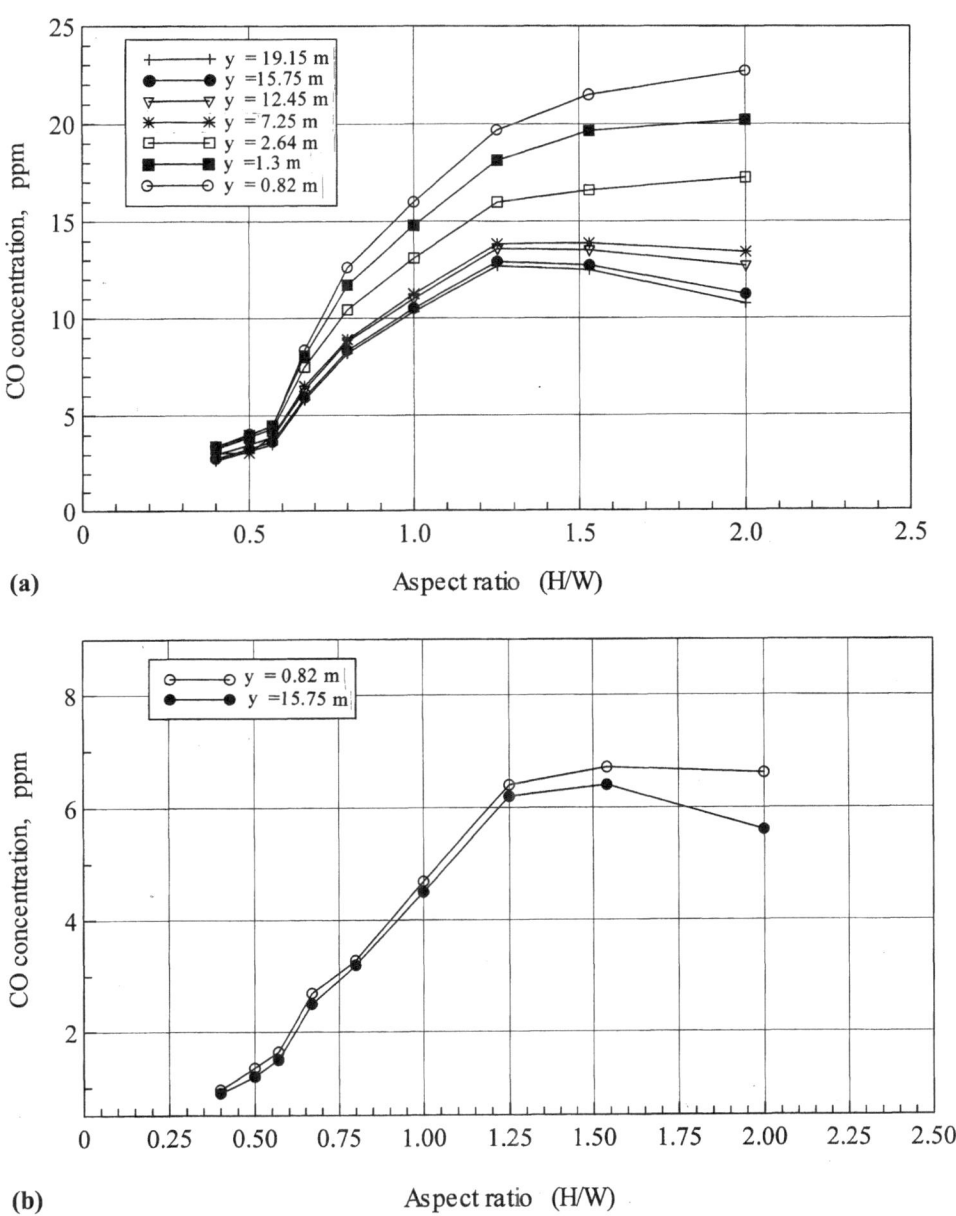

Figure 3. (a) Leeward face (L = 180, W = 20 m, incident velocity = 5 m s^{-1} and traffic flow 1000 vehicles per hour). (b) Windward face (L = 180, H = 20 m, incident velocity = 5 m s^{-1} and traffic flow 1000 vehicles per hour).

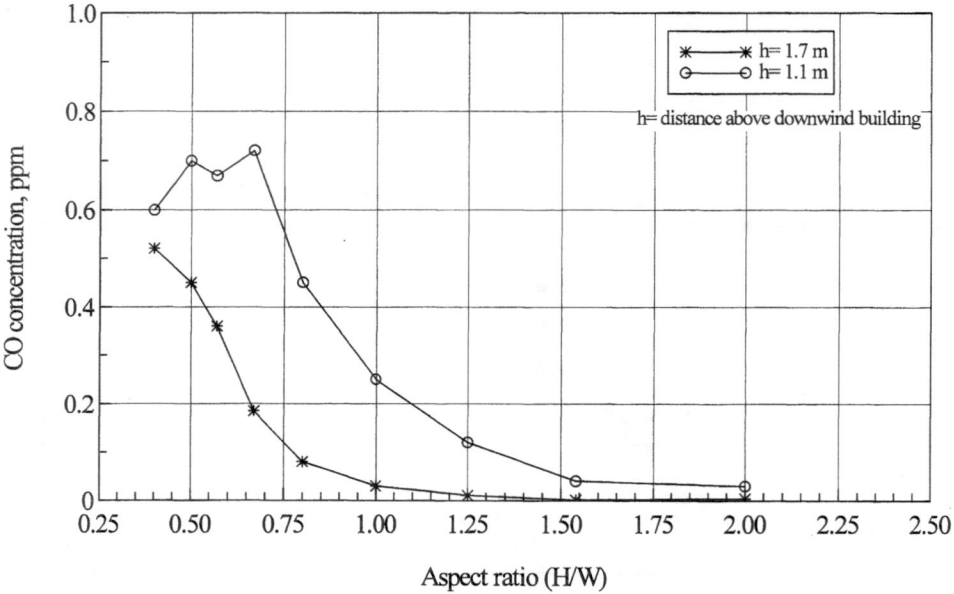

Figure 4. The effect of aspect ratio on pollutant concentrations above downwind buildings.

120 m, the concentrations were found to be relatively constant. The results show that reducing the street length also reduces street-level pollution at mid-canyon.

Also, the roof-level concentration above the downwind building for different street lengths are shown in Figure 6, indicating that mid-canyon pollutant concentration is approximately constant for all lengths greater than 100 m.

3.2. EFFECT OF WIND DIRECTION

Pollutant dispersion is also significantly affected by the variability in wind direction. For all cases simulated, the street geometry (L, W and H) are held to be constant and equal to 180, 25, 20 m, respectively, and the above building incident wind speed is held at 2 m s^{-1}. The wind directions, ϕ, ranged in 10° increments from perpendicular ($\phi = 0°$) to parallel ($\phi = 90°$) to the target street.

The initial investigations of the pollutant distributions were when the wind was blowing parallel to the street, as illustrated in Figure 7a. It was found that the CO concentrations along the street at the two sides were identical and characterized by small concentrations in the region of initial dispersion where the wind dilutes the pollutants while flushing them along the street. Also, by examination of calculated concentrations of CO across the street width, it was found that the concentrations were higher in the middle of the street as shown in Figure 7b.

For winds almost parallel to the street ($\phi = 80$ and 70°), the calculated pollutant concentrations at the leeward face were higher than at the windward face and are characterized by small concentrations at the region of the initial dispersion.

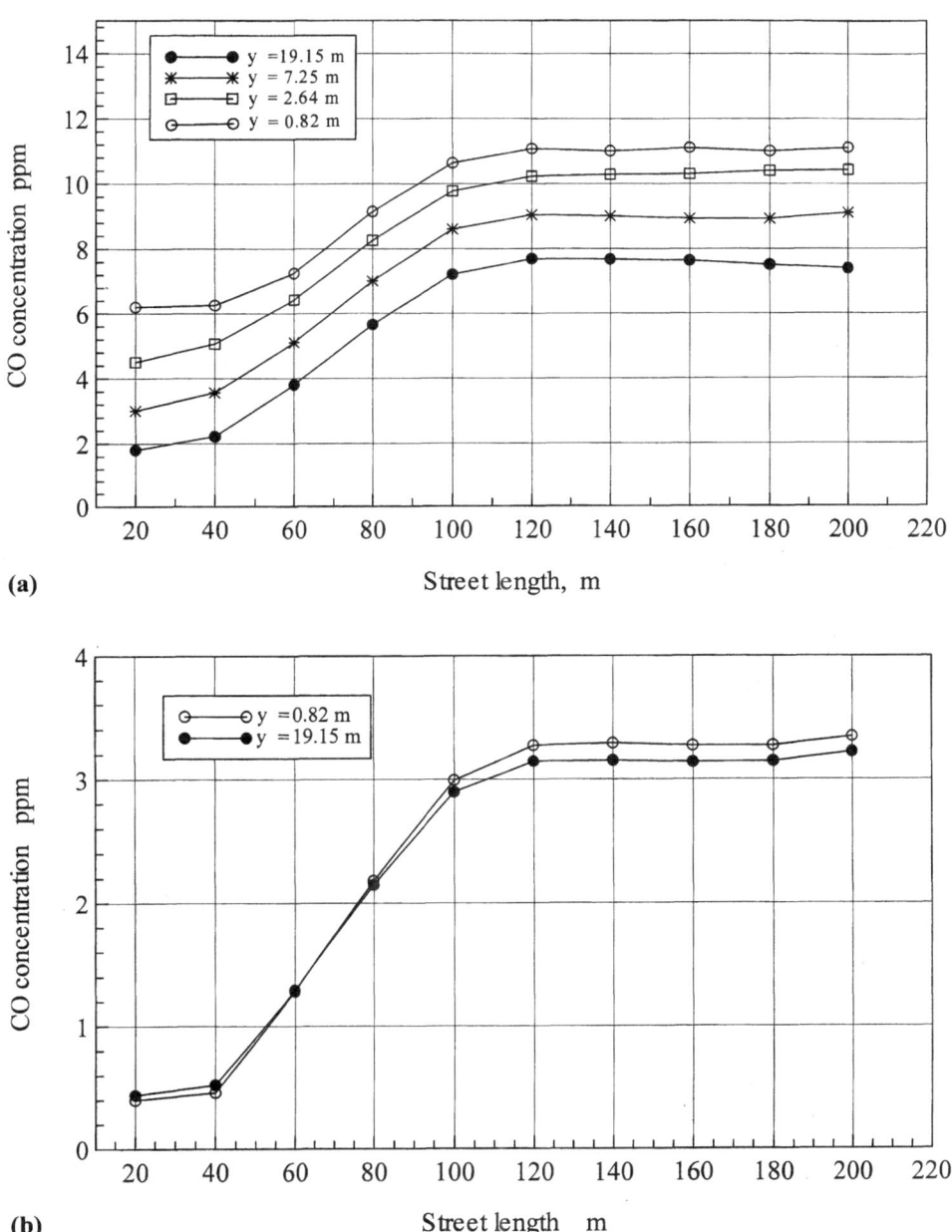

Figure 5. (a) The effect of street length on leeward face pollutant concentrations. (b) The effect of street length on windward face pollutant concentrations.

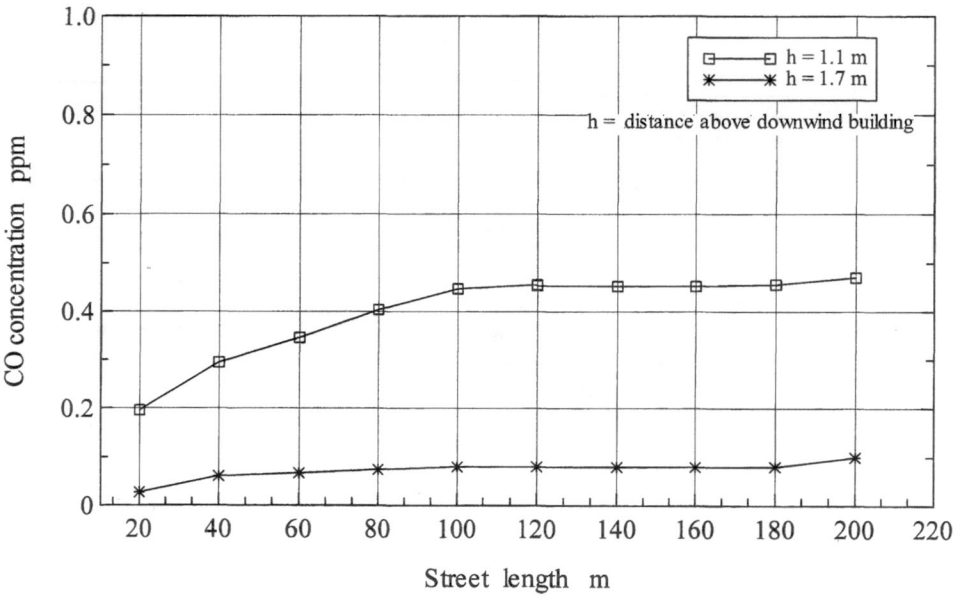

Figure 6. The effect of street length on pollutant concentrations above downwind buildings.

Maximum concentrations were found at the end of the street and these decrease with increasing height as shown in Figures 8a and b ($\phi = 80°$).

The situation is complicated when the wind angle, ϕ, is smaller, (between 60 and 10°) as shown in Figures 9a–b and 10a–b. It is clear that for winds with directions between (60–40°) the pollutant are flushed out at the end of the street and characterized with a minimum concentration at the input wind region. The maximum concentration occurs inside the street depending on the flow characteristics, which change when the wind angle changes. For winds more perpendicular to the street, the pollutants are distributed over a larger area of street and have higher values. That means that the amount of pollutant flushed out is small compared to the above-mentioned cases.

The results for oblique incidence of the incoming wind are tentative and require validation by field observation or physical simulation studies. There are few previous studies focusing on the effect of changing wind direction on distribution of pollutants within street canyons but Hoydysh and Dabberdt (1988) used a wind tunnel to investigate the pollutant concentration pattern at mid-block of a model street canyon. They reported the concentrations of pollutant as a function of wind angle for both leeward and windward faces and showed that the concentration at leeward face decreased from a maximum at $\phi = 0°$ to a minimum at about 30°. In the present study, the CO concentration at mid-block of the street canyon has been investigated as shown in Figures 11a and b. The results showed that the maximum concentration at leeward face occurred when the wind direction was perpendicular

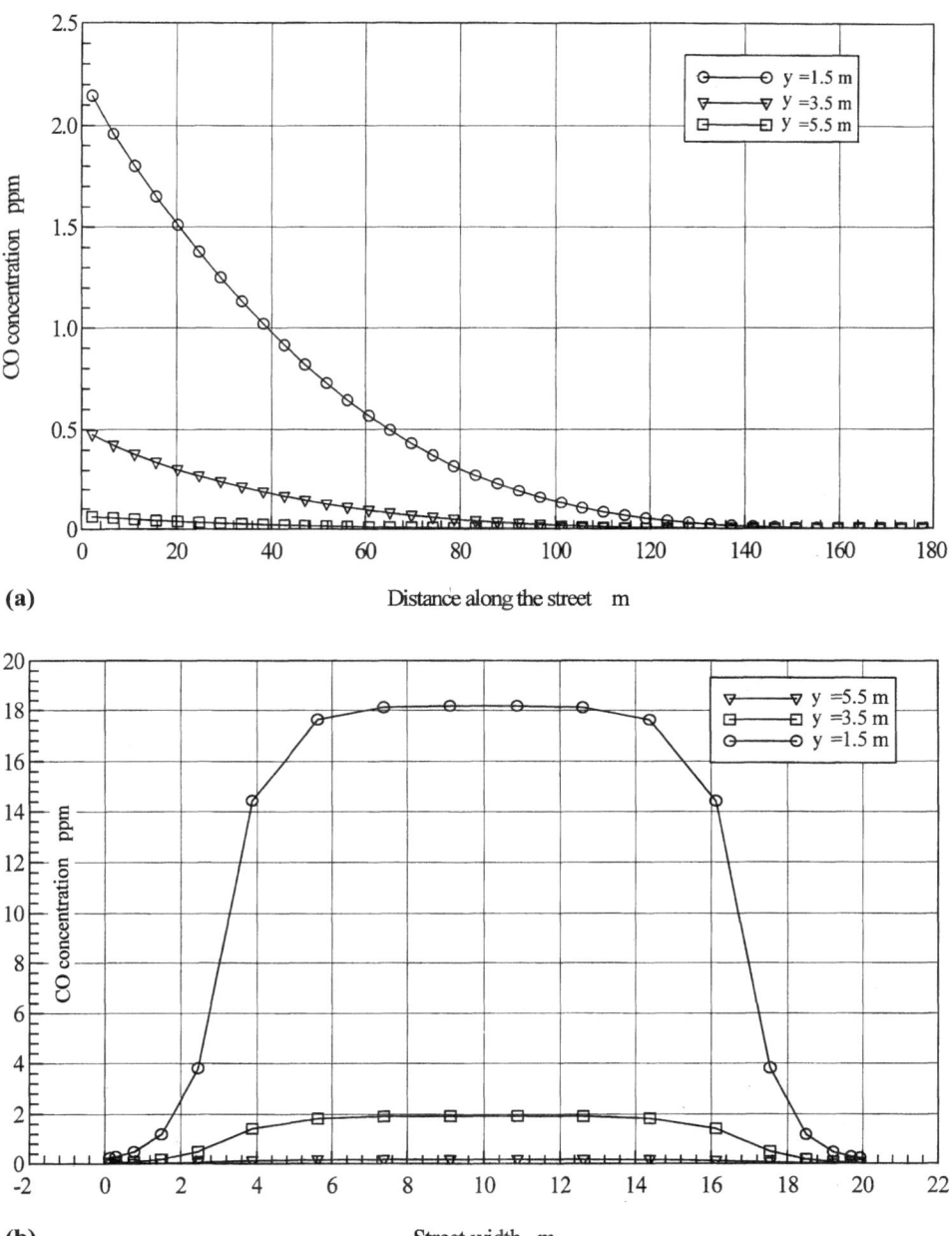

(a)

(b)

Figure 7. (a) CO for a vertical central longitudinal symmetry plane ($\phi = 90°$) wind right to left. (b) CO concentrations at the vertical mid-block symmetry plane ($\phi = 90°$).

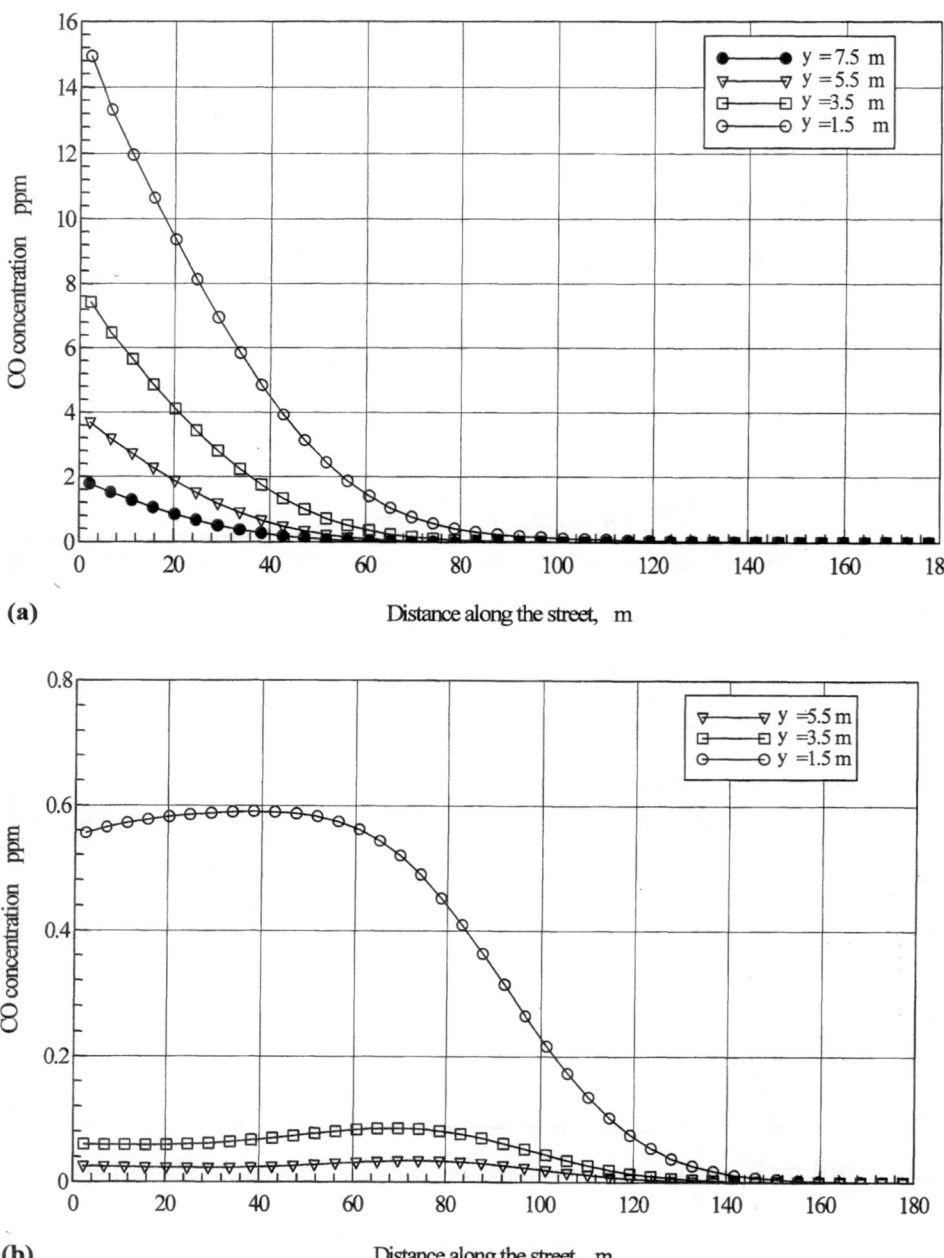

Figure 8. (a) Leeward face CO concentrations ($\phi = 80°$) wind right to left. (b) Windward face CO concentrations ($\phi = 80°$) wind right to left.

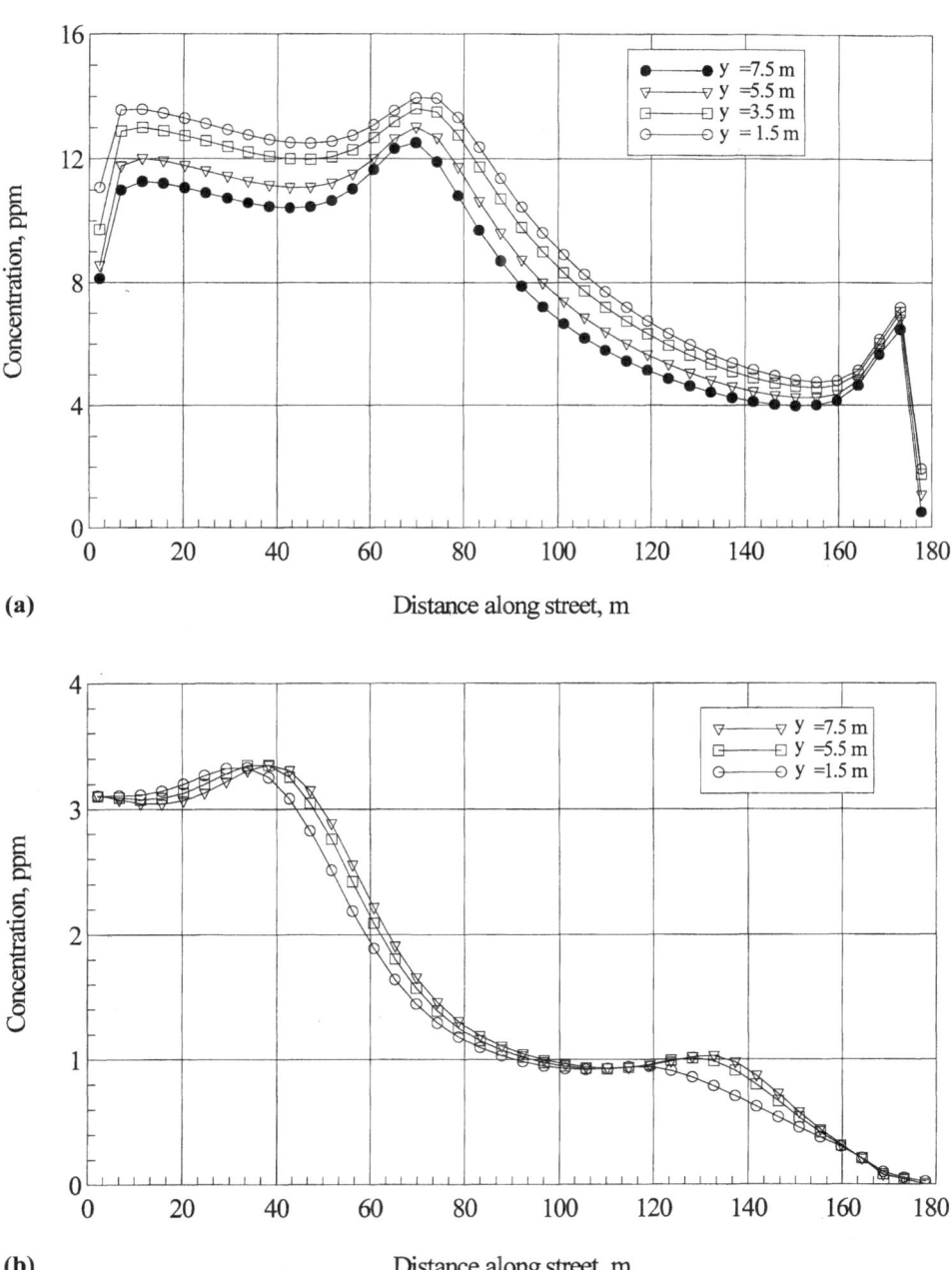

(a)

(b)

Figure 9. (a) Leeward face CO concentrations ($\phi = 20°$) wind right to left. (b) Windward face CO concentrations ($\phi = 20°$) wind right to left.

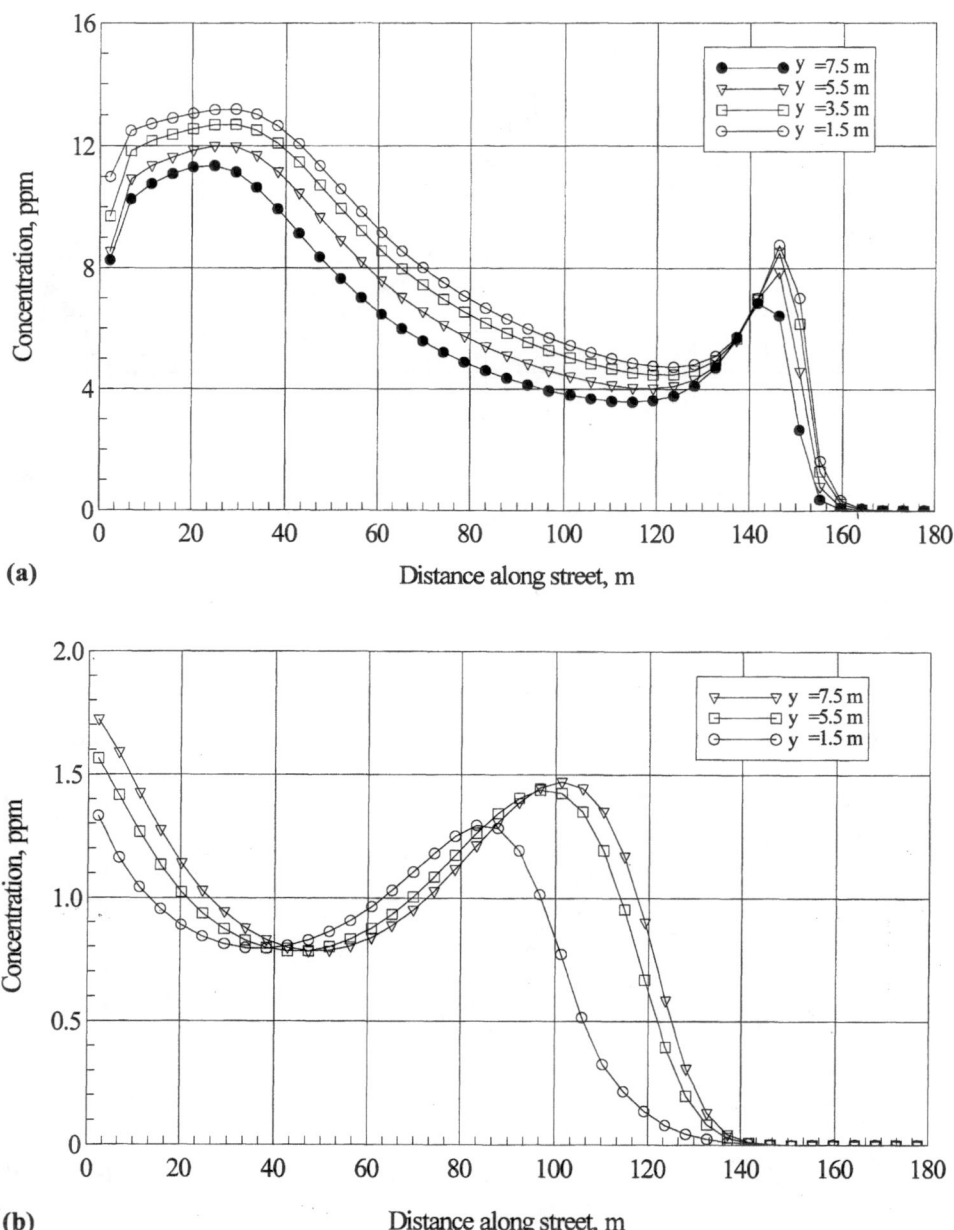

Figure 10. (a) Leeward face CO concentrations ($\phi = 40°$) wind right to left. (b) Windward face CO concentrations ($\phi = 40°$) wind right to left.

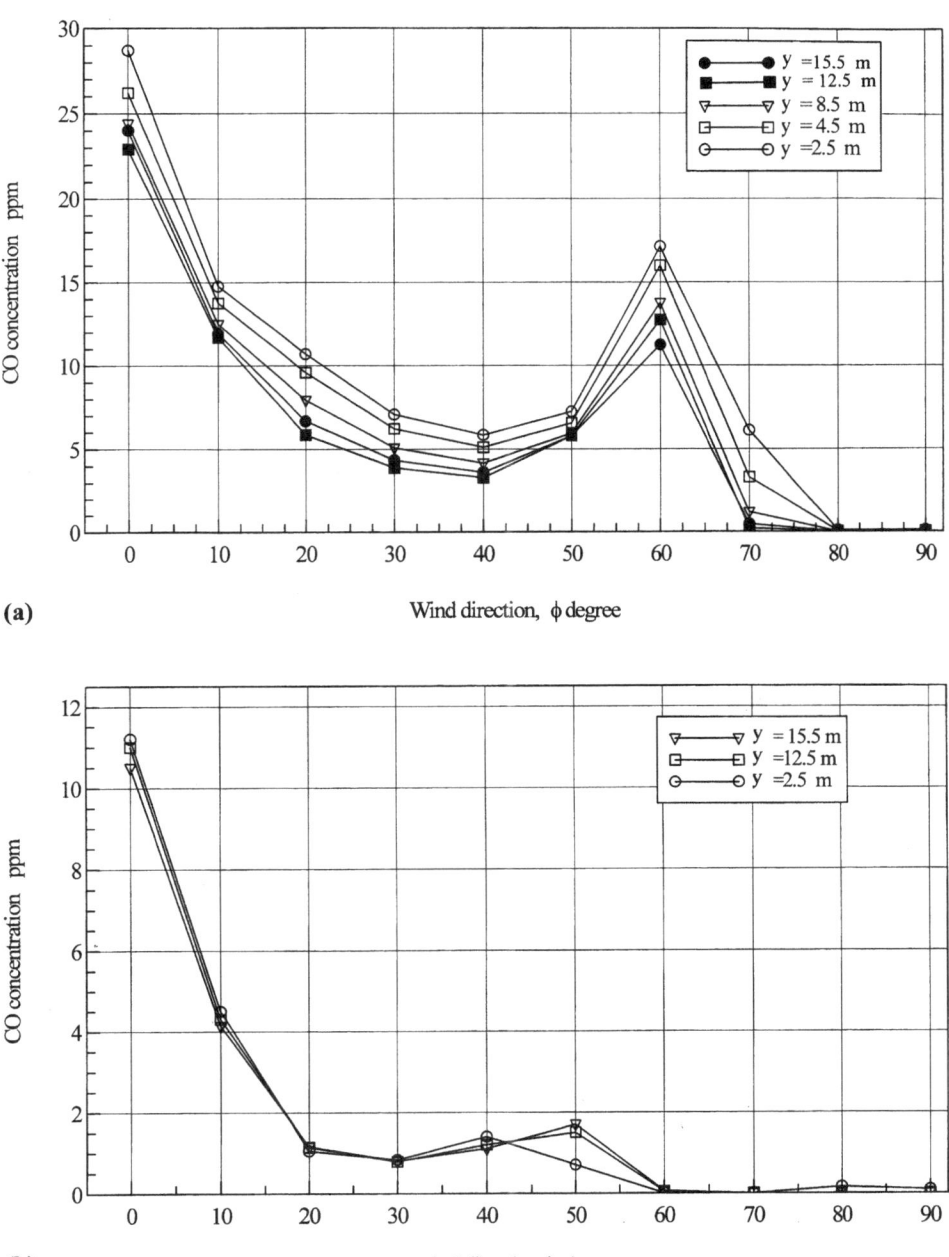

(a)

(b)

Figure 11. (a) Leeward face CO concentrations at mid-block as a function of wind direction. (b) Windward face CO concentrations at mid-block as a function of wind direction.

to the street (i.e. $\phi = 0°$). The concentration decreased with increasing wind direction to reach a minimum value at $\phi = 40°$ and then started to increase at $\phi = 60°$. At the windward face the concentration was found to have high values for $\phi = 0°$ and low values when the wind direction was parallel to the street. The concentration pattern obtained by Hoydysh and Dabberdt is quite similar to our present results except when the wind direction was parallel to the street.

4. Conclusions

In this article the PHOENICS CFD has been used to simulate the dispersion of air pollutants in a three-dimensional urban street canyon. The results show that the canyon geometry and wind direction can influence strongly the dispersion of pollutants. It should be stressed that these findings for oblique incidence require validation by field observations and wind tunnel studies. The k-ε turbulence model appears to be capable of simulating the dispersion of vehicular emissions within street canyons and the present results agree qualitatively and quantitatively with previous wind tunnel and field research findings at normal incidence (Hassan and Crowther, 1998b).

References

Arnfield, A. J. and Mills, G. M.: 1994, 'An analysis of the circulation characteristics and energy budget of a dry, asymmetric, east-west urban canyon. I. Circulation characteristics', *Int. J. Climatol.* **14**, 119–134.

Bardeschi, A., Colucci, A., Gianelle, V., Gnagnetti, M., Tamponi, M. and Tebaldi, G.: 1991, 'Analysis of the impact on air quality of motor vehicle traffic in the Milan urban area', *Atmos. Environ.* **25B**(3), 415–428.

Dabberdt, W. F. and Hoydysh, W. G.: 1991, 'Street canyon: Sensitivity to block shape and entrainment', *Atmos. Environ.* **25A**, 1143–1153.

Dabberdt, W. F., Shelar, E., Marimont, D. and Skinner, G.: 1981, 'Analysis, Experimental Study, and Evaluation of Control Measures for Air Flow and Air Quality on and near Highways', Federal Highway Administration, FHWA/RD-81/051, *Final Report*, SRI International, Menlo Park, CA 94025.

DePaul, F. T. and Sheih, C. M.: 1984, 'A Study of Pollutants Dispersion in an Urban Street Canyon', Argonne National Laboratory, *Technical Report*, ANL/ER-84-1,84.

DePaul, F. T. and Sheih, C. M.: 1986, 'Measurements of wind velocities in a street canyon', *Atmos. Environ.* **20**(3), 455–459.

Hassan, A. A. and Crowther, J. M.: 1998a, 'A simple model of pollutant concentrations in a street canyon', *Env. Mon. and Assessment* **52**, 269–280.

Hassan, A. A. and Crowther, J. M.: 1998b, 'Modelling of fluid flow and pollutant dispersion in a street canyon', *Env. Mon. and Assessment* **52**, 281–297.

Hoydysh, W. G. and Dabberdt, W. F.: 1988, 'Kinematics and dispersion characteristics of flow in asymmetric street canyons', *Atmos. Environ.* **22**, 2677–2689.

Hoydysh, W. G., Ogawa, Y and Griffiths, R. A.: 1974, 'A Scale Model Study of Dispersion of Pollution in Street Canyons', APCA Paper No. 74-157, *67th Annual Meeting of the Air Pollution Control Association, Denver, Colorado*, 9–13 June.

Hunter, L. J., Johnson, G. T. and Watson, L. D.: 1992, 'An investigation of three-dimensional characteristics of flow regimes within the urban canyon', *Atmos. Environ.* **26B**(4), 425–432.

Johnson, G. T., Hunter, L. J. and Arnfield, A. J.: 1990, 'Preliminary field test of an urban canyon wind flow model', *Energy and Building* **15–16**, 325–332.

Johnson, W. B., Ludwig, F. L., Dabberdt, W. F. and Allen R. J.: 1973, 'An urban diffusion simulation model for carbon monoxide', *J. Air Poll. Control Assoc.* **23**, 490–498.

Jounard, R.: 1987, *iEmissions Unitaires des Vehicules Legers*, Laboratoire Energies Nuisance – Case 24-F69675, Institut National de Recherche sur les Transports et leurs Securité, Bron Cedex, France.

Kennedy, I. M. and Kent, J. H.: 1977, 'Wind tunnel modelling of CO dispersal in city streets', *Atmos. Environ.* **11**, 541–547.

Launder, B. E. and Spalding, D. B.: 1972, *Lectures in Mathematical Models of Turbulence*, Academic Press, London and New York.

Lee, I. Y. and Park, H. M.: 1994, 'Parameterization of the pollutant transport and dispersion in urban street canyons', *Atmos. Environ.* **28**(14), 2343–2349.

Mohammadi, B. and Pironneau, O.: 1994, *Analysis of the k-Epsilon Turbulence Model*, John Wiley & Sons, New York.

Moriguchi, Y. and Uehara, K.: 1993, 'Numerical and experimental simulation of vehicle exhaust gas dispersion for complex urban roadways and their surroundings', *J. of Wind Eng. and Ind. Aerodyn.* **46–47**, 689–695.

Nakamura, Y. and Oke, T. R.: 1988, 'Wind, temperature and stability conditions in an east-west oriented urban canyon', *Atmos. Environ.* **22**, 2691–2700.

Nicholson, S. E.: 1975, 'A pollution model for street-level air', *Atmos. Environ.* **9**, 19–31.

Qin, Y. and Kot, S. C.: 1993, 'Dispersion of vehicular emission in street canyons, Guangahou City, South China (P.R.C.)'. *Atmos. Environ.* **27B**(3), 283–291.

Yamartino, R. J. and Gotz, W.: 1986, 'Development and evaluation of simple models for the flow, turbulence and pollutant concentration fields within an urban street canyon', *Atmos. Environ.* **20**, 2137–2156.

Wedding, J. B., Lombardi, D. and Cermak, J.: 1977, 'A wind tunnel study of gaseous pollutants in city street canyons', *J. Air Poll. Control Assoc.* **27**, 557–566.

ANALYSIS OF THE ST. PETERSBURG TRAFFIC DATA USING THE OSPM MODEL

ALEXANDER ZIV[1]*, RUWIM BERKOWICZ[2], EUGENE GENIKHOVICH[1], FINN PALMGREN[2] and EKATERINA YAKOVLEVA[1]

[1] *Department of Monitoring of Air Pollution, Main Geophysical Observatory, St. Petersburg, Russia;* [2] *Department of Atmospheric Environment, National Environmental Research Institute, Roskilde, Denmark*

(* *author for correspondence, e-mail: adz@main.mgo.rssi.ru, fax: +7 812 247 86 61*)

Abstract. Since October 1998 two DOAS instruments were installed at the level of the first floor and at the top of a building located in St. Petersburg at Pestelya Street. The collected data covers the time period of December 1998–March 2001, and include concentrations of benzene, toluene, NO and NO_2, ozone and SO_2. There is also an additional information about the traffic intensity and meteorological conditions. The results of the analysis of this data set, using the OSPM model, are presented here with the goal to understand the features of the air pollution dispersion in this street canyon and to analyse the information about the emission factors of the vehicles. In particular, the model results are used for the solution of the inverse problem of reconstructing the emission factors from measured concentrations. The results obtained indicate that most of the concentrations are well inside the Russian standards with the only exception of NO_2 (mean and 98-th percentile are equal to 57.8 and 119.2 μg m^{-3} for the street level). The same values for benzene are 18.5 and 62.6, respectively. Emission estimates show that there is a possibility that the NO_x and benzene basic emission factors recommended by the Russian national guidelines could result in overestimating the traffic emissions. These considerations are supplemented with the model sensitivity tests carried out in connection with the problem of predictability of NO_2 concentrations in the street canyon. Tests indicate that NO_2 concentrations are not very sensitive to NO_x emissions because of the usually low urban background ozone levels.

Keywords: air pollution, inverse dispersion calculations, OSPM model, street canyon

1. Introduction

In the framework of the Danish-Russian project 'Monitoring and analysis of the air pollution in St. Petersburg', two DOAS instruments were installed in St. Petersburg on Pestelya Street, which could be considered as a street canyon (Genikhovich *et al.*, 2000). In the current study we used data obtained during this project to investigate the air pollution in one of the streets in the downtown of St. Petersburg and to estimate the emission factors for different types of vehicles using Danish Operational Street Pollution Model (OSPM) (Berkowicz *et al.*, 1997a). Pestelya Street is oriented approximately from the East to West (the site map is shown in Figure 1). DOAS instruments are located in the street (2.5 m above the ground) and

Water, Air, and Soil Pollution: Focus **2:** 297–310, 2002.
© 2002 *Kluwer Academic Publishers.*

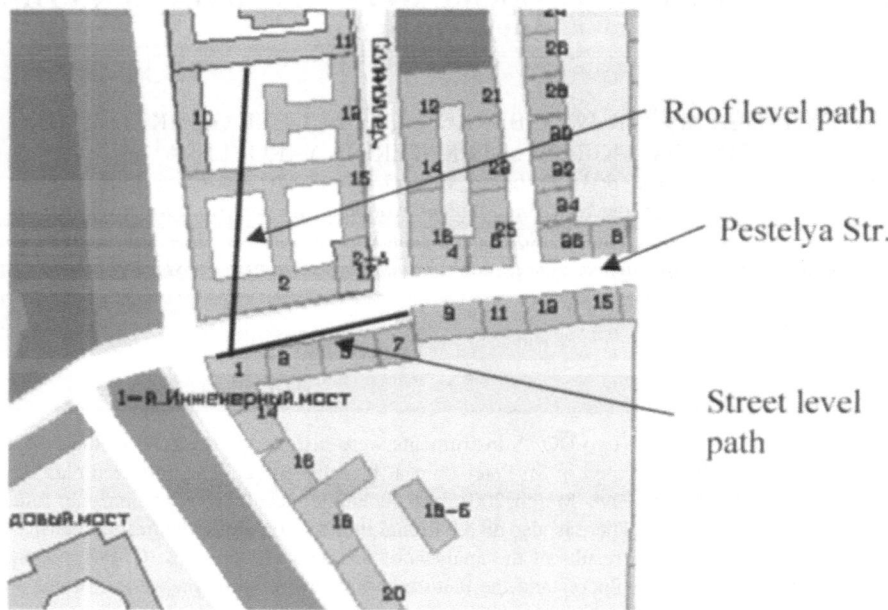

Figure 1. The map of the site where the instruments are located.

at the roof (15 m high) level. The distances from the instruments to the reflecting mirrors are 138 and 158 m, respectively. The street level measurement path goes along the southern side of the street and the upper-level path goes across the street and over the roofs of the surrounding buildings. A characteristic feature of the St. Petersburg experiment is, that it is one of the few where a DOAS instrument is used for the street level measurements during a rather long period. The collected data covers the time period of December 1998–March 2001, and included concentrations of benzene, toluene, NO, NO_2, ozone and SO_2. There is also an additional information about the traffic intensity measured manually for different types of vehicles. As it was expected, most of the vehicles on Pestelya Street were passenger cars. The diurnal variation of the number of vehicles of different types is shown in Figure 2 and corresponds to the weekly period in May–June, 1999. The additional measurements were carried out from time to time during the period of December 2000–February 2001, and showed that the traffic intensity in this winter period became 10–20% lower. It should be mentioned that approximately half of the vans, trucks and buses have gasoline engines. Only a small percentage of passenger cars is equipped by catalysts.

Continuos meteorological measurements at Pestelya Street started first in September 2000. The concentrations measured before this time were analysed using the data obtained from the meteorological mast located about 5 km from the DOAS instruments. Comparison of the wind directions and wind speeds at these two sites, measured during five-months period from September 2000, has shown

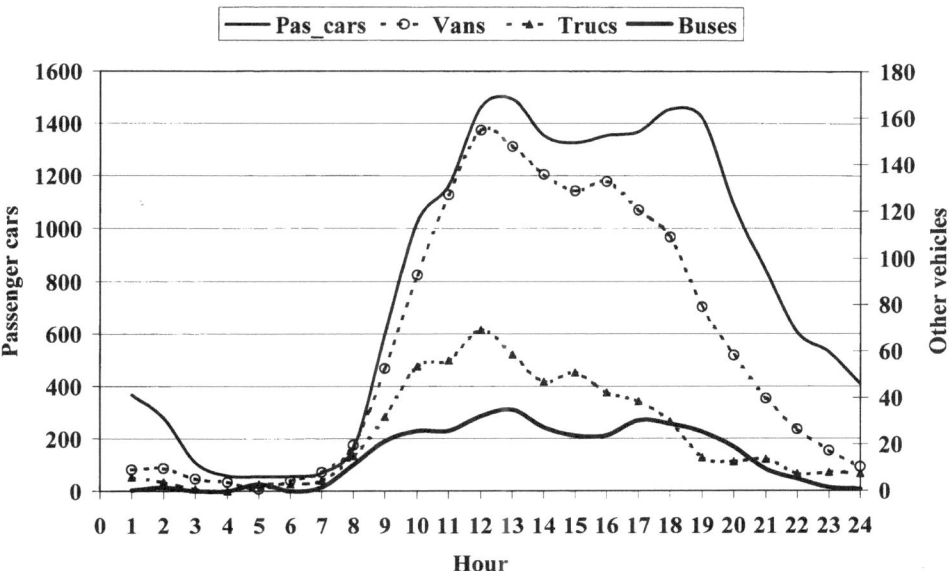

Figure 2. Traffic intensity (vehicles per hour) on Pestelya street.

that they are in a good agreement, but the factor of 0.8 was obtained from the regression analysis for correcting wind speed at the remote mast.

2. Statistical Results

The statistical analysis of the results of measurement campaigns in different cities is widely reported in papers (Sandroni *et al.*, 1995; Hargreaves *et al.*, 2000; Kourtidis *et al.*, 2000; Karppinen *et al.*, 2000; etc.). Depending on the main ideas of the investigation this analysis can be more or less detailed. In the present study we consider only the results of the general statistical analysis of the hourly averaged concentrations measured at the street and roof levels (see Table I). Most of the concentrations are well inside the Russian standards, both for long and short averaging times (means and 98 percentiles, respectively), with the only exception of NO_2, but even in this case the exceedance of standards is not higher than 1.45 times. In comparison with the proposed EU standards, however, concentrations of benzene are very high. The ratio of NO_2 and NO_x average values, which can be considered as measure of transformation $NO \rightarrow NO_2$, is different at the street (≈ 0.5) and roof (≈ 0.65) levels. The ratios of mean concentrations at two levels for benzene, toluene and NO_x are equal to 2.9, 2.6 and 1.7, respectively. One can notice that the ratio for NO_x is significantly less than for hydrocarbons. This fact could be attributed to the influence of all sources of NO_x located outside Pestelya Street. Notice that the variance coefficients in Table I change for different pollutants – they are higher for

TABLE I

The statistical characteristics of the observed concentrations (in $\mu g \ m^{-3}$)

	Mean	Median	Variance coeff.	Min	Max	98%	99.9%
Street level							
Benzene	18.5	13.8	0.82	−1.3	121.8	62.6	74.9
Toluene	66.6	45.3	0.97	−4.8	478.9	253.2	297.4
NO	40.9	28.7	0.93	−1.7	211.5	144.2	162.4
NO_2	57.8	55.5	0.44	5.6	242.6	119.2	127.7
Ozone	25.5	23.8	0.60	0.0	95.9	61.2	65.3
NO_x	116.7	98.3	0.65	5.1	420.9	313.9	342.3
SO_2	12.3	7.1	1.33	0.0	301.0	56.2	76.2
Roof level							
Benzene	6.3	5.0	0.84	−1.5	76.8	21.0	26.6
Toluene	25.5	19.4	0.96	−5.8	348.7	92.8	115.3
NO	16.3	10.2	1.10	−0.7	156.6	71.4	89.9
NO_2	43.8	40.9	0.46	3.8	221.6	93.7	105.0
Ozone	37.0	34.7	0.48	1.4	164.4	76.1	80.3
NO_x	67.2	56.3	0.63	6.8	374.3	185.9	214.7
SO_2	11.1	5.8	1.40	−1.1	237.0	54.8	74.1

hydrocarbons and SO_2. However, except ozone and NO, for each of the species, these variances are almost equal at both levels.

Relatively high cross-correlations (see Table II) between the concentrations at the street and roof levels and between the differences (DIFF) between both levels, was found for non-reactive pollutants benzene, toluene and NO_x. It confirms, in particular, that these pollutants are emitted mainly by traffic. The slightly lower correlations for differences in comparison with those for the street level are probably explained by the higher contribution of the noise in data, when we subtract the values at the roof level from those at the street level. It follows from Table I that the means for sulphur dioxide are almost the same for both levels. Though more accurate analysis (after removing of several outliers) shows, however, the correlation equals to 0.72 between $DIFF_{NO_x}$ and $DIFF_{SO_2}$. This indicates that SO_2 is partly emitted from vehicles in the street as well.

TABLE II

Cross-correlation between the traffic induced pollutants

		BEN			**BEN**			**BEN**	
		Street			Roof			DIFF	
Street	**TOL**	0.97	**TOL**						
	NOx	0.87	0.91	**NOx**					
Roof	**BEN**	0.76	0.78	0.68	**BEN**				
	TOL	0.74	0.79	0.71	0.93	**TOL**			
	NOx	0.73	0.78	0.80	0.80	0.87	**NOx**		
DIFF	**BEN**	0.96	0.92	0.84	0.58	0.58	0.61	**BEN**	
	TOL	0.94	0.96	0.86	0.65	0.64	0.65	0.95	**TOL**
	NOx	0.75	0.75	0.87	0.45	0.40	0.39	0.77	0.80

3. Emission Estimates

For the estimations of both total emissions and emission factors Danish OSPM model had been used previously (Palmgren et al., 1999). It is a street pollution model (Berkowicz et al., 1997a) based on a simplified description of flow and dispersion conditions in street canyons. Concentrations at any point inside the canyon are computed using a combination of a plume model for the direct contribution from street traffic, box model for the recirculating part of pollutants and three reactions (between NO, NO_2 and O_3) chemical model. As it was shown in several studies (Berkowicz et al., 1997b; Berkowicz, 1999; Kukkonen, 2001) this model usually gave very promising results. For the present study OSPM model has been adjusted in such a way that the output values became ones averaged over the measurement path. Only the western part of Pestelya Street, where the street level measurement path goes (see Figure 1), was considered for the modelling. Because this path is long enough, we assume that the contribution of traffic emissions from the adjoining streets (with very comparable traffic intensity) is negligible.

The seasonal variation of the total emissions from traffic in Pestelya Street has been estimated in the following way. For each of the 24 hours of the day for the four seasons separately, we first calculated the coefficients in the linear regression of measured differences of the concentrations between street and roof levels on modelled concentrations with unity emissions. Emission values for each season have been obtained then as averages of the corresponding 24 coefficients. The results for benzene and NO_x are presented in Figure 3. The levels of the benzene emission are much higher for the winter period and decrease during the rest of seasons. It is well known that the benzene emission intensity varies significantly with the speed of vehicles (Heeb et al., 2000). During the winter and early in spring the traffic

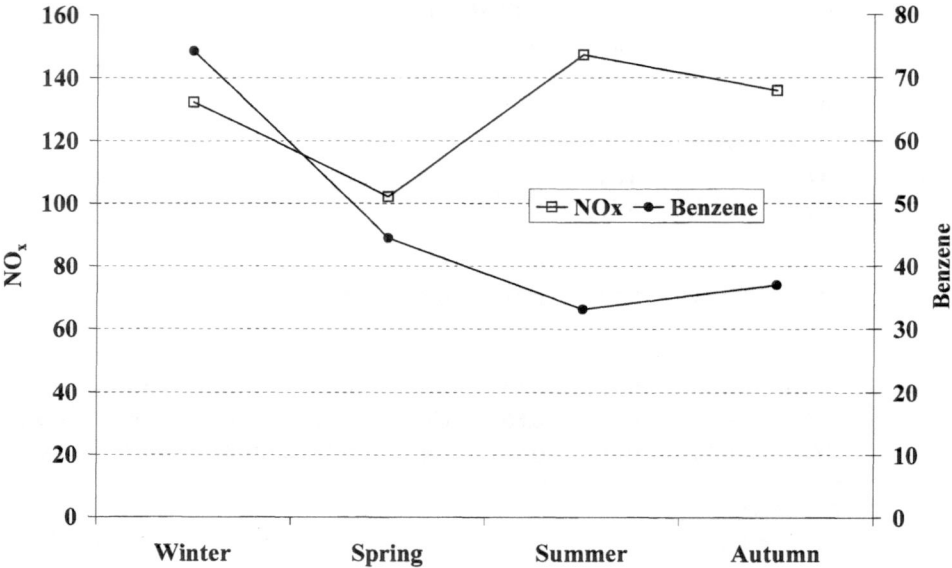

Figure 3. Seasonal variation of the averaged calculated emissions (μg m^{-1} sec^{-1}) for benzene and NO$_x$.

conditions often become much worse and the car fleet velocity correspondingly drops. It might be that this factor even overlaps the slight reduction in the car fleet intensity during this period. For NO$_x$ the variation of the emission levels is not so definite taking into account the influence of random errors. This distinction between NO$_x$ and benzene is in a good agreement with the above-mentioned higher coefficient of variance for the hydrocarbons in Table I.

The problem of obtaining the realistic values of the emission factors is widely discussed (KuKhlwein and Friedrich, 2000; Vogel *et al.*, 2000; Venkatram, 2000). The OSPM calculations were firstly carried out using the emission factors from the Russian guidelines (Milyaev *et al.*, 1999). The results show a relatively good agreement from the point of the correlation but significant discrepancy in scale. To derive the actual emission factors from the joint use of the modelled and measured concentrations we considered the approximation of the observed time series of differences D by the following sum of four terms:

$$D \approx E_c * T_c + E_v * T_v + E_b * T_b + E_t * T_t , \qquad (1)$$

where scalars E denote the emission factors (in g km^{-1}); time series T denote the traffic intensity multiplied by the speed correction factor, corresponding unit conversion coefficient and the model result for the unity emission; subscripts c, v, b, t correspond to the different types of vehicles (passenger cars, vans, buses and trucks).

TABLE III

Emission factors (g km^{-1}) for NO$_x$ and benzene

	Cars	Vans	Trucks	Buses
NO$_x$				
Reference value	1.8	2.9	6.45	6.65
Experimental	0.84	0.84	3.28	3.28
Benzene				
Reference value	0.63	0.45	0.91	0.98
Experimental	0.16	0.16	0.44	0.44

The emission factors should be obtained then from Equation (1) as regression coefficients. However, due to the high correlation between the fleet intensity of the different types of vehicles and small contribution of both light-duty and heavy-duty cars (see Figure 2), the errors in initial data induces the non-stable solutions. It means that the direct use of the linear regression could lead to significant errors in the results. In fact we consider the task, which is close to the variety of source apportioned problems (Henry, 1987; Hopke, 1988; Shen and Israel, 1989; Wang and Hopke, 1989). When the similar problem was considered in connection with the data from Jagtvej Street in Copenhagen (Palmgren et al., 1999), it was found, in particular, that for simplification of the task it was worthwhile to combine several categories of vehicles into one category.

In the present study we estimated the emission factors splitting all the vehicles into two groups only: one combining passenger cars and vans and second – trucks and buses. The estimation procedure was as follows. The mean squared modelling error, i.e. the difference between left-hand and right-hand parts of (1) (only two terms on the right), was minimised subject to the four constraints: both means and 98-th percentiles of the observed and modelled values should differ less than 10% and both emission factors should be positive. Because of the second constraint the solution was found by the direct enumeration. The results of the estimation of emission factors are presented in Table III (experimental estimates) together with those taken from Milyaev et al. (1999, reference values). The evident discrepancy needs a further investigation.

Figures 4 and 5 show the quintile–quintile plots both for the street level concentrations and the differences between street- and roof levels. It follows from these plots that the distributions of the modelled and measured concentrations are rather similar with slight model underestimation especially for the street level values.

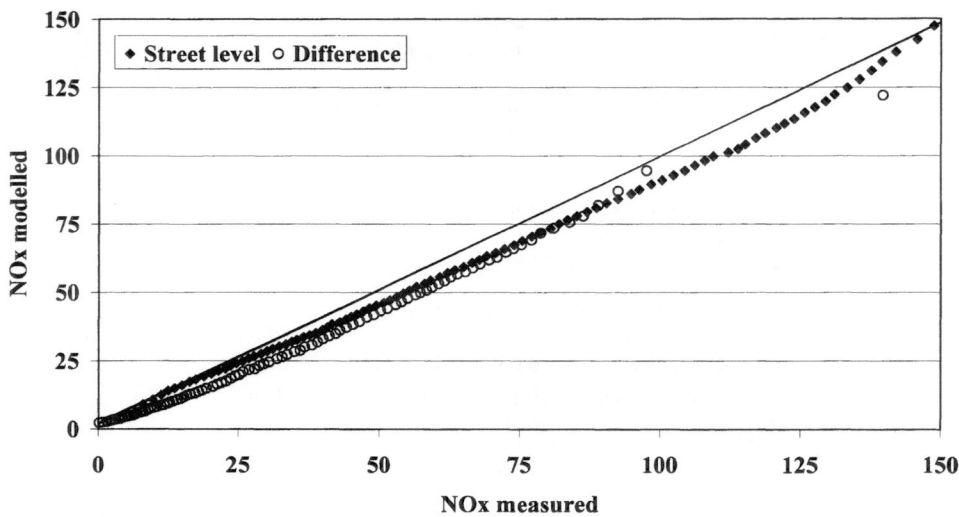

Figure 4. Quintile–quintile plot for NO_x.

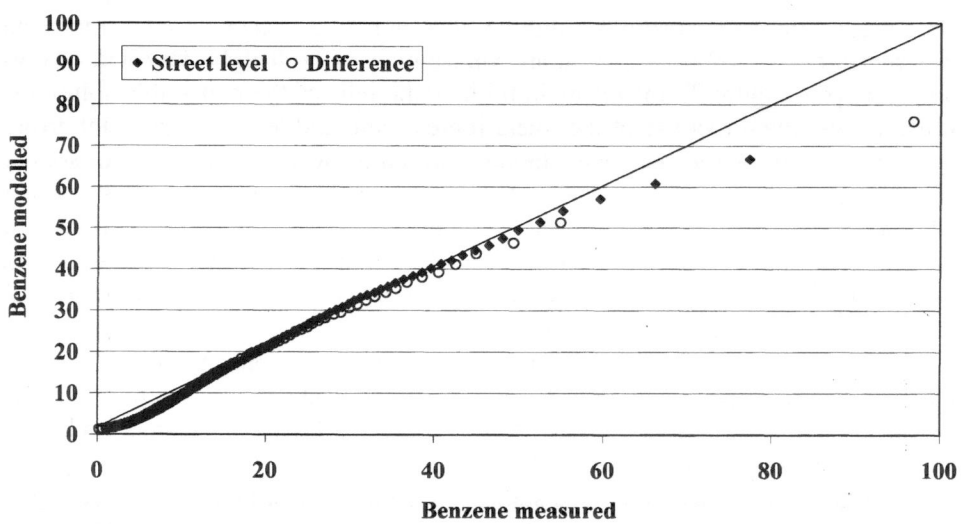

Figure 5. Quintile–quintile plot for benzene.

Obtained emission factors have been used further for the comparison of observed and modelled concentrations. As an example the scatterplot of NO_x is shown in Figure 6.

The equations for the trendlines with zero and non-zero intercept and corresponding correlation coefficients both for the street level concentrations and differences (DIFF), which are actually modelled, are presented in Table IV.

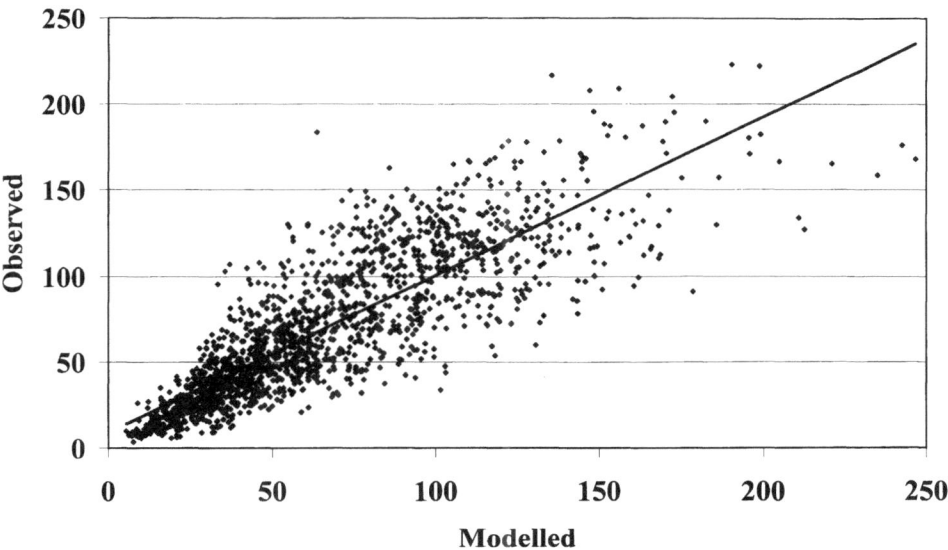

Figure 6. Observed versus modeled concentrations for NO_x.

TABLE IV

The trendlines for the scatterplots of observed versus modelled concentrations

Trendline type	NO_x		Benzene	
	Equation	Correlation	Equation	Correlation
Zero intercept	y = 1.02x	0.88	y = 1.1493x	0.81
Non-zero intercept	y = 0.91x + 10.71	0.89	y = 1.04x + 2.91	0.82

Trendline type	DIFF NO_x		DIFF Benzene	
	Equation	Correlation	Equation	Correlation
Zero intercept	y = 0.99x	0.62	y = 1.17x	0.67
Non-zero intercept	y = 0.76x + 11.45	0.68	y = 0.98x + 3.87	0.7

It can be seen that the agreement between the observed and modelled results is better in case of the street level. It is evident of course since, in the first case we add the background concentrations, which are measured at the roof level, to the calculated ones, and these background concentrations are not small and well correlate with the street level measurements (see Tables I and II).

We considered also the distribution of the modelled and observed concentrations (differences) over the wind directions for the two wind speed classes (U <

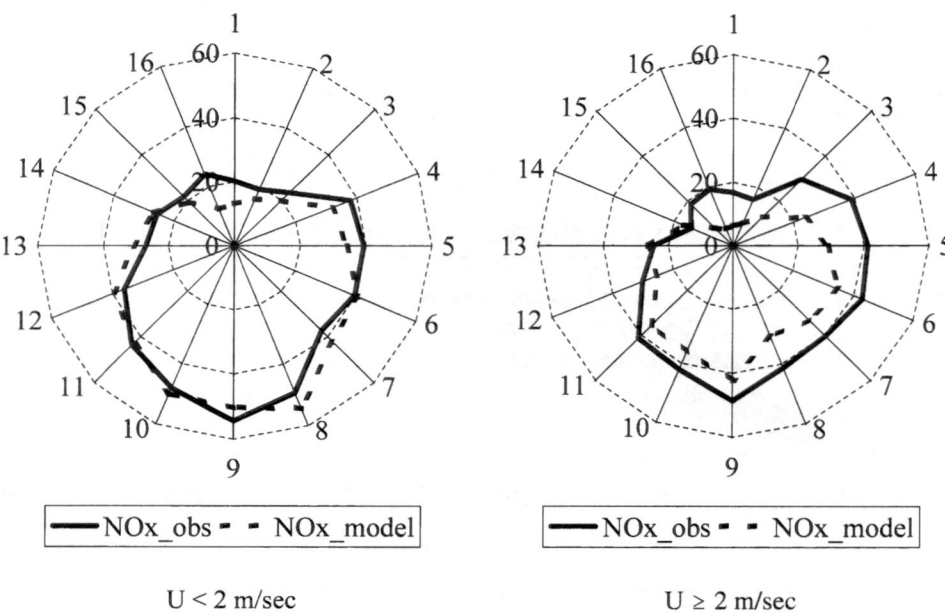

Figure 7. Distribution of the modelled and observed differences for NO_x over the wind directions and two ranges of the wind speed.

2 m s^{-1} and $U{\geq}2 \text{ m s}^{-1}$). For each class the data have been divided in accordance with 16 wind direction sectors (the middle point of the first one corresponds to the wind direction from the northern perpendicular to the street canyon). The average values for the each subset for both measured differences and modelling results are presented in Figure 7, in the form of the concentrations 'roses'. It can be seen that for the both classes of wind speed the shapes of the curves follows one another well enough. However, we can observe that the model results are in a better agreement with the measured values for the low wind speeds. At the right-hand plot the experimental data overestimates the modelled results. Probably it can be attributed to the fact that the calculations are too sensitive to the wind speed.

Figure 8 demonstrates for the street level the relations between NO_2 and NO_x concentrations (in $\mu g \text{ m}^{-3}$) and the ratio of these two concentrations. For the high concentrations this ratio drops, and it can be seen from the top plot that NO_2 concentrations become less dependent from NO_x. A simple chemical model is used in OSPM to calculate the concentrations of NO_2. In case of low urban background ozone levels (that is the case for the St. Petersburg conditions) calculated NO_2 concentrations are not very sensitive to NO_x emissions. To demonstrate it, NO_2 concentrations were calculated using two sets of emission factors for NO_x from Table III. There is an almost perfect agreement between these two calculation results. The corresponding slope of the regression line equals to 0.92 and the correlation coefficient to 0.97 (the corresponding slope for NO_x is 0.45). The

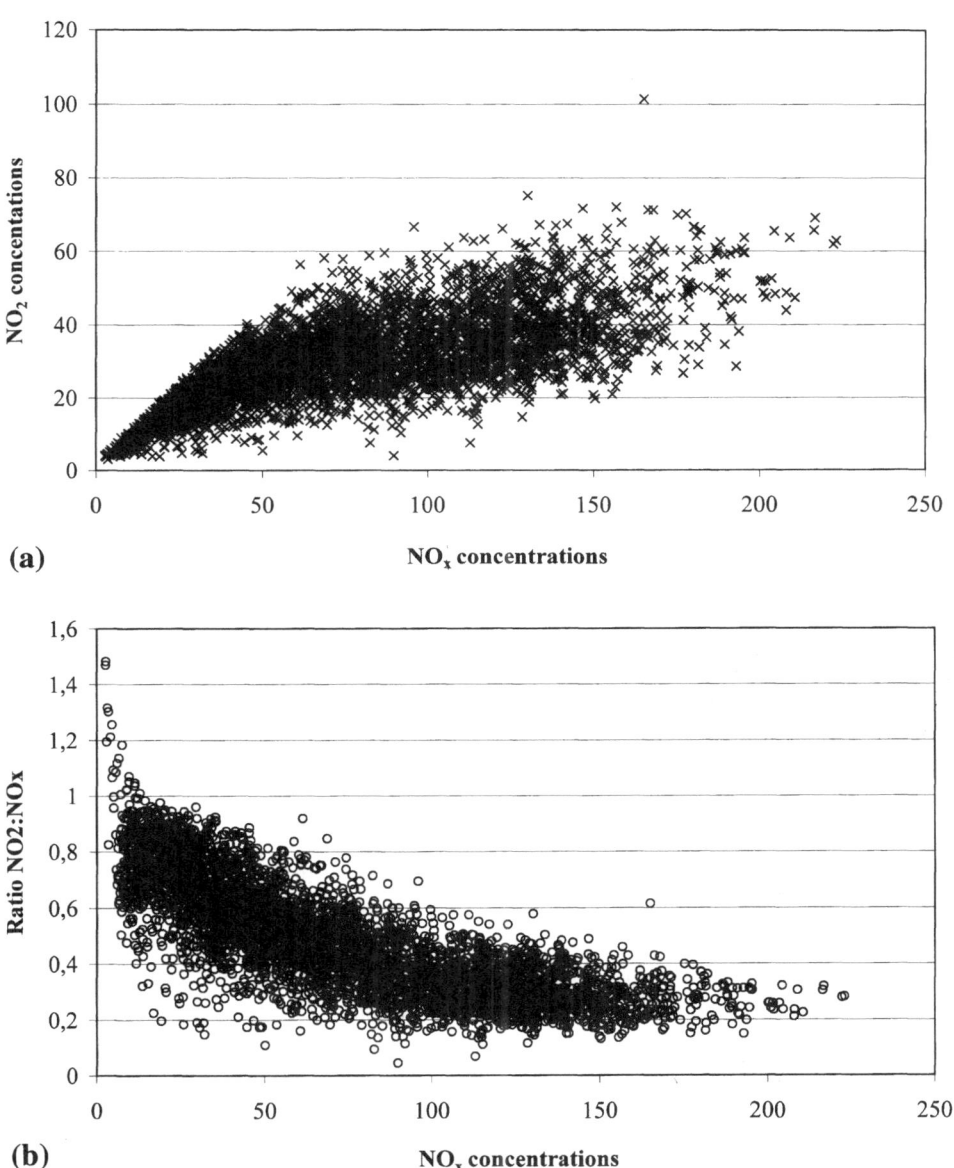

Figure 8. NO$_2$ versus NO$_x$ concentrations for the street level (a) and the corresponding ratio versus NO$_x$ (b).

conclusion, which can be made from these considerations, is that the emission estimations should not be carried out using NO_2 measurements.

4. Conclusions

Since December 1998 a measurement campaign has been carried out in the street canyon of Pestelya Street in the downtown of St. Petersburg. Concentrations of benzene, toluene, NO, NO_2, ozone and SO_2 are measured using DOAS instruments at both street- and roof levels. These experimental results together, with the relevant meteorological information and traffic intensity data, were used to investigate the main features of the air pollution in this street canyon, which is considered as typical one for St. Petersburg. From the joint analysis of concentrations measured at two levels it was found, in particular, that the concentrations of benzene, toluene and NO_x are well correlated and, therefore, definitely induced mainly by the traffic emissions. The comparison of the simple statistical characteristics of these pollutants at two levels showed, however, the differences, which could be attributed probably to that fact that in case of NO_x the contribution of the sources located outside the Pestelya Street is more significant. The evaluation of inter-seasonal variations of the traffic emissions using OSPM model indicated also some difference between NO_x and benzene. The emissions of benzene have the definite maximum during the winter time, which could likely be explained by the large sensitivity of the benzene emissions to the vehicles speed.

Runs of the OSPM with the reference values of the emission factors, accepted in Russia, does not show the satisfactory correlation of the modelled and observed concentrations, but the difference in their levels. The special technique was developed in this article and applied to the estimation of the emission factors. The results demonstrated a significant discrepancy with the above mentioned reference values, and this discrepancy should be specially investigated. Runs of the OSPM model, using obtained emission factors, showed that the probability distributions of calculated and measured concentrations are very close. The correlation coefficients for the scatterplots (observed versus modelled concentrations) of NO_x and benzene are 0.68 and 0.7, respectively. However, if we consider concentrations observed at the street level and the sum of modelled concentrations and those measured at the roof level, the correlation coefficients for the scatterplots are equal to 0.89 and 0.82. The model calculations indicated also that the NO_2 concentrations in St. Petersburg area were not very sensitive to NO_x emissions, because of the usually low urban background ozone levels.

All these results confirm that, in general, OSPM model reproduces the air pollution processes in the Pestelya Street canyon reasonably well. Moreover these results of application of OSPM are in a good agreement with those previously published by Berkowicz et al. (1997a, b), Palmgren et al. (1999) and Kukkonen et al. (2001). At the same time there are still some questions for the future investigation.

In particular – it is still an open question – why OSPM performs better for low wind speeds (less than 2 m s^{-1}). This model should be also tested on an 'independent' set of data, which were not used when the emission factors were determined.

Acknowledgements

We wish to express our great thanks to the Danish National Research Council, which supported this work. It was carried out as a part of the EUROTRAC-2 project (subproject SATURN), which provided an excellent platform for the international scientific co-operation. We are also very grateful to the Environmental Protection Agency of St. Petersburg, headed by Mr. A. S. Baev and also to his colleagues Ms. S. Gluhova, Mr. I. Chmilev, and Mr. A. Zaporozgez for the support of our study. They kindly provide us also with the information on meteorological observations in St. Petersburg. We would like to acknowledge the technical stuff of the institute 'Nevskgeologia' in St. Petersburg and, in particular, Mr. Yu. Pravdyuk for their help in maintenance of the instruments on Pestelya Street.

References

Berkowicz, R., Hertel, O., Larsen S., Sorensen, N. and Nielsen, M.: 1997a, *Modelling Traffic Pollution in Streets*, National Environmental Research Institute, Roskilde, Denmark, 52 pp.

Berkowicz, R., Hertel, O., Sorensen, N. and Michelsen, J.: 1997b, 'Modelling air pollution from traffic in urban areas', in R. Perkins and S. Belcher (eds), *Flow and Dispersion through Groups of Obstacles*, Clarendon Press, Oxford, pp. 121–141.

Berkowicz, R.: 1999, 'OSPM – Parameterised street pollution model', *Environmental Monitoring and Assessment*, accepted for publication.

Genikhovich, E., Ziv, A., Iakovleva, E., Palmgren, F. and Berkowicz, R.: 2000, 'Monitoring and Analysis of Air Pollution in St. Petersburg', A contribution to subproject SATURN. Eurotrac Newsletter, Vol. 22, p. 27.

Hargreaves, P., Leidi, A., Grubb, H., Howe, M. and Mugglestone, M.: 2000, 'Local and seasonal variations in atmospheric nitrogen dioxide levels at Rothamsted, U.K., and relationships with meteorological conditions', *Atmosph. Environ.* **34**, 843–853.

Heeb, N., Forss, A., Bach, C. and Mattrel, P.: 2000, 'Velocity-dependent emission factors of benzene, toluene and C 2-benzenes of a passenger car equipped with and without a regulated 3-way catalyst', *Atmosph. Environ.* **34**, 1123–1137.

Henry, R.: 1987, 'Current factor analysis receptor models are ill posed', *Atmosph. Environ.* **21**, 1815–1820.

Hopke, K.: 1988, Target transformation factor analysis as an aerosol mass apportionment method: A review and sensitivity study', *Atmosph. Environ.* **22**, 1777–1792.

Karppinen, A., Kukkonen, J., ElolaKhde, T., Konttinen, M., Koskentalo, T. and Rantakrans, E.: 2000, 'A modelling system for predicting urban air pollution: Model description and applications in the Helsinki metropolitan area', *Atmosph. Environ.* **34**, 3723–3733.

Kourtidis, K., Ioannis Ziomas, I., Christos Zerefos, C., Gousopoulos, A., Balis, D. and Tzoumaka, P.: 2000, 'Benzene and toluene levels measured with a commercial DOAS system in Thessaloniki, Greece', *Atmosph. Environ.* **34**, 1471–1480.

Kuhlwein, J. and Friedrich, R.: 2000, 'Uncertainties of modelling emissions from road transport', *Atmosph. Environ.* **34**, 4603–4610.

Kukkonen, J., Valkonen, E., Walden, J., Koskentalo, T., Aarnio, P., Karppinen, P., Berkowicz, R. and Kartastenpaa, R.: 2001, 'A measurement campaign in a street canyon in Helsinki and comparison of results with predictions of the OSPM model', *Atmosph. Environ.* **35**, 231–243.

Milyaev, V. B., Burenin, N. S., Shemyakov, P. M., Dvinyanina, O. V., Nesterova, N. N., Nahimovskaya, I. N., Lozhkin, V. N. and Volkodaeva, M. V.: 1999, *Guidelines for Determination of Traffic Emission Factors for the City Air Pollution Estimations*, St. Petersburg, 17 p.

Palmgren, F., Berkowicz, R., Ziv, A. and Hertel, O.: 1999, 'Emission estimates from the actual car fleet by air quality measurements in streets and street pollution models', *Sci. Total Environ.* **235**, 101–109.

Sandroni, S., Cerutti, C., Noriega, A. and Palmgren, F.: 1995, *Air Quality Measurements in Brussels (1993–1994)*, Published by the European Commission, Luxemburg, 40 p.

Shen, J. and Israel, G.: 1989, 'A receptor models using specific non-negative transformation technique for ambient aerosol', *Atmosph. Environ.* **23**, 2289–2298.

Venkatram, A.: 2000, 'A critique of empirical emission factor models: A case study of the AP-42 model for estimating PM_{10} emissions from paved roads', *Atmosph. Environ.* **34**, 1–11.

Vogel, B., Corsmeier, U., Vogel, H., Fiedler, F., KuKhlwein, J. and Friedrich, R.: 2000, 'Comparison of measured and calculated motorway emission data', *Atmosph. Environ.* **34** 2437–2450.

EXAMINATION OF TRAFFIC POLLUTION DISTRIBUTION IN A STREET CANYON USING THE NANTES'99 EXPERIMENTAL DATA AND COMPARISON WITH MODEL RESULTS

R. BERKOWICZ[1]*, M. KETZEL[1], G. VACHON[2], P. LOUKA[2], J.-M. ROSANT[2],
P. G. MESTAYER[2] and J.-F. SINI[2]

[1] *National Environmental Research Institute, Department of Atmospheric Environment, Roskilde, Denmark;* [2] *Laboratoire de Mécanique des Fluides UMR 6598 CNRS – Ecole Centrale de Nantes, Nantes Cedex 3, France*
(author for correspondence, e-mail: rb@dmu.dk, fax: +45 46301214)*

Abstract. During the Nantes'99 experiment, pollution concentrations, temperature, flow and turbulence conditions were measured at several locations in Rue de Strasbourg, Nantes, France. Traffic was measured by vehicle counters at different places within the street. Traffic speed was monitored as well. The measuring campaign was conducted in the period June–July 1999 but only data from a selected intensive observation period are used in this study. This period was selected to suit conditions required for study of the traffic produced turbulence and the thermal effects and is characterised by quite low wind speeds. The data are used here for examination of concentration distributions in the street. Measurements are compared to model results calculated by a simple parameterised model, the Operational Street Pollution Model (OSPM) and a 3-D CFD model MISKAM. Both models reproduce reasonably well the observed distribution of pollutants in the street. Due to predominantly low wind speed conditions, such effects as the traffic produced turbulence play a quite significant role. The model results provided by MISKAM are scaled using a velocity scale depending on the traffic produced turbulence. Application of a scaling velocity depending on wind speed only, provides unrealistic results.

Keywords: MISKAM, OSPM, traffic pollution, traffic produced turbulence, vertical distribution

1. Introduction

Traffic pollution in a street canyon is characterised by large temporal, horizontal and vertical variability, which is not only related to the diurnal variation in the traffic amount but is also influenced by the meteorological conditions. This has a great importance for e.g. evaluation of the urban population's exposure to the traffic pollution and therefore requires special attention. Very little is still known about the vertical distribution of the pollution in street canyons. Only a few experimental data exists and most of the data are from wind tunnel experiments (Pavageau *et al.*, 1997). Recently a new unique data set became available based on results from the Nantes'99 experiment. During this experiment, pollution concentrations, temperature, flow and turbulence conditions were measured at several locations in Rue de Strasbourg, Nantes, France. A detailed description of the experimental set-up

Water, Air, and Soil Pollution: Focus **2**: 311–324, 2002.

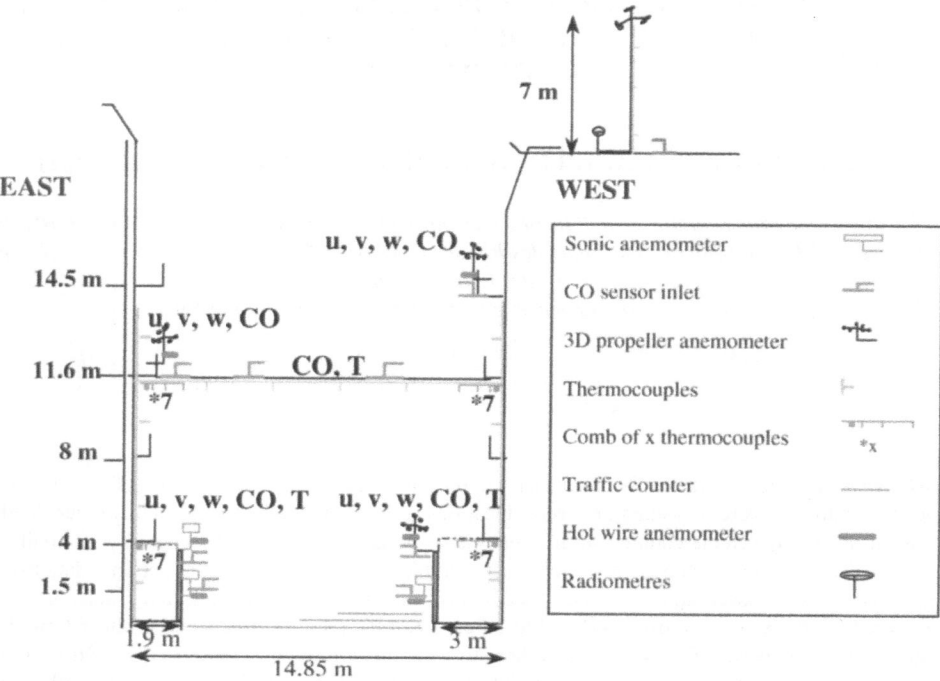

Figure 1. The experimental set-up of the measuring campaign in Rue de Strasbourg.

is given in Vachon *et al.* (1999). Analyses of the turbulence measurements in the street, with special emphasis on the dependence on traffic flow and thermal effects, are presented in Vachon *et al.* (2001) and Louka *et al.* (2001). In this article we focus on examination of the spatial distribution of traffic pollution and the effect the meteorology has on this variation.

The experimental data are compared with results from two models: the Operational Street Pollution Model (OSPM) (Berkowicz, 2000) and a 3-D CFD model MISKAM (Eichhorn, 1995). The models are not described in this article, and the readers are referred to the original publications. However, it should be emphasised that OSPM is a highly simplified, parameterised model, while the CFD model MISKAM is much more advanced and is believed to be suitable for detailed modelling of flow and dispersion conditions in urban streets. Some results of evaluation of several CFD models, incl. MISKAM, on wind-tunnel and field data are presented in Sahm *et al.* (2001) and Ketzel *et al.* (2001a).

2. The Site and the Experimental Set-up

Rue de Strasbourg is a 3-lane, one-way, highly trafficked street. The mean height of the buildings along the street is ca. 21 m and the width of the street is ca. 15 m, resembling a street canyon of the W/H ratio of ca. 0.7. The street orientation is 28° to West with respect to North. The experimental set-up is shown in Figure 1.

The components of flow and turbulence were measured by 3-D sonic and propeller anemometers at three levels on each side of the street. Meteorological parameters were also measured on a 7 m high roof mast, on the westerly side of the street.

Concentrations of CO were measured on both sides of the street at 1.5, 4 and ca. 12 m on the East side, and 1.5, 4 and ca. 16 m on the West side. Measurements were also available from an upper level location in the middle of the street, but they are not used in this study. Background concentrations were monitored at the roof location, the same place as the meteorological mast.

Traffic was measured by vehicle counters at different places within the street. Traffic speed was monitored as well.

3. The Data

The measuring campaign was conducted in the period June–July 1999 but only data from a selected intensive observation period are used in this study. This period was selected to suit conditions required for study of the traffic produced turbulence and the thermal effects (Vachon *et al.*, 2001; Louka *et al.*, 2001) and is characterised by quite low wind speeds. The frequency distribution of wind speed measured on the roof mast and the frequency distribution of wind directions are shown in Figure 2.

The available traffic measurements were used to construct an average diurnal traffic profile for the street (Figure 3a). The CO emissions calculated using the traffic flow data and vehicle emission factors are shown in Figure 3b. It should be noted here, that significant uncertainty must be attributed to the emission data. CO emissions are known to depend very much on driving conditions, the vehicle types and such factors as e.g. cold-start percentage. Only very rough estimates of these parameters were available for this study.

The data mentioned in this section, together with background concentrations measured at the roof location (see Figure 1) are used for the presented model calculations.

4. The Modelling Procedure

Due to the low wind speed conditions prevailing during the measuring campaign, some special adaptation of MISKAM modelling results was required.

MISKAM is a very CPU time-demanding model and calculations are usually done for selected wind directions (in this case for 12 directions with 30° interval) and one reference wind speed only. Assuming that both the concentrations and the

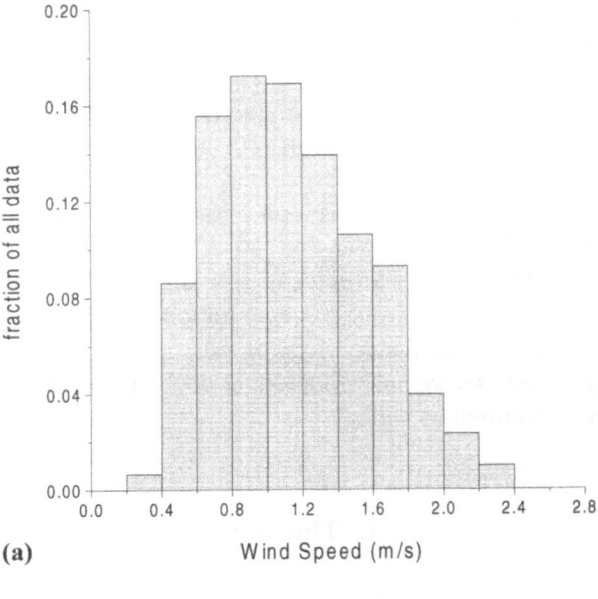

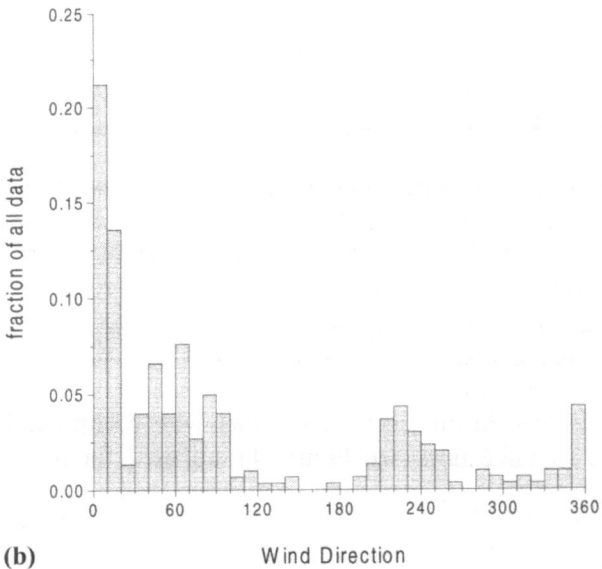

Figure 2. Frequency distribution of (a) wind speed and (b) wind direction as measured during the selected intensive observation period.

wind field data will scale with the reference wind speed, the model results can be given in terms of the non-dimensional concentration c^*,

$$c^* = c \cdot u_r \cdot S/Q ,$$

(1)

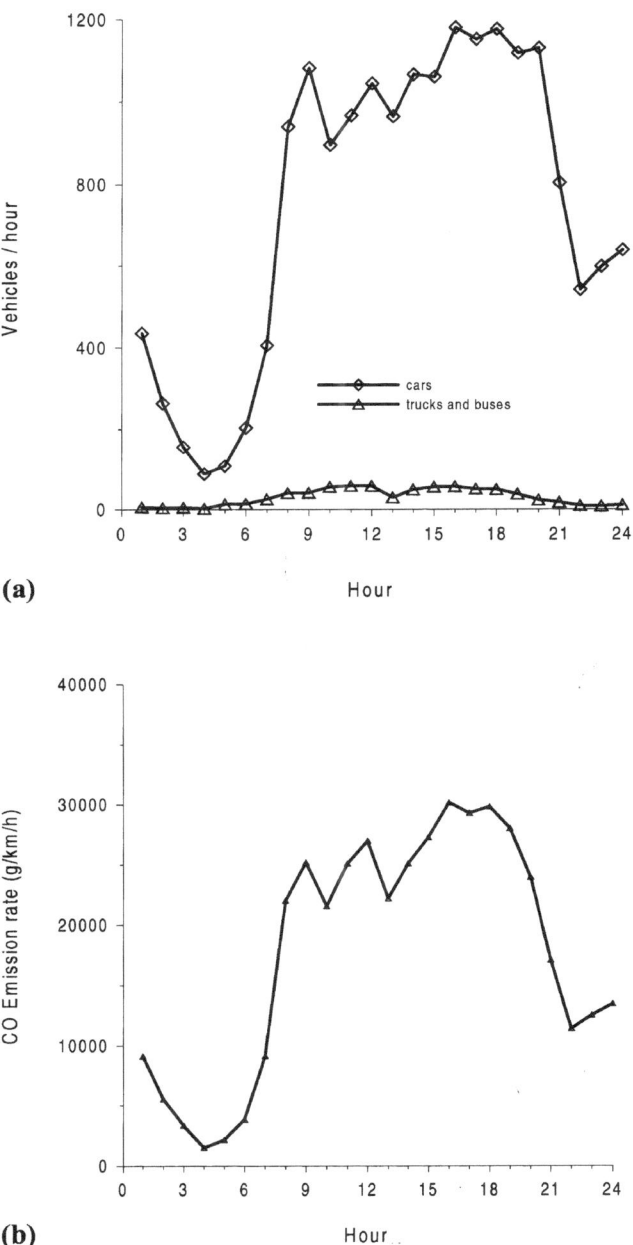

Figure 3. The average diurnal traffic profile (a) and the CO emission rate (b).

where u_r is a reference wind speed, Q is the emission rate and S is a length scale related to the street dimensions (here we use the height of the buildings).

Using Equation (1) for calculation of concentrations for any value of wind speed implies the assumption that the ambient wind is the only mechanism responsible for dispersion of pollution in the street. This is a reasonable approximation for higher wind speeds, but fails totally for lower wind speeds. It is believed that at low wind speed conditions, the main mechanism governing dilution of the car exhaust gases is the turbulence created by the traffic itself (Kastner-Klein et al., 2001; Vachon et al., 2001). However, the traffic produced turbulence is not directly included in the present version of MISKAM. Therefore, in order to account for this additional dilution mechanism, the concentrations within the street canyon, for a particular ambient wind speed u, are calculated here using the following formula:

$$c = c^* \cdot \frac{Q}{S \cdot \sqrt{u^2 + (\alpha \cdot \sigma_{\text{traf}})^2}} \ . \tag{2}$$

In Equation (2), σ_{traf} is the additional turbulence created by the traffic in the street and α is an empirical parameter, which depends only on the ambient wind direction. No provision was made here to account for the variation of the traffic produced turbulence with the height above the street. This means that the vertical concentration gradients predicted by MISKAM are not altered by this 'renormalisation' procedure. The traffic contribution to the turbulence in the street (the value of σ_{traf}) is calculated in the same way as in OSPM (Berkowicz, 2000). More details concerning the empirical parameter α, as well as some more test results of the procedure are presented in Ketzel et al. (2001b).

5. Results

Hourly averaged CO concentrations were calculated with OSPM for all the measuring points in the street. OSPM is a highly simplified model but previous model tests have proven its reasonably good performance. This study is however the first intensive test of the model using data from different heights in the street. Results of model comparison with measurements in the street are shown in Figure 4.

A similar comparison, but with results obtained with the 3-D numerical model MISKAM (version 4.1) using the procedure outlined in Section 4, is shown in Figure 5.

In each of the above figures, both the 1:1 line, as well as the best-fit line, are indicated too. The scattering around the 1:1 line is quite substantial, and especially for the lowest receptor heights. Some of the scatter can be attributed to the large uncertainties in determination of the hourly emissions of CO from the traffic in the street. However, the model deficiencies to reproduce the flow and dispersion conditions at the low wind speed conditions, should also be taken into account.

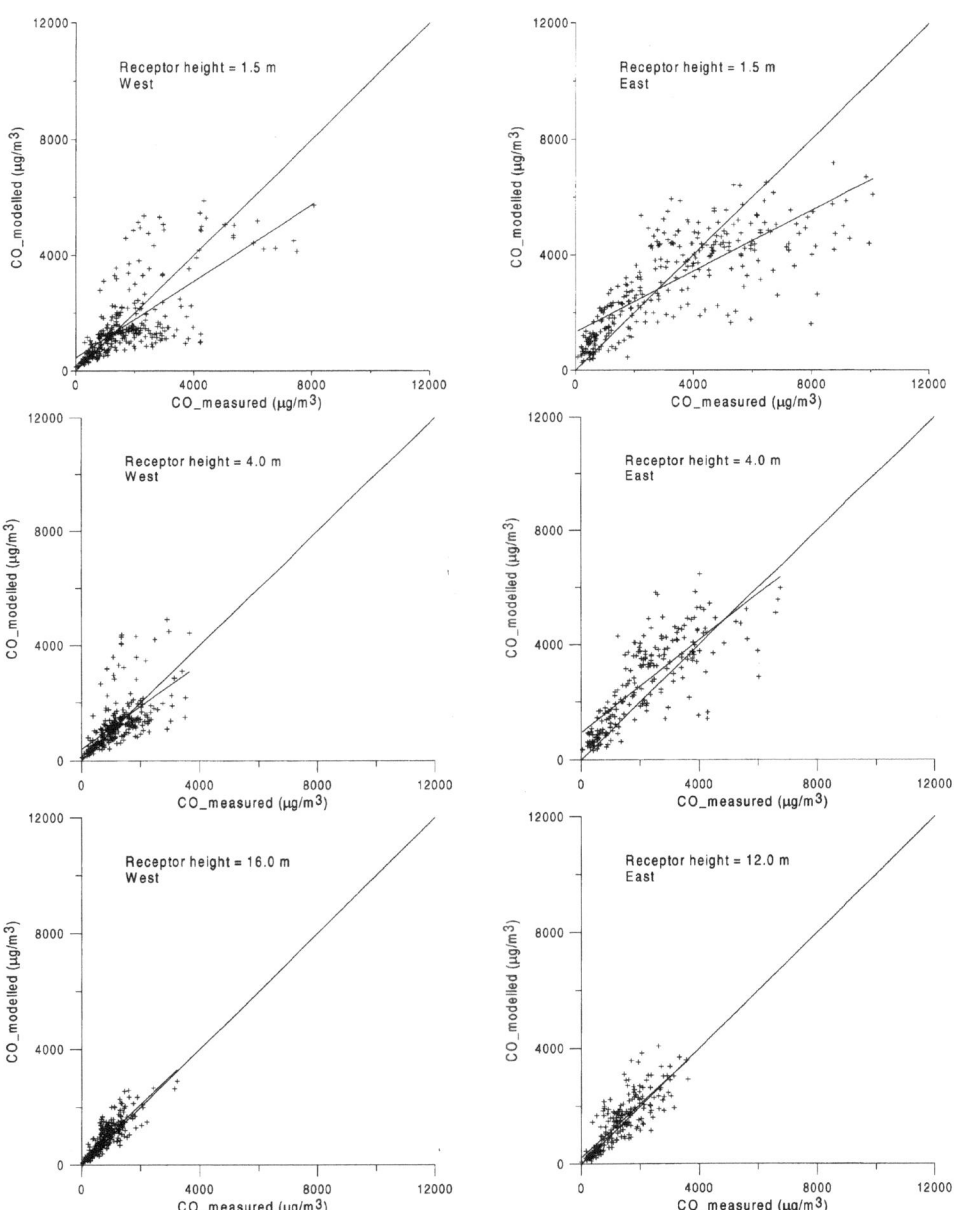

Figure 4. Comparison of OSPM results with CO measurements in the street. Location of the receptor points is indicated in the figures.

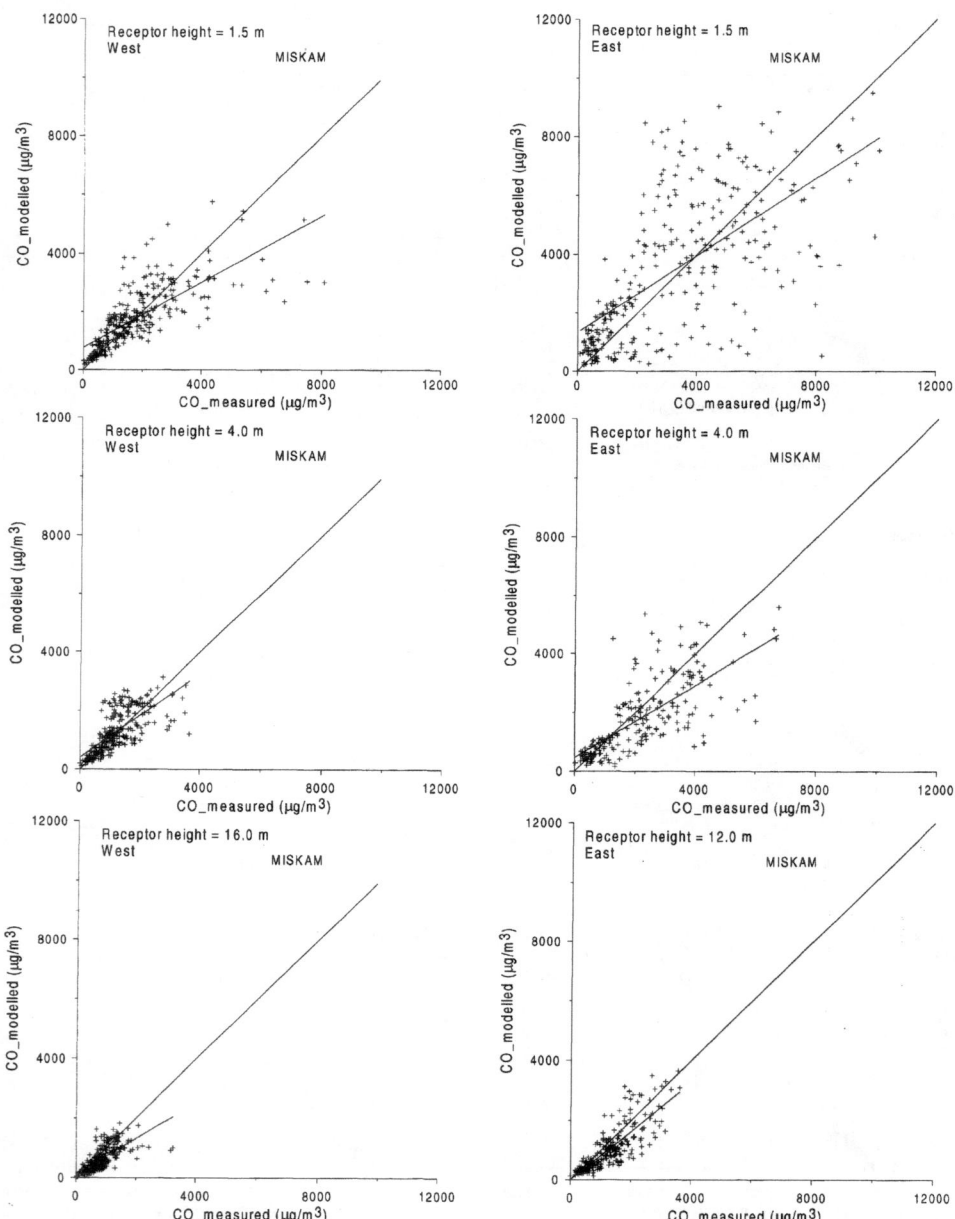

Figure 5. Comparison of MISKAM results with CO measurements in the street. Location of the receptor points is indicated in the figures.

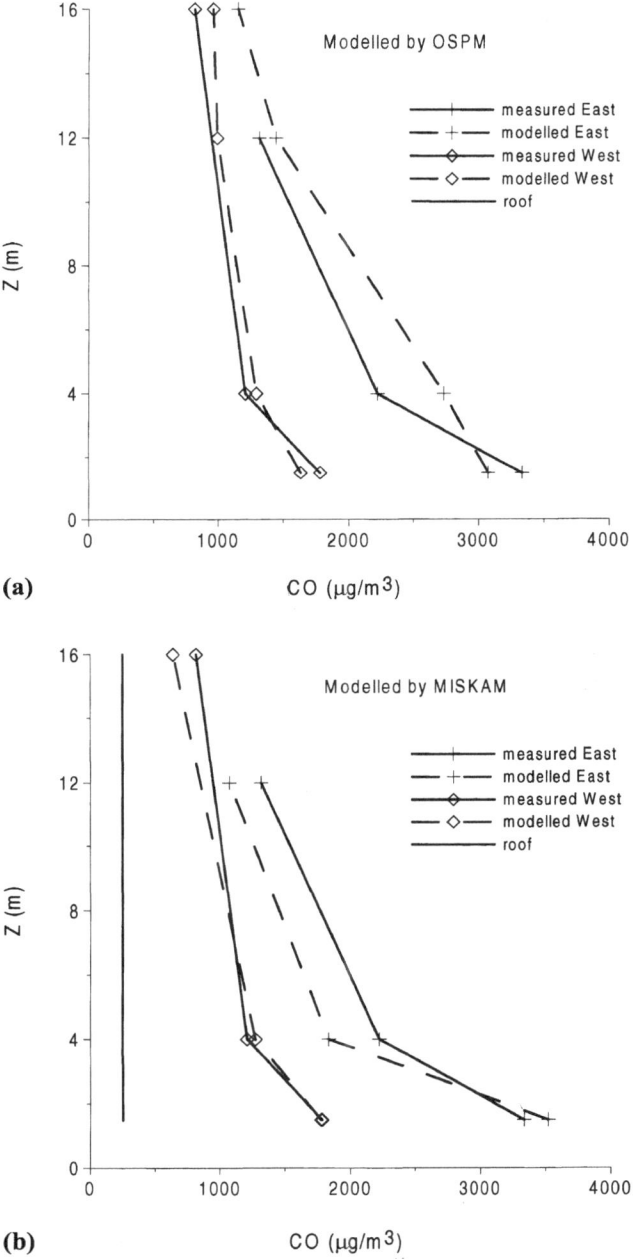

Figure 6. The vertical distribution of the measured and modelled concentrations. Results are shown both for (a) OSPM and for (b) MISKAM.

The vertical distribution of CO concentrations in the street is shown in Figure 6. The presented profiles represent average concentrations for each of the measuring locations. The averaging is made using all the available data. Background concentrations (roof monitor) are shown as well.

The average concentrations on the East side of the street are significantly higher than on the West side. This is the result of prevailing Easterly winds during the campaign (see Figure 2). For Easterly winds the East side of the street is leeward and receives higher concentrations than the windward West side. The variation with height is somewhat smaller on the West side (predominantly windward) than on the East side. This variation is quite well reproduced by MISKAM. OSPM underestimates the vertical gradient on the East side. The main reason for the underestimation of the vertical gradient by OSPM is the assumption on the initial vertical dispersion of vehicle exhausts. In OSPM this value is assumed to be 3 m. In MISKAM, the initial mixing is set to 2 m, which is the height of the first numerical grid. Smaller initial mixing height results in larger vertical gradients in the lowest level of the street.

The good agreement between the MISKAM results and the average vertical profiles shown in Figure 6 might however be somewhat fortuitous. The absolute values of the averaged concentrations are quite dependent on the above mentioned 'renormalisation' procedure. Taking into consideration that this procedure was parameterised using the same data as used for the present comparison, the good agreement for the average concentrations, in spite of the large scatter evident in Figure 5 for the single observations, is not surprising. However, the good performance of MISKAM concerning predictions of the magnitude of the vertical concentration gradients is still obvious.

The difference in the behaviour of pollutants on the leeward and the windward sides of the street is illustrated in Figure 7. Here, the dependence of concentrations (street contribution only) normalised by emissions is shown as function of wind speed. The wind directions are selected so, that the East side is always leeward and the night time hours with small emissions are excluded. The dependence on wind speed is less pronounced on the leeward side (the East side). This is again believed to be due to larger contribution of the traffic produced turbulence to the dispersion conditions on the leeward side. The scatter in the experimental data is, however, very large and the results must be taken with some caution.

6. Conclusions

The experimental data collected during the measuring campaign Nantes'99 in Rue de Strasbourg provide an excellent opportunity to study the traffic pollution distribution in a street canyon. The data represent conditions with predominantly low wind speeds and such effects, as traffic produced turbulence have a significant impact on the dispersion of pollutants in the street.

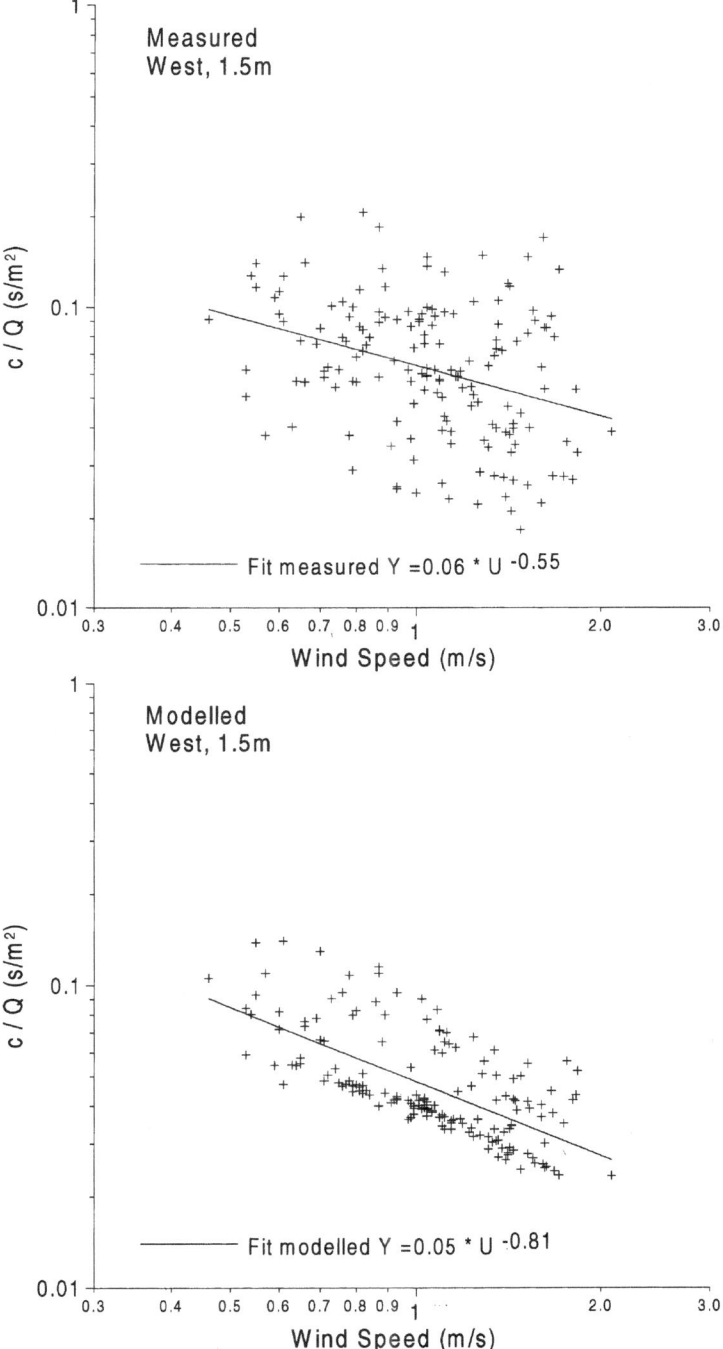

Figure 7a. The dependence of the windward (West) concentrations on wind speed. Both model results (OSPM) and measurements are shown.

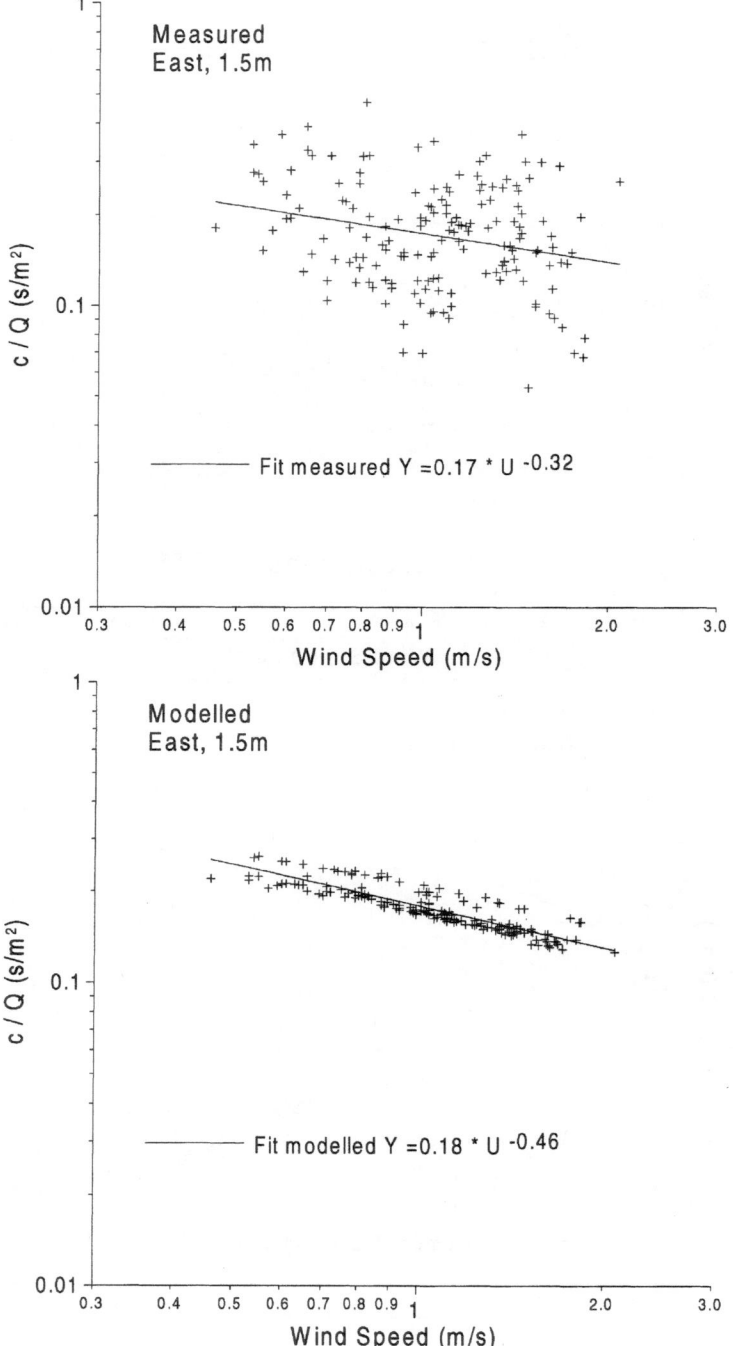

Figure 7b. The dependence of the leeward (East) concentrations on wind speed. Both model results (OSPM) and measurements are shown.

The concentration levels on the leeward side of the street are, as a rule, higher than on the windward side. Remarkable vertical gradients are observed and these gradients are more pronounced on the leeward side than on the windward side.

The experimental data are compared to model results using a simple parameterised model, OSPM, and a more advanced 3-D CFD model MISKAM. Both models reproduce reasonably well the observed distribution of pollutants in the street. The MISKAM model performs better in reproducing the vertical gradients. The OSPM results show less variation with height in the lowest 4 m than the observed data. This behaviour is believed to be due to the assumptions made about the initial dilution height of car exhaust gases.

A significant improvement of model results is achieved when a velocity scale based on the traffic produced turbulence is applied to calculate concentrations with the CFD model MISKAM. Scaling based on wind speed alone results in totally unrealistic concentration values.

It would be desirable to extend this study with data from the remaining period of the experimental campaign. Examination of pollution distribution under higher wind speed conditions will provide additional, valuable information for development and test of traffic pollution models.

Acknowledgements

The support provided by the European Commission within the framework of the TRAPOS Network (Contract Number ERBFMRXCT97-0105) is highly acknowledged.

References

Berkowicz, R.: 2000, 'OSPM – A parameterised street pollution model', *Environ. Monit. Assess.* **65**, 323–331.

Eichhorn, J.: 1995, 'Validation of a Microscale Pollution Dispersal Model', *Proceedings of the 21st International Meeting on Air Pollution Modeling and its Application*, Baltimore, Maryland, U.S.A.

Kastner-Klein, P., Berkowicz, R. and Fedorovich, E.: 2001, 'Evaluation of Scaling Concepts for Traffic-produced Turbulence based on Laboratory and Full-scale Concentration Measurements in Street Canyons', *The Third International Conference on Urban Air Quality*, 19–23 March 2001, Loutraki, Greece.

Ketzel, M., Louka, P., Sahm, P., Guilloteau, E., Sini, J.-F. and Moussiopoulos, N.: 2001a, 'Intercomparison of Numerical Urban Dispersion Models – Part II: Street Canyon in Hannover, Germany', *The Third International Conference on Urban Air Quality*, 19–23 March 2001, Loutraki, Greece.

Ketzel, M., Berkowicz, R., Flassak, T., Lohmeyer, A. and Kastner-Klein, P.: 2001b, 'Adaptation of Results from CFD-models and Wind-tunnels for Practical Traffic Pollution Modelling', *The 7th International Conference on Harmonisation within Atmospheric Dispersion Modelling for Regulatory Purposes*, May 2001, Italy.

Louka, P., Vachon, G., Sini, J.-F., Mestayer, P. G. and Rosant, J.-M.: 2001, 'Thermal Effects on the Airflow in a Street Canyon – Nantes'99 Experimental Results and Model Simulations', *The Third International Conference on Urban Air Quality*, 19–23 March 2001, Loutraki, Greece.

Pavageau, M., Rafailidis, S. and Schatzmann, M.: 1997, 'A comprehensive experimental databank for the verification of urban car emission dispersion models', *Int. J. Environ. Pollut.* **8**, 738–746.

Sahm, P., Louka, P., Ketzel, M., Guilloteau, E. and Sini, J.-F.: 2001, 'Intercomparison of Numerical Urban Dispersion Models – Part I: Street Canyon and Single Building Configurations', *The Third International Conference on Urban Air Quality*, 19–23 March 2001, Loutraki, Greece.

Vachon, G., Rosant, J.-M., Mestayer, P.G. and Sini, J.-F.: 1999, 'Measurements of Dynamic and Thermal Field in a Street Canyon, URBCAP Nantes'99', *Proceedings of the 6th Int. Conf. on Harmonisation within Atmospheric Dispersion Modelling for Regulatory Purposes*, 11–14 October, Rouen, France, Paper 124.

Vachon, G., Louka, P., Rosant, J.-M., Mestayer, P.G. and Sini, J.-F.: 2001, 'Measurements of Traffic-induced Turbulence within a Street Canyon during the Nantes'99 Experiment', *The Third International Conference on Urban Air Quality*, 19–23 March 2001, Loutraki, Greece.

STUDIES ON POLLUTANT DISPERSION FROM MOVING VEHICLES

A. G. VENETSANOS*, D. VLACHOGIANNIS, A. PAPADOPOULOS, J. G. BARTZIS
and S. ANDRONOPOULOS

National Centre for Scientific Research 'Demokritos', Environmental Research Laboratory, Institute of Nuclear Technology and Radiation Protection, Aghia Paraskevi, Attiki, Greece
(* author for correspondence, e-mail: venets@avra.ipta.demokritos.gr, fax: +30 1 6525004)

Abstract. Remote sensing instruments may be used to make measurements of emission levels from individual vehicles under real driving conditions. In designing such an instrument, that will identify the gross polluters, the CFD modelling approach has been used to study the effects of the parameters involved on the characteristics of the exhaust gas plume. The analysis was performed using the ADREA-HF code in open space, street canyon and tunnel environments, with sensitivity parameters the car speed and meteorological conditions. The calculations have been performed in a moving coordinate system, with the car and site geometry being fully resolved. A discussion is given on the effects of the various parameters affecting the design of the instrument.

Keywords: CFD modelling, gross polluters, moving vehicles, pollutant dispersion, remote sensing

1. Introduction

It is known (Crookell and Sinclair, 2001; Stedman, 1989) that 10% of all vehicles generate at least 50% of the CO emissions and unburned HC. Therefore, the identification of gross polluting vehicles and subsequently the requirement of repair or removal from service will have a significant impact on reducing the emission effects. The solution of periodic vehicle testing, at idle, suffers from not correlating with real behavior. Remote sensing instruments on the other hand offer the possibility to detect emission levels under real conditions, for example Bishop and Stedman (1996). In this study the objective was to apply CFD modeling to determine the impact of various factors that may affect the exhaust plume dispersion and hence the deployment of a remote sensing instrument.

2. Methodology

CFD calculations were performed for a typical passenger car for different driving environments, car speeds and meteorological conditions. It was assumed that the car belongs to the category of open loop catalytic gasoline cars, $1400 < cc < 2000$ (Ntziachristos *et al.*, 1999). Furthermore, assumed car dimensions were of

Water, Air, and Soil Pollution: Focus **2**: 325–337, 2002.
© 2002 *Kluwer Academic Publishers.*

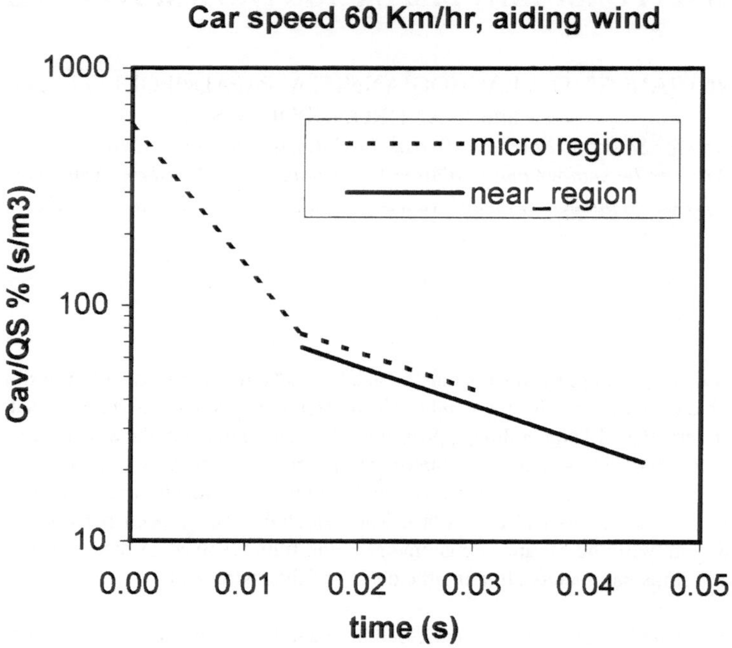

Car speed 60 Km/hr, aiding wind

Figure 1. Open road micro versus near range results.

$4.5 \times 1.8 \times 1.4$ m (length$\times$width$\times$height), with a 4 cm diameter exhaust pipe at 20 cm distance from ground.

Different driving environments were considered such as open road, tunnel and street canyon. For the open road case, we assumed a road of 10 m width, with a field on both sides of hydrodynamic roughness of 0.1 m. For the canyon, we assumed a street of 10 m width, with 2.5 m sidewalks on both sides of it and building height to canyon width ratio equal to 1. For the tunnel, we assumed a road of 7.4 m width and sidewalks on both sides of 0.85 m width each as well as a tunnel 6.6 m high.

The parametric analysis was performed for car speeds of 40 and 60 km hr^{-1} for all driving environments, 80 km hr^{-1} for tunnel and canyon and 120 km hr^{-1} for the open road.

The influence of the meteorological conditions for both the open space and canyon cases was considered, only with regards to the wind direction effect. Thus, the assumption was made on three wind directions, aiding, opposing and crosswind (with respect to the dispersion, i.e. aiding wind blowing opposite to the car moving direction). The undisturbed ambient wind velocity was assumed 5 m s^{-1} at 10 m height at neutral stability conditions. Ventilation effects in the tunnel were not taken into account.

The flow and dispersion calculations were performed using the ADREA-HF CFD code Bartzis (1991). The ADREA-HF code is a prognostic CFD model designed to predict atmospheric dispersion in complex terrain under various source

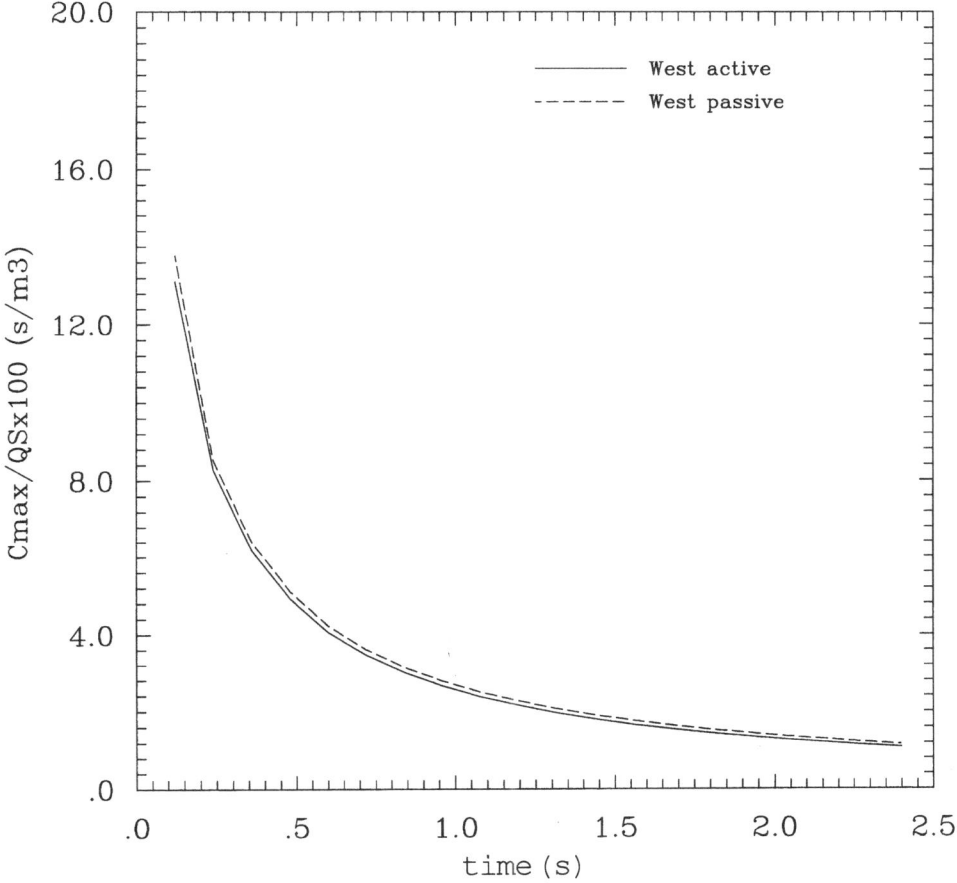

Figure 2. Open road near range results. Exhaust exit conditions effects (west = opposing).

and stability conditions. Validation studies of the code can be found in Androno-poulos *et al.* (1994), Bartzis *et al.* (1997), and Venetsanos *et al.* (1998, 2000).

The calculations were performed on a Cartesian moving grid at car speed, with car and site geometry fully resolved, using the DELTA-B geometrical pre-processor Venetsanos *et al.* (1995).

The open space and canyon runs were performed on a domain of $300 \times 60 \times 20$ m, in X, Y and Z directions, respectively. The grid was non-equidistant comprising $63 \times 39 \times 28$ cells in X, Y and Z directions, respectively. The minimum grid cells were 0.5, 0.45 and 0.2 m in X, Y, Z directions, selected close to ground and source. The maximum expansion ratio for the cells was 1.12. This domain was used to cover the 'near range' dispersion, i.e. 0.02–1.5 sec after car has passed.

The tunnel runs were performed on a domain of size $150 \times 10 \times 7$ m, in X, Y and Z directions, respectively. The grid was non-equidistant comprising $74 \times 47 \times 43$ cells in X, Y and Z directions, respectively. The minimum grid cells

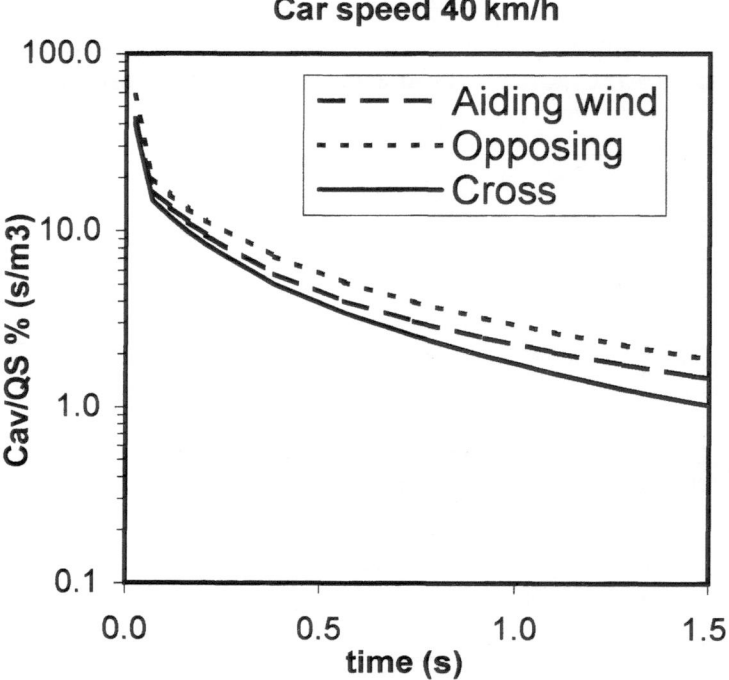

Figure 3. Open road near range results. Effect of ambient wind.

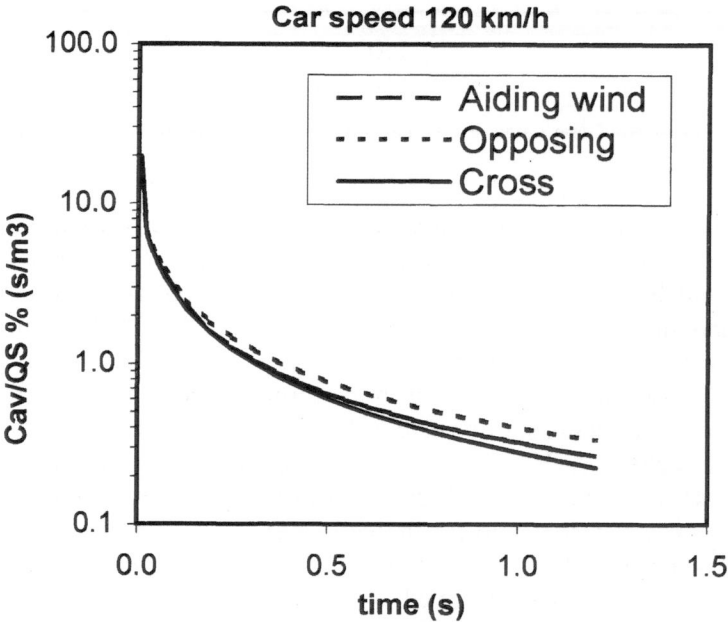

Figure 4. Open road near range results. Effect of ambient wind.

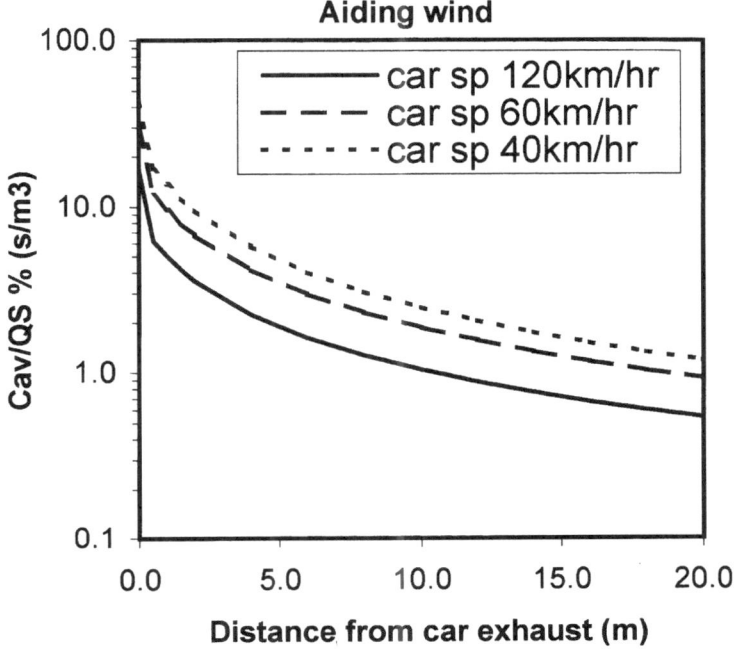

Figure 5. Open road near range results. Effect of car speed on average crossroad concentration.

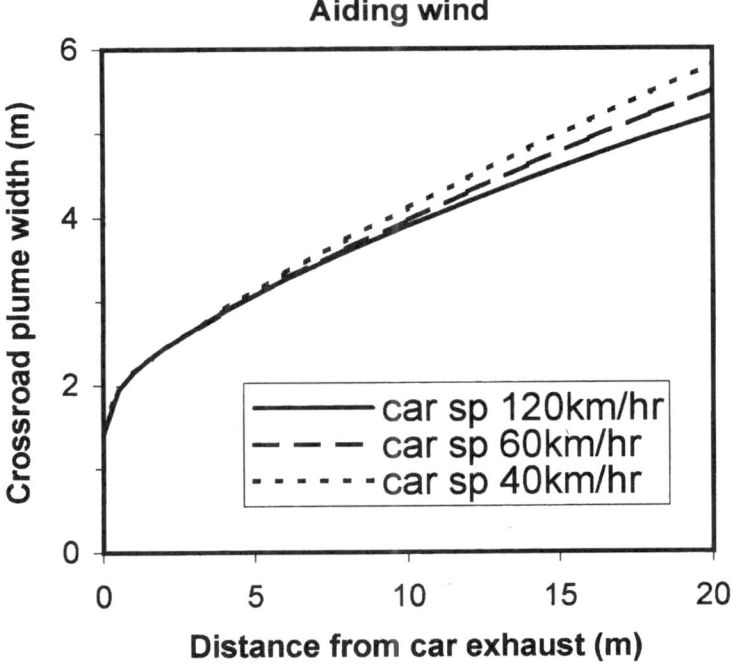

Figure 6. Open road near range results. Effect of car speed on plume width in the crossroad direction.

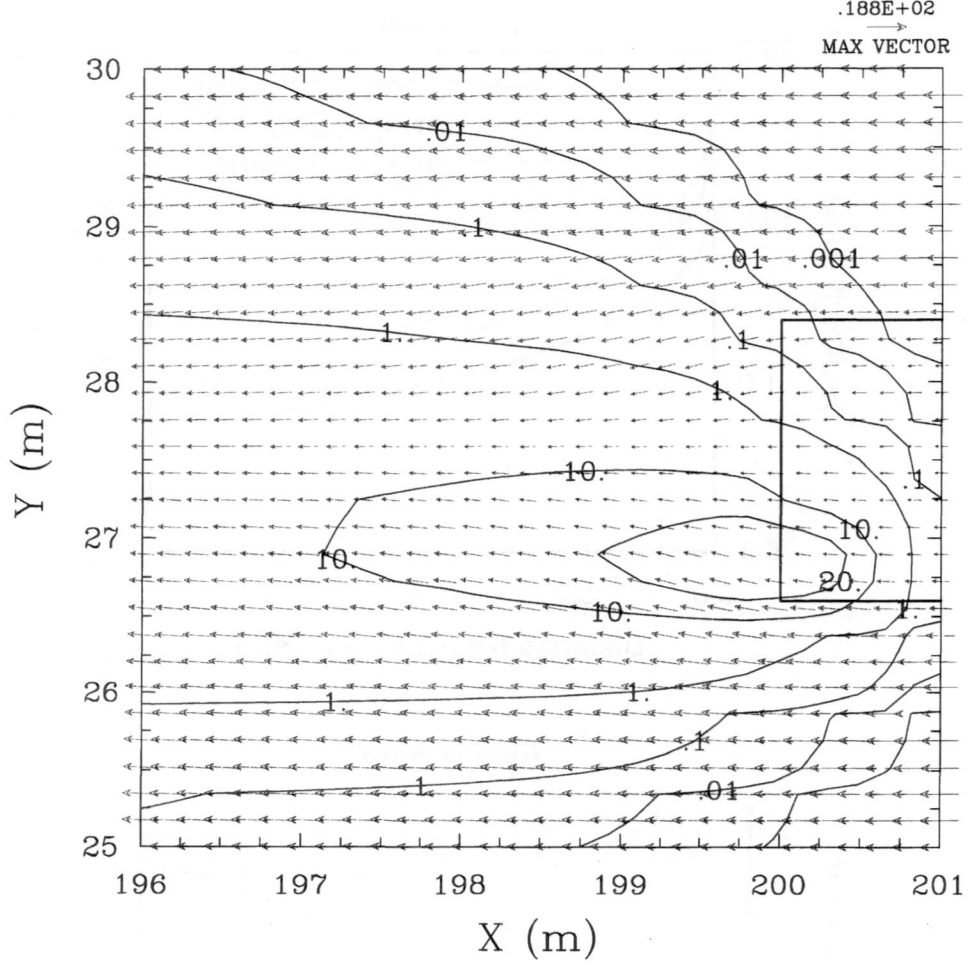

Figure 7. $C/Q_s\%$ contours and vectors on horizontal plane at source level for open road, 60 km hr^{-1} car speed and aiding wind.

were 0.25, 0.1 and 0.05 m in X, Y, Z directions, selected close to ground and source. The maximum expansion cell ratio was 1.12.

In all cases, turbulence was modelled using the default k-1 model of the ADREA-HF code, which is a one-equation model for the turbulent kinetic energy, with length scales calculated from algebraic relations, taking into account the distance from solid surfaces, the stability and the asymptotic pressure gradient. A detailed description of this model can be found in Andronopoulos *et al.* (1994).

Initial condition for the concentration was zero background. Computations were stopped when steady state was reached. In all the calculations performed, the pollutant dispersion was assumed passive, except for one case where 'real' exit condi-

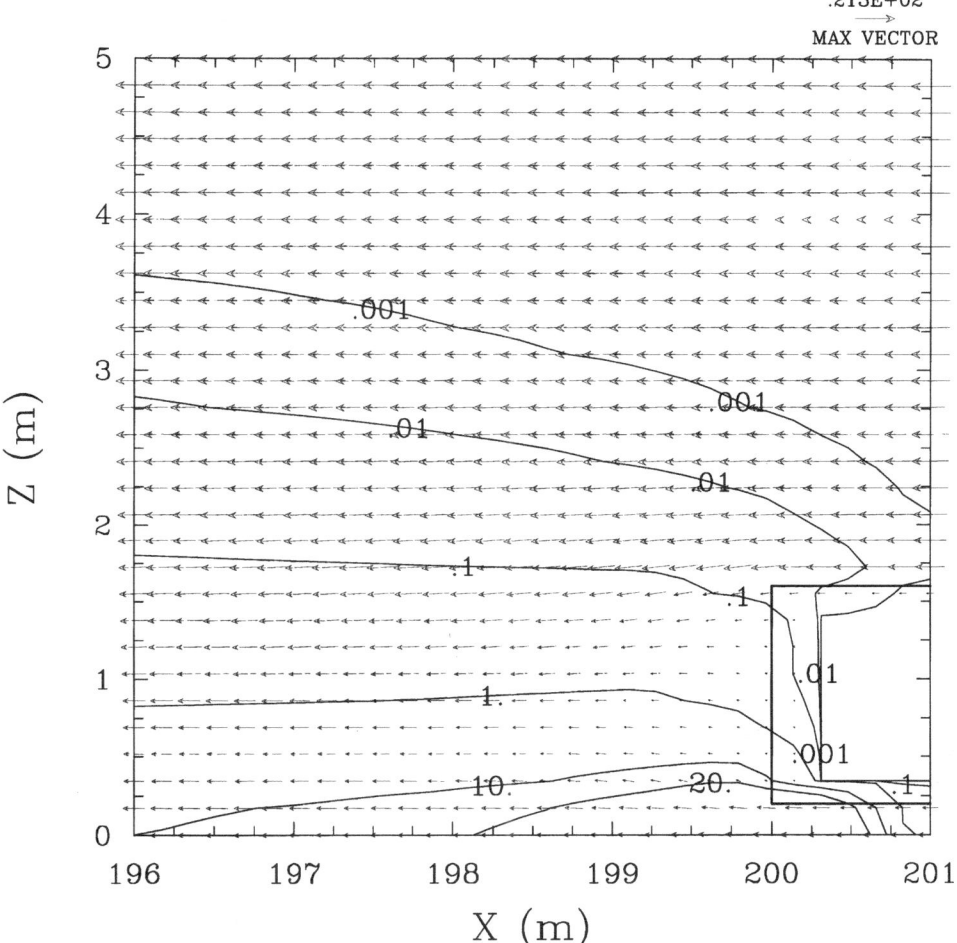

Figure 8. $C/Q_s\%$ contours and vectors on vertical plane at source level for open road, 60 km hr^{-1} car speed and aiding wind.

tions were assumed, to test the validity of the passive approach as explained below. For all passive runs, the assumed source emission was $Q_s = 1.5$ mg s^{-1}.

3. Results and Discussion

Before performing calculations in the near range it was necessary to establish that the applied computational procedure provides a match between near range and micro range (0.0–0.1 sec) solutions. The selected micro range domain was $2.0 \times 0.4 \times 0.6$ m around the source, with a minimum grid step of 0.01 m. Both near range and micro range calculations were performed for open road, with 60 km

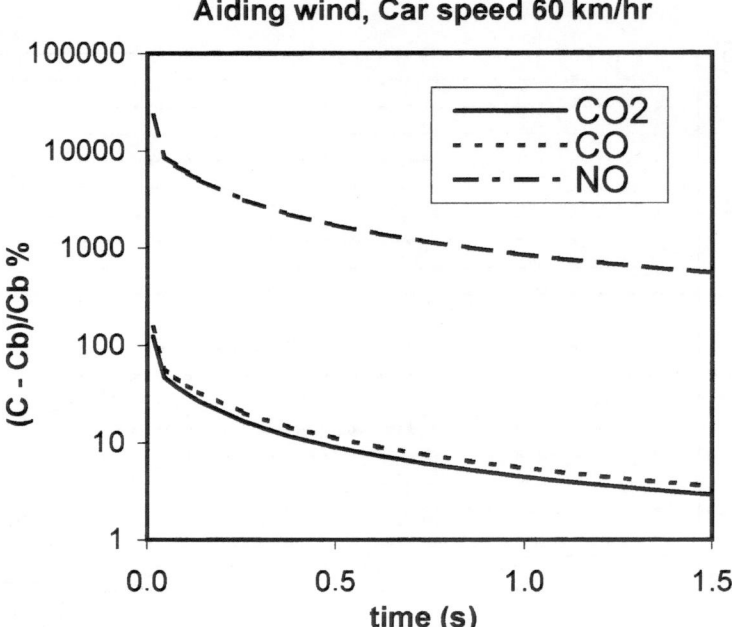

Figure 9. Open road, near range. Effect of pollutant.

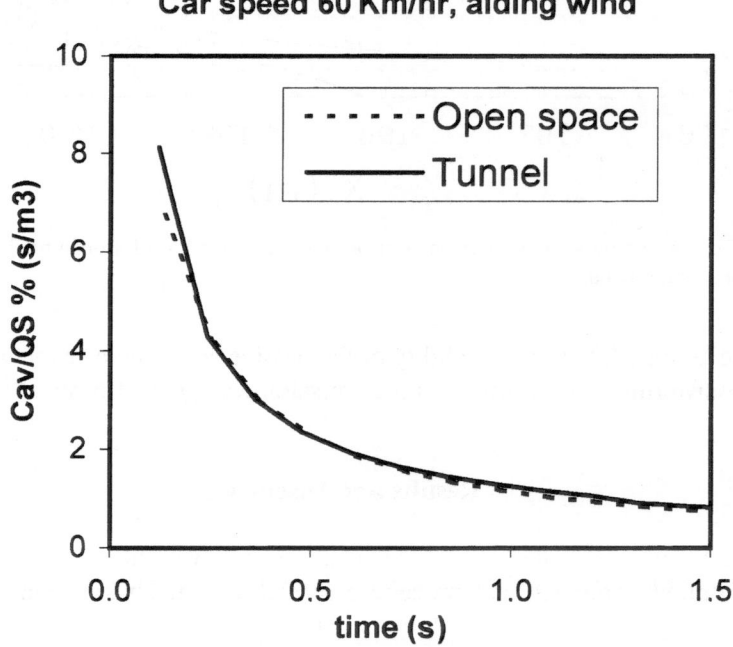

Figure 10. Open road versus tunnel near range results.

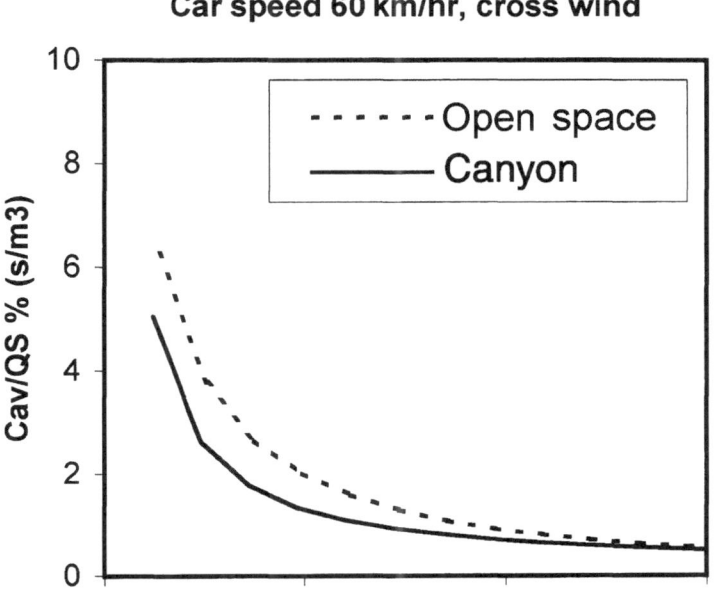

Figure 11. Open road versus canyon near range results.

hr^{-1} car speed and aiding wind. Figure 1 shows the predicted maximum crossroad concentration at the exhaust height, as function of time after the car has passed. It was concluded that the micro range and near range results were in very good agreement and more specifically, in the micro range the concentration fell by a factor of 100.

The effect of the exhaust exit conditions was investigated by performing open space, near range, passive and non-passive calculations for 60 km hr^{-1} car speed and opposing wind. For the non-passive run, an exhaust exit temperature of 373 K was assumed and the solution was provided taken also into account the mixture enthalpy conservation equation. Figure 2 shows the maximum crossroad concentration at the exhaust height, as function of time. It was observed that passive and non passive results were in good agreement.

The effect of the ambient wind direction on the average crossroad concentration at the source height for open space is small, as shown in Figures 3 and 4 and it was seen to decrease with increasing car speed. Crossroad averaging was performed assuming a Gaussian distribution for the crossroad concentration and estimating

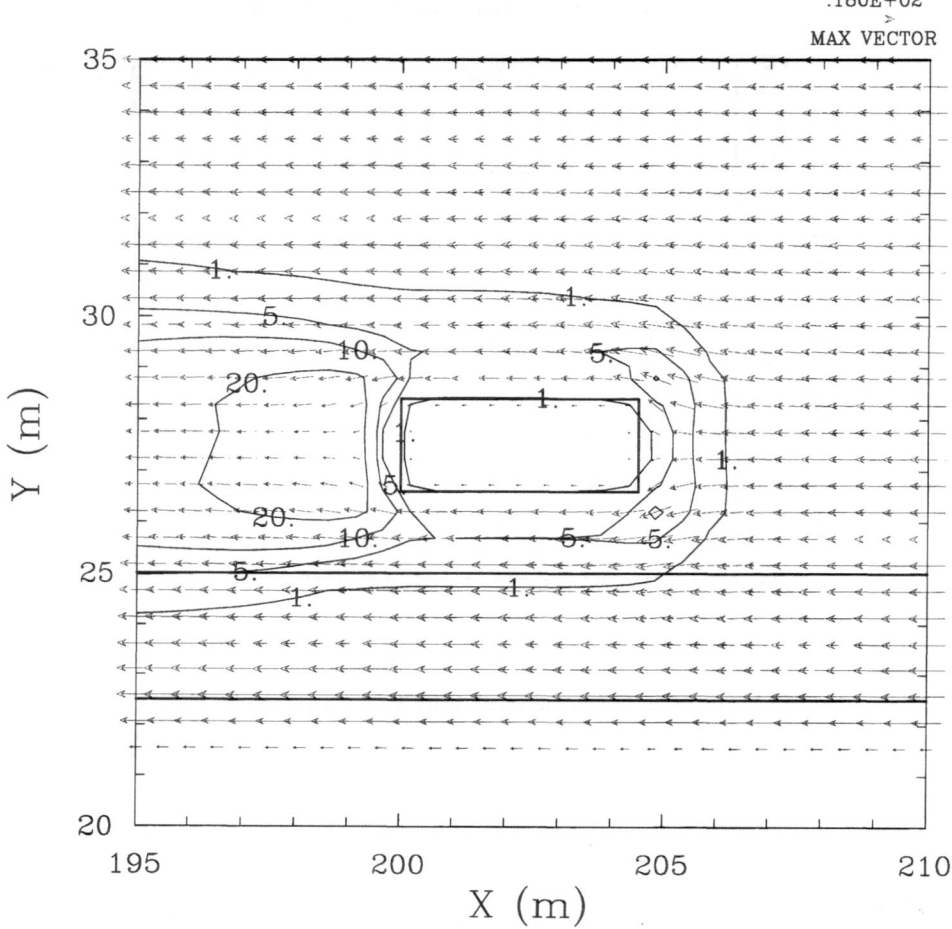

Figure 12. Canyon with cross wind and 60 km hr^{-1} car speed (Canyon y = [22.5, 37.5], buildings height 15 m, sidewalks y = [22.5, 25] and [35,37.5]). Shown is the turbulent kinetic energy contours and velocity field on horizontal plane, 0.5 m from ground.

the variance σ, average concentration C_{av} and cloud width w, from the following relations:

$$\sigma^2 = \frac{\int_{-\infty}^{\infty} C\,(y - y_0)^2\, d\,(y - y_0)}{\int_{-\infty}^{\infty} Cd\,(y - y_0)} \qquad C_{av} = \frac{\int_{-\infty}^{\infty} Cd\,(y - y_0)}{w} \qquad w = 4\sigma\,,$$

where y_0 is the crossroad location of the maximum.

Figures 3 and 4 show that the dilution in case of opposing wind is smaller than in the case of aiding. The reason for this is that the ambient air is moving with a higher

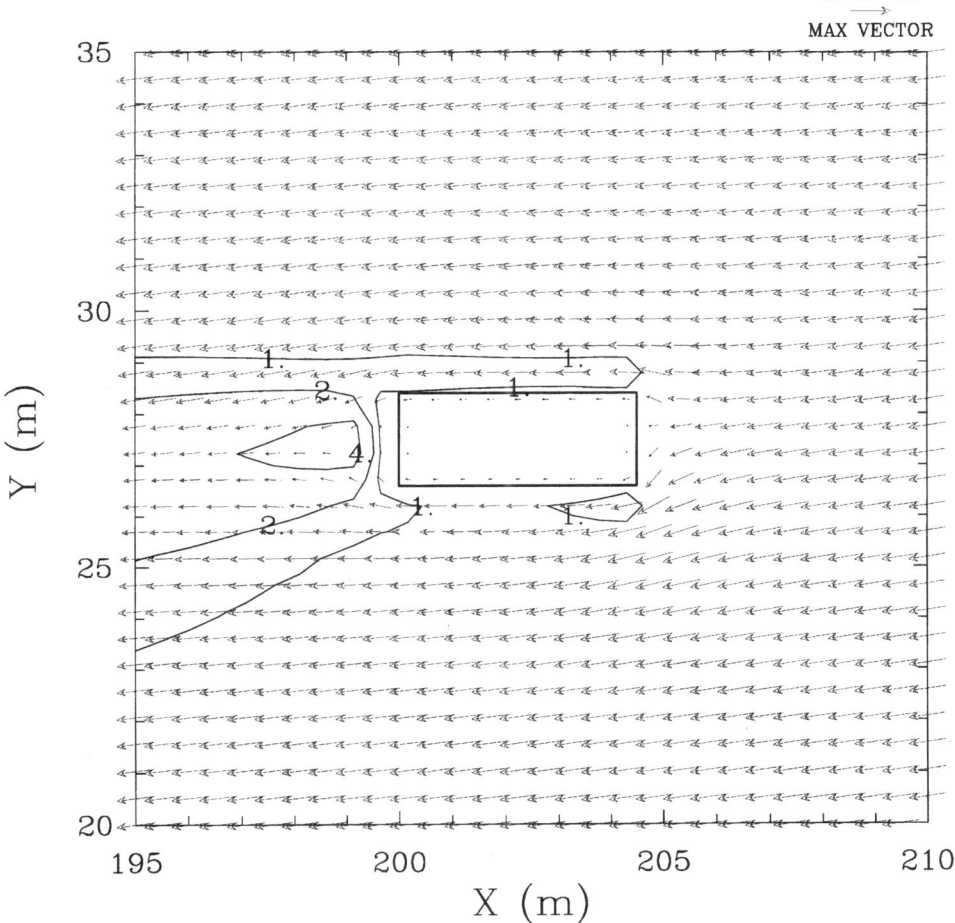

Figure 13. Open road with cross wind and 60 km hr^{-1} car speed. Shown is the turbulent kinetic energy contours and velocity field on horizontal plane, 0.5 m from ground.

velocity against the car for aiding wind leading to higher turbulence production and therefore to higher dilution. It can also be observed that the maximum dilution occurs under crosswind conditions. In this case, the turbulence levels are increased, due to the cross flow between ambient air moving normal to and against the car. Finally, it can also be deduced that the average concentrations fall with a factor of 30 in the near range.

The effect of car speed on crossroad averaged concentration and cloud width, for open space and aiding wind is significant, as shown in Figures 5 and 6. The dilution increases with car speed. This is due to higher turbulence production at higher car speeds. Similar results were obtained for the other wind directions. The predicted wind field in the moving system and the predicted concentration patterns at the rear of the car, on the horizontal and vertical planes through the exhaust are depicted in Figures 7 and 8, for the open space near range results, for car speed 60 km hr^{-1} and aiding wind. Shown also in the figure is the footprint of the car. The wind field is distorted due to the car presence. Also, there is a decrease in the velocity magnitude in the near wake. The cloud width given in Figure 6 can be qualitatively compared with the concentration patterns given in Figure 7.

The predicted CO_2, CO and NO dilution levels for open space, 60 km hr^{-1} and aiding wind are shown in Figure 9. Assumed background concentrations were $C_b(CO_2) = 376$ ppm, $C_b(CO) = 9.0$ ppm, $C_b(NO) = 0.02$ ppm. The exhaust pollutant concentration for CO_2 and CO was on the average of the order of 20% above background level while for NO, it was several times above the background level.

In general, the effect of driving environment was found to be small. Figure 10 shows the predicted average crossroad concentration at the source level for the tunnel and open space, for 60 km hr^{-1} car speed and aiding wind. In the open space and canyon results, the largest discrepancy was found for crosswind conditions, where the driving environment seemed to have an effect on concentrations (Figure 11). The dilution in the canyon case was calculated to be higher than in open space. This is attributed to higher car induced turbulence in the case of the canyon see Figure 12 for canyon and Figure 13 for open road.

4. Conclusions

The present analysis led to the following conclusions.

In the micro-range defined close to the exhaust exit, the maximum concentration drops approximately by a factor of 100 within a few tenths of a second. The horizontally averaged concentrations in the near- range (0.02–1.5 sec) fall on the average by a factor of 30.

The exit conditions of the exhaust pipe do not have a significant effect in the near-range.

The effect of ambient wind direction on plume dispersion is relatively small and decreases with increasing car speed.

The effect of vehicle speed is important in all wind directions and the exhaust plume dispersion increases with car speed.

The exhaust pollutant concentration for CO_2 and CO was found to be on the average of the order of 20% above background level while for NO, it was several times above the background level.

For the configurations of this study, no significant difference was observed between tunnel and open road cases. Small differences were found between open road and street canyon cases for aiding and opposing wind directions while a rather higher divergence in the results was calculated for cross wind conditions.

Acknowledgements

The authors would like to thank the European Commission for partially financing the present work, in the framework of the REVEAL project (Remote Measurement of Vehicle Emissions at low cost), contract No: GRD1-0657 of the GROWTH programme.

References

Andronopoulos, S., Bartzis, J. G., Würtz, J. and Asimakopoulos, D.: 1994, 'Modelling the effects of obstacles on the dispersion of denser than air gases', *J. Hazard. Mater.* **37**, 327−352.

Bartzis, J. G.: 1991, 'ADREA-HF: A Three-dimensional Finite Volume Code for Vapor Cloud Dispersion in Complex Terrain', *EUR Report 13580 EN*.

Bartzis, J. G., Venetsanos, A. G., Varvayanni, M., Andronopoulos, S., Davakis, S., Statharas, J., Catsaros, N. and Deligiannis, P., 1997, 'Wind Flow and Dispersion Modeling over Terrain of High Complexity', *Proceedings of the Fifth International Conference AIR-POLLUTION 97*, 16−18 September, Bologna, Italy.

Bishop, G. A. and Stedman, D. H.: 1996, 'Measuring the emissions of passing cars', *Acc. Chem. Res.* **29**, 489−495.

Crookell, A. and Sinclair, P.: 2001, 'Remote Analysis of Motor Vehicle Exhaust: The Elusive Emissions Factor', *3rd International Conference on Urban Air Quality*, 19−23 March, Loutraki, Greece.

Ntziachristos, L. and Samaras, Z., contributions from: Eggleston, S., Gorißen, N., Hassel, D., Hickman, A.-J., Joumard, R., Rijkeboer, R. and Zierock, K.-H.: 1999, 'COPERT III: Computer Program to Calculate Emissions from Road Transport. Methodology and Emission Factors', *Draft Report*, 1st ed. (July), European Environmental Agency, European Topic Centre on Air Emission, 83 pp.

Stedman, D. H.: 1989, 'Automobile carbon monoxide emission', *Environ. Sci. Technol.* **23**(2), 147−148.

Venetsanos, A. G., Catsaros, N., Würtz, J. and Bartzis, J. G.: 1995, 'The DELTA_B Code. A Computer Code for the Simulation of the Geometry of Three Dimensional Buildings. Code Structure and Users Manual', *EUR Report 16326 EN*.

Venetsanos, A. G., Andronopoulos, S., Statharas, J. and Bartzis, J. G.: 1998, 'Local Scale Dispersion Model Evaluation Exercise', *Proceedings of the Sixth International Conference AIR POLLUTION 98*, Genoa, Italy.

Venetsanos, A. G., Bartzis, J. G., Würtz, J. and Papailiou, D. D.: 2000, 'Comparative modeling of a passive release from an L-shaped building using one, two and three-dimensional dispersion models', *Internat. J. Environ. Pollut.* **14**(1−6), 324−333.

A NEW CONCEPT FOR AIR QUALITY MODELLING IN STREET CANYONS

C. MENSINK*, N. LEWYCKYJ and L. JANSSEN

VITO, Centre for Remote Sensing and Atmospheric Processes, Boeretang 200, Mol, Belgium
(* author for correspondence, e-mail: clemens.mensink@vito.be, fax: +32 14 322795)

Abstract. We present an analytical model that predicts concentrations in street canyons assuming a uniform distribution within a street, dimensioned by its length and width and the height of the surrounding built-up area. Using the Prandtl-Taylor hypothesis, the concentration in the street is determined from a mass flux balance between a horizontal advective flux, a turbulent diffusive vertical flux and a continuous road transport emission source. The model does not necessarily assume re-circulation of the flow in the street canyon, but rather considers the turbulent intermittency in the shear layer shed from the upwind roof level as the driving force. This concept is in agreement with recent measurements and observations. The model has been applied to compute benzene concentrations based on hourly emissions obtained for 1963 streets and road segments in the City of Antwerp, Belgium. The results are compared with diffusive sampler measurements carried out at 101 locations in several streets of Antwerp, during 4 periods of 5 days in 1998. When averaged over periods of 5 days, the calculated benzene concentrations show a very good agreement with the results obtained by the diffusive sampler measurements.

Keywords: air quality modelling, benzene, road transport emissions, street canyons, urban air pollution

1. Introduction

Various models exist to compute concentrations of pollutants inside a street canyon. On an expert level, complex CFD (computational fluid dynamics) models can be used to simulate in detail the transport of mass, momentum and heat inside the street (see for example Huang *et al.*, 2000). For routine evaluations of the impact of vehicle emissions on air quality in street canyons only a few models are available.

The 'Street' model as developed by Johnson *et al.* (1973) is an empirical model often used to obtain a first estimate of the concentration levels. The concentration is calculated as a function of the emission source strength, the height and the width of the street canyon, the wind speed at the roof level and some initial mixing height. Both leeward and windward side concentrations can be calculated. The effect of wind direction is absent in the model. In the Canyon Plume-Box Model (CPBM) developed by Yamartino and Wiegand (1986) the concentrations are computed by a combination of a plume model for the direct contributions from the traffic emissions and a box model for the re-circulating part of the pollutants. A similar method

Water, Air, and Soil Pollution: Focus **2**: 339–349, 2002.
© 2002 *Kluwer Academic Publishers.*

is applied in the Operational Street Pollution Model or OSPM model (Berkowicz, 1998). OSPM makes use of parameterisations based on extensive experimental data. Concentrations can be calculated on both sides of the street. The model needs as input the (varying) height of the buildings along the street, the width, length and angle of the street, meteorological data (wind speed and wind direction at 10 m above roof level) and background concentrations.

In this paper we present a newly developed analytical model that assumes a uniform concentration distribution over the street and is therefore called 'Street box' model, with the box dimensioned by the length and width of the street and the height of the surrounding built-up area. The concentration in the street is determined from a mass flux balance between a horizontal advective flux, a turbulent diffusive vertical flux and a continuous road transport emission source. In contrast with both the CPBM and OSPM model, the 'Street box' model does not necessarily assume re-circulation of the flow in the street canyon, but rather considers the turbulent intermittency in the shear layer shed from the upwind roof as the driving force. The turbulent diffusive flux is described using the Prandtl-Taylor hypothesis. The result is a vertical exchange of the pollutant over a characteristic length which is associated with a typical mixing length created by turbulent eddies shedding off at roof level, enhancing the exchange of mass and momentum.

This concept is supported by measurements and observations made by Louka et al. (2000). They found that the mean re-circulation within the street was much weaker than the unsteady turbulent fluctuations and concluded that the mean flow is merely a residual of an unsteady turbulent re-circulation that is coupled to the wind aloft through an unstable shear layer developing at roof level.

In Section 2 it is briefly shown how the model equation can be derived from the basic mass transport equations. In Section 3 it is shown how the model has been applied to compute benzene concentrations for a large number of streets in the City of Antwerp, Belgium. The results are compared with diffusive sampler measurements carried out at 101 street locations in Antwerp, during 4 periods of 5 days in 1998. This is discussed in Section 4.

2. Methodology

As mentioned in the introduction, inside the street box only horizontal convection along the street (x-direction) and vertical diffusion processes (z-direction) are considered, together with a continuous source term. Net contributions of horizontal turbulent fluxes are neglected as well as diffusion in horizontal directions. Through these assumptions, the change of concentration in a non-reactive flow can be expressed by:

$$\frac{\partial \bar{c}}{\partial t} = -\frac{\partial}{\partial x}\left(\overline{v_x c}\right) - \frac{\partial}{\partial z}\left(\overline{v_z' c'}\right) + D\frac{\partial^2 \bar{c}}{\partial z^2} + S, \tag{1}$$

where the concentration c (μg m^{-3}) has been replaced by a time-smoothed value $\bar{c}$ and a turbulent concentration fluctuation c'. The same is applied to velocity vector. The first term on the right hand side in Equation (1) represents the advective mass transport, the second term the mass transport due to turbulent fluctuations, and the third term the contribution due to laminar diffusion with coefficient D (m^2 s^{-1}). The vertical turbulent mass flux term in Equation (1) is approximated by applying the eddy diffusivity concept in analogy of Fick's law (Bird *et al.*, 1960):

$$\overline{v'_z c'} = -K \frac{\partial \bar{c}}{\partial z} \, . \tag{2}$$

For a turbulent free stream flow the eddy diffusivity K (m^2 s^{-1}) can be related to a charactristic length scale ℓ (m) and the free stream flow velocity gradient by applying the Prandtl-Taylor hypothesis (Hinze, 1987):

$$\overline{v'_z c'} = -\ell^2 \left| \frac{dU}{dz} \right| \frac{\partial \bar{c}}{\partial z} \, . \tag{3}$$

In the context of the street box, the characteristic length ℓ is associated with a typical mixing length or mixing length created by turbulent eddies shedding off at roof level. The velocity gradient over this mixing length is assumed to be constant and equal to the free stream velocity $U_\perp$ above the roof tops in the direction of the eddy shedding, i.e. perpendicular to the street direction, divided by the mixing length ℓ. Thus, conform Prandtl's mixing length theory, the eddy diffusion becomes equal to the product of a mixing length ℓ and some suitable velocity, expressed here by $U_\perp$:

$$K = \ell U_\perp \, . \tag{4}$$

Substitution of Equations (2) and (4) into Equation (1) yields:

$$\frac{\partial \bar{c}}{\partial t} = -\frac{\partial}{\partial x} (\overline{v_x c}) + (D + \ell U_\perp) \frac{\partial^2 \bar{c}}{\partial z^2} + S \, . \tag{5}$$

Equation (5) can be reformulated in terms of a mass flux balance assuming a steady state approach, i.e. no change in concentration during one hour. The mass flux balance is then given by:

$$F_{in} - F_{out} - G + S = 0 \, , \tag{6}$$

where the horizontal fluxes F (g s^{-1}) in Equation (6) are given by:

$$F_{in} = W \cdot H \cdot U_{\equiv} \cdot C_{in} \tag{7}$$

and

$$F_{out} = W \cdot H \cdot U_{\equiv} \cdot C_{out} \, . \tag{8}$$

In Equations (7) and (8), H is the height (m) and W the width (m) of the street canyon, $U_{\equiv}$ is the wind speed (m s^{-1}) parallel to the street and C_{in} and C_{out} are the in and outgoing concentrations, respectively. The vertical flux G (g s^{-1}) can be derived from Equation (5) using Equations (2) and (4). It is approximated by:

$$G = L \cdot W \cdot \overline{\left(v_z' c'\right)} = - L \cdot W \cdot (D + U_\perp \ell) \frac{\partial \overline{c}}{\partial z} \cong - L \cdot W \cdot (D + \ell U_\perp) \frac{C_b - C}{H}, \quad (9)$$

where C is the concentration to be calculated (μg m^{-3}), C_b is the background concentration and L the length (m) of the street canyon. Finally, the source term S (g s^{-1}) in Equation (6) is given by

$$S = Q \cdot L, \quad (10)$$

where Q is the emission source strength per unit length (μg m^{-1} s^{-1}). Equation (6) can be solved for several cases, depending on the assumptions made with regard to C_{in} and C_{out}. Here we assume that $C_{in} = C_b$ and that $C_{out} = C$, i.e. we focus on the contribution of the vehicle emissions in the street itself, assuming no additional fluxes from connecting street boxes. As a result we obtain:

$$C - C_b = \frac{Q}{U_{\equiv} \cdot \left(\frac{H}{L}\right) \cdot W + (D + \ell U_\perp) \cdot \left(\frac{W}{H}\right)}. \quad (11)$$

In Equation (11) the wind speed parallel to the street $U_{\equiv}$ is responsible for the 'ventilation' of the street box, whereas wind speed perpendicular to the street $U_\perp$ is responsible for the vertical exchange of the pollutant over a characteristic length ℓ. This characteristic length ℓ can be associated with a typical mixing length caused by turbulent eddies shedding off at roof level. The laminar diffusion coefficient D in Equation (11) can be neglected in most cases, although recently Copalle (1999) showed that at low wind velocity this diffusion can play a role. He suggests a value of $D = 1.5$ m^2 s^{-1}. The mixing length is set to $\ell = 1$ m. Equation (11) is implemented in such a way that for values of $U_\perp < 1.5$ m s^{-1}, $U_\perp$ is set to zero and a constant laminar diffusion is applied ($D = 1.5$ m^2 s^{-1}). When $U_\perp > 1.5$ m s^{-1}, the laminar diffusion is switched off. In this way, the concentration rise due to low wind speeds is limited by the laminar diffusion process, which takes over in this case, as described by Copalle (1999).

3. Model Application

The analytical model expressed by Equation (11) has been applied to calculate benzene concentrations in the City of Antwerp for 4 periods of 5 days in 1998. For these periods diffusive sampler measurements were carried out in 101 streets in Antwerp and at 4 regional background locations (Geyskens et al., 1999). The measurements were carried out in the framework of a European LIFE project (LIFE96ENV/IT/90) called MACBETH (Monitoring of Atmospheric Concentrations of Benzene in European Towns and Homes). The project included benzene measurements in 6 European cities (Antwerp, Copenhagen, Rouen, Murcia, Padova and Athens).

In the second European daughter directive (2000/69/EC) on ambient air quality assessment and management, limit values for the protection of human health are set at 5 μg m^{-3} for yearly averaged benzene concentrations. This limit value should be attained in 2010. As a consequence the ambient air quality for benzene needs to be assessed against this limit value. This assessment can be carried out by monitoring or by modelling, depending on the levels obtained in the zone or agglomeration that is considered. Since benzene has a low reactivity, with atmospheric half-lives of 5.7 days for reactions with OH radicals and 170 000 days for reactions with ozone (Verschueren, 1996), it is expected that benzene emissions and background concentrations resulting from atmospheric transport are the dominant constituents of benzene concentrations in the ambient air.

In order to provide a detailed picture of all benzene sources in and around the Antwerp area (20 × 20 km), an emission model was developed. This model consists of an urban transport emission model for Antwerp region, combined with a detailed benzene emission inventory of industrial point sources and area sources for spatial heating and domestic solvent use (with a resolution of 1 × 1 km). The results are collected in a GIS environment, from which the input for various air quality models is generated.

Benzene emissions from industrial sources are derived from the Flemish Emission Inventory (VMM, 1999). Since 1993, industrial companies are obliged to report their emissions on a yearly base. However, these registrations are only related to stack emissions. In addition, diffusive emissions form an important source of benzene. For 1998 the additional non-registered diffusive emissions were estimated to be 20%. Emissions for spatial heating and domestic solvent use are based on fuel consumption data and statistical data from a national inquiry on public housing.

The traffic emission model computes hourly benzene emissions coming from passenger cars, light duty vehicles, heavy duty vehicles and busses in the Antwerp area, in function of road type, vehicle type, fuel type, traffic volume, vehicle age, trip length distribution and the actual ambient temperature (Mensink et al., 2000). Cold start emissions and evaporation losses are included in the model. The emission calculations are based on an urban traffic flow model, actually used by the Antwerp City authorities for traffic management purposes. The urban traffic flow model contains a network with 1963 road segments and was implemented in a GIS environment. For each segment of the network the number of vehicles is predicted. The emission factors used in the model are derived from the COPERT-II methodology (Ahlvik et al., 1997) and are at present being adapted to the COPERT-III methodology.

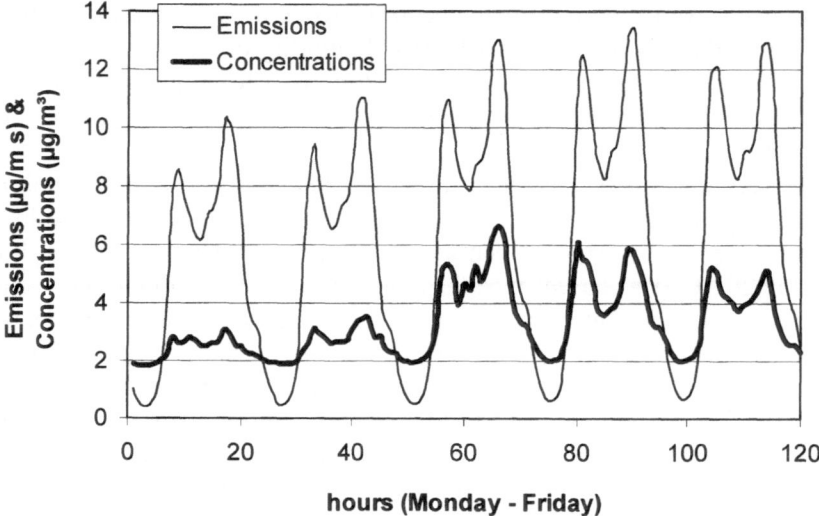

Figure 1. Benzene emissions (upper) and concentrations (lower) in Antwerp, 19–23 January 1998.

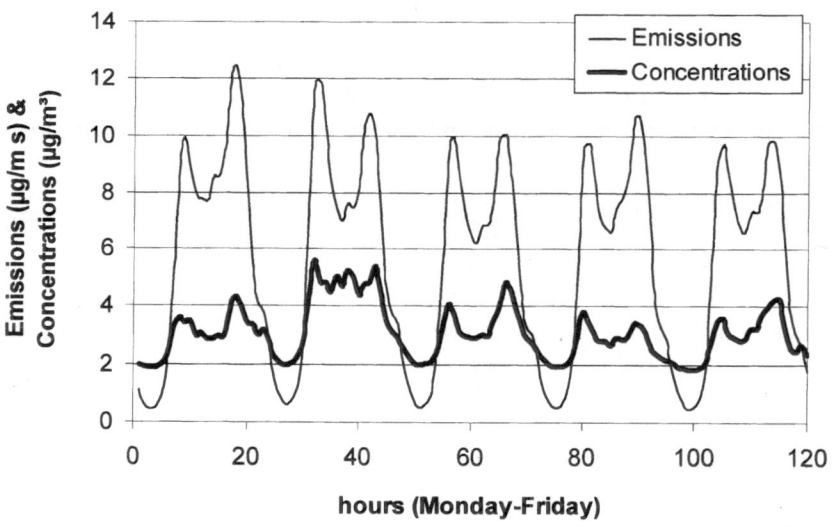

Figure 2. Benzene emissions (upper) and concentrations (lower) in Antwerp, 23–27 March 1998.

The benzene concentrations were calculated by applying Equation (11), assuming an average street aspect ratio (W/H) of 1. The hourly values for wind speed and wind direction were obtained from two meteorological towers located in the city. Wind speed at roof level was calculated from a wind profile described by a power law, with the exponent derived from the wind speed measured at heights of 30 and 153 m, respectively. The measured averaged regional background benzene concentrations were used to estimate C_b.

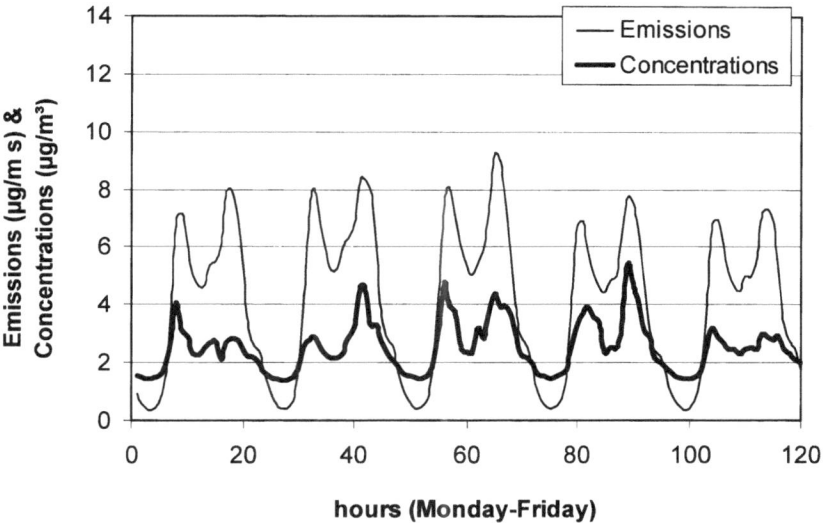

Figure 3. Benzene emissions (upper) and concentrations (lower) in Antwerp, 25–29 May 1998.

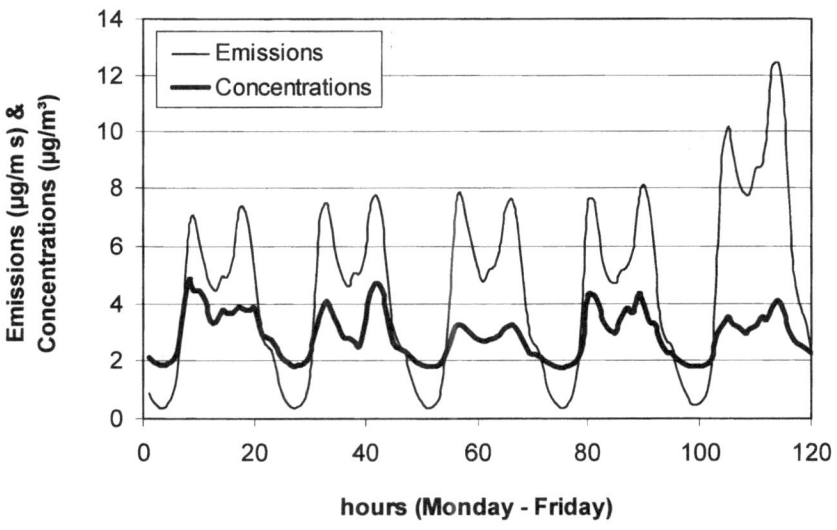

Figure 4. Benzene emissions (upper) and concentrations (lower) in Antwerp, 28 September–2 October 1998.

4. Results and Discussion

The benzene emissions and concentrations were calculated on an hourly base for 4 different periods, from Monday to Friday each:

(1) 19–23 January 1998;

TABLE I

Averaged measured and calculated benzene concentrations for four different periods of five days (Monday–Friday) in 1998 (N is the number of sample locations)

Period	Measured background concentration (N = 4)[a]	Measured concentration in the streets (N = 101)[a]	Measured standard deviation (N = 101)[a]	Calculated concentration in the streets (N = 1963)[a]
			$(\mu g\ m^{-3})$	
1) 19–23 January	1.8	3.3	±0.9	3.26
2) 23–27 March	1.8	3.1	±0.8	3.11
3) 25–29 May	1.3	2.6	±1.0	2.53
4) 28 September–2 October	1.7	3.0	±0.9	2.90

[a] N is the number of sample locations.

TABLE II

Averaged daily temperature (°C) as measured from Monday to Friday for four different periods

Period	Monday	Tuesday	Wednesday	Thursday	Friday
1) 19–23 January	5.9	4.4	1.0	−1.8	0.0
2) 23–27 March	5.4	4.5	8.3	8.0	10.7
3) 25–29 May	14.9	13.3	13.4	16.1	17.2
4) 28 September–2 October	15.8	15.9	15.3	14.7	8.3

(2) 23–27 March 1998;

(3) 25–29 May 1998;

(4) 28 September–2 October 1998.

Figures 1 to 4 show the calculated temporal evolution of benzene emissions and concentrations from Monday to Friday for each of these 4 periods. Since the diffusive samplers only provided long term averaged benzene concentrations over 5-day periods, the calculated hourly benzene concentrations cannot be compared directly. Table I shows the measured and calculated benzene concentrations for the 4 different periods as averaged over 5 days (Monday–Friday). Table II provides the daily averaged temperatures from Monday to Friday for these 4 periods. Table III shows the daily averaged wind speeds from Monday to Friday for the 4 periods.

TABLE III

Averaged daily wind speed ($m\ s^{-1}$) as measured at a height of 30 m. from Monday to Friday for four different periods

Period	Monday	Tuesday	Wednesday	Thursday	Friday
1) 19–23 January	8.1	6.2	2.5	3.1	3.6
2) 23–27 March	4.6	2.4	3.5	5.2	4.5
3) 25–29 May	3.2	3.3	2.3	2.2	2.7
4) 28 September–2 October	1.6	2.7	3.8	3.0	4.5

By comparing the Figures 1 to 4 one can clearly see the seasonal and daily changes in emission patterns. These changes are to some extend due to the variations in traffic volumes, but the main influence comes from the variations in temperature, as indicated in Table II. This temperature effect becomes very clear when comparing the emissions on 19–20 January to those on 21–23 January. After a considerable drop in temperature, the emission level increases, mainly because of the increased contributions from cold start emissions. Also on Friday 2 October the temperature effect is very clear.

Note that high emissions do not necessarily lead to high concentrations in the street. From Table III it can be seen that high concentrations in the streets are associated with low wind speed conditions. This becomes very clear by comparing the benzene concentrations and wind speeds on 19–20 January with those on 21–23 January. The wind speed effect is also remarkable on Monday 23 March compared to Tuesday 24 March, on Monday 28 September (low averaged wind speed) and on Friday 2 October (relatively high averaged wind speed). Averaged over the 5-day periods, the computed concentrations in Table I show a very good agreement with the measured benzene concentrations. Note however, that both measured and calculated concentrations are averaged over 5-day periods and over a large number of streets. In this way discrepancies for individual streets do not show up. A detailed comparison for individual streets is not carried out here, because of the momentary lack of detailed information on street and building dimensions. Because of the uniformity assumption in the box approach, concentration distributions inside the street cannot be considered, although such variations inside the street canyon are observed in reality. This should be born in mind when using the model concept and judging its validity. It makes Equation (11) more applicable and useful in a screening context, e.g. for evaluation and comparison of several streets in a city, rather than for a detailed examination of concentration variations inside the street itself.

Like in other regions in Northwest Europe, road transport emissions are the dominant source (75%) of benzene in Flanders, followed by spatial heating (12%),

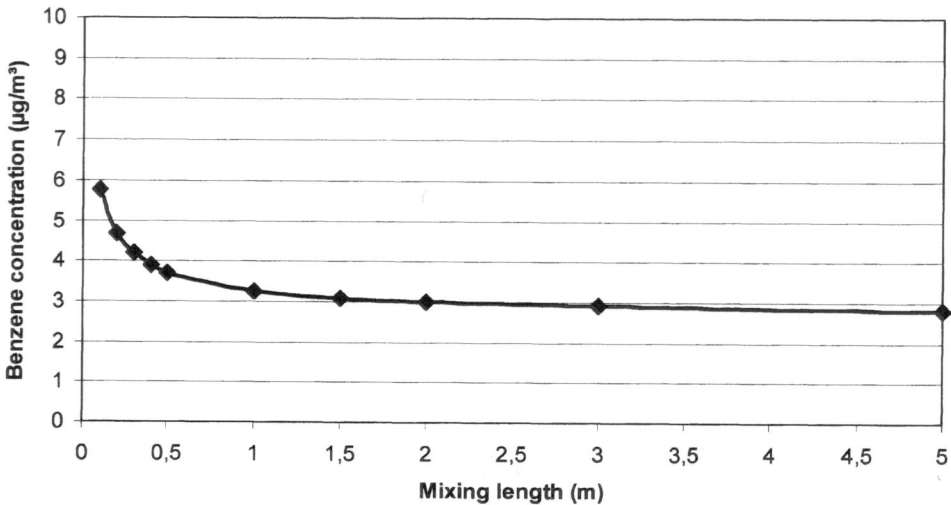

Figure 5. Sensitivity of the calculated benzene concentrations to variations in mixing length for the period 23–29 January 1998.

industrial activities (8%) and domestic solvent use (5%), as has been reported for 1998 (VMM, 1999). However many of the industrial sources are concentrated in the harbour area of Antwerp, especially emissions associated with oil refineries, fuel extraction, energy transformation and chemical process industry, resulting in relatively higher levels of benzene in this area. This industrial influence is confirmed by measurements from the monitoring stations near the harbour area, indicating ratios of benzene to the total BTEX of 17.6 to 20.7% (VMM, 1998), which is higher than values that are expected when no specific sources exist (11 to 17%). Outside the harbour area it was found that the concentrations within the streets are much less influenced by industrial sources. It was shown that these concentrations are directly related to the road transport emissions within the street itself.

A critical parameter in Equation (11) is the characteristic length or mixing length over which the turbulent mass exchange occurs. Its value was tuned and set to 1 m. Figure 5 shows the sensitivity of the results with respect to variations in mixing length for the period 23–29 January 1998. It was found that, except for very small values of l, the results are not very sensitive to this parameter. Larger values for ℓ are expected for unstable atmospheric conditions, whereas the value for ℓ becomes smaller for stable conditions and might even vanish for low wind speed conditions. In that case only the laminar diffusion will be the driving force for vertical mass exchange, as can be seen from Equation (5). More research is needed to identify this parameter and to link it (empirically) to the atmospheric stability conditions, for example through the vertical temperature gradient.

5. Conclusions

A new analytical model has been developed for the evaluation of air pollution in street canyons. In contrast with existing models, the model is based on the concept of a turbulent intermittent shear layer shed from the upwind roof as the driving force. The turbulent diffusive flux associated with this phenomenon was described by using the Prandtl-Taylor hypothesis. Calculated benzene concentrations based on hourly emissions obtained for 1963 road segments in Antwerp show a very good agreement with the results of diffusive sampler measurements when averaged over periods of 5 days. A streetwise comparison of measured and calculated benzene concentrations is needed to provide a more detailed comparison. This work is in progress.

References

Ahlvik, P., Eggleston, E., Gorissen, N., Hassel, D., Hickman, A.-J., Joumard, R., Ntziachristos, L., Rijkeboer, R., Samaras, Z. and Zierock, K.-H.: 1997, *COPERT II Computer Programme to Calculate Emissions from Road Transport, Methodology and Emission Factors*, 2nd ed., European Environmental Agency, European Topic Center on Air Emissions.

Berkowicz, R.: 1998, 'Street scale models', in J. Fenger, O. Hertel and F. Palmgren (eds), *Urban Air Pollution – European Aspects*, Kluwer Academic Publishers, Dordrecht, pp. 223–251.

Bird, R. B., Stewart, W. E. and Lightfoot, E. N.: 1960, *Transport Phenomena*, John Wiley & Sons, New York, 780 pp.

Coppalle, A.: 1999, 'A Street Canyon Model for Low Wind Speed Conditions', *Proc. of the 6th Int. Conf. on Harmonisation within Atmospheric Dispersion Modelling for Regulatory Purposes*, 1–14 October 1999, INSA de Rouen, France (CD-Rom).

Geyskens, F., Bormans, R., Lambrechts, M. and Goelen, E.: 1999, 'Measurement Campaign in Antwerp for Quantification of Personal Exposure to Representative Volatile Organic Compounds', *Vito Report 1999/DIA/R/048*, Mol, 168 pp. (in Dutch).

Hinze, O.: 1987, *Turbulence*, McGraw-Hill, New York, 790 pp.

Huang, H., Akutsu, Y., Arai, M. and Tamura, M.: 2000, 'A two-dimensional air quality model in an urban street canyon: Evaluation and sensitivity analysis', *Atmosph. Environ.* **34**, 689–698.

Johnson, W. B., Ludwig, F. L., Dabbert, W. F. and Allen, R. J.: 1973, 'An urban diffusion simulation model for carbon monoxide', *JAPCA* **23**, 490–498.

Louka, P., Belcher, S. E. and Harrison, R. G.: 2000, 'Coupling between airflow in streets and the well-developed boundary layer aloft', *Atmosph. Environ.* **34**, 2613–2621.

Mensink, C., De Vlieger, I. and Nys, J.: 2000, 'An urban transport emission model for the Antwerp area', *Atmosph. Environ.* **34**, 4595–4602.

Verschueren, K., 1996, *Handbook of Environmental Data on Organic Chemicals*, 3rd ed., Van Nostrand Reinhold, New York, 2064 pp.

Vlaamse Milieumaatschappij (VMM): 1998, *Volatile Organic Compounds in Ambient Air, Annual Report 1997*, Flemish Environmental Agency, Aalst, 26 pp. (in Dutch).

Vlaamse Milieumaatschappij (VMM): 1999, *Air Emission Report 1997–1998*, Flemish Environmental Agency, Aalst, 118 pp. (in Dutch).

Yamartino, R. J. and Wiegand, G.: 1986, 'Development and evaluation of simple models for flow, turbulence and pollutant concentration fields within an urban street canyon', *Atmosph. Environ.* **20**, 2137–2156.

THERMAL EFFECTS ON THE AIRFLOW IN A STREET CANYON – NANTES'99 EXPERIMENTAL RESULTS AND MODEL SIMULATIONS

P. LOUKA*, G. VACHON, J.-F. SINI, P. G. MESTAYER and J.-M. ROSANT

Laboratoire de Mécanique des Fluides UMR 6598 CNRS – Ecole Centrale de Nantes, Nantes Cedex 3, France

(* *author for correspondence, address: Laboratory of Heat Transfer and Environmental Engineering, Aristotle University Thessaloniki, Box 483, 54006 Thessaloniki, Greece, e-mail: petroula@aix.meng.auth.gr, fax: +30 31 996012*)

Abstract. The thermal effects on the airflow within a street canyon, which are produced by the variation of direct solar heating of the street sides and ground, are examined in this article. The investigation is based on the experimental results of the Nantes'99 campaign and numerical simulations performed with the Computational Fluid Dynamics (CFD) code CHENSI using the standard k-ε model. The Nantes'99 experimental campaign was performed in a North-to-South oriented central street canyon of Nantes, France. It was observed that a thin thermal layer develops locally within a few centimetres from the heated wall. It is anticipated that, the convective flow close to the windward wall, which was visualised during the experiment, carries air masses from the street level upwards, where normally cleaner air is transported. Consequently, thermal effects may be important for the air quality in the street. Based on the temperature and wind flow measurements, the flow and temperature fields were simulated first in two dimensions with the CFD code CHENSI. It was found that CHENSI overestimates the thermal effects on the canyon airflow showing the main re-circulation simulated in the isothermal case to change into two counter-rotating vortices after the inclusion of the heating of the windward wall. A reason for this overestimation is possibly the temperature wall function implemented for such thin thermal boundary layers in conjunction with the limitations in grid resolution.

Keywords: experimental campaign, heated wall, numerical simulations, street-canyon flow, thermal effects

1. Introduction

The airflow within a street canyon is dominated by aerodynamic, thermal or other effects such as traffic influences. Understanding these effects is crucial for improving the knowledge on the mechanisms that govern the air pollutant dispersion in urban streets. The aerodynamic effects, including building geometry and architecture as well as street canyon dimensions, have been extensively studied mainly with wind-tunnel experiments (e.g. Oke, 1987) and numerical models (e.g. Hunter *et al.*, 1992) and only with a few full-scale experiments (e.g. Louka *et al.*, 2000).

The different reception of direct solar radiation by the buildings' facades and street surface during the day may induce thermal effects on the airflow within a street canyon. It is expected that the thermal effects are relatively greater under low wind conditions and wind perpendicular to the street. Previous studies have mainly focused on the developed street-canyon flow under isothermal conditions, while the input of thermal effects on the airflow within a street canyon is insufficiently investigated numerically as well as experimentally. Sini *et al.* (1996) have studied these effects numerically using the CFD code CHENSI, within a model cavity and demonstrated that wall temperatures can largely influence the in-street flow and its vertical transport capabilities. In particular, the direct heating of the windward side of the street induces large modifications in the airflow changing the single vortex of the isothermal case to a double re-circulation system. On the other hand, the direct heating of the leeward side of the street led to the reinforcement of the isothermal vortex. However, further validation of these results is required. The two-dimensional airflow in a North-to-South oriented cavity was studied by Ca *et al.* (1995) with a Large Eddy Simulation model coupling the heat and mass transfer processes in the canyon. They found that for street canyons with small aspect ratios and strong ambient wind, the thermal convection due to the heating of the ground and walls was much lower than the mechanically produced turbulence, while for large aspect ratios or weak ambient wind thermal convection patterns were apparent. However, verification of the model with experimental results is required.

The temperature stratification during the day was the subject of a full-scale experiment within an East-to-West oriented street in Kyoto (Nakamura and Oke, 1988). It was observed that temperature on the North facade is greater than this of the South side during the day, while temperature differences between wall and air as large as 14 °C were measured close to the surface directly exposed to direct solar radiation. Nevertheless, only one velocity point-measurement was available and no further information on the structure of the dynamic field depending on the thermal stratification within the street canyon is available. Santamouris *et al.* (1999) have studied the thermal characteristics in a deep ($H/W = 2.5$) street canyon with Northwest-to-Southeast orientation in Athens under hot weather conditions. The experiment revealed that the surface temperature difference between the two street facades reaches 19 °C, while the airflow pattern for wind perpendicular to the street was characterised either by one main re-circulation or a double re-circulation system of two counter-rotating vortices. The presence of the double re-circulation was almost always associated with temperature stratification within the canyon that drives an upward air movement along the southwest wall and the lower layers of the canyon. Further work is necessary in order to understand and adequately represent in numerical models the thermal effects on the flow structure and pollutant dispersion within a street canyon.

The Nantes'99 experimental campaign aimed, among others, at providing an insight on the thermal effects on wind and CO concentration fields within a street canyon. The detailed database that was developed is a useful tool for the evaluation of numerical models that deal with this topic. Since previous numerical investigations (Sini *et al.*, 1996) suggested the largest impact of the thermal effects on the airflow when the windward side is heated, our interest in this article will be particularly focussed on this case.

2. Study Approach

The methodology adopted for investigating the thermal effects in a street canyon included the field experimental campaign of Nantes'99 and numerical modelling based on the experimental results.

2.1. THE NANTES'99 EXPERIMENTAL CAMPAIGN

The field experimental campaign was conducted during June and beginning of July 1999 in a street canyon in the city of Nantes, France. This summer month is usually characterised by high solar intensities and generally low wind-speeds, i.e. conditions expected to favour the enhancement of the thermal effects. Apart from studying these thermal effects on wind flow and pollution dispersion, the Nantes'99 campaign aimed at investigating the effects of traffic on the wind field developing within the street and the dispersion of vehicular exhausts discussed in Vachon *et al.* (2002).

Rue de Strasbourg (Figure 1a) is a three-lane one-way street situated in the centre of Nantes and is one of the most traffic congested streets of the city. The orientation of the street long axis is approximately North to South. It is an asymmetric street canyon with its west side slightly lower ($h_w = 19.4$ m) than its east side ($h_e = 22.8$ m). The aspect ratio of the width of the street, $W = 14.85$ m, over the mean height of the buildings, $H = 21$ m, is $H/W = 1.4$ implying that a main vortex develops within the street when the ambient wind is perpendicular to its axis (Oke, 1987). The experimental section was located midway between two crossroads which are 60 m apart. Detailed description of the experimental site and the available measurements can be found in Vachon (2001).

Measurements of surface and air temperature were performed using thermocouples. The surface temperature of the building walls was measured at four levels, while the air temperature was measured at similar levels and at seven horizontal positions at each of these levels as shown in Figure 1b. Mean wind and turbulence were measured with sonic and propeller anemometers at similar heights, while reference air temperature and wind were obtained at the roof of the west side of the street. Measurements from several days with low easterly winds perpendicular to the street and intense solar radiation were analysed. Consequently, for the particular days, the west and east sides of the street corresponded to the windward and leeward side of the street, respectively.

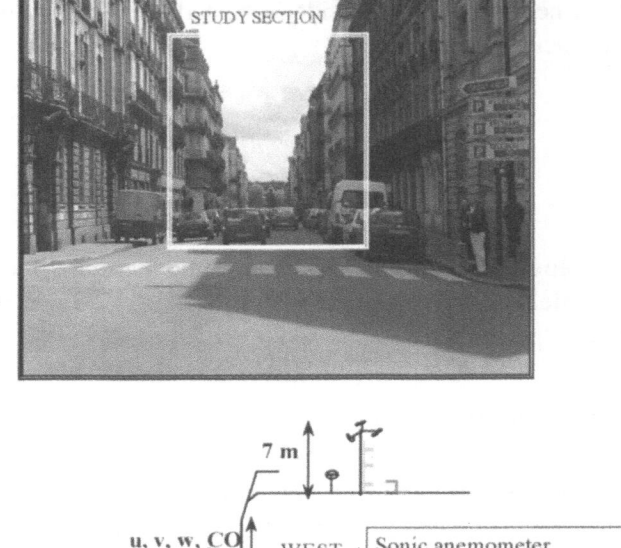

(a)

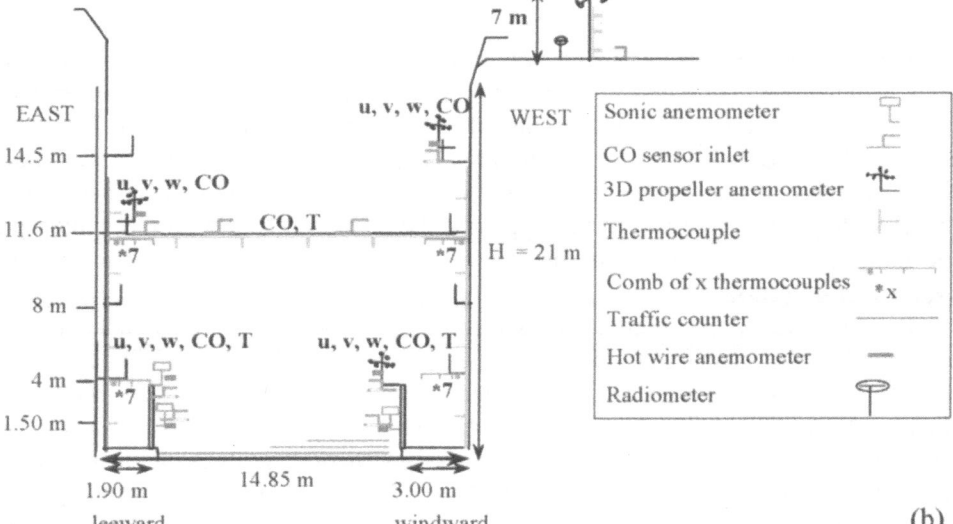

(b)

Figure 1. (a) The study section and (b) two-dimensional schematic representation of the measuring site and instruments involved.

2.2. NUMERICAL MODELLING

The CFD code CHENSI developed by the group Dynamique de l'Atmosphère Habitée of the Laboratoire de Méchanique des Fluides at Ecole Centrale de Nantes applies the standard k-ε model and was utilised for the numerical simulations. A detailed description of the model regarding the numerical scheme and the boundary conditions for wind and turbulence is available on the web site of the CFD working group of the TRAPOS network (http://www.dmu.dk/atmosphericenvironment/

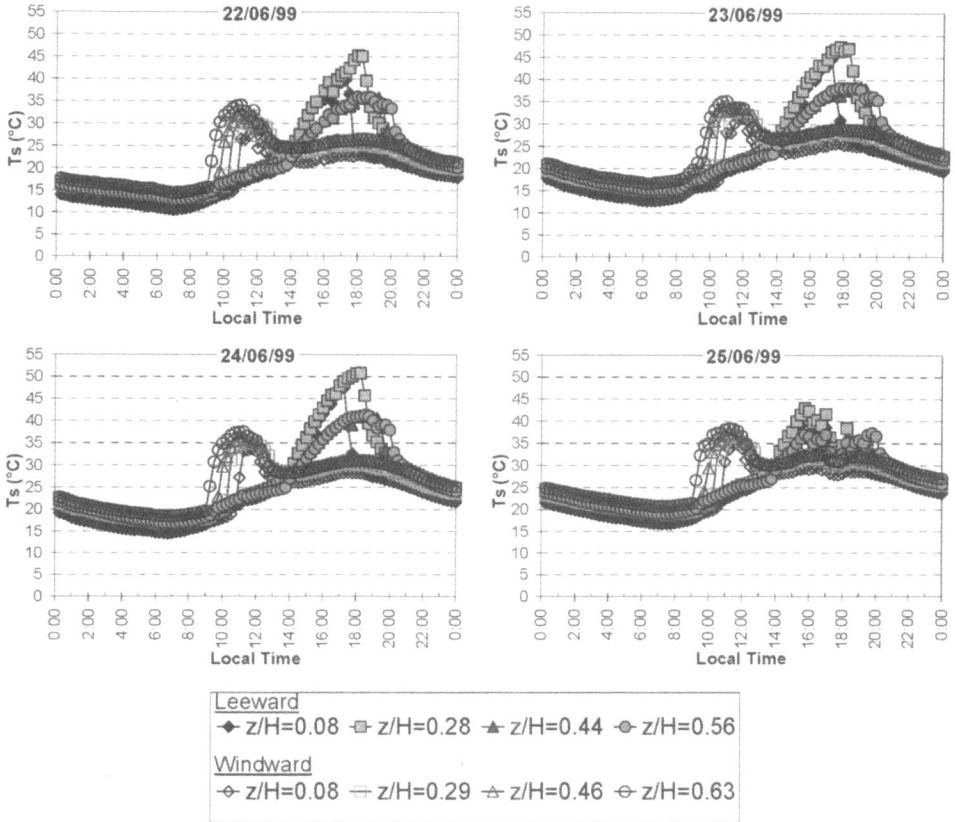

Figure 2. Diurnal wall temperature variation in Rue de Strasbourg during 22, 23, 24 and 25 June 1999.

trapos/cfd-wg.htm). Further description and discussion of the thermal boundary conditions applied in the code will be performed in the next sections.

3. Results

3.1. NANTES'99 EXPERIMENTAL RESULTS

Figure 2 shows the diurnal variation of the wall temperature at four different heights during four consecutive days of the experiment, namely 22, 23, 24 and 25 June 1999. It is clear that the North-to-South orientation of the street long axis results to the solar heating of its *west side* in the morning and progressively its *east side* in the afternoon, while a reduction of the east wall temperature during the afternoon of 25 June is attributed to cloud cover. It is worth noting that maximum wall temperature in the morning approaches 40 °C and in the afternoon exceeds 50 °C.

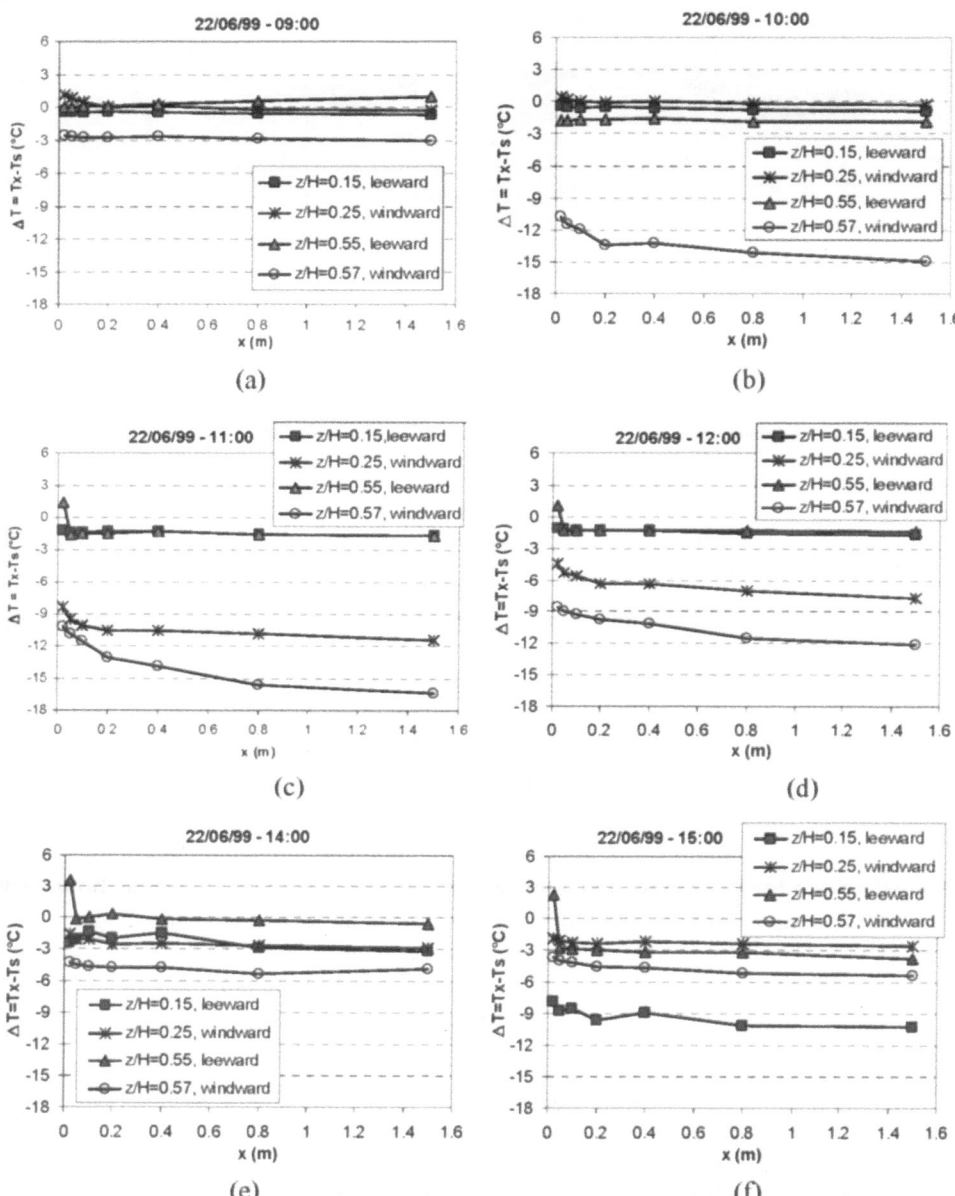

Figure 3. Measured temperature difference, ΔT, between air and walls versus distance from the walls at different levels at (a) 9:00, (b) 10:00, (c) 11:00, (d) 12:00, (e) 14:00 and (f) 15:00 local time.

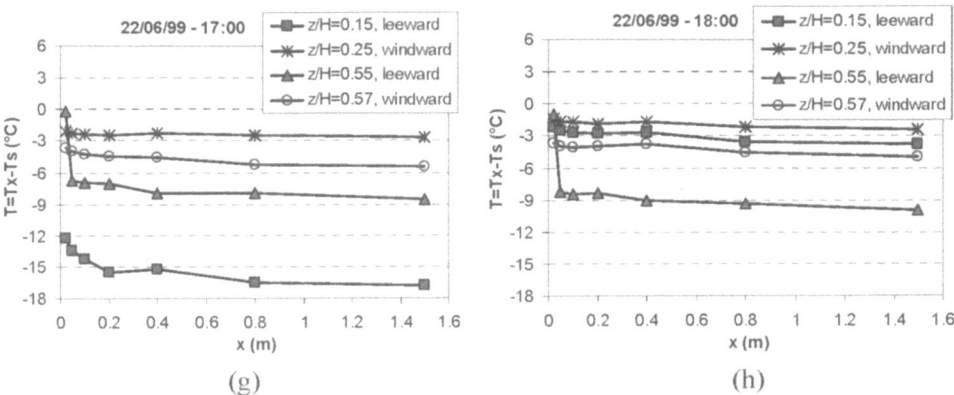

Figure 3. (continued) Measured temperature difference, ΔT, between air and walls versus distance from the walls at different levels at (g) 17:00 and (h) 18:00 local time.

Figure 3 demonstrates the variation of the temperature difference (ΔT) between air at the different x-locations (T_x) and adjacent building surface (T_s) on 22 June at different periods during the day. Similar patterns were observed also for the rest of the days. Early in the morning (9 a.m. local time), only the upper part of the windward side is directly heated by the sun leading to negligible ΔT at all positions apart from at $z/H = 0.57$ where ΔT is approximately constant along measuring path-x and equal to an absolute value of 3 °C (Figure 3a). An hour later the upper part of the windward side has been heated more and ΔT has increased reaching a maximum of 15 °C (Figure 3b). Within the next hour the sun has directly heated the lower levels of the west side with the wall temperature exceeding the air temperature by a maximum of approximately 18 °C at $z/H = 0.57$ (Figure 3c). At local noon similar behaviour is observed with temperature difference now being smaller than previously (Figure 3d), while at 2 p.m., that the sun has reached its zenith, ΔT as well as horizontal gradients have diminished (Figure 3e) due to the uniform heating of the canyon air. An hour later the lower part of the east (leeward) side of the street has been heated directly by the sun (Figure 3f) with ΔT changing almost uniformly horizontally from the wall. Large temperature difference is observed again at $z/H = 0.15$ at 5 p.m. with maximum ΔT approaching 18 °C (Figure 3g). Finally, at 6 p.m. only the upper part of the east side of the street is heated directly by the sun with maximum ΔT being approximately the same ($\sim$9 °C) at $z/H = 0.55$ (Figure 3h). Obviously, there is an experimental error for the first value of ΔT at $z/H = 0.55$ of the leeward side at all times.

The most important feature in Figure 3 is the development of a steep horizontal temperature gradient very close to the wall. During the morning hours when the windward wall is directly heated by the sun, already at 2 cm from the wall, the gradient is extremely high with a maximum of 10.7 °C/2 cm and it is still steep within the first 20 cm from the wall (maximum of 2.9 °C/20 cm at $z/H = 0.57$) implying the appearance of a thin thermal layer due to the direct solar heating of the

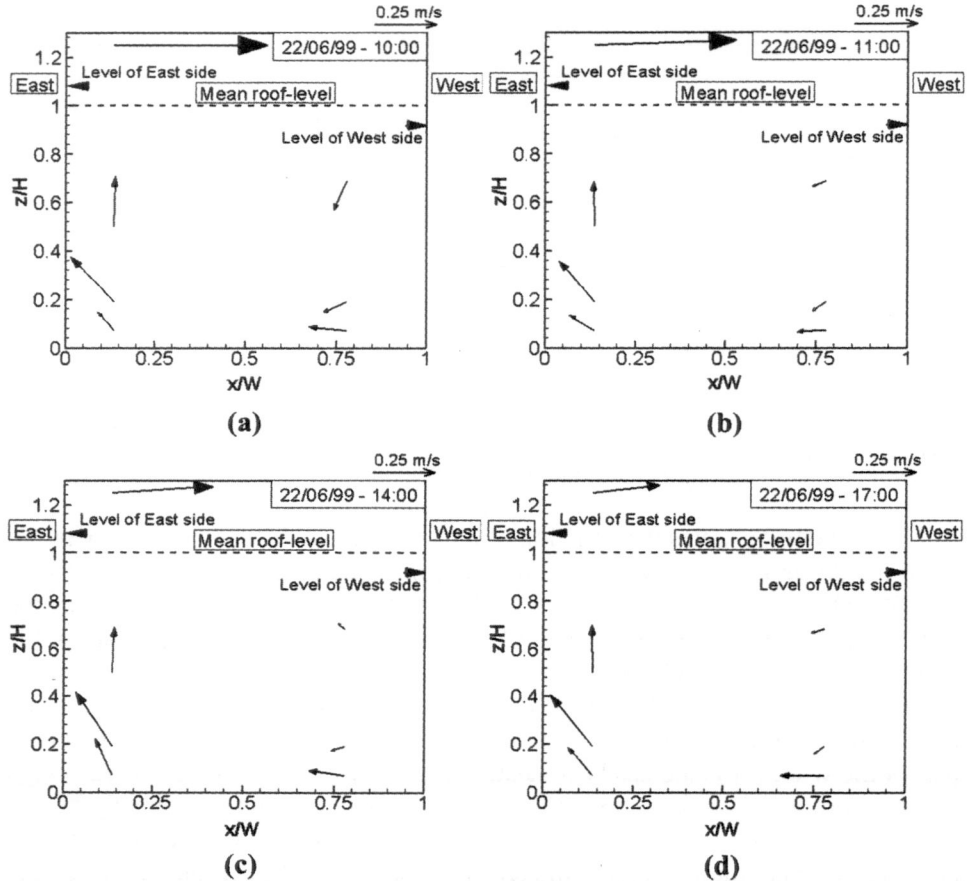

Figure 4. Two-dimensional wind speed measured on 22 June 1999 at (a) 10:00, (b) 11:00, (c) 14:00 and (d) 17:00 local time.

wall. At distances farther than 20 cm from the wall the gradient for ΔT is reduced to approximately 0.9–1.5 °C/130 cm (or 0.7–1.2 °C m^{-1}) as the temperature falls to its value in the main part of the canyon. Similarly, during the late afternoon when the leeward side receives direct solar radiation, a thin thermal layer close to the wall is observed; the maximum temperature difference in the first 2 cm reached the 12.3 °C at $z/H = 0.15$, while a gradient approximately equal to 3.2 °C/20 cm within 20 cm from the wall was observed.

These temperature differences, especially the maximum ones, lead to a strong buoyancy force close to the wall receiving direct solar radiation that may affect the transport of pollutants from the canyon to the layer aloft. As the horizontal gradients were observed to be steep very close to the walls up to 20 cm from it, it is suggested that the thermal effects on the airflow and dispersion of vehicular pollutants are greatest locally, close to the walls.

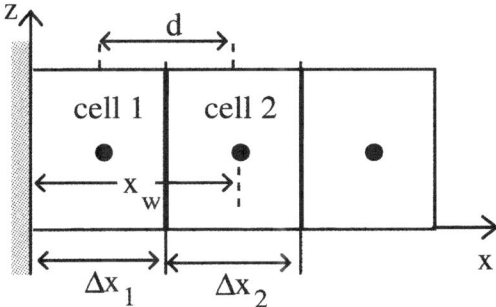

Diagram 2. Scheme of grid cells attached to a vertical wall as treated in CHENST.

The investigation of such thermal effects on the airflow was complicated due to the absence of wind speed measurements at the positions of temperature measurements as well as due to the combination with the traffic effect. A strong convection close to the walls was observed with the visualisation experiment carried out during the same period using non-buoyant helium balloons. The quantitative examination of the flow field using the available measurements was made with a two-dimensional projection of the velocity vectors as shown in Figure 4. A main re-circulation was suggested by the measurements, which was present during the whole day for wind perpendicular to the street. However, the wind-speed measurements were taken farther from the walls than the temperature measurements, therefore the available flow measurements cannot be used to extract conclusions on the thermal effects on the air motion up to 1.5 m from the walls.

The presence of a thin thermal layer close to the walls is an important characteristic of the microenvironment around the buildings and is difficult to be accurately represented by CFD codes. This aspect will be discussed in the next section.

3.2. NUMERICAL SIMULATIONS

The wall boundary condition used for temperature in CHENSI is a wall function based on temperature gradient normal to the wall (Sini *et al.*, 1996). This type of wall function for temperature has been validated for thick thermal boundary layers (Levi-Alvares, 1991).

Diagram 1 illustrates three grid cells attached to a vertical wall (shadowed area) as used in CHENSI. Cell 1 is where the wall conditions are applied, while cell 2 is where temperature is calculated.

The temperature at cell 1 is given by temperature at cell 2 from

$$T_1 = T_2 - A\,(T_2 - T_s)\ ,\tag{1}$$

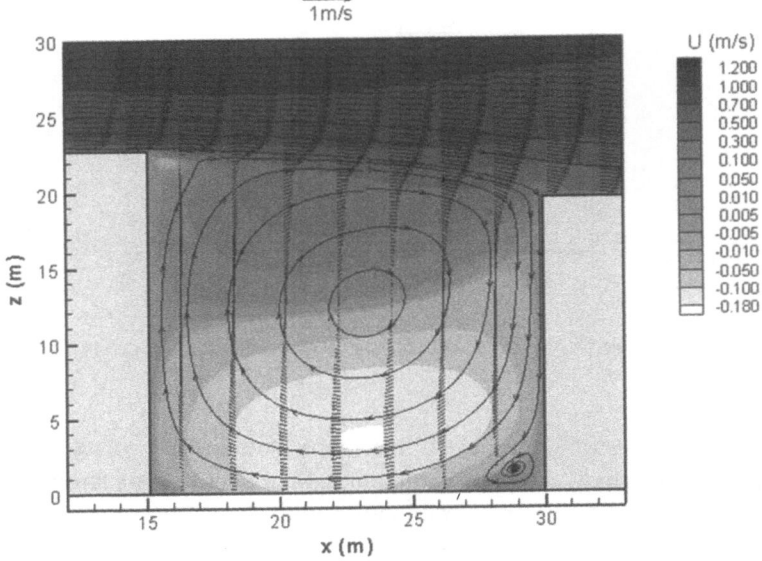

Figure 5. Simulated wind field without thermal effects (isothermal case).

where T_s is the fixed wall temperature and function A is given by

$$A = \frac{\kappa C_\mu^{1/4} d}{\ln\left(\frac{x_w}{z_o}\right) + \frac{P}{\kappa}} \frac{\sqrt{k}}{\nu_t} , \tag{2}$$

with κ the von Karman's constant ($=0.4$), $C_\mu = 0.09$, d the distance in the x-direction between the cell centres, z_0 the roughness length of the wall, x_w the distance from the wall, k the turbulent kinetic energy and ν_t the turbulent viscocity in cell 2. P is the widely used Jayatilleke (1969) parameter:

$$P = 9.24 \left[\left(\frac{Pr}{Pr_t}\right)^{0.75} - 1 \right] \left[1 + 0.28 \exp\left(-0.007\frac{Pr}{Pr_t}\right) \right] , \tag{3}$$

dependent of the mean and turbulent Prandtl number.

Initially, two-dimensional calculations were performed with size of the cells within the street equal to 24.75 cm. Due to the non-availability of measurements at the inlet of the calculated domain, in the particular case upstream the eastern side, the velocity and turbulent kinetic energy profiles were such that would lead to wind speed, U_{ref}, and turbulence values, k_{ref}, as close as possible to the measured ones at the reference position (roof of west building), i.e. $U_{ref} \sim 1 \text{ m s}^{-1}$ and $k_{ref} \sim 0.1 \text{ m}^2$ s^{-2}, for the selected days.

Initially, the dynamic field was calculated without any temperature effects (isothermal case). The features of the simulated flow field (Figure 5) are a main re-

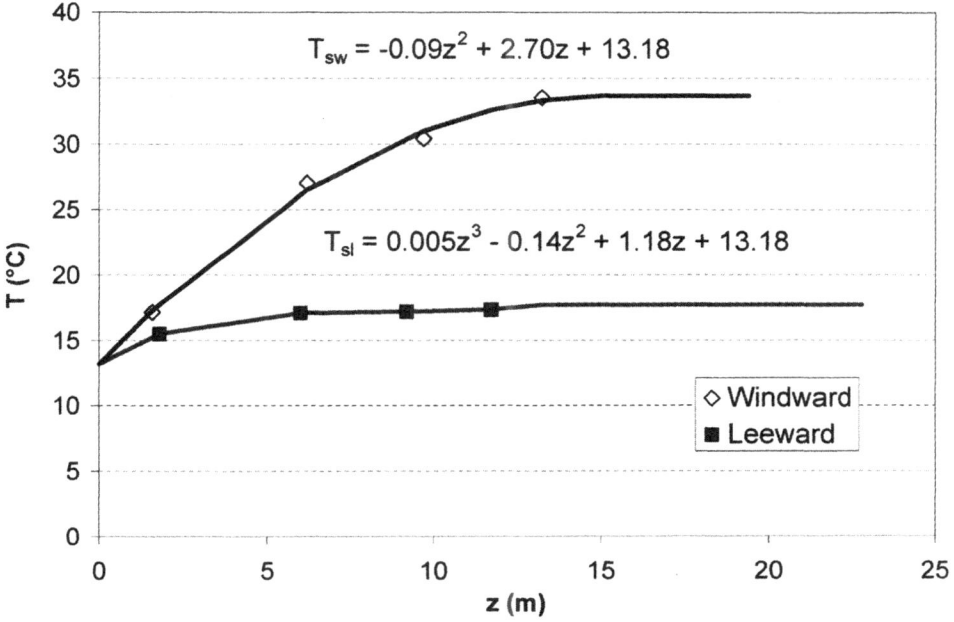

Figure 6. Measured leeward and windward wall temperature and fitted profiles.

circulation and a small secondary vortex at the windward corner at the floor of the street canyon.

As it was mentioned in the introduction, this article gives special emphasis to the case that the windward side of the street is heated in order to investigate the effect of the ascent of air masses induced by the heating of the wall on the main re-circulation of the isothermal case. For this reason the numerical simulation of that case will be presented here.

Fitted profiles on the surface temperature of the leeward and the windward walls at 11 a.m. provided the wall temperature values for the simulations with CHENSI. The profiles were extended for the whole height of the walls as shown in Figure 6. The simulation (Figure 7) shows that the heated windward wall has created a distinctive effect on the main flow, which is now characterised by two main counter-rotating vortices and a secondary smaller eddy at the leeward side at the bottom of the street. The re-circulation of the isothermal case has been suppressed at the top of the canyon, while a strong updraft close to the heated wall leads to transfer of air from the canyon floor to the roof-level.

It is apparent that CHENSI overestimates the thermal effects on the flow within Rue de Strasbourg leading to the prediction of a largely modified street flow. As it was observed during the experimental campaign, a very thin thermal boundary layer was present next to the walls having its largest intensity at 2 cm from the wall and being apparent up to 20 cm from the wall. Although the thermal law of the wall

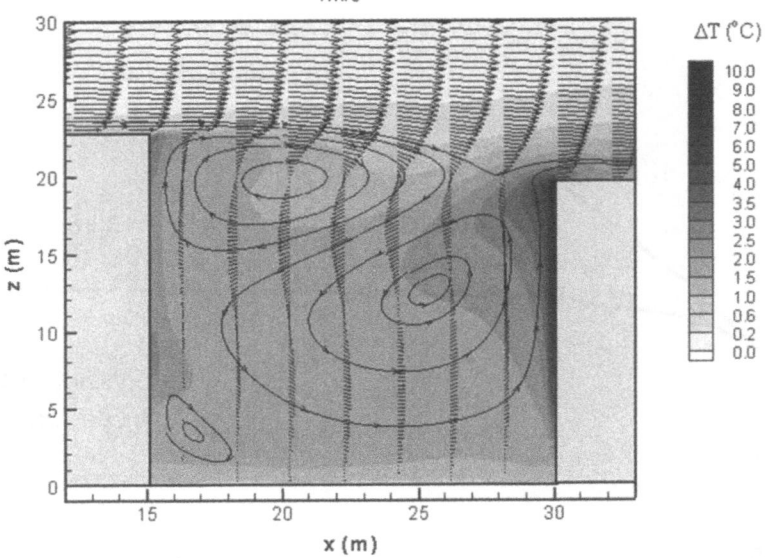

Figure 7. Simulated wind field with thermal effects and temperature difference isocontours.

implemented in CHENSI was shown to be good for thick thermal boundary layers (Levi-Alvares, 1991), for the present study of a thinner layer, the wall condition is used outside this thermal layer. Therefore, due to the wall function applied, making use of the temperature gradient normal to the wall and the restrictions to grid resolution (inevitable geometrical simplifications), CFD codes cannot resolve such thin thermal boundary layers. The question that should be answered is: 'How can thermal effects be represented integrally in such boundary layers at the scale of a building facade or roof?' For providing a solution to this problem, further investigation with new microscale experiments is required that may lead to further development and validation of CFD codes.

Similar thin thermal layers were also observed in the ENFLO wind tunnel investigating the effects of the uniformly heated windward wall of a cavity (Kovar-Panskus *et al.*, 2002). Finally, it is expected that under such low velocities within the street canyon the airflow is three-dimensional rather than two-dimensional as it was demonstrated by the same wind tunnel experiment.

4. Conclusions

This article presented the thermal effects on the airflow in Rue de Strasbourg as were observed during the Nantes'99 experimental campaign and predicted by the numerical simulations with the CFD code CHENSI.

The Nantes'99 experiment suggested that the effects of the heated walls on the street canyon flow dynamics have a small extent and could be significant in the close vicinity of the buildings. In particular, a steep horizontal temperature gradient was observed revealing a thin thermal layer extended up to 20 cm from the walls and having its greatest intensity at 2 cm from the walls. Farther from the walls the horizontal temperature gradient was smoother representing the cooling rate of the air away from the heated wall, as temperature reduces reaching its value at the main part of the canyon. As the flow in the wall boundaries carries air from the street level upwards, where normally cleaner air is transported from above, thermal effects may be important and therefore taken into account in practical models investigating air quality issues in street canyons.

In the case that the windward side is heated, for which more emphasis was given in this article, it is anticipated that the convective flow, observed close to the walls with the visualisation experiment, leads to introduction of colder air from below and thereby the formation of a secondary vortex close to the wall. However, the horizontal extent of this possible vortex is expected to be less than 3 m, i.e. will cover less than 20% of the total street width, as the airflow measurements obtained at that location did not suggest such behaviour.

The numerical simulation of this thermal case showed that the CFD code CHENSI overestimates the influence of the heated windward side leading to a system of two main counter-rotating vortices and a secondary smaller eddy at the leeward side at the bottom of the street. The possible reason for this overestimation is the implementation of a temperature wall function based on temperature gradients normal to the wall, in conjunction with the size of the grid cells close to the walls. Due to grid resolution restrictions the wall condition is used outside the thin thermal boundary layer observed. Therefore, it can be concluded that the 'standard' wall function cannot be used in the framework of obstacle resolving urban atmosphere because the thermal boundary layers that develop on the vertical surfaces are too thin to be resolved. Therefore, further study is required in order to understand the integrated effect of such thermal phenomena.

A more appropriate condition for modelling such flows would be the reformulation of the temperature wall conditions in terms of thermal fluxes based on the thermal balance of the walls. For this reason an experiment has been undertaken to investigate the flow and thermal field close to the near wall region. The analysis of these data is expected to give some further intake in the thermal effects on the airflow close to the walls.

In addition, it is to be expected that spanwise re-circulations associated with the essential three-dimensionality of the configuration will also affect the general flow within the street. Therefore, a three-dimensional simulation has commenced that will calculate the airflow within Rue de Strasbourg and in its neighbourhood taking also into account the complex geometry of the buildings and the effects of the side streets.

Apart from the obvious interest on air quality the investigation of thermal effects is also important for practical applications on building energy budgets associated with indoor ventilation and heating.

Acknowledgements

The present work was carried out within the TRAPOS network ('Optimisation of Modelling Methods for Traffic Pollution in Streets') in the framework of Training and Mobility of Researchers Programme (TMR contract ERNFMRXCT97-0105). The authors wish to acknowledge the support from the PRIMEQUAL-PREDIT (contrat INERIS 1998), the Agence de l'Environnement et de la Maîtrise de l'Energie (ADEME) and PSA Peugeot Citroën. We also acknowledge the computer support given by Institut de Développement et de Recherche pour l'Informatique Scientifique (IDRIS), France.

References

Ca, V. T., Asaeda, T., Ito, M. and Armfield, S.: 1995, 'Characteristics of wind field in a street canyon', *J. Wind Eng. Ind. Aerodyn.* **57**, 63–80.

Hunter, L. J., Johnson, G. T. and Watson, I. D.: 1992, 'An investigation of three-dimensional characteristics of flow regimes within the urban canyon', *Atmos. Env.* **26B**, 425–432.

Jayatilleke, C. L. V.: 1969, 'The influence of Prandtl number and surface roughness on the resistance of the laminar sublayer to momentum and heat transfer', *Prog. Heat Mass Trans.* **23**(6), 193–329.

Kovar-Panskus, A., Moulinneuf, L., Savory, E., Abdelqari, A., Sini, J.-F, Rosant, J.-M., Robins, A. and Toy, N.: 2002, 'A wind tunnel investigation of the influence of solar-induced wall-heating on the flow regime within a simulated urban street canyon', *Water, Air, and Soil Pollut.*, this issue.

Levi-Alvares, S.: 1991, 'Simulation Numérique des Ecoulements Urbains a l'Echelle d'une Rue a l'Aide d'un Modèle k-epsilon', *Thèse de Doctorat de l'Ecole Centrale de Nantes et de l'Université de Nantes*.

Louka, P., Belcher, S. E. and Harrison, R. G.: 2000, 'Coupling between airflow in streets and the well-developed boundary layer aloft', *Atmos. Env.* **34**, 2613–2621.

Nakamura, Y., and Oke, T. R.: 1988, 'Wind, temperature and stability conditions in an east-west oriented urban canyon', *Atmos. Env.* **22**, 2691–2700.

Oke., T. R.: 1987, *Boundary Layer Climates*, 2nd ed., Methuen, U.K.

Sini, J.-F., Anquetin, S. and Mestayer, P. G.: 1996, 'Pollutant dispersion and thermal effects in urban street canyons', *Atmos. Env.* **30**, 2659–2677.

Santamouris, M., Papanikolaou, N., Koronakis, I., Livada, I. and Asimakopoulos D.: 1999, 'Thermal and air flow characteristics in a deep pedestrian canyon under hot weather conditions', *Atmos. Env.* **33**, 4503–4521.

Vachon, G.: 2001, 'Transferts des Polluants des Sources Fixes et Mobiles dans la Canopée Urbaine: Evaluation Experimentale', Ph.D. Thesis, Ecole Centrale de Nantes, France.

Vachon, G., Louka, P., Rosant, J.-M., Mestayer, P. G. and Sini, J.-F.: 2002, 'Measurements of traffic-induced turbulence within a street canyon during the Nantes'99 experiment', *Water, Air, and Soil Pollut.*, this issue.

INFLUENCE OF GEOMETRY ON THE MEAN FLOW WITHIN URBAN STREET CANYONS – A COMPARISON OF WIND TUNNEL EXPERIMENTS AND NUMERICAL SIMULATIONS

A. KOVAR-PANSKUS[1], P. LOUKA[2], J.-F. SINI[2], E. SAVORY[3]*, M. CZECH[1],
A. ABDELQARI[2], P. G. MESTAYER[2] and N. TOY[1]

[1] *Fluids Research Centre, School of Engineering, University of Surrey, Guildford, Surrey, U.K.;*
[2] *Ecole Centrale de Nantes, Lab de Mecanique des Fluides, Nantes, France;* [3] *Department of
Mechanical and Materials Engineering, Faculty of Engineering, University of Western Ontario,
London, Ontario, Canada*
(* *author for correspondence, e-mail: esavory@eng.uwo.ca*)

Abstract. A comparison between numerical simulations and wind tunnel modelling has been performed to examine the variation with streamwise aspect ratio (width/height, W/H) of the mean flow patterns in a street canyon. For this purpose a two-dimensional (2-D) cavity was subjected to a thick turbulent boundary layer flow perpendicular to its principal axis. Five different test cases, W/H = 0.3, 0.5, 0.7, 1.0 and 2.0, have been studied experimentally with flow measurements taken using pulsed-wire anemometry. The results show that the skimming flow regime, with a large vortex in the canyon, occurred for all the cases investigated. For the cavities with W/H≤0.7 a weaker secondary circulation developed beneath the main vortex. The narrower the canyon, the smaller the wind speed close to the cavity ground, giving increasingly poor ventilation qualities. The corresponding numerical results were obtained with the Computational Fluid Dynamics (CFD) code CHENSI that uses the standard k-ε model. The intercomparison showed good agreement in terms of the gross features of the mean flow for all the geometries examined, although some detailed differences were observed.

Keywords: 2-D cavity, canyon, numerical simulation, pulsed-wire-anemometry, standard k-ε model, ventilation, vortex

Nomenclature

d	=	Displacement height (m);
H, W	=	Height and width of canyon (m);
k	=	Turbulent kinetic energy ($m^2\ s^{-2}$);
L	=	Spanwise length of canyon (m);
Re	=	Reynolds number, $U_{ref}\ H/\nu$;
U	=	Mean velocity in X direction ($m\ s^{-1}$);
U_{ref}	=	Freestream velocity ($m\ s^{-1}$);
U_*	=	Friction velocity ($m\ s^{-1}$);

Water, Air, and Soil Pollution: Focus **2**: 365–380, 2002.

u'	=	Streamwise turbulence intensity (m s^{-1});
v'	=	Lateral turbulence intensity (m s^{-1});
w'	=	Vertical turbulence intensity (m s^{-1});
X, Y, Z	=	Cartesian coordinates (m);
z_o	=	Roughness length (m);
δ	=	Boundary layer height (m);
ε	=	Dissipation (m^2 s^{-3});
ν	=	Kinematic viscosity of air (m^2 s^{-1}).

1. Introduction

Many different aspects of the wind flow within urban street canyons have been studied over the past few decades, including the influence of the flow pattern on the local pollutant transport. There have been many full-scale studies, such as De-Paul and Sheih (1986), which have provided descriptions of the canyon mean flow recirculation pattern. These investigations, with aspect ratios (Width (W)/Height (H)) of 0.5 to 1, showed that the main vortex shifts upwards within the canyon with decreasing canyon width (DePaul and Sheih, 1986). This also yields much higher pollutant concentrations in a canyon with W/H = 0.5 in comparison to W/H = 1 (Pavageau et al., 1996). For canyons with W/H = 1, the wind speed at the roof level was found to scale linearly with the strength of the vortex flow (Nakamura and Oke, 1988). Further wind tunnel studies have investigated the influence of roof shape on the distribution of pollutants within the canyon (Rafailidis, 1997; Kastner-Klein and Plate, 1999). Liedtke et al. (1999) studied the influence on pollutant dispersion of the geometrical resolution of the wind tunnel model used and found distinct differences in comparison to the modelled field site study of Berkowicz et al. (1996). Similarly, Leitl et al. (2001) studied the influence of wind tunnel model detail on pollutant dispersion in street canyons and found notable differences with different degrees of complexity of the model. Other research has clearly shown that the observed changes in flow pattern depended on both geometry and the interaction of different street canyons, for example at street intersections or T-junctions (Scaperdas, 2000). Extensive sets of wind tunnel measurement data around street intersections and buildings is available to researchers on the internet (MIHU, 2001). Wind tunnel studies of the flows in canyons with unequal wall heights have also been conducted (Dabberdt and Hoydysh, 1991). It was found that for a step-up canyon the vortex was significantly stronger than with a symmetrical canyon. In the step-down canyon the measured concentrations were slightly lower on the upwind side. This agrees with numerical simulations (Assimakopoulous et al., 2000), where the highest concentrations seemed to be trapped near the windward-wall.

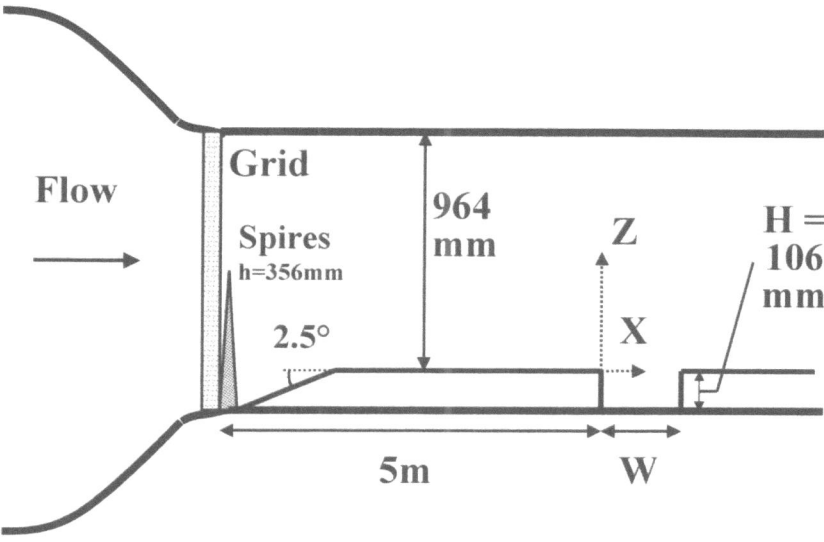

Figure 1. Diagrammatic layout of the wind tunnel and canyon model.

The flows associated with different canyon aspect ratios have been classified by Oke (1987). Since the main canyon vortex directly affects the local distribution of pollutants, the manner in which the flow and the associated turbulence quantities are dependent on the variation of canyon aspect ratio is of key importance. The present research was a study of the mean flow behaviour inside relatively narrow canyons to determine where transitions between different regimes occur. A nominally 2-D street canyon model was exposed to a flow normal to its principal axis and the aspect ratio systematically varied from W/H = 0.3 to 2.0. Detailed profiles of the three wind velocity components were obtained. Parallel to these experiments, the same regimes were numerically simulated with the standard k-ε model CHENSI (Sini *et al.*, 1996), run in 2-D to provide mean velocities and turbulent kinetic energy (TKE) for the streamwise (X) and vertical (Z) wind components. For these comparisons, the boundary layer mean velocity and turbulence profiles were used at the inlet of the calculated domain.

2. Details of Wind Tunnel Experiments

2.1. EXPERIMENTAL ARRANGEMENT

A nominally 2-D cavity of fixed depth (H = 106 mm) was mounted on the wall of a University of Surrey wind tunnel, with its principal axis normal to the wind direction, as shown in Figure 1. The working section dimensions are 1.37 m height × 1.07 m width × 9 m length, giving a canyon lateral aspect ratio of L/H = 12.9.

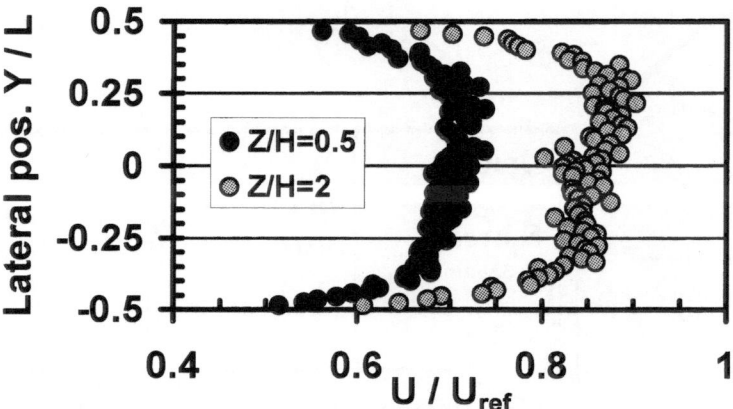

Figure 2. Lateral profiles of U component mean velocity in the boundary layer.

The streamwise cavity width (W) was varied to give 5 cases of W/H = 0.3, 0.5, 0.7, 1.0 and 2.0. A thick turbulent boundary layer was produced by a grid, a fence and vorticity generators at the entrance to the working section, whilst the ground plane and cavity walls were roughened using 'Lego' boards with distributed $4.5 \times 7.5 \times 15$ mm rectangular brick elements. The neutrally stratified approach conditions were; boundary layer height $\delta = 737$ mm ($\delta/H = 6.95$), roughness length $z_0 = 0.3$ mm, displacement height $d = 1$ mm and friction velocity $U_*/U_{ref} = 0.050$, based on the freestream velocity $U_{ref} = 8$ m s^{-1}. The freestream turbulence level, u'/U_{ref}, was 9% (at Z/H = 7), whilst the integral turbulence length scales were typically 2 H, 1.5 H and 1.0 H, in the X, Y and Z directions, respectively. The test Reynolds number (Re) was 5.6×10^4. The flow measurements were taken by pulsed-wire-anemometry, which simultaneously measures the flow speed and direction (Castro and Cheun, 1982). The sampling rate was 50 Hz and 10 000 samples were taken at each point. The measurement data were accurate to within ±10% for the mean values and ±15% for the stresses (Castro and Cheun, 1982).

The geometrical scaling of the simulated boundary layer is approximately 1:500, representing an unrealistic 53 m deep full-scale canyon. The resulting roughness length of 0.15 m is certainly not representative of an array of buildings of this height according to the literature (Theurer, 1994). The cavity depth fits a scale of 1:200, giving a building height of approximately 20 m and a scale factor mismatch of 2.5. It is known (Jensen, 1958), that H/z_0 is a key scaling parameter for modelling full-scale cases as too large a scale exposes the model to only small-scale turbulence (Hunt, 1981). However, a mismatch of 2–3 in scaling is acceptable for the present generic study.

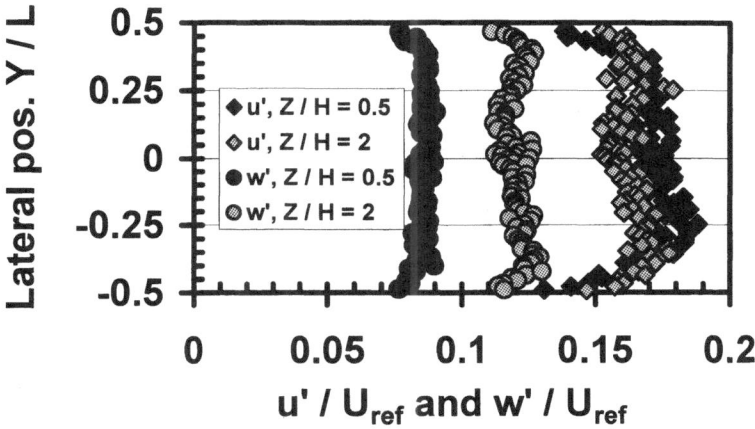

Figure 3. Lateral profiles of u' and w' turbulence intensity in the boundary layer.

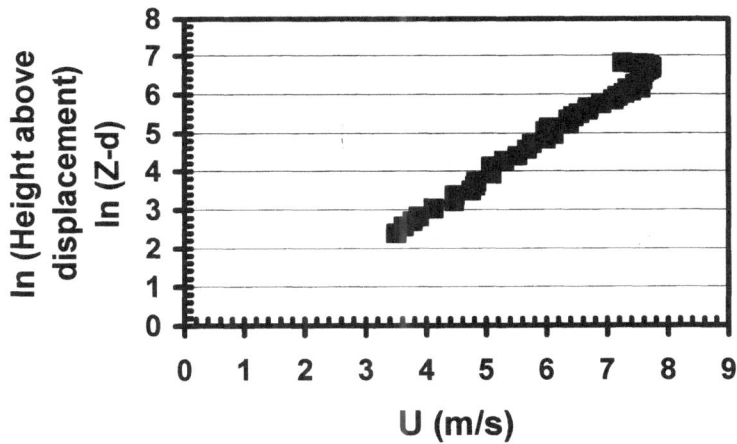

Figure 4. Boundary layer mean velocity profile before separation at cavity.

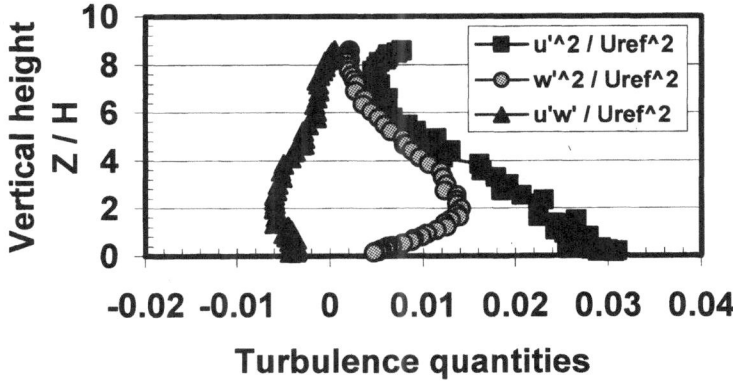

Figure 5. Boundary layer turbulence intensity profiles before separation at cavity.

2.2. APPROACH FLOW CONDITIONS

The lateral two-dimensionality in the approach flow and in the cavity was studied extensively to ensure good quality data for CFD validation purposes. This was examined at two heights upstream of the cavity where the boundary layer was already fully developed. The results show that the variation laterally across the boundary layer (Y/L) of the U component mean velocity is consistent to within $\pm6\%$ for both heights, Figure 2, whilst the corresponding variations of the u' and w' turbulence intensities are within ±8 and 6%, respectively, of the mean value, averaged across the cavity span, Figure 3. Two-dimensionality was clearly established for the central 60% of the spanwise length of the cavity (that is, approximately 7.8 H). The study included some spanwise traverses within the W/H = 1.0 cavity which showed that the U component was within $\pm5\%$ of the mean along each profile. The angle of deviation of the flow direction from the XZ plane (zero for an ideal 2-D flow), was in always less than 3°. The u' turbulence intensity was within $\pm3\%$ of the mean value along each profile. Hence, it may be concluded that the initial conditions were satisfactory for minimising the three-dimensionality effects. The mean velocity and turbulence profiles for the boundary layer, just before separation at the cavity, are shown in Figures 4 and 5, respectively. The small perturbations in the profiles at their outer edges are due to the boundary layer on the opposing side of the tunnel from the canyon model.

3. Details of the Numerical Simulations

The numerical simulations used the 3-D CFD code CHENSI (Sini *et al.*, 1996), which is based on the standard k-ε model. The turbulent kinetic energy k and its dissipation rate ε are calculated from the semi-empirical transport equations of Hanjàlic and Launder (1972) and the empirical modelling constants are given the most commonly used values for industrial flows (Launder and Spalding, 1974). CHENSI is based on the widely adopted 'Marker And Cell' (MAC) method (Hirt *et al.*, 1975), that uses a staggered grid and the solution of a Poisson equation for the pressure at every time-step. The method was initially developed for unsteady problems involving free surfaces; to allow the surface location to be determined as a function of time, markers (massless particles) were introduced into the flow. In CHENSI the method derived from the MAC approach has no markers, but the original staggered grid is retained (scalars located at the mesh centre and velocity components at the centre of the mesh faces). The iterative method for solving the pressure-velocity coupling is also kept. A finite volume method is applied on a non-uniform staggered grid and the three velocity components are defined at the centres of the cell faces while all other scalar dependent variables are located at the cell centres. The numerical procedure uses centred differences for the diffusion and source terms and an upwind-weighted scheme for the advection terms. At each time step, the continuity condition on mean velocities is directly satisfied by means of the artificial compressibility method described by Chorin (1967). This implicit iterative method simultaneously relaxes velocity and pressure fields in such a way that the Navier-Stokes solution complies with the continuity equation. The numerical procedure is explicit in time. The time-dependent problems are run from the pre-set initial state until steady state conditions are reached. Initial field values within the computation domain are chosen to hasten convergence. In the present case the initial wind speed and TKE field are derived from the experimental boundary conditions. The inlet logarithmic wind profiles are based on the measured z_o and U_* values, whilst the turbulence profiles of k and ε are built from measured values, with $k = (u'^2 + 2w'^2)$. The boundary conditions are specified in 'fictive' cells that border the computation domain. These facilitate the formulation of the boundary conditions and are artifacts for numerical convenience. The flow field is solved in the fluid (real) domain which 'sees' the boundary conditions as being the enveloping layer of fictive cells representing the external, uncalculated flow. The von Neumann zero normal derivative condition is used at the top and outflow free boundaries, whilst the near wall region is described by the wall function of Launder and Spalding (1974). The initial experimental conditions, together with the W/H = 2.0 data set, have recently been a test case for numerical model intercomparison within several European institutions (Sahm et al., 2002). This showed that CHENSI predicted the mean flow field to an accuracy as good as the other models.

4. Results and Discussion

4.1. MEAN PROFILES

The results presented here concern the mean flow patterns within the different canyons. Figure 6 shows the comparison between the experiments (symbols) and the CHENSI predictions (lines) of the U component mean velocity on five vertical profiles within the W/H = 1 cavity. A similar plot for the W/H = 2 case may be found in the paper of Sahm et al. (2002). The data sets collapse extremely well within the shear layer above the canyon, with the most upstream profile (X/W = 0.09) well predicted immediately below the shear layer. Through the cavity the agreement is good, although CHENSI tends to underpredict the magnitude of the positive and reversed flow velocities, as noted by Sahm et al. (2002). The discrepancy is greatest near the cavity base, probably due to the wall function used in the model. However, in regions where the velocity is low (e.g. for W/H = 0.7, not shown here) the CHENSI predictions agree well with the experiment. The differences between the experimental and predicted data near the wall may be significant when modelling traffic pollution. The bottom of the canyon is where traffic emissions take place and so correct modelling of the flow in this region may be of crucial importance in some canyon flow and dispersion scenarios.

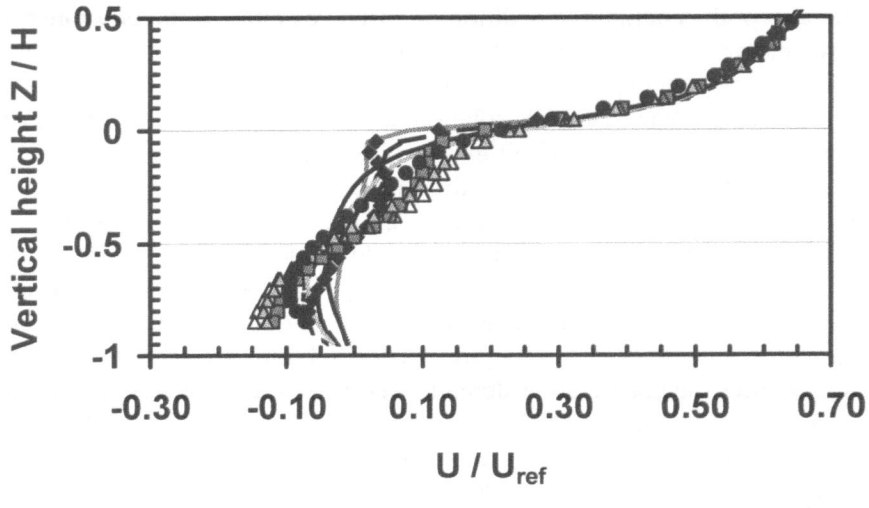

Figure 6. Comparison of U component velocity distributions in canyon, W/H = 1.

4.2. SKIMMING FLOW

Projected mean velocity vectors on the XZ centre-plane (Y = 0) for all the 5 cases, together with their streamtrace patterns, are shown in Figures 7 to 11. Each figure shows the experimental data plot followed by the CHENSI predictions. All the cases studied show the skimming flow regime, with the separating shear layer originating from the upstream cavity corner passing across the cavity opening with little deflection into the canyon. This agrees with Johnson and Hunter's (1999) field observations of a skimming flow regime up to an aspect ratio threshold of W/H = 2.5. Earlier numerical simulations gave a threshold value of W/H = 1.54 (Hunter *et al.*, 1992; Sini *et al.*, 1996).

4.3. VORTEX STRUCTURE

Both predictions and measurements show a dominant vortex within the canyon for W/H = 2.0 and 1.0, but only the W/H = 2.0 case also shows a clear secondary

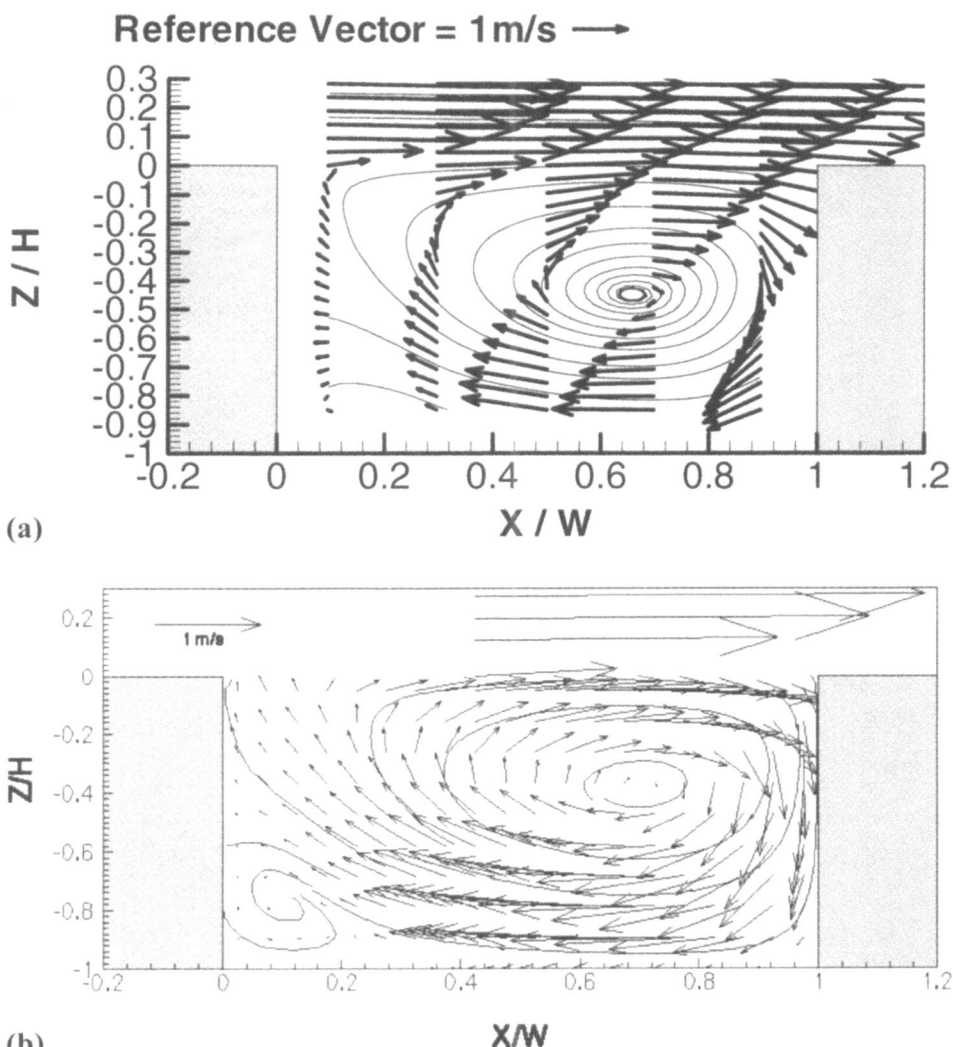

Figure 7. Mean velocity vectors and streamtraces on centre-plane for W/H = 2.0. (a) Experiments, (b) CHENSI.

vortex near the upstream wall. For a reduced aspect ratio, W/H = 0.7, the main vortex remains, but a recirculation region appears near the ground next to the *downstream* wall in both the experiments and predictions. In the W/H = 0.5 case the lower of these two vortices grows in size, although it remains considerably weaker than the upper recirculation. In the experiments the centre of the secondary vortex remains close to the downstream wall whereas the predictions show a more centred vortex. The same features, but less pronounced, are found for W/H = 0.3, where the experiments show almost zero velocity in the lowest third of the cavity, indicating

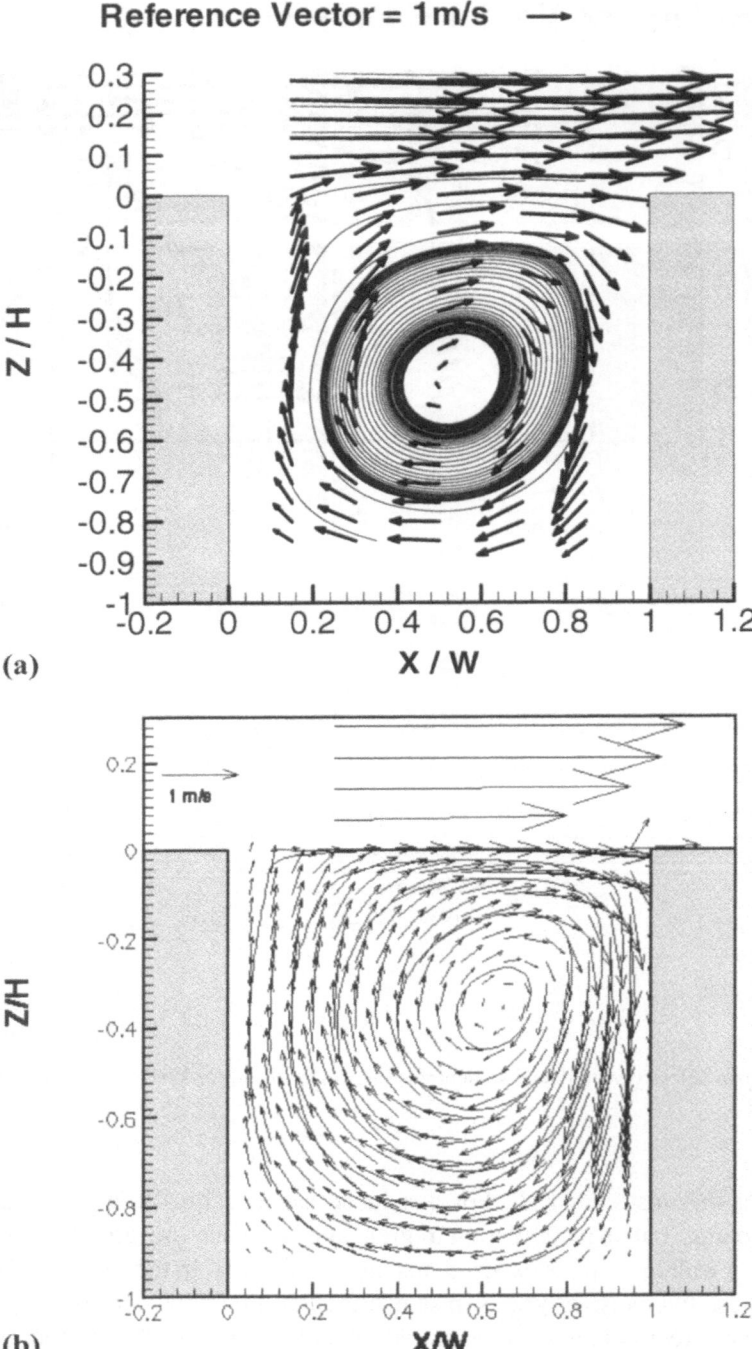

(a)

(b)

Figure 8. Mean velocity vectors and streamtraces on centre-plane for W/H = 1.0. (a) Experiments, (b) CHENSI.

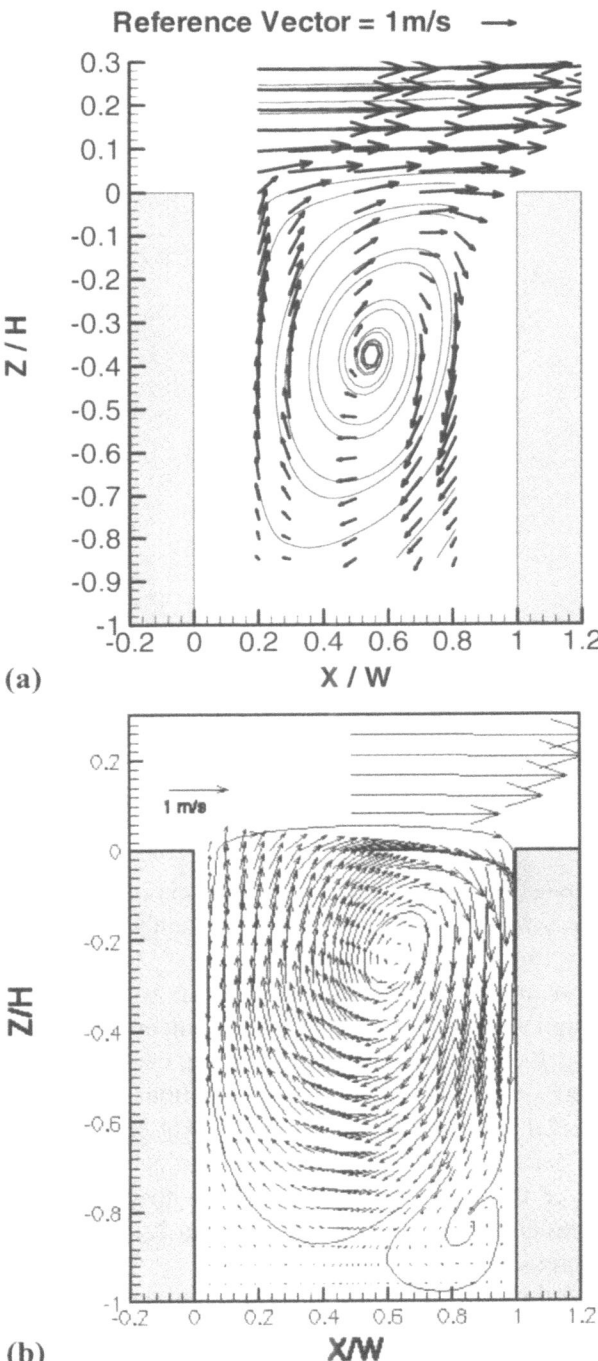

Figure 9. Mean velocity vectors and streamtraces on centre-plane for W/H = 0.7. (a) Experiments, (b) CHENSI.

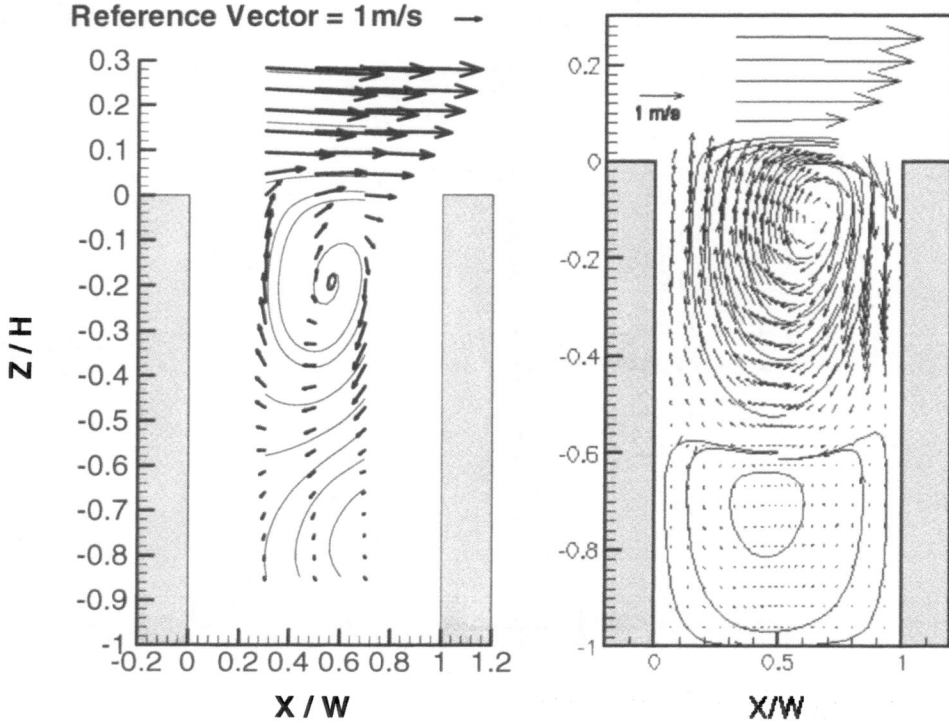

Figure 10. Mean velocity vectors and streamtraces on centre-plane for W/H = 0.5. (a) Experiments, (b) CHENSI.

a flow weakly connected to that above, giving very poor ventilation properties. The numerical predictions show 3 vortices immediately above each other, with the lowest vortex being the weakest of them.

The centre of the upper vortex in the simulations is always closer to the downstream wall and higher within the canyon when compared to the experiments. This is shown in Figure 12 by coordinates of the vortex centre (X_c, Z_c) presented as a plot of experiments versus predictions. The predictions are always 5–15% greater than the experimental values, indicating a tighter vortex in the upper downstream part of the cavity, perhaps due to a weaker diffusion of vorticity in the predictions. Whilst the height of the vortex centre increases with decreasing W/H, its downstream location remains static for W/H$\leq$1, at about X_c/W = 0.55 and 0.65 for the experiments and predictions, respectively.

It is known that the flow in regularly shaped canyons is strongly influenced by initial conditions, with the boundary layer friction velocity being a suitable normalising parameter for some aspects of the flow, such as wall pressure distributions and overall cavity drag force. Other recent work (Kovar-Panskus *et al.*, 2002), in which a different canyon with W/H = 1 was subjected to a boundary layer with U_*/U_{ref}

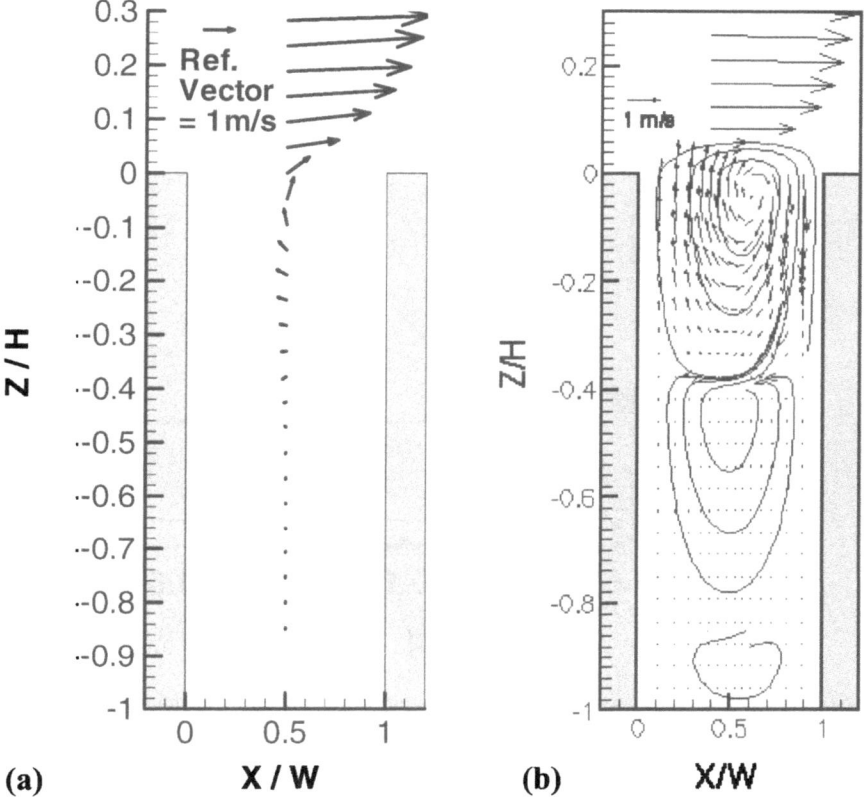

Figure 11. Mean velocity vectors and streamtraces on centre-plane for W/H = 0.3. (a) Experiments, (b) CHENSI.

= 0.070, gave a maximum reversed flow velocity near the ground of 0.19 U_{ref}. This is greater than the 0.15 U_{ref} obtained here for a lower friction velocity of U_*/U_{ref} = 0.050, indicating the role of the skin friction in driving the canyon flow. This second study gave a vortex centre location of X_c/W = 0.64, Z_c/H = –0.50, compared to 0.53 and –0.44, respectively, in the present study, again showing the sensitivity of the flow pattern to initial conditions.

5. Concluding Remarks

A wind tunnel investigation and a numerical study using the CFD code CHENSI have been conducted to investigate the flow regimes within a nominally 2-D cavity with W/H varied from 0.3 to 2.0. For all the cases under consideration the skimming flow regime has been observed, indicating that the transition to wake interference flow takes place at an aspect ratio of W/H > 2.0. The agreement between the wind tunnel measurements and the numerical study, using the same inlet bound-

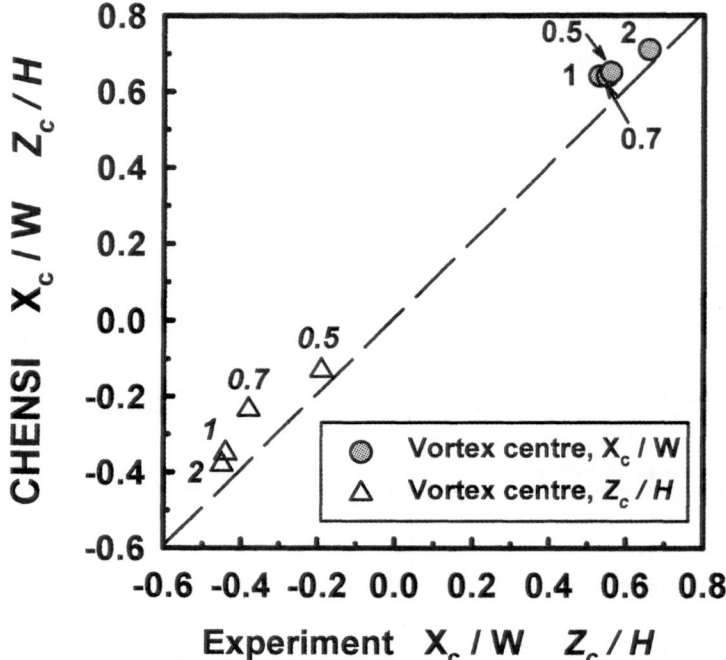

Figure 12. Comparison of the experimental and predicted location of the primary vortex centre (X_c, Z_c). Data points labelled by W/H, Z_c location in *italics*.

ary conditions was reasonably good, although CHENSI tended to underpredict the magnitudes of the velocities within the canyon. The main flow features, in terms of the large vortex structures, were predicted extremely well. However, the centre of the main vortex was 5–15% higher and more downstream in the cavity in the predictions than in the experiments, indicating a tighter and less diffuse vortex. Further work is required to clarify the influence of initial conditions on the flow regimes in canyon configurations and to achieve a suitable parameterisation of those conditions.

Acknowledgements

Thanks are due to the EU (TRAPOS project funding), Mr. T. Renouf and the Institut de Développement et de Recherche pour l'Informatique Scientifique (IDRIS), France.

References

Assimakopoulos, V., ApSimon, H., Sahm, P. and Moussiopoulos, N.: 2000, 'Effects of Street Canyon Geometry on the Dispersion Characteristics in Urban Areas', *Proceedings of the 16th IMACS World Congress*, 21–25 August 2000, Lausanne, Switzerland.

Berkowicz, R., Hertel, O., Larsen, S. E., Sorensen, N. N. and Nielsen, M.: 1997, *Modelling Traffic Pollution in Streets*, NERI, Roskilde, Denmark.

Castro, I. P. and Cheun, B. S.: 1982, 'The measurement of Reynolds stresses with a pulsed-wire anemometer', *J. Fluid Mech.* **118**, 41–58.

Dabberdt, W. F. and Hoydysh, W. G.: 1991, 'Street canyon dispersion: Sensitivity to block shape and entrainment', *Atmosph. Environ.* **25A**, 1143–1153.

DePaul, F. T. and Sheih, C. M.: 1986, 'Measurements of wind velocities in a street canyon', *Atmosph. Environ.* **20**, 555–559.

Hanjàlic, R. and Launder, B. E.: 1972, 'A Reynolds stress model of turbulence and its application to thin shear flows', *J. Fluid Mech.* **52**, 609–638.

Hirt, C. W., Nichols, B. D. and Romero, N. C.: 1975, 'Sola: A Numerical Solution for Transient Fluid Flows', *Report LA5852*, Los Alamos Laboratory, University of California, Los Alamos Scientific Laboratory, New Mexico 87544, U.S.A.

Hunt, A.: 1981, 'Scale Effects on Wind Tunnel Measurements of Surface Pressures on Model Buildings', *Proceedings of Colloquium 'Designing with the Wind*, CSTB, Nantes, France.

Hunter, L. J., Johnson, G. T. and Watson, I. D.: 1992, 'An investigation of three-dimensional characteristics of flow regimes within the urban canyon', *Atmosph. Environ.* **26B**, 425–432.

Jensen, M.: 1958, 'The model law phenomena in natural wind', *Ingenioren* **2**, 4.

Johnson, G. T. and Hunter, L. J.: 1999, 'Some insights into typical urban canyon airflows', *Atmosph. Environ.* **33**, 3991–3999.

Kastner-Klein, P. and Plate, E.J.: 1999, 'Wind tunnel study of concentration fields in street canyon', *Atmosph. Environ.* **33**, 3973–3979.

Kovar-Panskus, A., Moulinneuf, L., Savory, E., Abdelqari, A., Sini, J.-F., Rosant, J.-M., Robins, A. and Toy, N.: 2002, 'A wind tunnel investigation of the influence of solar-induced wall-heating on the flow regime within a simulated urban street canyon', *Water, Air, and Soil Pollut.*, this issue.

Launder, B. E. and Spalding, D. B.: 1974, 'The numerical computation of turbulent flows', *Comp. Meth. Appl. Mech. Eng.* **3**, 269–289.

Liedtke, J., Leitl, B. and Schatzmann, M.: 1999, 'Dispersion in a street canyon: comparison of wind tunnel experiments with field measurements', in P. M. Borrel and P. Borrel (eds), *Proc. of Eurotrac Symposium '98*, WIT Press, Southampton, pp. 806–810.

Leitl, B., Chauvet, C. and Schatzmann, M.: 2001, 'Effects of Geometrical Simplifications and Idealization on the Accuracy of Microscale Dispersion Modelling', *Extended abstract for the 3rd International Conference on Urban Air Quality*, 19–23 March 2001, Loutraki, Greece.

M.I.H.U.: 2001, Internet database, available at University of Hamburg, funded by the Umweltbundesamt: http://www.mi.uni-hamburg.de/cedval/categoryB.htm#B1-5.

Nakamura, Y. and Oke, T. R.: 1988, 'Wind, temperature and stability conditions in an E–W-oriented canyon', *Atmosph. Environ.* **22**, 2691–2700.

Oke, T. R.: 1987, *Boundary Layer Climates*, 2nd ed., Routledge, New York.

Pavageau, M., Rafailidis, S. and Schatzmann, M.: 1996, 'A Comprehensive Experimental Databank for the Verification of Urban Car Emission Dispersion Models', *Proceedings of the 4th Workshop on Harmonisation within Atmospheric Dispersion Modelling for Regulatory Purposes*, 6–9 May 1996, Ostende.

Rafailidis, S.: 1997, 'Influence of building area density and roof shape on the wind characteristics above a town', *Bound. Layer Meteor.* **85**, 255–271.

Sahm, P., Louka, P., Ketzel, M., Guilloteau E. and Sini, J. F.: 2002, 'Intercomparison of numerical urban dispersion models - Part I: Street canyon and single building configurations', *Water, Air and Soil Pollut.*, this issue.

Scaperdas, A.: 2000, 'Modelling Air Flow and Pollutant Dispersion at Urban Canyon Intersections', Ph.D. Thesis, University of London.

Sini, J.-F., Anquetin, S. and Mestayer, P.: 1996, 'Pollutant dispersion and thermal effects in urban street canyons', *Atmosph. Environ.* **30**, 2659–2677.

Theurer, W.: 1994, 'Ausbreitung bodennaher Emissionen in komplexen Bebauungen', *Mitteilungen, Heft 45*, Institut für Hydrologie und Wasserwirtschaft, Universität Karlsruhe, Germany.

SPATIAL VARIABILITY AND SOURCE-RECEPTOR RELATIONS AT A STREET INTERSECTION

ALAN ROBINS[1]*, ERIC SAVORY[1], ATHENA SCAPERDAS[2] and
DIMOKRATIS GRIGORIADIS[3]

[1] *School of Engineering, University of Surrey, Guildford, Surrey, U.K.;* [2] *WS Atkins Engineering Science Ltd, Woodecote Grove, Epsom, Surrey, U.K.;* [3] *NCSR Demokritos, 153 10 Aghia Paraskevi, Attiki, Greece*

(* author for correspondence, e-mail: a.robins@surrey.ac.uk, fax: +44 (0) 1483 259546)

Abstract. A wind tunnel study of dispersion at a simple urban intersection comprising two perpendicular streets is described. Concentration and flow field measurement were undertaken to determine the importance of the exchange of pollutants between the streets and to investigate source-receptor relationships at the intersection. The results showed that only in a symmetrical situation were exchanges negligible and that small departures from symmetry, brought about in the experiments through an off-set in the street alignment or a change of orientation relative to the wind, were sufficient to establish significant exchanges. The results also showed that significant structure appeared in the concentration fields in the streets as a result. Examples are shown where concentrations on one side of a street are entirely due to emissions from the perpendicular street, whereas on the opposite side concentrations depend on emission upwind in the same street as the receptor. The results imply that exchanges between street systems are likely to be the norm in practice and that the consequences of such exchanges are not confined to the immediate vicinity of the intersection.

Keywords: air quality, dispersion modelling, intersections, street-canyons

1. Introduction

The study of line sources in 'urban street canyons' has dominated short range dispersion research in urban areas. However, a number of studies (e.g. Hoydysh and Dabberdt, 1994; Scaperdas, 2000; Soulhac, 2000) have shown the importance of three dimensional effects at intersections, in particular the exchange of air, and hence pollutants, between the street systems involved. Air exchanges can be quantified through flow field measurements (or predictions) whereas pollutant transfers are most easily understood by using point sources at various locations within the street systems near the intersection.

Figure 1 is based on one of the cases investigated by wind tunnel simulation by Scaperdas (2000). It shows the measured flow interchanges at a simple intersection between two perpendicular streets. The approaching flow is along the 'x-street' and there is a lateral off-set at the intersection of $\Delta y = 0.6$ H, where H is the height (125 mm) of the four blocks used to define the intersection; the street width is also

Water, Air, and Soil Pollution: Focus **2:** 381–393, 2002.
© 2002 *Kluwer Academic Publishers.*

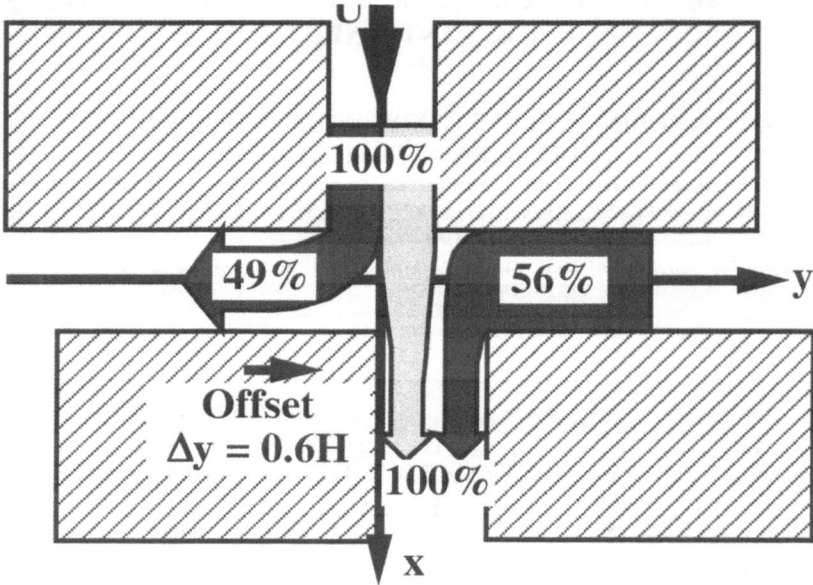

Figure 1. Flow interchanges at an intersection defined by four blocks of height H, with a 0.6 H off-set. Block size 6 H×4 H.

H in this case. The arrows and labels mark the volume flux exchanges relative to the volume flux in the upwind x-street (denoted as 100%). Clearly, a considerable flow of air passes into the y-street (in the section y < 0) from the upwind part of the x-street (x < 0). Such exchanges and the mechanisms associated with them lead to:

1. significantly larger pollutant source terms in the y-street than would otherwise be expected;
2. marked spatial structure in concentration distributions near the intersection; and
3. the development of high levels of concentration fluctuations.

particularly when the traffic flows in the x and y-streets are different. The objectives of the work discussed in this article were to explore these matters in further detail and to provide comprehensive sets of data for testing the predictive ability of numerical methods. Here, we concentrate on the former.

2. Methodology

The experiments were undertaken in the EnFlo Meteorological Wind Tunnel at the University of Surrey, which has a 20×3.5×1.5 m high working section and can be

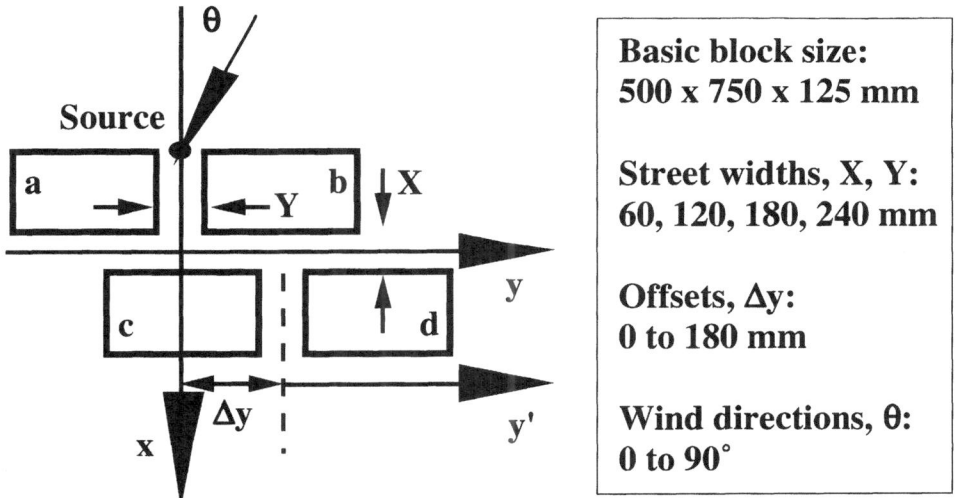

Figure 2. Schematic diagram of model arrangements including definitions of axes, orientation and off-set.

used to simulate a wide range of stable, neutral and unstable boundary layers. In this work only neutral flow was considered. The approach flow was a 1 m deep neutral boundary layer with friction velocity $u_*/U_{ref} = 0.055$, where U_{ref} is the wind speed at the boundary layer edge, and surface roughness length, $z_o = 1$ mm. The building blocks were 125 mm high and 750×500 mm in plan, arranged as shown in Figure 2, with off-sets from 0 to 1.5 H. Wind directions, as defined in the figure, from 0° to 90° were investigated. The experiments were carried out at reference speeds in the range from 2 to 2.5 ms^{-1}, and the characteristic Reynolds number, HU_{ref}/v, was of order 20 000.

The four blocks stood alone in the tunnel – they were not part of a 'larger' urban array. This configuration was chosen as a test case for CFD work because it provides particularly simple boundary conditions and is not excessively demanding on grid design. The block size ensures that the flow over the roof reattaches well upstream of the separation at the trailing edge, so that the flow above the street canyon is essentially horizontal. The ratio H/z_o was about 125, whereas figures in the range from 10 to 30 are more typical of urban areas. This is not a significant mis-match as concern rests in generic features of the flow within the street canyon, not in the structure of the boundary layer above the buildings. For similar reasons, no great significance attaches to the relatively shallow boundary layer used in the wind tunnel.

The distribution of mean concentration within the street network was measured with flame ionisation detector based instrumentation, trace gas being released from a (nominally) point source at ground level. Line source results were calculated from the point source data. A multi-channel sampling and analysis system was used for

TABLE I

Measured flow exchanges at the intersection, expressed as a percentage of the flow
from the upwind street, Q_1

Off-set	Orientation	Relative volume flux towards intersection (%)			
$\Delta y/H$	$\theta°$	Q_1 (x < 0)	Q_2 (x > 0)	Q_3 (y < 0)	Q_4 (y > 0)
0	0	100	−112	6	6
0.1	0	100	−110	−9	20
0.6	0	100	−100	−49	56
0	10	100	−95	−9	32

mean concentration measurement and two Cambustion Fast FIDs for fluctuation
measurement. Sources were positioned on the centre line of the x and y-streets. A
two component laser-Doppler anemometer was used to map the velocity field in
a limited number of cases. Quantitative measurements were preceded by comprehensive flow visualisation studies that were used to gain insight into the dispersion
processes and to guide the selection of cases for subsequent detailed study.

3. Results

All the results discussed in this section were measured at a height of H/6 above
street level; i.e. they describe street-level conditions.

3.1. EXCHANGES AT THE INTERSECTION

The exchanges between the streets at the intersection were calculated from measurements of the distribution of mean velocity over planes near the intersection that
were perpendicular to the streets. Table I lists the relative volume flux exchanges
for four arrangements, specified in terms of the off-set, Δy, and orientation, θ.
Fluxes are positive into the intersection and expressed as percentages of Q_1, the
flux from the upwind street, x < 0; Q_2 is the flux from the downwind street, x > 0,
Q_3 from y < 0 and Q_4 from y > 0. Any imbalance is due to experimental error and
vertical exchange.

In the first case, $\Delta y = \theta = 0$, there is no asymmetry to create secondary flow and
the weak flux from the y-street into the x-street is driven by mixing at the intersection. As will be seen later, this mean flow masks the fact that mixing transfers
some pollutant from the x-street into the y-streets. An off-set as small as 0.1H is
sufficient to create well defined secondary flows in the lateral streets, with significant transfer from the upwind x-street into the left lateral street (y < 0), and from
the right lateral (y > 0) into the downwind x-street. Increasing the offset to 0.6 H

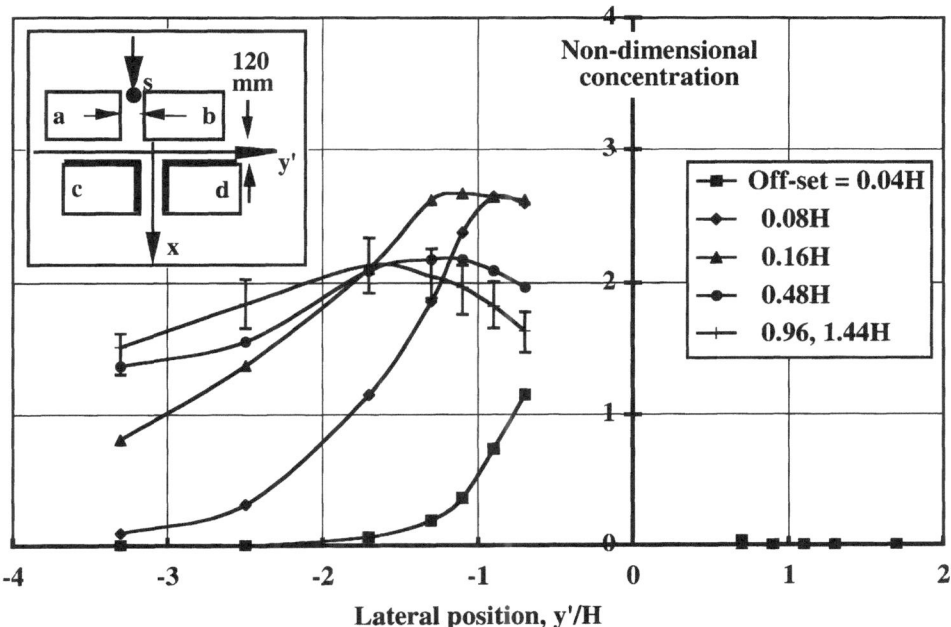

Figure 3. Non-dimensional concentration profiles along lateral streets at x = –0.5 H for $\theta = 0°$ and off-sets $\Delta y = 0.04$ H to 1.44 H; the source was located at the entrance to the upwind street and y′ is measured from the centre of the downwind street.

produces the results shown in Figure 1, with approximately 50% of the flow from the upwind street passing into the left lateral street and a somewhat larger flow from the right lateral into the downwind street. In circumstances like this the pollutant flux from the upwind street would be relatively large, because the mean flow is along the street, and would provide a substantial source term in the lateral street. The lateral street would generally be modelled as a street-canyon with wind normal to it, and clearly the flux from the intersection will completely change the character of the concentration field within the street. How far such changes penetrate into the lateral street is discussed later. Larger exchanges between the street systems are also observed in the case $\Delta y = 0$, $\theta = 10°$, though in this case there is a substantial imbalance in the horizontal fluxes. Whether this is due to experimental error or indicates a large vertical flux into the flow above the buildings is not known. Both this and the $\Delta y/H = 0.1$, $\theta = 0°$ case show that behaviour is very sensitive to small perturbations (as always exist in reality), a point returned to later.

3.2. CONCENTRATION DISTRIBUTIONS

Figure 3 shows measurements of mean concentration, C (0.5 H, y′, 0.16 H), along the downwind side (x = 0.5 H) of the lateral streets for a point source in the upwind street at x/H = –4; results for an off-set of 1.44 H were virtually identical to those

TABLE II

The four receptors chosen for the discussion of source-receptor relationships

Receptor location			Description
x/H	y/H	z/H	
1.38	−0.32	0.16	A pair on either side of the x-street, 0.9 H from the street entrance.
1.38	0.64	0.16	
0.48	−1.12	0.16	A pair on the downwind side of the y-street, 0.64 H from the entrance
0.48	1.04	0.16	to each segment.

shown for 0.96 H. Concentrations are plotted in non-dimensional form, CUH^2/Q (where U is the wind speed in the undisturbed flow at height H and Q is the source strength), as a function of position, y'/H, measured relative to the centre of the downwind street. Error bars are included for the 0.96 H off-set data and these are typical of all results. The rapid increase of penetration into the left lateral street ($y' < 0$) with increasing off-set is clear, whilst concentrations in the right lateral street ($y' > 0$) are zero. Maximum concentration levels are of the same magnitude as those near the intersection in the x-street. These results are entirely consistent with the flow patterns illustrated in Figure 1. Significant concentrations occurred throughout the lateral street (in these experiments of length 6 H), suggesting that nowhere within it were conditions attained for valid application of a standard street-canyon model with wind normal to the street.

Results for zero off-set have not been included in Figure 3. Measurement repeatability in this symmetrical arrangement proved to be much worse than usual and very sensitive to minor perturbations to the model set-up in the wind tunnel. Breaking the symmetry, even by as little as an off-set of 0.04 H, was sufficient to eliminate the high run-to-run variability and restore the expected level of repeatability. However, significant concentration levels were not measured beyond $|y'/H| = 1$, though this was the only case investigated where concentrations were observed in both segments of the lateral street.

3.3. SOURCE-RECEPTOR RELATIONSHIPS

Mean concentrations were measured at a fixed set of receptor locations, each approximately 1 to 1.5 H from the centre of the intersection, for a range of ground level point source positions along the x and y axes for the model arrangement $\Delta y/H = 0.16$, $\theta = 0°$. Four receptor locations, chosen to illustrate the characteristic features of dispersion behaviour at the intersection, will be used to the discuss source-receptor relationships. The selected receptors are described in Table II and the results are present in Figures 4 and 5.

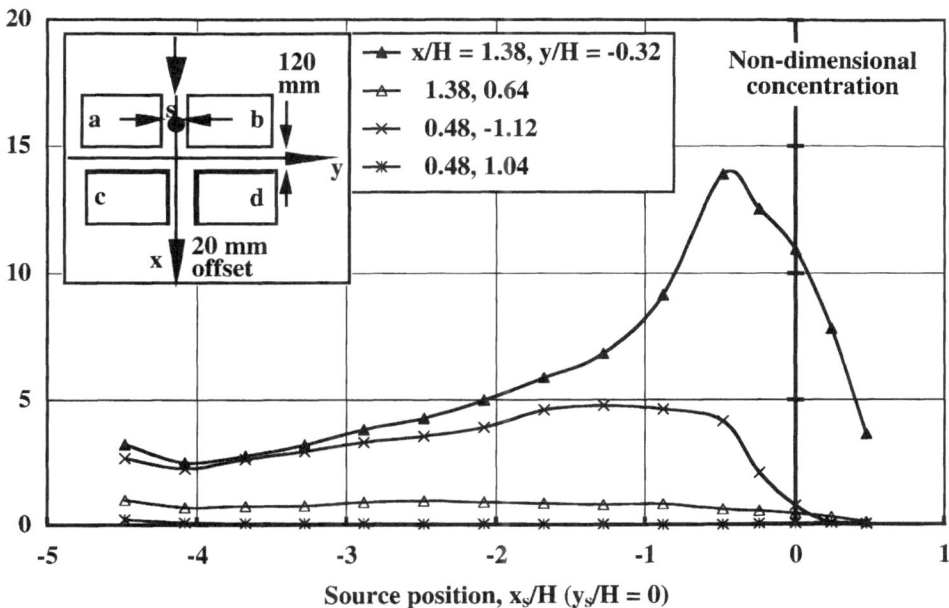

Figure 4. Receptor concentrations at four selected locations as a function of the source position in the x-street, $\Delta y/H = 0.16$, $\theta = 0°$.

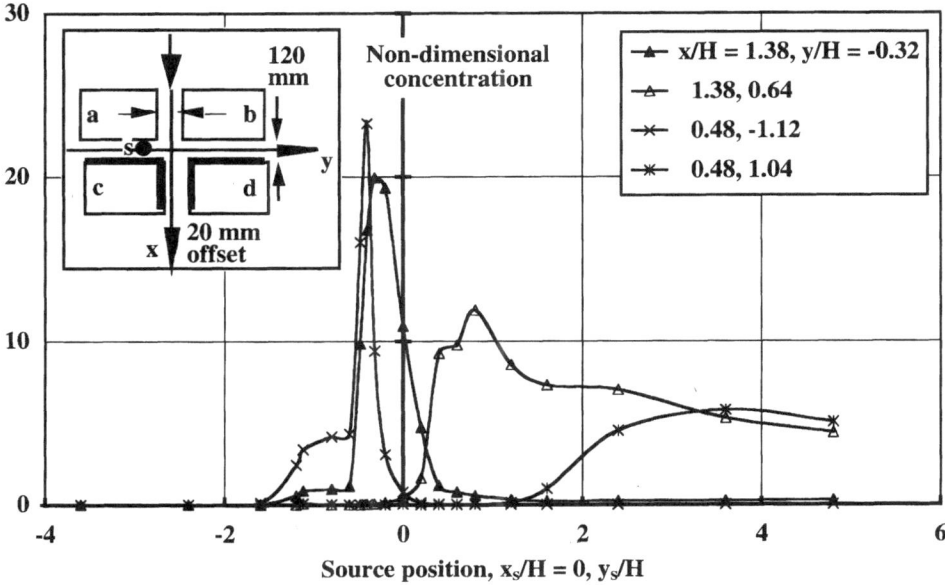

Figure 5. Receptor concentrations at four selected locations as a function of the source position in the y-street, $\Delta y/H = 0.16$, $\theta = 0°$.

Figure 4 treats sources in the x-street. Concentrations measured on either side of the downwind sector of the x-street (x/H = 1.38, y/H = –0.32 and 0.64) reveal considerable differences. The highest concentration shown occurs at the first of these locations when the source is at x_s/H≈–0.5. Thus in travelling approximately 2 H the plume has dispersed sufficiently to generate the maximum street-side concentration; plume spread and lateral displacement are both likely to be important in this process. The second of this pair of locations is affected by the clean air that comes from the right hand lateral street, as illustrated in Figure 1. The resulting cross-street structure persists far into the x-street and is little reduced at x = 4 H, the final receptor position studied. Concentrations in the left side of the lateral street are advected from the upwind x-street and decay quite slowly with fetch along the street. Concentrations fall as the source moves downstream from about the centre of the intersection because material then no longer reaches the receptor point by advection and diffusion across the street. Concentrations are zero in the other half of the lateral street since it is exposed solely to clean air flowing along it in the negative y direction.

Figure 5 treats the same receptors, but for sources located in the y-street. The pattern is now quite different, with sharp peaks arising at two of the locations when the source is near the intersection. These occur at the receptors at x/H = 1.38, y/H = –0.32 and 0.48, –1.12 when the source position is in the range from y_s/H≈0 to –0.8, indicating direct advection of the plume from source to receptor. The first of these receptors is only affected for a narrow range of source positions as might be excepted from the flow fields sketched in Figure 1. There is some evidence of dispersion counter to the overall mean flow direction in that concentrations are observed at this receptor even when the source is at y_s/H = –1. In contrast, the opposite receptor (x/H = 1.38, y/H = 0.64) is exposed to pollutant from all source positions in the right hand segment of the lateral street and no source positions in the left hand segment, again reflecting the flows sketched in Figure 1.

The sharpest peak in the concentration distributions occurs for the receptor in the right hand lateral street (x/H = 0.48, y/H = –1.12). No material from the other segment of the street reaches this position and there is no evidence of dispersion counter to the overall mean flow since concentrations fall to zero when y_s < y. This, together with the observations discussed above, suggests that y_s/H≈–1.2 is approximately the limiting source position from which material can disperse into the downwind x-street. Concentrations are only observed at the receptor in the other section of the lateral street, where y > 0, when the source is upstream of the receptor in terms of the secondary flow within the street.

3.4. LINE SOURCES

Line source results can be synthesised from the point source data to quantify the relative contributions from each street to concentrations at a given receptor. We assume that the concentration, C_{Ln}, at a given receptor due to emissions from the

TABLE III

Receptor positions for analysis of line source results

Receptor	x/H	y/H	Receptor	x/H	y/H
Lateral street at z/H = 1/6					
1	0.48	−2.12	4	0.48	−0.53
2	0.48	−1.12	5	0.48	0.84
3	0.48	−0.72	6	0.48	1.04
Longitudinal street at z/H = 1/6					
7	3.68	−0.32	12	0.68	0.64
8	2.48	−0.32	13	0.88	0.64
9	1.28	−0.32	14	1.28	0.64
10	0.88	−0.32	15	2.48	0.64
11	0.68	−0.32	16	3.68	0.64

'nth' element of the line source, length λ_n, can be written as follows in terms of the concentration, C_n, from a point source at the centre of the element:

$$C_{Ln}UH/q = (CUH^2/Q)_n(\lambda_n/H) \, ,$$

where q is the line source strength per unit length and Q the point source strength. Summing over all such elements gives the concentration due to the line source, C_L:

$$C_L UH/q = \Sigma_n (CUH^2/Q)_n(\lambda_n/H) \, .$$

In this way results can be derived for a line source along the longitudinal street from x/H = −4.48 to 0.48 and a second line source along the lateral street from y/H = −4.8 to 4.8. The section of each line source that provides the main contribution to the concentration levels at a given receptor can also be determined.

Figure 6 shows non-dimensional concentrations, $C_L UH/q$, at the 16 receptor positions specified in Table III, assuming equal source strengths for the two line sources. The contribution from each line source is shown separately. Receptors 1 to 4 are in the left lateral street (y < 0) and here both line sources contribute, with the x line source responsible for about 2/3 of the total near the intersection. By contrast, the two receptors, 5 and 6, in the other segment of the lateral street (y > 0) are only influenced by sources in the same segment because the mean flow is along the street towards the intersection. There is also a strong contrast between the two sides of the longitudinal street (x > 0). The left hand side (y < 0) receptors (7 to 11) are largely affected by the line source in the same street, with only about 1/4 of the total derived from the y line source, whereas concentrations on the right

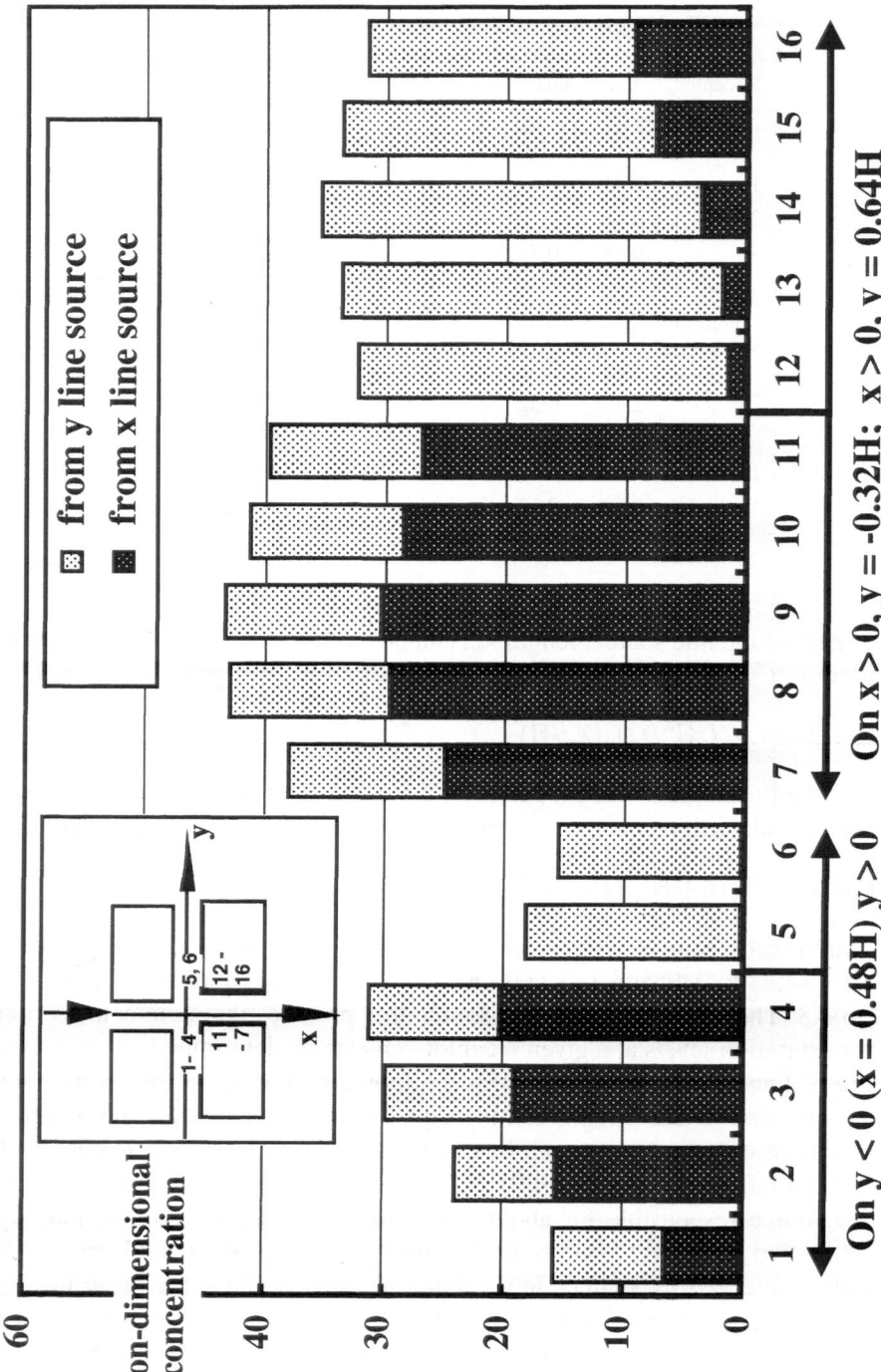

Figure 6. Contributions to pollutant levels at the 16 receptors for equal strength×and y line source emissions; Δy/H = 0.16, θ = 0°.

TABLE IV

The 'active' lengths of the line sources affecting each receptor

Receptor location			Longitudinal line source		Lateral line source	
x/H	y/H	z/H	x_{s10}/H	x_{s90}/H	y_{s10}/H	y_{s90}/H
1.28	−0.32	0.17	−3.2	−0.1	−0.5	0.4
1.28	0.64	0.17	−3.8	−0.6	0.7	3.5
0.48	−1.12	0.17	−3.6	−0.8	−1.0	−0.3
0.48	1.04	0.17	−[a]	−[a]	2.5	4.5

[a] No measurable concentrations due to longitudinal line source.

(receptors 12 to 16) are dominated by the y line source. The situation scarcely changes with increasing distance along the street, the only significant development being a gradual increase in the contribution from the x line source on the right hand side of the street.

A convenient measure of the 'active' length of a line source is the extent responsible for, say, 80% of its contribution to the mean concentration at a receptor. This can be specified by end points x_{s10}, and x_{s90} for the x line source (defined such that 10% contributions derive from $x_s < x_{s10}$ and $x_s > x_{s90}$) and y_{s10} and y_{s90} for the y line source. Table IV presents such active length data for the four receptors previously discussed (numbers 2, 6, 9 and 14 in Table III). The results contrast two cases with a relatively short active length (of order H), with two where the active length is of order 3 H long. The former correspond to the narrow peaks seen in Figure 5, which are associated with the direct advection of the plume from its source to a receptor. These are the only situations shown where concentrations would not increase were the line sources made more extensive (apart from the trivial case in Figure 4 where concentrations are zero for all source locations).

4. Discussion

The results clearly reveal the importance of the transfer of pollutants from one street to the other at an intersection and show that this becomes significant as soon as there are minor departures from symmetry in the geometry. The magnitude of the fluxes involved is very sensitive to the local geometry, in these experiments the magnitude of the off-set, and the wind direction. One implication is that such transfers are likely to be the norm at urban street intersections. The experiments also show that the consequences of such transfers are not confined to the immediate vicinity of the intersection and may penetrate far into the streets involved. The sensitivity to small changes in geometry is also associated with a high level of long time scale concentration fluctuations near the intersection, as was first observed during initial flow visualisation studies and later revealed in detail by fluctuation measurements (to be published at a later date). In practice, such fluctuations may be greatly enhanced by the unsteadiness in the source terms that results from the characteristics of the traffic and its movement. However, this does not affect the mean concentrations discussed in this article since all that is required for their evaluation is appropriate mean source terms.

Integral models can be readily written for flow and dispersion along the streets comprising the intersection. These can be designed so that they tend to a standard street-canyon model (e.g. Berkowicz et al., 1997) far from the intersection, though something simpler serves as an illustration. Define C and U_S as the average concentration and mean flow speed within a street canyon in a plane normal to the street, where U_S is driven by processes at the intersection and the external wind direction is perpendicular to the street. The conservation equation for the concentration is:

$$d/ds(CHWU_S) = q - CU_E W ,$$

where s is measured along the street from the intersection, q is the local line source strength, W the street width and U_E an entrainment velocity modelling mixing over the top of the canyon. The boundary condition is that $C = C_o$ at $s = 0$ and the solution, assuming U_S is constant, is:

$$C = C_o \exp\left(-U_E s/U_S H\right) + \{q/U_E W\} \{1 - \exp\left(-U_E s/U_S H\right)\} ,$$

and, with $L = HU_S/U_E$ defined as a decay length scale, this can be written as:

$$C = C_o \exp(-s/L) + \{q/U_E W\}\{1 - \exp(-s/L)\} .$$

In many of the situations studied the air flow into the lateral street was a significant fraction of that along the upwind longitudinal street; for the case shown in Figure 1 a reasonable estimate of U_S/U is 0.5. The entrainment velocity ratio, U_E/U, is likely to be of order 0.1 so that in this case the decay length scale for concentrations in the lateral street is of order $L \approx 5$ H. The standard local equilibrium result for a street canyon:

$$C = q/U_E W$$

is attained once $s/L \gg 1$, emphasising again that the consequences of the air exchanges between the street at the intersection persist far into the lateral street.

Integral modelling of this sort can be considerably refined but is of limited value until linked to algorithms for the boundary conditions, C_o and Q_o (WHU$_s$). The first can probably be estimated to an acceptable accuracy from integral modelling applied to the upwind street but Q_o, being very sensitive to the geometry and wind direction, can only be defined empirically. Nevertheless, this does offer the possibility of using wind tunnel or CFD modelling to determine the behaviour of Q_o and then some form of modified street-canyon model, along the lines discussed above, to undertake a full assessment of concentrations in the vicinity of the intersection.

5. Conclusions

The results show that significant exchanges occur at intersections between two streets and that, as a consequence, pollution emitted in one street can penetrate far into the other. A further consequence is the development of persistent structure in the concentration field within a street, with receptors on opposite sides being primarily affected by entirely separate sets of sources. The exchanges are driven by minor departures from symmetry in the geometry or orientation of the intersection and are likely to be the norm in practice. As the effects of the exchanges are not confined to the immediate vicinity of the intersection, they need to be addressed in the analysis of urban air quality, whether by calculation or monitoring. The successful inclusion of such effects in dispersion models will rest heavily on the provision of a realistic empirical description of the air exchanges at the intersection and their dependence on wind direction. Such information is likely to be site-specific. Detailed studies of dispersion processes at both generic and realistic intersections are required to develop further understanding of the processes involved and to pave the way for improved dispersion models. In all probability, such research will involve both experimentation and computations (CFD).

Further analysis of the results currently available will concentrate on the effect of wind direction on the concentration field and air exchanges so that a description like that presented above can be developed for all wind directions. Annual average concentration fields can then be investigated and their sensitivity to the line source strengths in each street and the form of the wind rose established.

Acknowledgements

This work was in part funded by the UK EPSRC and the European Community through the Large Scale Facilities section of the Training and Mobility of Researcher (TMR) program, contract ERBFMGECT980117.

References

Berkowicz, R., Hertel, O., Larsen, S. E., Sørensen, N. N. and Nielsen, M.: 1997, *Modelling Traffic Pollution in Streets*, National Environmental Research Institute, Roskilde, Denmark, 52 pp., ISBN 87-7772-307-4.

Hoydysh, W. G. and Dabberdt, W. F.: 1994, 'Concentration fields at urban intersections: Fluid modelling studies', *Atmos. Environ.* **28**(11) 1849–1860.

Scaperdas, A.-S.: 2000, 'Modelling Air Flow and Pollutant Dispersion at Urban Canyon Intersections', Ph.D. Thesis, Imperial College, University of London.

Soulhac, L.: 2000, 'Modèlisation de la Dispersion Atmosphérique a l'Interieur de la Canopée Urbaine', Ph.D. Thesis, Ecole Centrale de Lyon.

APPLICATION OF DISPERSION MODELLING FOR ANALYSIS OF PARTICLE POLLUTION SOURCES IN A STREET CANYON

O. LE BIHAN*, P. WåHLIN, M. KETZEL, F. PALMGREN and R. BERKOWICZ

National Environmental Research Institute, Department of Atmospheric Environment, Roskilde, Denmark

(* author for correspondence, e-mail: Olivier.Le-Bihan@ineris.fr, fax: +33 344556302)

Abstract. The dominating source of particles in urban air is road traffic. In terms of number concentration, its main contribution is within the range of ultrafine particles (Dp < 100 nm). The dispersion conditions, i.e. transport and dilution, of the submicrometer particles are expected to be like for gases and therefore the particle concentrations in a street canyon can be calculated using gaseous pollutants dispersion models. Such processes, like coagulation or condensation, are less important due to the short residence time within the street canyon environment. Two extensive measuring campaigns were conducted in the street Jagtvej in Copenhagen, Denmark. The particle size distributions were measured by a Differential Mobility Analyser (DMA) coupled to a particle counter, providing high time resolution data (1/2 hourly) on a continuous basis. Measurements of NO_x, CO and meteorological parameters were also available. The measured particle number concentrations, especially below 100 nm, reveal very similar dependence on the meteorological conditions as the NO_x concentrations. This underpins the conclusion that dilution properties are similar for particles and NO_x. For particle sizes over 100 nm, somewhat different behaviour is observed. This points toward existence of additional particle sources, not related to traffic emissions within the street canyon. A significant contribution is believed here to be attributed to long-range transport. The total particle emission from traffic, including daily variation and size distribution, has been calculated using the OSPM dispersion model. Results are in accordance with a previous analysis based on statistical modelling.

Keywords: dispersion modelling, OSPM, particle, particulate matter, street canyon, traffic pollution

1. Introduction

Particles in urban air are suspected to present some important adverse health effects (Künzli, 2000) with a special concern on ultrafine particles i.e. particles of a diameter less than ca. 100 nm (Kittelson and Watts, 2000).

The dominating source is road traffic. In terms of number concentration, its main contribution is within the range of ultrafine particles; exhaust pipe, both for diesel and petrol vehicles, is believed to be the main source of this mode (Wåhlin, 2001a, b).

In a street canyon, dilution is the main process affecting particle concentration after emission (Vignati, 1999). Such processes, like coagulation or condensation, are less important due to the short residence time within the street canyon envir-

Water, Air, and Soil Pollution: Focus **2**: 395–404, 2002.
© 2002 *Kluwer Academic Publishers.*

onment. The dispersion conditions, i.e. transport and dilution, of submicrometer particles are expected to be like for gases; therefore the absolute source contribution to particle pollution in a street canyon can be calculated using dispersion models developed for gaseous pollutants.

Two field campaigns have been carried out in Copenhagen in winter 1999 and winter 2000. They were analysed using statistical modelling (Wåhlin, 2001a, b). The dependence on meteorology was neutralised by using a ratio particle/gas, with NO_x and CO as tracers.

Firstly this paper presents a comparison of the dispersion properties of particles with those of NO_x. Secondly, it considers the analysis of these measurements by application of the Operational Street Pollution Model (OSPM) (Berkowicz, 1996).

2. Methodology

Two extensive measuring campaigns (January–March 1999 and January–March 2000) were conducted in the street Jagtvej in Copenhagen, Denmark. During these campaigns the particle size distributions were measured by a Differential Mobility Analyser (DMA) coupled with a particle counter (Wåhlin, 2001a, b). This system provided high time resolution data (1/2 hourly) on a continuous basis. Simultaneous measurements of NO_x and CO were available. The meteorological data were provided from a nearby meteorological mast.

3. Results and Discussion

3.1. DILUTION PROPERTIES

Pollution measurements performed at the street sites exhibit a pronounced dependence on meteorological conditions. The observed dependence of concentrations on wind is an indirect manifestation of formation of a wind vortex in the street canyon.

The direction of the Jagtvej Street is about 30° and the station is on the East Side; in consequence the clean part (windward side) of the vortex corresponds to direction 300° and the polluted part (leeward side) to direction 120°.

The NO_x concentration versus wind direction during both campaigns exhibits the typical patterns at the Jagtvej station: the effect of the clean sector is particularly evident (300°); the other part of the wind direction range, including the leeward side results in a broad maximum. Results for the winter 1999 campaign are plotted in Figure 1.

The correlation coefficient between NO_x and particles is very high for particle diameter below 100 nm. It decreases gradually with increasing particle diameter from 100 to 700 nm. As a consequence, we will consider successively particles below and over 100 nm.

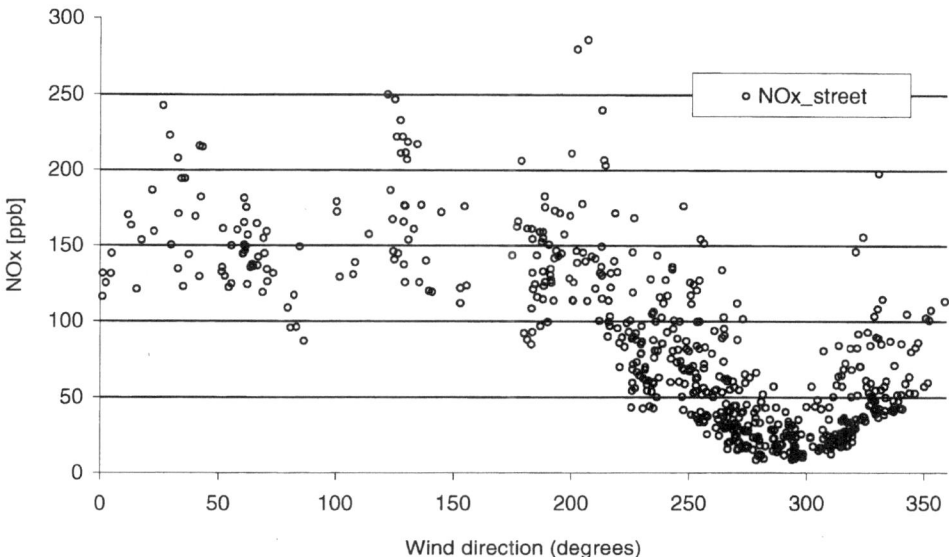

Figure 1. NO_x concentration versus wind direction – winter 1999 working days and hours – wind speed >2 m s^{-1}.

3.1.1. *Particle Diameter Below 100 nm*

The channel of the DMA centred around 22.5 nm is representative of the sub-100 nm mode. As shown in Figure 2, its coefficient of determination with the NO_x concentration is high ($R^2 = 0.8$).

Figure 3 presents a simultaneous comparison of NO_x and particle (channel 22.5 nm) dependence on wind direction. The mean slope from Figure 2 has been used to set-up the scales of the graph.

It appears that NO_x and channel 22.5 nm show a very similar shape. It suggests that sub-100 nm particles and NO_x have comparable dilution properties and originate from the same source.

3.1.2. *Particle diameter over 100 nm*

As indicated previously, the correlation between NO_x and over-100 nm particles is decreasing: for instance in winter 2000 the coefficient of determination R^2 drops from 0.8 at 100 nm to 0.3 at 700 nm. NO_x versus channel 550 nm (Figure 4) illustrates this situation. The scatter is here much larger than for the 22.5 nm particles (Figure 2). This behaviour can be explained by assuming that the particle concentration in this size range is a sum of several sources.

Firstly, these include traffic-related sources, namely exhaust pipe, and also mechanical sources such as tyres, brakes. Because they are traffic-related, they correlate with NO_x. However the relation with the mechanical sources is indirect because they do not produce any NO_x.

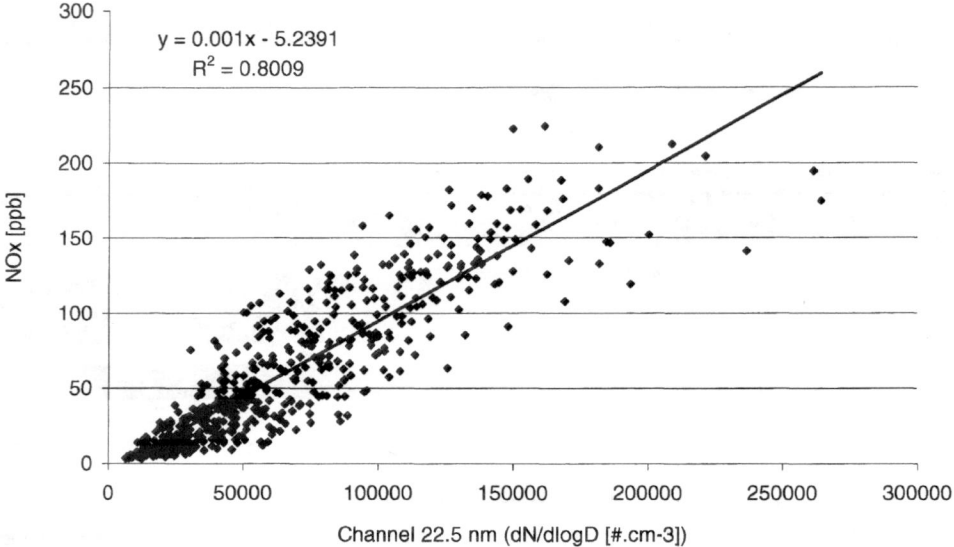

Figure 2. NO$_x$ versus channel 22.5 nm – winter 1999 working days and hours – wind speed >2 m s^{-1}.

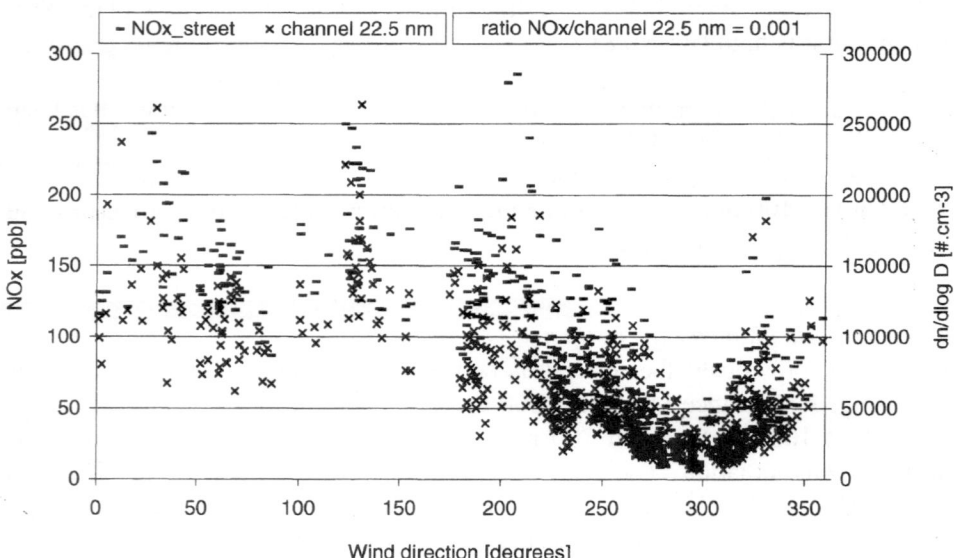

Figure 3. NO$_x$ and channel 22.5 nm versus wind direction – winter 1999 working days and hours – wind speed >2 m s^{-1}.

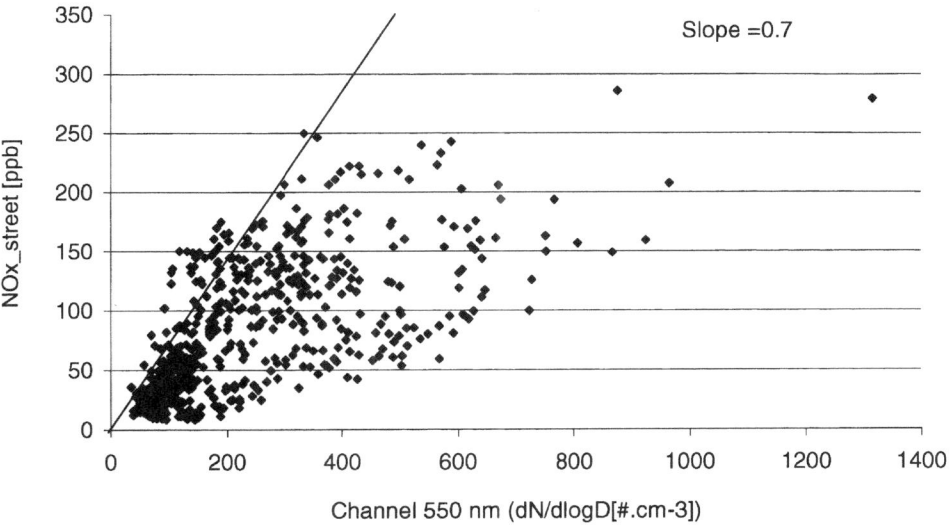

Figure 4. NO$_x$ versus channel 550 nm – winter 1999 working days and hours – wind speed >2 m s^{-1}.

Secondly, it includes non-traffic-related sources (at least not local street traffic), such as long range transport.

Because the NO$_x$ concentrations at a street location are totally dominated by emissions from the traffic, and contribution from other sources is marginal, it is possible to split these two groups using NO$_x$ as a tracer: the traffic-related component is illustrated by the solid line in Figure 4. The difference between the total particle concentration and the solid line can be explained by the additional sources.

The slope of the solid line (0.7, calculated manually) is used as a scaling ratio between NO$_x$ and channel 500 nm when plotting these parameters versus wind direction (Figure 5).

It appears that the lowest concentration of channel 550 nm fits with NO$_x$. Additional particle sources might be responsible for higher concentration for wind direction sectors for instance around 40°, 150° and 240°. This assumption is supported by the fact that the 230–250 degrees range is known to correspond to important long-range transport events; so in this case the enforcement of concentration would be explained.

Therefore the decrease of correlation between 100–700 nm particles and NO$_x$ could be explained by additional sources rather than by a variation of the dilution process.

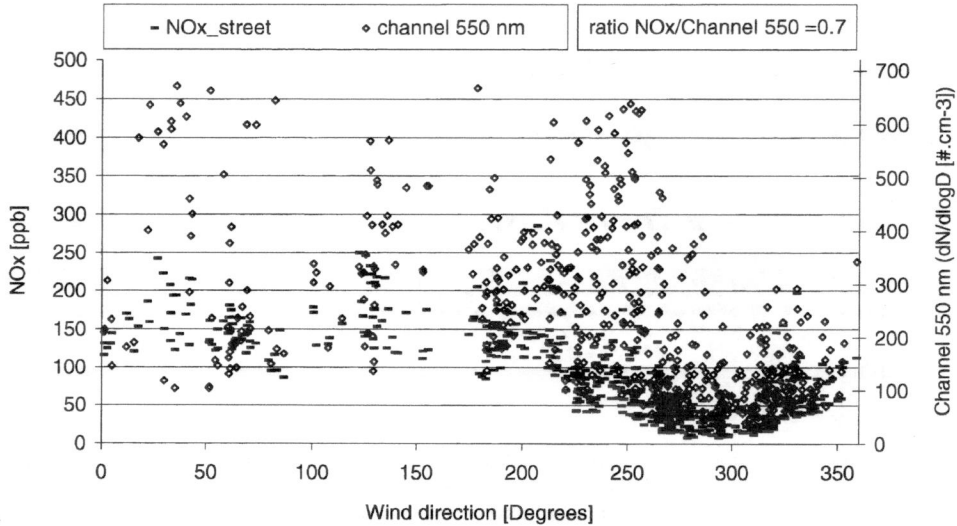

Figure 5. NO$_x$ and channel 550 nm versus wind direction – winter 1999 working days and hours – wind speed >2 m s^{-1}.

3.2. CALCULATION OF THE DILUTION

3.2.1. *The OSPM Model*

The Danish Operational Street Pollution Model (OSPM), has shown to give a satisfactory description of the air gas pollutant dispersion in urban street (Berkowicz, 1996).

The purpose of OSPM is to link a source of pollutant (traffic) and its concentration at any point of a street canyon. The principles behind this model, test results and limitations have been detailed previously (Berkowicz, 1996).

Using the meteorological data, the geometrical street configuration and the traffic conditions, the model calculates the dilution rate occurring during the movement from the source to the sample point. This relationship can be expressed as:

$$C = Q \cdot F + C_{\text{background}} , \tag{1}$$

where Q is the traffic emission rate and $C_{\text{background}}$ is the background contribution, i.e. contribution to the concentrations C from other sources than the traffic in the street. The factor F, and which is calculated by OSPM, provides the link between the source emissions and the concentrations in the street depending on the meteorological conditions. We will subsequently denote this factor as the dilution factor or dilution function.

On the opposite, it is possible to calculate the emission of pollutant (Q), from the measurements (C and $C_{\text{background}}$) and the dilution factor F provided by OSPM:

$$Q = \frac{C - C_{\text{background}}}{F} . \tag{2}$$

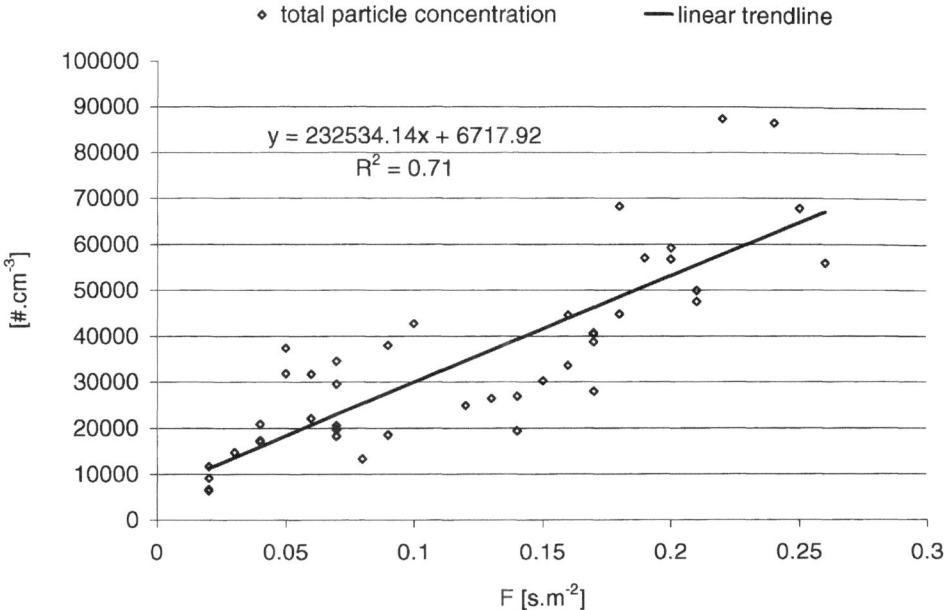

Figure 6. Total concentration of particles versus F – 11:00 winter 2000 – working days – wind speed >2 m s^{-1}.

3.2.2. *Application of OSPM to the Submicrometer Particles*

A prediction of the concentration caused by traffic needs a characterisation of the emission profiles. This could be provided by laboratory measurements. However some experimental artefacts exist affecting especially the ultrafine particles and make laboratory studies not representative for real world conditions (Kittelson and Watts, 2000).

As a consequence, our work will concern the assessment of the emission factor using field measurements.

The total emission rate Q is calculated using linear regression (e.g. Figure 6). It has to be noted that $C_{background}$ for particles has not been measured. The emission rates estimated by this method is used in the subsequent analyses.

3.2.3. *Results*

The total particle concentration is dominated by the sub-100 nm mode. Figure 7 gives a comparison between the daily variation of the total particle emission rate and the corresponding emission rate of NO$_x$. The results for particles are well correlated with NO$_x$ measurements and as a consequence with daily traffic pattern.

While the daily variations of NO$_x$ emissions are similar in winter 1999 and in winter 2000 (Figure 8), we observe a difference for particles (Figure 9). More precisely we observe a decrease of the concentration and a shift to the larger diameters (Figures 9 and 10). In accordance with the previous results obtained with statistical

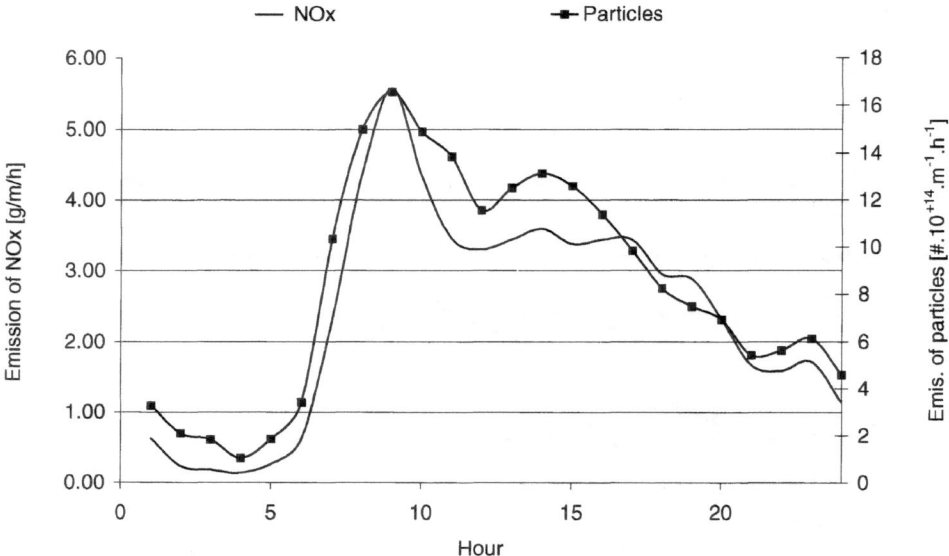

Figure 7. Daily total emission of NO$_x$ and particle winter 1999 – working days – wind speed >2 m s^{-1}.

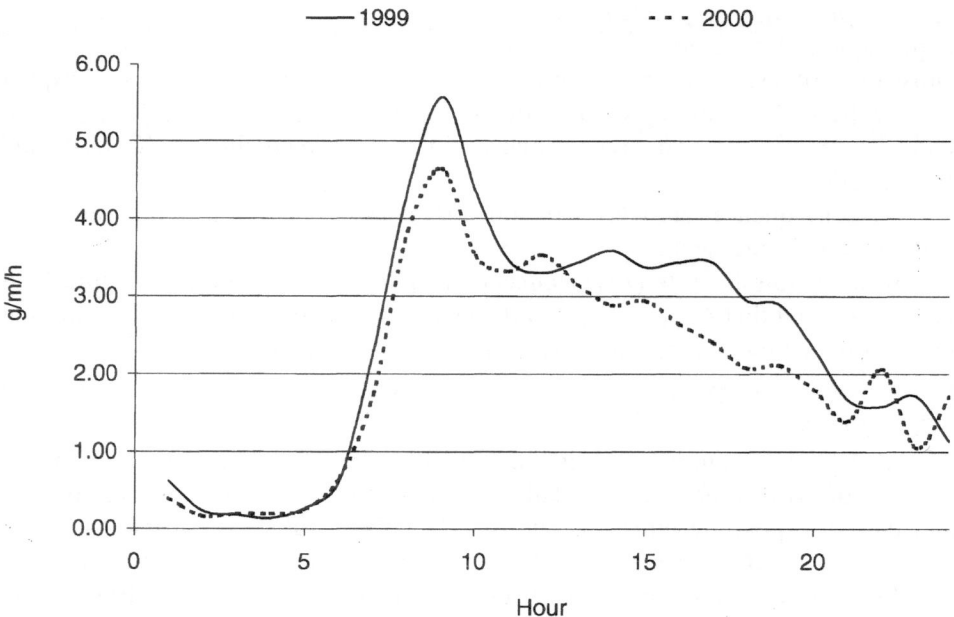

Figure 8. Daily total emission of NO$_x$ winter 1999 and 2000 – working days – wind speed >2 m s^{-1}.

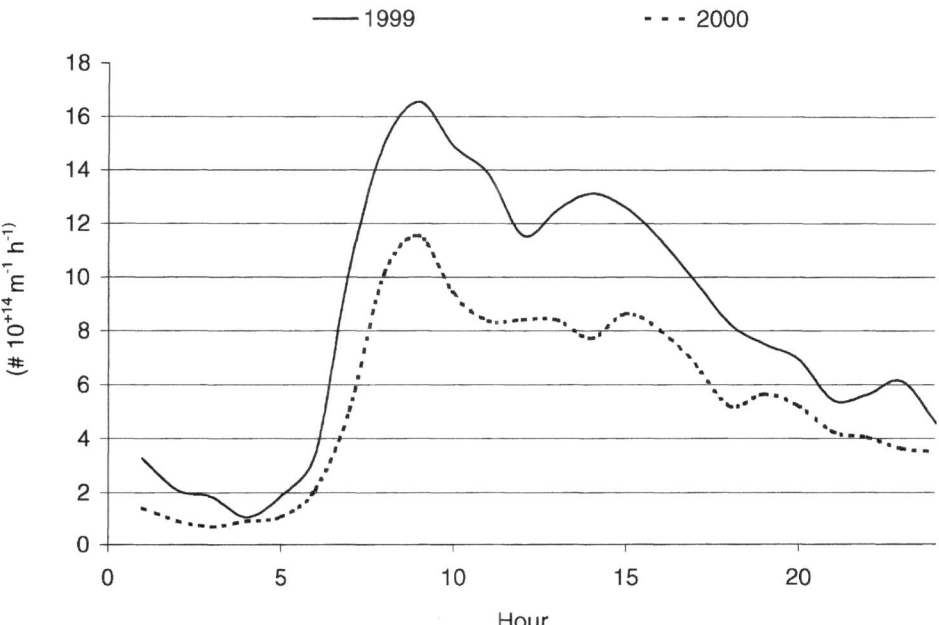

Figure 9. Daily total emission of particles winter 1999 and 2000 – working days – wind speed >2 m s^{-1}.

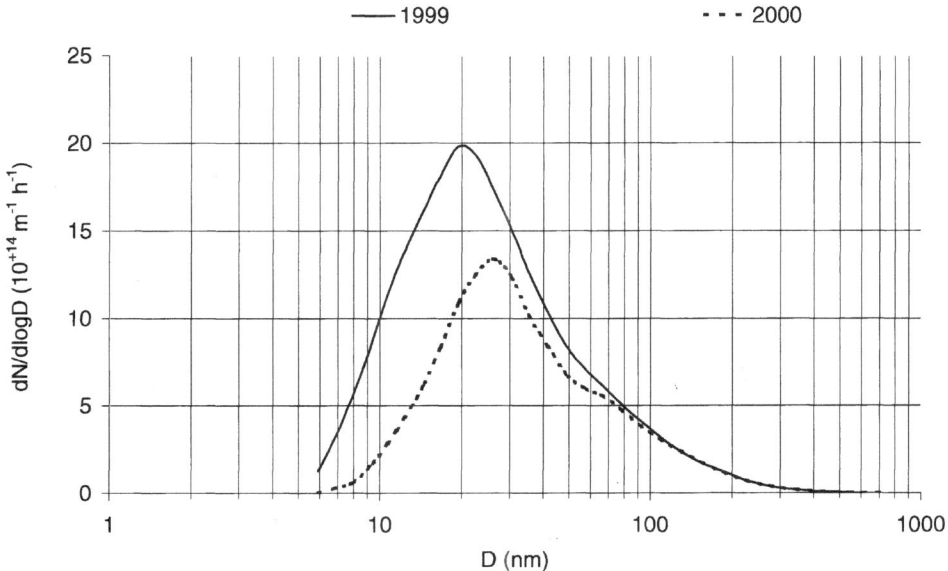

Figure 10. Size distribution of the total emission of particles – 11:00 winter 1999 and 2000 – working days – wind speed >2 m s^{-1}.

modelling (Wåhlin, 2001b), the most likely reason is the substantial decrease of the sulphur-content in the Danish diesel fuel in summer 1999.

4. Conclusions

Measurements of the concentration and size distribution of 10–700 nm particles at the Copenhagen Jagtvej station in winter 1999 and 2000 have been considered.

The dilution properties seem to be comparable for particles and NO_x, especially below 100 nm. Over 100 nm, some additional sources to NO_x-related traffic emission have to be taken into account, including long-range transport.

The total emission from traffic, including daily variation and size distribution, has been calculated using the OSPM dispersion model. Results are in accordance with the previous analysis based on statistical modelling.

We intend to use this method for detailed evaluation of particle emissions from traffic. The background concentration of particle will be characterised simultaneously to street measurements for that purpose. In addition, we will look for some tracers of the different possible sources.

Acknowledgements

The authors wish to express their appreciation for support from the European Commission's Training and Mobility of Researchers Programme (TMR) within the frameworks of the European Research Network on 'Optimisation of Modelling Methods for Traffic Pollution in Streets' (TRAPOS).

References

Berkowicz, R., Palgrem, F., Hertel, O. and Vignati, E.: 1996, 'Using measurements of air pollution in streets for evaluation of urban air quality-meteorological analysis and model calculations', *Sci. Total. Environ.* **189/190**, 259–265.

Kittelson, D. B. and Watts, W.: 2000, 'Nanoparticle Emissions from E.engines', *Third Joint ESF-NSF Symposium on 'Nanoparticles: Applications in Material Science and Environmental Science and Engineering'*, 6 September, Dublin.

Künzli *et al.*: 2000, 'Public-health impact of outdoor and traffic-related air pollution: A European assessment', *The Lancet* **356**, 2 September.

Vignati, E., Berkowicz, R., Palmgren, F., Lyck, E. and Hummelshøj, P.: 1999, 'Transformation of size distribution of emitted particles in streets', *The Sci. Total Environ.* **235**, 37–49.

Wåhlin, P., Palmgren, F. and Van Dingenen, R.: 2001a, 'Experimental studies of ultrafine particles in streets and the relationship to traffic', *Atmosph. Environ.* (in press).

Wåhlin, P., Palgrem, F. and Van Dingenen, R.: 2001b, 'Source Apportionment and Size Characterisation of Aerosol Particles Measured in a Copenhagen Street Canyon', *Third International Conference on Urban Air Quality*, 19–23 March 2001, Loutraki, Greece.

MODELLING OF FLOW AND POLLUTION DISPERSION IN A TWO DIMENSIONAL URBAN STREET CANYON

D. VLACHOGIANNIS*, S. RAFAILIDIS, J. G. BARTZIS, S. ANDRONOPOULOS and A. G. VENETSANOS

Environmental Research Laboratory, INTR-P, National Centre for Scientific Research 'Demokritos,' Aghia Paraskevi Attikis, Greece

(* author for correspondence, e-mail: mandy@avra.ipta.demokritos.gr, fax: +30 1 6525004)

Abstract. The wind-driven flow patterns and the dispersion of vehicle exhaust pollutants released at street level has been simulated with the three-dimensional (3-D) dispersion model ADREA-HF (Andronopoulos *et al.*, 1993), for idealised two-dimensional urban fetches occupied by buildings with slanted roofs. The simulation used oncoming atmospheric boundary layer characteristics corresponding to realistic above-town wind characteristics, as measured in reference wind tunnel experiments (Rafailidis, 1997). At that stage, analysis was limited to neutral stability conditions only. Firstly, the quality assurance of the numerical model was investigated in terms of the sensitivity to different grid allocations. The modelling results were corroborated by comparison with wind tunnel measurements in a similar two-dimensional domain (Pavageau *et al.*, 1997). The numerical modelling replicated well the high degree of non-uniformity in the dispersion field in the test street, and the results agreed satisfactorily with the experimental measurements. The reasons for the differences observed have been investigated. With the model thus validated, three different exhaust release scenarios have been tested, keeping the same overall emission rate but different spatial patterns of street-release. The effect of the different street-release scenarios was found to be only marginal, with the dispersion patterns on the sidewalls affected only locally, close to the street level.

Keywords: CFD modelling, dispersion, grid, source orientation, street geometry, urban canyon

1. Introduction

The continuing urbanisation in Europe means that environmental assessment moves closer to local scale, with main interest in near-field effects in urban street canyon situations. Emphasis is now placed on simulations of the wind flow and pollution dispersion at street level in-between the buildings. The pollutant concentrations within a street canyon depend mainly on traffic and car driving patterns, the street configuration, including neighbouring streets and the meteorological conditions (oncoming wind speed and wind direction, atmospheric stability, etc.). Practical and accurate numerical tools are thus required to cover relatively large town domains, to detect trends and evaluate reliably 'what–if' scenarios before implementation. Currently, uncertainties in dispersion modelling may rise from limitations in spatial grid resolution, domain size, boundary conditions, real topography simulation and lack of knowledge on exact source strength, orientation and distribution

Water, Air, and Soil Pollution: Focus **2**: 405–417, 2002.
© 2002 *Kluwer Academic Publishers.*

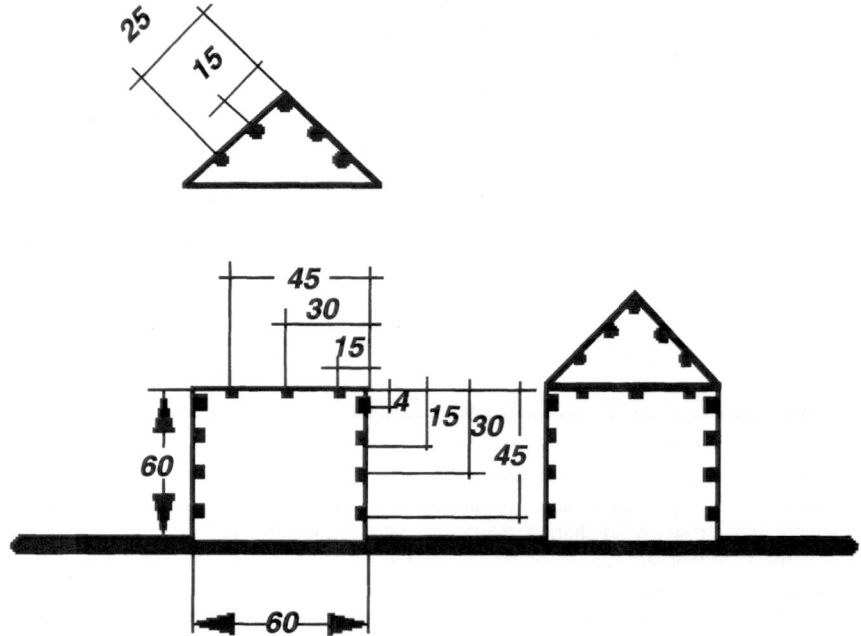

Figure 1. Positions of wind tunnel concentration measurements (numbered 1 to 18) on either side of the street canyon containing the source. Dimensions are in mm.

(Robins *et al.*, 2000). The aim of this work is to quality assure a model by investigating the model sensitivity to different grid allocations and subsequently study the influence of street canyon pollution dispersion of three different source release scenarios.

Experimental measurements of urban canyon pollution dispersion performed in the Blasius wind tunnel of Hamburg University have been used for a consequent modelling analysis (Rafailidis, 1997; Pavageau *et al.*, 1997). The experimental urban configuration consisted of a two-dimensional (2-D) span of identical parallel street canyons and buildings with slanted roofs, normal to the oncoming wind. The canyon aspect ratio was B/H = 1 (street width B to height of the building walls H). The building walls were 60 mm high. The apex of the slanted roofs in the models reached 90 mm above the floor (Figure 1). Slanted roofs had been placed on the five buildings upstream and two downstream of the test canyon. Further fifteen street canyons, surrounded by buildings with flat roofs, were also placed upwind. A line source was situated at the bottom of the test street canyon in the wind tunnel, ejecting outwards with velocity 0.012 ms^{-1}, a mixture of ethane and air horizontally in two opposite directions by placing a flat plate at a distance of 1 mm, thereby removing the vertical momentum. The source length was 10 mm and the height 1.2 mm. The source strength for ethane was equal to 2.6×10^{-7} m^3

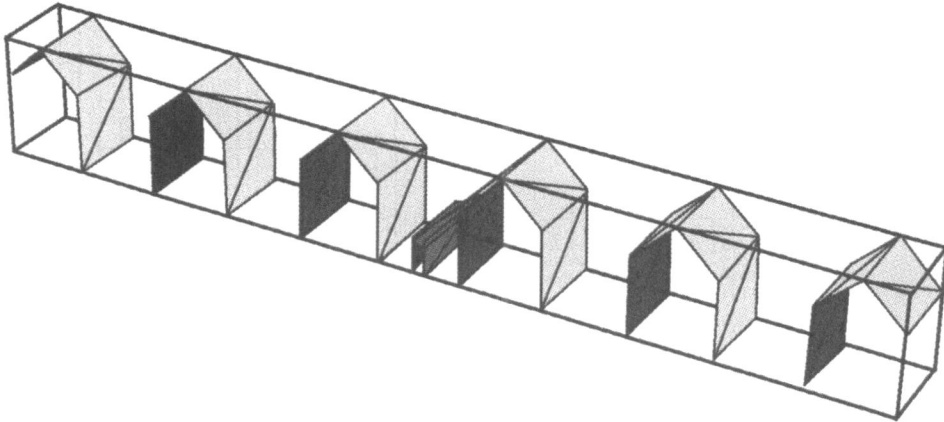

Figure 2. Urban model of parallel streets and buildings with slanted roofs and source in the test canyon. Aspect ratio assumed as B/H = 1. Dimensions in y-direction exaggerated.

s^{-1} and for air 2.6×10^{-5} m^3 s^{-1}. Further details on the boundary layer in the wind tunnel are given by Rafailidis (1997).

The 3-D CFD (Computational Fluid Dynamics) ADREA-HF model was employed for the simulation of wind flow and pollutant dispersion; this is a transient, non-hydrostatic, fully compressible transport code especially developed for dispersion of buoyant or passive gases over complex terrain in local scale. The fluid dynamics and thermodynamics are described by the air-gas mixture continuity, momentum and internal energy equations, whereas mass conservation of the pollutant substance is fulfilled through a separate mass transport equation. These equations are descritised, using the finite volume method with a staggered grid for the velocities. The method, which is based on the SIMPLER algorithm (Patankar, 1980), is fully implicit in time and uses the upwind scheme for the convective terms. From the turbulence closure schemes available in the code, the eddy viscosity/diffusivity concept is used to maintain the computation time and storage at realistic levels. The eddy viscosity is calculated using a 1-equation k-l model, solving a transport equation for the turbulent kinetic energy. A volume and surface porosity concept is used for the description of the complex geometry.

The geometry of the domain has been simulated using the DELTA-B pre-processor (Venetsanos *et al.*, 1995), applying a geometric scaling of 1000:3, with three buildings upstream and three downstream the test canyon holding the source (Figure 2). As shown by the results, this was sufficient to recreate the steady state, fully grown boundary layer conditions in the tunnel.

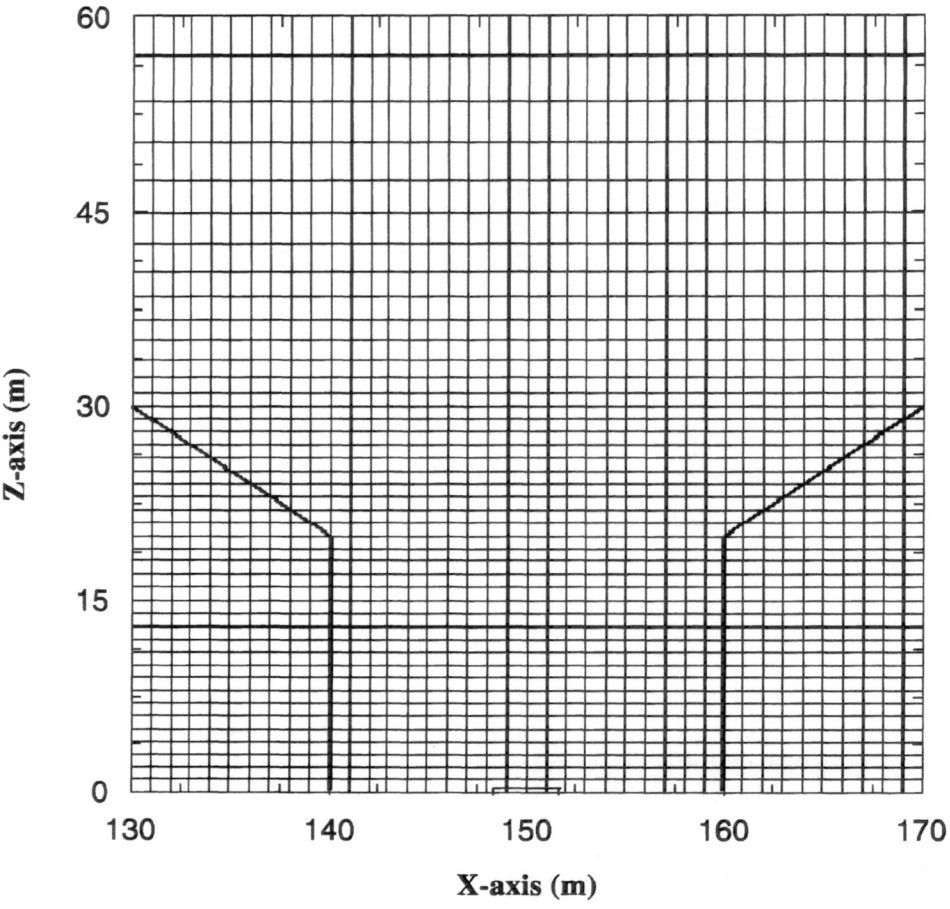

Figure 3. Section of the modelling domain in the test canyon shown for the base case of 260×61 cells; flow is from left to right.

2. Methodology

2.1. GRID SENSITIVITY

First of all, the sensitivity of the model predictions to different grid allocations was investigated for a 2-D domain with dimensions 260 m (horizontal) × 164 m (vertical). A base case was assigned as 2-D grid of 260 × 61 cells (Figure 3). Three other different grid configurations were constructed comprising 65 × 61, 130 × 61 and 520 × 61 cells. The grid spacing in the base case was $\Delta x = 1$ m in the horizontal direction while in the vertical, the minimum resolution was of $\Delta z(min)$ = 1 m, applied within the street canyons, increasing gradually above roof level to a maximum of $\Delta z(max) = 6.5$ m. This vertical resolution was kept the same for all

BL profile developing above urban fetch of slanted roofs

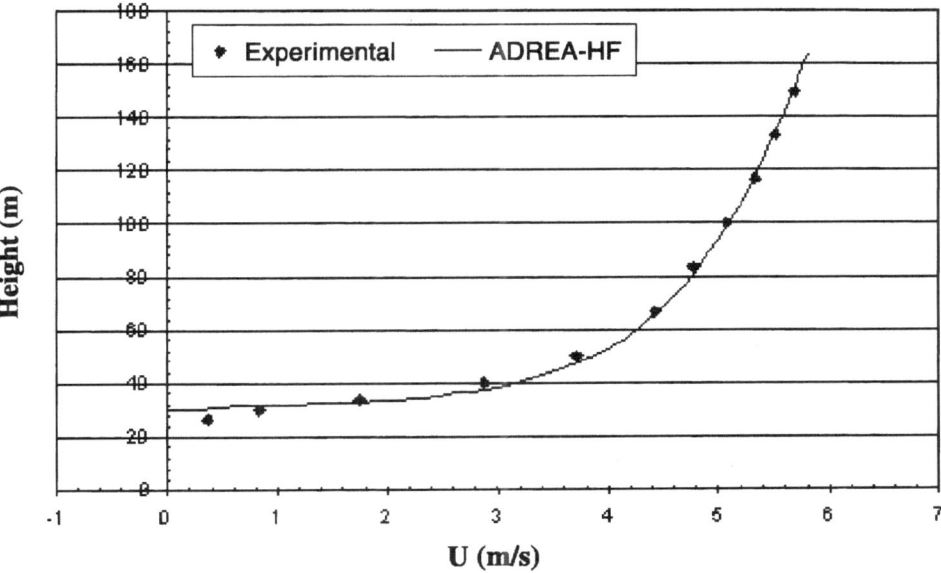

Figure 4. Comparison of the experimental and model calculated inflow boundary-layer (BL) profile.

grid configurations with only the horizontal resolution being altered from 4, 2 and 0.5 m (fine resolution) on the 65×61, 130×61 and 520×61 grids, respectively.

The source was modelled as a continuous line source placed in the centre of the test canyon, ejecting the gas horizontally in two opposite directions by imposing a horizontal obstacle of thickness 0.05 m and length 3.3 m, at a distance of 0.4 m from the ground. The jet exit velocity was equal to 0.012 ms^{-1} in the x-direction and the exit mass fraction was 0.01. These exit conditions corresponded exactly to the experimental measurements while the source dimensions were replicated applying the appropriate geometric scaling.

Simulations were performed with this specific source configuration using the ADREA-HF model for the base case as well as the three different grid allocations. For the model simulation, the domain upstream included an urban area where wind conditions prevailed above the average building height and also, the free stream wind profile was the same as that measured in the wind tunnel. With this approach to simulating the problem, the sudden appearance of obstacles against the undisturbed flow was avoided and hence, the developed boundary layer characteristics corresponded to realistic above town wind characteristics.

At the main inflow boundary of the 2-D simulation, the profile of the horizontal velocity was calculated by the model using the experimental free stream velocity value of $U\delta = 5.85$ ms^{-1} and the experimental measurements scaled-up to full scale, i.e. surface roughness length of $z_0 = 0.6$ m and displacement height above the ground of D = 30 m (Figure 4). The roughness for the street surface was estimated to correspond to the experiment at 0.1 m.

The modelling results were compared with wind tunnel pollutant concentration measurements in a similar two-dimensional domain (Pavageau *et al.*, 1997).

2.2. SENSITIVITY TO SOURCE RELEASE DIRECTION

Further to the grid sensitivity study, an investigation of the model sensitivity to different source release scenarios was carried out with regards to the direction of the jets. The emitted pollutant was neutrally buoyant. The momentum effects have been taken into account by modelling the source as a jet with an exit velocity. Thus, three exhaust release conditions were considered on the grid of 260×61 cells. Firstly, the source was modelled as two horizontal symmetric jets in opposite directions (base case) while for the other two scenarios, the same source strength has been assumed but with the source comprising one jet releasing in either the windward or leeward directions. The source strength remained the same as that used in grid sensitivity analysis, described in the previous section. The jet exit velocity was equal to 0.012 ms^{-1} for all source release scenarios. The same initial profile of the horizontal component of wind velocity was imposed at the inflow plane as shown in Figure 4.

3. Results and Discussion

3.1. GRID SENSITIVITY

The grid sensitivity analysis was carried out in order to verify the quality assurance of the model using as a reference the results from the fine resolution grid of $\Delta x = 0.5$ m with 520×61 cells. Pollutant concentrations calculated by the model at steady state conditions and at 407 sensor positions (equally spaced every 4 m in the horizontal direction and 1 m in the vertical direction) within the test canyon were plotted for the fine resolution grid versus those obtained from the other previously described grids (Figure 5). The calculated values using the different grid allocations have revealed that the horizontal resolution (for grid spacing $\Delta x \leq H/10$, where H is the building height) has little further effect on the accuracy of the simulation. Thus, the sensitivity of the model to different grid allocations has been proved to be small, an important conclusion for the quality of the model.

Based on this result, further model simulations were carried out only for the base case of 260×61 cells. Figure 6 shows the wind velocity field developed in the test canyon at steady state conditions. A circulating vortex formed symmetrically between the buildings at eaves level in agreement with wind tunnel observations (Rafailidis, 1997). At roof level, strong gradients of the wind velocity field developed, which consequently increased the turbulent mass transport and canyon ventilation.

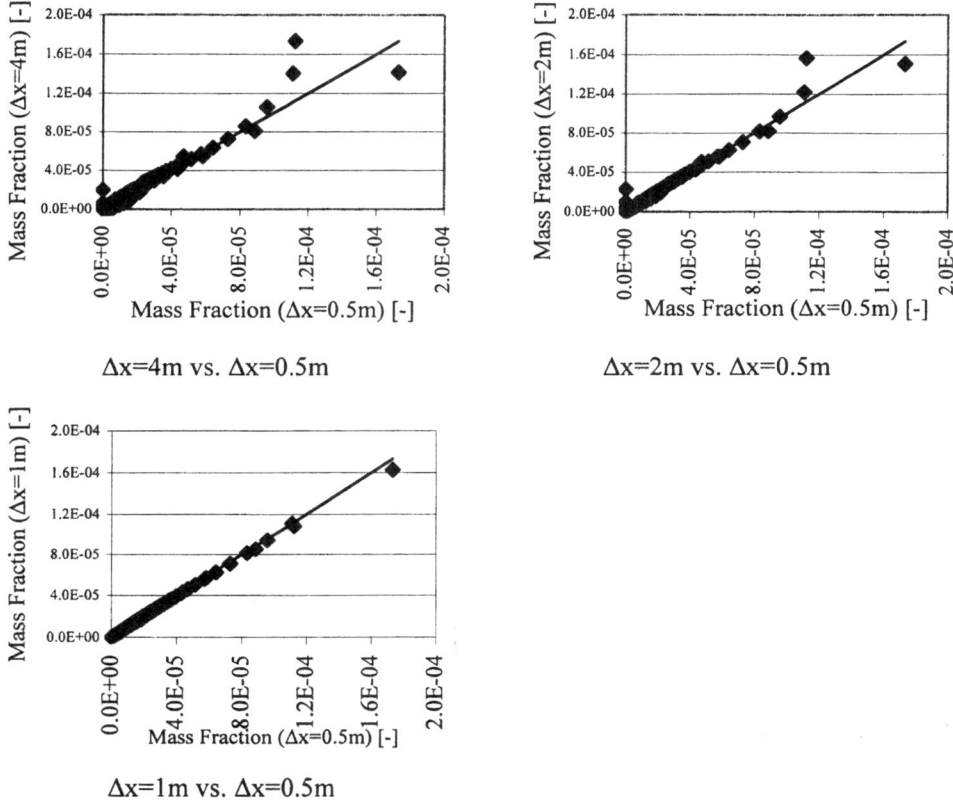

$\Delta x=4m$ vs. $\Delta x=0.5m$ $\Delta x=2m$ vs. $\Delta x=0.5m$

$\Delta x=1m$ vs. $\Delta x=0.5m$

Figure 5. Investigation of the influence of the grid step-size on concentration. Fine resolution case grid of $\Delta x = 0.5$ m (520 × 61cells) versus grids of $\Delta x = 4$ m (65 × 61 cells), $\Delta x = 2$ m (130 × 61 cells) and $\Delta x = 1$ m (260 × 61 cells).

The wind field in the whole computational domain is depicted in Figure 7. In this way the influence of the extent of the computational domain upstream and downstream of the test canyon on the results was studied. Apart from the test canyon where the release of the pollutants affects dispersion locally, at the street level close to the source, the wind speed distribution in the test canyon is similar to the ones in the other two upstream and downstream canyons. This helps to conclude that for the current case study, the computational domain that included two street canyons on either side of the test canyon allows sufficient development of the boundary layer above roof level and the subsequent steady state of the flow between the buildings.

To compare the model concentrations with experimental measurements, non-dimensional values of the calculated concentrations were determined using the following equation:

$$C^* = CU_\delta H/q,$$

(1)

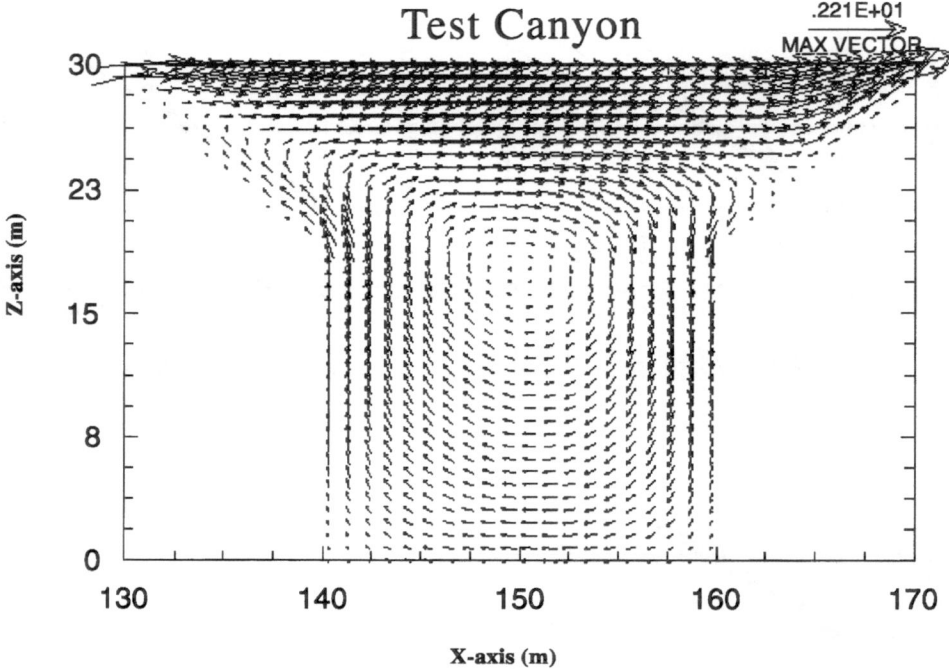

Figure 6. Vector plot of simulated wind field inside the test canyon in a plane normal to the street axis.

where C^* is the non-dimensional concentration, C (in ppm) is the calculated ethane tracer concentration, $U\delta$ (ms^{-1}) is the wind velocity measured in the free stream, H (in m) is the height of building, q (in m^3 s^{-1} m^{-1}) is the line source strength. Figure 8 presents a comparison between non-dimensional concentration values of the model results and the experimental measurements at 18 sensor positions, distributed on the sidewalls and roofs of the test canyon as shown in Figure 1. It can be seen clearly from this graph that the agreement between calculated concentrations and wind tunnel measurements is very good and any discrepancies found were calculated to be within the factor of two band. The factor of two on concentration predictions is well within the expected uncertainties derived from such modelling, as deduced from Robins *et al.* (2000), where the spread of the model results relative to wind tunnel data was found to be between a factor of 5 and 100.

3.2. SENSITIVITY TO SOURCE RELEASE DIRECTION RESULTS

To investigate the impact of different directions of the source release jets on the dispersion field within the test canyon, contours of the tracer mass fraction have been plotted for the three scenarios (Figure 9). It can be seen that the contour distribution does not significantly change except in the region at low level in the canyon close to the source. Furthermore, Figure 10 shows the model calculated

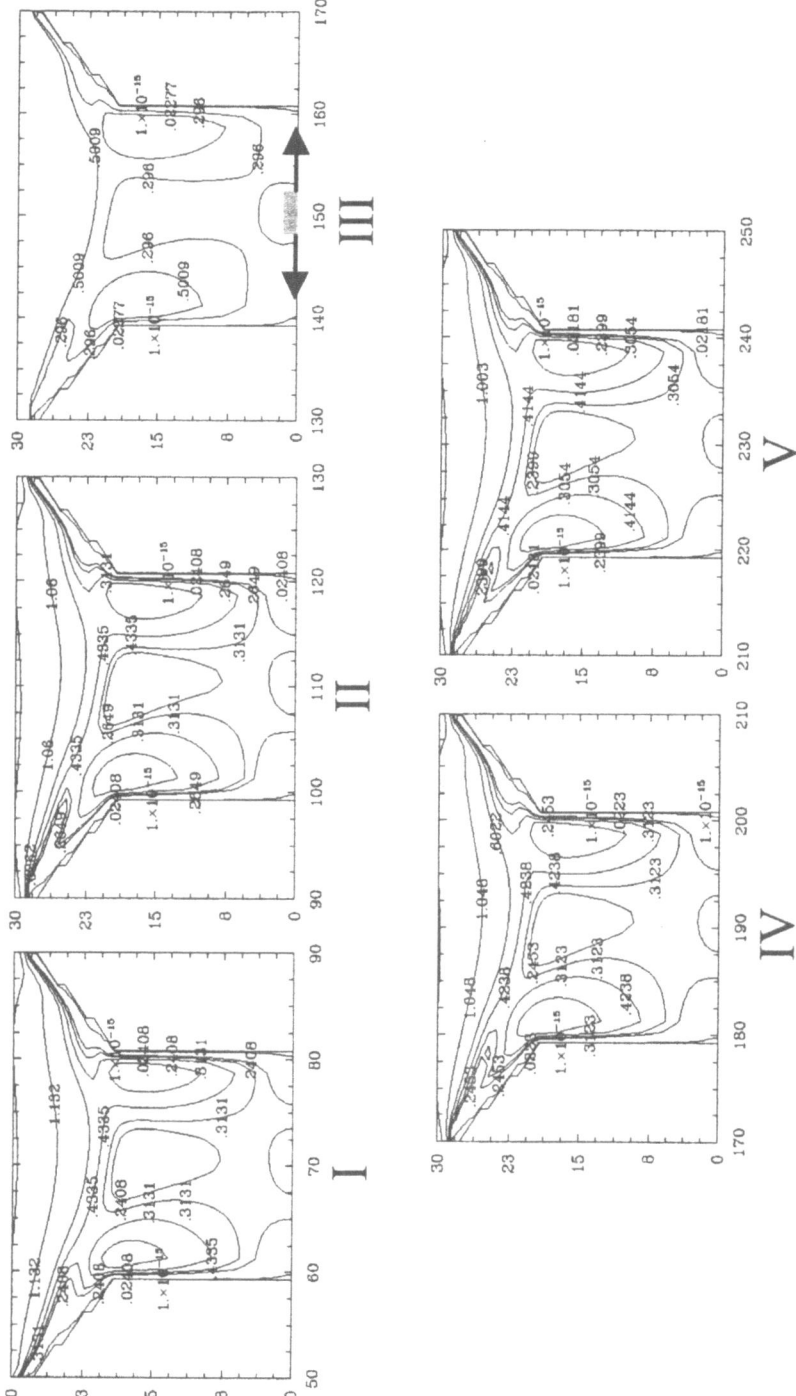

Figure 7. Comparison of the model calculated wind speed distribution in the street canyons upstream and downstream of the test canyon (base case).

Comparisons of model vs experimental results

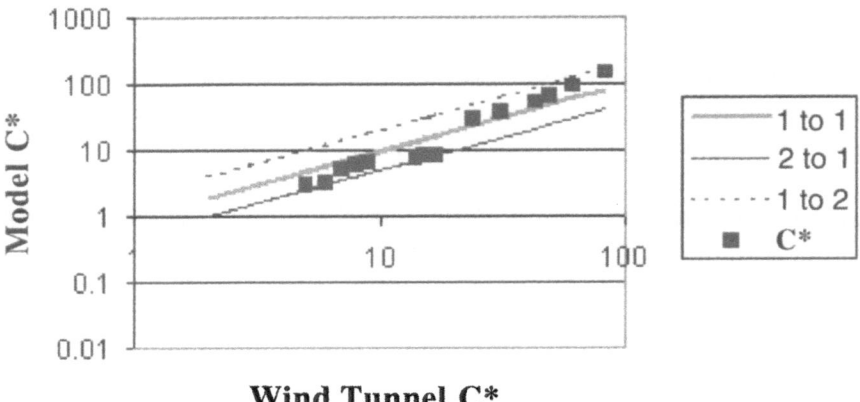

Figure 8. Comparison of non-dimensional concentration values of model against experimental results at the 18 measurement positions (Figure 1).

non-dimensional concentration values for the different release conditions at sensor positions of the test canyon on upwind and downwind roofs and building walls (see Figure 1 for sensor locations in the domain). This detailed analysis of the results showed that at the street level in the bottom of the test canyon, the concentrations differed more (though not significantly) on the leeward canyon side (sampling positions 1, 2, 3 and 4) than on the windward (sampling positions 10, 11, 12 and 13). As expected, the pollutant concentrations attain their highest values on the leeward side and these are more enhanced when the source is ejecting leeward, enriching the field driven by the vortex transport. In the case of two symmetric jets of opposite direction, the concentrations values are within the range defined by the other two release directions. However, the variations in the concentrations found across the three release cases are not significant hence, the effect of different street release conditions is not very important. Such a conclusion facilitates urban pollution management as not being influenced by, for instance, the different traffic distributions across the street lanes of an urban canyon with this specific geometry at neutral stability conditions (Figure 9).

4. Conclusions and Summary

A thorough numerical modelling investigation of the wind field and dispersion of pollutants within an urban street canyon has been carried out for neutral stability conditions corroborating the calculated results with experimental measurements from a wind tunnel. The comparisons of the model results with the experimental

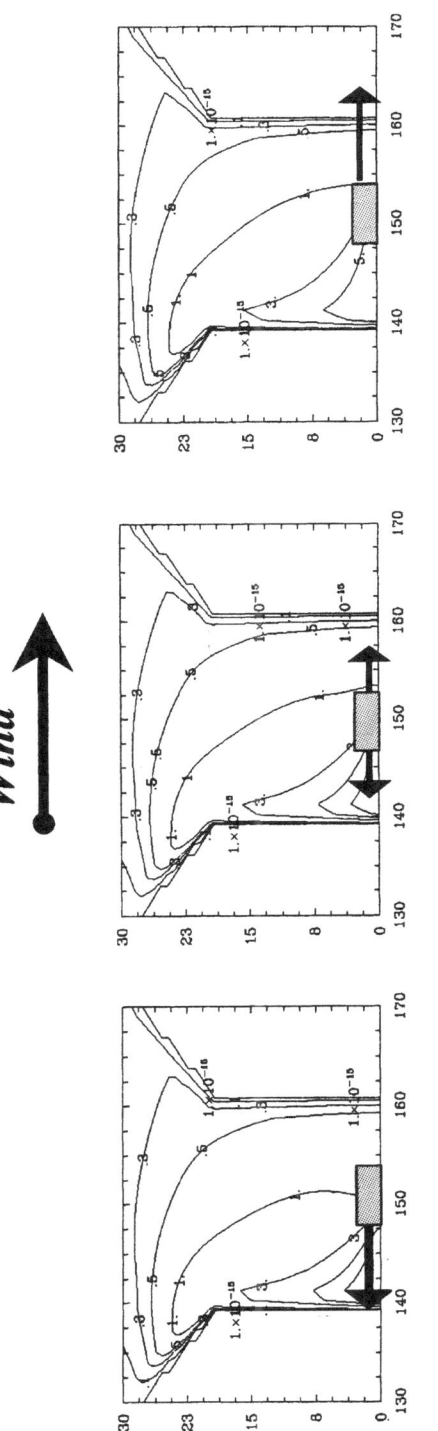

Figure 9. Comparison of model calculated distribution of tracer mass fraction in the test canyon for different directions of source jets while keeping the same source strength. Source dimensions exaggerated.

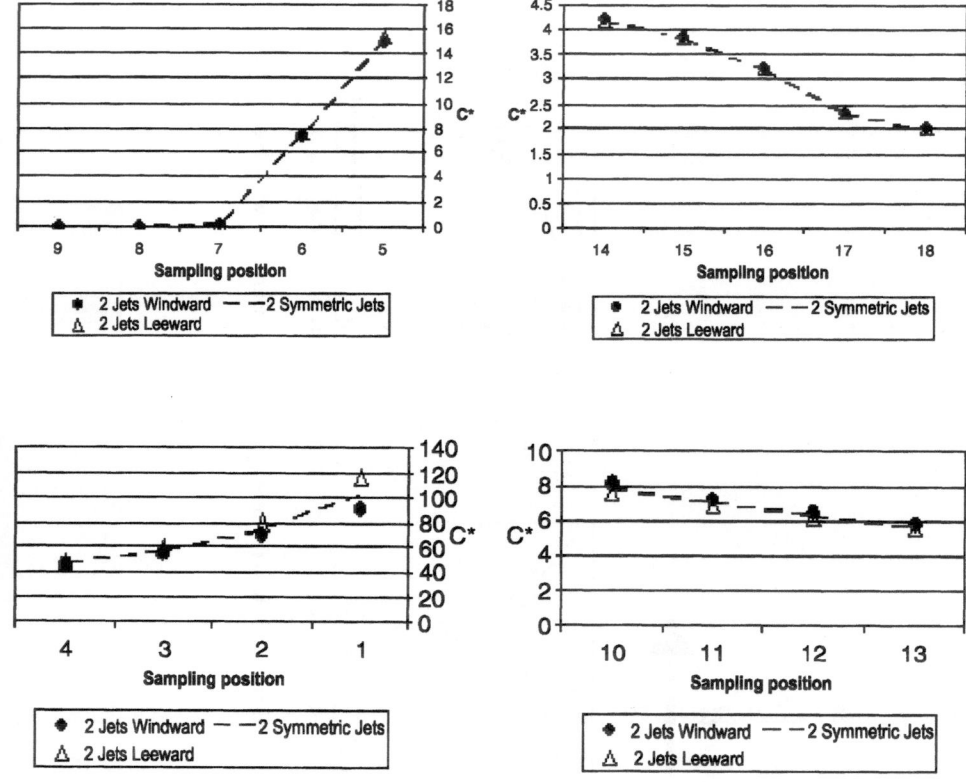

Figure 10. Investigation of the effect of different release conditions at test canyon sensor positions on upwind and downwind roofs and building walls as defined in Figure 1.

pollutant concentration data yielded a good agreement and any discrepancies found were within a factor of two. To quality assure the model (ADREA-HF), different grid refinements have been tested. The sensitivity of the model to these refinements proved to be minor, a positive indication of the quality of the model. For this study case, the grid horizontal resolution (for grid spacing less or equal to H/10, where H is the building height) was found to have little further effect on the accuracy of the simulation. The simulated wind field compared well to the wind tunnel field, with a vortex formed between the buildings, aiding the ventilation of the canyon from the pollutants emitted from the source. The different release cases showed that the wind fields and concentrations were largely independent of the direction of the source jets. The repercussions of this observation for urban pollution management could be assessed as not being influenced by, for instance, the different traffic distributions across the street lanes in a canyon of aspect ratio B/H = 1 surrounded by buildings with slanted roof shapes.

Furthermore, the influence of the extent of the computational domain upstream and downstream of the test canyon on the results was studied. Apart from the test canyon where the release of the pollutants affects dispersion locally, at the street level, the wind vector distribution in the test canyon is similar to those in the other two adjacent upstream and downstream canyons. Thus, for the present 2-D test case with the specific characteristics, the extent of the computational domain with a total of five urban street canyons is sufficient for the development of a well-formed flow of the atmospheric boundary layer above the roofs and subsequently, a steady state flow between the buildings.

References

Andronopoulos, S., Bartzis, J. G., Wurtz, J. and Asimakopoulos, D.: 1993, 'Simulation of the Thorney Island dense gas trial No. 8, using the code ADREA-HF', *Process Safety Progr.* **12**(1), 61−66.

Patankar, S. V.: 1980, *Numerical Heat Transfer and Fluid Flow*, Hemisphere, Washington.

Pavageau, M., Rafailidis, S. and Schatzmann, M.: 1997, 'A comprehensive experimental databank for the verification of urban car emission dispersion models', *Int. J. Environ. Pollut.* **8**(3−6), 738−746.

Rafailidis, S.: 1997, 'Influence of building areal density and roof shape on the wind characteristics above a town', *Bound.-Layer Meteorol.* **85**, 255−271.

Robins, A. G., Hall, R., Cowan, I. R., Bartzis, J. G. and Albergel, A.: 2000, 'Evaluating modelling uncertainty in CFD predictions of building affected dispersion', *Int. J. Environ. Pollut.* **14**(1−6), 52−64.

Venetsanos, A. G., Catsaros, N., Würtz, J. and Bartzis, J. G.: 1995, 'The DELTA_B Code. A Computer Code for the Simulation of the Geometry of Three-dimensional Buildings. Code Structure and Users Manual', *Report EUR 16326 EN*.

ESTIMATION OF CUMULATIVE FREQUENCY DISTRIBUTION FOR CARBON MONOXIDE CONCENTRATION FROM WIND-SPEED DATA, IN BUENOS AIRES (ARGENTINA)

NICOLÁS A. MAZZEO* and LAURA E. VENEGAS

National Scientific and Technological Research Council (CONICET), Department of Atmospheric Science and the Oceans, Faculty of Sciences, University of Buenos Aires, Ciudad Universitaria, Pab. 2., Buenos Aires, Argentina
(* author for correspondence, e-mail: mazzeo@at.fcen.uba.ar, fax: (+54 11) 4576 3356)

Abstract. In this article we apply and test a methodology to estimate cumulative frequency distribution for air pollutant concentration from wind-speed data. We use the inverse relationship after Simpson *et al.* (Atmospheric Environment, 19, 75–82, 1985) between the opposing percentile values in the statistical distributions for air pollutant concentrations and wind-speed data. This relationship is valid, irrespective of the statistical distributions of both variables, if an inverse relationship between them is also applicable. The available data are five years of 8-h average carbon monoxide concentration and 8-h mean wind-speed, observed in Buenos Aires (Argentina). The performance of the obtained empirical expressions in estimating cumulative frequency distributions for 8-h CO is statistically evaluated. The results show that it is possible to obtain an acceptable cumulative frequency distribution for 8-h CO concentration at the site if the cumulative frequency distribution for wind-speed is known. Q–Q plots show a good agreement between estimated and observed values. From our data, the mean relative error of the estimations was found to be as much as 8.0%.

Keywords: box models, carbon monoxide data, cumulative frequency distribution, exceedances of air quality standard, percentile analysis, urban air pollution

1. Introduction

Different urban air quality models are used to estimate concentrations of several pollutants resulting from urban emissions. The urban dispersion models range from simple empirical models to complex three-dimensional urban airshed models. Numerous studies have shown that simple modelling approaches may be used to estimate long term mean concentrations (see Arya, 1999).

Simple atmospheric dispersion models for urban areas which are usually referred to as box models, may be simplified to the following relationship between air pollutant concentration (C) and wind-speed (u) (Bencala and Seinfeld, 1976; Simpson *et al.*, 1983, 1985):

$$C = \frac{K}{u} \tag{1}$$

Water, Air, and Soil Pollution: Focus **2**: 419–432, 2002.
© 2002 *Kluwer Academic Publishers.*

where K depends on the area source strength and the height of the upper boundary of the box within pollution is completely mixed, atmospheric stability conditions and the characteristics of the pollutant. Equation (1) states that the average concentration in the box is mainly given by the local source strength, the wind-speed and the prevailing atmospheric stability conditions. Simpson *et al.* (1983, 1985) demonstrated that irrespective of the statistical distributions of air pollution data and wind-speed, if Equation (1) is applicable, then the cumulative frequency distributions $F(C)$ and $G(u)$ of air pollution data and wind-speed, respectively, are related by:

$$F(C) = 1 - G(u) , \tag{2}$$

where $F(C)$ is the probability of concentration to by greater than a given value C and $G(u)$ is the probability of wind-speed greater than a given value u. Equation (2) implies that the value of C corresponding to the p-percentile (C_p) in $F(C)$ and the value of u corresponding to the (*100-p*)-percentile, $u_{(100-p)}$ in $G(u)$, are related by (Simpson *et al.*, 1983):

$$C_p = \frac{K}{u_{(100-p)}} . \tag{3}$$

The objective of this article is to test the use of Equation (3) in estimating cumulative frequency distribution of 8-h CO concentrations from 8-h averaged wind-speed data, observed at one site in the city of Buenos Aires (Argentina). We have five years of data. Equation (3) is used to estimate K. We consider only one year of data at a time for calculations and the rest of the four years to test the results.

2. Distributions of Wind-speed and CO Concentration Data in Buenos Aires City

The city of Buenos Aires is situated on a flat terrain, it has an extension of 200 km² and a population of three million inhabitants. There are daily three million vehicles in the city, one million enters the city from the surroundings during working hours. Gasoline powered motor vehicles are the major sources of carbon monoxide (CO) in urban atmosphere. Measurements of carbon monoxide concentration are done by an automated continuous nondisperse infrared monitoring system sited in a dense populated area with intense activity during working hours. The available CO data include five years (1994–1998) of 8-h average concentration from 8 a.m. to 4 p.m. registered at one site (Venegas and Mazzeo, 2000). In order to assume no variation in emission rate (Q = constant), we consider 8-h CO concentration observed during working days only. January and February data are not included in the analysis because the number of vehicles decreases during these months due to summer holidays. The number of values for each year, included in the analysis

TABLE I

Linear relationship between K and the p-percentile ($p \leq 80\%$) and its correlation coefficient (R), values of coefficients a and b in Equation (4) ($K = a+bp$)

Year	Regression line	R
1994	$K_{94} = 30.128 + 0.124\ p$	0.828
1995	$K_{95} = 32.677 + 0.134\ p$	0.973
1996	$K_{96} = 27.617 + 0.161\ p$	0.993
1997	$K_{97} = 33.674 + 0.119\ p$	0.958
1998	$K_{98} = 24.385 + 0.320\ p$	0.990

is: 204 (1994), 206 (1995), 204 (1996), 197 (1997) and 194 (1998). Hourly wind-speed and wind direction are measured at a meteorological station of the National Weather Service of Argentina, located in the city.

2.1. WIND SPEED

The cumulative frequency distributions of 8-h mean wind-speed observed during 1994, 1995, 1996, 1997 and 1998 are included in Figures 1a, b, c, d and e, respectively. The plots for wind speed data indicate that, the straight-line approximation suggested by a log-normal distribution is quite good. For each data set, the geometric mean (μg) and the standard geometric deviation (Sg) are included in each plot. It can be seen, for example, that the frequency of 8-h mean wind-speed greater than 4 m s^{-1} is between 38% (1996) and 53% (1998).

2.2. CARBON MONOXIDE

Several authors (Bencala and Seinfeld, 1976; Simpson *et al.*, 1983) have shown that urban pollutant concentration can be nearly log-normally distributed. Figures 2a–e show the log-normal plot of cumulative frequency distributions for 8-h CO concentration registered at this site in Buenos Aires city, during 1994, 1995, 1996, 1997 and 1998, respectively. The geometric mean (μg) and the standard deviation (Sg) determined from each data set are also included in Figure 2. The lowest and the highest frequency of 8-h CO greater than 9 ppm (Air Quality Standard for CO, for an 8-h averaging period, recommended by the World Health Organisation (WHO, 1980) are 55% (for 1994) and 71% (for 1995), respectively.

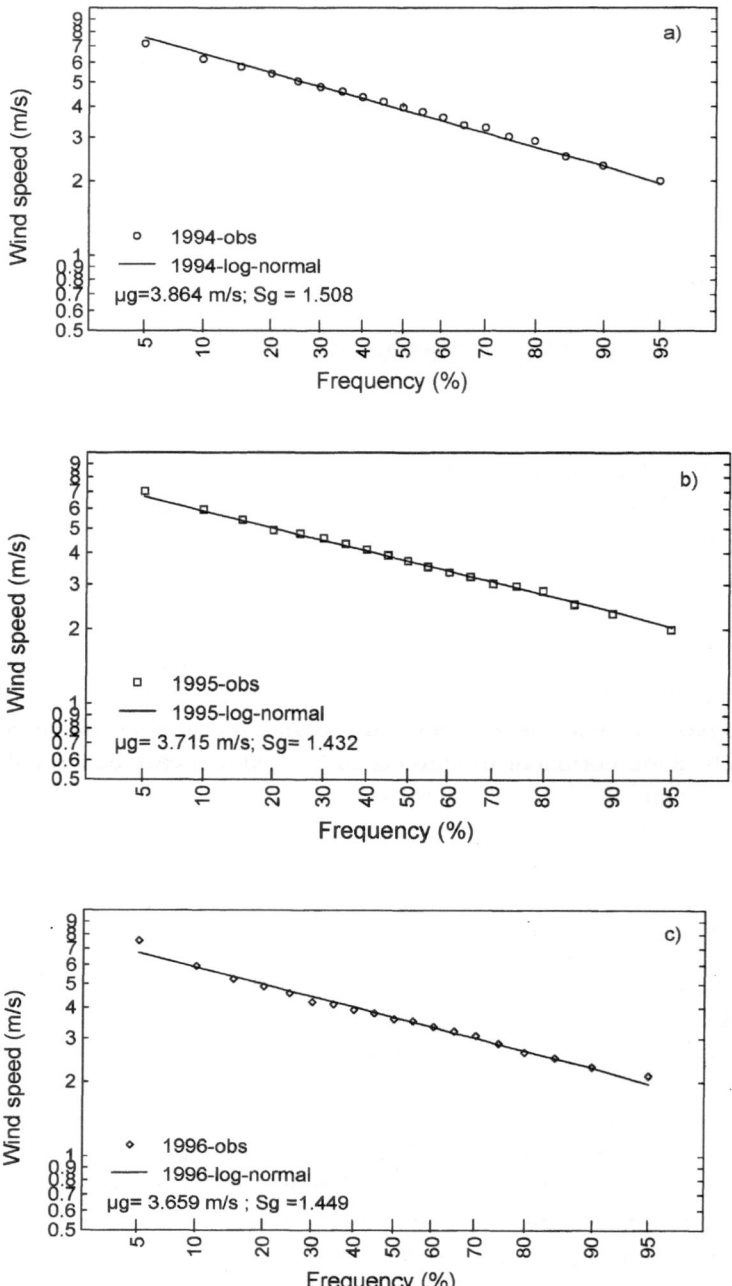

Figure 1. Cumulative frequency distribution for 8-h average wind-speed (from 8 a.m. to 4 p.m.) registered during the same days of 1994 to 1998 considered in CO data. The line represents the two-parametric log-normal distribution, μg is the geometric mean and Sg is the standard geometric deviation.

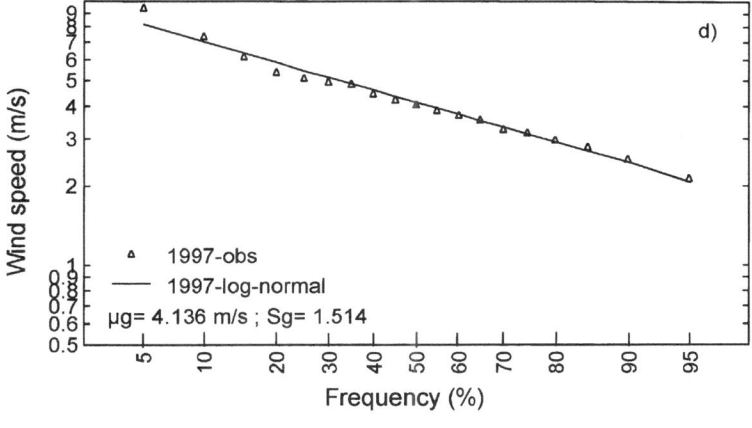

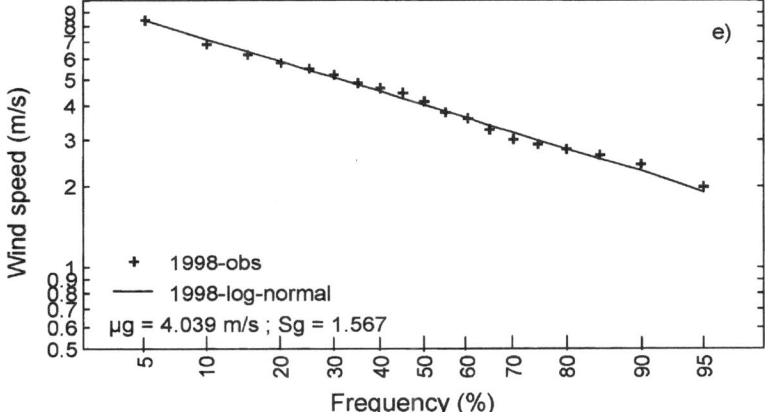

Figure 1. (continued)

3. Values for K

The values of $K = C_p u_{(100-p)}$ (from Equation (3)) were calculated for each percentile p varying from 5 to 95% every 5%. The variation of K with p-percentile for years 1994 to 1998 is shown in Figure 3. It can be seen that at percentiles smaller than 80%, relative bias between the values of K calculated for 1994, 1995, 1996 and 1997 is less than 20%. The coefficients a, b of the fitted regression lines of $K(p) = a + bp$, and its correlation coefficient (R) obtained from ordinary linear regression analysis considering $p \leq 80\%$, for 1994, 1995, 1996, 1997 and 1998, are included in Table I.

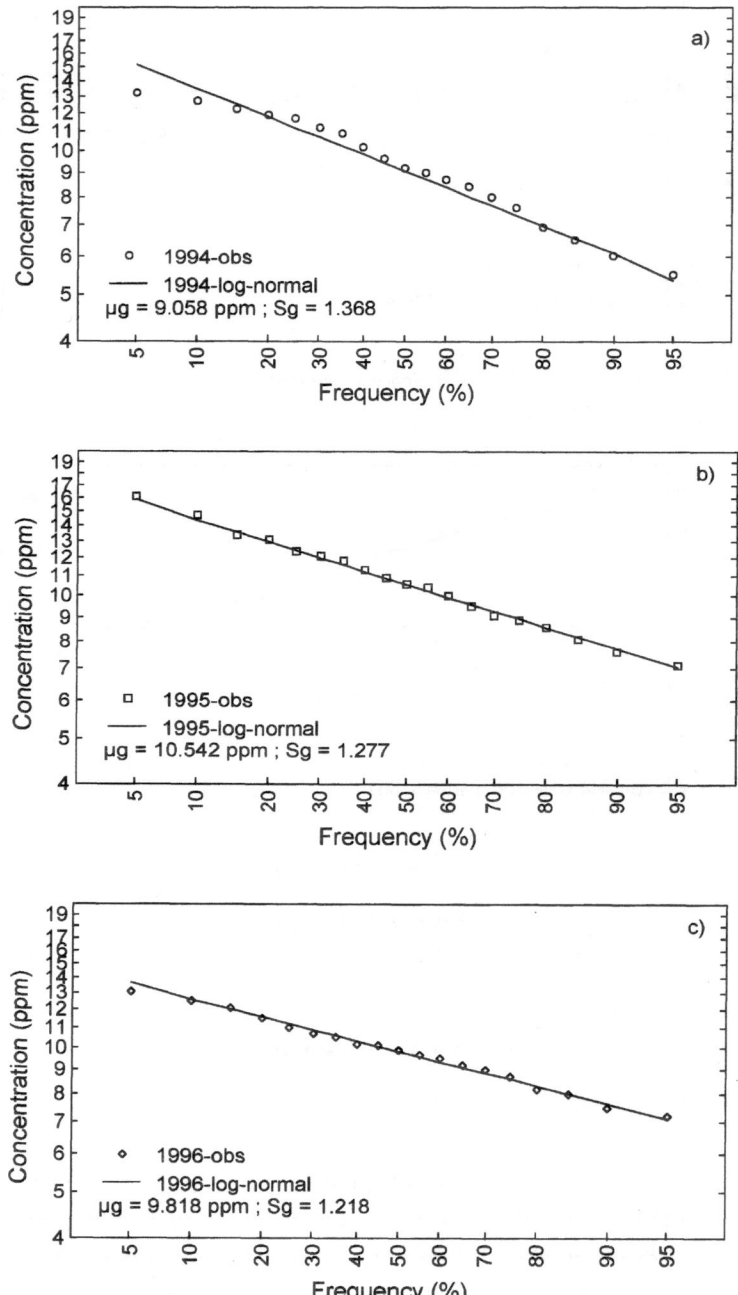

Figure 2. Cumulative frequency distribution for 8-h average CO concentration (from 8 a.m. to 4 p.m.) observed during the working days of 1994 to 1998. The line represents the two-parametric log-normal distribution, μg is the geometric mean and Sg is the standard geometric deviation.

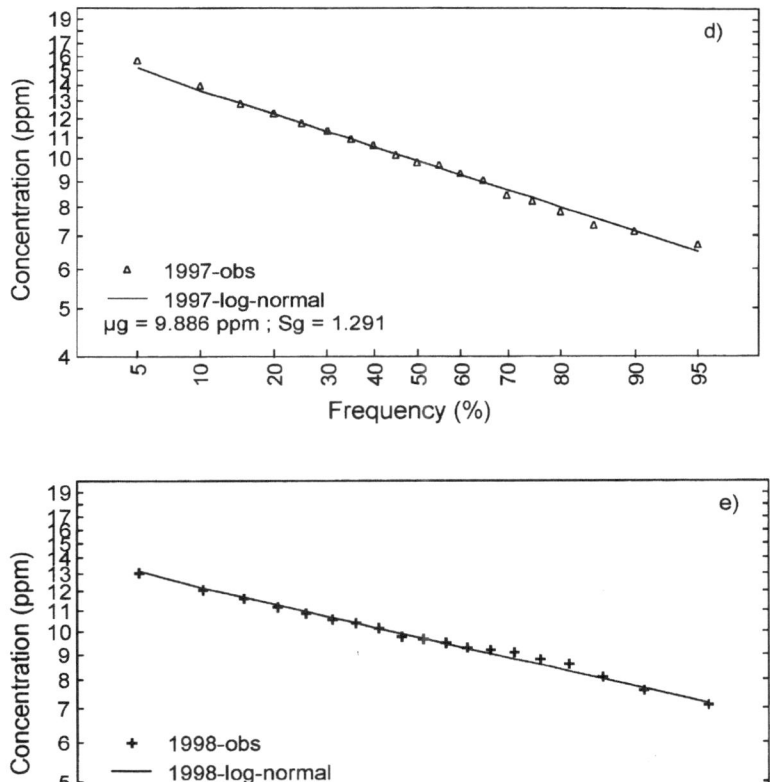

Figure 2. (continued)

Substituting the regression line defined by $K(p) = a + bp$, into Equation (3) gives:

$$C_p = \frac{a + bp}{u_{(100-p)}} .$$ (4)

The aim is to evaluate the performance of Equation (4), with coefficients a and b obtained for a given year (Table I), in estimating the cumulative frequency distribution (for $p \leq 80\%$) of 8-h CO concentration for the other four years, when the cumulative frequency distributions of mean wind-speed for those years are known. To illustrate, it may be useful to mention as an example, that we use K_{94} ($a = 30.128$ and $b = 0.124$, obtained from 1994 data) in Equation (4) to estimate the values of C_p for 1995, 1996, 1997 and 1998 from wind speed data of 1995, 1996, 1997 and 1998, respectively. Figures 4a–e show quantile-quantile plots of observed (C_p-obs)

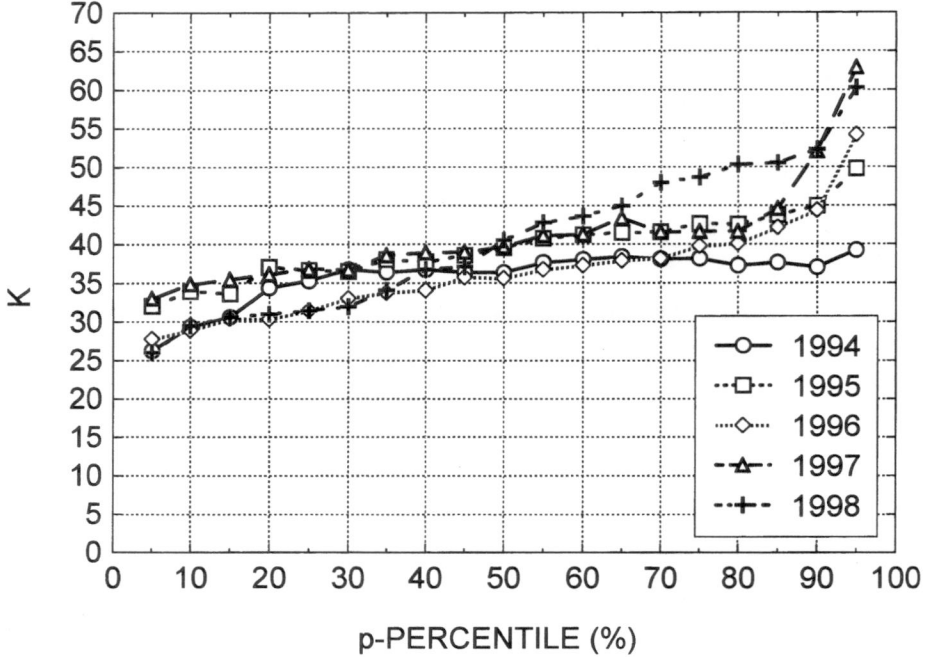

Figure 3. Variation of *K* with percentile for 1994 to 1998 computed from Equation (3).

and estimated (C_p-est) concentrations values considering the regression line for K_{94}, K_{95}, K_{96}, K_{97} and K_{98} in Equation (4), respectively. From plots on Figure 4, it can be seen that Figure 4a shows the best estimation of CO frequency distributions. This means that the use of expression K_{94} in Equation (4) gives better estimation of CO distributions for 1995, 1996, 1997 and 1998 than the use of K_{95} (Figure 4b) in Equation (4) to estimate CO distributions for 1994, 1996, 1997 and 1998, or to consider K_{96} (Figure 4c) in estimating CO distributions for 1994, 1995, 1997 and 1998, or to include K_{97} in Equation (4) (Figure 4d) to estimate CO distributions for 1994, 1995, 1996 and 1998, or using K_{98} (Figure 4e) to estimate CO distributions for 1994, 1995, 1996 and 1997.

Table II includes the statistical parameters (Hanna, 1993; Olesen, 1995) used in the statistical evaluation of the performance of Equation (4) in estimating p-percentile concentrations, with the different forms of $K (= a + bp)$ (Figures 4a–e).

Table III summarises the values of the statistical parameters included in Table II. Though results of Equation (4) considering the expression for K_{98} have the smallest bias, fractional bias (FB) and mean relative error (MRE), they show the lowest correlation coefficient (R) and the highest normalised mean square error (NMSE) and fractional bias (FS). These statistical measures and results in Figure 4e suggest that the performance of Equation (4) with K_{98} is relatively poor.

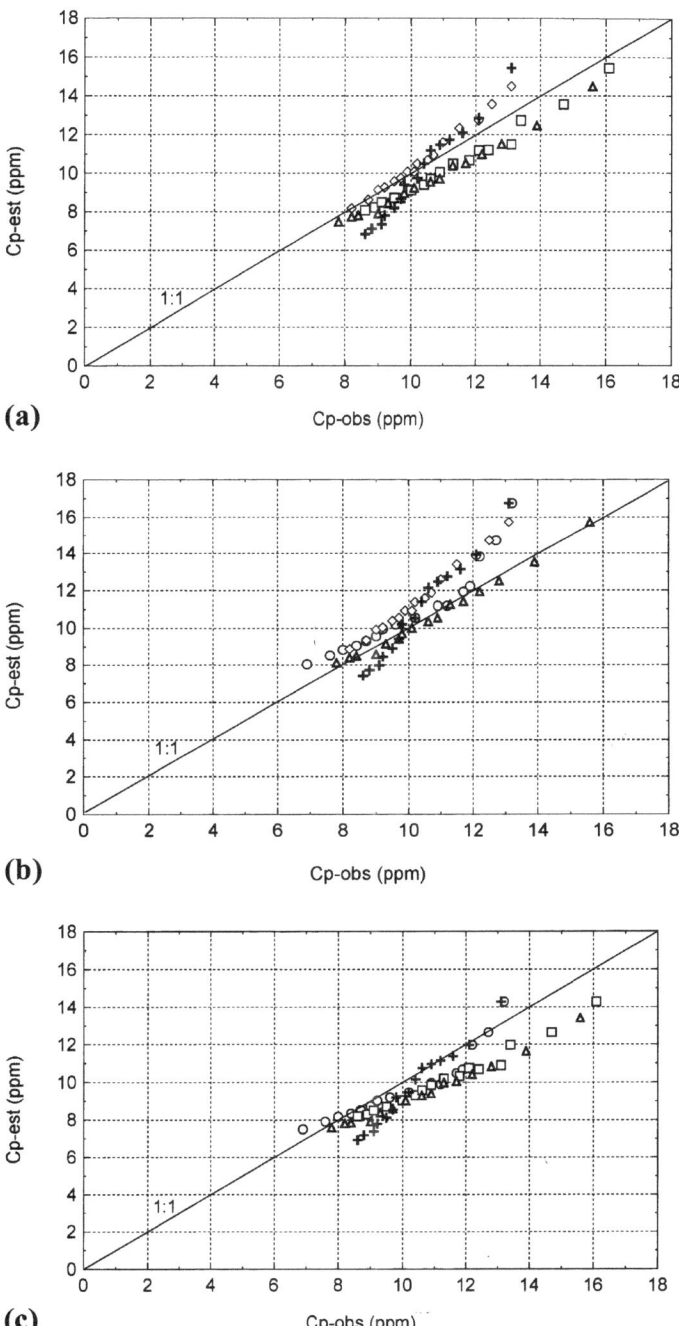

Figure 4a-c. Quantile-quantile plots of concentrations. Observed and estimated concentrations occurring at the percentile p ($p \leq 80\%$). Cp-est calculated using Equation (4) with: (a) K_{94}; (b) K_{95};
(c) K_{96}. ($\bigcirc$ 1994; $\square$ 1995; $\Diamond$ 1996; $\triangle$ 1997; $+$ 1998).

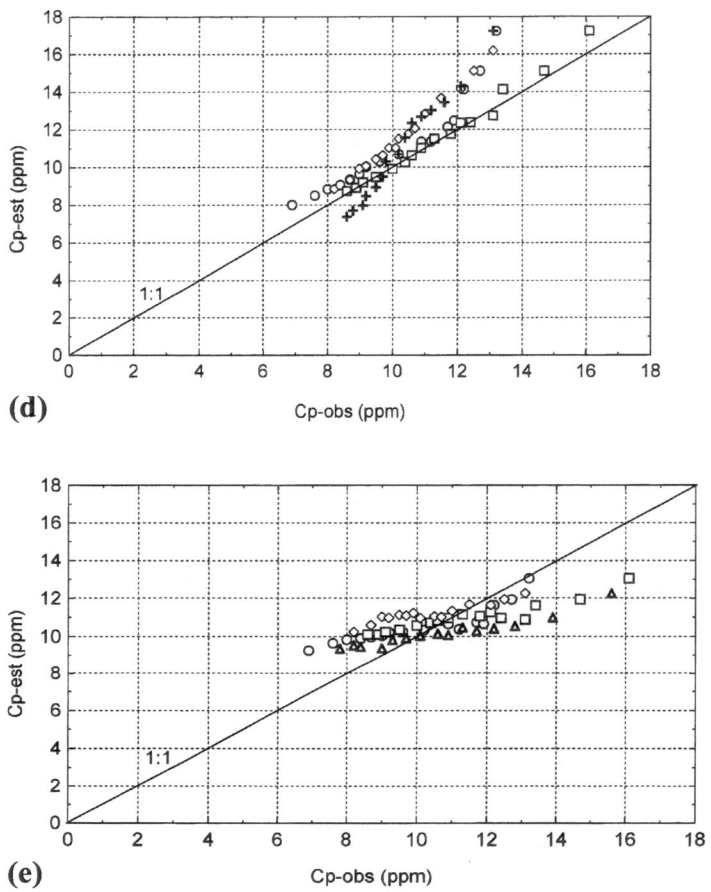

Figure 4d–e. Quantile-quantile plots of concentrations. Observed and estimated concentrations occurring at the percentile p ($p \leq 80\%$). Cp-est calculated using Equation (4) with: (d) K_{97}; (e) K_{98}. (○ 1994; □ 1995; ◊ 1996; △ 1997; + 1998).

The statistics shown in Table III reveal that performances of Equation (4) including K_{94}, K_{95}, K_{96}, or K_{97} are very similar. However, it appears that the results using K_{94} are slightly better. The results with K_{94} show the smallest bias, FB and MRE. Equation (4) with K_{94} gives a good estimation of the cumulative frequency distribution for concentrations of 1995, 1996, 1997 and 1998 with a mean relative error of +4.5%, indicating that C_p values are slightly underestimated.

Comparison of cumulative frequency distribution of 8-h CO concentration estimated using Equation (4) with K_{94} with the observed one for 1995, 1996, 1997 and 1998 are shown in Figure 5 (for $p \leq 80\%$).

Table IV shows the estimated (using $K = a + bp$ for 1994, in Equation (4)) and observed frequencies of 8-h CO concentration greater than 9 ppm for 1995, 1996,

TABLE II

Statistical parameters used in the statistical evaluation of the performance of Equation (4) in estimating C_p (C_{obs} is the observed value, C_{est} is the estimated value, the over bar indicates an average, σ_{Cobs} and σ_{Cest} are the standard deviations of observed and estimated quantities, respectively)

Bias	BIAS	$= \overline{(C_{obs} - C_{est})}$
Normalised mean square error	NMSE	$= \dfrac{\overline{(C_{obs} - C_{est})^2}}{\overline{C_{obs}}\,\overline{C_{est}}}$
Fractional bias	FB	$= \dfrac{\overline{C_{obs}} - \overline{C_{est}}}{0.5\left(\overline{C_{obs}} + \overline{C_{est}}\right)}$
Fractional variance	FS	$= \dfrac{\sigma_{Cobs} - \sigma_{Cest}}{0.5\left(\sigma_{Cobs} + \sigma_{Cest}\right)}$
Correlation coefficient	R	$= \dfrac{\overline{\left(C_{obs} - \overline{C_{obs}}\right)\left(C_{est} - \overline{C_{est}}\right)}}{\sigma_{Cobs}\sigma_{Cest}}$
Mean relative error	MRE	$= \overline{\left(\dfrac{C_{obs} - C_{est}}{C_{obs}}\right)} \times 100\%$

1997 and 1998. For the entire period 1995–1998, the relative error in estimating the frequency of 8-h CO concentration greater than 9 ppm is less than 11%.

TABLE III

Statistics for percentile concentrations (C_p) estimated using Equation (4) with the regression lines of K obtained for 1994 to 1998 (see Table I). NMSE: normalised mean square error; FB: fractional bias; FS: fractional variance; R: correlation coefficient; MRE: mean relative error (see Table II for definitions)

K in Equation (4)	BIAS	NMSE	FB	FS	R	MRE
K_{94}	0.453	0.0079	0.043	−0.134	0.915	+4.5
K_{95}	−0.638	0.0119	−0.060	−0.287	0.912	−5.8
K_{96}	0.873	0.0126	0.086	0.078	0.917	+8.0
K_{97}	−0.808	0.0146	−0.074	−0.320	0.917	−7.2
K_{98}	−0.065	0.0162	−0.006	0.797	0.796	−2.8

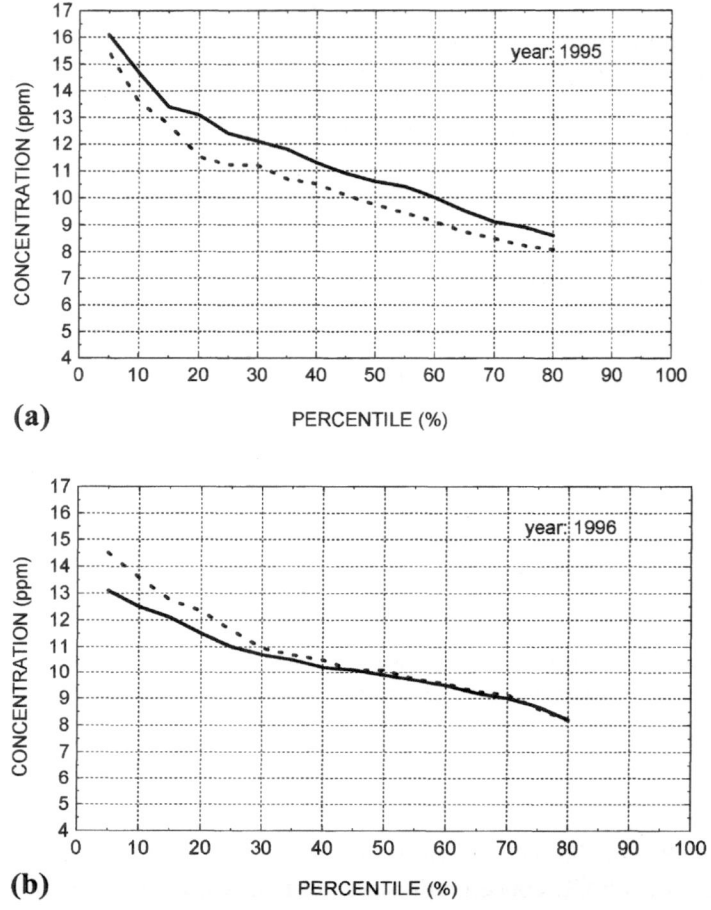

Figure 5. Observed and estimated frequency distribution for 8-h average CO valid for $p \leq 80\%$.
_____ observed; $- - - - - -$ Equation (4) with coefficients a and b of K_{94}.

4. Conclusions

The frequency distribution of an air pollutant concentration (C) is usually the result of complex phenomena. From data obtained at one site downtown in Buenos Aires city, we found that wind-speed and CO concentration are approximately log-normally distributed. Simpson *et al.* (1985) suggested that irrespective to the statistical distributions of air pollution data and wind speed, if $C = K/u$, then the C value corresponding to the p-percentile (C_p) and the u value corresponding to the $(100 - p)$ percentile (u_{100-p}) are related by $K = C_p u_{(100-p)}$. The equation of several straight lines between K and the p-percentile were obtained from cumulative frequency distributions for 8-h CO concentration and 8-h mean wind-speed data collected at this site during 1994 to 1998. The fitted curves were valid for $p \leq$

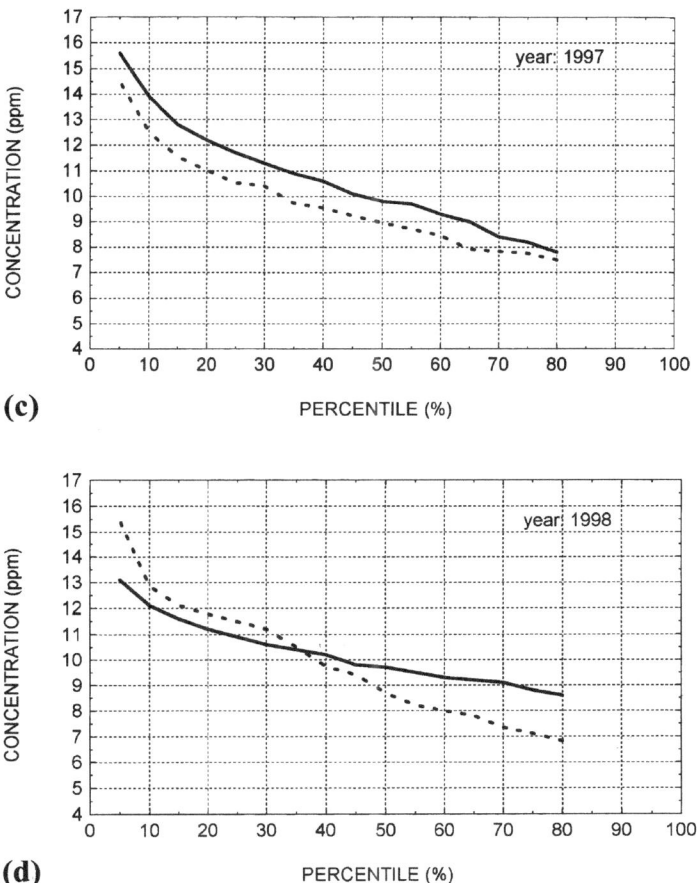

(c)

(d)

Figure 5. (continued)

TABLE IV

Observed and estimated frequency of 8-h CO concentrations greater than 9 ppm for 1995, 1996, 1997 and 1998. The estimations are obtained considering K_{94} in Equation (4) and using the statistical distribution for wind-speed of each year

	Frequency	
	$(C \geq 9 \text{ ppm})_{(obs)}$	$(C \geq 9 \text{ ppm})_{(est)}$
1995	71%	62%
1996	70%	72%
1997	64%	50%
1998	70%	48%

80%. From our data, the expression of $K(p)$ obtained form data observed during 1994, K_{94}, gives the best estimation of cumulative frequency distribution of 8-h CO concentration at the location, for 1995, 1996, 1997 and 1998. Statistical evaluation indicates that the use of K_{94} may slightly underestimate the concentration (C_p) occurring at percentile p, with a mean relative error of 4.5%. For the four-year period (1995–1998) the relative error in the estimation of frequency of 8-h CO concentration greater than 9 ppm is less than 11%. The expression of $K(p)$ is valid only for this site. However, if wind speed data are available, a new form of $K(p)$ can be obtained from one year of air pollution concentration at another site. Afterwards, if air pollution concentrations at that site are not available and the emission conditions remain within little variation, $K(p)$ can be used to obtain an estimation of the cumulative frequency distribution for air pollutant concentration, using wind speed data. It is also necessary to stress that the methodology described in this article is applicable at a city located in flat terrain, when only vehicle emissions are taken into account.

Acknowledgements

This work was supported by the University of Buenos Aires (X093) and the National Scientific and Technological Research Council (PIP 0424/98). Wind-data have kindly been supplied by the National Weather Service of Argentina.

References

Arya, S. P.: 1999, *Air Pollution Meteorology and Dispersion*, Oxford University Press, Inc., New York, 310 pp.

Bencala, K. E. and Seinfeld, J. H.: 1976, 'On frequency distributions of air pollutant concentrations', *Atmos. Environ.* **10**, 941–950.

Hanna, S. R.: 1993, 'Uncertainties in air quality model predictions', *Bound.-Layer Meteor.* **62**, 3–20.

Olesen, H. R.: 1995, 'The model validation exercise at Mol: Overview of results', *Internat. J. Environ. Pollut.* **5**, 761–784.

Simpson, R. W., Daly, N. J. and Jakeman, A. J.: 1983, 'The prediction of maximum air pollution concentrations for TSP and CO using Larsen's model and the ATDL model', *Atmos. Environ.* **17**, 2497–2503.

Simpson, R. W., Jakeman, A. J. and Daly, N. J.: 1985, 'The relationship between the ATDL model and the statistical distributions of wind-speed and pollution data', *Atmos. Environ.* **19**(1), 75–82.

Venegas, L. E. and Mazzeo, N. A.: 2000, 'Carbon monoxide concentrations in a street canyon of Buenos Aires City (Argentina)', *Environ. Monit. Assess.* **65**, 417–424.

WHO: 1980, *Environmental Health Criteria 13-Carbon Monoxide*, World Health Organisation, 135 pp.

AN EVALUATION OF DAUMOD MODEL IN ESTIMATING URBAN BACKGROUND CONCENTRATIONS

LAURA E. VENEGAS* and NICOLÁS A. MAZZEO

National Scientific and Technological Research Council (CONICET), Department of Atmospheric Science and the Oceans, Faculty of Sciences, University of Buenos Aires, Ciudad Universitaria, Pab. 2, Buenos Aires, Argentina

(* author for correspondence, e-mail: venegas@at.fcen.uba.ar, fax: (+54 11) 4576 3356)

Abstract. This article presents an evaluation of the performance of the urban atmospheric dispersion model (DAUMOD) in estimating nitrogen oxides (NO_x) background concentrations in Copenhagen. Estimations of hourly average (averaged over a year), mean daily and mean monthly concentrations of NO_x are compared with observed values for two years of data. The model slightly underestimates low hourly average values and overestimates high values. The cumulative frequency distribution of mean daily concentration obtained from model estimations is in good agreement with the obtained from observed data. We performed a statistical analysis to determine the agreement between estimated and observed concentration values. The results show that 95.8% of hourly average estimations, 86.8% of mean daily and 100% of monthly average concentrations are within a factor of two of the observed values. The normalised mean square error of predictions is +0.13 for hourly average estimations, +0.22 for mean daily values and +0.02 for monthly mean concentrations. The fractional bias values are: −0.049 for hourly mean estimations, −0.047 for mean daily values and −0.053 for monthly average estimations. The values of the statistical parameters allow us to consider that though estimations are lightly larger than the observed values, the model performance is acceptable.

Keywords: model performance evaluation, pollution concentrations, urban air quality predictions, urban background estimates, urban dispersion modelling

1. Introduction

The deteriorated state of air quality in many urban centres is well known to everybody. The air· pollution level in cities often is dominated by a countless number of small, not-well-determined emission sources. Within the framework of an urban air quality management system, atmospheric dispersion models provide a link between the source emissions and the ambient concentrations. In the urban environment, such models may consider two different scales: the urban scale that covers the whole domain of a city and the micro-scale that covers a single street canyon or up to a few city blocks. Dispersion models for the urban scale provide information about the urban background concentrations. These models usually combine most of the small sources that exist in an urban area, into larger area sources, and assume that emissions are uniform over that particular area. In the last decades a number of models have been developed to predict the dispersion

Water, Air, and Soil Pollution: Focus **2:** 433–443, 2002.
© 2002 *Kluwer Academic Publishers.*

of pollutants in urban atmospheres. These models range from simple empirical models to complex three-dimensional urban airshed models (see Arya, 1999).

Sometimes, in practical applications, the quality of input data, for example emission data, is not sufficient in order to justify application of the very complex numerical tools. In these cases, simple urban background pollution models become an alternative that can provide as good results as computations with more sophisticated models (Hanna, 1971; Berkowicz, 2000).

Since Lucas' (1958) early study, a number of simple urban air pollution models were proposed (Turner, 1964; Gifford and Hanna, 1973; Turner et al., 1985). In recent years, several authors have developed different modelling systems for predicting atmospheric dispersion of pollution in an urban area, some of them, applying the basic principles of the former simple models. Differences can be in how the area source summation is carried out, or how various meteorological and dispersion parameters are included, or if chemistry reactions and removal processes are taken into account or not. Descriptions, applications or validations of some of these models have been reported by de Haan et al. (1998), Härkönen et al. (1998), Luhana and Sokhi (1998), Berkowicz (2000), Brechler (2000), Owen et al. (2000), de Haan et al. (2001), Kousa et al. (2001).

The urban dispersion model DAUMOD is a simple non-gaussian model that estimates background air pollution concentrations in urban areas. It has been developed and reported over last decade (Mazzeo and Venegas, 1991; Venegas and Mazzeo, 1998). The earliest version of this model estimates mean annual concentrations using climatological data as input and assuming neutral atmospheric conditions. The results obtained from this former version of the model were very encouraging so some improvements were made. The actual version of DAUMOD model uses routine hourly meteorological data and it is run hourly. In this way, the hourly variation of emission rate can also be considered. This version also takes into account hourly atmospheric stability condition. It incorporates expressions to compute the parameters included in the model knowing the surface roughness length and the Monin-Obukhov length.

The objective of this article is to apply and to evaluate the last version of DAUMOD model. We apply DAUMOD to estimate urban background hourly nitrogen oxide (NO_x) concentrations in Copenhagen. In order to evaluate the performance of this version of the model, we compare estimated hourly average (averaged over a year), mean daily and mean monthly concentration with the values obtained from two years of hourly measurements of NO_x concentrations at an urban background station in Copenhagen.

2. Methodology

The urban air quality model (DAUMOD) is based on the equation of mass continuity. Steady state conditions are assumed. If the effluents are emitted continuously from surface level with source strength (Q) expressed as mass per unit area per unit time, the concentration ($C(x, z)$) satisfies the lower boundary condition:

$$K(z) \left. \frac{\partial C}{\partial z} \right|_{z=0} = -Q \,, \tag{1}$$

where x-axis is in the direction of the mean wind, z-axis is vertical and $K(z)$ is the vertical eddy diffusivity for contaminants. On the other hand, the top boundary of the model (h) coincides with the upper boundary of the plume of contaminants. Assuming that there is no transport of mass through the upper limit of the plume, the upper boundary condition is given by:

$$K(z) \left. \frac{\partial C}{\partial z} \right|_{z=h} = 0 \,. \tag{2}$$

In the development of the model, vertical variation of concentration at a given distance $[C(x, z)]$ is fitted to polynomial expressions (Mazzeo and Venegas, 1991):

$$C(x, z) = C(x, 0) \sum_{i=0}^{6} A_i \left(\frac{z}{h} \right)^i \,, \tag{3}$$

where coefficients A_i depend on atmospheric stability conditions.

The height of the top boundary (h) can be obtained integrating the continuity equation:

$$\int_0^x Q dx = \int_0^h u(z) C(x, z) dz \,, \tag{4}$$

with $C(x, z)$ given by Equation (3) and $u(z)$ is the diabatic wind profile (Arya, 1988). The variation of $h(x)$ obtained from Equation (4) can be fitted with great accuracy to potential functions given by:

$$\frac{h}{z_0} = a \left(\frac{x}{z_0} \right)^b \tag{5}$$

where z_0 is the surface roughness length and a and b depend on atmospheric stability (Mazzeo and Venegas, 1991). From Equation (1), considering the vertical eddy diffusivity profile depending on atmospheric stability (Arya, 1988) and Equation (3) the following expression is obtained

$$C(x, 0) = \frac{Q h(x)}{|A_1| k u^* z_0} \,, \tag{6}$$

where u^* is the friction velocity and k is the von Kármán's constant.

In an urban area, we may assume horizontal distribution of area sources with strength $Q(x)$, varying according to a typical grid pattern. Substituting Equation (5) in Equation (6) and assuming that each square grid emission strength is given by $Q_i (i = 0, 1, 2, \ldots, N)$, the ground level air pollution concentration distribution can be estimated by:

$$C(x, 0) = \frac{a \left[Q_0 x^b + \sum_{i=1}^{N} (Q_i - Q_{i-1})(x - x_i)^b \right]}{\left(|A_1| \, k z_0^b u_* \right)}. \tag{7}$$

The model is discussed elsewhere (Mazzeo and Venegas, 1991; Venegas and Mazzeo, 1998) though some improvements have been made that include the use of Monin-Obukhov length (L) to describe hourly atmospheric stability. The expressions of $a(z_0/L), b(z_0/L)$ and $A_1(z_0/L)$ included in the present version of DAUMOD are:

(i) $a(z_0/L)$,

$z_0/L < -10^{-4}$ a $= 3.618833 + 0.2369076 * \ln(|z_0/L|)$
$-10^{-4} \leq z_0/L \leq 10^{-4}$ a $= -384.73(z_0/L) + 1.4$
$10^{-4} < z_0/L$ a $= 0.6224632 + 7.37387 \times 10^{-5} / \ln[(z_0/L) + 1]$

(ii) $b(z_0/L)$,

$z_0/L < -10^{-4}$ b $= 0.5356147 + 0.0234187 * \ln[(|z_0/L|) + 0.01]$
$-10^{-4} \leq z_0/L \leq 10^{-4}$ b $= -130.0(z_0/L) + 0.415$
$10^{-4} < z_0/L$ b $= 0.5065736 - 1.196137 / \ln[2802.315 + 9/(z_0/L)]$

(iii) $|A_1|(z_0/L)$,

$z_0/L < -10^{-4}$ $|A_1|$ $= 9.254667 + 0.8043134 * \ln(|z_0/L|)$
$-10^{-4} \leq z_0/L \leq 10^{-4}$ $|A_1|$ $= -3853.31 + 1.461$
$10^{-4} < z_0/L$ $|A_1|$ $= 0.05478233 + 0.0001021171 / \ln((z_0/L) + 1)$.

This version of DAUMOD can be run using hourly meteorological data as input. To evaluate the performance of DAUMOD model in estimating NO_x background concentrations in Copenhagen, we used a typical diurnal variation of NO_x emission data for this city subdivided into a grid net with a resolution of 2 km×2 km and covering most of the Great Copenhagen. Hourly meteorological data (wind-speed and direction, global radiation, air temperature) and measured NO_x concentration values are taken at an urban background station located on the roof of a building at the centre of the grid. In addition, the contribution from regional sources or regional background is obtained from the measured concentrations at a rural monitoring station located at about 25 km west of Copenhagen. Two years (1994 and 1995) of

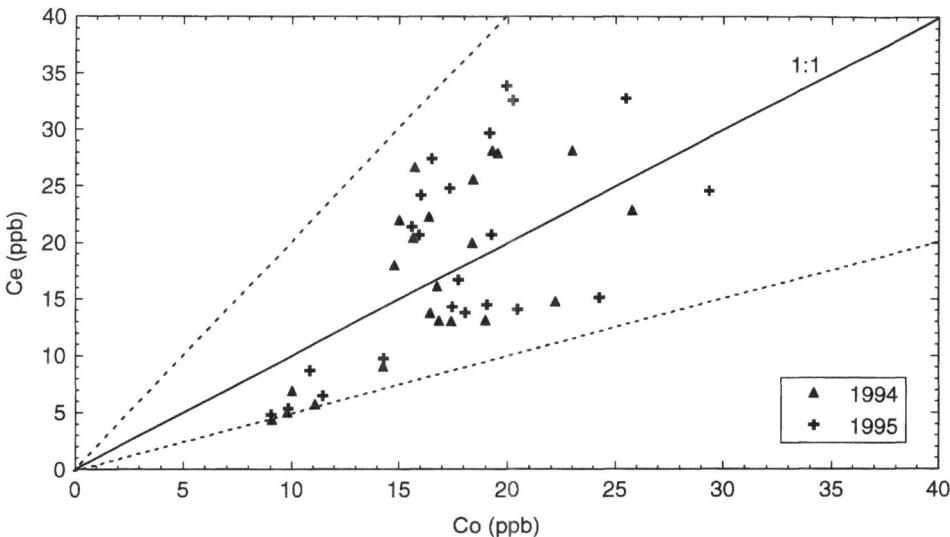

Figure 1. Scatter plot of hourly average NO$_x$ concentrations (Ce: estimated; Co: observed). The lines 1:1/2, 1:1 and 1:2 are included.

data are used for computations, kindly made available to us by Dr. R. Berkowicz from the National Environmental Research Institute (Roskilde, Denmark).

The 'observed' value results after the elimination of regional pollution contribution from measured concentration at the city. If wind blows from the western side of the city, regional contribution is given by measured concentrations at the rural station. If wind blows from any other sector, no regional contribution is assumed.

3. Results and Discussion

3.1. HOURLY AVERAGE CONCENTRATIONS

The comparison of hourly average (averaged over a year) values of NO$_x$ concentrations estimated by the model (Ce) with 'observed' (Co) values is shown in Figure 1 for the two years. The model underestimates the low values and slightly overestimates the high ones. The following statistical parameters (Hanna *et al.*, 1991; Olesen,

TABLE I

Statistical measures obtained for DAUMOD hourly average estimations when comparing with observations

Model	mean	sigma	bias	nmse	cor	fa2	fb	fs
All observations (n = 48)								
OBS	16.70	4.62	–	–	–	–	–	–
DAUMOD	17.53	8.45	–0.84	0.13	0.691	0.958	–0.049	–0.587
Mean hourly 1994 data (n = 24)								
OBS	16.20	4.26	–	–	–	–	–	–
DAUMOD	16.63	7.81	–0.44	0.12	0.722	0.917	–0.027	–0.588
Mean hourly 1995 data (n = 24)								
OBS	17.20	4.90	–	–	–	–	–	–
DAUMOD	18.43	8.95	–1.23	0.15	0.661	1.000	–0.069	–0.586

n = Number of estimations, mean (ppb), sigma (ppb), bias (ppb), nmse: normalised mean square error; cor: correlation coefficient; fa2: fraction within a factor of two; fb: fractional bias; fs: fractional variance.

1995) concerning the evaluation of model performance when compared against the observations are included in Table I.

$$-\text{bias} = \overline{(C_o - C_e)}$$

$$-\text{normalised mean square error (nmse)} = \frac{\overline{(C_o - C_e)^2}}{\overline{C_o}\,\overline{C_e}}$$

$$-\text{correlation coefficient (cor)} = \frac{\overline{(C_o - C_o)(C_e - C_e)}}{\sigma_{Co}\sigma_{Ce}}$$

$$-\text{fa2 is the fraction of data for which} \quad 0.5 \le Ce/Co \le 2.0$$

$$-\text{fractional bias (fb)} = \frac{\overline{C_o} - \overline{C_e}}{0.5(\overline{C_o} + \overline{C_e})}$$

$$-\text{fractional variance (fs)} = \frac{\sigma_{Co} - \sigma_{Ce}}{0.5(\sigma_{Co} + \sigma_{Ce})}$$

where σ_{Co} and σ_{Ce} are the standard deviations of observed and estimated quantities, respectively, and the over bar indicates an average.

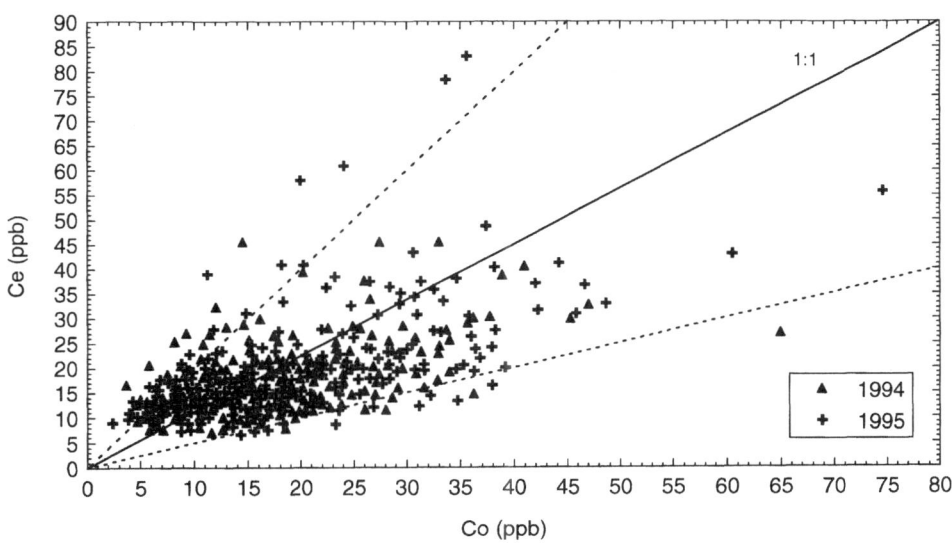

Figure 2. Scatter plot of mean daily NO$_x$ concentrations (Ce: estimated; Co: observed). The lines 1:1/2, 1:1 and 1:2 are included.

3.2. MEAN DAILY CONCENTRATIONS

Mean daily NO$_x$ concentrations are obtained from hourly estimations. Figure 2 includes the scatter plot of mean daily concentrations estimated for 1994 and 1995. The statistical measures obtained from the comparison between estimated and 'observed' values are included in Table II. Mean values of estimated and 'observed' data are slightly different. The bias and fractional bias (both negatives) indicate a tendency towards overprediction. However, considering all data set (680 values), 86.8% of the estimations are within a factor of two the observed values. Figure 3 displays the quantile-quantile plot of mean daily concentrations of NO$_x$ obtained for both years. In this case, the model overestimates low mean daily concentrations and underestimates high values.

3.3. AVERAGE MONTHLY CONCENTRATIONS

Average monthly NO$_x$ concentrations are computed from hourly values. Figure 4 shows the results for both years. Table III includes the statistical measures obtained from these data. Statistics reveal a tendency for the model to slightly overestimate the observed average monthly concentrations. However, all of the estimations are within a factor of two the observed values. The normalised mean square error and the fractional bias show low values.

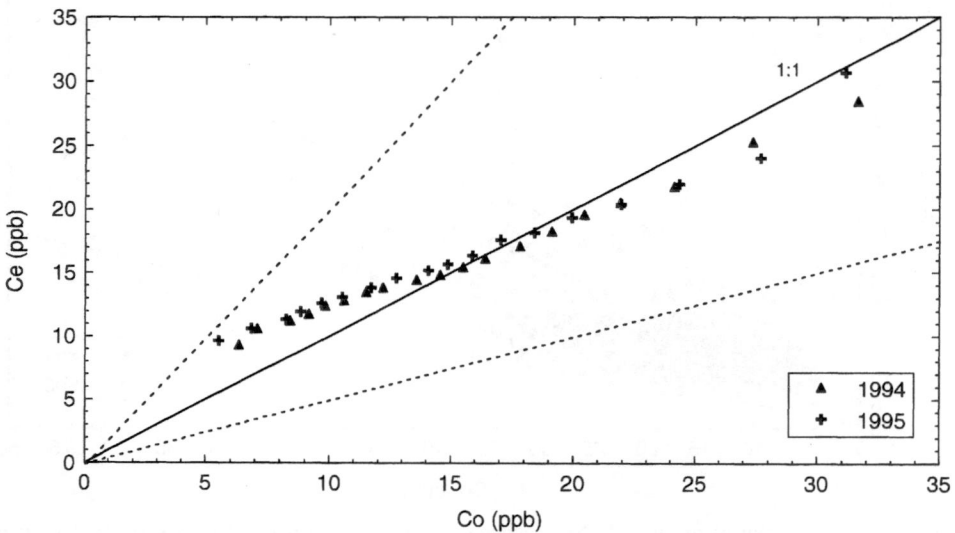

Figure 3. Quantile-quantile plot of mean daily NO$_x$ concentrations (Ce: estimated; Co: observed). The lines 1:1/2, 1:1 and 1:2 are included.

TABLE II

Statistical measures obtained for DAUMOD mean daily estimations when comparing with observations

Model	mean	sigma	bias	nmse	cor	fa2	fb	fs
All observations (n = 680)								
OBS	16.63	9.20	–	–	–	–	–	–
DAUMOD	17.42	8.19	–0.79	0.22	0.589	0.868	–0.047	0.116
Daily 1994 data (n = 346)								
OBS	16.11	8.28	–	–	–	–	–	–
DAUMOD	16.52	6.34	–0.41	0.20	0.529	0.884	–0.025	0.265
Daily 1995 data (n = 334)								
OBS	17.16	10.04	–	–	–	–	–	–
DAUMOD	18.35	9.66	–1.19	0.24	0.625	0.850	–0.067	0.039

n = Number of estimations, mean (ppb), sigma (ppb), bias (ppb), nmse: normalised mean square error; cor: correlation coefficient; fa2: fraction within a factor of two; fb: fractional bias; fs: fractional variance.

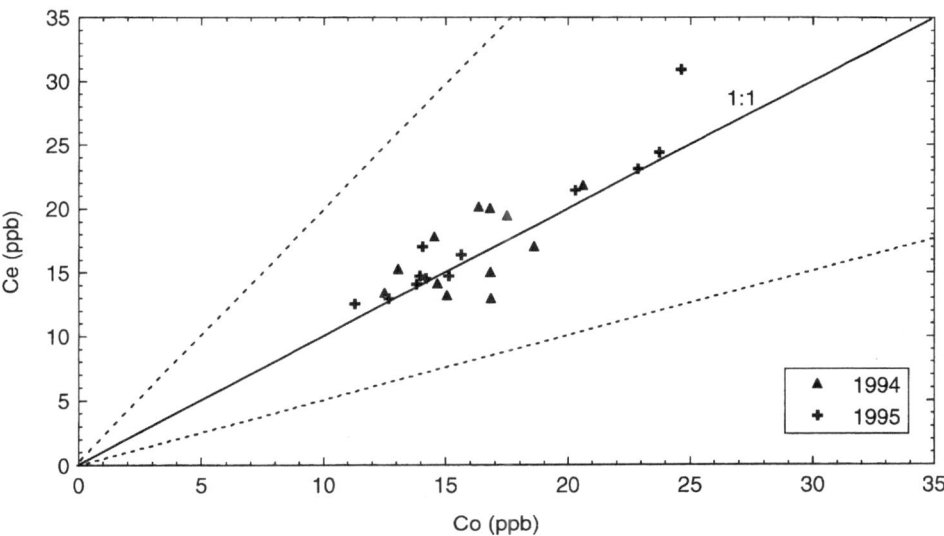

Figure 4. Scatter plot of average monthly NO$_x$ concentrations (Ce: estimated; Co: observed). The lines 1:1/2, 1:1 and 1:2 are included.

TABLE III

Statistical measures obtained for DAUMOD average monthy estimations when comparing with observations

Model	mean	sigma	bias	nmse	cor	fa2	fb	fs
All observations (n = 24)								
OBS	16.49	3.55	–	–	–	–	–	–
DAUMOD	17.38	4.44	–0.89	0.02	0.885	1.000	–0.053	–0.223
Monthly 1994 data (n = 12)								
OBS	16.12	2.19	–	–	–	–	–	–
DAUMOD	16.69	2.98	–0.57	0.02	0.620	1.000	–0.035	–0.307
Monthly 1995 data (n = 12)								
OBS	16.86	4.48	–	–	–	–	–	–
DAUMOD	18.08	5.43	–1.21	0.01	0.957	1.000	–0.070	–0.191

n = Number of estimations, mean (ppb), sigma (ppb), bias (ppb), nmse: normalised mean square error; cor: correlation coefficient; fa2: fraction within a factor of two; fb: fractional bias; fs: fractional variance.

4. Conclusions

The present version of the urban air pollution model (DAUMOD) can be used to estimate hourly concentrations using hourly meteorological and emission data as input. From the evaluation of model performance in estimating NO_x concentration in Copenhagen during 1994–1995, we found:

- the fractional bias of the estimated hourly average values (fb = –0.049) indicates the overall tendency of the model to slightly overestimate these values,
- the scatter plot shows that the model underestimates low hourly average concentrations and overestimates the high values,
- cumulative frequency distribution of mean daily concentration obtained from model estimations is in good agreement with the obtained from observed data,
- the estimations of mean daily concentrations are acceptable, as 86.8% of them are within a factor of two of the observed values,
- the fractional bias of the estimated and measured 24-hr averaged concentration values is –0.047 and the normalised mean square error is +0.22,
- estimated average monthly concentrations of NO_x are in good agreement with the observed values as all estimations are within a factor of two of the observed ones and the normalised mean square error of estimations and observations is very low (= +0.02).

The performance of DAUMOD model in estimating hourly average and mean daily concentrations is good and it improves when estimating monthly average values.

DAUMOD is a simple non-gaussian urban dispersion model that can be used to estimate horizontal distributions of urban background concentrations of pollutants emitted from area sources. This model may be a useful tool to be incorporated in an urban air quality management system.

Acknowledgements

The authors would like to thank Dr. Ruwim Berkowicz for kindly providing the data sets used in this study and also for his valuable comments.

This work has been supported by UBACYT-Project X093 and CONICET – PIP 0424/98.

References

Arya, S. P.: 1988, *Introduction to Micrometeorology*, Academic Press, Inc., San Diego, California, 303 pp.
Arya, S. P.: 1999, *Air Pollution Meteorology and Dispersion*, Oxford University Press, New York, 310 pp.

Berkowicz, R.: 2000, 'A simple model for urban background pollution', *Environ. Monit. Assess.* **65**, 259–267.

Brechler, J.: 2000, 'Model assessment of air-pollution in Prague', *Environ. Monit. Assess.* **65**, 269–276.

de Haan, P., Rotach, M. W. and Werfeli, M.: 1998, 'Extension of an operational short-range dispersion model for applications in an urban environment', *Int. J. Veh. Des.* **20**, 105–114.

de Haan, P., Rotach, M. W. and Werfeli, M.: 2001, 'Modification of an operational dispersion model for urban applications', *J. Appl. Meteorol.* **40**, 864–879.

Gifford, F. A. and Hanna, S. R.: 1973, 'Modeling urban air pollution', *Atmos. Environ.* **7**, 131–136.

Hanna, S. R.: 1971, 'A simple method of calculating dispersion from urban area sources', *J. Air Pollut. Control Ass.* **21**, 774–777.

Hanna, S., Strimaitis, D. and Chang, J.: 1991, *User's Guide for Software for Evaluating Hazardous Gas Dispersion Models*, Sigma Research Corp., 71 pp.

Härkönen, J., Kukkonen, J., Valkonen, E. and Karppinen, A.,: 1998, 'The influence of vehicle emission characteristics and meteorological conditions on urban NO_2 concentration, *Int. J. Veh. Des.* **20**, 125–130.

Kousa, A., Kukkonen, J., Karppinen, A., Aarnio, P. and Koskentalo, T.: 2001, 'Statistical and diagnostic evaluation of a new-generation urban dispersion modelling system against an extensive dataset in the Helsinki area', *Atmos. Environ.* **35**, 4617–4628.

Lucas, D. H.: 1958, 'The atmospheric pollution of cities', *Int. J. Air Pollut.* **1**, 71–86.

Luhana, L. and Sokhi, R.: 1998, 'Application of the Box Model (PEARL) to Simulate the Seasonal Air Quality in Urban Centres', in *Proceedings of the 11th World Clean Air Congress,* IUAPPA, Vol. 1, pp. 2G-1-1–6.

Mazzeo, N. A. and Venegas, L. E.: 1991, 'Air pollution model for an urban area', *Atmos. Resear.* **26**, 165–179.

Olesen, H. R.: 1995, 'The model validation exercise at Mol: Overview of results', *Int. J. Environ. Pollut.* **5**, 761–784.

Owen, B., Edmunds, H. A., Carruthers, D. J. and Singles, R. J.: 2000, 'Prediction of total oxides of nitrogen and nitrogen dioxide concentration in a large urban area using a new generation urban scale dispersion model with integral chemistry model', *Atmos. Environ.* **34**, 397–406.

Turner, D. B.: 1964, 'A diffusion model for an urban area', *J. Appl. Meteorol.* **3**, 83–91.

Turner, D. B., Irwin, J. S. and Busse, A. D.: 1985, 'Comparison of RAM model estimates with 1976 St. Louis RAPS measurements of sulfur oxide', *Atmos. Environ.* **19**, 247–253.

Venegas, L. E. and Mazzeo, N. A.: 1998, 'Urban Diffusion-Deposition Air Pollution Model', in *Proceedings of the 11th World Clean Air Congress*, IUAPPA, Vol. 2, pp. 6E-4-1–6.

CFD PREDICTION OF AIR QUALITY IN THE AREA BETWEEN TWO CITY VEHICULAR TUNNELS CONSIDERING MOVING CARS

J. KATOLICKÝ, J. POSPÍŠIL and M. JÍCHA*

Brno University of Technology, Faculty of Mechanical Engineering, Department of Thermodynamics and Environmental Engineering, Technicka 2, Brno, Czech Republic
(* author for correspondence, e-mail: jicha@dt.fme.vutbr.cz, fax: +420 5 4114 3365)

Abstract. Concentration fields of different pollutants that spread outside two road tunnels predicted with a CFD code will be presented. The solution domain represents the city area located between two tunnel outlets – tunnel Strahov and tunnel Mrazovka in Prague. The vicinity of both tunnels is a heavily built up area with tall buildings forming typical street canyons. The CFD modelling predicts the situation after the tunnel Mrazovka will be finished and traffic will increase considerably between both tunnels. Namely, an interest was given to the prediction of dispersion of emissions leaving both tunnel and the area touched by the traffic. For the CFD predictions, a method previously developed for moving vehicles was used. The method uses combination of Eulerian and Lagrangian approaches to moving objects and is capable of modeling different speeds and traffic rates of cars as well as traffic-induced turbulence. Influence of several meteorological parameters was studied, namely wind speed and direction and traffic parameters, like traffic rates and speed of cars. The method separates contributions from different sources to the total concentration field, namely from background, tunnel outlet and roadway. Results are presented in the form of horizontal and vertical concentration fields of NO_x.

Keywords: CFD modelling, Eulerian-Lagrangian approach, moving vehicles, pollutants dispersion, urban air quality, vehicular tunnels

1. Introduction

The system of air quality control in big cities must be capable of providing information about short-term peak levels of pollution as well as about city background level. Peak levels are often directly connected with pollution originating from traffic. Despite their short-distance reach, knowledge of the peak values and their dispersion is important as they most frequently overcome threshold values.

1.1. LOCAL GEOMETRY

To correctly describe situation around traffic constructions like vehicular tunnels, large city crossroads and/or street canyons, relations between sources and receptors must be established. In this procedure, traffic plays a significant role. Ignoring traffic would neglect one of the most important phenomena that influences mixing processes in the proximity of traffic paths within canopy layer, namely in the

 Water, Air, and Soil Pollution: Focus **2**: 445–457, 2002.

situations of very low wind speed. Moving vehicles intensify both micro- and large-scale-mixing processes in the environment by inducing turbulence and enhancing advection by entraining masses of air in the direction of vehicle motion.

Dispersion of pollutants originating from traffic is rather short-distance and it is obvious that the actual geometry of the adjacent area plays an important role. The canopy layer is often strongly disturbed by different constructions that influence the flow field in the vicinity of traffic paths and by the shape of terrain that gives rise to a variety of recirculation zones and vortexes. When in the inner city, buildings and other obstacles may influence the level of local concentrations by more than one order. In such situations it is essential to establish the influence of different parameters on the pollutants distribution and to assess their relative contributions. The principal parameters can be classified as

- geometrical, i.e. the shape and type of surface (slopes, forests, road in a valley or vice verse in the open terrain), width and height of street canyons, shape of building roofs and building fronts and different design of sidewalks (trees along the roadway, etc.);
- traffic related, i.e. number of traffic lanes, one- or two-way traffic, traffic rate, and speed and type of cars;
- meteorological related (including climate and season conditions).

1.2. METEOROLOGY AND DISPERSION

Meteorological conditions play very significant role. It is mainly the speed and direction of the wind, solar irradiation of side walls and bottom of the canyon and thermal stratification of the atmosphere. The latter is probably less important in street canyons as the temperature distribution in the street canyon more reflects energy sources and sinks having origin in buildings, cars and solar radiation in an individual canyon. Thermal stratification rather plays a role of an *agent* that either enhances or suppresses ventilation of the canyon in the upward direction in the situation of very low wind speed. In the past some papers, (e.g. Hider *et al.*, 1997), deal with pollutant dispersion in the wake behind the vehicle but do not take account of dispersion in the traffic lane where the wake structure is certainly different from that for an individual car. Some other papers, (e.g. Kastner-Klein, private communication), make comparison between wind tunnel measurements and CFD calculations but this latter taking account only of traffic induced turbulence.

1.3. PRESENT STUDY

In the present contribution the authors focus on traffic and its impact on the pollutants distribution around two city vehicular tunnel outlets. A model based on Eulerian-Lagrangian approach to moving objects has been developed and integrated into a commercial CFD code StarCD. The subroutine is based on 3-D Eulerian-Lagrangian approach to moving objects and is capable of taking into account the traffic induced flow rate and turbulence (Jicha et al., 2000). The method has a significant advantage of separating contributions from different sources to a total concentration field of different pollutants. In such a way we are able to clearly determine what is the impact of tunnel outlets, line sources from vehicles moving along roadways and what is the background. The method has been applied to prediction of concentration field of CO and NO_x in the city area located between two vehicular tunnel outlets – tunnel Strahov and tunnel Mrazovka in Prague.

Predictions were done for different traffic situations at both tunnel portals (different traffic rate and speed) and different wind direction and speed. Also the 'no-wind' conditions were modelled and one situation in which the ventilation system inside the tunnel Mrazovka is out of order which results in the highest amount of polluted air leaving the tunnel outlet into the adjacent area. Results show a very different area affected by emissions leaving the tunnel outlets and emissions from the road line sources under different conditions and can help local authorities to assess the interaction between tunnels and surrounding.

2. Mathematical Formulation and Solution Procedure

The set of equations for the conservation of mass, momentum, energy and a chemical species (pollutants concentration) is solved for the steady, incompressible turbulent flow. Three momentum equations for three velocity components u_i ($i = 1, 2, 3$) in three directions x_1, x_2, x_3 in Cartesian-tensor notation can be written in the form:

$$\frac{\partial}{\partial x_j} \left(\rho u_j u_i \right) = \frac{\partial}{\partial x_j} \left(\mu_{eff} \frac{\partial u_i}{\partial x_j} \right) - \frac{\partial P_{eff}}{\partial x_i} + S_\phi + S_\phi^p , \tag{1}$$

where ρ is the fluid density, μ_{eff} is effective viscosity (sum of molecular and turbulent viscosity), P_{eff} is the pressure defined as *head*, i.e. $P_{eff} = p + \rho g h$. Other conservation equations can be written with a general variable ϕ in the form:

$$\frac{\partial}{\partial x_j} \left(\rho u_j \phi \right) = \frac{\partial}{\partial x_j} \left(\Gamma_{eff} \frac{\partial \phi}{\partial x_j} \right) + S_\phi + S_\phi^P , \tag{2}$$

where Γ_{eff} is effective diffusion coefficient, variable ϕ substitutes temperature (or enthalpy) and/or chemical species in appropriate conservation equations and equals unity in the mass conservation equation. Source term S_ϕ represents additional forces in the momentum equations and sources of enthalpy and/or pollutants in the other conservation equations. Additional source term S_ϕ^p (superscript p stands for *particles*) in both equations results from the interaction between moving vehicles and the ambient air. In the momentum equations this term represents the momentum transported from the discrete moving objects into the ambient air. The interaction is treated using a modified Particle-Source-In-Cell (PSIC) technique

TABLE I

Variants of wind and traffic

Wind direction	Wind velocity (m s^{-1})	Traffic rates (car/hour/lane)	Car speed (km hr^{-1})
East, west[a]	0.2, 2.0	1500, 3000, 5000	40, 70, 90

[a] For the west wind with 2.0 m s^{-1} the situation with ventilation system of the tunnel Mrazovka out of order was also modelled.

by (Crowe et al., 1977). The set of Equations (1) and (2) is solved using control volume method. Turbulence is treated using standard $k - \varepsilon$ model of turbulence (Launder and Spalding, 1972) with additional source of kinetic energy of turbulence k that results from additional generation of turbulence by moving cars (Jicha et al., 2000). Calculation of the additional source term S_ϕ^p is based on the solution of Lagrangian equation for moving vehicles:

$$m_p \frac{d\vec{U}_p}{dt} = \frac{1}{2} \rho_\infty C_D A_p \left| \vec{U}_\infty - \vec{U}_p \right| \left(\vec{U}_\infty - \vec{U}_p \right) , \qquad (3)$$

where m_p, $\vec{U}_p$ and A_p are mass, velocity and cross section of moving objects, respectively, ρ_∞ is density of air, C_D is drag coefficient, $\vec{U}_\infty$ is velocity of the ambient air and t is time. Equation (3) is integrated over an appropriate time step within each control volume through which cars move. The body of cars is split into several i-streaks each moving with the same velocity. The cars are set an initial velocity when entering a particular control volume and the change in velocity due to drag force is calculated using the Equation (3) for each i-streak. At the outlet from the control volume, which the car has just crossed, the velocity is reset to its original value. The momentum source term is then calculated from the change of momentum using the equation (for details Jicha et al., 2000):

$$S_{phi}^p = \sum_i S_{phi,i}^p = \sum_i \frac{\Delta H_i}{\Delta V} = \sum_i \frac{1}{\Delta V} = \dot{m}_{p,i} \left(\vec{U}_{p,i,out} - \vec{U}_{p,i,in} \right) , \qquad (4)$$

where ΔH is momentum change, ΔV is the control volume, $\dot{m}_{p,i}$ is mass flux of appropriate i-streak calculated from traffic rate and subscripts in and out refer to inlet and outlet, respectively from a particular control volume.

2.1. GEOMETRY OF THE SOLUTION DOMAIN

The solution domain represents the city area located between two vehicular tunnel outlets–tunnel Strahov and tunnel Mrazovka in Prague. The modeled area has dimensions approximately 600 m×600 m, the height of the canopy layer is approximately 300 m above the highest point in the area. The tunnel outlets are situated

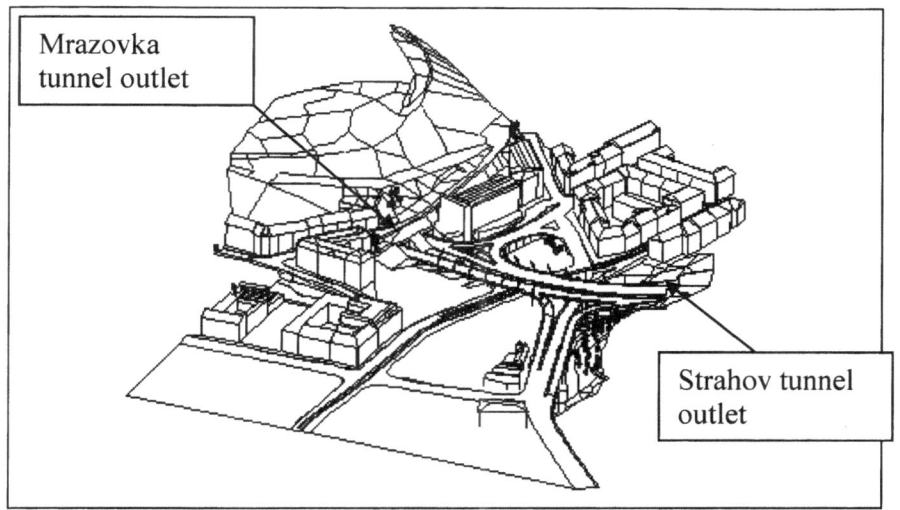

Figure 1. Schematic view of the area between two tunnel outlets.

between two hills. At the root of one, the outlet of the tunnel Strahov (already in service) is located, at the root of the other, the outlet of the tunnel Mrazovka (currently under construction) is located. The tunnel outlets are linked with two-lane dual carriageway and two ramps. Schematic view is shown in Figure 1. Several variants of wind direction and velocity and traffic rate and car speed have been modeled (see Table I).

2.2. BOUNDARY CONDITIONS AND SOLUTION PROCEDURE

The lateral faces of the solution domain were assigned either inlet or outlet conditions. Velocity at the inlet faces was decomposed into two horizontal components, opposite sides were assigned outlet boundary conditions. The upper face was assigned pressure boundary condition. The tunnel outlet and inlet were specified as internal positive and negative local sources of momentum with specified velocity.

To obtain concentration fields, sources of passive scalar representing the contribution from tunnel were specified in both tunnel outlets (as only the impact of tunnel outlets on the surrounding area was a concern, the procedure omitted contributions from line sources and background). The value of passive scalar at the outlets was set to 1.0. The actual CO and NO_x concentrations were re-calculated from the emission factors for the specific traffic fleet and speed of cars using PIARC methodology.

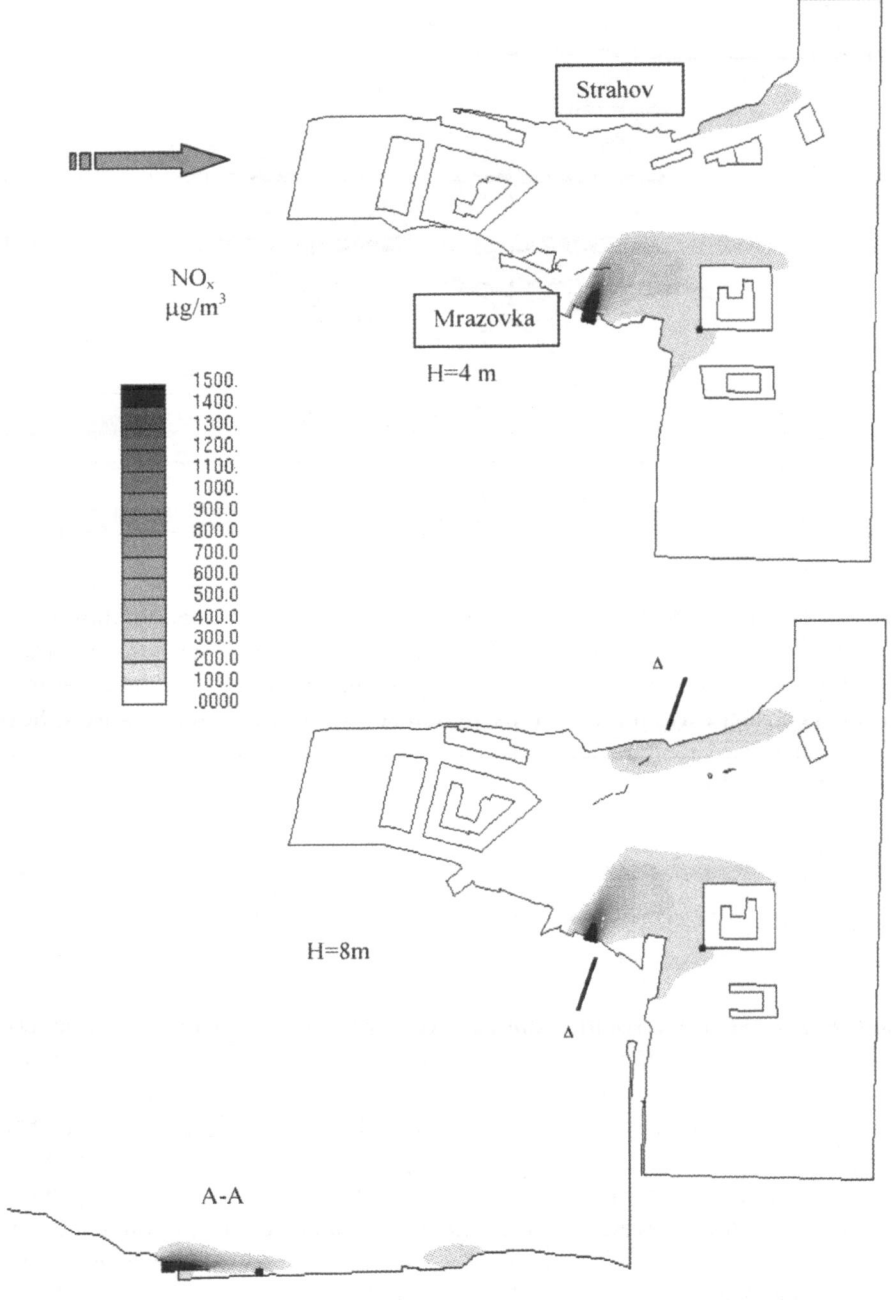

Figure 2. NO$_x$ concentrations: West wind with velocity of 2 m s^{-1}, traffic rate 1500 car/hour/lane, speed of cars 70 km hr^{-1}.

3. Results

In Figures 2 to 6 there are examples of the predictions for the case of west and east wind with velocity of 0.2 and 2 m s^{-1} and for traffic rate of 1500 car/hour/lane and 3000 car/hour/lane with the speed of 70 km hr^{-1}. The figures show the concentration field of NO$_x$ at two altitudes, namely at 4 and 8 m above the reference point in the solution domain (black dot in the lower left corner of one building). Both altitudes also correspond to the outlet of the Mrazovka tunnel. The altitude of 4 m corresponds with the level of the roadway, the altitude of 8 m corresponds with the level below the tunnel ceiling. The vertical cross section A–A is always the sectional view drawn in the direction of maximum concentrations from the tunnel outlet at the altitude of 8 m. The prediction was done without taking account of emissions originating from traffic on the roadway between both tunnel outlets. The momentum of moving vehicles along the roadway was considered using the Eulerian-Lagrangian method described earlier. The reason for this procedure was to see clearly the impact of polluted air penetrating from the tunnels into the surrounding area. From the same reason, the background values of concentrations were assumed zero. The main purpose of the study was to predict how large will be the area affected by the tunnel outlet under various meteorological and traffic conditions and how far the moving vehicles extract the emissions from the tunnel and/or keep them on the roadway, respectively. *(EIA study required that no emissions leave the outlet of tunnel Mrazovka after it is finished in order not to increase the ground level concentration of pollutants. So the tunnel ventilation system was designed to fulfil the requirements and the goal of the current study was to show whether the goal was reached.)* According to the authors of this article, to achieve zero emissions leaving the tunnel might be an impractical goal.

Figures 2 to 5 show results in the case of regular operation of the ventilation system inside the tunnel Mrazovka. In this case no emissions should penetrate outside the tunnel, as the EIA study required. Figure 6 shows the case when the ventilation system is out of order.

Let's first compare the influence of traffic rate on the pollutants dispersion – Figures 2 and 3. Both figures show results for the West wind with velocity of 2 m s^{-1}, speed of cars 70 km hr^{-1}, and traffic rate is 1500 car/hour/lane (Figure 2) and 3000 car/hour/lane (Figure 3). We can see that the higher traffic rate 3000 car/hour/lane demonstrates several effects:

- emissions are extracted from both tunnel outlets (Mrazovka and Strahov) into a much distant area;
- pollutants are kept on the roadway for a longer distance before they are swept down by the wind.

The impact of the wind velocity can be seen from Figures 2 and 4. Under a very low wind velocity (almost no wind conditions) of 0.2 m s^{-1} we can observe that

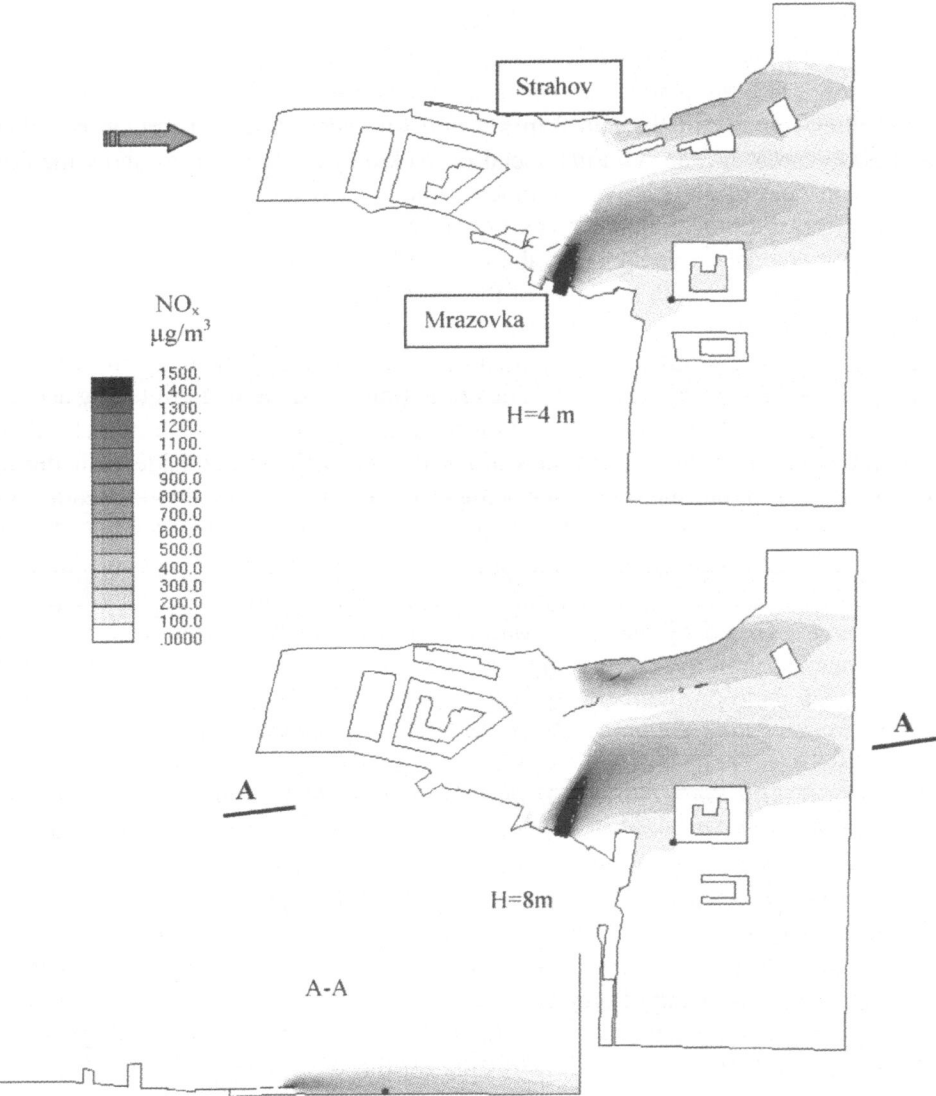

Figure 3. NO$_x$ concentrations: West wind with velocity of 2 m s^{-1}, traffic rate 3000 car/hour/lane, speed of cars 70 km hr^{-1}.

pollutants leaving the tunnel outlets:

- spread into a very large area;
- high concentrations are conserved for a long distance along the roadway due to high momentum of moving cars compared to momentum of wind;

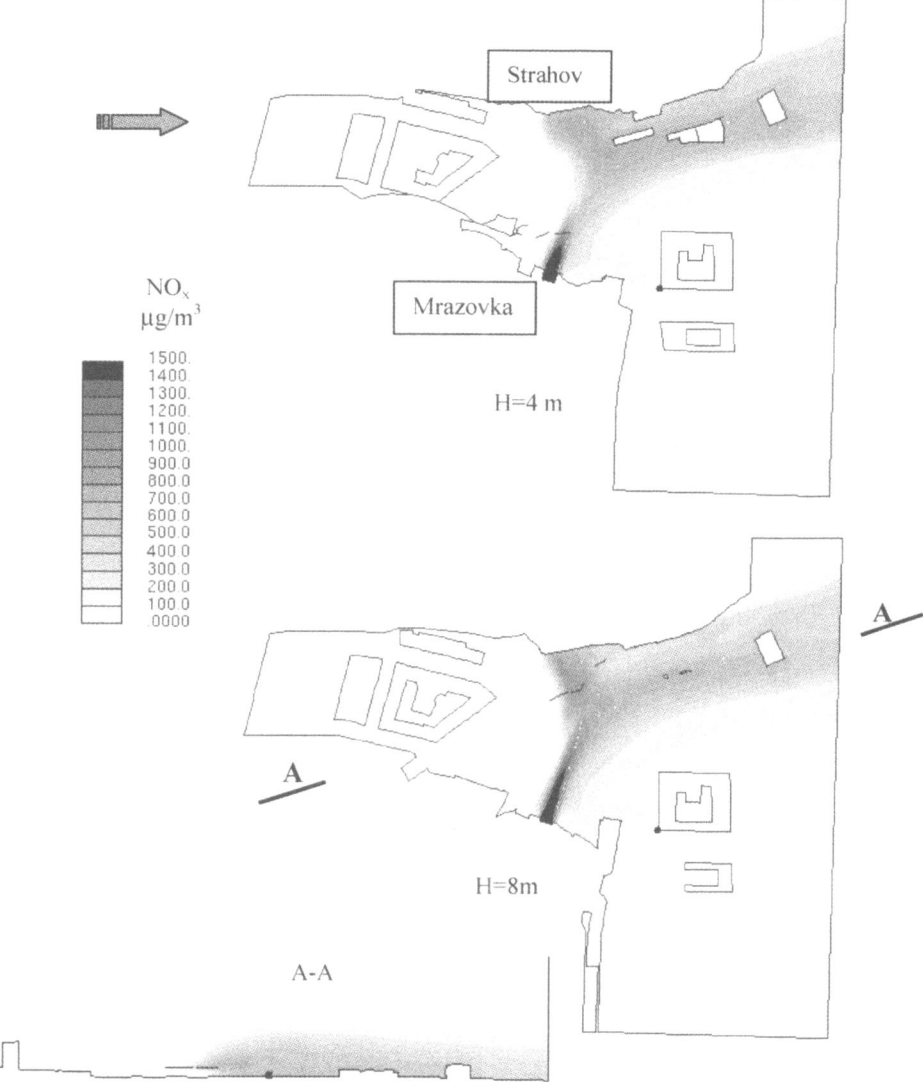

Figure 4. NO$_x$ concentrations: West wind with velocity of 0.2 m s^{-1}, traffic rate 1500 car/hour/lane, speed of cars 70 km hr^{-1}.

- concentration field is much less affected by the wind and does not follow strongly its direction as it is the case showed in Figure 2;
- The plume raises much higher.

A study was also done to see what happens if the ventilation system of the tunnel fails. The results are in Figure 5. The traffic rate was 1500 car/hour/lane and a

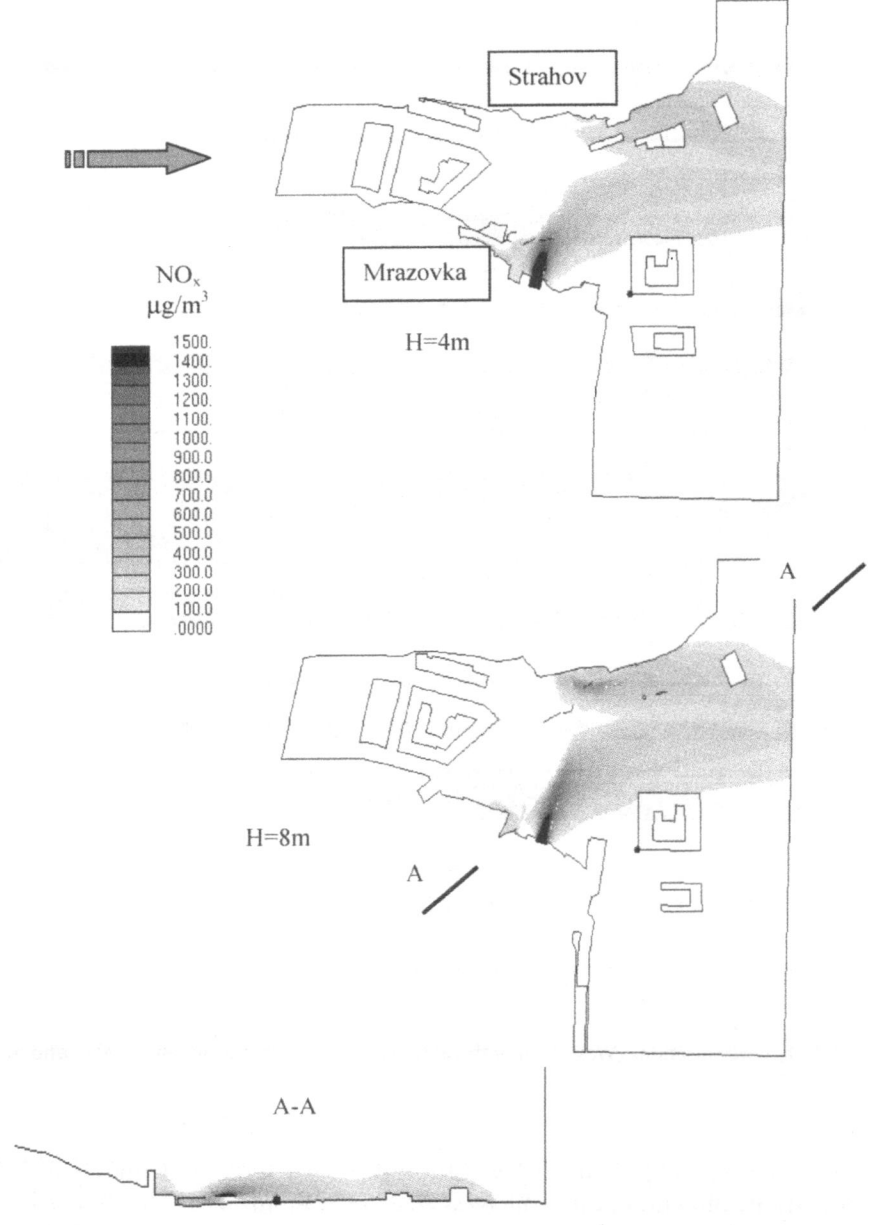

Figure 5. NO$_x$ concentrations: West wind with velocity of 2 m s^{-1}, traffic rate 1500 car/hour/lane, speed of cars 70 km hr^{-1}, tunnel ventilation system out of order.

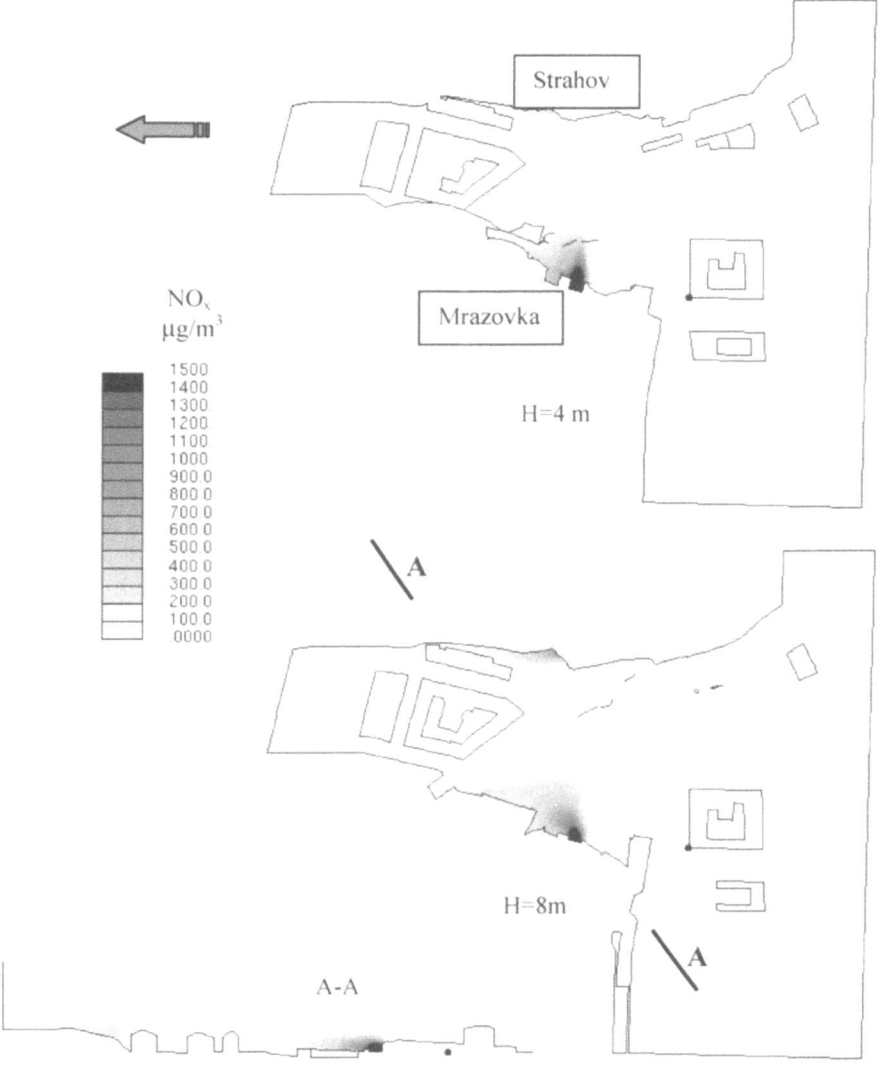

Figure 6. NO$_x$ concentrations: East wind with velocity of 2 m s^{-1}, traffic rate 1500 car/hour/lane, speed of cars 70 km hr^{-1}.

speed of cars 70 km hr^{-1}. We can see that the plume, extends into a much larger area due to higher velocity of air leaving the tunnel (in this case the 'tunnel air' is not plugged by the ventilation system that generates opposite force to the vehicles direction). The concentration field corresponds to a situation when the tunnel ventilation system is in service and traffic rate is 3000 car/hour/lane and speed of cars 90 km hr^{-1} (not shown here).

In Figure 6 the variant with East wind is shown. We can see how the geometry of the terrain affects the pollutants dispersion. The built up area in the west direction is very different from the east side. In the direction west of tunnel outlets the area contracts and wind velocity increases. The plume penetrating from the tunnel outlet is very intensively diluted and concentrations very rapidly drop.

4. Conclusions

From the results of modelling we can conclude that both, speed of cars and traffic rate considerably influence the dispersion of pollutants that spread outside the tunnel outlets. Higher is the speed of cars and/or traffic rate, larger is the area affected by tunnel emissions (due to limited space we do not present figures comparing the concentration fields for different car speed). Nevertheless, increasing the traffic rate has more pronounced effect than increasing the speed of cars (e.g. the affected area is much larger for doubled traffic rate than for doubled speed of cars). Moving cars create a strong dynamic effect that helps emissions keep in the direction of roadway. Namely in the situation of a very low wind speed. In such situation, the pollutants cover the largest area of all and also we can see that pollutants are kept for a long time in the course of road to where they are entrained by moving cars.

Except the traffic parameters, also configuration of the surrounding terrain has a significant impact on the concentration fields. When the terrain locally causes an increase in wind velocity (e.g. it contracts) pollutants are more diluted and concentrations decrease (the case of East wind). In the opposite situation (West wind) the dispersion of pollutants is much larger as the terrain open wider

For the given tunnel ventilation system the results did not confirm that no emissions leave the tunnel. The momentum of moving vehicles extracts emissions from the tunnel even when a strong opposite plug effect is created inside the tunnel.

Acknowledgement

This work is a part of Eurotrac-2 subproject SATURN OE32 /EU1489 and COST 715.80 and was supported by the Ministry of Education of the Czech republic and the Brno University of Technology Research Plan MSM 262100001.

References

Crowe, G. T., Sharma, M. P. and Stock, D. E.: 1977, 'The particle-source-in-cell model for gas-droplet flows', *J. Fluid Eng.* **99**, 325–332.
Hider, Z., Hibberd, S. and Baker, C. J.: 1997, 'Modelling particulate dispersion in the wake of a road vehicle', *J. Wind Eng. Ind. Aerodyn.* **67/68**, 733–744.

Jícha, M., Katolický, J. and Pospišil, J.: 2000, 'Dispersion of pollutants in street canyon under traffic induced flow and turbulence', *Environ. Monit. Assess.* **65**, 343–351.

Launder, B. E. and Spalding, D. B.: 1972, *Mathematical Models of Turbulence*, Academic Press, New York.

ANALYSIS OF 3-D URBAN DATABASES WITH RESPECT TO POLLUTION DISPERSION FOR A NUMBER OF EUROPEAN AND AMERICAN CITIES

C. RATTI[1], S. DI SABATINO[2,4]*, R. BRITTER[2], M. BROWN[3], F. CATON[5] and
S. BURIAN[6]

[1] *University of Cambridge, Department of Architecture, Martin Centre for Architectural and Urban Studies, Cambridge, U.K.;* [2] *University of Cambridge, Department of Engineering, Cambridge, U.K.;* [3] *Los Alamos National Laboratory, Energy and Environmental Analysis Group TSA-4, Los Alamos, U.S.A.;* [4] *Cambridge Environmental Research Consultants Ltd, Cambridge, U.K.;* [5] *Laboratorie d'Energétique et de Mécanique Théorique et Appliquée CNRS UMR 7563, Vandoeuvre Cedex, France;* [6] *University of Arkansas, Department of Civil Engineering, 4190 Bell Engineering Center, Fayetteville, AR, U.S.A.*
(* *author for correspondence, e-mail: silvana@cerc.co.uk, fax: +44 (0) 1223 357 492*)

Abstract. Dispersion models require as input various geometrical parameters to calculate the flow field and dispersion characteristics in the urban environment. As a result of recent advances in digital photogrammetry and remote sensing, databases of the actual 3-D geometry of city centre areas are now increasingly available. In this work we outline a procedure to reduce this large amount of data to a structured input for urban pollution dispersion models, i.e. to extract the important flow and dispersion parameters from the urban databases. Based on a review of the scientific literature, we have identified a number of parameters relevant to the modelling of pollution dispersion and atmospheric flows in urban areas. These parameters are: the plan and frontal area densities, the plan and frontal area density as a function of height, the distribution of heights, their standard deviation, the aerodynamic roughness length and the sky view factor. These parameters are obtained by analysing urban Digital Elevation Models (DEMs) which are regularly spaced grids of elevation values. Examples of the parameters calculated from high-resolution databases (with pixel size of about 1 m) for three European (London, Toulouse and Berlin) and two North American (Salt Lake City and Los Angeles) cities are presented and discussed. The calculated aerodynamic roughness length was smaller for the European cities than for the North American ones. A multiplicative correction factor κ to the aerodynamic roughness length is proposed to include the effect of the variability of the building heights.

Keywords: aerodynamic parameters, Digital Elevation Models (DEMs), image processing, pollution dispersion, urban morphometry, urban roughness

1. Introduction

In the context of pollutant dispersion modelling the knowledge of the aerodynamic roughness length z_0 is of prime importance. The derivation of this aerodynamic property and others such as the zero-plane displacement height z_d from morphometry is quite recent. Cities are characterised by large values of z_0 and z_d. These properties directly influence the surface shear stress, the scales and intensity of turbulence, the depth of the roughness sublayer, the wind speed and the shape of the wind speed profile and the type of flow found within the urban canopy layer. Accurate knowledge of the aerodynamic characteristics of cities is vital to describe, model and forecast the behaviour of urban winds, turbulence and the dispersion of pollutants at all scales.

The calculation of z_0 and z_d, however, is not straightforward. The classical way to estimate them in open flat terrain is based on the measurements of wind speed profiles from a tall mast or, less accurately, on the inference from published aerodynamic roughness values for similar terrain elsewhere (Davenport, 1960; Davenport et al., 2000). Both methods, however, are very difficult to apply to cities, due to the considerable height where wind measurements should be taken (two to three times the height of the average building height) and to the irregularities of urban texture.

A promising alternative that has become available in recent years, due to increased computing resources and the availability of high-resolution 3-D databases for urban areas, is based on the calculation of z_0 and z_d from the analysis and measure of the city geometry (urban morphometry). Grimmond and Oke (1999) discuss this method extensively; they used many available formulas and compared results with field measurements.

Urban morphometry provides a new range of parameters that can easily be calculated in urban areas and used as input for meso-scale and urban dispersion models. This article reviews a number of them and shows how they could be calculated from urban Digital Elevation Models (DEM) using image-processing techniques. It builds on the recent work by Ratti et al. 2000, extending the number of city case studies from London, Toulouse, Berlin to Salt Lake City and Los Angeles (see Figure 1).

2. Methodology and Data Set Description

The DEM is a digital image of a city, where each pixel has a grey-level proportional to the height of the buildings. It contains a full 3-D description of the urban surface on a 2-dimensional support (the image).

High resolution DEMs in urban areas are becoming increasingly available. The cost is still high but it is expected to decrease as the demand on such data is rapidly increasing.

The Los Angeles building data set shown in Figure 1e, for instance, is a commercial product by Aerotopia and contains building footprints and rooftop elevation information. Its resolution is 2 m horizontal and 1 m vertical and is indexed to universal transverse mercator (UTM) coordinates. The London, Berlin, Toulouse and Salt Lake City building data sets were produced in-house using satellite and high resolution aerial photographs (Müller et al., 1999). They have a very high resolution in plan (6 inches/pixel in Salt Lake City), although a lower one in elevation:

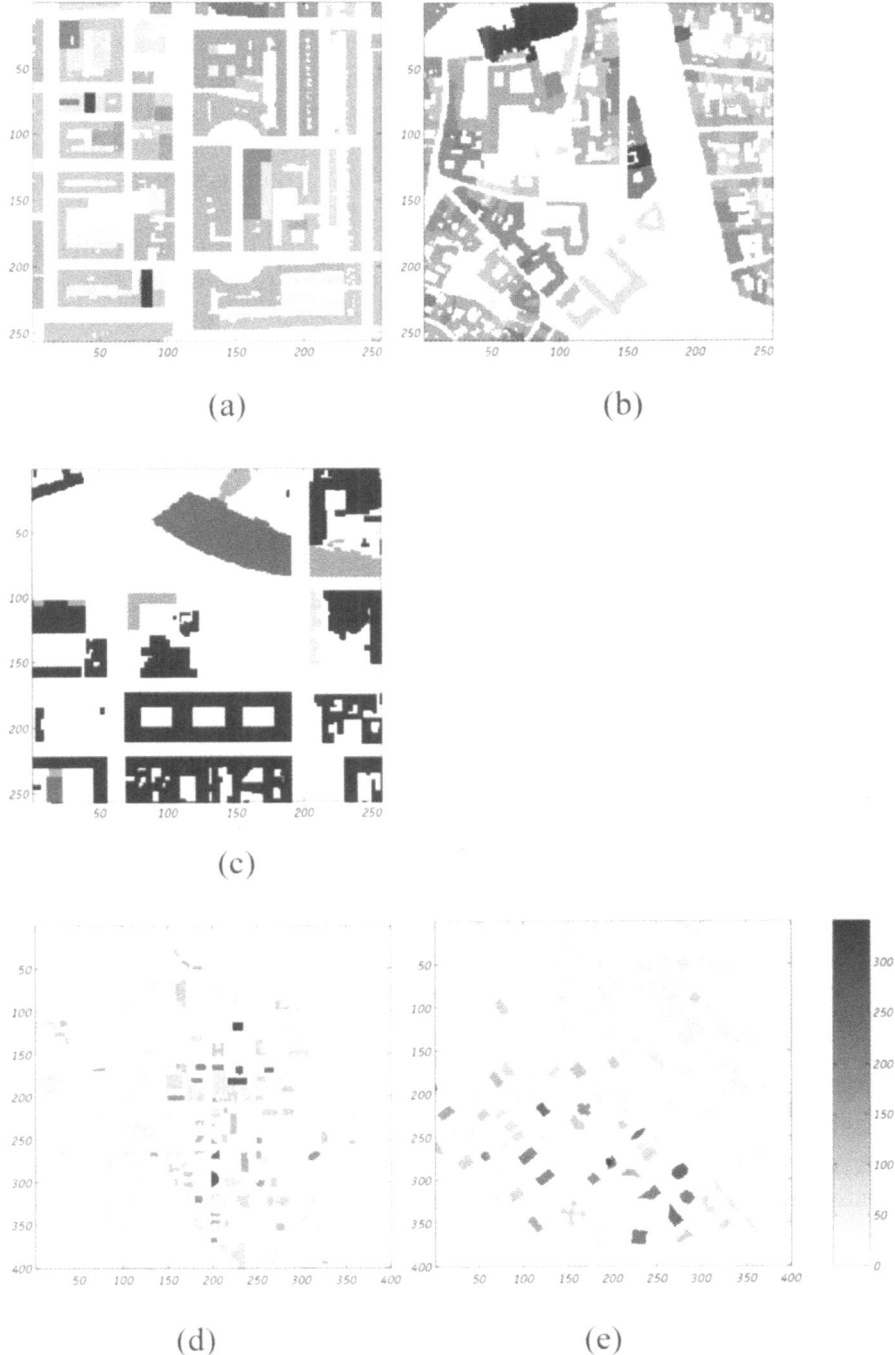

Figure 1. In an urban DEM, building height is proportional to the level of grey. The shaded bar refers to the Los Angeles DEM (e), where the maximum height is $h_{max} = 341$ m. In London (a) $h_{max} = 40$ m, Toulouse (b) $h_{max} = 32$ m, Berlin (c) $h_{max} = 21$ m, Salt Lake City (d) $h_{max} = 98$ m.

building heights were estimated by counting storeys from photographs and during visits to the city, and therefore have an uncertainty of a storey (∼3.5 m).

The Los Angeles and Salt Lake City DEMs, which cover areas of 2 and 3.5 km^2, respectively, encompass the downtown regions with small pockets of adjacent high-density residential, industrial and commercial landuses. The European DEMs which are approximately 0.2 km^2 each, describe only the central urban areas, although they are representative of larger portions of the cities.

Urban DEMs can be analysed with image processing techniques using simple packages like the Matlab Image Processing Toolbox. A review of this method, which has similarities with raster GIS analysis, is contained in Ratti and Richens (1999). The orientation of facades, the amount of solar radiation falling on the city, the Fourier and Radon transforms, the estimate of energy consumption in buildings, the travelling-time in the street network, etc., can be calculated. Other parameters related to flow and pollutant dispersion in the urban environment are reviewed below in detail.

3. Parameters Related to the Urban Flow Field and Pollutant Dispersion

A number of different parameterisation schemes are used in meso-scale models to approximate the effects of the urban canopy on the flow field. At a minimum level, urban landuse information is needed to estimate the aerodynamic roughness length and the surface energy balance. More complex urban canopy parameterisations (e.g., Sorbjan and Uliasz, 1982; Brown and Williams, 1998; Ca et al., 1999) require morphological information cross-correlated with landuse, average building height, plan area density, and building frontal area density. For example, the frontal area density is used in the momentum equations as part of the drag force term. Another important parameter is the sky-view factor, which can be used to determine the long-wave energy flux into and out of the urban canopy.

The following parameters have been calculated on the DEM using image processing techniques:

(1) The built to total area ratio (λ_P) at ground level and also its variation with height.
(2) The average building height ($\overline{H}$), the average of building heights squared ($\overline{H^2}$), the standard deviation of building heights (σ_H) and the average building height (z_H) where each building is weighted with its frontal area:

$$z_H = \frac{\sum \text{Height of the buildings} * \text{Frontal area of the buildings}}{\text{Total lot area}}$$

The parameter z_H is a function of the wind direction.

(3) The aerodynamic roughness length z_0 and the zero-plane displacement height z_d. They can be calculated from the above parameters plus the frontal area density λ_F. We have used the formulas from Macdonald *et al.* (1998) as these performed well in the comparison by Grimmond and Oke (1999):

$$\lambda_F = \frac{\sum \text{Frontal area of the buildings}}{\text{Total lot area}}$$

$$\frac{z_0}{H} = \left[1 - \frac{z_d}{H}\right] \exp\left[-\left[\frac{0.5\beta c_D \lambda_F}{k^2}\left[1 - \frac{z_d}{H}\right]\right]^{-0.5}\right]$$

$$\frac{z_d}{H} = 1 + \alpha^{-\lambda_P}(\lambda_P - 1) \ ,$$

with $\alpha = 4.43$, $\beta = 1.0$, $k = 0.4$, $c_D \approx 1$.
All those parameters (z_0, z_d, λ_F) are functions of the wind direction.

(4) A correction κ to z_0 to account for the height variability of the buildings. Based on our interpretation of results from Hall *et al.* (1996), the z_0 values calculated with the formulas above should be corrected by a factor κ where

$$\kappa = \left(1 + 4\frac{\sigma_H}{H}\right) \ .$$

(5) The sky view factor from the streets to the sky. This parameter can be calculated on the DEMs as explained in Ratti and Richens (1999). Its average value ψ_{sky} can be used to predict the maximum heat island intensity $\Delta T_{\text{max urban-rural}}$, using a formula by Oke (1981).

4. Results and Discussion

Our results are summarised in Figures 2–4 and in Table I. In the case studies considered here, North American cities show a smaller λ_P (built to non-built area ratio), a greater maximum height and also a larger variability of building heights than with the European cities. This is in agreement with what is qualitatively expected but here we have quantified these factors.

Looking at the calculated values of the aerodynamic roughness length in Table I we see that they are smaller for the European cities than for Salt Lake City and Los Angeles. According to Hanna *et al.* (2001) a value of z_0 around 1 m could be used for screening purposes for most cities. From Table I we see that z_0 has an unusually low value for London and an unusually high value for Los Angeles.

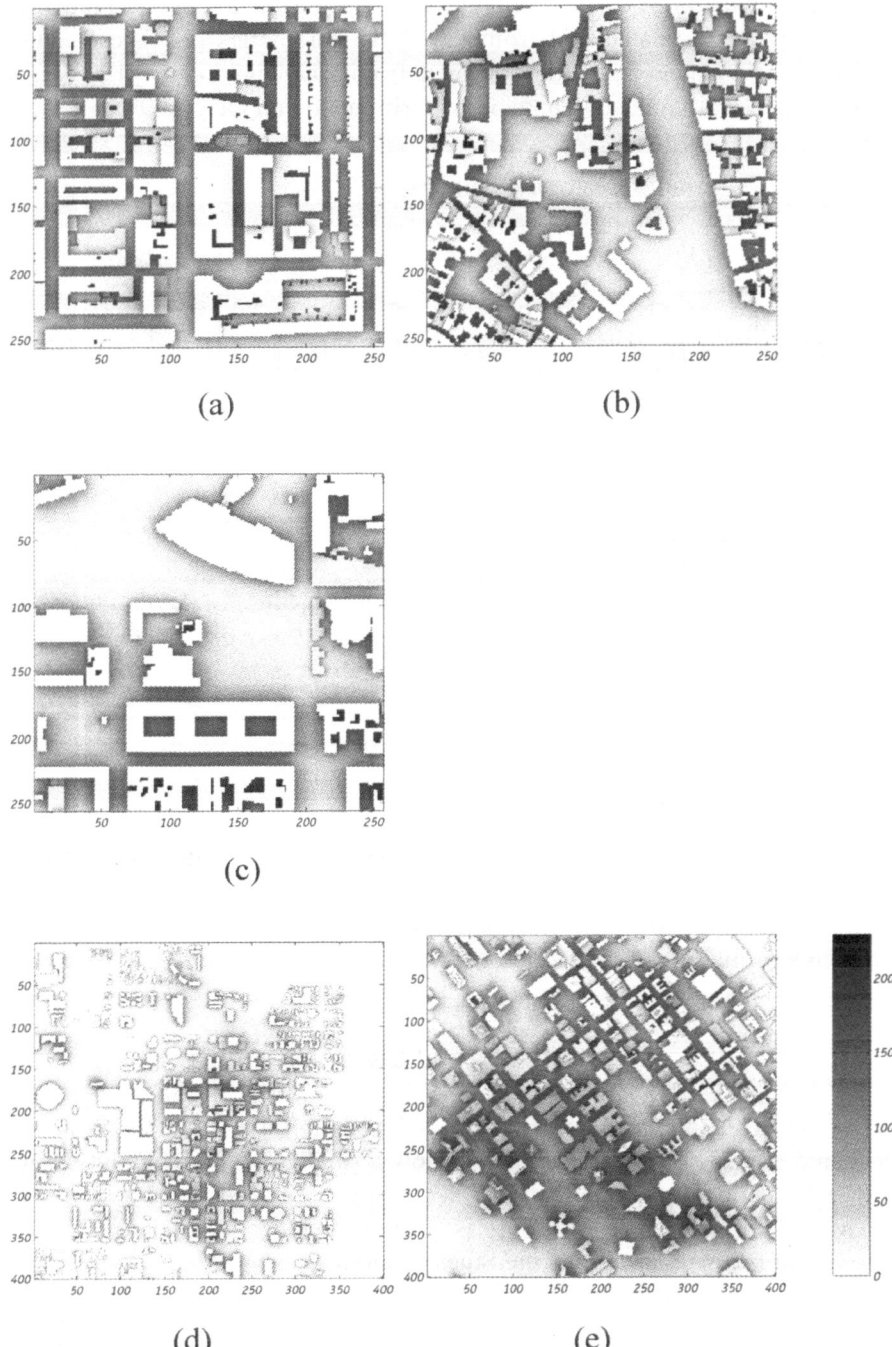

Figure 2. Sky view factors in London (a), Toulouse (b), Berlin (c), Salt Lake City (d) and Los Angeles (e).

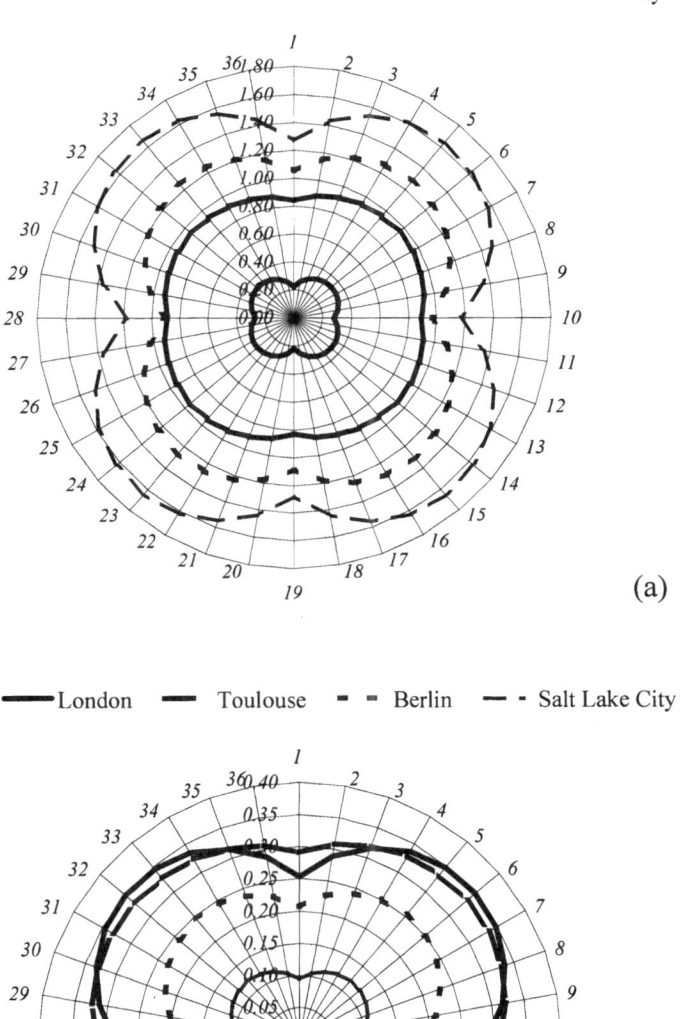

Figure 3. Variation of aerodynamic roughness length z_0 (a) and λ_F (b) with azimuth, respectively (Los Angeles data have been omitted).

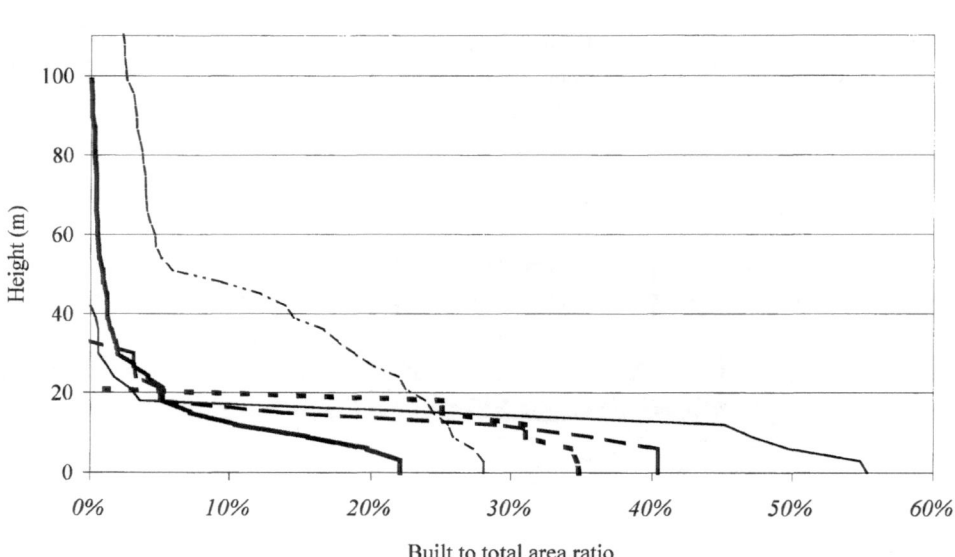

Figure 4. Built to non-built area ratio in percentage (c) at different heights (Los Angeles data have been truncated).

We have calculated the aerodynamic roughness length based on the formulas by Macdonald *et al.* (1998) shown earlier; in addition we have recalculated z_0 using the same formulas but replacing the standard average building height $(\overline{H})$ with z_H, i.e. the average building height weighted with the frontal area of the buildings.

Results in Table I show that z_H and $\overline{H}$ values are similar for the European data sets while they differ greatly for the North American data sets. Thus, the alternative use of z_H was not significant for the European cities; we obtained only a slightly larger z_0 value using z_H than we have obtained using $\overline{H}$. The use of z_H did not seem to be appropriate for Los Angeles; using $\overline{H}$ has still produced an unusually large value of z_0. There must be other factors that account for the unusually small value for London and the unusually large number for Los Angeles.

The small value for London arises because the formulas by Macdonald *et al.* (1998) that we have used imply $z_0 \to 0$ for $\lambda_P \to 1$ and London has the largest λ_P of the five cities analysed. The formulas we have used for z_0 do not account for the building height variability that is expected to have a large influence on the aerodynamic roughness length values, particularly for those cases with a large λ_P. Physically we expect cities with a large λ_P to be very sensitive to building height variability.

The height variability has been estimated by comparing the ratio of the standard deviation of building heights to their average height i.e. $\sigma_H/\overline{H}$. From Table I we see that the North American cities have approximately twice the height variability

TABLE I

Calculated parameters for London, Toulouse, Berlin, Salt Lake City and Los Angeles

	London	Toulouse	Berlin	Salt Lake City	Los Angeles
λ_P, built to total area ratio	0.55	0.40	0.35	0.22	0.28
$\overline{H}$, average of building heights (m)	13.6	15.3	18.6	16.3	51.3
$\overline{H^2}$, average of the heights squared (m²)	211	270	364	464	5289
σ_H, standard deviation of the heights (m)	5.0	6.1	4.3	14.1	51.5
$\sigma_H/\overline{H}$ building height variability	0.37	0.40	0.23	0.87	1.00
σ_H/z_H building height variability	0.34	0.38	0.22	0.54	0.50
$z_H/\overline{H}$ building height ratio	1.09	1.05	1.07	1.60	2.01
z_H, average of the heights weighted with frontal area (also averaged all azimuth) (m)	14.8	16.1	19.9	26.0	103.0
λ_F, frontal area density (average all azimuth)	0.32	0.32	0.23	0.11	0.45
z_d, zero-plane displacement height (average all azimuth) (m)	11.9	10.9	12.1	11.4	54.3
z_0, roughness length (average all azimuth) (m)	0.29	0.89	1.08	0.94	7.2
z_0, calculated using z_H instead of $\overline{H}$, (also averaged all azimuth) (m)	0.30	0.92	1.18	1.50	14.36
κ, roughness length correction factor	2.47	2.59	1.92	4.48	5.02
ψ_{sky}, average view factor from the streets to the sky	0.529	0.646	0.720	0.866	0.602

of the European cities. This large difference between the two sets of data, however, may just be an artefact due to different coverage areas being represented by the sets of data i.e. 2 and 3.5 km² for Los Angeles and Salt Lake City against 0.2 km² for the European cities. From Table I, we see that $\sigma_H/\overline{H}$ and σ_H/z_H (where z_H is the average building height weighted with the frontal area of the buildings) values are similar for the European cities while they differ greatly for the North American cities. This means that the European data sets represent portions of cities with more homogeneous building shapes than the North American ones, possibly a further reflection of the different coverage areas.

To include the building height variability in the calculation of z_0 a multiplicative correction κ must be used. Our preliminary estimate of κ, based on our interpretation of results by Hall *et al.* (1996) was shown in the previous section. Using this formula we have obtained a correction to z_0 which seems appropriate for the

European data sets; by applying the correction to those cases we obtain values of z_0 around 1 to 2 m. Our use of κ is less satisfactory for Salt Lake City and Los Angeles as the application of the correction to those data sets would lead to very large values for z_0. The formula for κ is not a function of λ_P and we expect a λ_P dependence to be included. The refinement to the formula for κ, to link together the influence of the height variability and the building packing density, is currently under consideration.

In summary, the height variability formula is probably satisfactory for the z_0 values typical of the European cities. It is likely to produce over predictions for the λ_P of the North American cities. Additionally, and probably more importantly, some of the spatial inhomogeneities over the large coverage area of the North American data sets may not actually be contributing to z_0.

Despite the greater height of buildings in North American cities, the average view factor from streets to the sky is comparable with that in Europe, due to the lower built to non-built area ratio λ_P of the North American cities. Further work is currently in progress to determine and compare not only the average values of the view factors and the other urban parameters, but also their spatial variation.

5. Conclusions

Digital Elevation Models (DEMs) contain information on the urban morphometry. They provide data that can be analysed in a structured way to extract useful parameters linked to the structure of the urban canopy and the flow and pollutant dispersion characteristics within and above the urban canopy. Those parameters may be used as inputs for meso-scale and urban pollution dispersion models. In this article we have identified some of relevant parameters for this purpose and outlined a method to extract those from DEMs.

Data sets for three European cities (London, Toulouse and Berlin) and two North American cities (Salt Lake City and Los Angeles) have been analysed.

Our results showed that North American cities have a smaller λ_P, a greater maximum building height and also a larger variability of building heights than the European cities.

The calculated aerodynamic roughness length was smaller for the European cities than for the North American ones. A multiplicative correction factor κ to z_0 has been proposed to include the effect of the variability of the building heights. This correction does not explicitly account for variations in building packing density. The proposed formula is probably satisfactory for the z_0 values typical of the European cities that have, in our case, a large value of λ_P, while some refinement is required for its applicability to cases with smaller λ_P values.

Acknowledgements

Carlo Ratti and Silvana Di Sabatino acknowledge support from the EU-projects Precis and TMR-TRAPOS.

References

Brown, M. and Williams, M.: 1998, 'An Urban Canopy Parameterization for Mesoscale Meteorological Models', *2nd AMS Urban Env. Symp.*, Albuquerque, NM, U.S.A.

Ca, V., Asaeda, T., and Ashie, Y.: 1999, 'Development of a numerical model for the evaluation of the urban thermal environment', *J. Wind Eng. Ind. Aerodyn.* **81**, 181–196.

Davenport, A.: 1960, 'Rationale for determining design wind velocities', *J. Struct. Div. Am. Soc. Civ. Eng.* **86**, 39–68.

Davenport, A., Grimmond, C., Oke, T. and Wieringa, J.: 2000, 'The Revised Davenport Roughness Classification for Cities and Sheltered Country', *3rd AMS Urban Env. Symp.*, Davis, CA, U.S.A.

Grimmond, C. and Oke, T.: 1999, 'Aerodynamic properties of urban areas derived from analysis of surface form', *J. Appl. Meteorol.* **38**, 1261–1292.

Hall, D. J., Macdonald, R., Walker, S. and Spanton, A. M.: 1996, 'Measurements of Dispersion within Simulated Urban Arrays – A Small Scale Wind Tunnel Study', *BRE Client Report CR 178/96*, Building Research Establishment (BRE).

Hanna, S., Britter, R. and Franzese, P.: 2001, *The Effect of Roughness Obstacle on Flow and Dispersion at Industrial Facilities*, Hanna Consultants, WE, U.S.A., Project 132.

Macdonald, R. W., Griffiths, R. F. and Hall, D. J.: 1998, 'An improved method for estimation of surface roughness of obstacle arrays', *Atmos. Environ.* **32**, 1857–1864.

Müller, C., Burian, S. and Brown, M.: 1999, 'Urban Database Issues', LA-UR-99-5874.

Oke, T.: 1981, 'Canyon geometry and the nocturnal urban heat island: Comparison of scale model and field observation', *J. Climat.* **1**, 237–254.

Ratti, C. and Richens, P.: 1999, 'Urban Texture Analysis with Image Processing Techniques', *Proc. CAADFutures99*, Atlanta, GE.

Ratti, C. F., Di Sabatino, S., Caton, F. and Britter, R. E.: 2000, 'Morphological Parameters for Urban Dispersion Models', *3rd AMS Urban Env. Symp.*, Davis, CA.

Sorbjan, Z. and Uliasz, M.: 1982, 'Some numerical urban boundary-layer studies', *Bound.-Layer Meteorol.* **22**, 481–502.

OZONE EPISODE IN THE MILAN METROPOLITAN AREA: A COMPARATIVE EVALUATION OF UAM-V AND CALGRID MODELS

F. POLLA MATTIOT[1]*, C. GARIAZZO[1], P. BUTTINI[1], G. CARIZI[2], A. LEVY[2] and E. REBESCO[3]

[1] *EniTecnologie S.p.A, MonterotondoScalo, Italy;* [2] *Aquater Loc. Miralbello, San Lorenzo in Campo, Italy;* [3] *Agip Petroli CREA, S. Donato, Milanese, Italy*
(author for correspondence, e-mail: fpolla@enitecnologie.eni.it, fax: 39 6 90 673 263)*

Abstract. Data from an ozone episode (2–5 June, 1998) in the Milan metropolitan area were used for an application of two photochemical grid models: UAM-V and CALGRID. To assure a fair comparison, the models were run on the same domain and grid size, with same source emission inputs, CALMET diagnostic meteorology, and initial and boundary conditions taken from air quality data and literature values. Hourly emissions were derived from the AutoOil-II programme inventory except for on-road mobile source emissions; a new traffic emission inventory, based on both COPERT II methodology and road classification has been developed. NO_x and O_3 concentration results were compared to local network monitoring data. Results indicate that both models predict the highest ozone values along the north-east direction and are able to reproduce the ozone daytime trend though differences can be found between the two models on ozone spatial distribution. Average normalised bias for both models is about 50%, peak daily ozone concentrations are underestimated, with simulated peak shapes broader than the observed ones and a temporal shift between the two models. Night-time concentration levels of pollutants were not successfully reproduced due to an incorrect parameterisation of vertical turbulence calling for further work.

Keywords: emission inventory, model comparison, model evaluation, photochemical grid models, urban ozone

1. Introduction

The Milan metropolitan area located in the North of Italy, includes the city of Milan and a large territory surrounding it which is highly populated. It also hosts working commuters from several Provinces like Bergamo, Como, Pavia, Piacenza and Varese. Inside this area there are about 600 km of highways and 1700 km of extra urban roads. The surrounding mountains with elevations up to 2000 m and the Po valley also characterise this basin. All these features make this area an ideal site to study the traffic effects on ozone and other related primary and secondary pollutants. This was in fact the ultimate goal of our study which intended to investigate the effect on air quality of detailed traffic emission inventory, addressing further studies to modifications on VOC-NO_x emitted by vehicles.

In Italy, the Minister of Environment (DM 16/05/96), according to EC Directive 92/72, defined ozone thresholds for health protection (110 μg m^{-3} for 8 hr average

◢◣ *Water, Air, and Soil Pollution: Focus* 2: 471–486, 2002.

concentration) and for vegetation protection (200 μg m^{-3} for 1 hr average concentration). Both thresholds were exceeded in Lombardia Region many times during 1998: more than 6700 cases over 110 μg m^{-3} and about 2000 hours in the second situation were recorded.

Photochemical grid models have been largely used to study ozone episodes (Kumar et al., 1994; Ziomas et al., 1998; Hanna et al., 1996; Jiang et al., 1998; Mayes et al., 1996). Among photochemical models, CALGRID (California Air Resources Board Airshed Model, Yamartino et al., 1989, 1992) and UAM (Urban Airshed Model, SAI, 1995) have been extensively used as regulatory models; a combined approach of both models has been suggested by different authors. The main difference between the two models is the chemical scheme used: SAPRC (Carter, 1990) and Carbon bond IV (Gery et al., 1988). Silibello et al. (1998) studied the Milan metropolitan area using the CALMET-CALGRID modelling system. The present work updates the study to 1998 and focuses on traffic sources using a different approach to estimate the vehicle emissions in the Milan urban area.

The CALGRID and UAM-V photochemical grid models were applied on the same domain 100 × 100 km^2 wide. The meteorological field produced by CALMET was provided to both models together with the temporally and spatially disaggregated emission inventory. Results coming from both models were then evaluated and compared to ozone, NO and NO$_2$ monitoring data.

2. Simulation Models

Photochemical grid models are generally used to study ozone episodes. They calculate concentrations of pollutants by simulating the physical and chemical processes in the atmosphere. The basis of these models is the atmospheric diffusion equation which represents a mass balance that includes all of the relevant emissions, transport, diffusion, chemical reactions, and removal processes in mathematical terms. The three dimensional structure of these models needs a large number of inputs such as: speciated and spatially/temporally disaggregated emission inventory, full three dimensional meteorological fields, initial and boundary conditions. Two Eulerian photochemical grid models, CALGRID and UAM-V, were used in the present work to reproduce the O$_3$, NO and NO$_2$ concentration field. The CALGRID model incorporates the SAPRC chemical mechanism where hydrocarbons with similar functional group and reactivities are lumped together. The UAM-V model uses an extended version of the Carbon Bond IV (CB-IV) mechanism representing different types of carbon bonds in hydrocarbons. The reactions included in the two mechanisms for inorganic species are nearly identical while organic species cannot directly be compared except when treated explicitly (Dodge, 2000).

In order to provide the meteorological field to the photochemical models the diagnostic meteorological model CALMET has been used. The wind field is obtained in two steps: in step one a first guess wind velocity is adjusted for local effects by empirical formulas (Liu *et al.*, 1980; Scire *et al.*, 1997; Allwine *et al.*, 1985). In step two a weighted averaging of step one wind field and available wind measurements is performed and a divergence minimisation procedure (Goodin, 1980) is applied to the three-dimensional wind field to assure overall mass conservation.

While the CALMET meteorological output can be directly provided to CAL-GRID, some adaptations must be done for the UAM-V meteorological input. It must be pointed out that UAM-V has been designed for use with prognostic rather than diagnostic meteorological models, therefore meteorological inputs for UAM-V are possibly not entirely suitable and not directly comparable to what used by CALGRID. Specifically the vertical turbulence exchange coefficients, the 3-D humidity and pressure files must be calculated. The former ones have been estimated applying a CALGRID algorithm which uses the friction velocity and Monin-Obukhov length calculated by CALMET as suggested by Hanna (Hanna *et al.*, 1996). The specific humidity has been derived using a modified version of a UAM-V meteorological pre-processors which simply interpolates surface and upper air measurements, spatially and temporally. The 3-D pressure field has been calculated using a theoretical equation (Pielke, 1984) based on surface measurements and the CALMET generated temperature field.

The model region consists of a 50×50 grid domain with 2 km horizontal resolution, centred on the city of Milan. Vertically, a stretched grid has been applied up to 4000 m, with the first level located 20 m above the ground.

3. Input Data for Base Runs

3.1. AIR QUALITY DATA

Numerous studies were conducted on Lombardia Region because dangerous ozone exceedances had been detected and many chemical data from monitoring network had been available. In addition the so-called PIPAPO (*Pi*anura *Pa*dana *P*roduzione di *O*zono: Ozone production in the Po basin) experimental monitoring campaign was recently promoted by an EU project in order to perform measurements of photochemical smog compounds and precursors at ground and upper level in this area.

Among PIPAPO and local network monitoring stations (about 40 in the domain), eight monitoring stations, spread over all the area, were chosen as reference. The selected stations are: Bergamo, Como, Milan (Viale Marche, EniTecnologie mobile laboratory), Pavia (four urban stations), Brugherio, Landriano, Legnano and Merate.

The observations show high ozone values (over 100 ppb) over all the selected stations during 2–5 June, 1998.

In this work many urban and suburban stations were used to determine initial conditions. Constant values in space and time were assumed for boundary conditions for almost all compounds except for ozone concentrations which have been

TABLE I

Boundary concentrations assumed for the two models

UAM-V species (ppb)		CALGRID species (ppb)	
Lateral boundary conditions		Lateral boundary conditions	
NO	1	NO	1
NO_2	3	NO_2	3
O_3	30 (2–3 June)	O_3	30 (2–3 June)
	40 (4–5 June)		40 (4–5 June)
OLE	0.09	OLE1	0.5
		OLE2	0.5
PAR	4	ALK1	1
		ALK2	0.5
TOL	0.05	ARO1	0.5
XYL	0.03	ARO2	0.5
FORM	0.5	HCHO	0.5
ALD2	0.3	CCHO	0.5
		RCHO	0.5
ETH	0.5	ETHE	0.5
CRES	0.5	CRES	0.5
ISOP	0.5	OLE3	0.5
PAN	0.5	PAN	0.5
CO	0.5	CO	0.5
HNO_3	1	HNO_3	1

assumed to increase from 30 to 40 ppb in the last two days of simulation. These values, reported in Table I, are taken from the quoted study on the same area (Silibello *et al.*, 1998), which presented a chemical speciation according to SAPRC. As the chemical mechanism of the two models is different, the VOC Table I can only be compared as a whole.

3.2. METEOROLOGICAL DATA

Inside the study area, seventeen surface stations and two vertical soundings (Figure 1) were selected. As the domain topography is characterised by the presence of plain, hill and mountain, the surface stations were selected in order to cover these areas inside the domain. Hourly wind speed and direction, temperature, humidity and pressure data were used as surface data input. The meteorological upper soundings were collected from the Linate Airport (data every 6 hr) and from a mobile meteorological station sited at Seregno (hourly data; PIPAPO campaign).

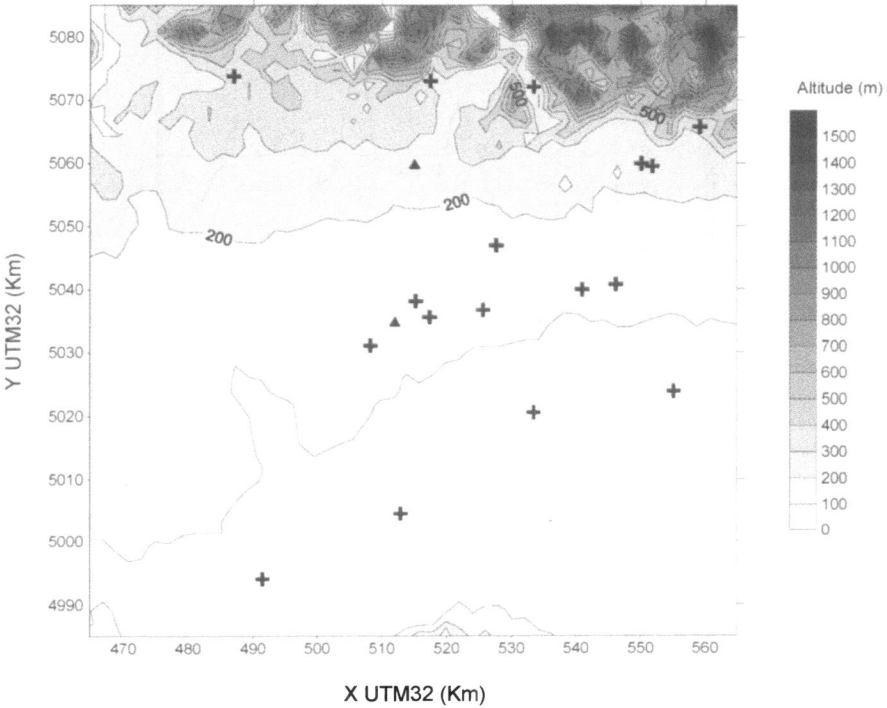

Figure 1. Study area with orography and siting of the meteorological measurements stations (+ surface and ▲ upper air sounding).

In the days 2–5 June 1998, the region was crossed by a ridge of high pressure and low wind speed blowing from North Africa towards N–NE; this particular synoptic condition promoted the development of the photochemical cycle.

3.3. EMISSIONS INVENTORY

The emission inventory is made up of selected inventories coming from other projects and based on different methodologies. The inventory includes CO, NO_x and VOC. In terms of emission sources this inventory deals with macro-sources such as: total area sources, total traffic sources and other area sources. According to another inventory (Silibello *et al.*, 1998) point sources account for less than 10 and 3% for NO_x and VOC, respectively. As the available data emission were not updated, their uncertainty is very high and relative abundance is quite small, compared to that coming from traffic sources, point sources were not included in this study. The other non-traffic emission inventories were derived from the AutoOil-II Programme (AOIIP) (Skouloudis, 1999) which updated, at the year 1995, the European emission inventory based on CORINAIR. According to AOIIP classification, total area sources (TAS) include combustion in manufacturing in-

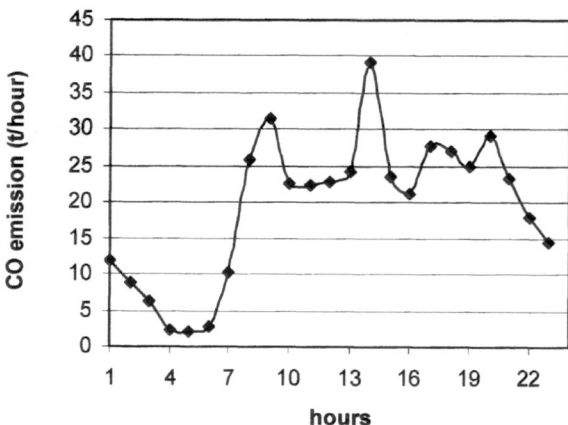

Figure 2. Average daily distribution of CO traffic emission calculated by Tesea for the city of Milan (i.e. temporal profile).

dustry, solvent and other product use, extraction and distribution of fossil and other fuels, combustion in energy and transformation small industries, and non-industrial combustion plants (during winter). Other mobile sources, agriculture and natural emissions are merged into the source classified as other area sources (OAS). AOIIP data were supplied spatially disaggregated as t yr^{-1} $cell^{-1}$.

In order to provide the appropriate detail in time and space, the inventory has to be disaggregated to hourly emissions for cell and organic compounds must be speciated according to the chemical scheme used by each model. This task requires a proper temporal modulation and a chemical profile for each source. The temporal disaggregation was deduced by Simoni *et al.* (1998). In particular the June emission represents 8 and 12% of yearly emission for TAS and OAS, respectively. Hourly data were derived by an averaged emission temporal profile referred to an industrial combustion process for TAS and to a natural emission for OAS.

Organic compound profiles, according to SAPRC lumped species, were taken from literature (Silibello *et al.*, 1998) referring to typical industrial and biogenic source. There is no univocal correspondence to Carbon Bond species, therefore CB species were forced to reproduce a similar industrial profile, as previously described for boundary condition.

Traffic emission inventory was estimated using the COPERT II methodology (Ntziachristos *et al.*, 1999), through a software named 'Tesea'. Outside the city of Milan, Italian average traffic parameters were used (vehicle average speed, share of mileage driven on different road classes, etc.), whereas fleet composition was derived from registered fleet. Inside the city of Milan, a more detailed traffic description was instead adopted: roads were divided into six categories according to their structure and typical traffic conditions. A road measurement campaign was performed by evaluating the main traffic parameters (speed and traffic fluxes) of

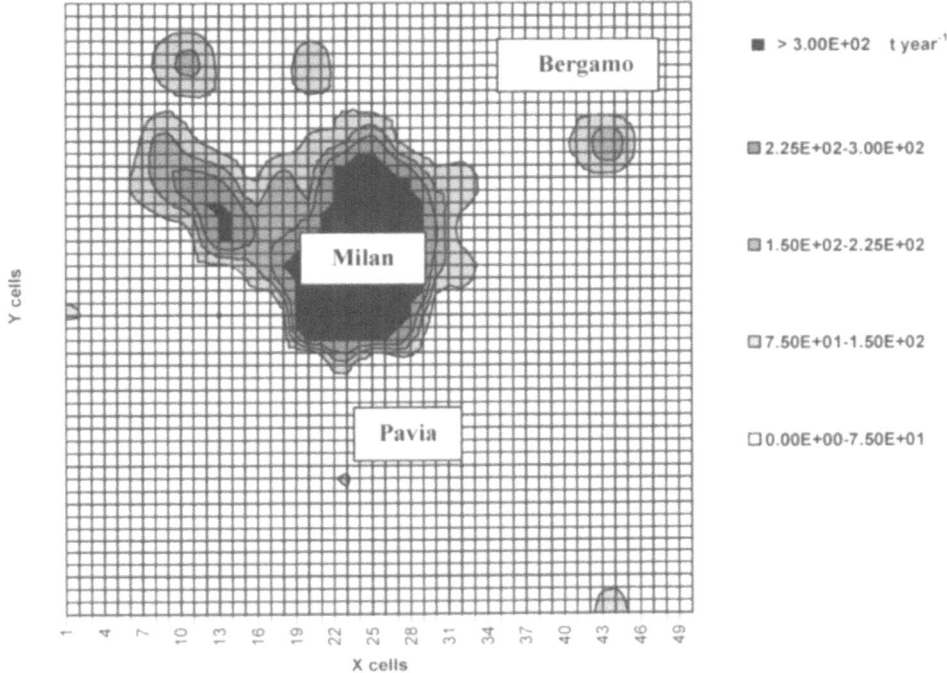

Figure 3. Map of the CO vehicular emissions (t yr^{-1}) over the domain (Tesea output). Scale is from 0 to 300 t yr^{-1}, with an interval of 75, except for last level that includes emissions from 300 to 2500 t yr^{-1}.

each category during week and week-end days. Car emissions were calculated for each road category and then added in every cell of the grid according to the kilometres of each type of road included in the cell. Hourly disaggregation for the traffic source was supplied by those measurements. Based on this temporal disaggregation, the daily CO traffic emission is reported in Figure 2 where midday peaks are shown. Organic compound speciation was performed matching COPERT profiles for different type of vehicles into both SAPRC and CB modelling species.

The CO yearly traffic emission map is reported in Figure 3 which shows the high emission density within the city of Milan. Daily emissions for the whole area according to source type are reported in Table II. From the comparison between the Tesea and similar (Silibello *et al.*, 1998) vehicular traffic inventories (Table III) the CO scenarios appear to be comparable: in fact a reduction of 60% from 1990 to 1998 might be reasonable. If VOC and NO$_x$ are expected to decrease as CO, they would have reduced by 50–60%. As the ratio between the two traffic inventories is only 30% for VOC and NO$_x$, it is possible that those pollutants have been overestimated in our calculations.

TABLE II

Daily emission inventory for the whole simulated area according to source type and chemical species

Compound	Total area sources (t day^{-1})	Vehicular traffic source (t day^{-1})	Other area source (t day^{-1})
CO	41	1249	131
NO$_x$	27	289	28
VOC	121	305	116

TABLE III

Ratio between TESEA and Silibello's vehicular emission inventories on the simulated area according to chemical species

	CO	NO$_x$	VOC
Tesea/Silibello's	0.4	0.7	0.7

4. Simulation Results

4.1. METEOROLOGICAL SIMULATION

The meteorological field for 2–5 June 1998, has been reconstructed by CALMET in the study area, using both the observed meteorological parameters and the geophysical features of the terrain. In order to check the reconstruction of the wind field, results were compared with measurements of the PIPAPO campaign. This phase made possible to tune those parameters which affect the CALMET capability to explain the main features of the local atmospheric circulation. On overall the simulated wind field fairly reproduces the existing wind regimes, though domain limits seem to affect the field close to the Alps.

Regarding those variables where no observed data are available, some remarks must be pointed out. Mixing heights calculated by CALMET range from 100 to 2600 m; at night-time 100 m mixing height, referred to stable conditions, is generally predicted. This affects the Kz evaluation needed for UAM-V. The algorithm has shown an incorrect prediction of nocturnal Kz which is fixed to a minimum constant value that does not represent the usual night-time conditions where heat island enhances vertical mixing. Therefore for urban cells a parabolic-like behaviour was assumed for Kz from 9 p.m to 6 a.m. (with a minimum value of 0.1 m^2 s^{-1}).

4.2. MODEL EVALUATION OF BASE-CASE PREDICTIONS

The two models produced hourly 3-D concentration fields of O_3, NO, NO_2 for the modelled period. In order to closely compare the results, surface maps and concentration time series at selected monitoring points were extracted.

Simulated surface O_3 maps at a selected time (4 p.m., 4 June) are shown in Figures 4a and b for UAM-V and CALGRID, respectively. Both models reveal maximum concentration levels downwind along the N–NE direction, although UAM-V results have lower values. Looking at the spatial distribution around the city of Milan, CALGRID exhibits a depletion zone that starting from the southern boundary limit reaches the city with concentration values about 60 ppb. UAM-V shows similar concentration values in the city of Milan but the depletion area is limited to the city. The main reason for this difference on spatial distribution could be ascribed to meteorological aspects as discussed in other UAM-V/CALGRID comparison studies (Jiang *et al.*, 1998).

In Figure 5, the time series of observed and predicted surface concentrations of O_3 at some of the selected monitoring stations are shown. Observed ozone concentration values at the monitoring stations, generally show the highest ozone concentration (over 80 ppb) at the last two days of simulation. Both models are able to reproduce the day-night cycle of ozone but maximum values are, on average, underestimated. In general the comparison between UAM-V and CALGRID simulation results, revealed a temporal shift and a peak shape broader than the observed daytime peak.

Focusing on the urban station located within the city of Milan (Viale Marche), a specific behaviour stands up for UAM-V results related to the peak shapes at noon. In fact UAM-V simulation shows a double peak which is anti-correlated with the Milan measured hourly traffic emission profile as shown in Figure 2. This feature is not present in the CALGRID results in this station, suggesting that it could be ascribed to the different chemical scheme dynamics of the two models. The same effects are not present on stations outside Milan because a different temporal disaggregation of traffic emission inventory was used for areas outside Milan as previously described.

Pavia station, due to its position, upwind to the advected ozone plume produced by the city of Milan, has a key role to evaluate how much boundary conditions affect the results. Observed ozone concentrations at Pavia are continuously increasing during the simulated period, addressing to a developing regional ozone episode which transports ozone inside the domain. This effect is only partially revealed by the simulations, suggesting that both larger domain and temporal variable boundary conditions can increase the accuracy of the results. As far as Merate station results are concerned, both models predict the observed peak quite well on the second simulation day while on the third UAM-V largely underestimates the maximum value.

Night-time observed ozone concentrations levels and episodic peaks are in gen-

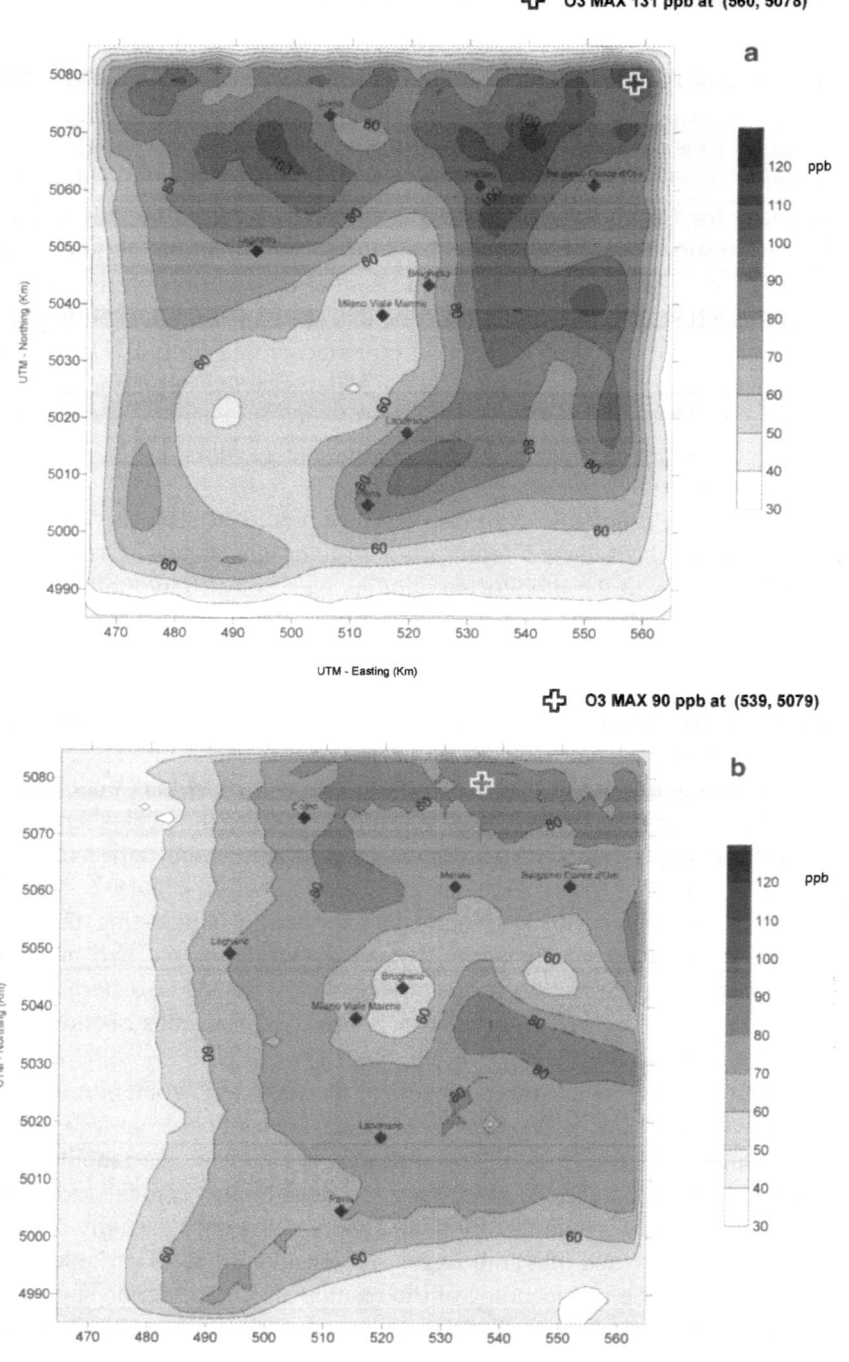

Figure 4. UAM-V (a) and CALGRID (b) ozone surface maps, 4 June at 4 p.m. ♦ Chemical selected monitoring stations.

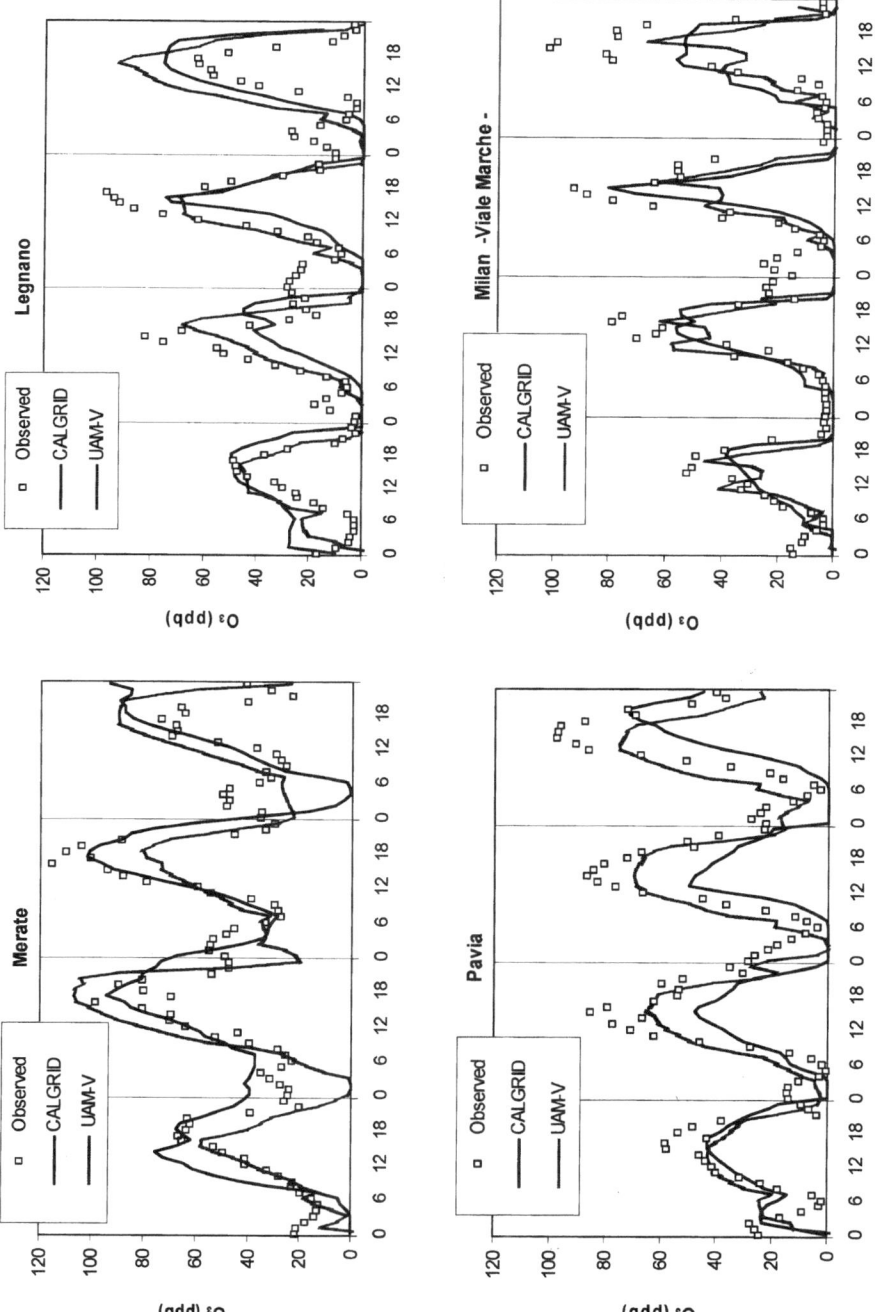

Figure 5. Ozone time series. Simulated and observed data are compared in some of the selected monitoring stations, for the entire episode.

eral not well reproduced by both models. Simulated results underestimate the ozone concentrations during night-time at urban stations like Milan-Viale Marche and Pavia. At Merate station both models are in agreement with high ozone levels on the third simulated night, while on second and fourth nights they have an opposite behaviour.

In order to evaluate the agreement between observed and predicted ozone concentrations, statistical indices can be used. Among them we selected daytime normalised bias D and time unpaired highest prediction accuracy Au defined as follows:

$$D = \frac{1}{N} \sum_{i=1}^{N} \frac{Cp(x_i, t) - Co(x_i, t)}{Co(x_i, t)} \qquad t = 7, 20$$

$$Au = \frac{1}{N} \sum_{1}^{N} \frac{Cp(x_i, t'')\text{max} - Co(x_i, t')\text{max}}{Co(x_i, t')\text{max}},$$

where $Co(x_i, t)$ and $Cp(x_i, t)$ are the observed and predicted O_3 concentration values at monitoring station i for hour t, N is the number of the selected monitoring stations, $Co(x_i, t')\text{max}$ is the daily maximum 1 hr observed concentration at monitoring station i and $Cp(x_i, t'')\text{max}$ is the daily maximum 1 hr predicted concentration over the station surface grid cell.

In Table IV the statistical measures are reported. The negative Au values confirm that the daily maximum ozone concentrations are underestimated by both models although D values indicate that, taking into account the whole peak, the models tend to overpredict. The latter result can be ascribed to the simulated peak shape which is broader than the observed one, as previously described. Statistical evaluations made by another comparison study (Jiang *et al.*, 1998) confirm general underprediction of peak ozone concentrations (-28 and -13% of Au for CALGRID and UAM-V, respectively).

Ozone underestimation suggests further investigation on boundary conditions. It is well known that boundary conditions play, in many applications, a dominant role in producing peak ozone concentration (Kumar *et al.*, 1994; Silibello *et al.*, 1998). Emission inventory accuracy must also be taken into account to improve ozone predictions, especially for areas outside the city of Milan where a less detailed traffic description was adopted.

All stations located inside urban areas are highly influenced by direct traffic emissions and good agreement can be noticed for daytime concentration of ozone and to a certain extent for NO_2 and NO. In particular at Viale Marche station (Figure 6) both models are able to reproduce the low NO values at midday and the early morning NO peak due to emissions and meteorological factors. At the same site NO_2 average concentration, over the simulated period, is reproduced with small differences by the two models.

At night-time large discrepancies for the NO and NO_2 observed values are manifest. Unrealistic peaks are shown by both models and that could be ascribed

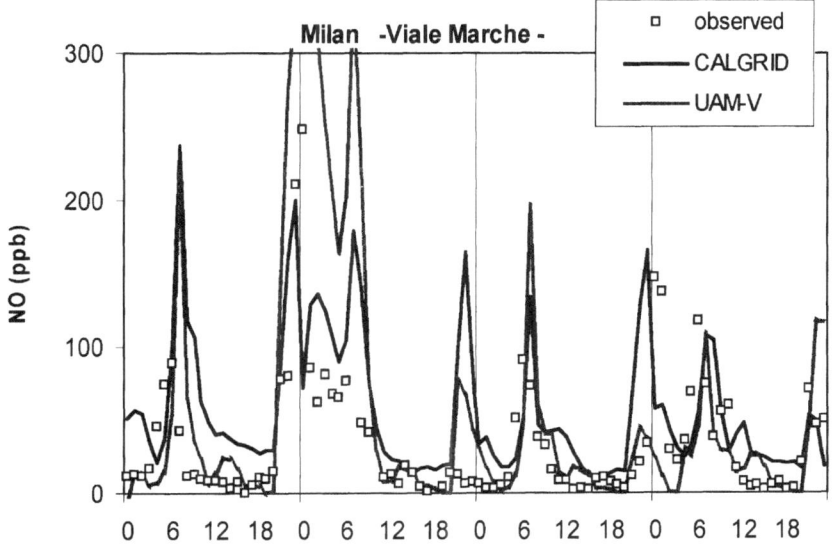

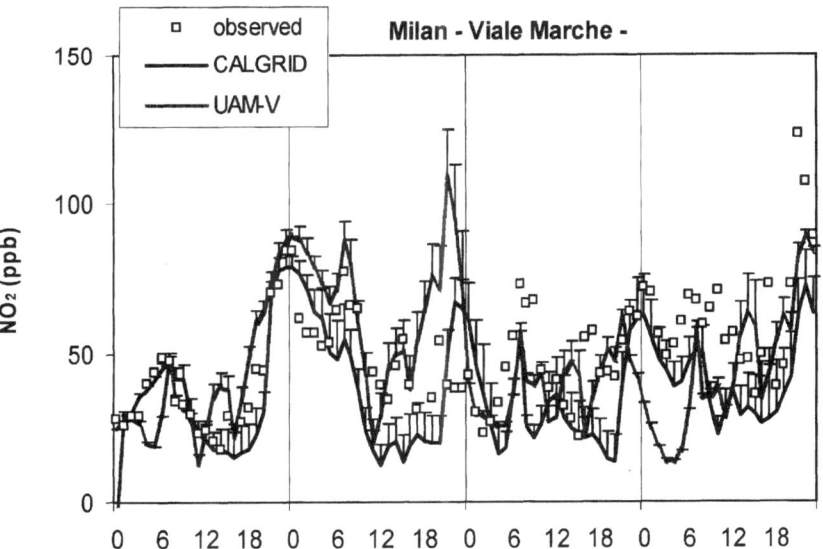

Figure 6. NO and NO$_2$ time series. Simulated and observed data at Milan Viale Marche monitoring station. Bars are added to NO$_2$ simulated data representing the sum of HNO$_3$ and PAN predicted concentrations. In fact these species are measured as if they were NO$_2$ by the chemiluminescent NO$_x$ monitors in current use (Winner and Cass, 1999).

TABLE IV

Statistical evaluation of models performance on reproducing daytime ozone peaks and maximum values in selected monitoring stations

Ozone	D	Au
UAM-V		
2 June	23%	−11%
3 June	48%	−6%
4 June	21%	−23%
5 June	112%	1%
CALGRID		
2 June	71%	8%
3 June	53%	−5%
4 June	20%	−21%
5 June	48%	2%

to an incorrect parameterisation of the vertical dispersion in urban areas, that suppresses the dilution of the pollutant emitted in the lower layers.

5. Conclusions and Recommendations

The first comparison of the application of UAM-V and CALGRID models on an Italian area has been performed for the ozone episode registered in the Lombardia region over 2–5 June, 1998. Furthermore a novel traffic emission inventory has been designed and constructed in order to provide updated input for the photochemical models. Meteorological and 3-D concentration fields have been produced for the modelled domain.

The comparison study has shown that both models are able to consistently reproduce the ozone daytime trend nevertheless the accuracy is far from satisfactory. Average normalised bias for both models is about 50% and peak values are underestimated up to 23%. Both models predict the highest ozone concentrations along the NE direction.

Small differences can be ascribed to the different chemical scheme adopted by the models, in particular UAM-V is more sensitive to emission temporal patterns.

Deviations from observed values for predicted pollutants were registered over-night. That evoked difficulties in meteorological description and matching of variables like night-time vertical exchange coefficients.

The results obtained from both models direct to a better determination of boundary conditions. The option of the enlargement of the simulation domain, which should at least cover the entire Lombardia Region, and the use of nesting techniques could solve this limitation.

It must be pointed out that, in order to make the comparison study, the UAM-V model has been used without the new features available for this version respect to the previous one (UAM-IV) especially the prognostic meteorological model and the nesting capabilities. The use of these features could probably enhance UAM-V performances but would have carried on different working conditions for the two models, preventing comparison.

References

Allwine, K. J. and Whiteman, C. D.: 1985, *MELSAR: A Mesoscale Air Quality Model for Complex Terrain, Vol. 1, Overview, Technical Description and User's Guide*, Pacific Northwest Laboratory, Richland, Washington.

Carter, W.: 1990, 'A detailed mechanism for the gas-phase atmospheric reactions of organic compounds', *Atmosph. Environ.* **24A**, 481–518.

Dodge, M. C.: 2000, 'Chemical oxidant mechanism for air quality modelling: Critical review', *Atmosph. Environ.* **34**, 2103–2130.

Gery, M. W., Whitten, G. Z. and Killus, J. P.: 1988, *Development and Testing of the CBM-IV for Urban and Regional Modelling*, EPA/600/3-88/012, U.S. Environmental Protection Agency, Research Triangle Park, NC.

Goodin, W. R., McRae, G. J. and Seinfeld, J. H.: 1980, 'An objective analysis technique for constructing three-dimensional urban scale wind fields', *J. Appl. Meteorol.* **19**, 98–108.

Hanna, S., Moore, G. and Fernau, M.: 1996, 'Evaluation of photochemical grid models using data from the Lake Michigan ozone study', *Atmosph. Environ.* **30**, 3265–3279.

Jiang, W., Hedley, M. and Singleton, D.: 1998, 'Comparison of the MC2/CALGRID and SAIMM/UAM-V photochemical modelling systems in the Lower Fraser Valley, British Columbia', *Atmosph. Environ.* **32**, 2969–2980.

Jiang, W., Singleton, D. L., Hedley, M. and McLaren, R.: 1997, 'Sensitivity of ozone concentrations to VOC and NO_x emissions in the Canadian Lower Fraser Valley', *Atmosph. Environ.* **31**, 627–638.

Kumar, N., Russell, A., Tesche, T. and McNally, D.: 1994, 'Evaluation of CALGRID using two different ozone episodes and comparison to UAM results', *Atmosph. Environ.* **28**, 2823–2845.

Liu, M. K. and Yocke, M. A.: 1980, 'Siting of wind turbine generators in complex terrain', *J. of Energy* **4**, 10–16.

Mayes, P., Moore, G. and Schulman, L.: 1996, 'Comparison of CALGRID, UAM-IV and UAM-V Ozone Predictions in New England', *89th AWMA Annual Meeting and Exhibition*, Nashville, 23–28 June, 96-TA23A.01.

Ntziachristos, L. and Samaras, Z.: 1999, 'COPERT III Computer Programme to Calculate Emissions from Road Transport', *Final Draft Report EEA*.

Pielke, R. A.: 1984, *Mesoscale Meteorological Modeling*, Academic Press, Orlando.

SAI: 1995, *User's Guide to the Variable-Grid Urban Airshed Model (UAM-V), SYSAPP-95/027* ICF Kaiser, System Applications International, CA, U.S.A.

Scire, J. S. and Robe, F. R.: 1997, 'Fine-scale Application of the CALMET Meteorological Model to a Complex Terrain Site. Paper 97-A1313, *AWMA 90th Annual Meeting and Exhibition*, 8–13 June, Toronto, Ontario, Canada.

Silibello, C., Calori, G., Brusasca, G., Catenacci, G. and Finzi, G.: 1998, 'Application of a photochemical grid model to Milan metropolitan area', *Atmosph. Environ.* **32**, 2055–2038.

Simoni, P. and Angelino, E.: 1998, 'La predisposizione degli inventari di emissione per lo studio dell'inquinamento fotochimico', *Ingegneria Ambientale* **27**(1–2), 16–23.

Skouloudis, A.: 1999, 'Joint Research Center ISPRA' (personal communication).

Winner, D. A. and Cass, G. R.: 1999, 'Modeling the long-term frequency distribution of regional ozone concentrations', *Atmosph. Environ.* **33**, 431–451.

Yamartino, R. J., Scire, J. S., Carmichael, G. R. and Chang, Y. S.: 1992, 'The CALGRID: Mesoscale photochemical grid model, I – Model formulation', *Atmosph. Environ.* **26A**, 1493–1512.

Yamartino, R. J., Scire, J. S., Hanna, S. R., Carmichael, G. R. and Chang, Y. S.: 1989, *CALGRID: A Mesoscale Photochemical Grid Model*, Vol. 1, 2.

Ziomas, I., Tzoumaka, P., Balis, D., Melas, D., Zerefos, C. and Klemm, O.: 1998, 'Ozone episodes in Athens, Greece. A modelling approach using data from the MEDCAPHOT-TRACE', *Atmosph. Environ.* **32**, 2313–2321.

INFLUENCE OF TRAFFIC EMISSIONS ESTIMATION VARIABILITY ON URBAN AIR QUALITY MODELLING

C. BORREGO, O. TCHEPEL*, A. MONTEIRO, N. BARROS and A. MIRANDA

Department of Environment and Planning, University of Aveiro, Aveiro, Portugal
(* author for correspondence, e-mail: oxana@dao.ua.pt, fax +351 234 429290)

Abstract. The main objective of this work is to analyse how uncertainties in emission data of nitrogen oxides (NO_x) and volatile organic compounds (VOC), originated from road traffic, influence the model prediction of ozone (O_3) concentration fields. Different methods to estimate emissions were applied and results were compared in order to obtain their variability. Based on these data, different emission scenarios were compiled for each pollutant considering the minimum and the maximum values of the estimated emission range. These scenarios were used as input to the MAR-IV mesoscale modelling system. Simulations have been performed for a summer day in the Northern Region of Portugal. The different approaches to estimate NO_x and VOC traffic emissions show a significant variability of absolute values and of their spatial distribution. Comparison of modelling results obtained from the two scenarios presents a dissimilarity of 37% for ozone concentration fields as a response of the system to a variation in the input emission data of 63% for NO_x and 59% for VOC. Far beyond all difficulties and approximations, the developed methodology to build up an emission data base shows to be consistent and an useful tool in order to turn applicable an air quality model. Nevertheless, the sensitivity of the model to input data should be considered when it is used as a decision support tool.

Keywords: emission inventory, emission uncertainties, mesoscale modelling, photochemical pollution, road traffic emissions

1. Introduction

Dispersion and photochemical models are widely used as a tool in air quality management to predict the concentration fields of different pollutants. Therefore, errors of model prediction are of great concern. There are several sources of error integrated in modelling output mainly related to: (i) incorrect or incomplete implementation of physical and/or chemical mechanism, (ii) the stochastic nature of the phenomena to be simulated, and (iii) input information (Britter, 1994). The last component, the input data error, can have a significant contribution to the total model error. In particular, the emission data are identified as one of predominant error source in air quality modelling because of high uncertainties in their estimation (Ebel *et al.*, 1997).

There are two distinct approaches to calculate atmospheric emissions with the spatial and temporal resolution required by air quality models (Baldasano, 1998). The first one, 'top-down' approach, concerns the disaggregation of the existing

Water, Air, and Soil Pollution: Focus **2**: 487–499, 2002.

data by applying statistical indicators. The second method consists on the calculation of emissions based on local dataset together with emission factors and is referred as 'bottom-up' approach. Since this approach involves detailed information for emissions estimation, results would be more accurate from bottom-up methodology than after disaggregation process. However, this methodology is time-consuming, requires the availability of large amount of data and involves the application of some specific tools such as Geographical Information Systems. Due to these factors, the top-down approach is still widely used for compiling the emissions as input to air quality models (Palacios et al., 2001). It is an important task to compare the emission results from different disaggregation methods and approaches and to estimate their range of variability.

The main objective of this work is to analyse how uncertainties in emission data of nitrogen oxides (NO_x) and volatile organic compounds (VOC), originated from road traffic, influence the model prediction of ozone (O_3) concentration fields.

2. Methodology

The adopted methodology is clearly related with two main subjects of the work: traffic emission estimation and photochemical production modelling. Firstly, different methods to estimate emissions were applied and results were compared in order to obtain their variability. The emission values were used by the photochemical modelling system, aiming to investigate the influence of the uncertainties associated to the generation of emission data on air quality modelling results.

2.1. EMISSION DATA

The most recent Portuguese atmospheric emissions inventory is based on the CORINAIR methodology (CORINAIR, 1994) and reports to emissions at national level. In order to obtain the resolution required by air quality models, the emission data have to be disaggregated.

The statistical indicators used for data disaggregation are a crucial point and can significantly affect the final results. There is no general rule to select an appropriate factor for disaggregation of emissions coming from road traffic. Therefore, several statistical parameters are considered, taking into account their advantages and disadvantages. In present work, the following indicators have been used:

(i) Gasoline consumption. The detailed information about gasoline consumption is available for each municipal unit; however, other fuel types used by road transport are ignored.
(ii) Gasoline and diesel consumption. In this approach, the two principal fuel types used by vehicles are considered. The diesel consumption data represent total values for each municipality and do not distinguish between transport and other activities and this is the important disadvantage of the method.

(iii) Total vehicles miles travelled. These data reflect not only fuel consumption, but also the vehicle split for each municipality. The main limitation of this approach is that the referred values are results from calculations and not statistical information.

Each of these parameters was separately applied to disaggregate national emission data to the municipality level. As a next step, the emissions for municipal level (NUT IV), obtained on the basis of the different disaggregation factors, were processed applying Census data in order to obtain sub-municipal (NUT V) resolution and then transformed to the regular matrix required by the model. A comparison of the emission values coming from this approach is discussed below.

The bottom-up methodology was also considered to estimate road traffic emissions of NO_x and VOC for the study area. In the present approach, main roads were processed as line sources and emission calculations were based on daily mean traffic measured at several points. Emissions factors used in calculations are derived from the COPERT III methodology (COPERT III, 2000). These emission factors depend on several parameters, such as vehicle types and age, vehicle speed, etc. The application of this methodology has been described in more detail in Borrego et al. (1999a).

2.2. MODELLING SYSTEM

In this study, a version of the Systems Applications International Mesoscale Model (SAIMM) (Kessler and Douglas, 1992), a prognostic 3-D meteorological model, was used to generate the meteorological inputs needed by the photochemical Urban Airshed Model (UAM – CB IV) (U.S. EPA, 1990). These models were integrated in a numerical system named MAR IV (Modelo Atmosférico Regional) (Barros, 1999), especially developed to be applied to coastal regions. The MAR-IV system was evaluated with experimental data and its performance was also compared with other models (Barros, 1999; Borrego et al., 2000). In both cases, the system shown to be consistent, giving realistic results.

The applications of the MAR-IV system were performed over a modelling domain of 200×140 km^2 with a horizontal grid resolution of 5×5 km^2. The domain is located in the Northern part of Portugal including Oporto and other urban areas (Figure 1).

Due to the significant role played by the sea-breeze circulations on the study region, a summer day (20 August 1997) characterized by a typical summer synoptical forcing was selected to perform the numerical simulations. Several emission scenarios were defined in order to study the model sensitivity to the variability in emission input data. The methodology to construct these scenarios is presented in the next section.

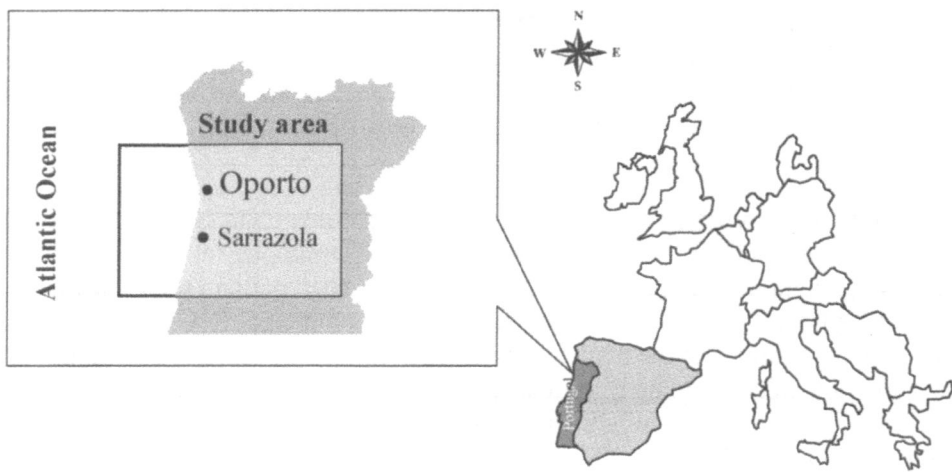

Figure 1. The study area used for mesoscale modelling.

3. Results and Discussion

Following the described methodology, emission results from the application of the different approaches and modelling results will be separately presented and discussed.

3.1. EMISSION DATA

Five matrixes of NO_x and VOC road traffic emissions estimated from different approaches were analysed. These matrixes correspond to: A – disaggregation based on gasoline consumption data; B – disaggregation based on consumption of gasoline and diesel data; C – total vehicle miles travelled; D – updated 1990 inventory data applying growth factors to an already disaggregated emission inventory (Borrego *et al.*, 1999b); and E – emissions estimated by the bottom-up methodology. Table I presents the total NO_x and VOC emissions calculated, for the entire simulation domain, by the five methods.

The emission values for the selected domain have a significant difference not only in their spatial distribution, but also in the total value as will be discuss below. To evaluate the emission estimations it is important to provide their uncertainty. Two main sources of uncertainties could be identified (IPCC, 1997): emission factors and activity data. In the scope of this work only uncertainty of activity data was analysed, due to the absence of the information needed to conclude the same analysis for the emission factors.

As presented in Table I, the total values of NO_x and VOC for the study area are relatively low for bottom-up estimation in comparison with other matrixes. The same tendency was observed in previous works when data comparison was performed for the Region of Lisbon (Borrego *et al.*, 1999a). One possible cause for

TABLE I

Comparison of total traffic emission values obtained by different methodologies

Pollutant	Top-down				Bottom-up
$(t \, yr^{-1})$	A	B	C	D	E
NO_x	68200	69534	68497	62722	48957
VOC	46945	47869	47153	43950	24967

these results is the incompleteness of the road network considered in estimations. In order to compute the data incompleteness, the bottom-up approach used for road traffic emission as described above, was also applied for estimation of fuel consumption. The results were compared with statistical information on fuel consumption within the study area. A significant difference between the estimated and statistical values was found for diesel consumption, since the industrial consumption of diesel is also contemplated in the statistical report. However, the difference between estimated and statistical values for gasoline consumption corresponds only to 14%. This value could be used to characterise the incompleteness of the activity data and consequently as an indication of resulting emissions uncertainties. It is important to stress that this parameter does not represent the overall uncertainty of the emission estimations, because the error arising from emission factors is not included. Nevertheless, the difference in total values presented in Table I reveals a divergence much higher than 14% for both NO_x and VOC, which can be related to the uncertainties of emission factors.

To provide the input to the photochemical model, two emission scenarios were compiled on the basis of maximum and minimum values, for each cell, obtained from the five approaches above described. Figure 2 presents the spatial distribution of minimum (a) and maximum (b) values of VOC emission calculated for each grid cell. In order to characterise the data variability, a maximum range between the two scenarios (600 kg hr^{-1} for NO_x and 255 kg hr^{-1} for VOC) and a standard deviation of the differences (63% for NO_x and 59% for VOC) were calculated.

As can be observed in Figure 2, spatial distribution of the emission values is well correlated with the road network. In the case of the minimum scenario, emissions are limited by the bottom-up approach (in road absence the emission is '0'). In the case of the maximum scenario, the bottom-up estimations are also dominant due to the highest values associated to the line sources while the top-down methodology emissions are estimated for the area sources and consequently represent more disperse low values.

To preserve the consistency of ozone concentration prediction by the photochemical model, all non-road emissions including anthropogenic and biogenic sour-

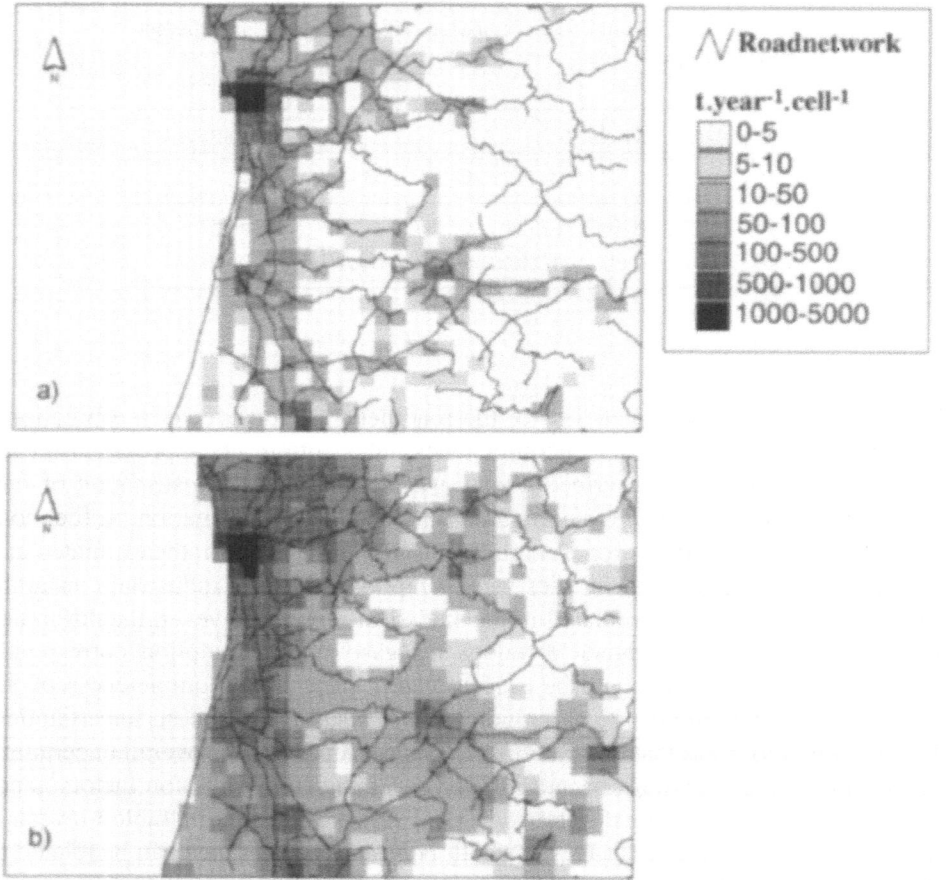

Figure 2. (a) Minimum and (b) maximum scenarios for VOC traffic emissions used in photochemical model.

ces were also considered as input to the modelling system. The methodology for compilation of non-road emissions for the study area is described in Borrego *et al.* (1999b).

3.2. MODELLING SYSTEM

It is possible to determine the contribution of input emission data for model prediction by applying the numerical system for the several emission scenarios. Analyses of the modelling results were performed, comparing the number of grids cells exposed to ozone concentration that exceed the defined limits: 180 μg m^{-3} (warning threshold of 92/72 EEC Directive), 240 μg m^{-3} (proposed alert threshold of COM (2000) 613 final) and 360 μg m^{-3} (alert threshold of 92/72 EEC Directive) (Table II).

TABLE II

Comparison of results obtained through the air quality model simulations

	Number of cells with O_3 concentrations higher than		
	$180 \ \mu g \ m^{-3}$	$240 \ \mu g \ m^{-3}$	$360 \ \mu g \ m^{-3}$
Minimum scenario	18	0	0
Maximum scenario	77	4	0

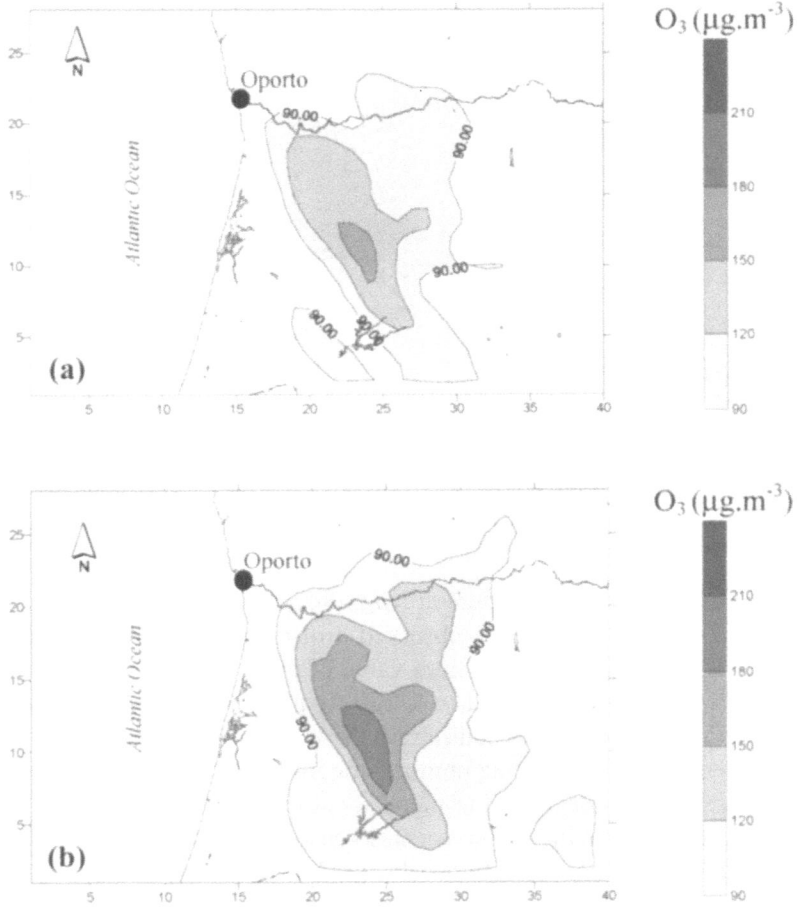

Figure 3. Simulation results of the photochemical model for the (a) minimum and (b) maximum emission scenarios for O_3 concentrations, at 15 UTC.

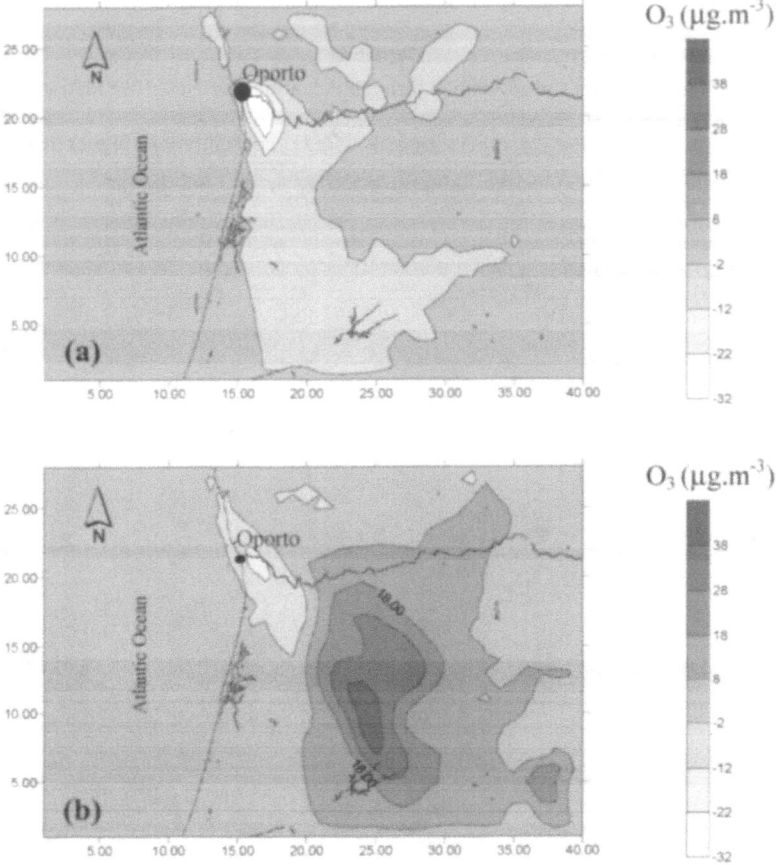

Figure 4. Spatial representation of the difference between the maximum and minimum scenarios for O$_3$ concentration fields at (a) 10 UTC and at (b) 15 UTC.

In Table II, and as expected, there are significant divergences in the magnitude of data values. In Figure 3 can be observed a discrepancy for ozone model prediction at 15 UTC showing a maximum value of circa 150 μg m^{-3} for one of the scenarios and concentration values above 200 μg m^{-3} for the other one. During the simulation period, the number of cells above the warning threshold in the maximum scenario exceeds about 4 times the number of cells above the warning threshold in the minimum scenario. Besides that there are some cells exceeding the proposal alert threshold only in the maximum scenario case. Nevertheless, the spatial distribution of surface ozone concentration fields predicted by the photochemical model for the minimum and the maximum scenarios shows a similar pattern as confirmed by Figure 3. Additionally, a standard deviation of 37% for ozone concentration differences has been estimated. This model prediction discrepancy is due to the input data variation used in the scenarios compilation.

Figure 4 shows a spatial representation of the difference between the maximum and the minimum scenarios for ozone concentration fields at 10 UTC (a) and 15 UTC (b). In the early morning, freshly emitted NO, linked to the traffic rush hours emissions, locally scavenge ozone (Borrell and Borrell, 2000). This titration process works as an ozone sink and can be the explanation of the verified large ozone level depression (above -32 μg m^{-3}), in particular around the Oporto urban area, as presented in Figure 4a. In the afternoon (Figure 4b), the mechanism is still active but less extensive. Nevertheless, downwind from the Oporto urban area, the rise in radiation allows the increase of ozone formation, which is more effective in the case of maximum emission scenario, and explains the observed positive difference.

In Figure 5, the daily variations of the difference in O$_3$ concentration and NO$_x$ emissions between the maximum and minimum scenarios for all surface domain cells are represented. During night, in absence of radiation, the amount of precursors emission plays a very small role; the deposition process and the night atmospheric chemistry scavenge limit the ozone atmospheric concentration (Zannetti, 1990). These processes are more effective in the maximum scenario (higher NO emissions) and explain the presence of negative values in the graph of ozone concentration difference between both scenarios during more intense traffic hours (Figure 5a). The number of negative values decreases during the day and is confined to the more intense traffic emissions. Nevertheless, the traffic emissions also contribute for downwind ozone formation, and very high positive differences (nearly 60 μg m^{-3}) can be observed during the day and the rush traffic hours, linked to NO$_x$ emissions amount (Figure 5b).

In Figure 6, the daily variation of the VOC/NO$_x$ concentration ratio for all surface domain cells for the maximum scenario is presented. Results for the minimum scenario case have a similar daily distribution. As can be observed, there is a clear difference between the diurnal and night patterns of ratio values. During the night and early morning, there is a significant number of cells with a relatively low ratio mainly as a consequence of nocturne chemistry of ozone consumption, which is consistent with the results observed in Figure 4. In the afternoon period, a significant part of the domain presents relatively high ratio values that induce the ozone production verified in the simulation results. The highest values of VOC/NO$_x$ ratio correspond to the cells in remote domain areas that are characterized by a considerable forest area, which is an important source of VOC biogenic emissions. In almost absence of anthropogenic NO$_x$ emissions in these remote areas, the high biogenic VOC emissions cause the rise of the numerical ratio value.

In order to verify the performance of the MAR-IV system for this particular application, a simple point-to-point comparison has been done for a single air quality station (Sarrazola) with available data for the simulation day (Figure 7). The MAR-IV system has a quite good representation of the reality for this particular station. Nevertheless, after sunset, the ozone concentration predicted by the Model System rapidly decreases in opposition to the ozone level recorded by the air

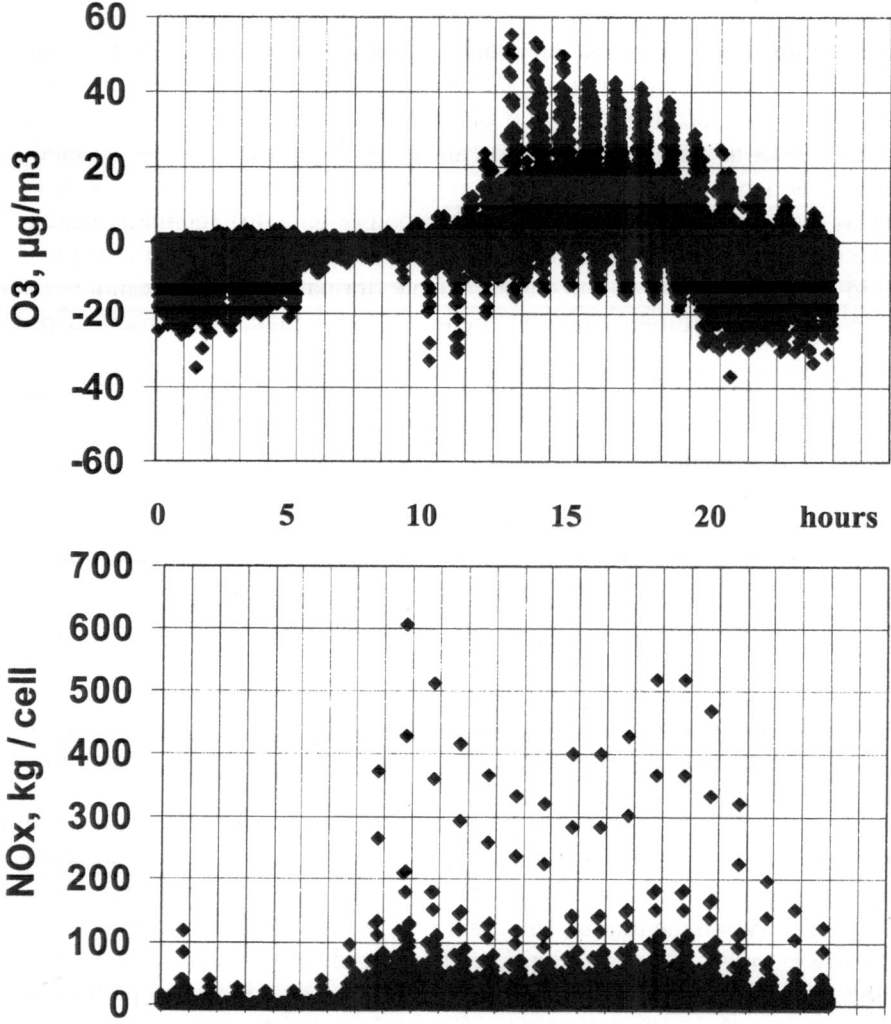

Figure 5. Daily variations of the difference in O_3 concentration (a) and NO_x emissions (b) between the maximum and minimum scenarios, for all superficial domain cells.

quality station. Another important point is that a very little difference was verified for the ozone predicted by the different scenarios. This fact can be explained by the remote position of the air quality station, which is not under direct influence of traffic emissions and of the downwind plume of the most important urban areas and roads.

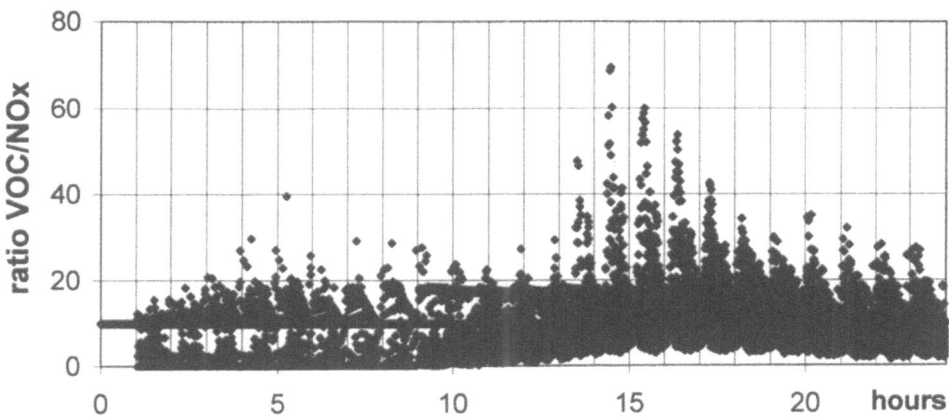

Figure 6. Daily variation of the VOC/NO$_x$ concentration ratio for all surface domain cells for the maximum scenario.

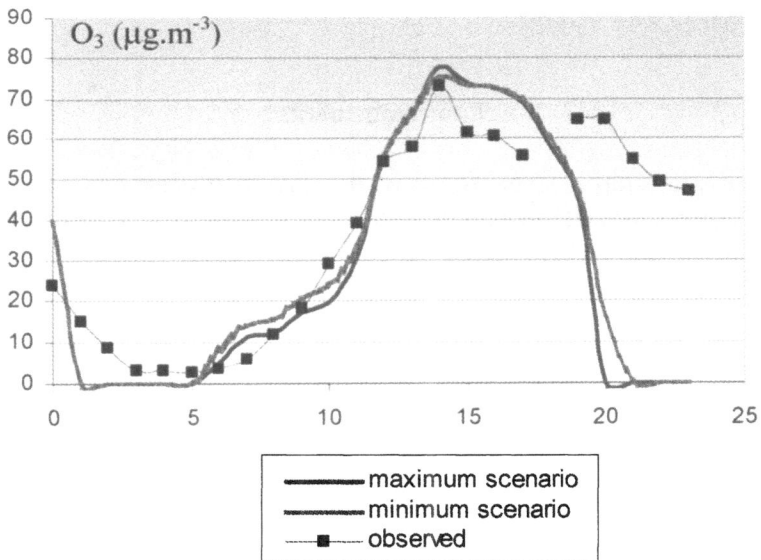

Figure 7. Comparison of the simulation results with observed data at Sarrazola station.

4. Conclusions

Different techniques of emission data generation with high spatial and temporal resolutions have been described and compared. This comparison demonstrates that there is a considerable divergence between the applied methodologies regarding the total absolute domain values (reaching 20 000 t yr^{-1} for both considered pollutants) and their spatial distribution. Aiming to measure the uncertainties of emission estimations, the data incompleteness has been calculated and corresponds to 14% for the study domain. The range of variability and the standard deviation for NO_x and VOC differences were also studied and considered for scenarios construction.

Analyses of the photochemical modelling results show that the variability in the input emission data influences significantly the air pollutant concentrations. A variation in the input emission data of 63% for NO_x and 59% for VOC produces a dissimilarity of 37% for ozone concentration fields. However, validation of the results based on a remote station observations do not reflect the uncertainties of model predictions for the overall domain due to the insignificant variability in emission estimation for the cells around the station and to their placement out of the urban plume influence. A comparison between results from the simulations considering the maximum and minimum scenarios indicates an increase of grid exposure to concentrations above the Directive warning threshold. Although all difficulties and approximations, the developed methodology to build up an emission data base shows to be consistent and an useful tool in order to turn applicable the air quality model. Nevertheless, the sensitivity of the model to input data should be considered when used as a decision support tool.

Acknowledgements

This work was developed in the scope of the CZCM (Centre of Coastal Zones and Sea) research unit and the authors would like to express their gratitude to the National Science and Technology Foundation for the Ph.D. Grant of O. Tchepel. Part of the research is included in the Subproject SATURN (EUROTRAC).

References

Baldasano, J.: 1998, *Guidelines and Formulation of an Upgrade Source Emission Model for Atmospheric Pollutants*, Air Pollution Emissions Inventory, CMP.

Barros, N.: 1999, 'Poluição Atmosférica por Foto-oxidantes: O Ozono Troposférico na Região de Lisboa', Ph.D. Thesis, Department of Environment, University of Aveiro, Portugal, 227 pp.

Borrego, C., Tchepel, O., Barros, N. and Miranda, A. I.: 1999a, 'Impact of road traffic emissions on air quality of the Lisbon region', *Atmosph. Environ.* **34**(27), 4683–4690.

Borrego, C., Barros, N., Lopes, M., Conceição, M., Valinhas, M. J., Tchepel, O., Ferreira, C., Coutinho, M. and Lemos, S.: 1999b, 'Emission inventory for simulation and validation of meso-scale models', in P. M. Borrel and P. Borrel (eds), *Proceedings of EUROTRAC Symposium '98*, WITPress, Southampton, England, pp. 715–719.

Borrego, C., Barros, N., Miranda, A. I., Carvalho, A. C. and Valinhas, M. J.: 2000, 'Validation of two photochemical numerical systems under complex mesoscale circulations', in S. Gryning and E. Batchvarova (eds), *Proceedings of the 23rd NATO-CCMS International Technical Meeting on Air Pollution Modelling and its Application*, Varna, Bulgaria, 28B2, October 1998, pp. 597–604.

Borrel, P. and Borrell, P. M.: 2000, 'Transport and chemical transformation of pollutants in the troposphere', in *Transport and Chemical Transformation of Pollutants in the Troposphere*, Vol. 1, EUROTRAC Special Report, Springer, Germany, 474 pp.

Britter, R. E.: 1994, 'The Evaluation of Technical Models used for Major-accident Hazard Installations', *Report EUR 14774*, European Communities-Commission, Luxembourg, 42 pp.

COPERT III: 2000, *Methodology and Emission Factors*, (Version 2.1), European Environment Agency, 86 pp.

CORINAIR: 1994, *Technical Annexes*, Vol. 2, Default Emission Factors Handbook, European Commission, EUR 1286/2 EN.

Ebel, A, Friedrich, R. and Rodhe, H.: 1997, 'Tropospheric modelling and emisson estimation', in *Transport and Chemical Transformation of Pollutants in the Troposphere*, Vol. 7, EUROTRAC Special Report, Springer, Germany, 440 pp.

IPCC: 1997, 'Greenhouse Gas Inventory Workbook, Vol. 2', in J. T. Houghton, L. G. Meira Filho, B. Lim, K. Tréanton, I. Mamaty, Y. Bonduki, D. J. Griggs and B. A. Callander (eds).

Kessler, R. C. and Douglas, S. G.: 1992, 'User's Guide to the Systems Applications International Mesoscale Model', *SYSAPP-92/001*, San Rafael, California.

Palacios, M., Martín, F. and Cabal, H.: 2001, 'Methodologies for estimating disaggregated anthropogenic emissions – Application to a coastal Mediterranean region of Spain', *J. Air Waste Manage. Assoc.* **51**, 642–657.

U.S. Environmental Protection Agency: 1990, *User's Guide for the Urban Airshed Model*, Research Triangle Park, N.C., Office of Air Quality Planning and Standards, EPA Report Nos. EPA-450/4-90-007A.

Zannetti, P.: 1990, *Air Pollution Modeling – Theories, Computational Methods and Available Software*, Van Nostrand Reinhold, New York, 444 pp.

AIR POLLUTION MODELLING AT AN URBAN SCALE – RUSSIAN EXPERIENCE AND PROBLEMS

E. L. GENIKHOVICH*, I. G. GRACHEVA, R. I. ONIKUL and E. N. FILATOVA

Department of Monitoring of Air Pollution, Main Geophysical Observatory, 194021 St. Petersburg, Russia

(* author for correspondence, e-mail: ego@main.mgo.rssi.ru, fax: +7-812-247-8661)

Abstract. Urban concentration fields are extremely inhomogeneous and their gradients are very high. It results in specific problems when modelling air pollution at the urban scale that are analysed in this paper. Some examples of misrepresentation of the urban concentration field, computed with the use of the source-receptor type models, are given here. They can help in understanding the nature of the problems we face and reveal deficiencies of the computational technologies in use. Filtering of the computed fields is suggested as an instrument for improving their quality. Its efficiency is proved on the test runs. An algorithm for refinement of the filtered fields is introduced in this paper.

Keywords: air pollution, concentration, field, filtering, modelling, refinement, urban

1. Introduction

The list of Russian national ambient air quality standards includes limit values of short-term (20 min) and long-term (say, one year) 'maximum permissible' concentrations (MPC) for more than two thousand different species. However a list of all air pollutants, systematically monitored in one or more of Russian cities, consists of less than fifty species. It explains the strong emphasis that the Russian national guidelines put on the air pollution modelling. In particular, dispersion calculations should be carried out in the urban scale when setting up emission standards for sources located in a city, designing urban road networks and so on. The computed fields should be mapped in the urban scale to cover the whole city or, at least, its districts. A non-Gaussian multiple-source national regulatory dispersion model OND-86 (see Genikhovich, 1995) is used in Russia in these calculations. This model, as well as Gaussian ones, is based on certain source-receptor relations.

Unlike the Gaussian models, OND-86 directly predicts the upper-limit concentration fields corresponding to the 98th percentile of PDF of concentrations (for the sake of brevity referred further as 'majorant' concentration field), but the technology of application of these models in the urban scale is more or less the same. When using them, the city area is usually covered with a regular grid mesh so that the concentration fields are 'sampled' at regular receptor points. Speaking about big cities, we assume here that the mapping area is about several tens by several tens kilometres. To map such an area, one cannot use the spatial resolution with a grid step D less than about several hundred meters, and the values of concentrations between the receptor points are reconstructed using some of the interpolation techniques. The quality of such a mapping strongly depends on the 'smoothness' of the interpolated field. It is known from the general theory that the larger are the gradients of these fields the more are the errors of interpolation. Unfortunately, the urban concentration fields are extremely irregular reflecting the complex structure of sources of pollution in the city, including low ones, and rapid changes of concentration, especially across the plumes. The gradients are the highest for the receptor points located in the vicinity of ground-level sources, in particular, roads. An additional 'noise' comes from perturbations in the wind flow and turbulence generated by buildings, street canyons and so on. In particular, the distance between two sidewalks in the street canyon is not resolved in the urban-scale dispersion models but the concentrations at these two sidewalks could differ as much as in three times. These small-scale features cannot be resolved with the step D due to 'loss' of the small-scale details of the field on the grid mesh.

In a sense, the sampling of computed urban concentrations at regular grid points which are arbitrarily located relative to the main roads and other sources resembles the sampling of stochastic time series. In particular, the effect of aliasing, i.e., confounding of frequencies, which is well known in the theory of time series (see Blackman and Tukey, 1958), is responsible for a significant fraction of the errors in computed fields. Here, the wave number k_Δ corresponding to the grid step Δ plays the role of the Nyquist frequency. Wave numbers larger than k_Δ cannot be correctly resolved on the grid. They, however, generate artificial oscillations on this grid with wave numbers less than k_Δ, which, actually, are nothing else but errors.

The existing technologies of dispersion calculations in urban areas usually neglect these features of the concentration fields and, in particular, do not account for the 'sampling effects'. As a result, the concentrations computed at the regularly spaced grid points do not represent the air pollution in corresponding grid cells. Consequently, the errors in isolines drawn with the use of the computed values could be significant. The computational technologies aimed to reduce these errors for the source-receptor type dispersion models are discussed in this paper. To make this discussion more focused, the errors related to physical deficiencies of the models, incompleteness of the model description of governing processes and errors in the input data are neglected here.

2. Misrepresentation of the Urban Concentration Fields

The effect of misrepresentation of the urban concentration fields could be understood better on calculations for the following model city. Let us consider four long evenly spaced identical parallel roads, the distance between the neighbouring roads being 2245 m. The origin of the rectangular computational domain is located on

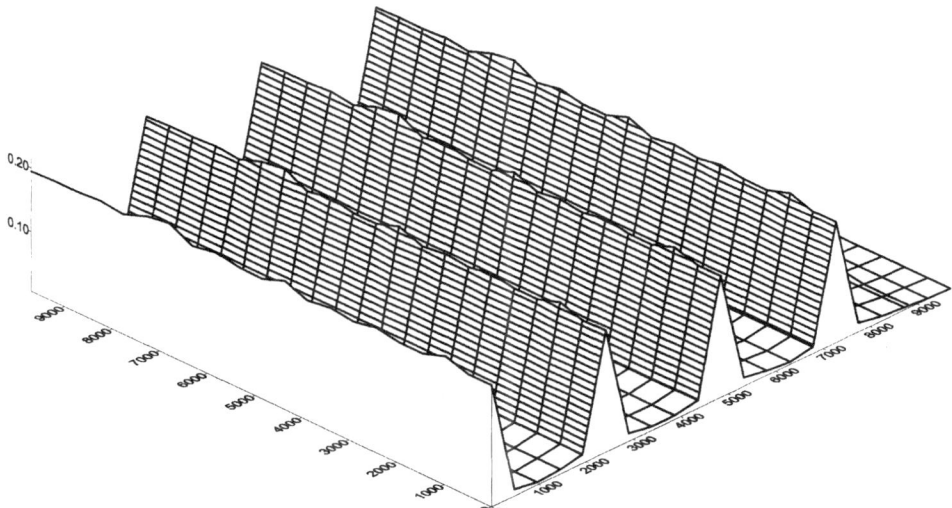

Figure 1. Majorant concentration field calculated for four roads ($\alpha = 0$; $\Delta = 485$ m).

the 'left-hand' road, and α is the angle between the y-axis of the domain and the roads. For the sake of simplicity the roads are represented here as the line sources, and concentrations are calculated numerically as integrals along these straight lines from corresponding concentrations from the point sources.

The majorant concentration field for such a layout calculated with $\alpha = 0$ and the grid step $\Delta = 485$ m along both, x- and y- directions is shown in Figure 1. Here, four 'ridges' correspond to locations of the roads. Small fluctuations in the values of concentrations along these ridges can be attributed to errors in the numerical integration. Corresponding isolines are shown in Figure 2. As one could expect, they are practically parallel to the roads. Numerical tests have confirmed that the same pattern of isolines is generated when doubling the resolution (i.e., when $\Delta = 242.5$ m; in this case the ridges have the same height but their width is decreased). It should be noted here that this and all other maps and graphs are mapped with the use of the procedure of interpolation known as kriging (in meteorological applications this procedure is known as optimal interpolation and its description can be found for example in Gandin, 1963); additional tests were used to prove that the results are not sensitive to interpolation techniques.

One could expect that the pattern of the majorant concentration field should not depend on orientation of the computational domain. These expectations are not supported by the results of computations with $\alpha = 14°$ shown in Figure 3 (the field there was computed with $\Delta = 500$ m). Instead of the ridges, one can see here a kind of a harrow turned upside down. The spikes are attributed to the receptor points located on the roads; they are separated one from another by the areas of low concentrations corresponding to receptor points off the roads. When doubling the resolution, the number of the spikes increases twofold but the pattern of the

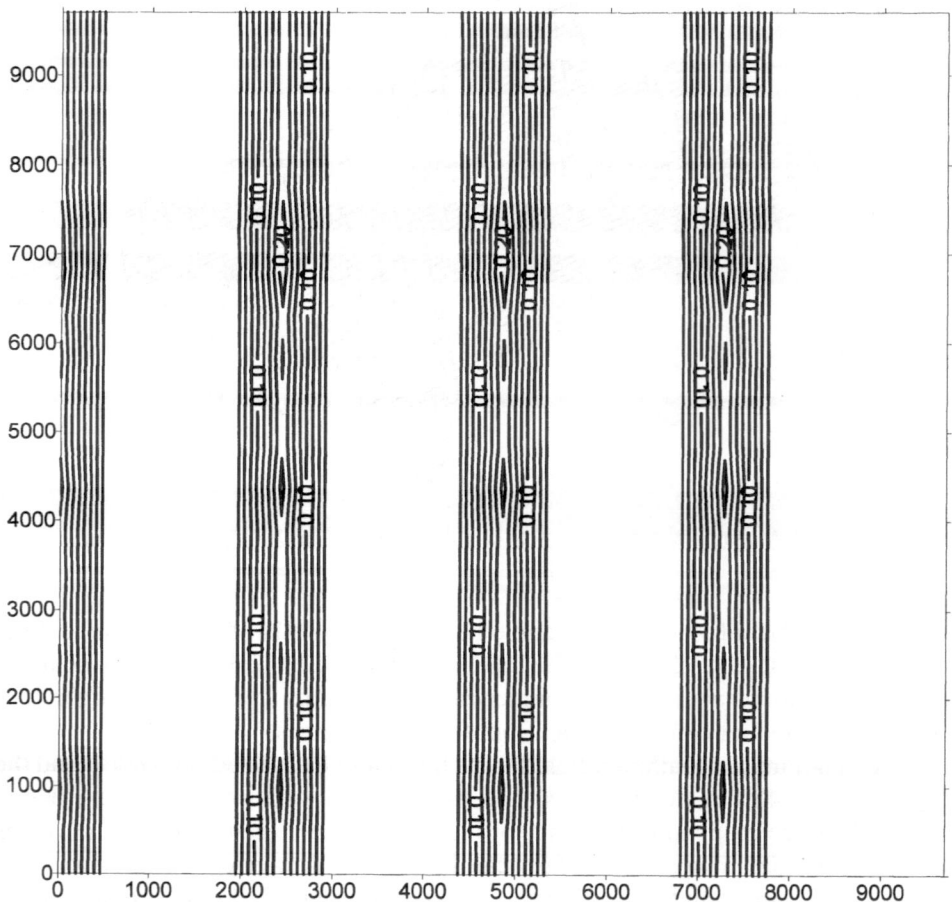

Figure 2. Map of concentrations corresponding to Figure 1. Isolines are at increment of 0.025.

field does not improved too much. It is especially noticeable, if one compares the concentration maps shown in Figures 2 and 4 (they were drawn using the same set of values which define the isolines).

3. Filtering the Concentration Fields

It has been mentioned in this paper that the misrepresentation of the urban concentration field mimics the effect of aliasing. To eliminate this problem, therefore, the oscillations with wave numbers larger than k_Δ should be filtered before mapping the computed field. The simplest way to do it is to average the initial source-receptor relationships over the length scale Δ. In this work, crosswind and along-wind distributions from OND-86 were averaged separately over the same

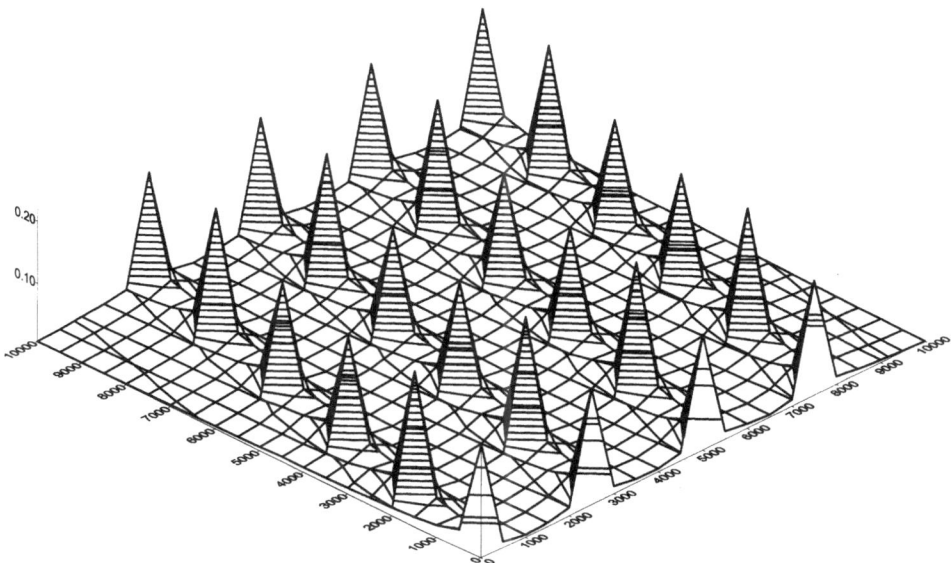

Figure 3. Computed field ($\alpha = 14°$; $\Delta = 500$ m).

Δ (assuming, therefore, equal steps along x- and y-directions). To avoid writing corresponding bulky formulae, one can illustrate this approach on Gaussian models. In this case the crosswind normal distribution $G_y[y;\sigma_y(x)] = \exp[-y^2/2\,\sigma_y^2(x)]/[\sigma_y(x)\sqrt{2\pi}]$ should be, for example, replaced with $G_{y,\Delta}$ which is defined as follows:

$$G_{y,\Delta} = \frac{1}{\Delta\sqrt{2\pi}} \int_{y-0.5\Delta}^{y+0.5\Delta} \frac{1}{\sigma_y(x)} \exp\left[-\frac{s^2}{2\mathrm{sigma}_y^2(x)}\right] ds. \tag{1}$$

A similar expression can be also introduced for $G_{z,\Delta}$.

In case of $\alpha = 0$, the pattern of the filtered, i.e., averaged over the cell, concentration field is essentially the same as those shown in Figures 1 and 2 but maximum values of concentrations are about 50% less (because spatial averaging reduces the peaks in concentrations) and the ridges are wider. More striking difference could be seen when comparing Figure 3 and 5. When doubling the resolution, the pattern of the 'averaged' concentration field presented in Figure 6 resembles those in Figure 2 even closer.

Additional computations of NOx majorant concentration fields were carried out for the real Russian city of Pskov. The emission data in use reflected main sources of NO_x including major roads. Resulting local (non-averaged) and filtered (averaged) fields are shown in Figures 7 and 8. It is evident that the filtered field (Figure 8) reproduces the pattern of the roads better than the local one (Figure 7); thus, spatial filtering helps reduce artefacts generated by a grid of receptors interacting with steep concentration gradients.

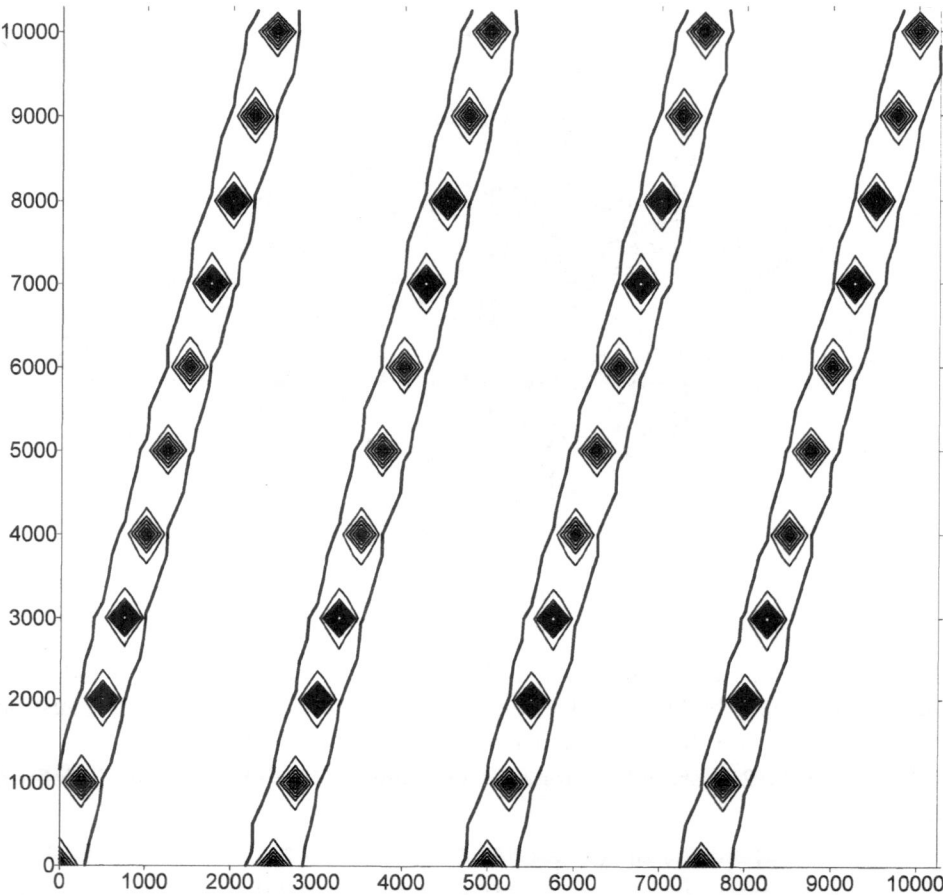

Figure 4. Map of concentrations corresponding to Figure 3. Isolines are at increment of 0.025.

4. Local Refinement of the Field of Concentration

A standard way for testing sensitivity of the computed field to the spatial resolution is to compare the computed values of concentration at the same grid point using the steps Δ and $\Delta/2$. If the results do not differ significantly, they could be considered as a 'final' estimate of the value of concentration at the given receptor point. If the difference is significant, the computed filtered field should be refined in the vicinity of this receptor point using the local-scale dispersion models. A corresponding algorithm will be discussed in this section. It should be noted first that, when calculating the filtered fields, there is no need to account for effects of buildings, structures and street canyons because they correspond to the scales smaller than the resolution of the model (they could be indirectly taken into account by the use of proper values of the roughness length and 'porosity' of the area covered

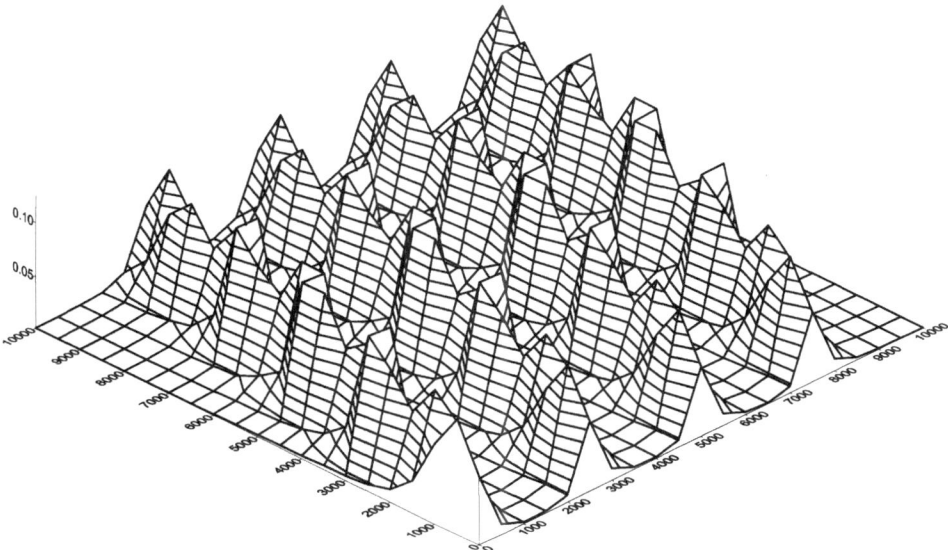

Figure 5. Computed filtered field ($\alpha = 14°$; $\Delta = 500$ m).

with roughness elements, displacement height and other input parameters). On the contrary, in the local-scale models these effects should be directly accounted for.

An algorithm of calculation of the refined concentration field could be described as follows:

1. Run calculations of the filtered concentration with grid steps Δ and $\Delta/2$.
2. Find the receptor points where concentrations are sensitive to variations of the grid step (i.e)., the differences are significant).
3. Find the sources that are responsible for major input into concentration at such a receptor point and sensitive to these variations ('sensitive sources').
4. Run the local-scale dispersion model for a cell surrounding the given receptor point, which is considered as a computation domain, with account for terrain and building effects on dispersion from the sensitive sources.
5. Repeat such a procedure for all receptor points where concentrations are sensitive to variations of the grid step
   A computer program based on this algorithm is now under development.

5. Discussion

In connection with the problem of assimilating the data of monitoring of urban air pollution, we have suggested to decompose the urban field of concentrations into a sum of components corresponding to different spatial length scales (see Genikhovich *et al.*, 2000 b; 2001c). In a sense, a similar idea is exploited in the paper presented here too. The filtered field of concentration, which corresponds

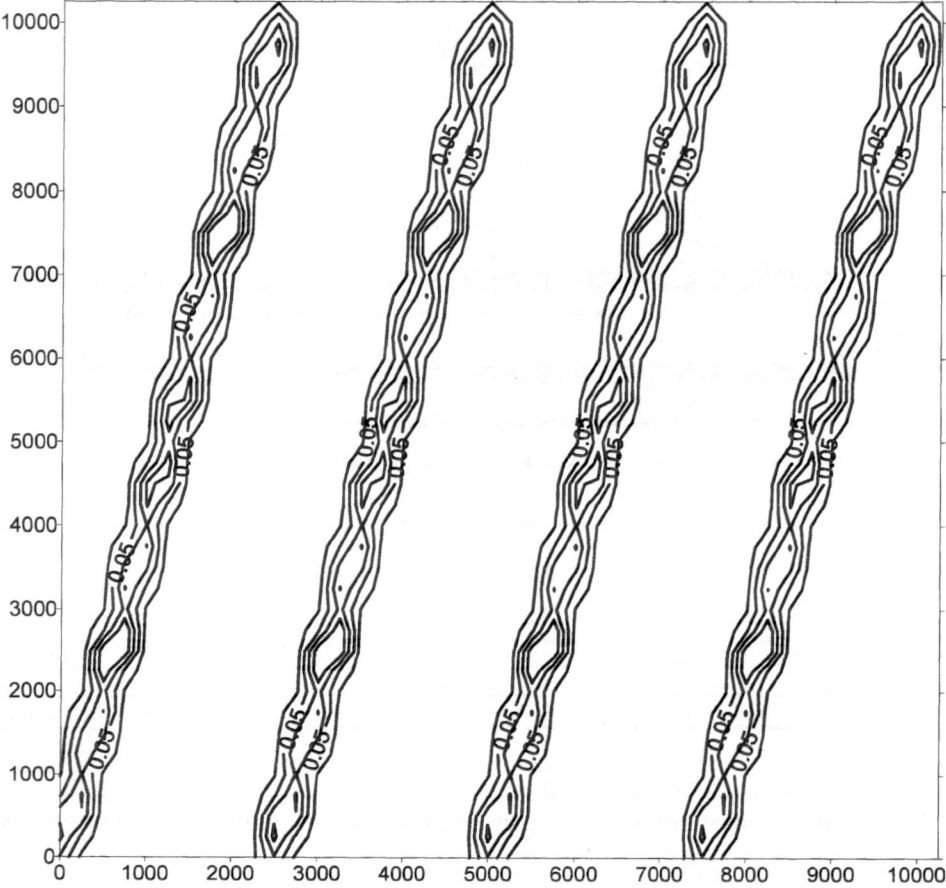

Figure 6. Map of concentrations corresponding to Figure 5. Isolines are at increment of 0.025.

to comparatively large length scales, is used here to map the air pollution in the whole city or its districts, and the local-scale features are simulated without such a filtration. In this case, the urban dispersion modelling should be carried out in two steps: first of them generates the filtered pattern of concentrations over the city to be used for mapping the urban air pollution, and the second one adds the local features.

The less smooth is the computed concentration field the more efficient is the two-step approach introduced in this paper. Nevertheless, it could be beneficial to reduce the "energy" of the small-scale fluctuations before applying this computational procedure. It should be noted in this connection that in many cases actual (corresponding to certain meteorological situations) concentration fields are used only as intermediate results to evaluate either majorant fields of concentration, which correspond to given, for example 98-th, upper percentiles of PDF of concentrations, or long-term averaged, for example mean annual, concentrations.

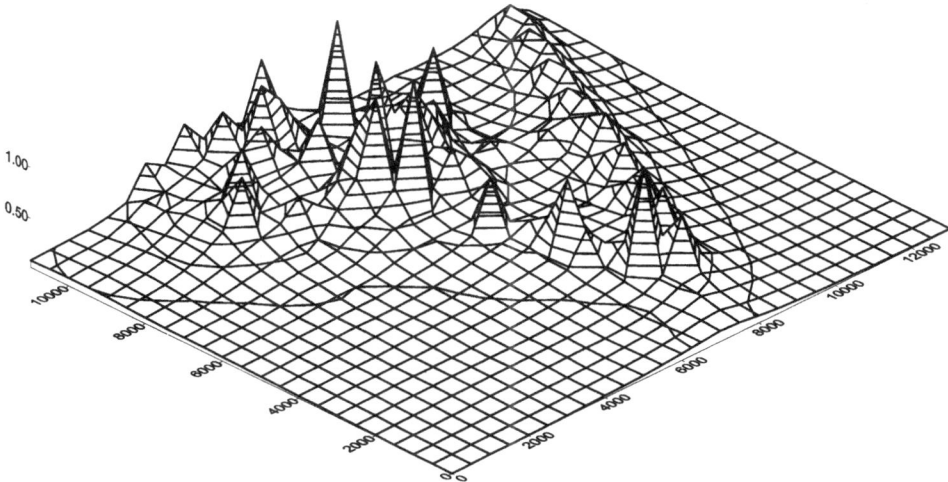

Figure 7. Local NO$_x$ concentration field for the city of Pskov ($\Delta = 500$ m).

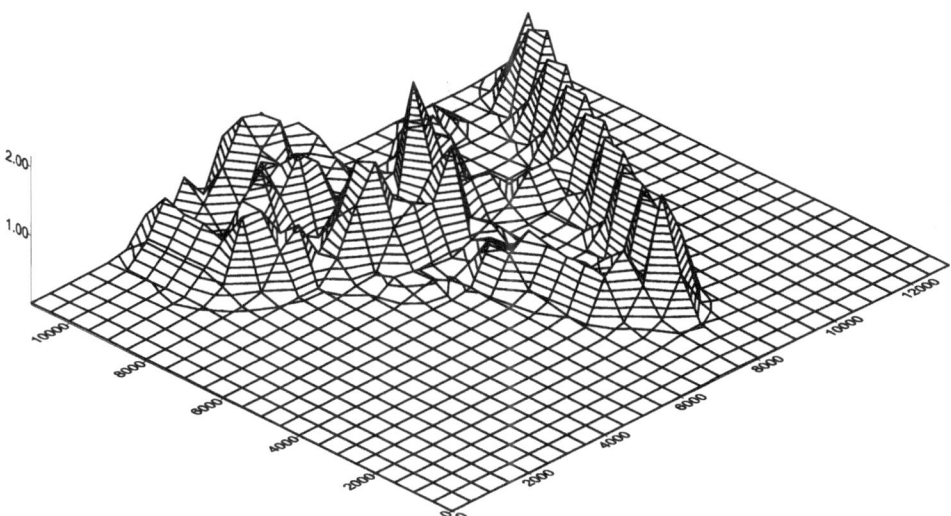

Figure 8. Spatially filtered NO$_x$ concentration field for the city of Pskov ($\Delta = 500$ m).

Both, majorant and mean annual concentration fields for a single point source are much smoother than a narrow plume, which represents the actual concentrations. Consequently, the urban majorant and mean concentration fields are smoother than the actual ones. It was taken into account when developing the regulatory dispersion model OND-86 (Genikhovich, 1995). This non-Gaussian model directly predicts the majorant concentration field. Similarly, the non-Gaussian model MEAN (Genikhovich *et al.*, 1998, 2000a) directly, i.e. without computing the actual concentrations, predicts the mean annual fields.

The approach we suggested actually corrects the difference in technologies of application of source-receptor models and numerical ones to urban-scale dispersion calculations. The applicability of numerical grid models in urban dispersion calculations was questioned by Schatzmann *et al.* (1999) on the ground of high sensitivity of urban concentrations to comparatively small-scale details of geometry of buildings. It should be mentioned, however, that only the problem of reproducing the small-scale features of the concentration fields was analysed in their paper. These features cannot be accounted for properly when mapping the air pollution in the urban scale. Filtered (cell-averaged) fields of concentration are generated 'automatically' when using numerical models based on grid-and final-element methods. From this point of view, the numerical models are completely appropriate for the use in calculations of the urban-scale air pollution. However it should be mentioned that, due to the high level of inhomogeneity of the urban concentration fields, the cell-averaged concentration is more-or-less as bad a representative of all values of concentrations in a cell, as the local one is. In this connection, many questions related to validation of urban-scale dispersion grid models, interpretation of the results of their application and even the results of the instrumental monitoring of the air pollution in cities should be reconsidered. The authors will present corresponding analyses in their following publications. Effects of blocking up the grid cell volume inside the urban canopy by buildings should be taken also into account in the numerical dispersion models, which do not resolve obstacles. They were discussed by Genikhovich *et al.* (2001 a,b) who suggested to use numerical grid models based on equations of hydrodynamics and turbulent diffusion in porous media.

6. Conclusion

It was shown in this paper that the 'random sampling' of values of modelled concentrations at the regular grid of receptor points results in erroneous maps of the concentration field because of significant variability of this field and its high gradients. The corresponding errors could be so large that they can drastically change the pattern of the field. In a sense, the problem of computation of the urban concentration field can be considered as a stiff one (see for example Fedorenko, 1994; Dekker and Verver, 1984). In other words, it is characterized by very large values of the ratio of the scales that should be resolved by the model. Indeed, the largest length scale to be resolved in computation is the city diameter (say, 20 km), and the smaller scale to be accounted for should be not more than the characteristic width of the street (say, 20 m). The ratio of these two scales is 1000 that is the number of the grid steps required in x- as well as in y-direction to resolve at least major features of the field. For two-dimensional computational domains it corresponds to the 'stiffness number' of 10^6. Therefore, special precautions should be taken to assure the quality of the computed concentration fields. It was demonstrated here that filtering of the computed fields (their averaging over the grid cell) could significantly reduce the level of errors and improve the quality of mapping the urban air pollution. A new algorithm of refinement of the filtered field was introduced in this paper too. Taken together, the filtering and refinement constitute a two-step algorithm that could be applied in practical computations.

Presented results and discussion can be considered as a part of the efforts to develop more reliable technologies of modelling and mapping the air pollution in cities. In its turn, it is a part of the broader problem of development of more efficient techniques of computational monitoring of the urban air pollution. In connection with the latest EU framework directives, one could expect that in the near future computational as well as hybrid monitoring (the last one being based on combination of instrumental and computational monitoring) will be more actively used in practical applications.

Acknowledgement

This work was supported by the Russian Foundation for Basic Research under the grant No 98–05–65606.

References

Blackman, R. B. and Tukey, J. W.: 1958, *The Measurement of Power Spectra from the Point of View of Communications Engineering*, Dover, New York, 190 pp.

Fedorenko, R. P.: 1994, *Introduction into Computational Physics*, Institute of Physics and Technology Publishers, Moscow, 526 pp. (in Russian).

Gandin, L. S.: 1963, *Objective Analysis of Meteorological Fields, Hydrometeoizdat*, Leningrad, 1963, 288 pp. (in Russian).

Genikhovich, E. L.: 1995, 'Practical applications of regulatory diffusion models in Russia', *Inter. J. Envir. Pollut.* **4–5**, No 4–6, 530–537.

Genikhovich, E. L, Berlyand, M. E., Gracheva, I. G., Eliseev, V. S., Ziv, A. D., Onikul, R. I., Filatova, E. N., Khurshudyan, L. H., Chicherin, S. S. and Iakovleva, E. A.: 1998, 'Operational model for calculation of long-term averaged concentrations', *Trudy GGO (Transactions of the Main Geophys. Observ.)* **549**, 11–31 (in Russian).

Genikhovich, E. L., Gracheva, I. G., Groisman, P. Ya. and Khurshudyan, L. G.: (2000a), 'A new Russian regulatory dispersion model MEAN for calculation of mean annual concentrations and its meteorological preprocessor', *Inter. J. Environ. Pollut.* **14**, Nos. 1–6, 443–452.

Genikhovich, E. L., Filatova, E. N. and Ziv, A. D.: 2000b, 'A method for mapping the air pollution in cities with combined use of measured and calculated concentrations', *Inter. J. Environ. Pollut.* (in press).

Genikhovich, E., Gracheva, I. and Filatova, E.: 2001a, 'Modeling of Urban Air Pollution: Principles and Problems', Preprints 25th NATO/CCMS International Technical Meeting on Air Pollution Modelling and Its Appl., Louvain-la-Neuve, Belgium, pp. 189–196.

Genikhovich, E., Gracheva, I., Filatova, E., Iakovleva, E. and Ziv, A.: 2001b, 'Characteristic Features of the Urban Field of Concentrations: An Experimental and Theoretical Study', Preprints 3rd International Symposium on Environmental Hydraulics, Tempe, Arizona, 8 p.

Genikhovich, E. L., Filatova, E. N. and Ziv, A. D.: 2001c, 'Adaptive dispersion modelling and its ap-
    plications to integrated assessment and hybrid monitoring of air pollution', in S.-E. Gryning and
    F. Schiermeier (eds), *Air Pollution Modelling and Its Application XIV*, Kluwer Academic/Plenum
    Publishers, NY, 475–480.
Dekker, K. and Verver, J. G.: 1984, 'Stability of Runge-Kutta Methods for Stiff Nonlinear Differential
    Equations', North-Holland, Amsterdam, 332 pp.
Schatzmann, M., Liedtke, J. and Leitl, B.: 1999, 'Dispersion Models for Urban Applications – A
    Critical Assessment of the Present 'State of Application', in *Neuere Entwicklungen bei der Mes-
    sung und Beurteilung der Luftqualitaet, VDI-Kommission Reinhaltung der Luft*, Bericht 1443,
    pp. 99–115.

CRITICAL RECONSIDERATION OF PHASE SPACE EMBEDDING AND LOCAL NON-PARAMETRIC PREDICTION OF OZONE TIME SERIES

P. HAASE, U. SCHLINK* and M. RICHTER

UFZ – Department of Human Exposure Research and Epidemiology, Centre for Environmental Research Leipzig-Halle, Permoserstr. 15, 04318 Leipzig, Germany
(author for correspondence, e-mail: schlink@expo.ufz.de, fax: ++49(341) 235-2288)*

Abstract. Phase space prediction is a feature selection method which tries to exploit non-linear dynamics of an underlying system. We describe and offer a critical reconsideration of this approach, discuss questions of whether non-linear methods are justified by the data, and apply them to ozone time series from single locations. Our main objectives are to obtain air quality forecasts in order to provide public health warnings and to provide an insight into the dynamics of the underlying system. Interestingly, comparable linear data sets (surrogates) have very similar structure and give similar prediction accuracy to that of the ozone data. In this instance there does not appear to be any advantage to applying the phase space approach to univariate time series.

Keywords: chaos, embedding, local non-parametric prediction, missing value reconstruction, non-linear dynamics, ozone concentrations, surrogates

1. Introduction

Recently researchers have outlined the application of non-linear techniques, e.g. those inspired by chaos theory, to ozone data (Raga *et al.*, 1996; Li *et al.*, 1994; Chen *et al.*, 1998; Kocak *et al.*, 2000). However, before applying such methods to real time series data we should first ask whether such advanced techniques are justified by the data. We might expect or even know that there are some non-linear components in the system but this does not prove that non-linearity is also reflected in a specific signal we measure from that system (Kantz and Schreiber, 2000). The method of surrogate data has become a very popular tool to address such questions, although there are some limitations and caveats. Based on ozone data of Berlin, Cincinnati and Siracusa we discuss certain problems in the application of non-linear methods, i.e. strong autocorrelation, periodicity and possibly violated assumptions, like non-stationarity. For comparison we also apply non-parametric techniques to a time series generated by the non-linear Lorenz system that was originally developed in an atmospheric context.

Water, Air, and Soil Pollution: Focus **2:** 513–524, 2002.
© 2002 *Kluwer Academic Publishers.*

2. Methodology

2.1. MISSING DATA RECONSTRUCTION

Missing data reconstruction is an important problem when dealing with real time series. Missing values occasionally occur because of the calibration or failure of automatic monitoring stations. The phase space method requires a continuous time series. For length of gaps greater than 20 h an interpolation based on the Kalman filter which utilises the spectral characteristics of the data series can be used (Schlink et al., 1997). Spectral analysis of ozone time series identified a low frequency part, and 12 and 24 h seasonal parts. A random walk (RW) model was used for the estimation of the trend and an integrated random walk (IRW) model was used for the dynamic harmonic regression of the amplitude of the seasonal parts. The Kalman acts as a low-pass for the trend and as a band-pass for the seasonals. Both the cut-off frequency and the band-width can be tuned by parameters. This way interpolation was made as a pre-processing step prior to phase space embedding and prediction.

2.2. LINEAR ANALYSIS OF TIME SERIES

Linear Gaussian models have dominated the development of time series models for the past decades. They incorporate lagged relationships and a correlation structure. In principle, all linear processes can be modelled with arbitrary accuracy by the well known autoregressive (AR) model. In an AR-model the present outcome is a linear combination of past observations, plus additive Gaussian white noise. The simple linear associations between the variables may be however inadequate to capture interactions and non-linearities in the ozone response. In addition the model order might be extremely large. AR-models can serve as reference model when testing for non-linearity. Since real data are often not Gaussian distributed one usually assumes that a non-linear transformation distorts the output of the Gaussian random process. Such non-linearities are called static, in contrast to non-linearities in a dynamic sense.

2.3. PHASE SPACE EMBEDDING AND LOCAL NON-PARAMETRIC PREDICTION

A time series (x_t) is embedded into a m-dimensional phase space using m-dimensional delay vectors $x_t = (x_{t-(m-1)\tau} \ldots x_{t-\tau} \, x_t)$ where τ is the time delay. Delay vectors represent the variables of the system. The phase space is spanned by these variables and describes the set of all possible states. The first zero crossing of the autocorrelation function gives a good estimate of the delay time τ. An optimal embedding dimension m can be obtained by estimation of the correlation dimension (Grassberger and Procaccia, 1983), false nearest neighbour statistics (Kennel et al., 1992) or phase space prediction (Sugihara and May, 1990). Prediction is done by searching for all neighbours of the point to be forecasted. The average of their

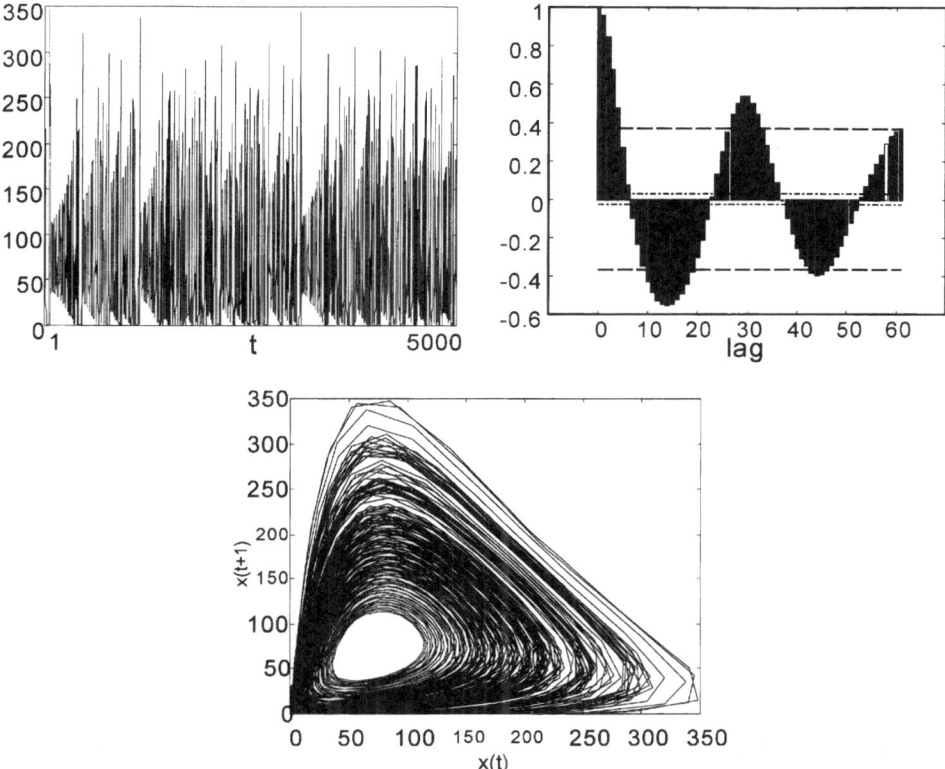

Figure 1. Time series plot, autocorrelation function and two-dimensional delay embedding for generated Lorenz data.

future values gives the predicted value (Hegger *et al.*, 1999). The search for nearest neighbours is based on the maximum norm. The quality of the forecast is measured by the (relative) root mean square error (RMSE).

2.4. PRECONDITIONS

Phase space embedding gives a proper reconstruction of the non-linear dynamics of a deterministic system if the embedding dimension is chosen sufficiently large (Takens, 1981). Whether the assumption of dynamic non-linearity is met or not can be tested by techniques using surrogate data which have been advocated by several authors (e.g. Theiler *et al.*, 1992). The main idea of the method of surrogate data is to compare a non-linear statistics, e.g. error of non-linear forecast, for an ensemble of surrogate data with the value obtained for the original data. Surrogates retain certain linear properties of the original data but are otherwise random. The null hypothesis of a possibly non-linearly transformed Gaussian linear process is rejected if the minimal forecast error for the surrogates is larger than the error for the original data.

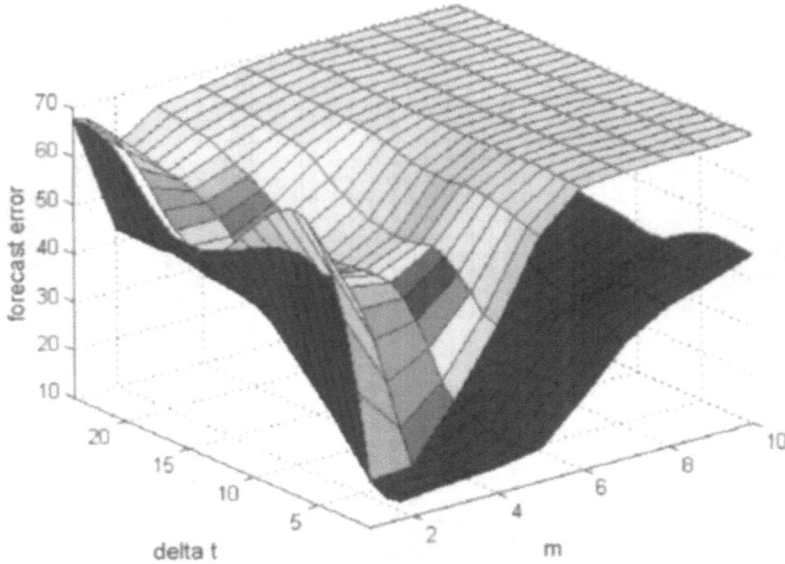

Figure 2. Non-linear forecast errors for Lorenz data (black surface) and an ensemble of surrogates (bright surface corresponds to minimum prediction error). 19 surrogate data sets were generated to ensure 95% confidence for the one-sided test.

Another assumption for non-linear time series analysis is stationarity. Most authors just assume stationarity in a widest sense looking at running means and standard deviations (e.g. Chen *et al*). In a non-linear setting this is however certainly inadequate. Here we have to assume that all quantities which are relevant for the dynamics of the system are constant over time. Unfortunately, a quantitative test for stationarity does not exist. We studied stationarity not only by calculation of first and second order moments but also by using recurrence plots. This is for example done by dividing the ozone time series into a number of segments and calculating the cross-forecast errors between the different segments. Visual inspection of error variations can be used to examine stationarity.

3. Results and Discussion

Our purpose is to study whether ozone data meet the preconditions for non-linear time series prediction. For comparison we also apply some of the methods to data generated by the non-linear Lorenz system.

3.1. LORENZ DATA

Figure 1 shows a realisation of the x-variable of length 5000 generated by the non-linear Lorenz system (upper left panel) (Lorenz, 1963), the autocorrelation

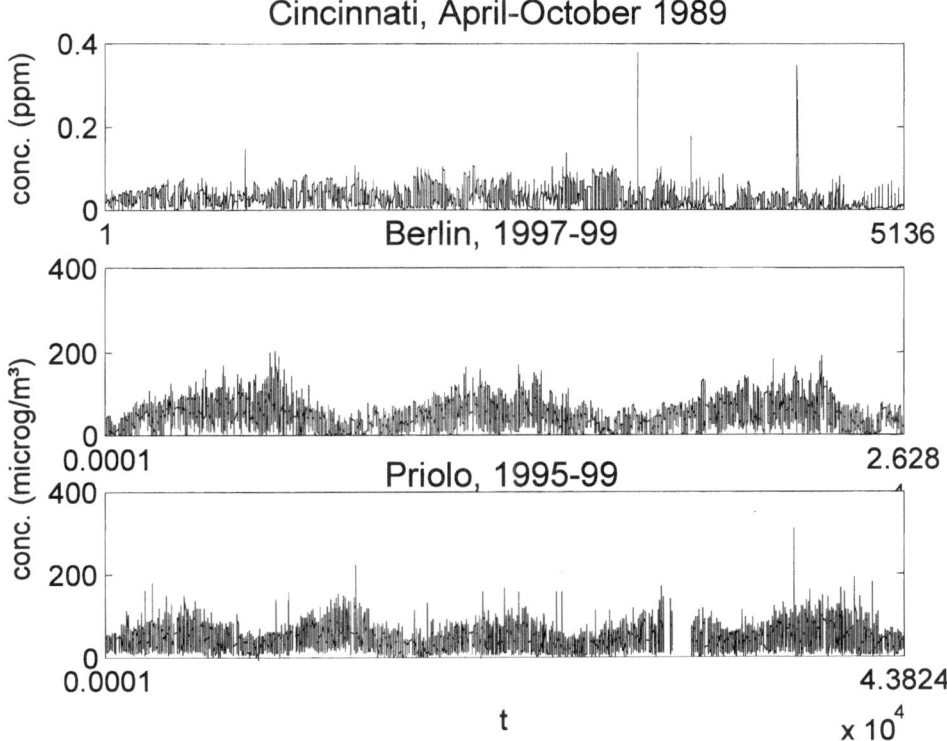

Figure 3. Hourly average values of ozone data from a) Cincinnati, April to October 1989, b) Berlin, 1997 to 1999 and c) Priolo, 1995 to 1999.

function (upper right panel) and a two-dimensional delay embedding (lower panel). The values were squared because the squared Lorenz time series shows similar properties like the ozone data, for example periodicity and strong autocorrelation.

The first zero of the autocorrelation function and two-dimensional delay embedding suggest the value $\tau \approx 5$. False nearest neighbour statistics show that an embedding dimension of three is sufficient.

While embedding dimension or prediction time is fixed in surrogate tests given in the literature, here we create a three-dimensional representation. Figure 2 shows the result of the surrogate test for embedding dimensions up to 10 and prediction times up to 24.

The null hypothesis is rejected if the minimal forecast error for the surrogates is larger than the error for the original data, i.e. for given embedding dimension m and prediction time $s = \Delta t$ the bright surface lies above the black one. This occurs in 238 cases, leaving two cases where the minimal forecast error for the surrogates is smaller than the error for the original data. However in these two cases the embedding dimension was one, which is not large enough. Thus, the test is able to identify the non-linear dynamics of Lorenz data.

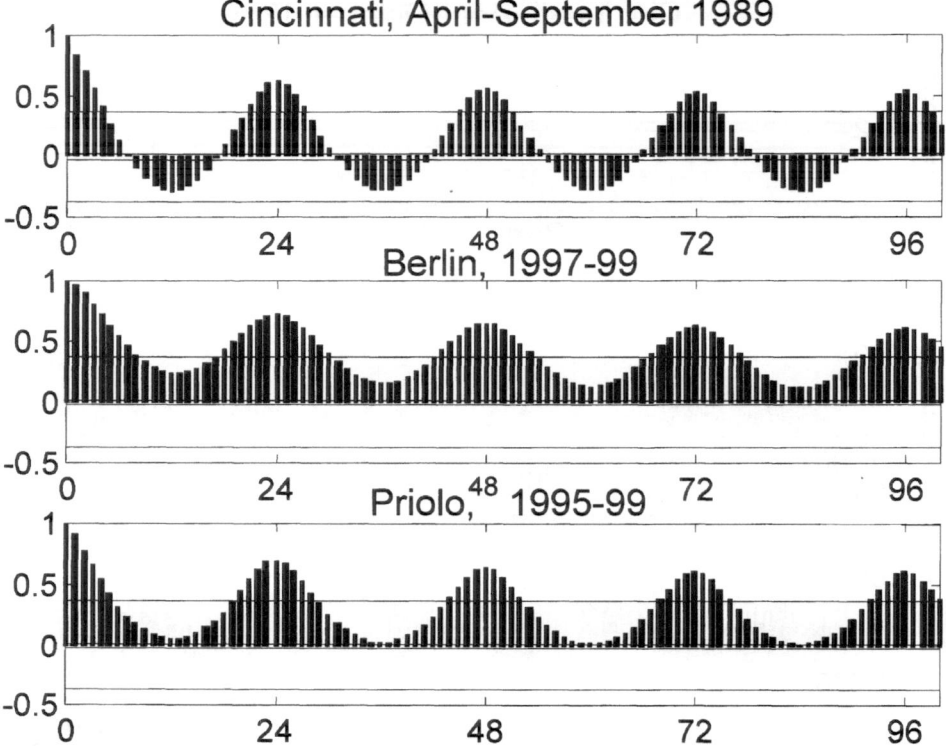

Figure 4. Autocorrelation functions for hourly average values of ozone data from a) Cincinnati, April to October 1989, b) Berlin, 1997 to 1999 and c) Priolo, 1995 to 1999 for lags up to 100.

3.2. OZONE DATA

Our ozone concentrations were collected from a local monitoring station in Berlin-Marienfelde (Germany) from 1997 to 1999 and Siracusa-Priolo (Italy) from 1995 to 1999. Data from Cincinnati were provided by Chen (1998). They were collected from a State and Local Air Monitoring Station in Cincinnati by the Department of Environmental Services, Hamilton County, Ohio from April to October 1989.

Figure 3 shows hourly average values of the Cincinnati, Berlin and Priolo time series the length of which are 5136, 26280 and 43824, respectively. Looking at the time series of Cincinnati we can see four extreme peaks. Such peaks are also observable in the Priolo time series. In contrast ozone concentrations from Berlin look relatively regular.

For Berlin and Priolo the annual cycle is very obvious. Cincinnati data are only given for the summer season which makes it impossible to detect an annual cycle.

Visual inspection of the autocorrelation functions (see Figure 4) reveals a constant period of 24 hours for each of the data sets. However if examining daily maximums and the time between them it is clear that this is not a periodicity in a

Priolo, 1995-99

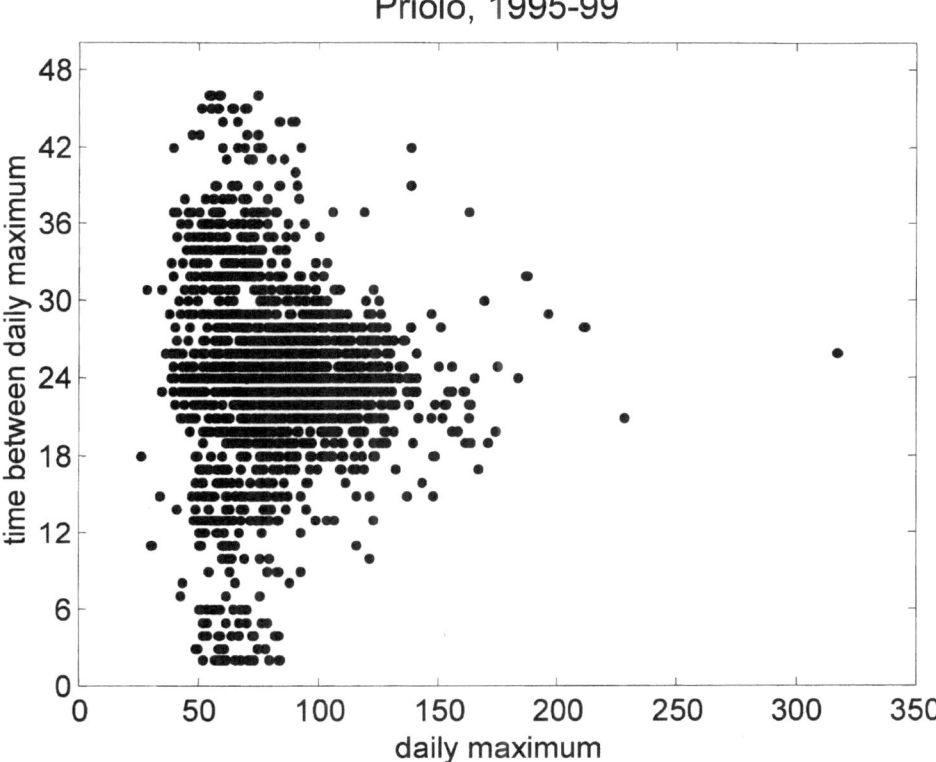

Figure 5. Plot of the time between local maximums against the maximums, Priolo, 1995 to 1999. The majority of the data clusters around the period 24.

mathematical sense, i.e. that the time series repeats itself after 24 h. As you can see from Figure 5 most of the points cluster around the period 24 but there are some outliers. The first zero of the auto-correlation function and the two-dimensional delay embedding yield delay 5, 9 and 6, respectively for ozone time series from Cincinnati, Berlin and Priolo. Interestingly, the autocorrelation function for Cincinnati data scatters around zero while the correlation for Berlin and Priolo time series is always greater than zero. The reason for this lies in the length of the time series and the corresponding mean value, respectively. Restriction of ozone time series from Berlin and Priolo to the summer season, say April to September 1999, gives a similar autocorrelation function to that of Cincinnati time series. Although ozone data seem to follow a Gaussian distribution the data fail a test for Gaussianity, thus ruling out a Gaussian linear stochastic process as their source.

As we can see from Figure 6 mean and standard deviation are quite constant over time if enough data points are used for estimation (here for Berlin ozone data). However, cross-forecast errors between different segments of the time series

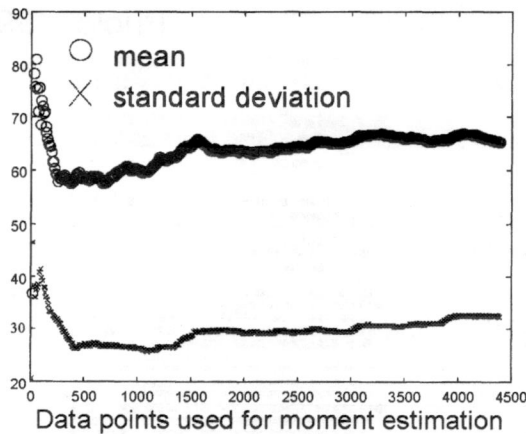

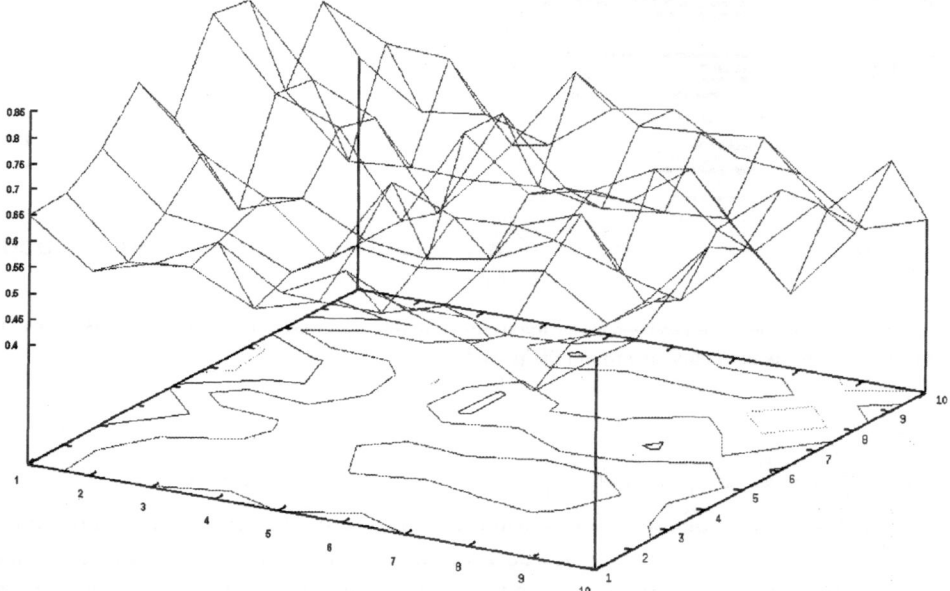

Figure 6. Plot of the a) first and second order moments and b) meta-recurrence plot for Berlin ozone data. x and y axes correspond to the different segments of the time series, z to the relative forecast error.

are very unstable suggesting non-stationarity (see also Haase *et al.*, 2001). We have got a quite similar result examining ozone concentrations for Cincinnati and Priolo.

Since the nature of the data set is still not so clear, we now use a surrogate test. Here we have confined ourselves to the period between April and September since ground level ozone only arises during warm days with high solar radiation. To overcome periodicity artefacts we have chosen a subinterval of the data such that the end points do match as closely as possible. We have then created amplitude

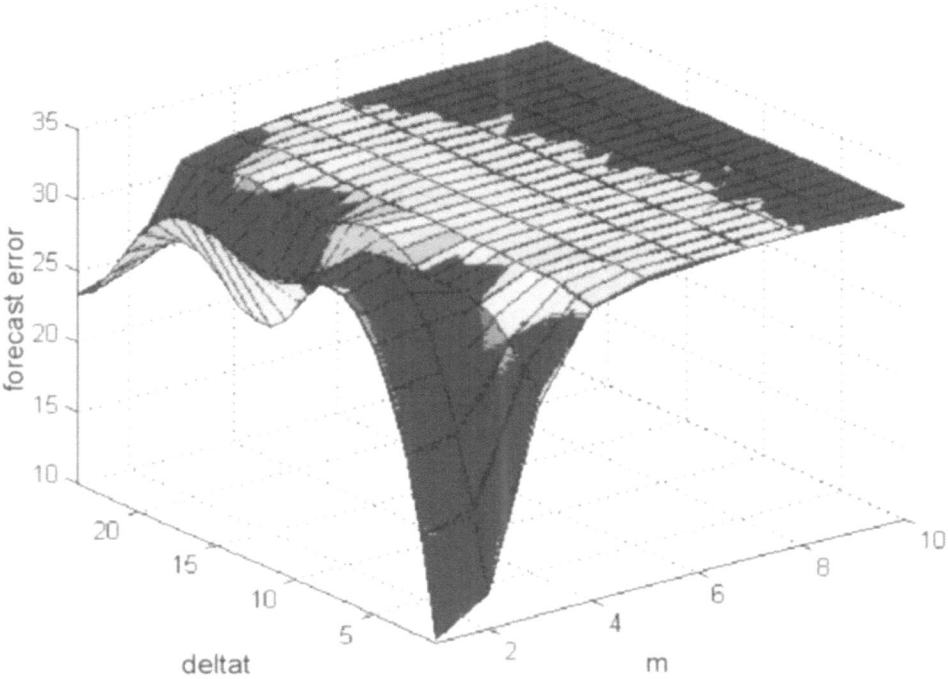

Figure 7. Same as Figure 2 but for ozone data from Berlin, April to September 1999.

adjusted surrogates i.e. conserving both the linear correlation and the distribution (Schreiber and Schmitz, 2000). Visual inspection of the original data and these surrogates in the time domain and in a two-dimensional delay embedding gives a rather good agreement. Comparing an ensemble of these amplitude-adjusted surrogate data sets with the original data by their predictability gives Figure 7 (Berlin ozone data). The hypothesis that the ozone data from Berlin is a non-linear transformation of a linear stochastic process is rejected for embedding dimensions 5 and 6 for all prediction times. The rejection could be due to non-linearity but also to non-stationarity. It is not rejected for embedding dimensions 9 and 10 for all prediction times. For all other embedding dimensions it is sometimes rejected or not. The differences between original data and surrogates are very small. The same analysis was done for Siracusa and Cincinnati ozone data and led to similar results.

In general we cannot reject the nullhypothesis of a linear probably non-linearly transformed Gaussian process. This is what we have observed also for other European sites. The minimum prediction error for an ensemble of surrogates is almost equal to the error for the observed data. We could not find an example which shows clear superiority of the observations in terms of prediction error.

Finally we have tested how the proposed phase space method predicts the future under different conditions. First of all we tried to obtain forecasts for hourly

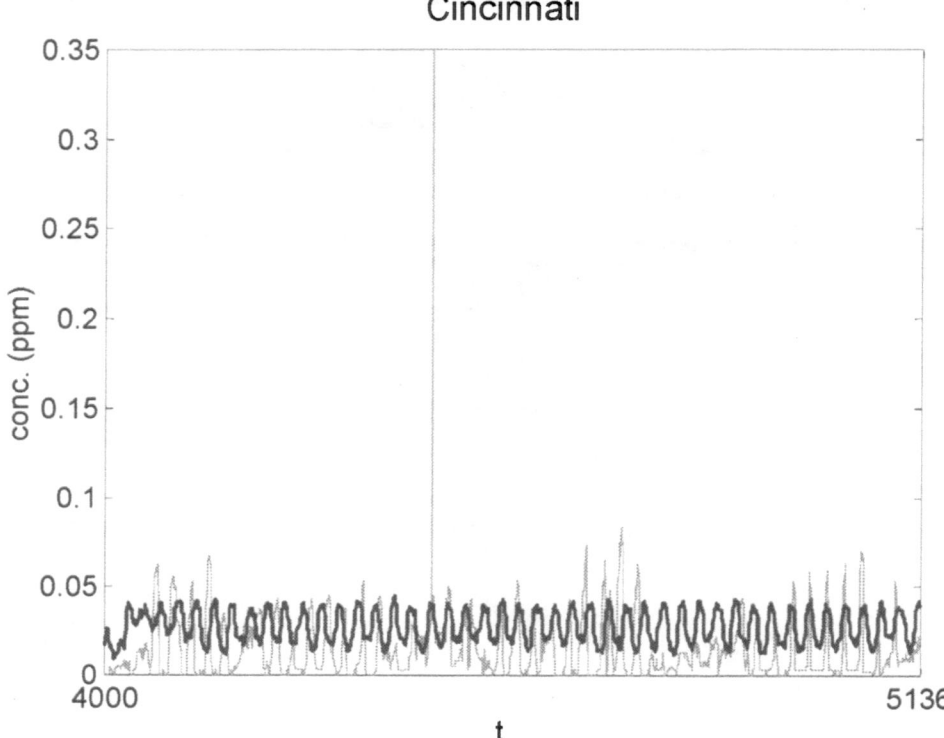

Figure 8. Cross forecasts based on the first 4000 hourly average values from Cincinnati. The black line corresponds to the predictions, the grey one to the observations (m = 8).

average values of Cincinnati using only the first 4000 data points. The algorithm is able to identify the period of 24 h (compare with Figure 8) or to capture the trend of the data but not the variability of the time series. The embedding dimension was chosen according to Chen *et al.* (1998).

One step predictions based on previous hourly average values which are shown in Figure 9 are however quite good. They will be getting worse if a larger prediction horizon is chosen.

4. Conclusions

In contrast to the result of the surrogate test for Lorenz data, non-linearity of ozone data is not obvious. If there is any non-linearity in the ozone time series it may be very weak. Probably almost all of the expected non-linearity discussed in the literature so far can just be found in the interrelation between various variables.

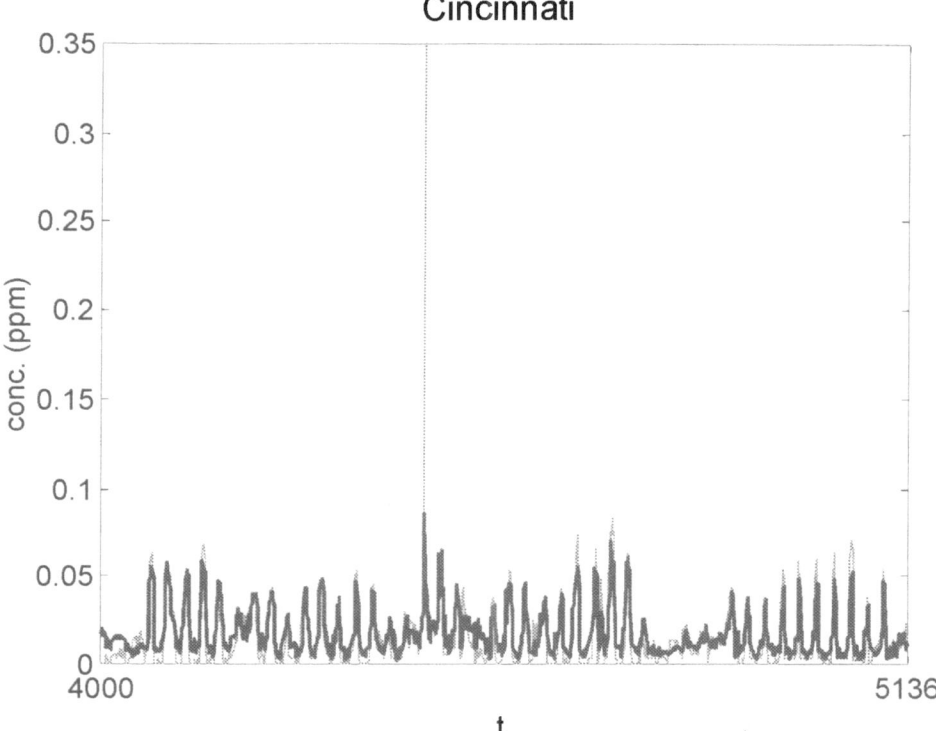

Figure 9. One step (short term) predictions using previous hourly average values for ozone time series from Cincinnati. The black line corresponds to the predictions, the grey one to the observations.

If the database is dynamically updated then the proposed prediction method gives excellent short term forecasts. If the training set is fixed the method was so far just able to capture the trend or the periodicity of the time series.

Prediction in phase space is attractive because it provides a basis to reconstruct a system using just one observed variable of the system and it is also easily extendable to a multivariate approach. A big caveat however is that it requires a time series without missing values.

Acknowledgements

Part of the work was done in the frame of the EUROTRAC-2 subproject SATURN. The research on non-linear statistical modelling techniques is organised by the APPETISE project and funded by the European Commission (No. IST-99-11764). We are grateful to Christian Merkwirth, University of Goettingen for providing the C routines which are also available from www.dpi.physik.uni-goettingen.de/tstool/index.html. In addition we made use of the TISEAN package which is available in the internet too: www.mpiks-dresden.mpg.de/~tisean/TISEAN_2.1. We also like to thank JL Chen for providing the data for Cincinnati. Ozone data from Berlin were provided by the 'Senatsverwaltung für Stadtentwicklung, Umwelt und Technologie', Berlin. Siracusa data is available from the APPETISE database.

References

Chen, J. L., Islam, S. and Biswas, P.: 1998, 'Nonlinear dynamics of hourly ozone concentrations: nonparametric short term prediction', *Atmos. Environ.* **32**(11), 1839–1848.

Grassberger, P. and Procaccia I.: 1983, 'Characterization of strange attractors', *Phys. Rev. Lett.* **50**(5), 346–349.

Haase, P., Schlink, U. and Richter M.: 2001, 'Non-Parametric Short-Term Prediction of Ozone Concentration in Berlin: Preconditions and Justification', *Proceedings of the International Conference on Air Pollution Modelling and Simulation*, Paris, France.

Hegger, R., Kantz, H. and Schreiber, T.: 1999, 'Practical implementation of nonlinear time series methods: The TISEAN package', *Chaos* **9**, 413.

Kantz, H. and Schreiber, T.: 2000, *Nonlinear Time Series Analysis*, Cambridge University Press, Cambridge, 304 pp.

Kennel, M. B., Brown R. and Abarbanel, H. D. I.: 1992, 'Determining embedding dimension for phase-space reconstruction using a geometrical construction', *Phys. Rev. A* **45**(6), 3403–3411.

Kocak, K., Saylan, L. and Sen, O.: 2000, 'Nonlinear time series prediction of O_3 concentration in Istanbul', *Atmos. Environ.* **34**, 1267–1271.

Li, I. F., Biswas, P. and Islam, S.: 1994, 'Estimation of the dominant degrees of freedom for air pollutant concentration data: applications to ozone measurements', *Atmos. Environ.* **28**(9), 1707–1714.

Lorenz, E. N.: 1963, 'Deterministic nonperiodic flow', *J. Atmos. Sci.* **20**, 130–141.

Raga, G. B. and Moyne, L. Le: 1996, 'On the nature of air pollution dynamics in Mexico City – I. Nonlinear analysis', *Atmos. Environ.* **30**(23), 3987–3993.

Schlink, U., Herbarth, O. and Tetzlaff, G.: 1997, 'A component time-series model for SO_2 data: Forecasting, interpretation and modification', *Atmos. Environ.* **31**(9), 1285–1295.

Schreiber, T. and Schmitz, A.: 2000, 'Surrogate time series', *Physica D* **142**, 346.

Sugihara, G. and May R. M.: 1990, 'Nonlinear forecasting as a way of distinguishing chaos from measurement error in time series', *Nature* **344**, 734–741.

Takens, F.: 1981, 'Detecting strange attractors in turbulence', in D. A. Rand and L. S. Young (eds), *Dynamical Systems and Turbulence*, Lecture Notes in Mathematics 898, Springer, Berlin, Germany, pp. 366–381.

Theiler, J., Eubank, S., Longtin, A., Galdrikian, B. and Farmer, J. D.: 1992, 'Testing for nonlinearity in time series: The method of surrogate data', *Physica D* **58**, 77–94.

NUMERICAL PREDICTION OF DISPERSION CHARACTERISTICS IN AN URBAN AREA BASED ON GRID REFINEMENT AND VARIOUS TURBULENCE MODELS

G. THEODORIDIS[1]*, V. KARAGIANNIS[2] and D. VALOUGEORGIS[2]

[1] *AIAS Research Ltd, Thessaloniki, Greece;* [2] *Department of Mechanical and Industrial Engineering, University of Thessaly, Volos, Greece*

(* *author for correspondence, e-mail: georgios@aias.com.gr, fax: +30 310402060)*

Abstract. A detailed simulation of the Goettinger Strasse pollutant dispersion problem is performed using the CFD code CFX-TASCflow for different wind directions. Two turbulence models, the $k - \varepsilon$ and the RSM are adopted on three grid refinement levels. Besides the typical reference grid implemented by the TRAPOS group, two different grid resolutions are introduced. The first refinement is in the whole street canyon region on the x—y level, while the second one is local in all three directions. Validation of all involved computational schemes is performed based on relative available experimental data. The computed velocity fields and concentration contours imply that the typical reference grid is a suitable choice for the velocity fields, while local grid refinement in all three directions in a small region containing the receptor is required to upgrade the pollutant concentration results with modest additional computational effort. Finally the RSM model resulted in smaller concentration levels. The $k - \varepsilon$ model compared to the RSM seems a more appropriate choice to solve this particular problem.

Keywords: CFD, grid dependence, turbulence modeling, urban air quality

1. Introduction

The dispersion of pollutants in urban scale is dominated by modifications of the atmospheric flow caused by buildings. Down-wash phenomena and increased local turbulence strongly influence the mean flow field and the diffusion parameters. Early studies (Yamartino and Wiegard, 1986; Depaul and Sheih, 1986) with simple urban canyons, subject to perpendicular flow, have provided some description of the mean flow recirculation pattern. The influence of the aspect ratio (width/height) on the main vortex has been examined and it has been found that it is shifting upwards with decreasing canyon width (Depaul and Sheih, 1986), resulting into significantly higher pollutant concentrations in canyons with low aspect ratios (Pavageau *et al.*, 1996). Further experimental studies have investigated the influence of different roof shapes on the distribution of pollutants (Rafailidis, 1997; Kastner-Klein and Plate, 1999). The effect of different wind tunnel geometrical resolution and details on the dispersion of pollutants, has been examined by Liedtke *et al.* (1999) and Leitl *et al.* (2001). They found significant differences depending upon

Water, Air, and Soil Pollution: Focus **2:** 525–539, 2002.
© 2002 *Kluwer Academic Publishers.*

the complexity and the details of the experimental model. In addition they have noticed large discrepancies comparing their results with numerical simulations (Berkowitcz *et al.*, 1997; Ketzel *et al.*, 1999).

It has become common evidence that the observed changes in flow patterns and pollutant concentrations highly depend on the detailed geometry of the street canyon (Scaperdas, 2000). It is necessary to introduce the so-called obstacle resolving or microscale models. Although a wide variety of such models have been recently developed still further calibration and validation is needed (Schatzmann *et al.*, 1999). The introduced uncertainties include typical computational problems (e.g. truncation error of the numerical scheme, implementation of boundary conditions, grid resolution, etc.). More specific difficulties are related with turbulence modeling and dispersion of pollutants in regions with high concentration gradients. Most of the existing studies are based on the typical $k - \varepsilon$ model (Sini *et al.*, 1996), which seems to be adequate when isolated buildings or simplified street canyons are simulated. When the street canyon becomes more complex it has been suggested by Kim and Boysan (1999) that more advanced models such as the RSM (Launder *et al.*, 1975; Speziale *et al.*, 1991) must be implemented. In addition Schatzmann *et al.* (1999) claim that special attention is needed in areas with high concentration gradients such as the regions close to the sources. As a result all implemented computational models must be carefully tested against reliable experimental data, before their application to real practical situations.

Recognising the importance of model validation for solving pollutant dispersion problems in urban areas, the computational group of the DGXII-TMR, TRAPOS project initiated and executed a series of benchmarking exercises, which included comparisons between experimental and numerical results. The most complex among them is the Goettinger Strasse case, in the city of Hanover. For this particular problem, field measurements (NLÖ, 1995) and laboratory data (Liedtke and Schatzmann, 1999) were available for pollutant concentration at one receptor point close to the traffic pollution sources and for meteorological data over the roof of the highest building. Very recently, Chauvet *et al.* (2001) have performed additional measurements for the same configuration providing additional data for wind velocity within the street canyon. They are also stressing out the need for comprehensive wind-tunnel data under carefully controlled conditions, before numerical models could be validated. The Goettinger Strasse case has also been simulated numerically with the aid of various microscale models by Ketzel *et al.* (2001) using one relatively coarse grid and the standard $k - \varepsilon$ model.

In this article an attempt is made to investigate some of the above issues and to provide specific answers. In particular the well-known and commonly used $k - \varepsilon$ model is compared to the more advanced and time demanding Reynolds Stress transport Model (RSM) in order to examine the influence of turbulence anisotropy to the dispersion characteristics in urban areas. Moreover several grid resolutions are implemented to study the effect of grid refinement on the produced velocity and concentration fields. Schatzmann *et al.* (1999), claimed that, when the receptor and source-containing-cells are adjacent, the concentration results at the receptor are strongly grid dependent. For that reason, besides the typical reference grid implemented by the TRAPOS group, two different grid resolutions are introduced. The first refinement is in the whole street canyon region on the x−y level, while the second one is local in all three directions.

2. Computational Methodology

Computational work is based on the CFX-TASCflow package, which is a general purpose CFD analysis system using a flexible multi-block curvilinear grid system. Within CFX-TASCflow, the conservation equations for mass, momentum, and scalar quantities like temperature, turbulent kinetic energy and any number of species are solved in curvilinear co-ordinates. The numerical solution is based on second-order in time and space discretisation, applied on a co-located grid arrangement. The discrete momentum and continuity equations are solved with a coupled elliptic solver. An efficient algebraic multi-grid solution technique is adopted, giving a practically constant rate of convergence, regardless the level of the grid refinement. A detailed description of the computational method can be found in Raw *et al.* (1989).

Turbulent diffusion can be described with the standard two-equation $k - \varepsilon$ turbulence, or the RSM, which unlike the standard $k - \varepsilon$ accounts for anisotropy of turbulence by solving six additional transport equations for the Reynolds stresses. Both models use the wall function approach to model near-wall viscous effects (Launder and Spalding, 1974). This approach assumes the universality of a logarithmic velocity profile in the near wall region and relies on the validity of near-wall turbulent equilibrium.

3. Grid Resolution and Boundary Conditions

The test case considered in this study corresponds to a complex urban area located in the Goettinger Strasse in Hanover. Laboratory data are available for pollutant concentration at one receptor location (see Figure 1) and vertical wind velocity and turbulence kinetic energy profiles over the anemometer (Liedtke and Schatzmann, 1999). The computational model in the present work is an exact replica of the experimental set up. The specifications of the case under consideration, including inflow boundary conditions and aerodynamic roughness, have been obtained from the www database (http://www.dmu.dk/AtmosphericEnvironment/trapos/cfd-wg.htm) of the DGXII-TMR TRAPOS project. Given a wind direction ϕ, the corresponding Dirichlet inflow boundary conditions are applied at two of the four lateral bound-

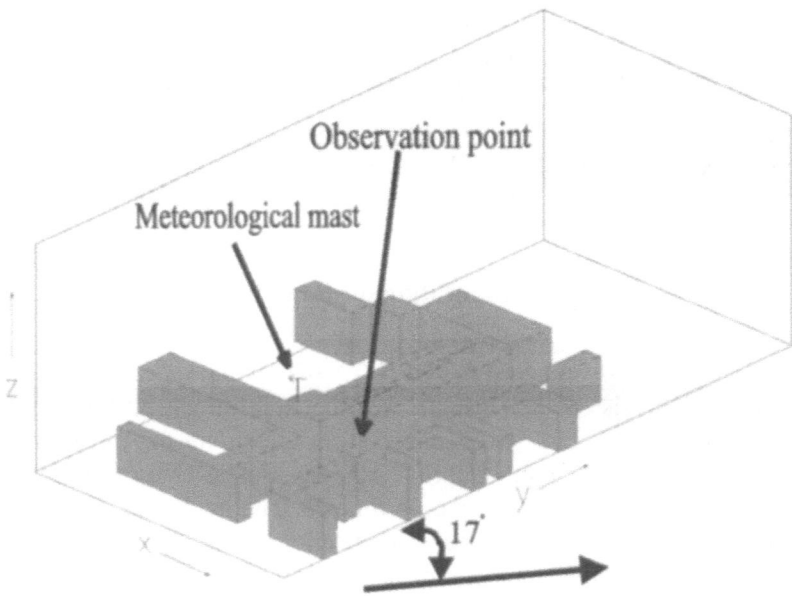

Figure 1. Geometrical configuration of urban area.

aries, while typical outflow conditions are applied at the remaining two. More specifically at the inflow boundaries the relationships

$$\frac{V}{u_*} = \frac{1}{\kappa} \ln\left(\frac{z}{z_0}\right) \; , \quad k = \frac{u_*}{\sqrt{c_\mu}} \; , \quad \varepsilon = \frac{u_*}{\kappa z} \; ,$$

are used, where V is the horizontal wind speed, z is the distance from the ground, $u_* = 0.526$ m s^{-1} is the friction velocity, $z_0 = 0.05$ m is the aerodynamic roughness and the empirical constants κ, c_μ take the values of 0.41 and 0.09, respectively. All walls are considered as rough no-slip walls with an aerodynamic roughness of $z_0 = 0.01$ m. At the upper boundary a symmetry condition of vanishing gradients with respect to z and vanishing normal velocity component w is applied. To justify the choice of the symmetry condition the distance between the lower and upper boundaries is taken equal to twelve times the average building height. Appropriate sources with a total strength of Q (Vol s^{-1}) are applied within the Goettinger Strasse, simulating the four-lane traffic emissions as shown in Figure 2.

Initially numerical simulations are performed on the reference numerical grid of $60 \times 73 \times 25$ points proposed by the CFD working group of the DGXII-TMR TRAPOS project (see Figure 2). Then a grid refinement (grid embedding) is applied in the x and y directions in the area bounded by the dashed frame, up to the roof of the highest building ($z = 30$ m), as shown in Figure 2. It is expected that since the receptor is located very close to the sources, the whole area adjacent to the sources and the receptor is a high concentration gradient region. By tripling the

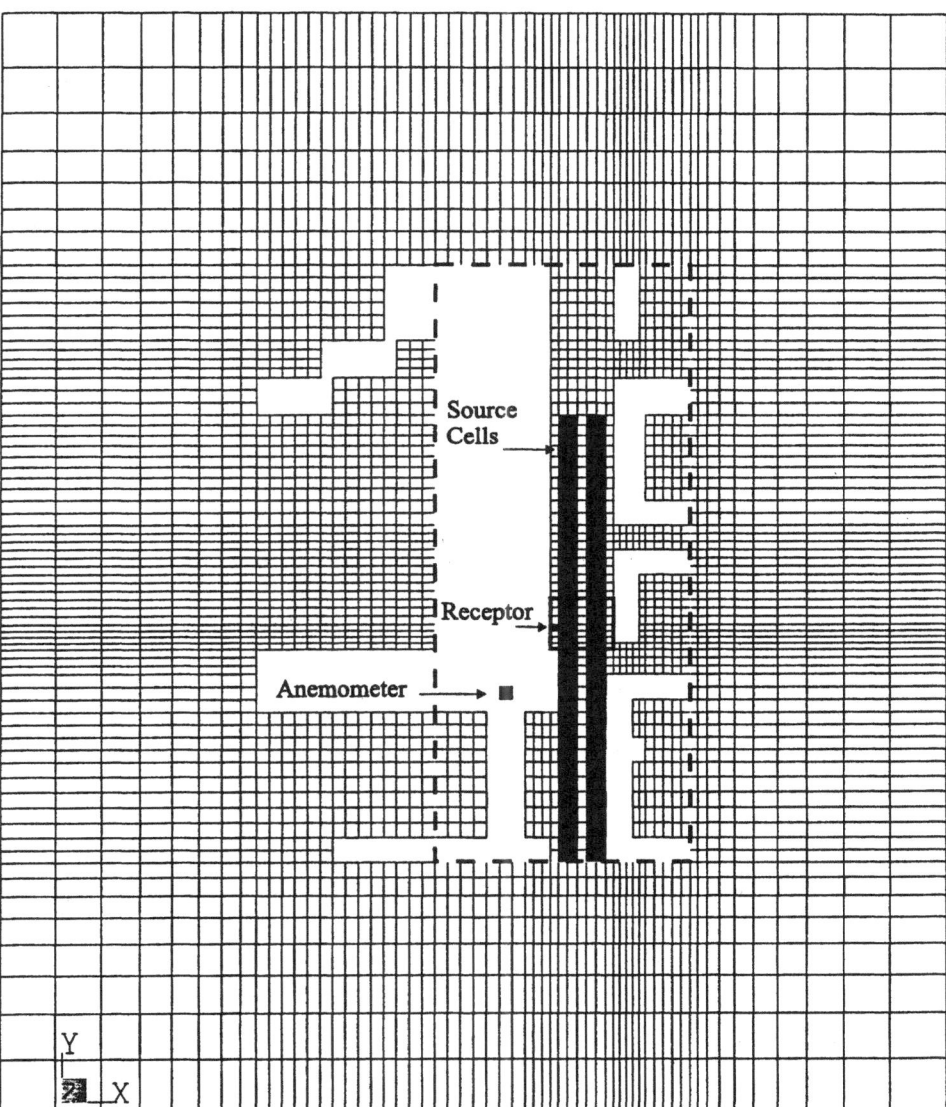

Figure 2. Reference grid and embedding regions at ground level. The grid is also refined by a factor of three in the horizontal direction (dashed line) and in all three directions close to the sensor location (solid line).

number of nodes in the x and y directions the relative coarse set up of the reference grid is refined and the grid dependency of the velocity and concentration fields is studied. This yields $82 \times 169 \times 12$ additional grid points in the refined region. Finally in order to examine the influence of the volume of the source-containing-cells on the computed concentration profiles (Schatzmann *et al.*, 1999), the near receptor region (area bounded by the blue frame in Figure 2 until the height of $z = 17.5$ m) is refined in all three directions. As a result $21 \times 21 \times 21$ grid points are added to the reference grid.

The computed concentration field c (Vol/Vol)is non-dimensionalised using the relationship $c^* = \frac{cV_{ref}H}{Q/L}$, where $V_{ref} = 10$ m s^{-1} is the reference wind speed velocity at a height of 100 m, $H = 20$ m is the average building height, and $L = 180$ m is the length of the line sources.

4. Results and Discussion

In this section first a comparison of our numerical results with experimental data is attempted and then the detailed velocity and concentration fields are given at selected positions within the street canyon, based on all three employed grid resolutions and two turbulence models. It is well known that, depending on the wind direction, minor geometrical modifications may result to significant changes in pollutant concentrations. Thus the validation of the present numerical work is based on the experimental data obtained from the simplified structure of the wind tunnel model with closed gateway wing (Schatzmann *et al.*, 1999). The geometric configuration of our virtual model, as mentioned before, is an exact replica of this experimental set up.

A first comparison between computed and measured velocity components and turbulence kinetic energy in the region above the anemometer is given in Figure 3, for two characteristic wind directions ($\phi = 0°$ and $\phi = 180°$). The computational results are obtained using the reference grid and both turbulence models under investigation. As it is seen the u and w profiles are slightly underestimated and overestimated respectively, while the v profile is in excellent agreement compared to the measured corresponding quantities for both turbulence models. The computed turbulent kinetic energy curves match perfectly the measured ones for the $k - \varepsilon$ turbulence model, while the RSM based results are slightly underdetermined. In the region however, very close to the roof of the highest building, where the experimental data show high levels of k there is an underestimation of the computed results for both models. Based on these results of the anemometer area, it seems that the $k - \varepsilon$ model produces superior results compared to the RSM results. Of course the overall agreement between computed and measured quantities is encouraging and remains unchanged for all wind directions tested between 0° and 360° with intervals of 10°.

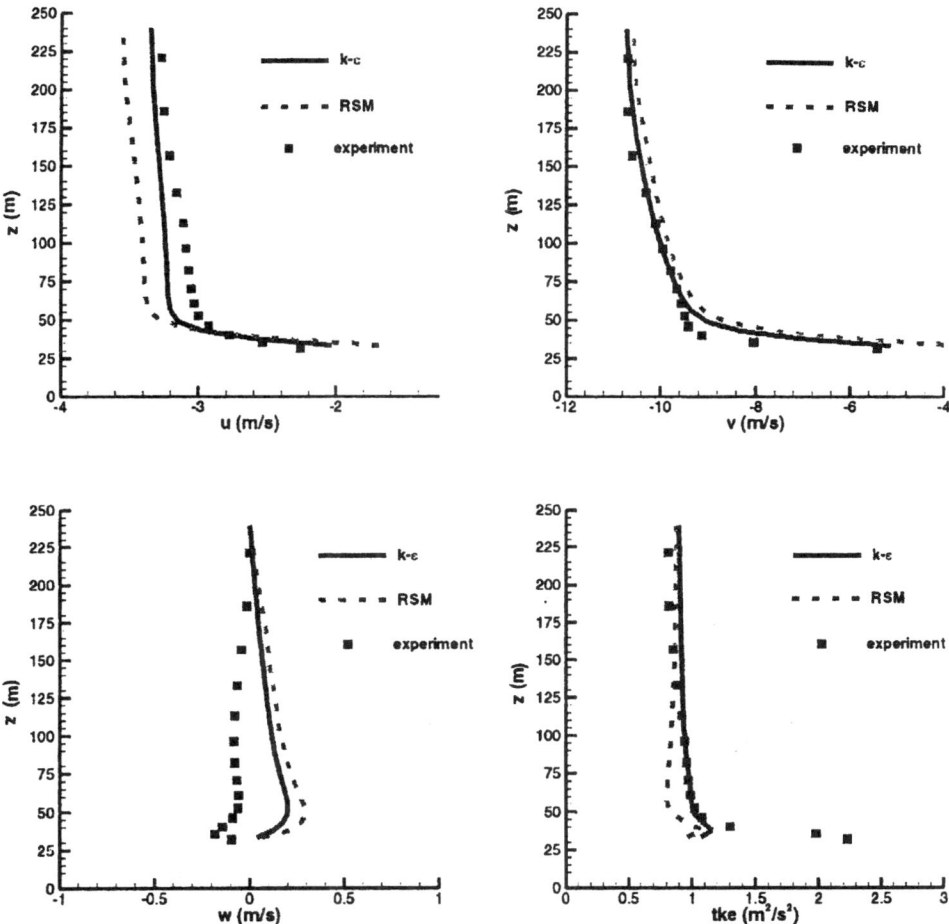

Figure 3a. Comparison between computed and measured wind components and turbulence kinetic energy at the region above the anemometer, for $\phi = 0°$.

Next in Figure 4 the computed concentrations at the monitoring point are compared to the corresponding measured ones for the complete range of the wind impingement angle. The results based on the $k - \varepsilon$ model are given for all three grid resolutions under consideration (i.e. the reference, the horizontaly refined and the grid with local refinement in all three directions), while the RSM results are shown only for the reference grid. It is seen that the shape and the peaks of the wind tunnel curve are reproduced and in most cases good qualitative and quantitative agreement between computational results and measurements is achieved. It is also obvious however that in certain wind direction angles the agreement is poor. In particular for $\phi = 100\pm10°$ and $\phi = 220\pm10°$ the departure between computed and measured concentrations is significant. This discrepancy may be due to the

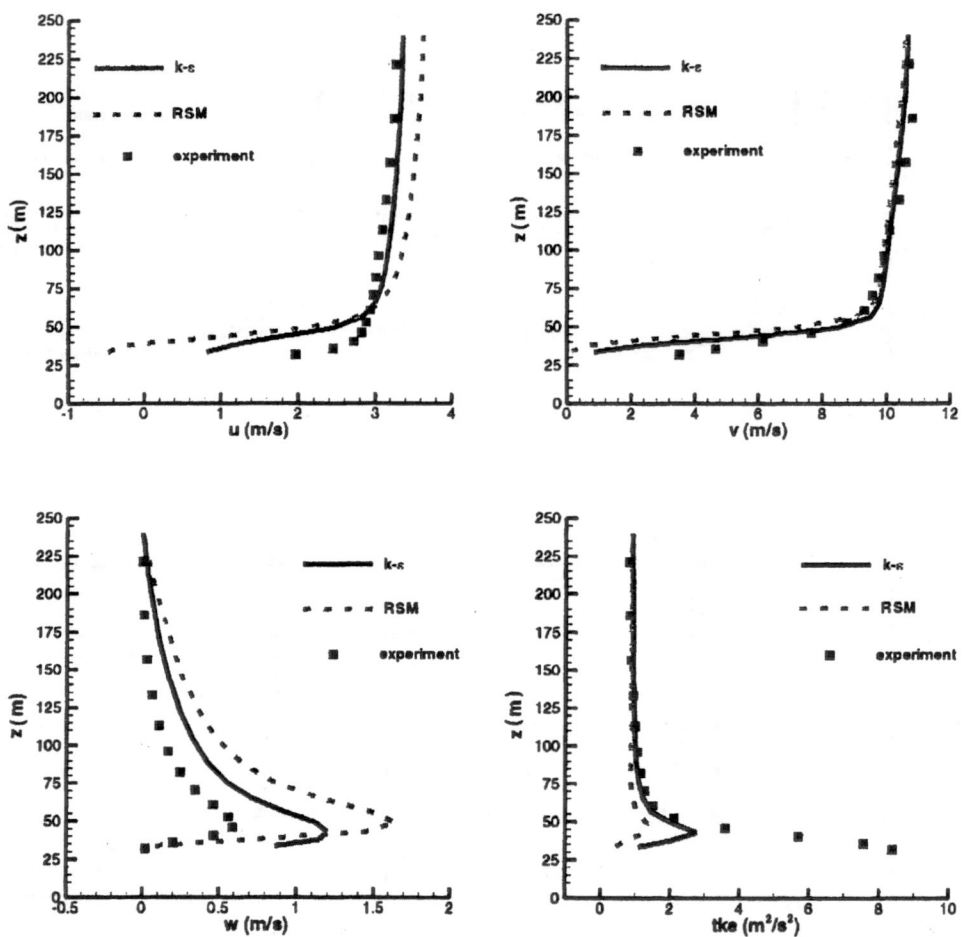

Figure 3b. Comparison between computed and measured wind components and turbulence kinetic energy at the region above the anemometer, for $\phi = 180°$.

fact that in the aforementioned wind directions the area around the receptor is characterised by high concentration gradients. As a consequence the uncertainties for both numerical and experimental work are increased. For example, small probe placement errors in the experiments would result in different probe readings. As reported earlier by Schatzmann *et al.* (1999) error bars of the order of ±30%, due to the size of the averaging time interval, should be considered in the particular wind tunnel results.

The results between the reference and the horizontal grid refinement are very close to each other (relative differences are less than 20%) indicating that no significant improvement is achieved by refining the grid in the horizontal direction only. It is emphasised that in this case the additional computational effort and required

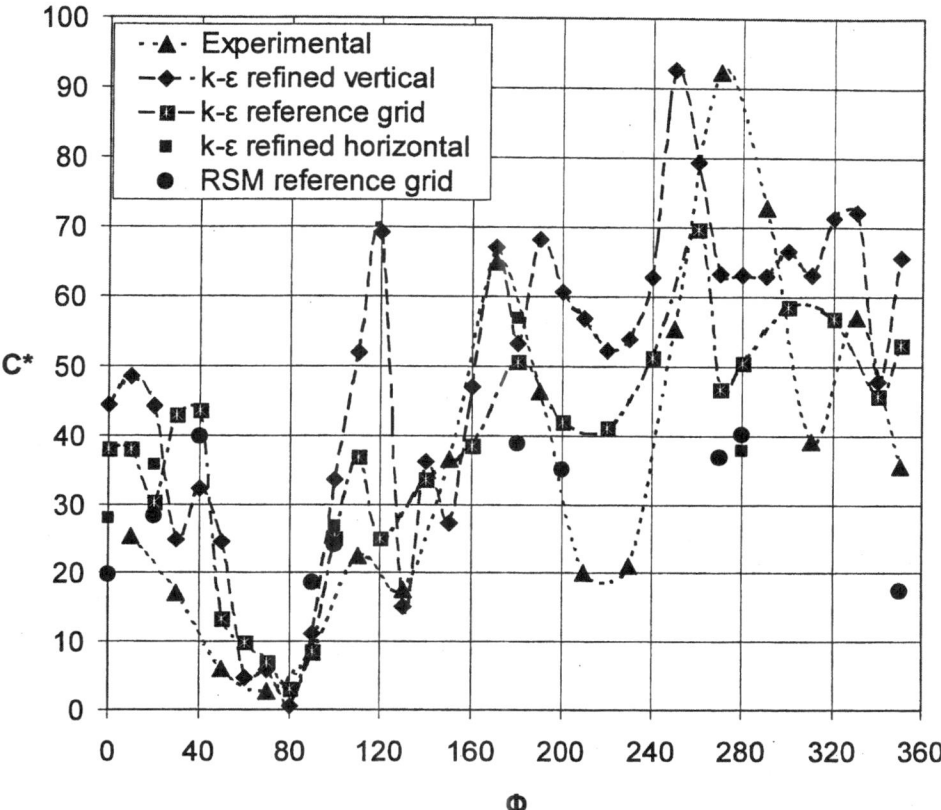

Figure 4. Comparison of the computed non-dimensional concentrations at the monitoring point for both turbulence models and all computational grids with experimental data.

memory space are both increased by a factor 4. On the other hand a properly selected local refinement of the grid in all three directions leads to quite different computed receptor concentrations. These concentrations compared with the ones obtained with the other two computational grids are overall in better agreement with the measured concentrations. This is not the case again for the specific wind directions with $\phi = 120°$ and $\phi = 220°$, where the measured concentrations are overestimated by a factor of about three. As mentioned before there is some significant difference between computed results with the different cell sizes but this departure is not proportional to the refinement of the source grid cells as it is speculated in previous work (Schatzmann *et al.*, 1999). This is an interesting matter, which will be discussed later in the article when the complete concentration profiles are given. Also in the same Figure 4, it is shown that the concentrations obtained with the RSM model and the reference grid are very close to the corresponding concentrations obtained with the $k - \varepsilon$ model. This is an indication that, at least for this particular problem under investigation, implementation of the RSM, which in general is a more advanced and reliable model than the $k - \varepsilon$, does not lead to more accurate results. Probably large eddy simulation models, able to directly simulate the large structure of atmospheric turbulence, may be the only remaining option when the required computational power is available. Finally it is noticed that in the local grid refinement case the required computational time and space compared to the corresponding ones with the reference grid are increased only about 20%. At this stage it seems appropriate to notice that in order to validate further the existing numerical models in an accurate and consistent manner more experimental data are required. In order to propose with confidence the best suited model for the particular problem it is essential to have complete experimental sets of data instead of the presently available isolated experimental values at one or more locations.

We turn now our attention to the detailed velocity and concentration prediction within the street canyon. In all following work the presented results are for $\phi = 0°$ and they are representative for all other wind directions. In Figure 5 the horizontal velocity vector fields are shown at $z = 3$ m for the $k - \varepsilon$ on the reference and the horizontally refined grids (left and middle panels, respectively) and the RSM model on the reference grid (right panel). For reasons of clarity, velocity vectors are only plotted every second grid location in both x and y directions. It can be seen that the $k - \varepsilon$ results on both grids correspond to very similar flow patterns. It is noticed that although the wind direction is almost parallel to the longitudinal axis of the main street, the flow is not channeling inside the Goettinger Strasse, due to the fact that wind is coming also from the three normal streets on the right side of the street canyon. Actually several recirculation regions are developed in both the reference and the horizontally refined grids. This is clearly indicated at the center panel of Figure 5. It is concluded that the reference grid is a suitable grid with grid independent numerical results for the velocity fields. This statement is valid only when second order accurate schemes are adopted for the advection terms, as in the case of the CFX-TASCflow. The wind field results inside the street canyon for the $k - \varepsilon$ and the RSM are quite similar, indicating that there is no justification to select a more complex and time consuming turbulence model, such as the RSM, over a conventional type model, such as the $k - \varepsilon$. This conclusion remains invariant for all wind impingement angles tested. As a result the RSM has not been tested any further with the other two grids implemented in the present work.

As stated by Schatzmann et al. (1999) in most numerical models the initial concentration (the concentration in the source containing grid cells) is solely dependent on the actual choice of the grid due to the fact that the pollutant flux is uniformly distributed over the source grid cells. Thus decreasing the volume of these source grid cells by a certain factor (or in other words increasing accordingly the grid resolution in the z direction) results into an increase of the initial concentration by the same factor. As a consequence, it is expected that the concentration of the grid cells adjacent to the source will be controlled, to a large degree, by the actual choice of the grid dimension.

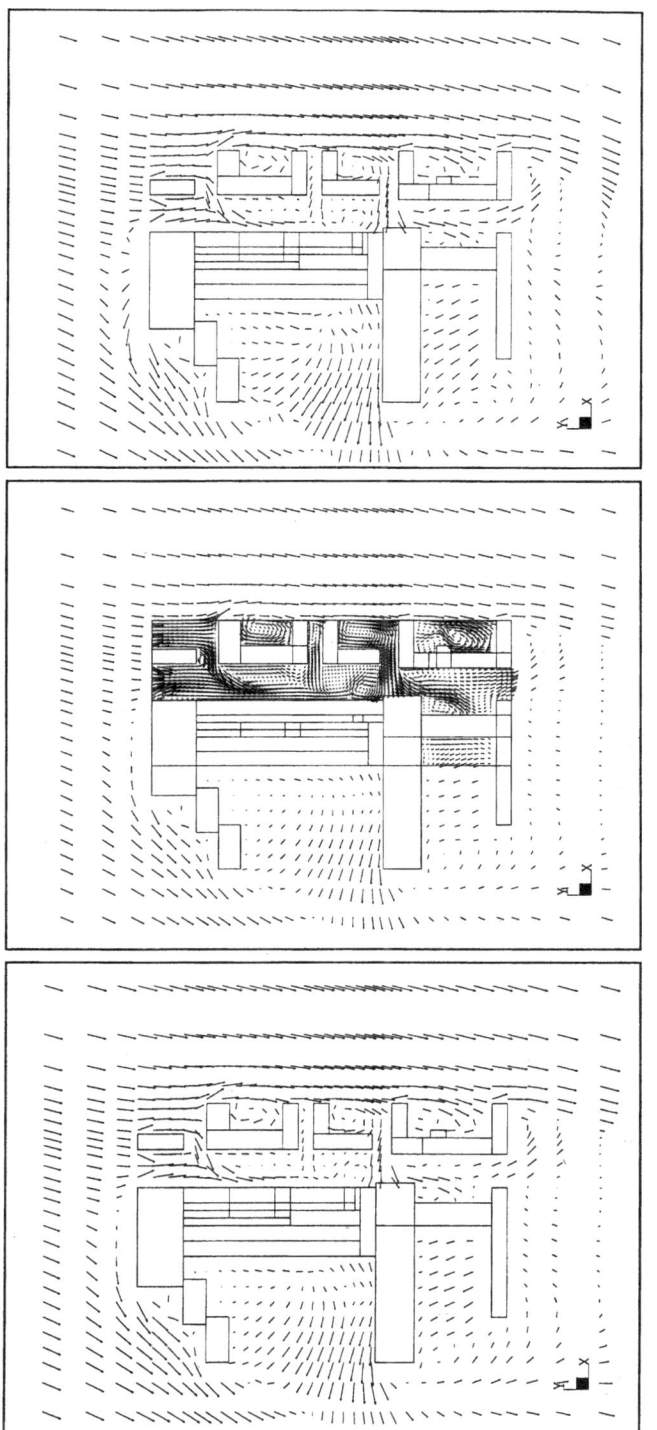

Figure 5. Computed velocity fields at $z = 3$ m and $\phi = 0°$: reference grid with $k - \varepsilon$ model (left); horizontally refined grid with $k - \varepsilon$ model (middle); reference grid with RSM model (right).

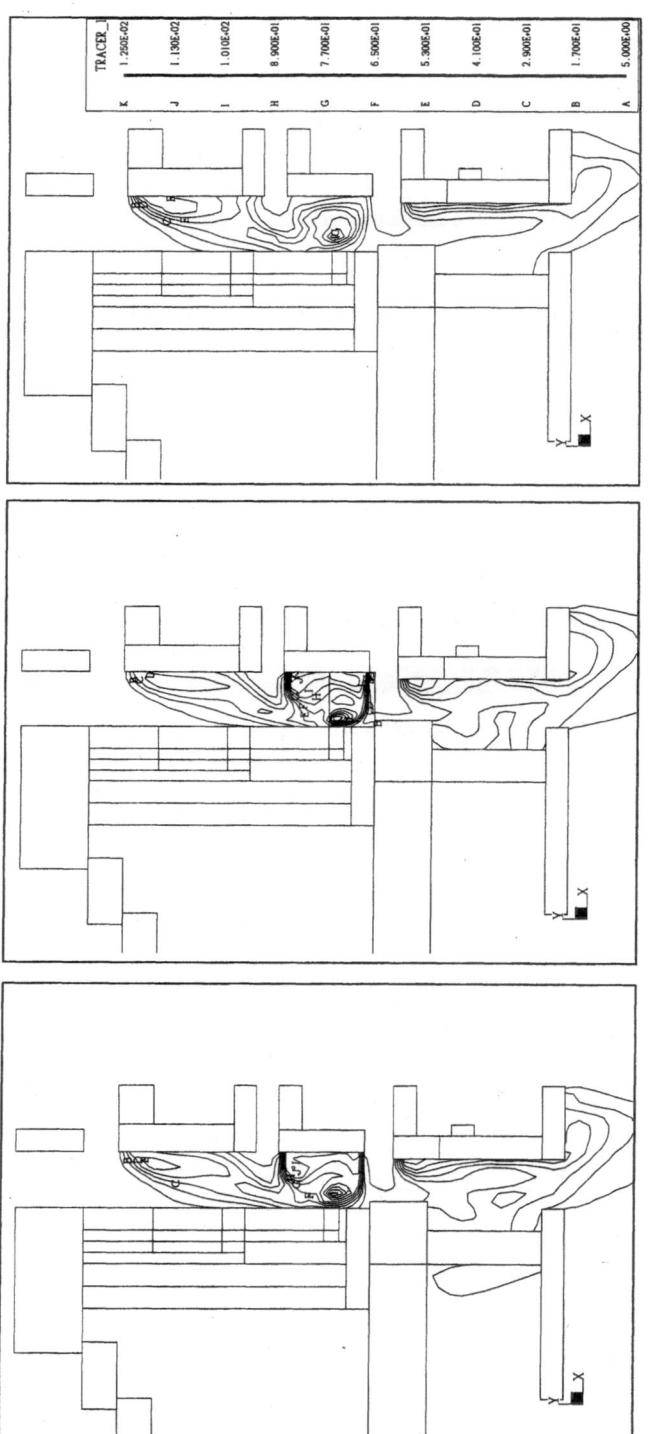

Figure 6. Computed concentration contours at z = 3 m and $\phi = 0°$: reference grid with $k - \varepsilon$ model (left); locally refined grid in all 3-directions with $k - \varepsilon$ model (middle); reference grid with RSM model (right).

As mentioned before the wind field is found to be insensitive to the horizontal grid refinement. Therefore the above interesting concern is tested by adopting a local grid refinement in all three directions only in the high gradient concentration region close to the receptor (see Figure 2). In the resulting grid, produced by tripling the number of points locally, there is one and a half-cell between the receptor and the sources on both the x and z directions. It is emphasised that in previous work by other investigators, related to the Goettinger Strasse, the receptor was always placed at the center of the first boundary cell adjacent to the source street lines. In Figure 6 the concentration contours are indicated for the $k-\varepsilon$ model on the reference and the locally refined grids (left and middle panels), as well as the RSM model on the reference grid (right panel). These concentration fields correspond to two of the velocity vector fields shown in Figure 5 (left and right panels). As it is seen the results with the standard $k-\varepsilon$ including the horizontally refined grid case which is not shown here, are quite similar. Maximum concentrations as high as $c^* = 125$ are experienced, located in the aforementioned regions of recirculating flow (see also Figure 5). More important, although the initial concentration at the source-containing-cells has been increased by a factor of three the corresponding receptor concentration has been increased, compared to that of the reference grid, only by 17% (see Figure 4). It is concluded that at the height of $z \geq 3$ m, the numerical results are practically grid independent. The concentration results at the receptor however are more sensible when local grid refinement in all three directions is applied.

Finally the concentration field, which is based on the reference grid and the RSM is shown in the right panel of Figure 6. Although the velocity fields inside the street canyon region are very similar for the two turbulence models, it seems that the RSM model tends to overpredict the eddy diffusivity, resulting in an overall underestimation of the concentration of pollutants. It is noticed that the $k-\varepsilon$ results based for the reference grid are in excellent agreement with recent computational results by Ketzel *et al.* (2001). As far as the authors are aware of there no available results in the literature to compare with for the case of the RSM and more dense grid resolutions.

5. Conclusions

A detailed simulation of the Goettinger Strasse pollutant dispersion problem is performed using the CFD code CFX-TASCflow for different wind directions. Two turbulence models, the $k-\varepsilon$ and the RSM are adopted and three grid refinement levels are used. Validation of all involved computational schemes is performed based on relative available experimental data. Complete velocity vector fields and concentration contours are reported.

The velocity fields are found to be similar for all implemented grids and turbulence models. This implies that the so-called 'reference grid' is a suitable choice with grid independent numerical results for the velocity fields, when second order accurate schemes are adopted for the advection terms. From the other hand, the so-called 'local refined' grid in all three directions in the region containing the receptor is the most suitable way to upgrade the pollutant concentration results with modest additional computational effort. Finally the RSM model resulted in smaller concentration levels. The $k - \varepsilon$ model seems a more appropriate choice to solve this particular problem compared to the RSM. Although the above conclusions are related to the specific problem under consideration general guidelines may be obtained, since the present problem is the most complex one in a series of test cases performed for model validation (Ketzel *et al.*, 2001).

It is important to notice that at this stage of the investigation is essential to have complete experimental sets of data instead of the presently available isolated experimental values at a very limited number of locations. Assuming that these experimental results are available the present work may be extended to validate existing computational models in a more integrated manner.

Acknowledgements

The specifications of the case studied were prepared by the CFD working group of the DGXII-TMR TRAPOS project. Availability of the specification and measurement data is highly acknowledged.

References

Berkowicz, R., Hertel, O., Larsen, S. E., Sorensen, N. N. and Nielsen, M.: 1997, *Modelling Traffic Pollution in Streets*, National Environment Research Institute, Roskilde, Denmark.

Chauvet, C., Leitl, B. and Schatzmann, M.: 2001, 'High Resolution Measurements in an Idealised Street Canyon', *The 3rd International Conference on Urban Air Quality*, Loutraki, Greece, March 2001.

DePaul, F. T. and Sheih, C. M.: 1986, 'Measurements of wind velocities in a street canyon', *Atmosph. Environ.* **20**, 555–559.

Kastner-Klein, P. and Plate, E. J.: 1999, 'Wind tunnel study of concentration fields in street canyon', *Atmosph. Environ.* **33**, 3973–3979.

Ketzel, M., Berkowicz, R. and Lohmeyer, A.: 'Dispersion of Traffic Emissions in Street Canyons – Comparison of European Numerical Models with Each Other as Well as with Results from Wind Tunnel and Field Measurements', *The 3rd International Conference on Urban Air Quality*, Madrid, March 1999.

Ketzel, M., Louka, P. Sahm, P. Guilloteau, E., Sini, J. F. and Moussiopoulos, N.: 2001, 'Intercomparison of Numerical Urban Dispersion Models – Part II: Street Canyon in Hanover, Germany', *The 3rd International Conference on Urban Air Quality*, Loutraki, Greece, March 2001.

Kim, S. E. and Boysan F.: 1999, 'Application of CFD to environmental flows', *J. Wind. Eng.* **81**, 145–158.

Launder, B. E., and Spalding, D. B.: 1974, 'The numerical computation of turbulent flows', *Comput. Meths. Appl. Mech. Eng.* **3**, 269–289.

Launder, B. E., Reece, G. J. and Rodi, W.: 1975, 'Progress in the development of a Reynolds-stress turbulence closure', *J. Fluid Mech.* **68**, 537–566.

Leitl, B., Chauvet, C. and Schatzmann, M.: 2001, 'Effects of Geometrical Simplifications and Idealization on the Accuracy of Microscale Dispersion Modeling', *3rd International Conference on Urban Air Quality*, Loutraki, Greece, March 2001.

Liedtke, J. and Schatzmann, M.: 1999, 'Autoabgasausbreitung in Straßenschluchten – Vergleich von Windkanal- und Naturmessungen an einem konkreten Fallbeispiel. Projekt Europäisches Forschungszentrum für Maßnahmen zur Luftreinhaltung. Forschungszentrum Karlsruhe', *Schlußbericht PEF 296001* (in German).

Liedtke, J., Leitl, B. and Schatzmann, M.: 1999, 'Dispersion in a street canyon: Comparison of wind tunnel experiments with field measurements', in P. M. Borrel and P. Borrel (eds), *Proc. of Eurotrac Symposium 98*, WIT Press, Southampton 1999, pp. 806–810.

NLÖ: 1995, 'Lufthygienisches Überwachungssystem Niedersachsen-Standortbeschreibung der NLÖ Stationen', Bericht, Niedersächsisches Landesamt für Ökologie, Göttinger Strasse 14, 30449 Hannover (in German).

Pavageau, M., Rafailidis, S. and Schatzmann, M.: 1996, 'A Comprehensive Experimental Databank for the Verification of Urban Car Emission Dispersion Models', *Proceedings of the 4th Workshop on Harmonisation within Atmospheric Dispersion Modeling for Regulatory Purposes*, Ostende, May 1996.

Rafailidis, S.: 1997, 'Influence of building area density and roof shape on the wind characteristics above a town', *Bound.-Layer Meteor.* **85**, 255–271.

Raw, M. J., Galpin, P. F. and Hutchinson, B. R.: 1989, 'A collocated finite-volume method for solving the Navier-Stokes equations for incompressible and compressible flows in turbomachinery: Results and applications', *Canadian Aeronautics and Space Journal* **35**, 189–196.

Scaperdas, A.: 2000, 'Modelling Air Flow and Pollutant Dispersion at Urban Canyon Intersections', Ph.D. Thesis, University of London.

Schatzmann, M., Liedtke, J. and Leitl, B.: 1999, 'Dispersion Models for Urban Applications. A Critical Assessment of the Present State of Application', *Internationale Kolloquium' Messung und Beurteilung der Luftqualität*, Heidelberg, April 1999.

Sini, J. F., Anquetin, S. and Mestayer, P.: 1996, 'Pollutant dispersion and thermal effects in urban street canyons', *Atmosph. Environ.* **30**, 15, 2659–2677.

Speziale, C. G., Sarkar, S. and Gatski, T. B.: 1991, 'Modeling the pressure-strain correlation of turbulence: An invarient dynamical systems approach', *J. Fluid Mech.* **227**, 245–272.

Yamartino, R. J. and Wiegand, G.: 1986, 'Development and evaluation of simple models for the flow, turbulence and pollutant concentration fields within an urban street canyon', *Atmosph. Environ.* **20**, 2137–2156.

PHYSICAL MODELLING OF URBAN ROUGHNESS USING ARRAYS OF REGULAR ROUGHNESS ELEMENTS

R. W. MACDONALD*, S. CARTER SCHOFIELD and P. R. SLAWSON

Department of Mechanical Engineering, University of Waterloo, Waterloo, Ontario, Canada
(* author for correspondence, e-mail: rwmacdon@uwaterloo.ca, fax: +1 519 888 6197)

Abstract. The aerodynamic behaviour of large urban agglomerations must be represented in increasingly greater detail, as large-scale numerical weather prediction and air pollution dispersion models are refined. The present study provides detailed measurements of the flow field in regular arrays of obstacles to obtain representative data on mean flow and turbulence statistics in urban-type areas. Obstacle arrays consisting of simple cubes and flat plate roughness commonly used in boundary layer simulations were placed in a simulated atmospheric boundary layer flow in a hydraulic flume. The scale factor was about 1:200 based on the obstacle height (50 mm). The results show no appreciable 'constant stress' region in the internal boundary layer above the buildings, since in any finite-length test array the boundary layer is always developing. However, if the RMS turbulence components are scaled by the local values of the shear stress, then there seems to be a universal scaling, with $\sigma_u/u_* = 2.1$, $\sigma_v/u_* = 1.65$ and $\sigma_w/u_* = 1.2$. This greatly simplifies the parameterization of the first order turbulence statistics in obstacle arrays. It was also observed during the experiments that, compared to results in the cube arrays, the turbulence kinetic energy and the Reynolds stresses were almost doubled in the flat plate roughness arrays.

Keywords: aerodynamic roughness, hydraulic flume, physical modelling, turbulence, urban boundary layer

1. Introduction

The flow field around a single bluff obstacle is quite complex, with regions of separated flow, concentrated vorticity, and large shear (Hosker, 1984; Meroney, 1982). The aerodynamic interaction of large groups of buildings introduces further complicating flow features such as wake interference and skimming flow (Oke, 1987). Often only the simplest parameterizations of the urban surface are used in large-scale numerical weather prediction (NWP) and air pollution dispersion models (e.g., Best, 2000; Martilli *et al.*, 2000). As these NWP codes and urban pollution dispersion models are further refined, they will require increasingly sophisticated methods (parameterizations) for incorporating the effects of large groups of buildings and obstacles on the wind flow (e.g. Brown and Williams, 1998; Brown and Spore, 2000).

In attempting to characterize the flow over urban areas, it is not practical to resolve the flow around individual obstacles. As a result, methods have been developed to parameterize the mean boundary layer flow in terms of a finite number

Water, Air, and Soil Pollution: Focus **2**: 541–554, 2002.
© 2002 *Kluwer Academic Publishers.*

of relatively simple parameters: the surface shear stress ($\tau_0 = \rho u_*^2$), the aerodynamic roughness (z_0), and the displacement height (d). In an equilibrium boundary layer flow the mean wind speed can be expressed in terms of these parameters according to the log-law profile,

$$u(z) = \frac{u_*}{\kappa} \ln \left(\frac{z - d}{z_0} \right) . \tag{1}$$

The aerodynamic parameters z_0 and d cannot be measured directly, but can be estimated from the mean building height (H), width (W), inter-obstacle spacing (S_y, S_x) and other geometrical variables (Figure 1). Various models are now available to do this (Bottema, 1997; Macdonald et al., 1998; Grimmond and Oke, 1999), however the available methods do not address the further problem of parameterizing the first-order turbulence statistics of the flow. In addition, the models assume the existence of a well-developed, equilibrium boundary layer above the buildings, which is only true for effectively infinite, homogeneous fetch. In practice, most urban areas are quite 'patchy' and at best we can only expect limited homogeneous fetch in areas where we need to estimate the overhead mean velocity profile. Only the lowest part of the internal boundary layer structure above the rooftops, with a depth of about 1% of the total fetch, can be expected to satisfy the equilibrium conditions which are assumed in the derivation of Equation (1) (Wieringa, 1993).

For the purpose of validating urban boundary layer parameterizations, the availability of high quality, full-scale measurements of mean flow, turbulence and shear stress around large groups of buildings in generic arrangements is fairly limited. The available data are usually only representative of a very specific area, and do not reveal clearly the influence of the various descriptors required for specifying the geometry of an obstacle array. Physical modelling studies are a very effective way of obtaining fluid dynamic data for controlled variations of the surface obstacles. In the present study detailed measurements of flow and turbulence quantities were made in regular arrays of cubes and two-dimensional 'billboard' obstacles at 1:200 scale. The latter obstacles offer an extreme example of sharp-edged obstacles.

2. Experimental Methodology

The experiments were conducted in a 12.6 m long hydraulic flume which has a 1.2 m × 1.2 m cross section with a 2.4 m long test section enclosed in tempered glass. A combination of turbulence spires and upstream surface roughness was used to generate a turbulent atmospheric boundary layer (ABL) flow with a mean surface roughness of $z_0 = 1.3$ mm. The mean profile could also be fitted with a simple power law:

$$\frac{u(z)}{u_H} = \left(\frac{z}{H} \right)^{0.26} , \tag{2}$$

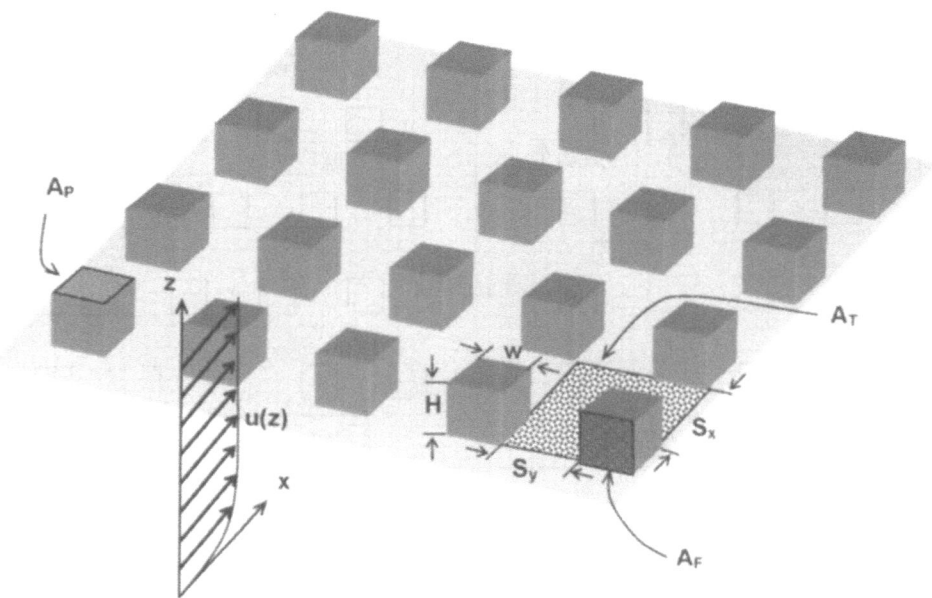

Figure 1. Definition sketch for flow over a generic obstacle array consisting of uniform blocks of height H and width W, with cross-stream spacing S_y and alongwind spacing S_x. The frontal packing density is $\lambda_f = A_F/A_T$ and plan area density is $\lambda_p = A_P/A_T$.

where H = 50 mm was the height of the typical obstacle studied. The profile exponent of 0.26 is typical of flow over rough, urban-like, terrain. Detailed specific-ations of the flow field, including turbulence information, are found in the original report (Macdonald *et al.*, 2000).

Regular arrays of obstacles, consisting of up to 24 rows, were placed in the test section of the hydraulic flume to simulate built-up areas. The surface obstacles consisted of 50 mm aluminum blocks or 50 mm sections of aluminum angle that were arranged in simple staggered and square (in-line) arrays. Cube arrays at three packing densities were tested, covering the range of isolated roughness flow ($\lambda_f =$ 0.0625), wake interference flow ($\lambda_f = 0.16$) and skimming flow ($\lambda_f = 0.44$). The definitions of the packing density and other geometrical parameters used are shown schematically in Figure 1.

For comparison with the arrays of solid cube obstacles, arrays of two-dimensional flat plates (billboard roughness) were also tested, but only at the single packing density $\lambda_f = 0.16$. During all of the experiments a reference upstream flow speed of $u_H = 50.5$ mm s^{-1}, measured at the height of the obstacles, was maintained. Figure 2 shows some of the obstacle arrays tested, the features of which are also summarized in Table I.

Each velocity profile in the arrays consisted of point measurements made at 24 heights between $z = 0$ and 400 mm using a SonTek^{TM} acoustic Doppler ve-

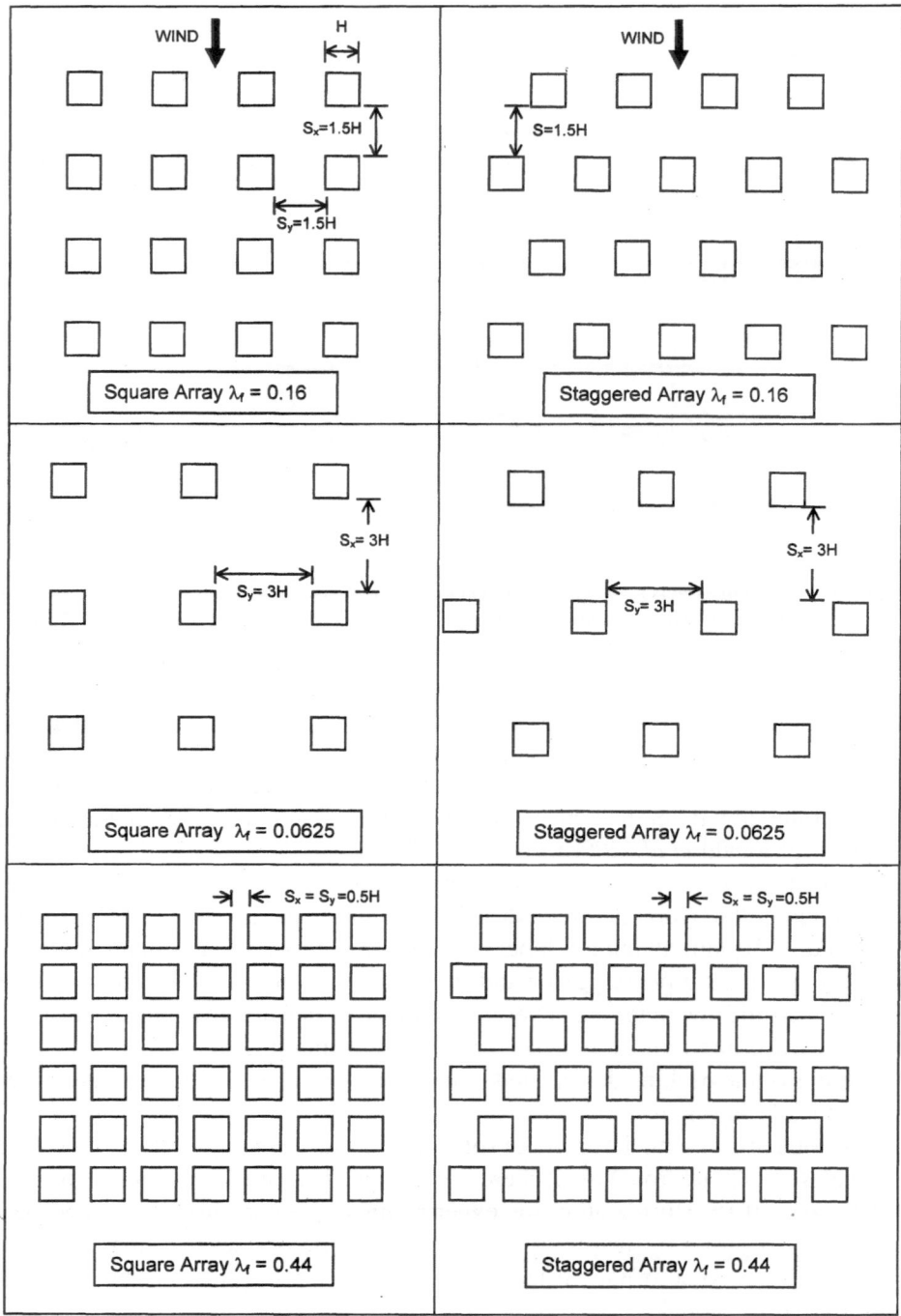

Figure 2. Schematic of the cube arrays tested in the hydraulic flume.

TABLE I
Obstacle array configurations tested in the hydraulic flume

Obstacle type	Frontal area density (λ_f)	Obstacle center-to-center spacing	Number of rows (total fetch)
Cubes	0.0625	S/H = 4.0	N = 14 (x/H = 43)
Cubes	0.16	S/H = 2.5	N = 21 (x/H = 50)
Cubes	0.44	S/H = 1.5	N = 16 (x/H = 23.5)
Billboards	0.16	S/H = 2.5	N = 21 (x/H = 50)

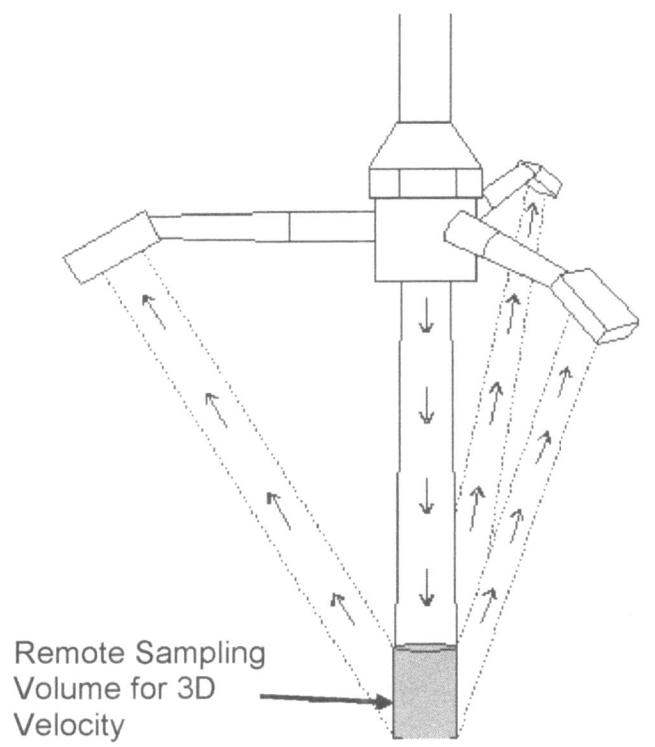

Remote Sampling
Volume for 3D
Velocity

Figure 3. Schematic of control volume for Sontek Micro-ADV probe.

locimeter (ADV) operating at 20 Hz. This instrument uses the Doppler shift of a 16 MHz acoustic pulse to measure the three components of the fluid velocity. The acoustic pulse is reflected off of tiny particles transported by the flow and it is sampled by an array of three receivers focused on a small control volume (Figure 3). A two-minute sampling period was found to give stable, representative values of the mean velocity and turbulence variances.

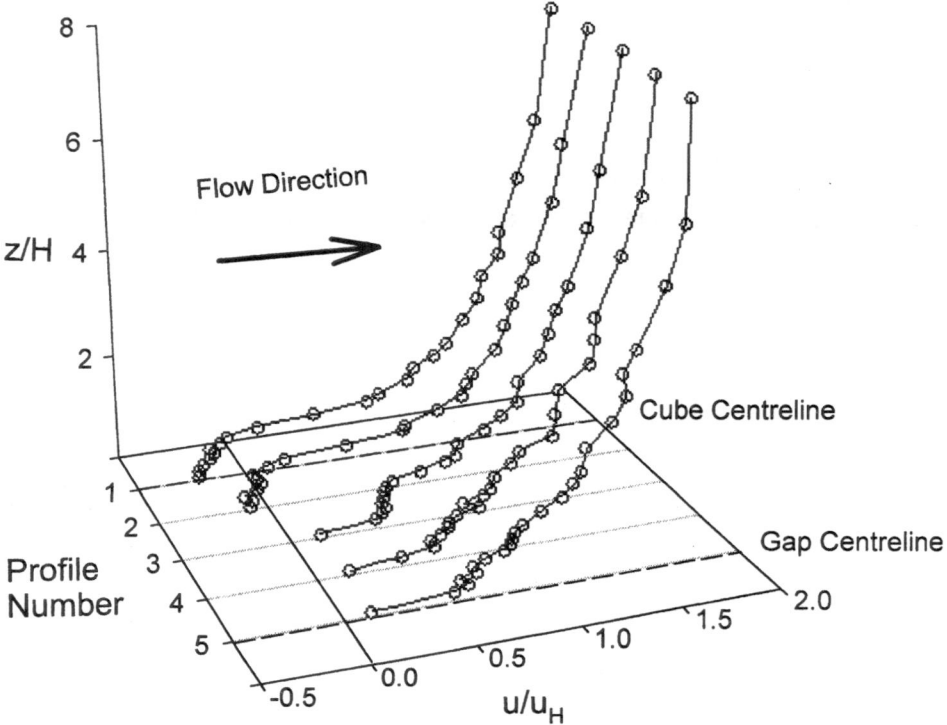

Figure 4. Plot of individual velocity profiles measured at five different positions behind the ninth row of the square cube array with a packing density $\lambda_f = 0.16$.

Fluid velocities measured at a point in an obstacle array are very sensitive to proximity to the obstacles and show much spatial variability. In order to remove the large spatial variability in the measured velocity profiles, and to provide a representative mean profile, the results from the experiments were spatially averaged. In this procedure, the profiles measured at several cross-stream locations behind a given row in the obstacle array are averaged to give the mean profile. Because of the regularity of the obstacle arrays, this could be done with relatively few individual profiles (typically five). Figure 4 shows five separate profiles measured behind the ninth row in the $\lambda_f = 0.16$ cube array. The presence of a recirculation cavity is indicated by the negative mean velocities in the lower part of the profile measured directly behind the cube. However, after the spatial averaging, this type of information is 'smeared' out. The resulting data set discussed below includes spatially averaged profiles of mean velocity, turbulence variances (σ_u, σ_v, σ_w), and Reynolds stresses ($\tau = -\overline{u'w'}$) between $z = 0$ and $z = 8H$.

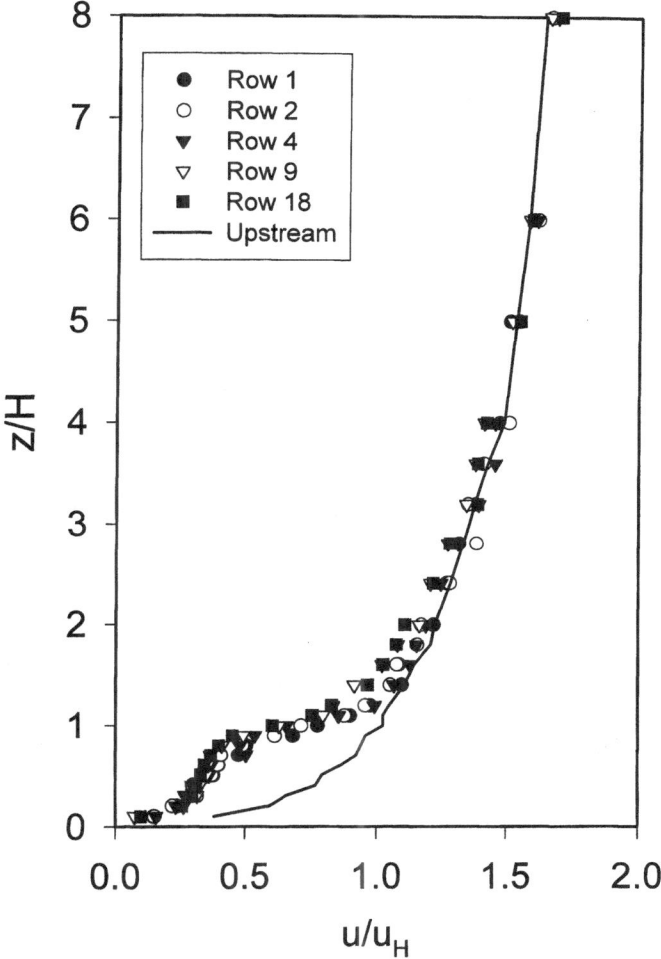

Figure 5. Mean velocity profiles measured behind rows 1, 2, 4, 9 and 18 of the square geometry cube array with a packing density of $\lambda_f = 0.16$.

3. Results

The mean velocity profiles measured behind Rows 1, 2, 4, 9 and 18 in the $\lambda_f = 0.16$ cube array (square configuration) are shown in Figure 5. Each of these profiles is a spatially averaged result, which includes measurements in the wake of the obstacles as well as in the streamwise gaps between the obstacles. The upstream profile is shown as a solid line in the figure. When the flow enters the obstacle array, there is a rapid adjustment to an equilibrium profile within the obstacle canopy ($z < H$), and a much slower diffusion of the momentum deficit upwards into the internal boundary layer. The height of this internal boundary layer was approximately 5H

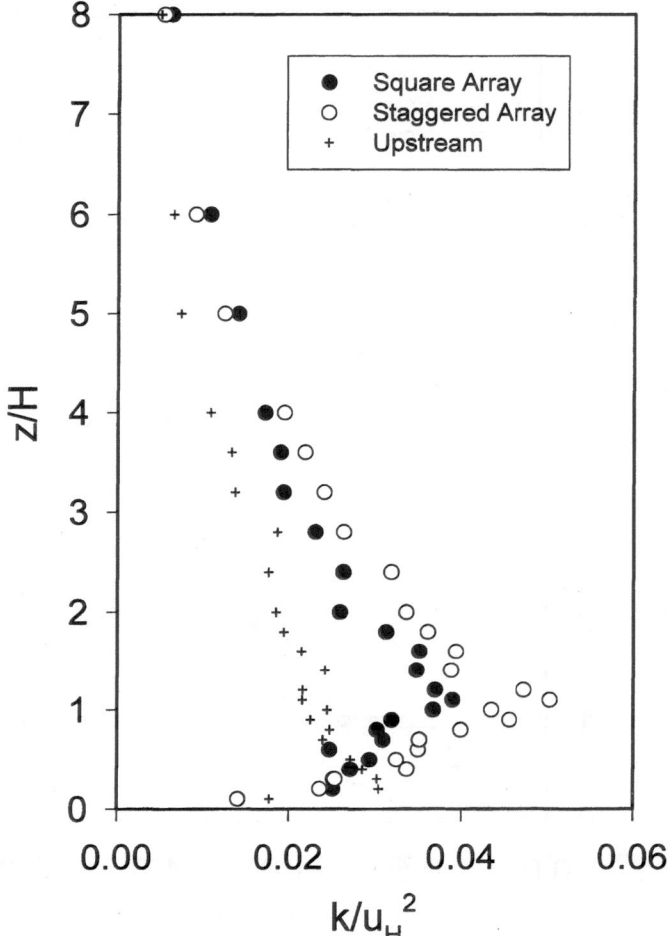

Figure 6. Turbulence kinetic energy profiles measured behind row 18 of the $\lambda_f = 0.16$ cube array in both a square and staggered configuration.

at the 18th row (x = 44H) of this array, i.e., about 10% of the fetch. In the densely packed $\lambda_f = 0.44$ array (results not shown), the mean flow in the canopy was much reduced, relative to the $\lambda_f = 0.16$ array. This was due to the onset of street canyon flow, and the appearance of a stable mean vortex between the rows in the $\lambda_f = 0.44$ array.

Figure 6 shows the turbulence kinetic energy (TKE) profiles measured behind the 18th row of the $\lambda_f = 0.16$ cube arrays. The TKE is defined as:

$$k = \frac{1}{2}\left(\sigma_u^2 + \sigma_v^2 + \sigma_w^2\right) .$$ (3)

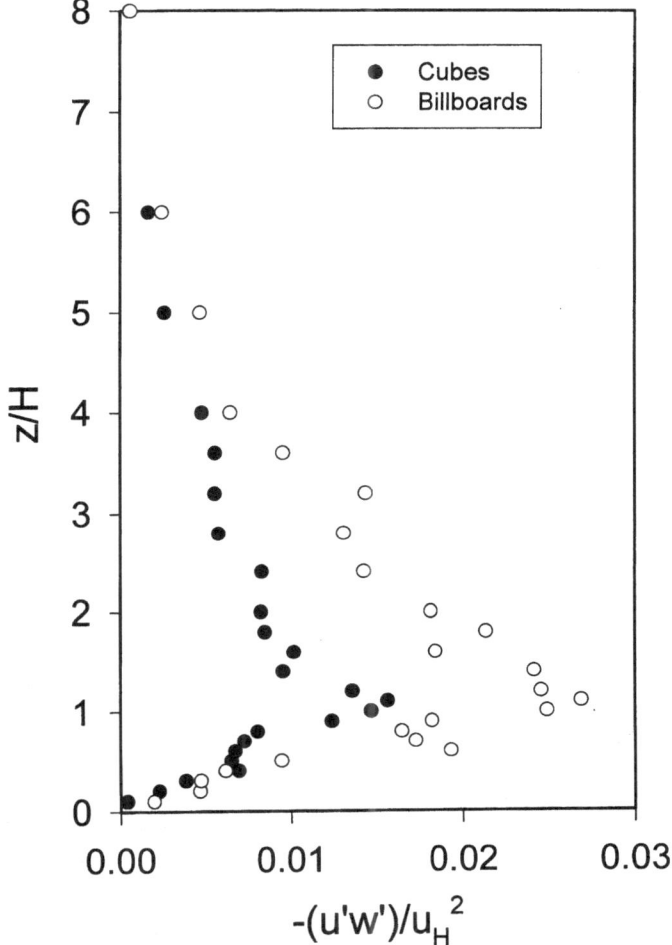

Figure 7. Reynolds stress profiles measured behind the 18th row of the $\lambda_f = 0.16$ arrays of cubes and flat-plate roughness in a staggered geometry.

This figure shows that the TKE is about 20% higher in the staggered arrays. This increased turbulence energy was found above the staggered arrays at all three packing densities ($\lambda_f = 0.0625$, 0.16, and 0.44), and is due to the greater aerodynamic exposure of the obstacles in the staggered configuration. This observation correlates with the slightly larger z_0 value for the staggered arrays (Macdonald *et al.*, 1998).

In all cases, at the top of the internal boundary layer above the arrays, the TKE profiles blended into the background turbulence levels in the approach flow. This blending occurs above $z = 6H$ in the profile shown in Figure 6.

Figure 7 shows the Reynolds stress profiles measured in the $\lambda_f = 0.16$ arrays of billboards and cubes. In each case the maximum shear stress was measured at the

TABLE II

Aerodynamic roughness and friction velocity in the various obstacle arrays

Frontal area density (λ_f)	Array type	Roughness height (z_0/H)	Friction velocity (u_*/u_H)
0.0625	Cubes, square	0.087	0.083
0.0625	Cubes, staggered	0.11	0.089
0.16	Cubes, square	0.11	0.089
0.16	Cubes, staggered	0.16	0.10
0.44	Cubes, square	0.022	0.075
0.44	Cubes, staggered	0.055	0.077
0.16	Billboards, square	0.32	0.13
0.16	Billboards, staggered	0.33	0.15

top of the obstacles and the $-\overline{u'w'}$ stress decreased significantly in the roughness sublayer between $z = H$ and $z = 2H$, so that it was not possible to identify a 'constant stress layer' above the arrays. The reference surface shear stress $\tau_0 = \rho u_*^2$ was therefore obtained by averaging the Reynolds stress in the layer between $z = H$ and $z = 2H$.

As shown in Figure 7, the shear stress in the billboard array was a factor of two larger than in the cube array. This translates into a friction velocity u_* about 50% larger. The roughness height z_0 was also much larger in the billboard arrays. Table II summarizes al the z_0 and u_* results obtained from the experiments.

Figure 8 shows the nondimensionalized RMS turbulence components in the internal boundary layer above the $\lambda_f = 0.16$ array. To nondimensionalize these data the friction velocity u_* used was based on the local Reynolds stress (evaluated at each height). These plots show that the nondimensional turbulence ratios above the obstacles are approximately constant in the internal boundary layer flow between $z = H$ and $z = 5H$. Similar results were found above all the arrays. In addition, the correlation coefficient,

$$R_{uw} = \frac{-\overline{u'w'}}{\sigma_u \sigma_w} \tag{4}$$

was also constant. Table III summarizes the results for the different obstacle array types.

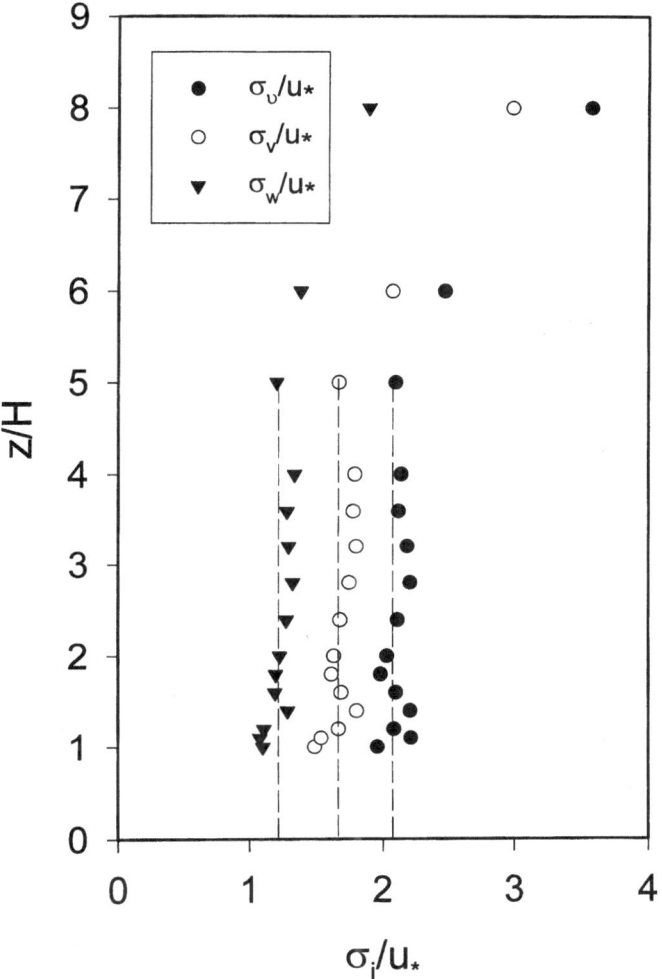

Figure 8. Non-dimensional RMS turbulence components measured behind row 18 of $\lambda_f = 0.16$ cube array in a square configuration.

TABLE III

Nondimensional turbulence quantities in the various arrays

Array type	σ_u/u_*	σ_v/u_*	σ_w/u_*	R_{uw}
Cubes – average of all array types	2.1	1.65	1.2	0.41
Billboards – average of $\lambda_f = 0.16$ arrays	2.0	1.6	1.2	0.45

4. Discussion

The z_0 results in Table II confirm previous findings by Macdonald *et al.* (1998) where it was also observed that the staggered arrays were aerodynamically rougher than square arrays. As a function of obstacle array packing density, the magnitude of z_0 reaches a peak at about $\lambda_f = 0.16$, but is much reduced in the densely packed $\lambda_f = 0.44$ array, due to the onset of skimming flow. In a real urban array, the dramatic decrease in z_0 for the densely packed ($\lambda_f = 0.44$) array would probably not occur, since the large height variability and irregular nature of the real urban landscape would tend to prevent skimming flow and the onset of such low values of z_0.

An interesting observation from these experiments is that the TKE and shear stress associated with the billboard arrays is much larger – they are typically twice as rough as in the equivalent cube arrays. This shows that frontal area density cannot be the only parameter that affects the value of z_0. For example, Lettau's (1969) expression:

$$\frac{z_0}{H} = 0.5\lambda_f , \tag{5}$$

can only be an approximation to the aerodynamic roughness of obstacle arrays over a restricted range of λ_f. Much more sophisticated parameterizations are required to accurately predict the aerodynamic roughness of real urban arrays (Grimmond and Oke, 1999; Hanna *et al.*, 2001).

The nondimensional turbulence ratios in Table III are similar to the standard scaling ratios for ABL flow over relatively smooth terrain which are suggested in many meteorology publications. If one defines the constants A, B, and C according to:

$$A = \frac{\sigma_u}{u_*}; \qquad B = \frac{\sigma_v}{u_*}; \qquad C = \frac{\sigma_w}{u_*}, \tag{6}$$

then most full-scale atmospheric studies over rural terrain yield average values $A = 2.4$, $B = 1.9$ and $C = 1.25$ (e.g., Panofsky and Dutton, 1984). Lower values for A and B are typically measured in wind tunnels, however, because in the full scale ABL the horizontal components of the wind variances typically include energy not due to mechanical shearing. The present results yield $A = 2.1$, $B = 1.65$ and $C = 1.2$ for scaling in the internal boundary layer above the rough obstacle arrays, based on local (z-dependent) values of u_*. Similar scaling rules for full-scale urban boundary layer data have been found by Roth (2000), and suggest that the urban scaling is similar to the rural scaling.

The usefulness of the scaling ratios in Table III comes from the fact that one only has to measure one of the turbulence components such as σ_u, which can be done with relatively simple instrumentation, in order to obtain estimates for the rest of the data (σ_v, σ_w, and u_*). This can lead to a large reduction in experimental effort. The present data also suggest that the value of the structure constant C_μ used in $k - \varepsilon$ modelling of the rough boundary layer,

$$C_\mu = \left(\frac{\tau_0}{k}\right)^2 , \tag{7}$$

should be between 0.054 and 0.063, which is somewhat less than the standard value of 0.09 used in modelling wall flows. An even lower value of $C_\mu = 0.033$ was proposed by Raithby *et al.* (1987) for CFD modelling of an atmospheric boundary layer flow over a hill, based upon the standard scaling values of $A = 2.4$, $B = 1.9$ and $C = 1.25$.

5. Conclusions

The turbulence results obtained here are valid for boundary layer flows over rough obstacle arrays with a range of packing density covering wake interference and skimming flow, and are thus applicable to most urban construction. The much larger TKE and shear stress in the flow above billboard obstacles shows the importance of including the sharp features such as peaked roofs on wind tunnel models in order to generate the necessary turbulence levels. These results also confirm that 2-D, flat plate roughness elements generate the maximum z_0 effect.

Perhaps the most important conclusion from these experiments is the observation that, despite the absence of a measurable constant stress layer in the flow above the arrays, when the RMS turbulence components are scaled with the local u_* data, they yield the standard scaling ratios $A = 2.1$, $B = 1.65$ and $C = 1.2$, for σ_u, σ_v and σ_w, respectively. This provides an important quality control test for numerical simulations of flow over large obstacle arrays and in fact the archived data set obtained in this study is currently being used in validating CFD codes and for testing out simple parameterizations of the urban canopy flow for use in dispersion models (Coirier, 2001).

Acknowledgements

The work described herein was funded by George Mason University under contract number 2943101 and is part of the Coordinated Hazardous Atmospheric Release Modeling (CHARM) project managed by Dr. Steven Hanna. Shelley Carter was funded in part by a research assistantship from grants awarded to R. Macdonald and P. Slawson by the Natural Sciences and Engineering Research Council of Canada.

References

Best, M. J.: 2000, 'Can we Represent Urban Areas in Operational Numerical Weather Prediction Models?', *Proceedings of the AMS Third Symposium on the Urban Environment*, Davis, CA, U.S.A., 14–18 August 2000.

Brown, M. J. and Williams, M. D.: 1998, 'An Urban Canopy Parameterization for Mesoscale Meteorological Models', *Proceedings of the AMS Second Urban Environment Symposium*, Albuquerque, NM, U.S.A., 2–7 November 1998.

Brown, M. J. and Spore, J.: 2000, 'Computational Fluid Dynamics Modeling Used to Assess Mesoscale Urban Canopy Drag and Turbulence Parameterizations', *Proceedings of the AMS Third Symposium on the Urban Environment*, Davis, CA, U.S.A., 14–18 August 2000.

Bottema, M.: 1997, 'Urban roughness modelling in relation to pollutant dispersion', *Atmosph. Environ.* **31**, 3059–3075.

Coirier, W. J.: 2000, *Development of a High Fidelity PC-based Simulator for Modeling the Atmospheric Transport and Dispersion of Nuclear, Chemical, Biological and Radiological Substances in Urban Areas*, CFD Research Corporation Final Report, Huntsville, AL, U.S.A., 56 pp.

Grimmond, C. S. B. and Oke, T. R.: 1999, 'Aerodynamic properties of urban areas derived from analysis of surface form', *J. Appl. Meteorol.* **38**, 1262–1292.

Hanna, S. R. and Britter, R. E.: 2001, *Wind Flow and Vapor Cloud Dispersion at Industrial and Urban Sites*, Draft Manual for American Institute of Chemical Engineers (AIChE) Center for Chemical Process Safety (CCPS), 136 pp.

Hosker, R. P.: 1984, 'Flow and diffusion near obstacles', in D. Randerson (ed.), *Atmospheric Science and Power Production*, U.S. Dept. of Energy Report No. TIC-27601, pp. 241–326.

Lettau, H.: 1969, 'Note on aerodynamic roughness-parameter estimation on the basis of roughness-element description', *J. Appl. Meteorol.* **8**, 828–832.

Macdonald, R. W., Griffiths, R. F. and Hall, D. J.: 1998, 'An improved method for estimation of surface roughness of obstacle arrays', *Atmosph. Environ.* **32**, 1857–1864.

Macdonald, R. W., Carter, S. and Slawson, P. R.: 2000, 'Measurements of Mean Velocity and Turbulence Statistics in Simple Obstacle Arrays at 1:200 Scale', University of Waterloo, *Thermal Fluids Report 2000-1*, Waterloo, ON, Canada, 130 pp.

Martilli, A., Clappier, A. and Rotach, M. W.: 2000, 'A Parameterization of Urban Effects for Mesoscale Models', *Proceedings of the AMS Third Symposium on the Urban Environment*, Davis, CA, U.S.A., 14–18 August 2000.

Meroney, R. M.: 1982, 'Turbulent diffusion near buildings', in E. J. Plate (ed.), *Engineering Meteorology*, Elsevier Scientific, Amsterdam, pp. 481–526.

Oke, T. R.: 1987, *Boundary Layer Climates*, Routledge Publishers, London.

Panofsky, H. A. and Dutton, J. A.: 1984, *Atmospheric Turbulence*, John Wiley & Sons, New York, p. 160.

Raithby, G. D., Stubley, G. D. and Taylor, P. A.: 1987, 'The Askervein Hill project: A finite control volume prediction of three-dimensional flows over the hill', *Bound.-Layer Meteorol.* **26**, 247–267.

Roth, M.: 2000, 'Review of atmospheric turbulence over cities', *Q. J. R. Meteorol. Soc.* **126**, 941–990.

Wieringa, J.: 1993, 'Representative roughness parameters for homogeneous terrain', *Bound.-Layer Meteorol.* **63**, 323–363.

A WIND TUNNEL INVESTIGATION OF THE INFLUENCE OF SOLAR-INDUCED WALL-HEATING ON THE FLOW REGIME WITHIN A SIMULATED URBAN STREET CANYON

A. KOVAR-PANSKUS[1], L. MOULINNEUF[2], E. SAVORY[3]*, A. ABDELQARI[2],
J.-F. SINI[2], J.-M. ROSANT[2], A. ROBINS[1] and N. TOY[1]

[1] *Fluids Research Centre, School of Engineering, University of Surrey, Guildford, Surrey, U.K.;*
[2] *Ecole Centrale de Nantes, Laboratoire de Mécanique des Fluides, Nantes, Cedex 3, France;*
[3] *Department of Mechanical and Materials Engineering, Faculty of Engineering, University of Western Ontario, London, Ontario, Canada*
(* author for correspondence, e-mail: e.savory@eng.uwo.ca, fax: +44 (0) 1483 450984)

Abstract. A wind tunnel study has been undertaken to assess the influence of solar-induced wall heating on the airflow pattern within a street canyon under low-speed wind conditions. This flow is normally dominated by large-scale vortical motion, such that the wind moves downwards at the downstream wall. In the present work the aim has been to examine whether the buoyancy forces generated at this wall by solar-induced heating are of sufficient strength to oppose the downward inertial forces and, thereby, change the canyon flow pattern. Such changes will also influence the dispersion of pollutants within the street. In the experiments the windward-facing wall of a canyon has been uniformly heated to simulate the effect of solar radiation. Four different test cases, representing different degrees of buoyancy (defined by a test Froude number, Fr), have been examined using a simple, 2-D, square-section canyon model in a wind tunnel. For reference purposes, the neutral case (no wall heating), has also been studied. The approach flow boundary layer conditions have been well defined, with the wind normal to the main canyon axis, and measurements have been taken of canyon wall and air temperatures and profiles of mean velocities and turbulence intensities. Analysis of the results shows clear differences in the flow patterns. As Fr decreases from the neutral case there are reductions of up to 50% in the magnitudes of the reverse flow velocities near the ground and in the upward motion near the upstream wall. A marked transition occurs at $Fr \approx 1$, where the single dominant vortex, existing at higher Fr values, weakens and moves upwards whilst a lower region of relatively stagnant flow appears. This transition had previously been observed in numerical model predictions but at a Fr at least an order of magnitude higher.

Keywords: buoyancy, cavity flow, Froude number, street canyon, thermal effects

Nomenclature

d	=	Displacement height	(m)
Fr	=	Froude number	(–)
H, W	=	Height and width of canyon	(m)
k	=	Turbulent kinetic energy	(m^2 s^{-2})
L	=	Spanwise canyon length	(m)

Water, Air, and Soil Pollution: Focus **2:** 555–571, 2002.
© 2002 *Kluwer Academic Publishers.*

Re	=	Reynolds number, U_{ref} H/ν	
T	=	Air temperature	(°C)
T_w	=	Temperature of heated wall	(°C)
T_{ref}	=	Freestream air temperature	(°C)
$\overline{U}, \overline{W}$	=	Mean velocity components	(m s^{-1})
U_{ref}	=	Freestream velocity	(m s^{-1})
u', w'	=	Rms velocity fluctuations	(m s^{-1})
u_*	=	Friction velocity	(m s^{-1})
Vel	=	Velocity vector, $(\overline{U}^2 + \overline{W}^2)^{1/2}$	(m s^{-1})
X, Y, Z	=	Co-ordinate system	(m)
z_0	=	Roughness length	(m)
δ	=	Boundary layer thickness	(m)
ν	=	Kinematic viscosity of air	(m^2 s^{-1})

1. Introduction

The wind flow field within urban areas has been studied for several decades, with several investigations of the main flow patterns occurring within street canyons (Albrecht, 1933; followed by Georgii et al., 1967; Ludwig and Dabberdt, 1972; DePaul and Sheih, 1986). All these researchers undertook full-scale measurements in a street canyon, under conditions of the wind perpendicular to the main canyon axis, and observed the main vortex that tends to be established under these wind conditions. Further research focussed on the dependence of the flow patterns and also the dispersion of pollutants on the aspect ratio of the canyon (streamwise width/depth), different roof shapes (Rafailidis, 1997), and the effect of traffic (Vachon, 2000).

Another topic of interest, especially under conditions of low wind speed, is the influence of wall heating in street canyons due to solar radiation incident on one or more walls during the course of a day. Full-scale measurements have been conducted to gain knowledge about the influence of wind speed, temperature, and stability conditions on the canyon flow regime (Nakamura and Oke, 1988; Vachon, 2000), whilst wind tunnel studies investigating the same problems have been conducted by Ogawa et al. (1981), and Uehara et al. (1997a, 2000). In addition, the dispersion of pollutants has been investigated within street canyons in stratified urban canopy layers (Uehara et al., 1997b). However, the influence of different buoyancy effects due to canyon wall heating has not previously been assessed through wind tunnel simulations under carefully controlled conditions, because it is a challenging problem to match the necessary similarity laws for velocity and temperature via an appropriate Froude number. The effect of wall heating on the flow and pollution dispersion characteristics within the canyon is likely to be

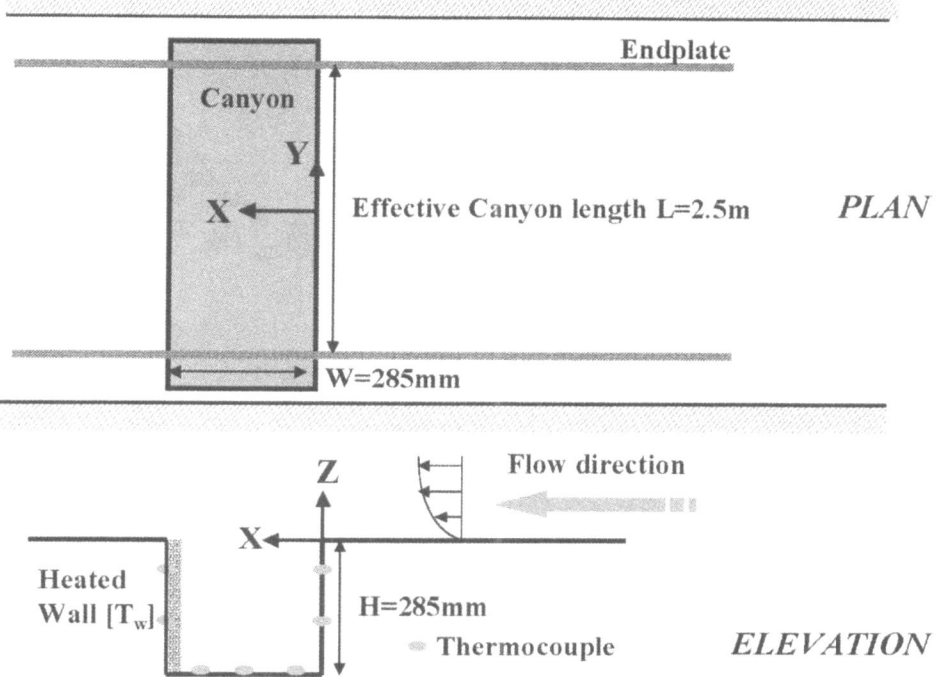

Figure 1. Diagrammatic layout of wind tunnel and canyon model.

greatest when the windward-facing wall is heated. This is because the buoyancy forces generated at that wall by the heating directly oppose the downward inertial forces in the same region that are associated with the dominant canyon vortex. It is anticipated, from previous numerical predictions (Mestayer *et al.*, 1995), that, under certain conditions, the buoyancy forces will be large enough to disrupt the canyon vortex to form a different regime with consequent effects on the local flow and pollutant dispersion characteristics.

This article reports on an initial examination of this complex problem, studying the effect of Froude number on the main flow features in a simple, yet not too unrealistic, 2-D canyon in order to determine the transition conditions between any different regimes and whether there is a threshold Froude number above which such thermal effects may be neglected. A second aim of the work is to provide a detailed data set for developing computational fluid dynamics (CFD) models for predicting these flow regimes.

2. Experimental Details

A comprehensive study was conducted in the EnFlo Laboratory wind tunnel of the Fluids Research Centre at the University of Surrey. A nominally two-dimensional

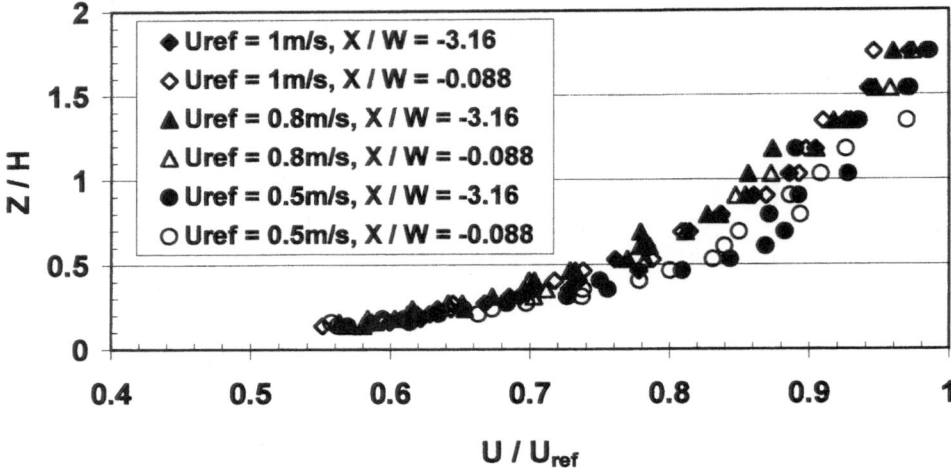

Figure 2. Boundary layer mean velocity profiles for different windspeeds.

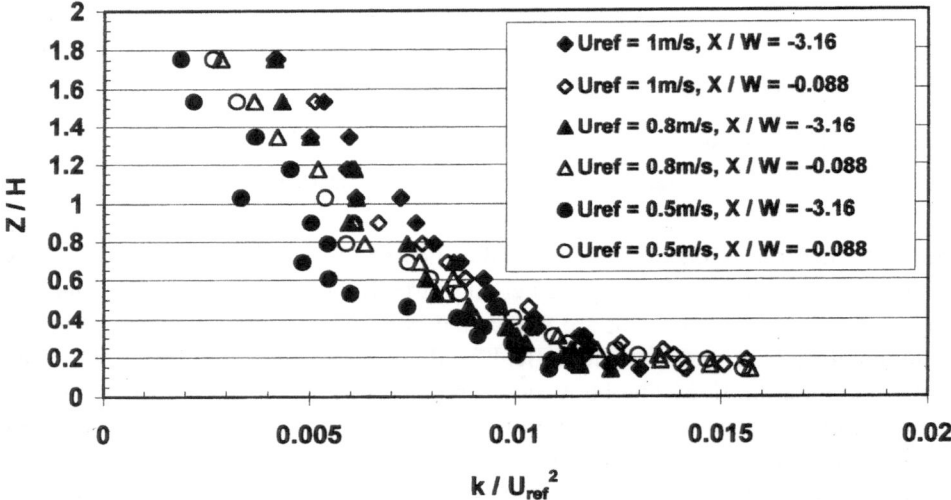

Figure 3. Boundary layer turbulent kinetic energy profiles for different windspeeds.

(2-D), 285 mm square section (W/H = 1) cavity of 3 m length was installed with its principal axis perpendicular to the oncoming flow (Figure 1). The stratification of the approach flow was kept neutral, whereas inside the cavity different buoyancy conditions were simulated by heating of the windward-facing wall. The boundary layer was simulated using vorticity generators and low density, flat plate roughness elements, with the resulting parameters of boundary layer height $\delta = 1$ m, displacement height $d = 0$ mm, and roughness length z_0 of between 1 and 1.6 mm, depending on the flow Reynolds number. The friction velocity u_*/U_{ref} varied between 0.064 and 0.070 over the freestream velocity range of 0.5 to 1 m s^{-1}. This

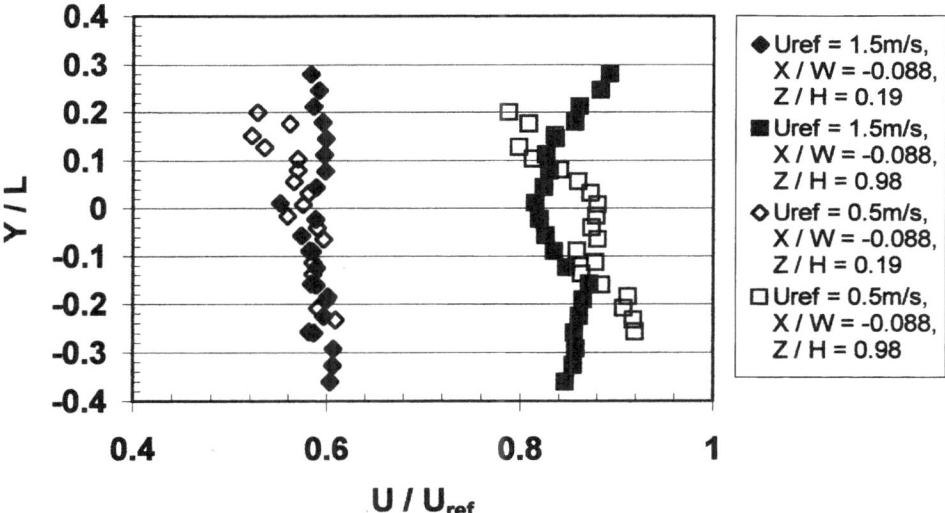

Figure 4. Spanwise variation of mean velocity in the boundary layer.

wind speed range gave canyon Reynolds numbers from 9.5×10^3 to 1.9×10^4. The boundary layer profiles of mean velocity $(\overline{U})$ and turbulent kinetic energy (k) for this range of freestream speeds are shown in Figures 2 and 3, respectively, for two different locations upstream of the canyon, illustrating how the boundary layer was fully developed in all cases.

The velocity measurements were carried out with a two-component Laser-Doppler-Anemometer, with an accuracy of $\pm 3.5\%$ for the mean values and $\pm 4\%$ for the turbulence quantities. This includes the accuracy of the measurement system and the statistical error, as well as the repeatability of the measurements. The sampling frequency, dependent on the local flow seeding, was always at least 40 Hz inside the cavity (usually about 100 Hz) and typically 30 000 samples were taken per measurement point. The flow temperature measurements were taken with thermocouples (K-Type) and a Platinum-Resistance-Thermometer (PRT), with an accuracy of within $\pm 1.2\,^\circ$ for the thermocouples and $\pm 0.5\,^\circ$C for the PRT.

The two-dimensionality both before and inside the cavity was established by the use of end plates that also enabled Reynolds number independence for the flow at velocities at or above 0.5 m s^{-1}. These end plates had a height of 0.8 m and they extended for distances of 4.5 and 2.4 m upstream and downstream of the canyon, respectively. Preliminary optimisation experiments showed that a spanwise spacing of L = 2.5 m between the end plates gave the highest degree of flow two-dimensionality. Hence, this setting was used in all the subsequent measurements, giving an effective spanwise aspect ratio of L/W = 8.8. Whilst this ratio is rather small for a nominally two-dimensional model it was considered that, for the purposes of this initial study, the most important factor was to optimise the quality of

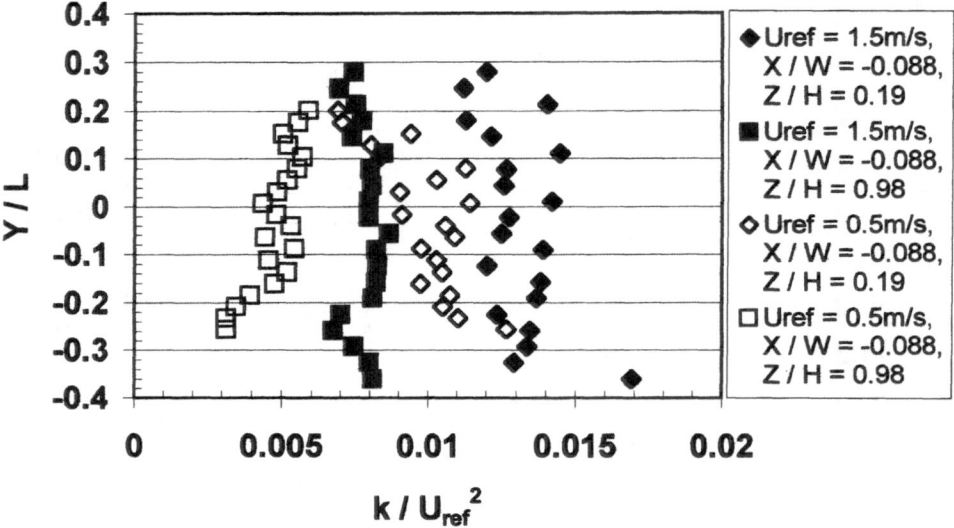

Figure 5. Spanwise variation of turbulent kinetic energy in the boundary layer.

the approach flow. At the two reference wind speeds examined, of 0.5 and 1.5 m s^{-1}, the lateral uniformity of the approach flow mean streamwise velocity and turbulent kinetic energy were within ±5 and ±10%, respectively, of their mean values measured over the central ±20% of the cavity length about the geometrical centre-plane, as illustrated in Figures 4 and 5.

The windward-facing wall was heated by heating mats, with the temperature being controlled via measurements from embedded thermocouples, whilst the other faces were cooled to ambient temperature. Although solar-induced wall heating is a complex issue, in this initial study only the differential temperature between the air and the heated wall was considered. A reasonable value for this is 5 °C (Sini *et al.*, 1996), although it will clearly vary with canyon geometry, cloud cover and time of day, for example. The thermal effect is only of potential importance at low wind speeds, certainly below 3 m s^{-1} when the 'urban dome' regime is dominating. Hence, a simple canyon of 20 m height has been modelled subjected to full-scale wind speeds of 1–3 m s^{-1}. This gives Froude numbers of 0.27 to 2. In order to match this range the temperatures on the heated wall were established at either ambient (neutral case), 80 or 120 °C with the variation over the whole heated wall being within ±5 °C of the mean value. The three different test velocities of 0.5, 0.8 and 1.0 m s^{-1} (measured by a reference sonic anemometer located at a height of 1 m) resulted in Froude numbers ranging from 0.27 to 2.03 for the heated wall cases (the neutral case giving an infinite Froude number). Here, the Froude number has been defined as $Fr = U_{ref}^2/(gH(T_w - T_{ref})/T_{ref})$, where T_w denotes the heated wall temperature, T_{ref} the ambient, freestream reference temperature (all temperatures in this expression are absolute values in Kelvin) and g the acceleration due to grav-

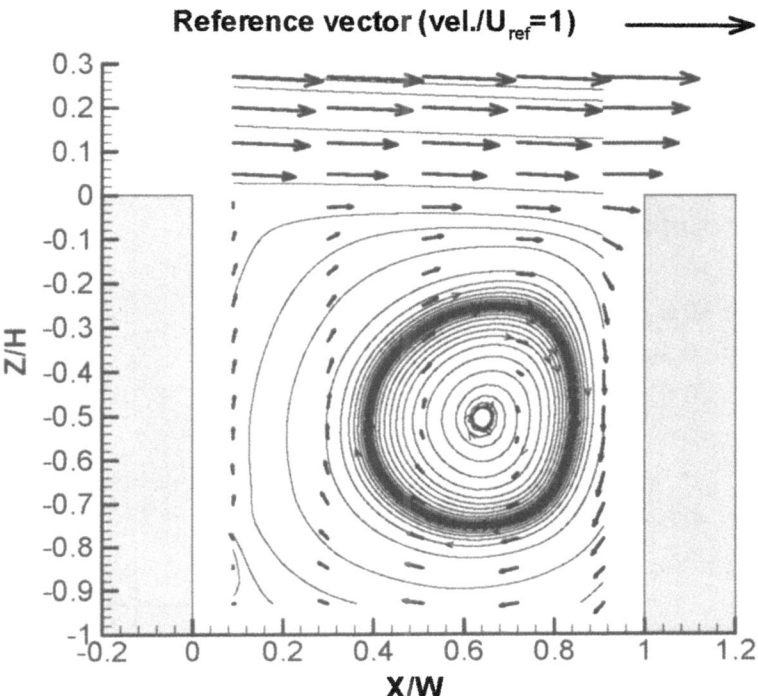

Figure 6. Projected mean velocity vectors and streampaths on canyon centre-plane. Neutral case, $U_{ref} = 1 \text{ m s}^{-1}$.

ity. Five vertical profiles of mean velocity and turbulence measurements were taken at the same streamwise locations, on the geometrical centre-plane, for all the different Froude number cases. Each profile commenced at 20 mm above the canyon base and extended into the shear layer where the data could be compared with the oncoming flow boundary layer conditions. At each position the time-histories of the streamwise (U) and vertical (W) velocity components were measured and the time-averaged velocity and turbulence statistics computed. Restrictions on the orientation of the LDA probe within the canyon flow regime prohibited the measurement of the third, lateral component of velocity (although, for completeness, these data were taken in the boundary layer). Hence, the turbulent kinetic energy (TKE) within the canyon was estimated as $k = 1/2(u'^2 + 2w'^2)$.

3. Results and Discussion

The projected mean velocity vectors on the centre-plane of the canyon, together with streampaths computed from these velocity distributions, are shown in Figures 6 to 10 for the neutral case through to the lowest Fr value. The figures for the heated wall cases also show the mean air temperature distributions. The influence

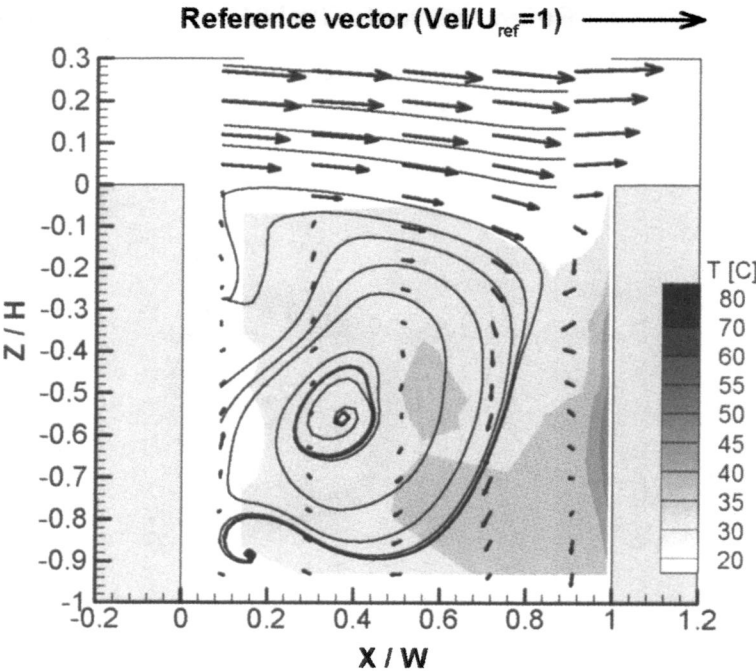

Figure 7. Projected mean velocity vectors, mean temperatures and streampaths on canyon centre-plane. $Fr = 2.03$, $U_{ref} = 1$ m s^{-1}, $T_w = 80\ ^\circ$C.

of the temperature due to the wall heating seems, in general, to be very small at the higher Fr cases examined. The overall feature of a single main vortex in the cavity remains for Fr down to 1.17, although the centre of the vortex tends to move slightly downwards and upstream with decreasing Fr value. For the lowest two Fr values a flow in the freestream direction, associated with a very weak secondary vortex, is found near the base of the canyon. This transition from a flow regime with single dominant vortex to one with a weaker secondary vortex close to the ground, that appears to take place at a Fr of the order of 1, is qualitatively, but not quantitatively, similar to the 2-D numerical predictions, using the code CHENSI (Sini et al., 1996), for a geometrically similar cavity (W/H = 1) but with different approach flow conditions (Mestayer et al., 1995). However, in their study this transition took place at a Fr value at least an order of magnitude higher. Further work is required, including direct comparisons of experiments and 3-D predictions of exactly the same test case, in order to explain these differences. However, it is to be expected that a 2-D prediction, with precisely defined and uniform heated wall temperature distributions, will yield different results from an experiment where there is variability in the wall heating and spanwise recirculations associated with the essential three-dimensionality of the configuration.

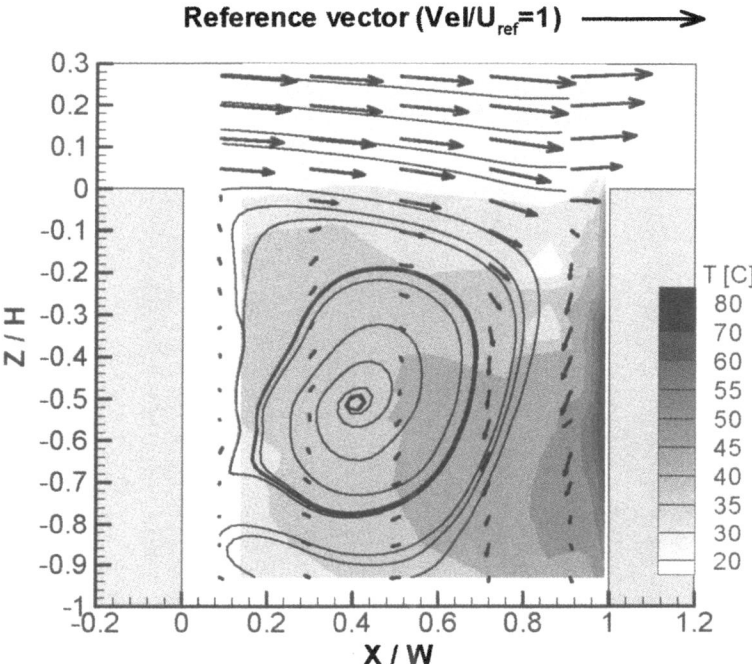

Figure 8. Projected mean velocity vectors, mean temperatures and streampaths on canyon centre-plane. $Fr = 1.17$, $U_{ref} = 1$ m s^{-1}, $T_w = 120$ °C.

In all the cases examined there does not seem to be any evidence of an updraft close to the heated wall, even for the very lowest Fr regime. Since the profile nearest to the heated surface was taken at a distance of 0.09 W (25 mm) from the wall it is probable that there is a very thin layer close to the wall within which an updraft occurs. As part of a recent field experiment undertaken in a street in Nantes, France (Louka *et al.*, 2002), some flow visualisation was carried out using neutrally buoyant, helium-filled balloons (Vachon *et al.*, 1999). For wind directions perpendicular to the street axis balloons released at street level tended to remain within 5 m of the ground with no obvious tendency towards upwards or downwards motion, suggesting that they were trapped within the weak secondary vortex. However, those balloons that approached very close to the heated wall, so that they were almost touching the surface, always rose upwards along that wall and into the upper levels of the canyon.

In the present work the temperature distributions inside the canyon display some degree of similarity, except very close to the heated wall. Figures 11 to 14 show enlarged plots of the temperature distributions close to the heated wall for the range of Fr examined, in order to show more clearly the differences between the cases. The vertical location of the maximum temperature in the profile measured closest to the wall (0.014 W, or 4 mm, from the wall) changes significantly with Fr, from

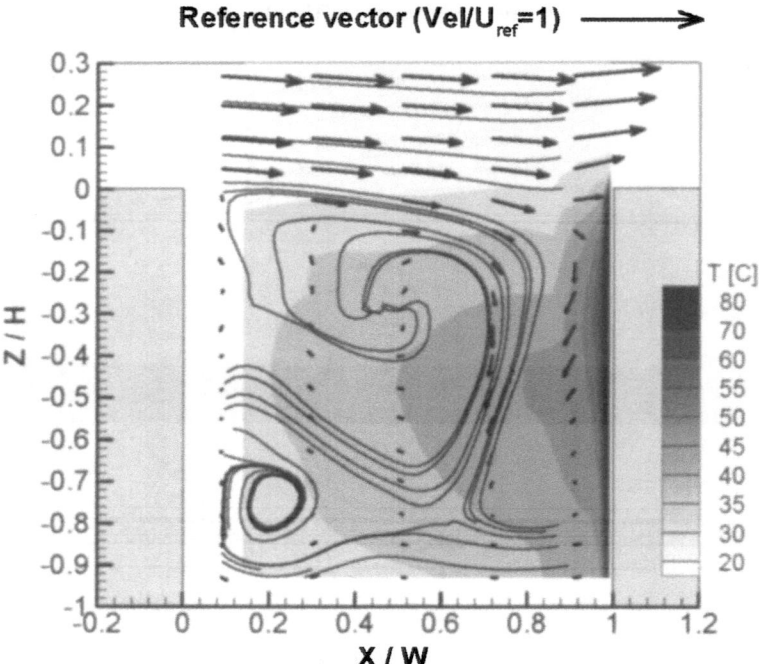

Figure 9. Projected mean velocity vectors, mean temperatures and streampaths on canyon centre-plane. $Fr = 0.73$, $U_{ref} = 0.8$ m s^{-1}, $T_w = 120$ °C.

a position of 0.40 H above the canyon base at $Fr = 2.03$ and then 0.47 H at $Fr = 1.17$ to 0.82H at $Fr = 0.73$ followed by 0.75 H at $Fr = 0.27$. Thus, as with the mean velocity data, there appears to be a transition at $Fr \approx 1$. The value of the maximum temperature in each of these near-wall profiles increases with a decrease in Fr down to $Fr = 0.73$ and then decreases again at the lowest Fr examined. The profile shapes for the monotonic decrease of air temperature with distance from the heated wall are similar for all the Fr values, giving a thermal boundary layer thickness of the order of 0.2 W, at those heights corresponding to the maximum near-wall temperature. Examining these plots, with those presented in Figures 6 to 10, clearly shows that the increase in height of the hottest region close to the heated wall is directly related to the weakening of the downwash associated with the dominant vortex. In addition, as the wall heating influence increases (decreasing Fr) and the main vortex weakens, along with the evolution of a region of stagnant flow beneath, the temperatures in the most upstream part of the canyon decrease. This indicates a reduced convection of the heat produced from the heated wall into this region.

Figures 15 to 19 show the distributions of TKE within and above the canyon for the different test conditions. It may be seen that for all the heated wall cases there is a significant increase in TKE near the heated wall. Given that the overall

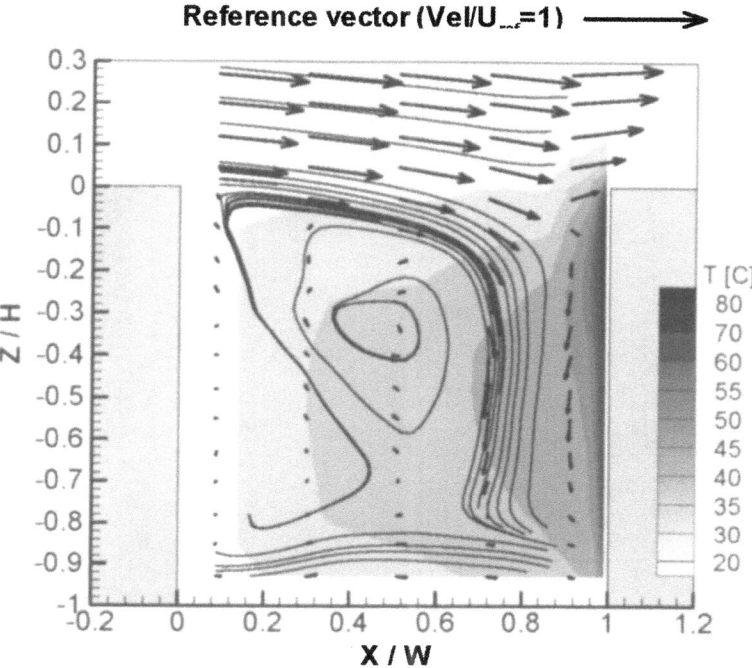

Figure 10. Projected mean velocity vectors, mean temperatures and streampaths on canyon centre-plane. $Fr = 0.27$, $U_{ref} = 0.5$ m s^{-1}, $T_w = 120\,°C$.

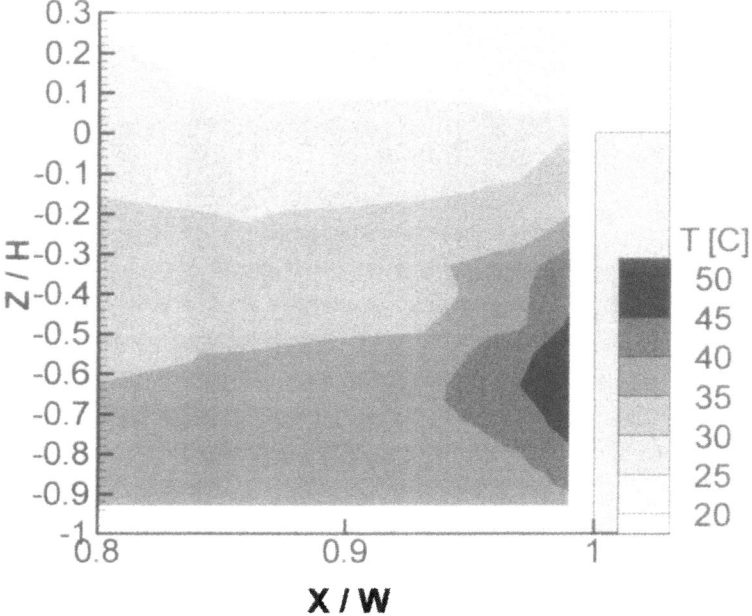

Figure 11. Detail of temperature distribution near heated wall for $Fr = 2.03$.

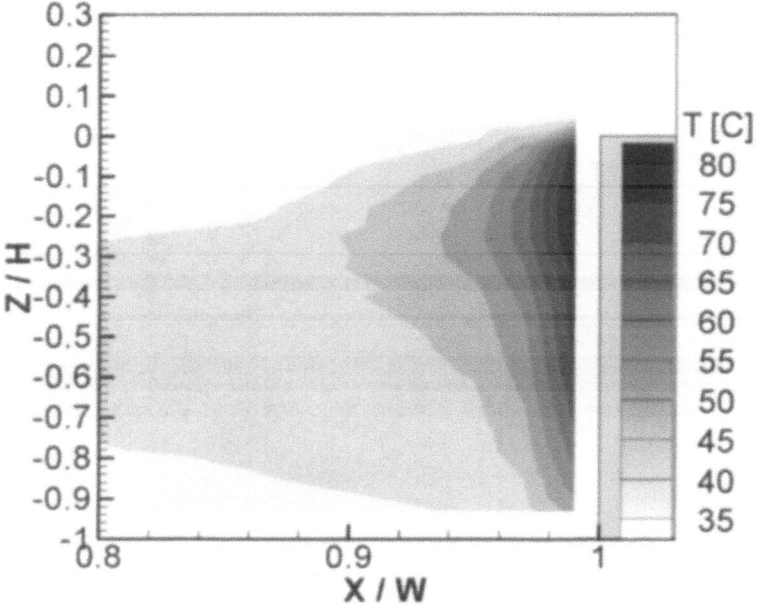

Figure 12. Detail of temperature distribution near heated wall for $Fr = 1.17$.

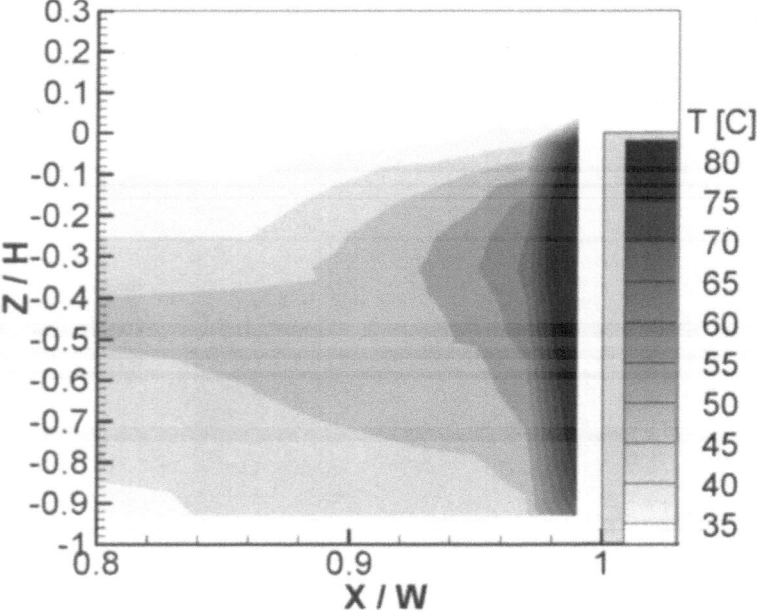

Figure 13. Detail of temperature distribution near heated wall for $Fr = 0.73$.

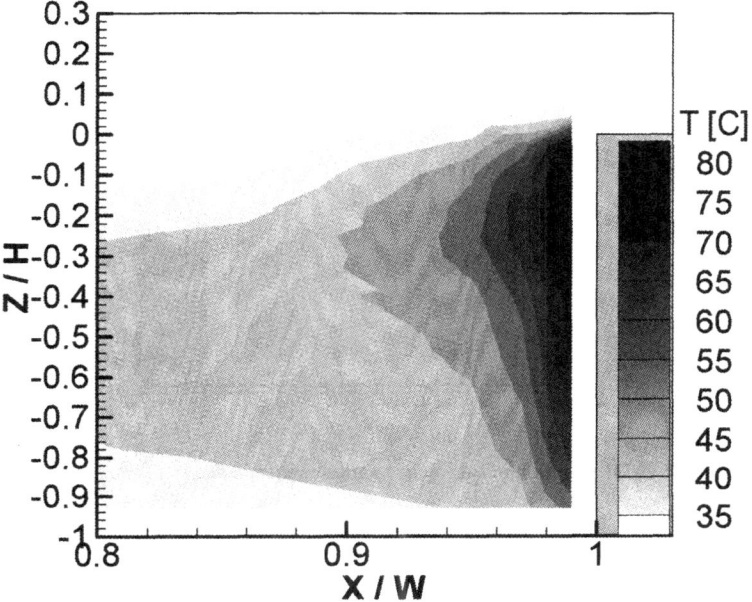

Figure 14. Detail of temperature distribution near heated wall for $Fr = 0.27$.

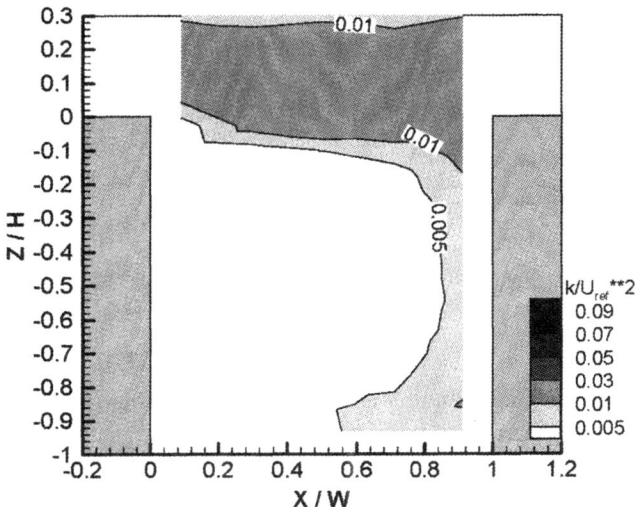

Figure 15. Distribution of turbulent kinetic energy in canyon for neutral case.

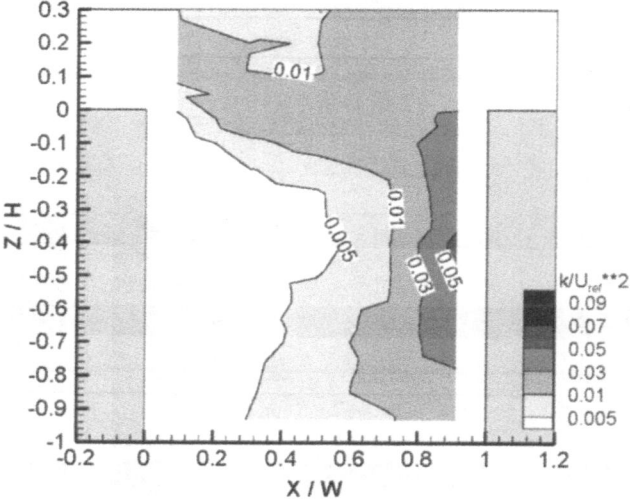

Figure 16. Distribution of turbulent kinetic energy in canyon for $Fr = 2.03$.

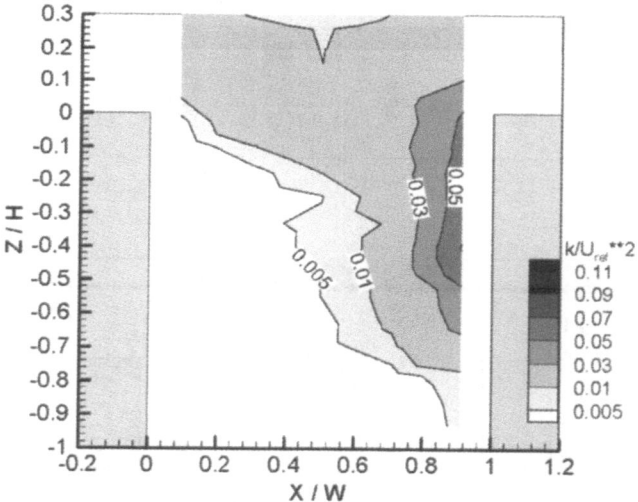

Figure 17. Distribution of turbulent kinetic energy in canyon for $Fr = 1.17$.

accuracy of the TKE measurements is $\pm12\%$ the increase in TKE with decrease in Fr is always greater than this experimental uncertainty; neutral case ($Fr = \infty$), $k = 0.007\pm0.0008$; $Fr = 2.03$, $k = 0.03\pm0.004$; $Fr = 1.17$, $k = 0.05\pm0.006$; $Fr = 0.73$, $k = 0.07\pm0.008$; $Fr = 0.27$, $k = 0.11\pm0.013$. These are the maximum values measured in the profile closest to the wall, taken from each plot. Since there is probably insufficient shear in the mean flow in this region to produce such high TKE production, the primary source of production may be the high thermal

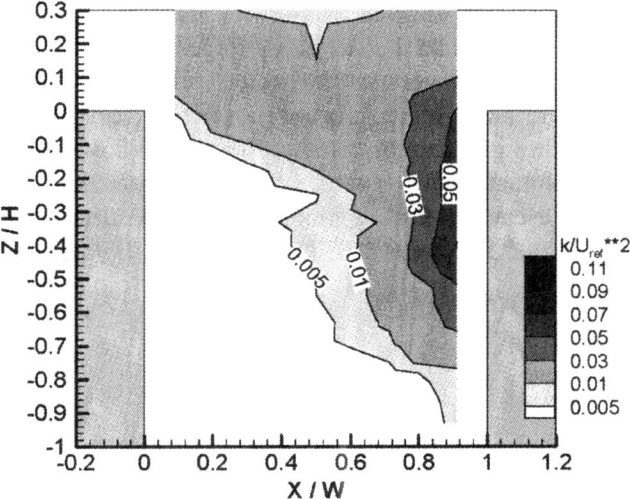

Figure 18. Distribution of turbulent kinetic energy in canyon for $Fr = 0.73$.

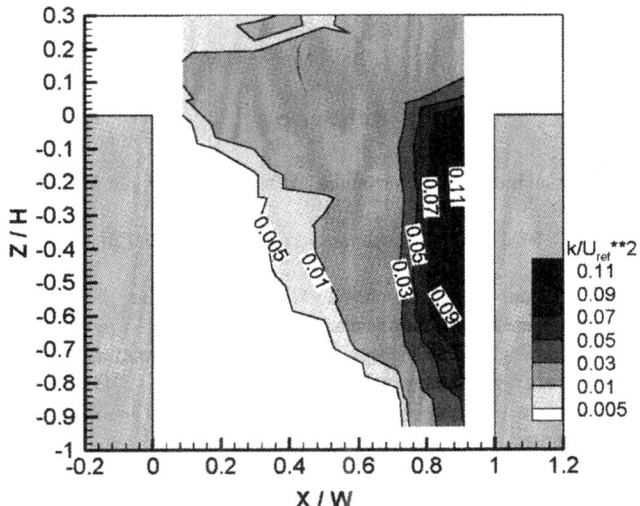

Figure 19. Distribution of turbulent kinetic energy in canyon for $Fr = 0.27$.

gradients near the wall. However, further measurements much closer to the heated wall would be required to verify the source of TKE production.

4. Concluding Remarks

The results from the present study indicate that the heating of the windward-facing wall does appear to have some influence on the generation of a very weak secondary flow close to the ground of the canyon at very low Froude numbers. However, so far there is little evidence that the buoyancy forces induce a widespread upward motion, except in a very thin layer near the heated wall, as also noted from field experiments in Nantes, France. Hence, it is not possible to clearly state that the effect of wall heating will be significant in terms of the canyon flow field and the motion and dispersion of pollutants. Further work is planned to examine the possible three-dimensionality of the canyon flow regime with wall heating, together with 3-D simulation, based on the CFD code CHENSI (Sini *et al.*, 1996), for predicting such flows.

Acknowledgements

The authors gratefully acknowledge funding from the EU within the TMR project TRAPOS (Contract ERBFMRXCT97-0105). The authors should also like to thank Dr. P. Hayden and Mr. T. Lawton from EnFlo and Mr. T. Renouf, Dr. P. Louka, Dr. P. G. Mestayer, and Mr. X. Mestayer from ECN for their contributions to this research.

References

Albrecht, F.: 1933, 'Untersuchungen der vertikalen Luftzirkulation in der Grossstadt', *Meteorologische Zeitung* **50**, 93–98.
DePaul, F. T. and Sheih, C. M.: 1986, 'Measurements of wind velocities in a street canyon', *Atmosph. Environ.* **20**, 555–559.
Georgii, H. W., Busch, E. and Weber, E.: 1967, 'Untersuchung über die zeitliche und räumliche Verteilung der Immissions-Konzentration des Kohlenmonoxids in Frankfurt am Main', Berichte des Instituts für Meteorologie und Geophysik der Universität Frankfurt am Main, 11.
Louka, P., Vachon, G., Sini, J.-F., Mestayer, P. G. and Rosant, J.-M.: 2002, 'Thermal effects on the airflow in a street canyon – Nantes'99 experimental results and model simulations', *Water, Air and Soil Pollut.*, this issue.
Ludwig, F. L. and Dabberdt, W. K.: 1972, 'Evaluation of the APRAC-1A Urban Diffusion Model for Carbon Monoxide', *Final Report*, Coordinating Research Council contract CAPA-3-68 (1–69), NTIS No PD 2210819.
Mestayer, P. G., Sini, J.-F. and Jobert, M.: 1995, 'Simulation of the Wall Temperature Influence on Flows and Dispersion within Street Canyons', in *Air Pollution '95, Vol. 1: Turbulence and Diffusion*, Porto Carras, Greece, September, pp. 109–106.
Nakamura, Y. and Oke, T. R.: 1988, 'Wind, temperature and stability condition in an east-west-oriented urban canyon', *Atmosph. Environ.* **22**, 2691–2700.
Ogawa, Y., Diosey, P. G., Uehara, K. and Ueda, H.: 1981, 'A wind tunnel for studying the effects of thermal stratification in the atmosphere', *Atmosph. Environ.* **15**, 807–821.
Sini, J.-F., Anquetin, S. and Mestayer, P.: 1996, 'Pollutant dispersion and thermal effects in urban street canyons', *Atmosph. Environ.* **30**, 2659–2677.
Uehara, K., Murakami, S., Oikawa, S. and Wakamatsu, S.: 1997a, 'Wind tunnel tests on stratified flow fields around urban street canyons by LDV', *J. Architect., Plann. Environ. Engineer. (Trans. of AIJ)* **492**, 39–46.

Uehara, K., Murakami, S., Oikawa, S. and Wakamatsu, S.: 1997b, 'Wind tunnel test of concentration fields around street canyons within the stratified urban canopy layer', *J. Architect., Plann.
Environ. Engineer. (Trans. of AIJ)* **499**, 9–16.

Uehara, K., Murakami, S., Oikawa, S. and Wakamatsu, S.: 2000, 'Wind tunnel experiments on how
thermal stratification affects flow in and above urban street canyons', *Atmosph. Environ.* **34**,
1553–1562.

Vachon, G., Rosant, J.-M., Mestayer, P. G. and Sini, J.-F.: 1999, 'Measurements of Dynamic and
Thermal Field in a Street Canyon. URBCAP Nantes'99', in *6th International Conference on Har-
monisation within Atmospheric Dispersion Modelling for Regulatory Purposes*, Rouen, France,
October 1999, 12 pp.

Vachon, G.: 'Nantes'99', Internet-publication on http://www.ec-nantes.fr/Fr/Recherche/
DEE/Dah/gaelle.htm.

A SENSITIVITY ANALYSIS OF OZONE FORMATION TO AMBIENT AIR COMPOSITION BY MEANS OF PHOTOCHEMICAL MODELS

STEFANO CORDINER[1], RENATO BACIOCCHI[2*] and MARINA ATTINÀ[2]

[1] *Dipartimento di Ingegneria Meccanica, Via di Tor Vergata 110;* [2] *Dipartimento di Scienze e Tecnologie Chimiche, Via della Ricerca Scientifica, Università degli Studi di Roma 'Tor Vergata', Roma, Italy*

(* *author for correspondence, e-mail: baciocchi@stc.uniroma2.it, fax: +39-0672594328*)

Abstract. The prediction of the atmospheric concentration of ozone and of other photochemical pollutants represents one of the major scientific challenges in the study of urban air quality. To this aim, photochemical models are widely applied as support tools for regulatory purposes, but their results are known to be affected by several uncertainties, in boundary conditions, geographical and meteorological data and the chemical composition of emission and ambient air as well. Also the mechanism selected to describe the urban chemistry may lead to large differences in the model results. In this work, the influence of ambient air composition on the results of photochemical model has been studied using a simple Lagrangian trajectory model coupled with different chemical sub-models (CB-IV and a simplified version of EMEP). To this aim a parametric sensitivity study of ozone with respect to its precursors' concentration was performed with reference to the same meteorological conditions and by simulating different scenarios characterized by specific emitted pollutants concentration. Rather than analyzing sensitivity by changing initial composition, an analytical local sensitivity concept has been applied; this allowed evaluation the influence of initial concentration on the predicted ozone concentration and to discriminate between VOC-sensitive and NOx-sensitive kinetic regimes. Finally, these results have been successfully compared with those obtained applying the concept of photochemical indicators (i.e. O_3/NO_y and H_2O_2/HNO_3 ratios).

Keywords: emissions, ozone, photochemical indicators, photochemical pollution, sensitivity analysis

1. Introduction

Photochemical oxidants, as ozone, PAN and other species, are formed in the lower atmosphere by the solar irradiation of mixtures of hydrocarbons and nitrogen oxides and the subsequent reactions with the short lived radicals which are formed. The general principles involved in the formation of these photochemical pollutants are fairly well known and have been described in a wide number of computer models (CB-IV/UAM (Gery *et al.*, 1989), EMEP (Van Loon, 1994), SAPRC/CALGRID (Carter, 1990). These models have been intensively used and one of the objectives of the simulations has been the assessment of their predictive performances (Jiang *et al.*, 1998; Polla Mattiot, 2001) and their role in helping to define air quality strategies (Volta *et al.*, 1999). However, the detailed description of photochemical

Water, Air, and Soil Pollution: Focus **2:** 573–585, 2002.

activity is complicated by the different photochemical behavior of the wide range of hydrocarbons (VOC), which are emitted both by anthropogenic and natural sources (Carter, 1989). Besides, application of photochemical models in supporting pollution control policies is still limited by other factors, i.e. model boundary conditions, meteorological and geographical data (Sillmann, 1999). Also the chemistry description used in the model may affect the predictions, depending on the criteria chosen for the parameterization that is still required to reduce computational times to reasonable values (Aumont et al., 1996, Andersson-Sköld et al., 1999). Nevertheless, application of photochemical models provided a better understanding of the urban troposphere behavior.

It has been demonstrated that some global parameters, such as the VOC/NOx concentration ratio, have a fundamental influence on photochemical pollution (Seinfeld and Pandis, 1998), determining whether the ozone formation process is sensitive to changes in VOC or NOx concentration. Thus, efforts in analyzing the effect of all these influence variables give the opportunity to develop a specific sensitivity, which could be fundamental while using model results as an important decision support tool in urban air quality policy. This sensitivity of photochemical model results to changes in VOC or NOx concentration has been often determined by repeating the model base case with reduced levels of NOx and VOC (Sillmann et al., 1997, Sillmann, 1999). Finally, in order to discriminate between VOC and NOx sensitive conditions, in recent literature it has been proposed to use the ratios between concentrations of some selected chemical species, the so-called Photochemical Indicators (such as H_2O_2/HNO_3 or O_3/NOy) (Sillmann et al., 1997).

In the present work, the concept of local analytical sensitivity (Dickinson and Gelinas, 1976), already widely applied in chemical kinetics (Varma et al., 1999), has been used for a systematic sensitivity study. The basic idea of the local sensitivity concept is to determine a relationship between the behavior of the real system and one or more parameters. The sensitivity value which can be determined in this way quantifies such relationship. In the present paper such an approach has been used to evaluate the influence of ambient air composition on the simulation results of photochemical smog episodes with emissions from internal combustion engines. The latter have been chosen due to their fundamental role in determining the urban air quality but also to their particular influence on the ozone formation processes, related to the presence of a large number of hydrocarbons with different ozone formation potentials. To study this influence a simple Lagrangian box-model was used, run under fixed meteorological conditions and coupled with two different chemical mechanisms. The local sensitivity approach has been also applied to discriminate between VOC-sensitive and NOx-sensitive kinetic regimes, and the results obtained have been compared with those obtained applying the concept of photochemical indicators.

2. Modelling

2.1. MODEL FORMULATION

Model calculations have been performed by means of a standard trajectory model based on the solution of the following equation system:

$$\frac{\partial}{\partial t} C_i = \frac{\partial}{\partial z} \left(K_{zz} \frac{\partial C_i}{\partial z} \right) + R_i + E_i \tag{1}$$

where C_i is the gas-phase concentration of the ith species among a total number of n, K_{zz} is the eddy diffusivity in the z direction, R_i is the rate of formation of the i-th species by chemical reactions and E_i is the rate of emission of the i-th species from sources. This set of equations is applied to describe the time evolution of species concentration of a air parcel moving with the local wind velocity and which develops from the ground level to the height of the mixing layer. It has to be coupled with a proper set of boundary conditions which defines the flux of species through the parcel top and bottom boundaries.

Two chemical sub-models have been used to describe the chemical transformations of ozone precursors emitted by stationary and mobile sources, namely the CB-IV mechanism and a simplified and highly parameterized version of the EMEP chemical model. These models address different requirements; being the CB-IV a detailed model of the environmental chemistry (32 species and 81 reactions), it is used as a predictive tool in the evaluation of urban acute pollution episode whereas the simplified version of the EMEP, still retaining the essential features of the urban chemistry, is characterized by a smaller number of species (15) and reactions (15) and can be usefully applied for those applications which require a large number of evaluations in a short time (what-if analysis, etc.).

The two different models have then been used to simulate a 48 h long pollution episode described by a unique meteorological scenario (typical summer conditions with low wind velocity and fixed boundary layer height) and different initial and boundary conditions. The vertical turbulent eddy diffusivity profile K_{zz} has been chosen as for moderately stable conditions (Businger and Arya, 1974). The emissions sources have been modeled as characteristics of an urban site with an high traffic concentration.

2.2. ANALYTICAL SENSITIVITY

As previously mentioned, from an analytical point of view the evolution of ambient pollutant concentration in the Lagrangian model can be described by a system of ordinary differential equations which can be expressed in the general form:

$$\frac{\partial y_i}{\partial t} = f_i(y_1, y_2, ..., y_N, t, C) \tag{2}$$

This admits solutions $y_i = (y_1, y_2, ..., y_N)$, that are function of time and of the parameter C which, for instance, can be the initial value of variable y_k [$C \equiv y_k(0)$].

The analysis of sensitivity of the solution y with respect to the parameter C requires the introduction of a set of new variables $Z_i (i = 1, ..., N)$:

$$Z_i = \frac{\partial y_i}{\partial C} \tag{3}$$

These new variables are the solution of a new set of equations

$$\frac{\partial Z_i}{\partial t} = \frac{\partial}{\partial C} \left(\frac{\partial y_i}{\partial t} \right) = \frac{\partial}{\partial C} \{ f_i [y (C, t), C, t] \} = \frac{\partial f_i}{\partial C} + \sum_{j=1,N} \frac{\partial f_i}{\partial y_j} \cdot \frac{\partial y_j}{\partial C} \tag{4}$$

which may be rewritten in a more compact form as $\frac{\partial Z}{\partial t} = \frac{\partial f}{\partial t} + [J]Z$ where
- $Z = (Z_1, Z_2, ..., Z_N)$
- $f = (f_1, f_2, ..., f_N)$ and
- $[J]$ is the Jacobian of the original differential equation system.

The two systems of equations may be solved together and the solution also supplies, besides the actual concentration of the species y_i, the sensitivity of the selected variables to the input parameters and/or other influence factors. In particular, in the present study, the sensitivity $\frac{d[O_3]}{dY_i}$ of ambient ozone concentration with respect to initial concentration of the emitted pollutants Y_i has been investigated to assess the importance of this input parameter on the simulated result.

3. Results and Discussion

As previously mentioned, emissions from mobile sources represent the most important origin of primary pollutants in urban environments and thus they also greatly affect the extent of photochemical pollution. Photochemical simulations of urban air quality evolution require, as input data, the emissions profile from mobile sources given by properly defined inventories, or eventually estimated through accurate models (Alexopoulos et al., 1993). Nevertheless, the internal combustion engines exhaust gases composition is a complex function of a large number of engine design and operating parameters and in the first part of the study a detailed experimental analysis has been performed by means of the experimental analysis/sampling system whose set-up is shown in Figure 1. The engine used for experimental analysis (a Ford 2.0 L 16 V Zetec spark ignition engine used on the Mondeo series) has been chosen as representative of latest generation spark ignition engines and is equipped with different technical solution required to control the gaseous emissions. Besides a last generation catalytic converter, the engine adopts a number of technical solutions aimed to shorten the warm-up period time length and thus to effectively reduce pollutant emissions also during low temperature operating conditions. This is particularly important in urban sites where drive

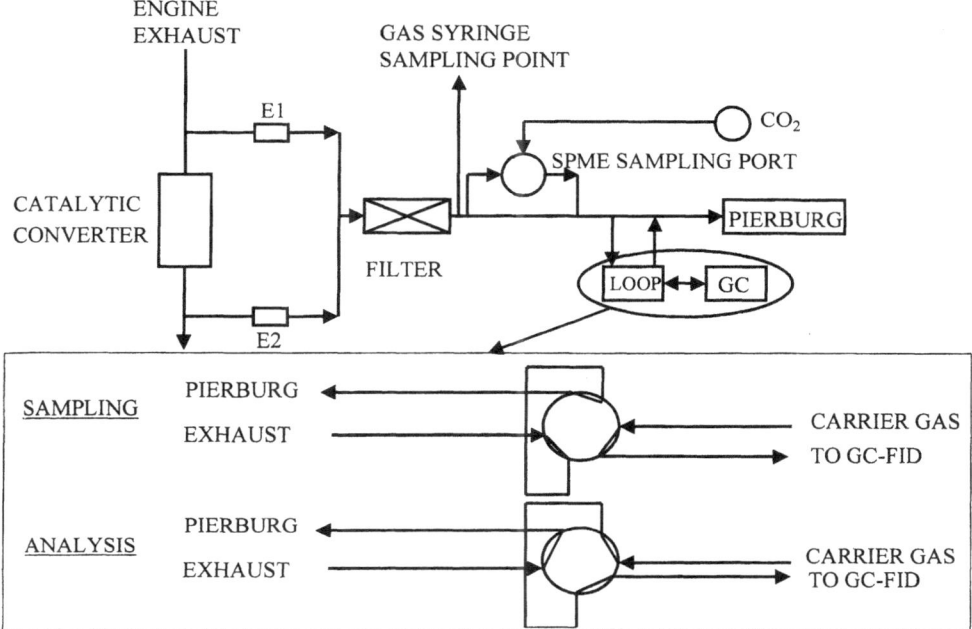

Figure 1. Experimental set-up of the sampling/analysis system of a Ford Mondeo 2000 engine's exhausts.

characteristics during traffic jams do not allow the engine to reach its optimum operating temperature.

The main objective of the experimental analysis has been then the characterization of engine emissions as a function of the operating conditions both in terms of quantity and detailed chemical composition. Furthermore the effect of the catalytic converter on the exhaust composition has been evaluated by comparing the upstream and down-stream chemical composition by means of the experimental set-up shown in Figure 1. Total emitted NO_x, CO and HC have been determined with a Pierburg AMA 100 analysis system by means of Chemiluminescence, Non-dispersive infrared and Flame ionization techniques respectively, whereas the organic fraction of the emissions has been speciated with either an on-line and off-line analysis. Off-line analysis have been performed with the Solid Phase Microextraction (SPME) sampling technique, using a Poliacrylate fiber, coupled to GC-MS performed on a VG Quattro instrument equipped with a Supelco Petrocol capillary column.

Typical chromatograms of the emissions' organic fraction, obtained with the SPME sampling technique coupled to GC-MS analysis, are shown in Figure 2 for a fixed operating condition of the engine (2000 rpm, medium load). Namely, the chromatogram shown in Figure 2a is relative to the raw emissions, whereas that shown in Figure 2b is relative to tailpipe emissions. Total HC concentration of

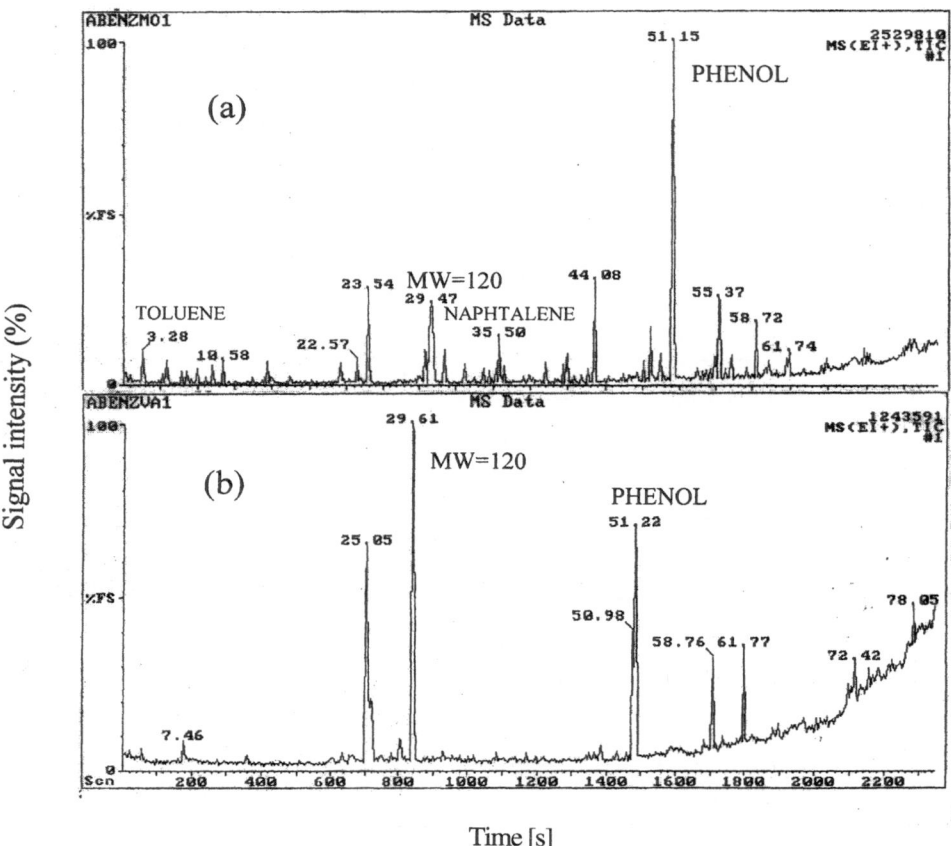

Figure 2. Fingerprint of the (a) raw and (b) tailpipe engine's emissions obtained applying Solid Phase Microextraction (SPME) coupled with capillary Gas Chromatography.

raw emissions was 400 ppm and was reduced to 40 ppm by the catalytic converter; this reduction was clearly not the same for all the hydrocarbons, as clearly shown comparing Figure 2a and Figure 2b. For instance, naphtalene and toluene are completely removed, since they are found in raw (Figure 2a) but not in tailpipe emissions (Figure 2b); on the other hand, substituted aromatics with Molecular weight, MW = 120, characterized by an higher reactivity to ozone formation, are removed with a much lower efficiency, and still represent an important fraction of total HC in tailpipe emissions.

The analysis of engines emissions allowed to identify the presence at high concentrations of different aromatic compounds, such as toluene, xylenes and other substituted aromatics; these compounds also have very high MIR (Maximum Incremental Reactivity) values, i.e. 2.7 for toluene and 8.2 for xylene (Carter 1989), thus having a great impact on ozone formation. For this reason, the sensitivity analysis study was focalized on some of these organic species, comparing their behavior with respect to ozone formation to that exhibited by other organics typically found in internal combustion engines emissions, such as paraffines and olefines.

The influence of ambient air composition on ozone concentration has been examined by calculating the sensitivity of the model solution with respect to the initial concentration of the chemical species included in the selected chemical mechanism (CB-IV or simplified EMEP), in accordance with the procedure discussed in Section 2. It has then been possible to calculate sensitivity coefficients for ozone with respect to selected species and some of the results obtained are reported in the following. In Figure 3 the sensitivity to different inorganic compounds is analyzed for both the implemented chemical mechanisms, with respect to the same scenario described in section 2.1. It can be seen that the results obtained with the two mechanisms are fairly in agreement, even if the CB-IV results look generally more sensitive to variations in initial composition. Besides, it is worth pointing out that sensitivity of ozone with respect to both nitrogen oxide and nitrogen dioxide is negative through all the simulation time (24 h). The results of the same simulation run, reported in terms of sensitivity with respect to selected organic species, reported in Figure 4, confirm the above observations, since CB-IV model shows again higher sensitivity values with respect to the highly simplified version of the EMEP. All sensitivities are in this case positive, confirming that through all the simulation the kinetic regime was clearly VOC-controlled. Finally, it can be noticed that the highest sensitivity values have been obtained for substituted aromatics, such as xylene; gradually lower sensitivities have been observed for aromatic compounds, olefins and paraffins, in agreement with the incremental reactivity scale developed by Carter (1989). Starting from these results, the local analytical sensitivity approach has also been applied to discriminate between VOC-sensitive and NO_x-sensitive kinetic regimes, and compared with the information obtained by using the concept of photochemical indicators.

To this aim, a set of 72 h simulations was performed, where the initial concentrations of all primary pollutants was set to zero and a sinusoidal emission profile for VOC and NOx, simulating rush hour conditions, was provided; VOC emissions were the same in all model runs, whereas the concentration of both NO and NO_2 in emissions (see Caption of Figure 5) was different in each run, in order to achieve different controlling regimes. The results of the base case, reported in Figure 5a, show that a VOC-sensitive regime was maintained through the whole simulation; the photochemical indicators values were observed to be very low, the H_2O_2/HNO_3 ratio was zero and the O_3/NO_y ratio was always below 0.2. The results of the simulations performed with the same input data, but with NO_x emission concentrations 300 and 3 times lower than the base case, are reported in Figures 5b and 5c, respectively. By looking at these figures, it is possible to find a satisfactory agreement between the sensitivity analysis and the photochemical indicators. Namely, a NO_x-sensitive regime was maintained through all the simulations whose results are reported in Figure 5b, while the values of both photochemical indicators (H_2O_2/HNO_3 ratio and O_3/NO_y ratio) were always high. Besides, the

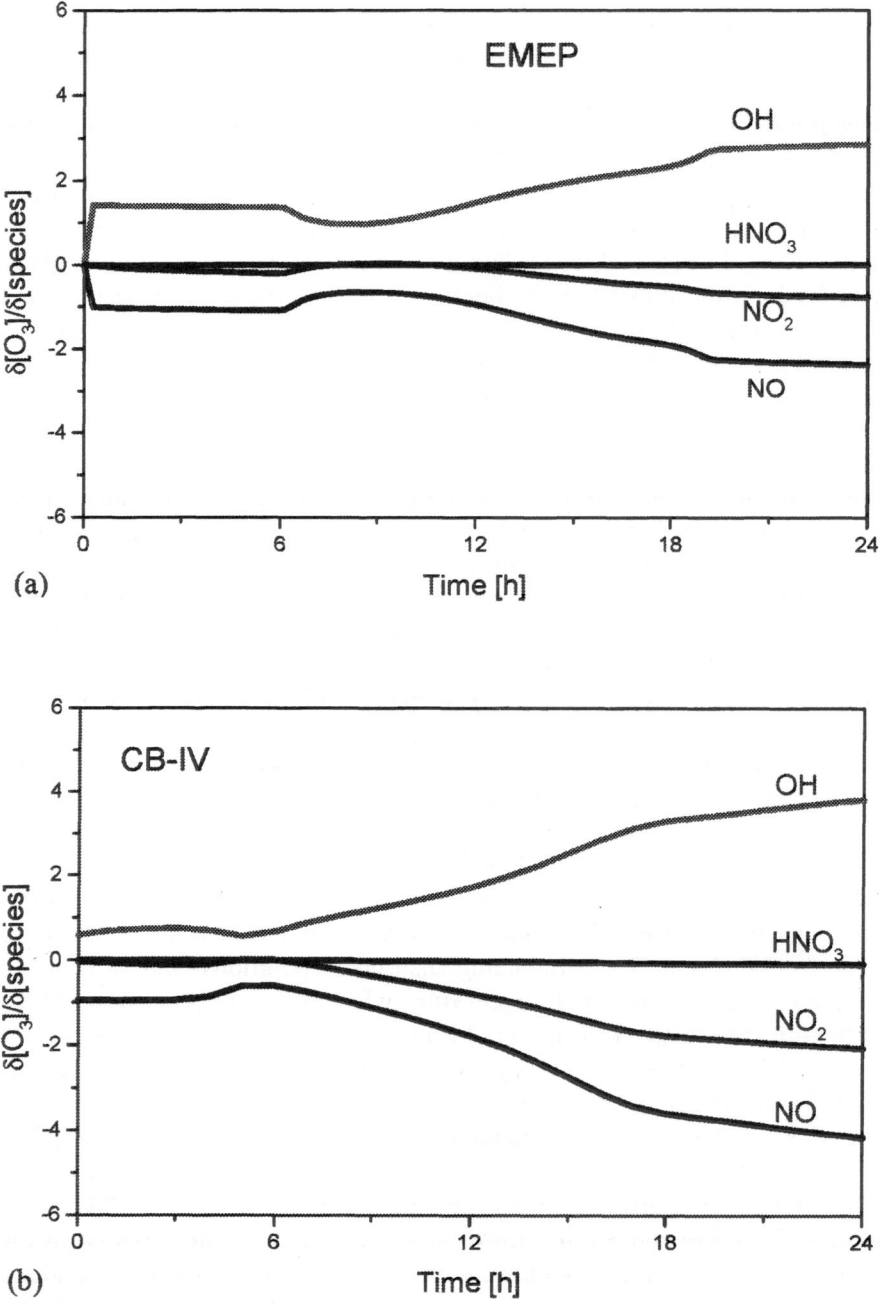

Figure 3. Sensitivity of ozone formation to inorganic species during a smog episode character-ized by VOC-limited chemistry obtained applying (a) simplified EMEP and (b) CB-IV chemical mechanisms.

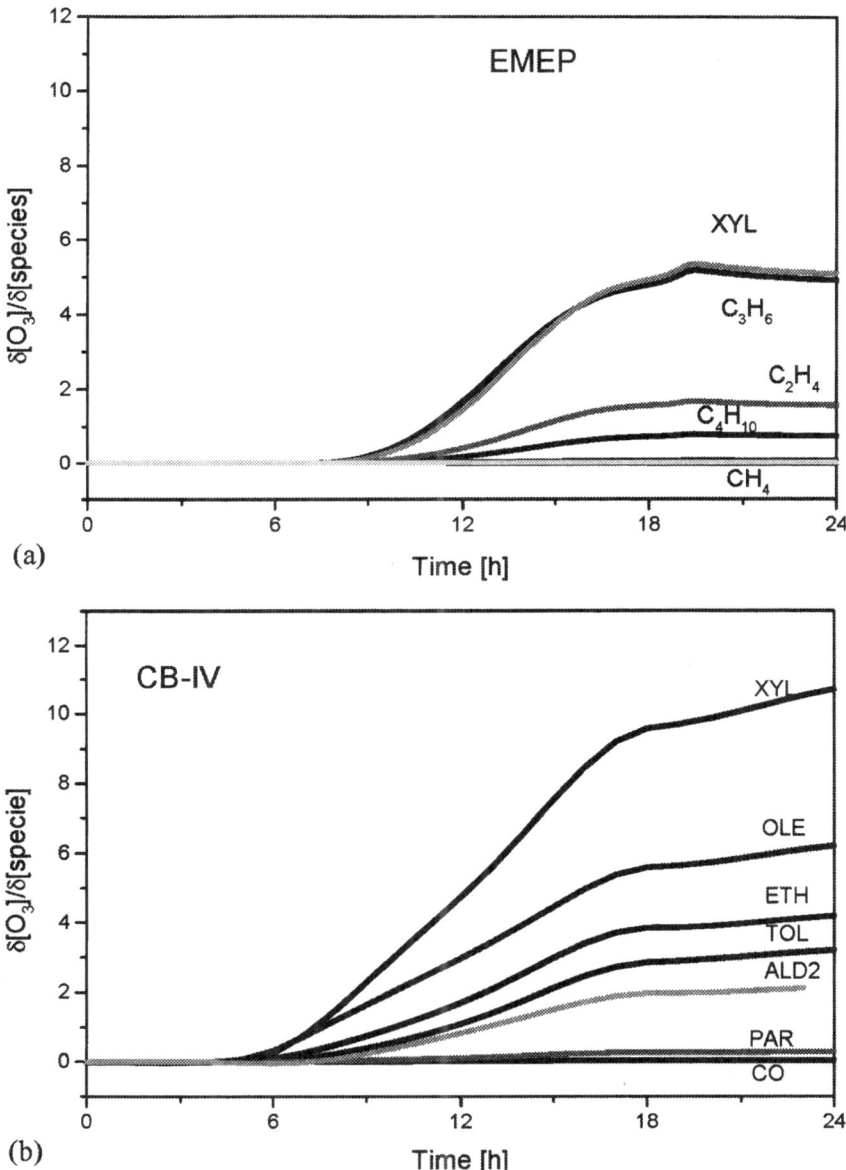

Figure 4. Sensitivity of ozone formation to different VOCs during a smog episode character-ized by VOC-limited chemistry obtained applying (a) simplified EMEP and (b) CB-IV chemical mechanisms.

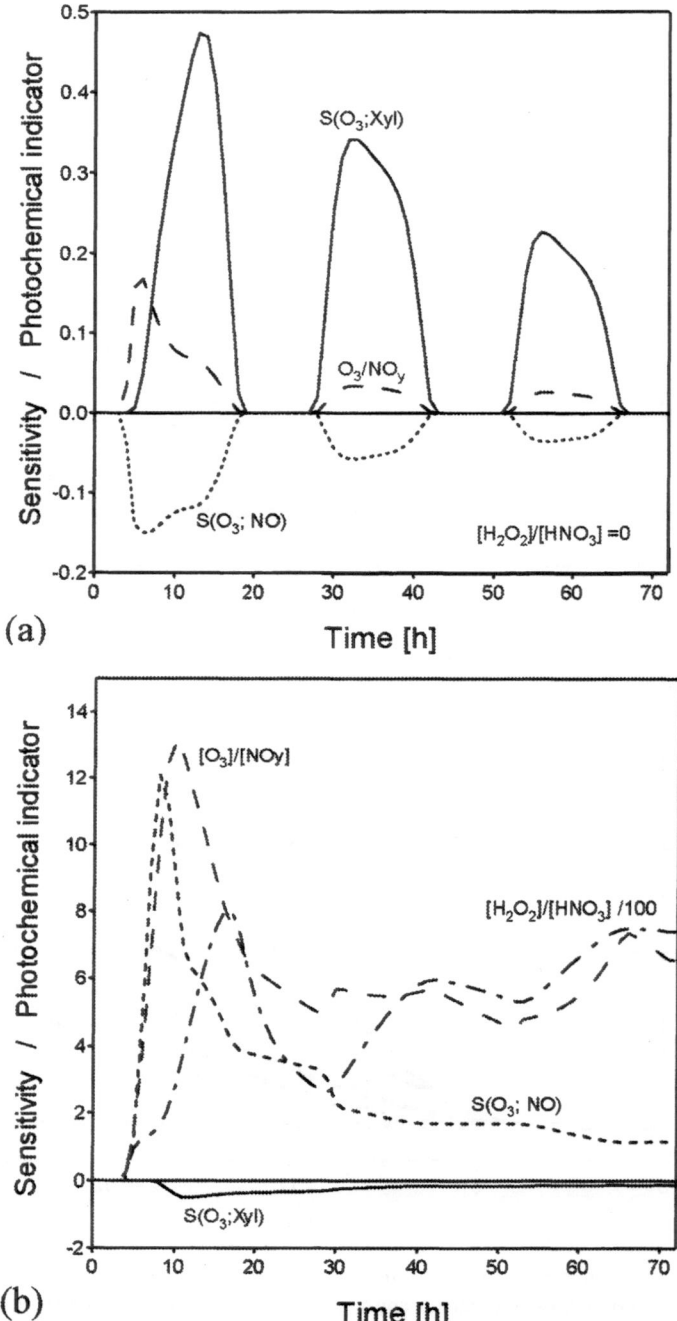

Figure 5. Sensitivity of ozone to Xylene and NO and values of two photochemical indicators (H_2O_2/HNO_3) and (O_3/NO_y) during a three day smog episode with initial concentration of pollutants set to zero and constant VOC emissions simulated using the CB-IV chemical mechanisms. (a) base case: NO_x concentration increase due to emissions ranging during the simulation with time-dependent sinusoidal law between 0 and 1.0 ppb per minute; (b) NO_x emissions 300 times lower than the base case; (c) NO_x emissions 3 times lower than the base case.

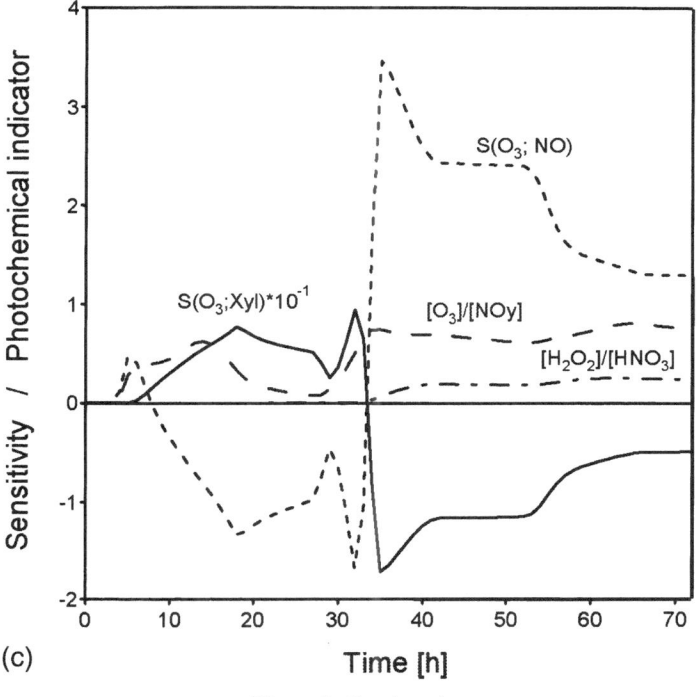

(c)

Figure 5. Continued.

switch between the initially VOC-sensitive regime and the NO_x-sensitive regime, which is clearly observed in Figure 5c, was accompanied by a sudden change of the photochemical indicators values, especially as far as the H_2O_2/HNO_3 ratio is concerned.

4. Concluding Remarks

The concept of analytical sensitivity has been used as a tool for assessing the influence of uncertainties in ambient air composition on photochemical model results. Namely, local analytical sensitivity studies have been performed on the results of a Lagrangian box model coupled with different chemical mechanisms evaluating the relative effect of the selected chemical species. An application of the proposed method was used to build criteria to discriminate between VOC-sensitive and NO_x-sensitive kinetic regimes. Such criteria were compared to analogous ones, based on the application of the so-called photochemical indicators. As far as the H_2O_2/HNO_3 ratio was considered, a very good agreement between the two criteria was observed, whereas a less clear correlation was found for the O_3/NO_y ratio. This difference may be attributed to the fact that the former photochemical indicator is directly related to the possible termination reactions of the photochemical cycle (Sillmann,

1999); besides, it is worth noting that the box model used in this work did not account for deposition processes, which may strongly influence the concentration of both nitric acid and hydrogen peroxide in the troposphere. Finally, it may be concluded that the analytical sensitivity approach may represent a more efficient and mathematically based tool for those sensitivity studies that are currently performed with time-consuming and cumbersome procedures, based on the comparison of model results obtained with different emissions profiles or initial conditions.

Acknowledgements

This work is dedicated to the memory of Prof. Marina Attinà, whose love for research and academy we will always bring with us.

References

Alexopoulos, A., Assimacopoulos, D. and Mitsoulis, E.: 1993, 'Model for traffic emissions estimation', *Atmos. Environ.* **4**, 435–446.

Andersson-Sköld, Y. and Simpson D.: 1999, 'Comparison of the chemical schemes of the EMEP MSC-W and IVL photochemical trajectory models', *Atmos. Environ.* **33**, 1111–1129.

Businger, J. A. and Ayra, S. P. S.: 1974, 'Height of the mixed layer in the stably stratified planetary boundary layer', *Adv. Geophys.* **18A**, 73–92.

Aumont, B., Jaecker-Voirol, A., Martin, B. and Toupance G.: 1996, 'Tests of some reduction hypotheses made in photochemical mechanisms', *Atmos. Environ.* **12**, 2061–2077.

Carter, W. P. L.: 1989, 'Development of ozone reactivity scales for volatile organic compounds', *J. Air Waste Manage. Assoc.* **44**, 881–889.

Carter, W. P. L.: 1990, 'A detailed mechanism for the gas-phase atmospheric reactions of organic compunds', *Atmos. Environ.* **24A**, 481–518.

Dickinson, R. P. and Gelinas, R. J.: 1976, 'Sensitivity analysis of ordinary differential equation systems – A direct method', *J. Computat. Phys.* **21**, 123–143.

Gery, M. W., Whitten, G. Z., Killus, J. P. and Dodge, M. C.: 1989, 'A photochemical kinetics mechanism for urban and regional scale computer modelling', *J. Geophys. Res.* **94**, 12925–12956.

Jiang, W, Hedley, M. and Singleton, D. L.: 1998, 'Comparison of the MC2/CALGRID and SAIMM/UAM-V photochemical modelling systems in the lower Fraser Valley, British Columbia', *Atmos. Environ.* **32**, 2969–2980.

van Loon, M.: 1994, 'Numerical Smog Prediction I: The Physical and Chemical Model', CWI Reports and Notes – Numerical Mathematics, Report 11., CWI, The Netherlands.

Polla Mattiot, F., Buttini, P., Gariazzo, C., Carizi, I., Levy A. and Rebesco, E.: 2001, 'Ozone Episode in Milan Metropolitan Area- A UAM-V/CALGRID Modelling Approach', *The Third International Conference on Urban Air Quality*, Loutraki, Greece, 19–23 March 2001.

Seinfeld, J. H. and Pandis, S. N.:1998, *Atmospheric Chemistry and Physics. From Air Pollution to Climate Change*, John Wiley and Sons, New York, pp. 300.

Sillmann, S, He, D., Cardellino, C. and Imhoff, R. E.: 1997, 'The use of photochemical indicators to evaluate ozone-NO_x-hydrocarbons sensitivity: case studies from Atlanta, New York and Los Angeles', *J. Air Waste Manage. Assoc.* **47**, 1030–1040.

Sillmann, S.: 1999, 'The relation between ozone, NO_x and hydrocarbons in urban and polluted rural environments', *Atmos. Environ.* **33**, 1821–1845.

Varma, A., Morbidelli, M. and Wu, H.: 1999, *Parametric Sensitivity in Chemical Systems*, Cambridge University Press.

Volta, M. and Finzi, G.: 1999, 'Evaluation of EU road traffic abatement strategies in Northern Italy by a photochemical modelling system', *Eurotrac Newsl.* **21**.

INTERCOMPARISON OF NUMERICAL URBAN DISPERSION MODELS – PART I: STREET CANYON AND SINGLE BUILDING CONFIGURATIONS

P. SAHM[1], P. LOUKA[1,2]*, M. KETZEL[3], E. GUILLOTEAU[4] and J.-F. SINI[2]

[1] *Laboratory of Heat Transfer and Environmental Engineering, Aristotle University Thessaloniki, Thessaloniki, Greece;* [2] *Fluid Mechanics Laboratory, Ecole Centrale de Nantes, Nantes, France;* [3] *Department of Atmospheric Environment, National Environmental Research Institute, Roskilde, Denmark;* [4] *Institute of Hydromechanics, University of Karlsruhe, Karlsruhe, Germany*
(* author for correspondence, e-mail: petroula@aix.meng.auth.gr)

Abstract. Microscale Computational Fluid Dynamics (CFD) models have become an efficient and common simulation tool for assessment and prediction of air quality in urban areas. The proper validation of such a model is a crucial prerequisite for its practical application. Within the framework of the European research network TRAPOS a working group on computational fluid dynamics modelling was established and model intercomparison exercises were launched. Different Computational Fluid Dynamics Codes were applied for simulating the wind flow and pollutant concentration patterns in several test cases. The aim of the present model intercomparison is (1) to assess and allocate the source of differences that appear when different CFD codes using the same turbulence model are applied to well defined test cases and (2) to improve the knowledge base for model development and application. Throughout the series of model applications covering manifold urban configurations, the overall agreement between the various models and experimental data is fair. In spite of quantitative differences between the various numerical results, the models are capable of reproducing the flow patterns and dispersion characteristics observed in urban areas but they show significant differences for the turbulent kinetic energy field that controls the dispersion of pollutants.

Keywords: atmospheric dispersion modelling, computational fluid dynamics, dispersion near buildings, microscale, model intercomparison, model validation, street canyon, traffic pollution

1. Introduction

Traffic pollution in urban areas has become the major concern for public health, since emissions from non-traffic sources have been constantly reduced. The aggregation of human activities, especially in areas of insufficient ventilation, has often led to pollution levels much higher than the limits set by the World Health Organisation. The building aggregates act as artificial obstacles to the wind flow and cause stagnant conditions in the city, even for relatively high ambient wind conditions. A typical configuration is the so-called street canyon, formed along a street in densely built urban areas.

Over the last years Computational Fluid Dynamics (CFD) using the Reynolds averaged Navier-Stokes (RANS) equations has become a standard simulation tool

Water, Air, and Soil Pollution: Focus **2:** 587–601, 2002.
© 2002 *Kluwer Academic Publishers.*

for the analysis, investigation and prediction of atmospheric flow and pollutant dispersion. The fundamental problem of CFD simulations lies in the physical difficulties of modelling the effects of turbulence. Other major issues are the accuracy in the spatial discretisation of complex urban geometries, the numerical procedure applied, the boundary conditions and physical properties selected and the validation of the models.

The research topic of the European Commission funded TRAPOS network concerns the improvement and optimisation of the methods that are used for modelling of traffic pollution in streets. Based on evaluation of presently available models and study of results from field and laboratory experiments, the major deficits in the description of important processes have been identified and addressed. The study included measure-ments in streets, wind tunnel experiments, and advanced numerical modelling and data evaluation. The main objectives of the project are to improve (a) the performance of models in the case of critical meteorological conditions; (b) their ability to deal with different street architectures and (c) the description of the chemical and physical conversion processes and treatment of new, not previously addressed pollutants. The numerical models used within the TRAPOS network comprise five advanced CFD models, i.e. CFX-TASCflow (Raw et al., 1989), CHENSI (Sini et al., 1996), CHENSI-2 (Guilloteau, 1999), MIMO (Ehrhard et al., 2000) and MISKAM (Eichhorn, 1989), for the numerical simulation of the three-dimensional flow field and the dispersion of pollutants in the microscale. Through close co-operation between the network teams an extensive CFD model evaluation study was organised. Several test cases have been defined, ranging from a single cavity case and a simple 3D case, to real case exercises. For all cases very comprehensive wind tunnel measurements were available. Additionally an extensive field data set exists for the real case. This paper, which is the first in a sequence of two, is focussed on the intercomparison of the models in the cases of a single cavity and a surface mounted cube. The intercomparison on a real case (i.e. a street canyon in Hanover, Germany) is described in part II paper (Ketzel et al., 2001).

2. Model Description

CFX-TASCflow may be used for simulating fluid flow, heat and mass transfer and fast chemistry in complex geometries. The code uses arbitrary curvilinear, body-fitted, multi-block, structured, non-staggered grids, a strong conservative form of the governing differential equations, a first-order accurate backward fully implicit scheme in time, a second-order bounded scheme for the spatial discretisation of advection terms, a coupled solution procedure for momentum and continuity equations, an algebraic multi-grid method for the solution of the sets of the algebraic difference equations and the standard k-ε turbulence model with wall functions. A detailed description of the computational method can be found in Raw *et al.* (1989) and Raw (1994).

CHENSI is a family of fully three-dimensional CFD codes developed by the group Dynamics of Inhabited Atmosphere of the Fluid Mechanics Laboratory at Ecole Centrale de Nantes. These codes are used for simulations of flow, heat transfer and passive scalar dispersion on staggered non uniform Cartesian grids. Their main applications are in-streets wind flows and traffic pollutant dispersion at local scales. They solve unsteady incompressible RANS equations with Boussinesq approximation and different k-ε turbulence closure models. The boundary condition used on solid surfaces is the wall function. Diffusion terms use the central difference scheme. The time scheme is explicit and first order accurate. CHENSI (Sini *et al.*, 1996) uses finite difference method and an upwind-weighted scheme for advection while CHENSI-2 (Guilloteau, 1999) uses finite volume method and a hybrid advection scheme.

MIMO is a prognostic microscale model which allows describing the air motion near complex building structures (Ehrhard *et al.*, 2000). Within MIMO, the Reynolds averaged conservation equations for mass, momentum and energy are solved together with additional transport equations for scalar quantities such as potential temperature, turbulent kinetic energy and specific humidity. A staggered grid arrangement is used and coordinate transformation is applied to allow non-equidistant mesh size in all three dimensions in order to achieve a high resolution near the ground and obstacles. Conservation properties are fully preserved within the discrete model equations. The discrete pressure equations are solved with a fast elliptic solver in conjunction with a generalized conjugate gradient method. The Reynolds stresses and turbulent fluxes of scalar quantities can be calculated by several linear and nonlinear turbulence models.

The model MISKAM consists of a 3-dimensional non-hydrostatic flow model and an Eulerian dispersion model (Eichhorn, 1989) and uses the k-ε closure. The physical basis is the complete 3-dimensional equations of motion of the flow field and the advection-diffusion equation to determine the concentrations of substances with neutral density. The calculated result is the stationary flow and pressure field, diffusion coefficients and the concentration field in an area of typically 500×500 m (100×100 cells or more, non equidistant grid).

A detailed description of the models is available on the websites of the TRAPOS network[*], and on the Model Documentation System (MDS) built by the European Topic Centre on Air Quality (ETC-AQ).

3. Single Cavity Case

The single cavity case was defined as the simplest two-dimensional case to investigate the performance of the codes in reproducing the flow field between buildings.

---

[*] http://www.dmu.dk/AtmosphericEnvironment/trapos/cfd-wg.htm

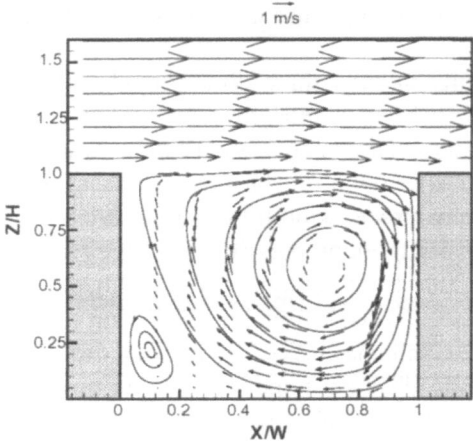

Figure 1. The flow field within and above the cavity as it was reproduced by CHENSI.

The experimental database for this case was established in the wind tunnel of the University of Surrey (Kovar-Panskus *et al.*, 2001). This database was developed for different cavity dimensions and the experiments aimed at assessing the effect of the cavity aspect ratio, i.e. width of cavity, W, over its depth, H, W/H, on the transformation of the flow from the one regime to the other and consequently on the dispersion of pollutants within a 'real' street. The first experimental case chosen to be studied by the numerical codes was a single cavity with aspect ratio W/H equal to 2 ($W = 0.212$ m and $H = 0.106$ m) (Figure 1). Vertical profiles of the mean wind field (u and w components), and the turbulent kinetic energy, k, were measured 27 mm upstream of the cavity and were specified as the input data for the models. In particular, the profiles fitted to the experimental values used as the inflow conditions for all codes were:

$$U(\text{m/s}) = 1.0776\text{In}(z - 0.106) + 8.1327, \ W\,(\text{m/s}) = 0$$

$$k(\text{m}^2/\text{s}^2) = 0.2768(z - 0.106)^4 + 10.028(z - 0.106)^3 - 14.431\,(z - 0.106)^2$$

$$+ \, 4{,}2727(z - 0.106) + 0.8716,$$

while the dissipation of the turbulent kinetic energy was considered having its standard formulation for the boundary layer:

$$\varepsilon(\text{m}^2/\text{s}^3) = u_*{}^3/(\kappa\,(z - 0.106)),$$

where u_*, the friction velocity, calculated from the measurements was equal to 0.4 m.s^{-1} and κ, the von Karman's constant equal to 0.4. Similar measurements of u, w and k were also performed within and above the cavity at positions $x/W \approx 0.1$, 0.3, 0.5, 0.7, and 0.9. All five models used the same domain and grid sizes.

All models were able to reproduce well the general flow pattern observed within the cavity. The flow within the cavity is dominated by a main re-circulation, while

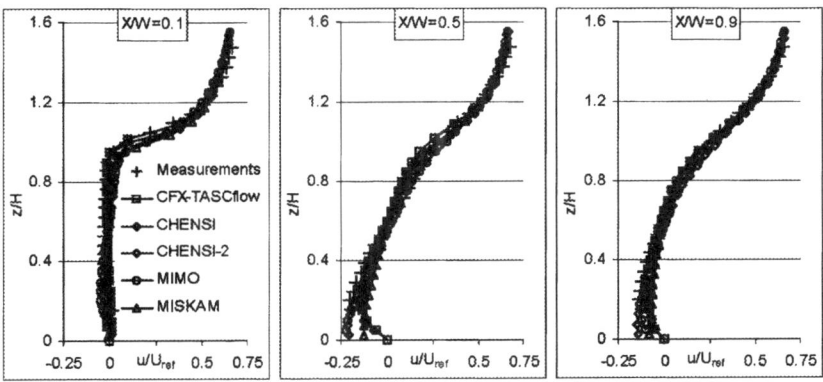

Figure 2. Comparison of the profile of the u-component normalised with the free-stream velocity ($U_{ref} = 8$ m s^{-1}) at $X/W = 0.1, 0.5$, and 0.9.

a secondary vortex rotating in the opposite direction is present at the leeward side of the cavity close to the ground (Figure 1). The re-circulation predicted by the models is characterised by mean velocities ranging between -2 and 2 m.s^{-1} and the secondary vortex by velocities varying between -0.15 and 0.15 m s^{-1} The comparison between the numerical results and the measurements at the different positions within the cavity shows a fairly good agreement especially for the mean wind field. Figure 2 shows the vertical profile of the u-component at $X/W = 0.1$, 0.5 and 0.9. The models have captured overall both the shape of the profile and roughly the magnitude of the wind speed.

Nevertheless, a focussed examination of the profiles close to the solid boundaries (building walls and ground) shows that the models divert in their predictions at these specific locations. The detailed examination of the source code showed that the origin of this difference is mainly due to the different implementation of the wall functions while differences in the numerical schemes may also be important. This conclusion was also reached after simulating a simple boundary layer in equilibrium over a rough flat plate (not shown here). This case was designed to investigate the effect of the different implementation of the wall function by the different codes. For this reason, fine grid was used, the advection terms were considered as negligible (consequently differences due to the numerical scheme used by the codes were negligible) and hence the discrete implementation of the wall boundary conditions was responsible for the observed differences in the airflow resulted.

It is generally observed, in Figure 2, that CHENSI and CFX-TASCflow predict very similar velocity values close to the walls, while CHENSI-2 and MIMO following the same wall-function implementation represent the second group of models that calculate very similar velocities. Due to the implementation of the advection scheme and boundary conditions on solid surfaces, MISKAM is mainly dedic-

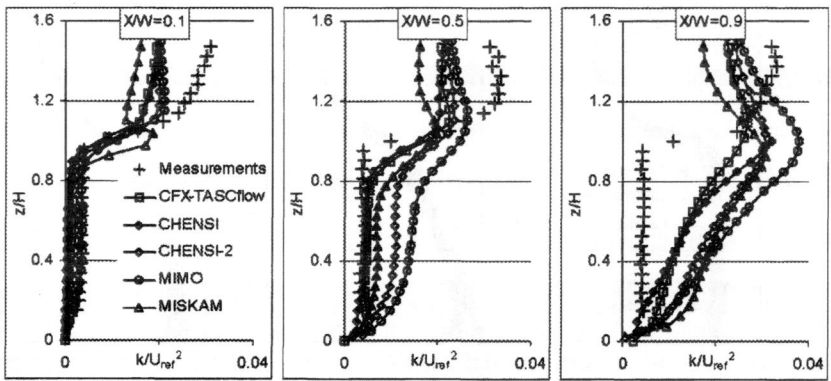

Figure 3. Comparison of the profile of the normalised turbulent kinetic energy observed in the wind tunnel with that predicted by the numerical models at X/W = 0.1, 0.5, and 0.9.

ated in simulating real-site flows therefore is probably less accurate in estimating small-scale flow patterns.

It has been shown that CHENSI generally underestimates the velocity close to solid boundaries in locations of stronger flow within a cavity (Kovar-Panskus *et al.*, 2001). This is also obvious here; CHENSI accompanied also by CFX-TASCflow predicts similar velocity values with the measured ones at X/W = 0.1, while at X/W = 0.5 and 0.9 the discrepancies from the measured values are larger (Figure 2). On the contrary, CHENSI-2 and MIMO show better agreement with the data at all examined positions. Therefore, it is observed that at very low local velocity conditions, the effect of the different wall-function implementation by the codes on the calculated velocities is small. As CHENSI-2 and MIMO showed better agreement with the data, it is suggested that the implementation of the wall-function applied by these two models is more appropriate for similar studies in cavities. It is stressed here that the models results are considered as grid independent as several tests have been carried out. Moreover, the size of the cells adjacent to the wall is restricted due to the value of the roughness length of the rough surface. Well above the cavity model results and data are exactly matched as the effect of the cavity on the airflow is small and the flow is in equilibrium with the surface below. The same conclusion was reached when simulating a simple boundary layer in equilibrium with a rough surface (not shown here).

The investigation of the turbulent kinetic energy showed that k is influenced by the implementation of the wall function in a greater extent than the mean velocity as this quantity involves more model parameters and numerics that may vary from model to model (cf. Figure 3). However, similar arguments are also applied here; the model agreement is observed to be better at locations of weaker than stronger airflow, although the measurements do not actually show that k changes greatly along the cavity. The overestimation of the turbulent kinetic energy by the models within the cavity may be explained if considering the turbulent kinetic energy budget:

$$\bar{U}\underbrace{\frac{\partial k}{\partial x}}_{Advection} = \underbrace{-\overline{U'W'}\frac{\partial \bar{U}}{\partial z}}_{Shear\ production} \underbrace{-(\overline{U'^2} - \overline{W'^2})\frac{\partial \bar{U}}{\partial x}}_{Streamwise\ production} \underbrace{-\frac{\partial(\overline{k'W'})}{\partial z}}_{Turbulent\ transport} \underbrace{-\varepsilon}_{Dissipation} ,$$

where buoyancy has been neglected and it has been assumed that the flow is steady and two-dimensional and that pressure transport terms are typically much lower than the turbulent transport term (Louka *et al.*, 2000). In order to understand the overprediction of k by the models we will focus our discussion only on the two production terms of the equation. We estimate the change in these production terms within the shear layer from the upwind side to the middle of the cavity. As it was measured and predicted by the models $\bar{U}$ increases along the cavity and therefore. $\frac{\partial \bar{U}}{\partial x} > 0$.Full-scale measurements in a street canyon (Louka *et al.*, 2000) have shown that $-(\overline{U'^2} - \overline{W'^2})$ is always negative, therefore, the streamwise production term is negative within the cavity leading to suppressing turbulence. In addition the produced turbulence is transported from the roof-level downwards into the cavity and advected by the main vortex. The streamwise production term is not taken into account in standard k-ε models, which assume that turbulence is isotropic and hence overpredict the turbulent kinetic energy production in cavities.

The small discrepancies in the calculated velocity and turbulent kinetic energy close to walls and ground among the models influence their prediction of pollutant dispersion within the cavity. Considering also the differences in the dispersion routines and the implementation of the source conditions the difference in the predicted pollution concentration among the models becomes large. Figure 4 illustrates the normalised concentration $c^* = (CU_{ref} H_{ref})/(Q/L)$ along the path ABCD within the cavity. C is the concentration, U_{ref} the velocity at H_{ref}, Q is the source strength and L its length. The source was placed between the positions $x/W = 0.425$ and 0.6 at the bottom of the cavity ($z/H = 0.05$). CHENSI and CFX-TASCflow predict the maximum concentration at the same location on part B, while for CHENSI-2 and MIMO the maximum concentration appears just on the leeward wall (end of part A of the path). As it was mentioned previously MISKAM is less accurate in estimating small-scale flow patterns and dispersion characteristics.

4. Flow Around a Wall-Mounted Cube

The wall-mounted cube case was defined as the geometrically simplest three-dimensional case to investigate the performance of the codes in reproducing the flow field around an idealised building. Although simple in the geometrical setup the corresponding flow is very complex with sharp pressure gradients, streamline curvature

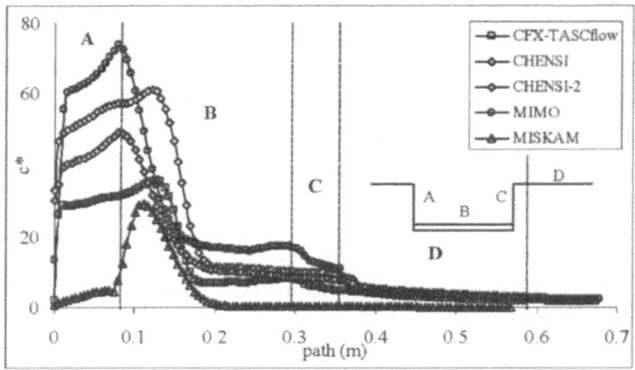

Figure 4. Comparison of the normalised concentration c^* along the path ABCD (A and C correspond to 75% of the cavity depth, while B and D are equal to its width).

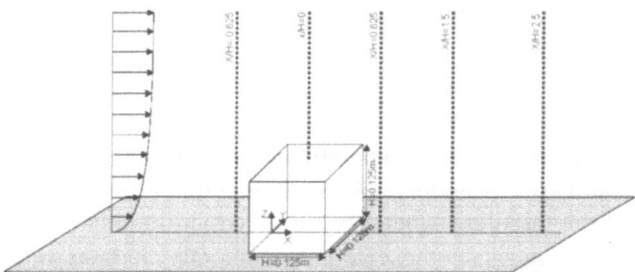

Figure 5. Experimental set-up of the wall-mounted cube; locations of the measuring positions in the centre plane.

and multiple, unsteady separation regions. Hence, this type of flow represents an excellent challenge for numerical models. An analogous experiment was carried out in the BLASIUS wind tunnel at the Meteorological Institute of Hamburg University at a scale of 1:200 and is included in the high-quality-assured CEDVAL database (Leitl *et al.*, 1999)*. A wooden cube (125 × 125 × 125 mm in model scale) was used to simulate an idealised model building of 25 m height (full scale). The cube was mounted in the centre of the wind tunnel test section.

The vertical profile of the mean wind field was measured upstream of the obstacle (i.e. at $X/H = -8$) and was used to derive u_* and z_0 and subsequently the logarithmic law was applied to obtain the vertical profile of u at inflow (Figure 5). For the turbulent kinetic energy k and the dissipation rate ε the approximations $k = u^{*2}/\sqrt{(c_\mu)}$ with $c_\mu = 0.09$ and $\varepsilon = u^{*3}/(\kappa \cdot z)$ were used to specify the input data for the models. All five models used the same computational domain of dimension 21H × 11H × 8H resolved on a grid with 96 × 96 × 48 grid points stretched away from the walls. The cube is represented by 16^3 grid points equidistantly distributed in the x- and y-direction.

* http://www.mi.uni-hamburg.de/cedval/introduc.htm

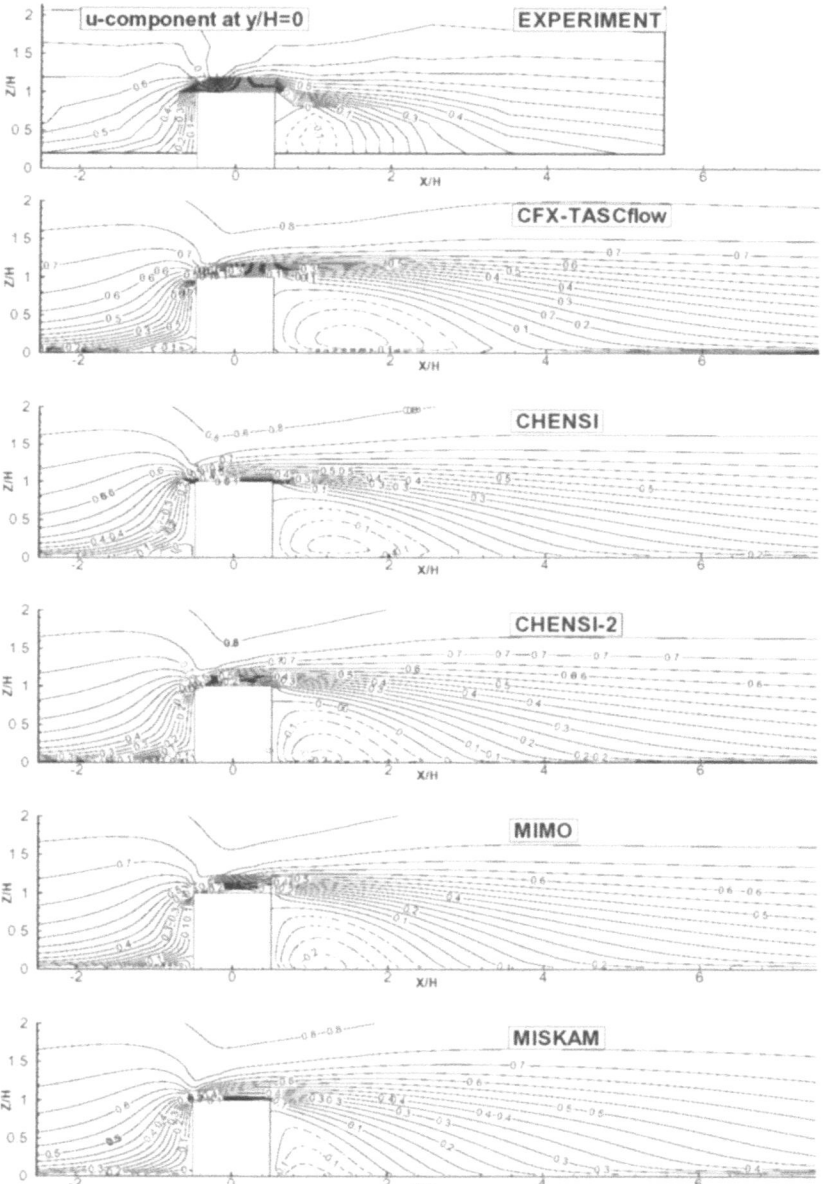

Figure 6. Vertical cross section of the dimensionless u-component normalised with the free-stream velocity ($U_{ref} = 6$ m s^{-1}).

TABLE I

Characteristic lengths of the flow field around
the cube (stagnation point Z_S (at $X/H = -0.63$),
separation point X_F (at $Z/H = 0.1$), reattachment
point X_R (at $Z/H = 0.1$))

Model	Z_S/H	X_F/H	X_R/H
Experiment	0.64	−0.88	1.50
CFX-TASCflow	0.68	−0.90	3.20
CHENSI	0.60	−0.94	2.89
CHENSI-2	0.68	−0.88	2.62
MIMO	0.61	−0.89	2.61
MISKAM	0.58	(-0.98)[a]	1.95

[a] Estimated at $Z/H = 0.03$ since MISKAM predicts a recirculation only in the lowest layer.

Figure 6 shows the vertical cross section of the dimensionless u-component
normalised with the free-stream velocity ($U_{ref} = 6$ m.s^{-1}) as observed in the wind
tunnel and predicted by the numerical models in the centre plane of the flow. In
agreement with the findings of the experiment, all models reproduce the following
typical flow pattern: the oncoming flow exhibits an impingement region at the
windward side of the obstacle. When approaching the cube the flow separates due
to increasing pressure leading to the development of a main horseshoe vortex wrap-
ping around the cube. At the upper leeward edge of the obstacle the flow separates
again and leads to an extended lee vortex formed in the cavity zone immediately
behind the cube which interacts with the horseshoe vortex. Table I summarises the
characteristic lengths of the flow field around the cube as predicted by the models
and derived from the measurements. The calculated stagnation point Z_S/H on the
upwind edge of the cube varies from 0.58 to 0.68. Considering that a value of 0.64
was derived from the measurements, all models predict the height of the stagnation
point satisfactorily. All models except MISKAM predict the upwind separation
length X_F/H in good agreement with the measurements. The upwind vortex calculated by MISKAM is limited to the lowermost computational layer, while the other
models show an upwind vortex in the two lower layers, (see also Figures 6 and 7,
profile at $X/H = -0.625$). The upwind vortex length X_F/H estimated for MISKAM
from the lowermost layer (−0.98) is close to the predictions of the other codes.
This upwind vortex is very sensitive to the differences in the implementation of
the boundary conditions on solid surfaces, since it is influenced by the windward
facing wall of the cube and the ground surface and is additionally characterised by
strong horizontal and vertical gradients.

All models overpredict the leeward re-attachment length X_R/H. The extent of the
lee vortex in the cavity behind the cube calculated by CFX-TASCflow is larger than

that of the other models. This can be attributed to the fact that in CFX-TASCflow less turbulent kinetic energy is produced in the impingement region and therefore less turbulent kinetic energy is transported around the cube leading to a lower level of turbulent kinetic energy in the wake region (cf. Figure 9).

Figure 7 shows the vertical profile of the u-component at different positions $(X/H) = -1.5, -0.625, 0, 0.625, 1.5$ and 2.5) in the centre plane of the flow. The agreement between measured and computed data at $X/H = -1.5$ is excellent; hence it can be argued that the computations simulated the exact experimental conditions prevailing upwind the cube. The comparison between the numerical results and the measurements at the different positions X/H shows a good agreement for locations above the cube (at dimensionless heights $Z/H > 1$). In agreement with observations, all models predict large gradients of the u-component for the lee vortex in the cavity zone behind the cube ($X/H = 0.625$). The airflow within this cavity zone predicted by the models is characterised by mean velocities ranging between -1 and -1.5 m.s^{-1}. Close to the observed reattachment point in the centre plane ($X/H = 1.5$) all models compute a negative velocity close to the surface indicating that this position is predicted to be still far inside the cavity zone, thus the models overestimate the reattachment length X_R (cf. Table I).

Except from CHENSI, all models perform similar upwind the obstacle; once more the underestimation of the velocity close to the obstacle by CHENSI indicates that the method of implementing the wall functions in this model is probably the main reason for the deviation. Largest differences among the model predictions can be found in the cavity zone behind the cube: Compared to the other models, MISKAM does not capture in full the lee vortex in the cavity zone, while CFX-TASCflow results show the most extended cavity zone. As already observed in the single cavity case, throughout the whole test case, CHENSI-2 and MIMO predictions lead to similar results.

Figure 8 shows the profile of the vertical wind component at different positions $(X/H = -1.5, -0.625, 0, 0.625, 1.5$ and 2.5) in the centre plane of the flow. The agreement between measured and computed data at $X/H = -1.5$ is fair. Above the cube next to its top, the (negative) w-component is underestimated with all models indicating that the models fail to correctly predict the re-circulation bubble above the cube. Nonetheless, this is consistent with published results of computations using the standard k-ε model in conjunction with wall functions which often do not show a detachment of the flow (re-circulation bubble) on top of the cube (cf. Lakehal and Rodi, 1997). This is associated with the overproduction of turbulent kinetic energy in the stagnation area upwind the cube (cf. Figure 9) leading to an overestimation of the eddy viscosity on the top of the cube. This results in an enhanced exchange of momentum, which counteracts a flow separation. Close to the leeward side of the cube the vertical wind component is overpredicted.

Figure 9 shows the vertical profile of the turbulent kinetic energy k at different positions ($X/H = -1.5, -0.625, 0, 0.625, 1.5$ and 2.5) in the centre plane of the flow. The models fail to predict the turbulent kinetic energy profile in the impinging

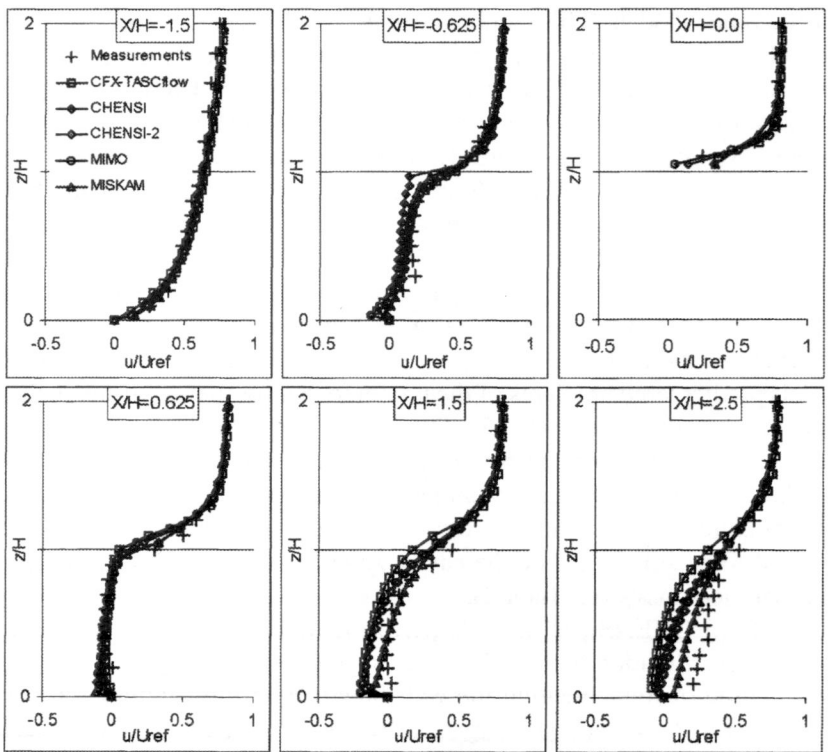

Figure 7. Comparison of the dimensionless profile of the u-component normalised with the free-stream velocity ($U_{ref} = 6$ m s^{-1}) at various distances.

region (at $X/H = -0.625$) and the strong gradient of the turbulent kinetic energy on the top of the obstacle. As already mentioned, this is a deficiency of the applied standard k-ε model which assumes the Reynolds stresses to be proportional to the local deformation of the flow (C_μ = constant only valid if production of k is equal to the dissipation ε (local equilibrium)) and in addition an isentropic eddy viscosity ν_t. The differences among the numerical results for k, once more, emphasise the importance of the different implementation of the wall function by the codes.

5. Conclusions

Improvement and optimisation of the methods used in practical application of traffic pollution models for air quality impact studies is one of the final goals of the TRAPOS network. Within the network CFD model intercomparison exercises were organised for model evaluation. Four different modelling groups participated in those exercises with their respective models. Several cases, e.g. the airflow within a single cavity and around a single cube, which have been employed within the

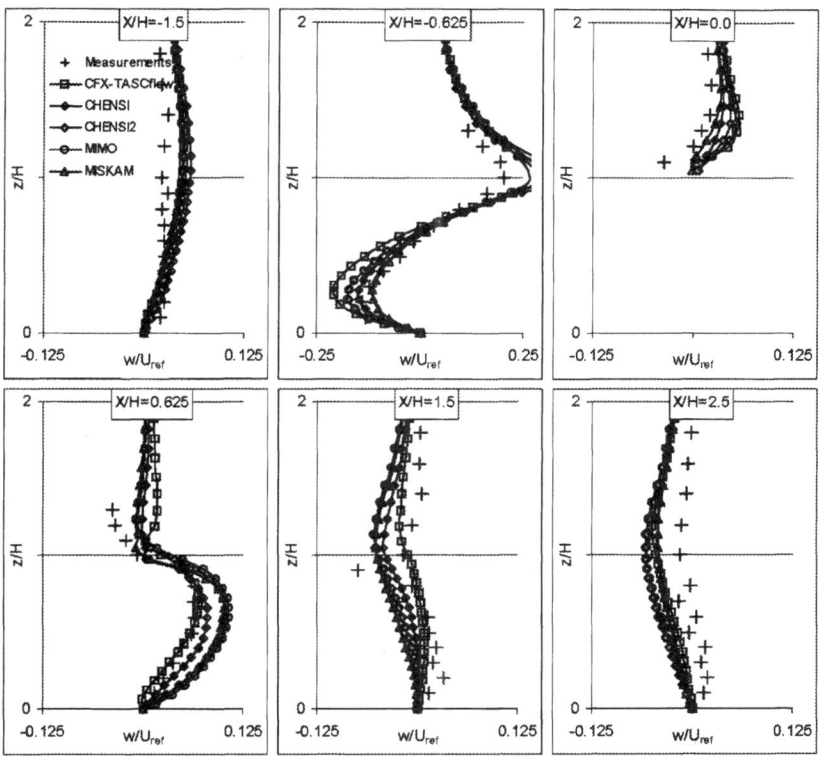

Figure 8. Comparison of the dimensionless profile of the w-component normalised with the free-stream velocity ($U_{ref} = 6$ m s^{-1}) at various distances.

model intercomparison exercise, indicate the applicability of the numerical models, in general, for predicting the main features of the airflow. The work presented here has pointed out that even though the case set up was relatively simple, discrepancies among model results were far to be negligible at individual locations. This was shown to be particularly true from the inter-comparison of the turbulent kinetic energy. The use of the standard k-ε turbulence model was found to be the principal factor being responsible for the deviation between measurements and model predictions. A detailed examination of the model predictions close to the solid boundaries suggested that the discrete implementation of the wall-function influences the calculated velocity values at locations where the local flow is strong. The most frequent discrete forms for the wall function applied in CFD codes have been used by the models of the present intercomparison exercise, therefore the results of this paper may be considered as representative for most of the CFD codes employing the unique law of the wall. The wind flow and pollutant dispersion within complex building structures are investigated by the part II paper (Ketzel *et al.*, 2001) giving also recommendations for the practical use of urban dispersion models.

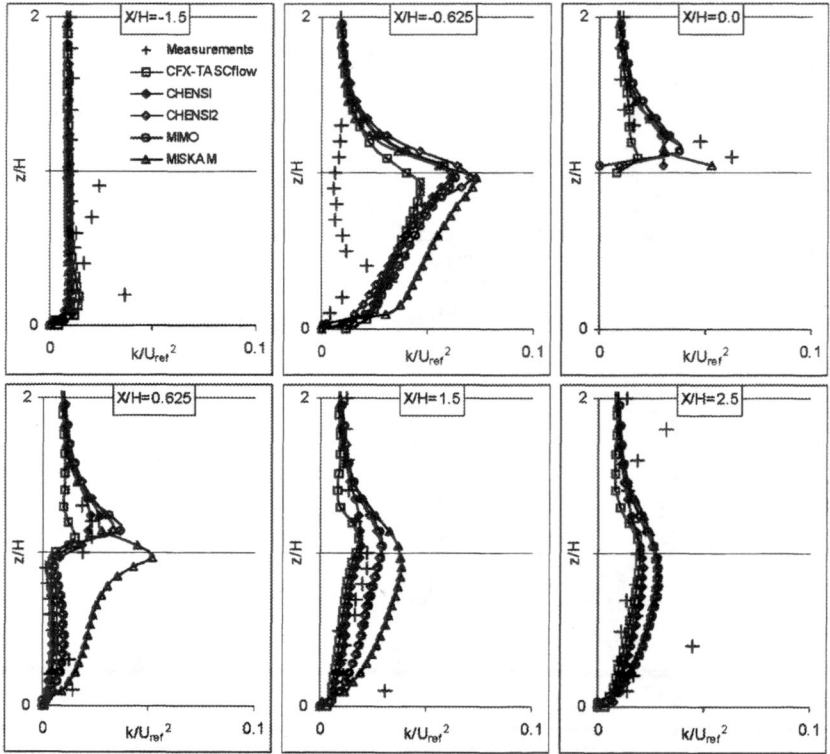

Figure 9. Comparison of the dimensionless profile of the normalised turbulent kinetic energy at various distances.

Acknowledgements

The presented work was carried out in the frame of the research network TRA-POS operated in the framework of EC's Training and Mobility of Researchers Programme. The authors acknowledge the provision of experimental data by Drs Savory and Kovar-Panskus, University of Surrey and Drs Leitl and Schatzmann, University of Hamburg. Computer support was given to the French team by the Institut de Développement et de Recherche pour l'Informatique Scientifique – IDRIS – (CNRS), Orsay, France.

References

Ehrhard, J., Khatib, I. A., Winkler, C., Kunz, R. Moussiopoulos, N. and Ernst, G.: 2000, 'The microscale model MIMO: Development and assessment', *J. Wind Engin. Indust. Aerodyn.* **85**, 163–176.

Eichhorn, J.: 1989, 'Entwicklung und Anwendung eines dreidimensionalen mikroskaligen Stadtklima-Modells', Dissertation, Universität Mainz, Germany.

Guilloteau E.: 1999, 'Modélisation des Sols Urbains pour les Simulations de l'Atmosphère aux Échelles Sub-Meso', PhD thesis, University of Nantes – Ecole Centrale de Nantes.

Ketzel, M., Louka, P., Sahm, P., Guilloteau, E.., Sini, J.-F. and Moussiopoulos, N.: 2001, 'Intercomparison of numerical urban dispersion models – Part II: Street canyon in Hannover, Germany', *Water, Air, and Soil Pollut.*, this issue.

Kovar-Panskus, A., Louka, P., Mestayer, P.G., Savory, E., Sini, J.-F. and Toy, N.: 2001, 'Influence of geometry on the flow and turbulence characteristics within urban street canyons – Intercomparison of wind tunnel experiments and numerical simulations', *Water, Air, and Soil Pollut.* (submitted).

Lakehal, D. and Rodi, W.: 1997, 'Calculation of the flow past a surface-mounted cube with two-layer turbulence models', *J. Wind Eng. Ind. Aerodyn.* **67**(68), 65–78.

Leitl, B. and Schatzmann, M.: 1999, 'Generation of high resolution reference data for the validation of micro-scale models', in *Recent Developments in Measurement and Assessment of Air Pollution*, VDI-Kommission Reinhaltung der Luft, Bericht 1443, ISBN 3-18-091443-2, pp. 647–656.

Louka, P., Belcher, S. E. and Harrison, R. G.: 2000, 'Coupling between airflow in streets and the well-developed boundary layer aloft', *Atmos. Env.* **34**, 2613–2621.

Raw, M. J.: 1994, 'A Coupled Algebraic Multigrid Method for the 3D Navier-Stockes Equations', *Proceedings of the 10th GAMM-Seminar*, Kiel, January 14-16, 1994.

Raw, M. J., Galpin, P. F. and Hutchinson, B. R.: 1989, 'A collocated finite-volume method for solving the Navier-Stokes equations for incompressible and compressible flows in turbomachinery: Results and applications', *Can. Aeron. Space J.* **35**, 189–196.

Sini, J.-F., Anquetin, S. and Mestayer, P. G.: 1996, 'Pollutant dispersion and thermal effects in urban street canyons', *Atmos. Environ.* **30**, 2659–2677.

INTERCOMPARISON OF NUMERICAL URBAN DISPERSION MODELS – PART II: STREET CANYON IN HANNOVER, GERMANY

M. KETZEL[1]*, P. LOUKA[2], P. SAHM[3], E. GUILLOTEAU[4], J.-F. SINI[2] and N. MOUSSIOPOULOS[3]

[1] Department of Atmospheric Environment, National Environmental Research Institute, Roskilde, Denmark; [2] Fluid Mechanics Laboratory, Ecole Centrale de Nantes, Nantes, France; [3] Laboratory of Heat Transfer and Environmental Engineering, Aristotle University Thessaloniki, Greece; [4] Institute of Hydromechanics, University of Karlsruhe, Karlsruhe, Germany
(* author for correspondence, e-mail: mke@dmu.dk, fax: +45 4630 1214)

Abstract. Microscale computational fluid dynamics (CFD) models developed in different European countries were applied to well defined test cases comprising a variety of 2 and 3 dimensional configurations for which measurements from wind tunnel or field studies were available. This paper presents the results of five CFD codes employing the widely used 'standard k-ε-model' (CHENSI, CHENSI-2, MIMO, MISKAM, TASCflow) for a street canyon in Hannover, Germany (Göttinger Strasse). Firstly the characteristics of the flow field predicted by the different codes are compared with high spatial resolution wind tunnel measurements; secondly the calculated concentration fields are compared with field and wind tunnel data. Both agreements (e.g. for the general flow and concentration fields) and disagreements (e.g. for the level of concentration) are observed in the comparison. The discussion aims at explaining the differences along with giving some suggestions to CFD model users on how to calculate such complex flows, but also to experimentalists on where concentration measurements should be taken in order to be more representative for a whole street and to avoid distinct local effects.

Keywords: atmospheric dispersion modelling, computational fluid dynamics, dispersion near buildings, microscale, model intercomparison, model validation, street canyon, traffic pollution

1. Introduction

Microscale computational fluid dynamics (CFD) models have become a useful and popular tool for assessment and prediction of air quality in urban areas. The proper validation of such a model is a crucial prerequisite for its practical application. Within the framework of the European research network TRAPOS a working group on computational fluid dynamics modelling was established and model intercomparison exercises were launched*. Several test cases comprising a variety of 2 and 3 dimensional configurations were defined and detailed descriptions of the grid set-up and the boundary and inflow conditions are available on the Internet. For all test cases measurements from wind tunnel or field studies were available. This

* http://www.dmu.dk/AtmosphericEnvironment/Trapos/cfd-wg.htm

Water, Air, and Soil Pollution: Focus **2**: 603–613, 2002.
© 2002 *Kluwer Academic Publishers.*

paper, which is the second in a sequence of two, presents the results of five models (CHENSI, CHENSI-2, MIMO, MISKAM, and TASCflow) for a street canyon (Göttinger Strasse) in Hannover, Germany. The intercomparison of the models in the cases of a single cavity and a surface mounted cube are subject of part I (Sahm et al., 2002).

2. Methodology

2.1. MODELS

The CFD codes applied in this exercise are frequently used by the different institutions participating in this comparison exercise. A more detailed description of the codes can be found in (Sahm et al., 2002). All codes employ the widely used 'standard k-ε -model' for the turbulence closure but different implementation of the boundary conditions and different numerical schemes are used.

2.2. FIELD AND LABORATORY DATA SET

A comprehensive field data set was obtained from measurements in Göttinger Strasse in Hannover. The State Environmental Agency of Lower Saxony operates a permanent monitoring station in this four-lane street canyon with a traffic load of ca. 30000 vehicles/day (NLÖ., 1993). The width of the canyon is 25 m and buildings on both sides of the street are ca. 20 m high.

In addition to street and background concentrations and meteorological data from a 10 m mast on top of a nearby building also traffic counts are available.

The data set was chosen for its completeness and quality as was shown in previous studies (Schädler et al., 1996; Ketzel et al., 2000). Another advantage of this field data set is the additional availability of wind tunnel measurements both for the concentration at the receptor point and for the flow field inside the street canyon, performed at the University of Hamburg (Liedke et al., 1998; Chauvet et al., 2002).

2.3. TEST CASE

Figures 1 and 2 show the central part of the modelling domain with the built-up area of 170 × 240 m. The average building height is ca. 20 m, the meteorological mast and the background station are placed on the roof of the highest building (30 m high). The dimensions of the full domain, including a margin without obstacles, are 370 × 440 × 240 m resolved by 59 × 72 × 24 grid boxes. This comparatively coarse grid is typical for many practical applications and was chosen as a compromise of resolution and calculation effort to make it possible to run different wind directions for most of the codes. The grid, the inflow and boundary conditions were exactly defined and used by all codes. The test case description can be downloaded

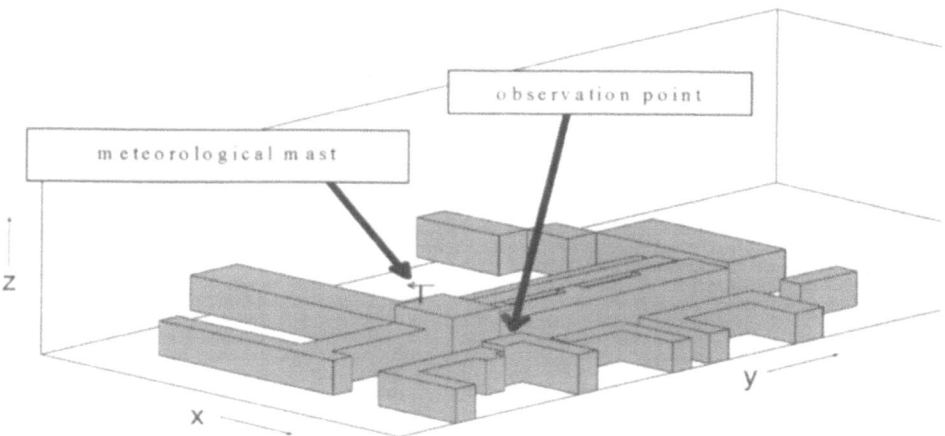

Figure 1. Central part of the modelling domain. The position of the meteorological mast and the point of the concentration measurements are indicated in the sketch.

from the webpage of the TRAPOS CFD working group[*] and used also for other numerical models. As shown in Figures 1 and 2, the building, which is used for the meteorological measurements, is much higher than the surrounding ones and it is shifted towards the street axis reducing the street width. The neighbouring building southwards contains a large opening (gateway, not shown in the pictures). On the opposite street side a series of side streets cause openings in the building front. Consequently the observation point is placed in a very complex situation, close to several irregularities in the building configuration. This has to be kept in mind when interpreting the model results.

3. Results and Discussion

Figure 3 shows the flow field for approaching flow from 240° in a horizontal plane 10 m above ground level as it was calculated by the five CFD codes and measured in the wind tunnel. The general flow pattern – the strong flow parallel to the street in its southern part and the inflow from the side street in middle position – are well reproduced by all the codes. For the details of the flow we observe large differences between the different codes and between codes and wind tunnel results. The wind tunnel results show a vortex with vertical axis in the central part of the street between the two side streets and south of this the flow has no vortex structure. CHENSI and MISKAM fail to reproduce the vortex between the two side streets, while TASCflow, MIMO and CHENSI-2 produce this vortex, but show an additional second vortex towards south, which is not present in the wind tunnel data.

[*] http://www.dmu.dk/AtmosphericEnvironment/Trapos/cfd-wg.htm

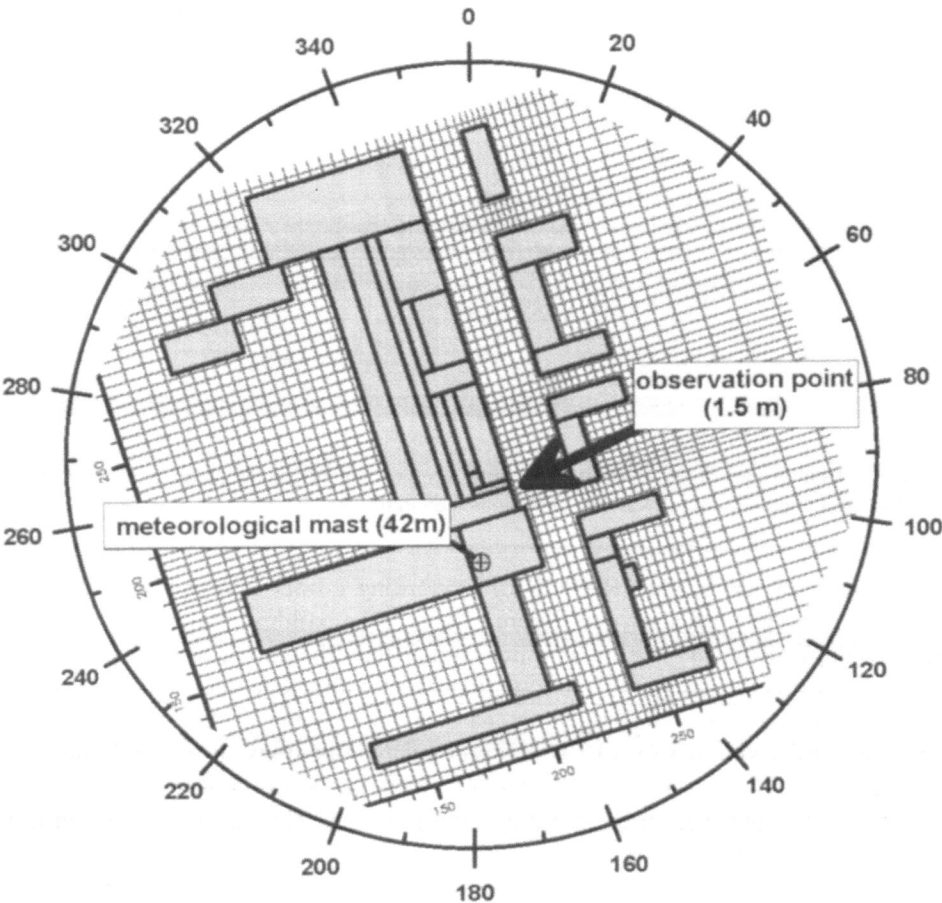

Figure 2. Top view on the modelling area. The grey areas indicate buildings. The mesh shows the resolution of the grid. The circular axis gives the direction with respect to North as it is used for the approaching flow direction in the following figures.

Figure 4 shows the situation for approaching flow from 260°. Also for this case the general flow pattern – the strong flow parallel to the street in its northern part and the vortex with vertical axis at the south end of the street – are well reproduced by all the codes. Close to the observation point, the wind tunnel experiment showed an area with very low wind speeds and a vortex with vertical axis as it was also reproduced by most of the codes. However the exact position of the vortex inside the horizontal plane as well as the flow rate differ from code to code. This will consequently have dramatic effects on the calculated pollutant concentration fields, as shown for this wind direction later (Figure 6).

For wind direction 280° (see Figure 5) we observe a similar picture of agreement in the general flow pattern and discrepancies in the details. The vortex at the

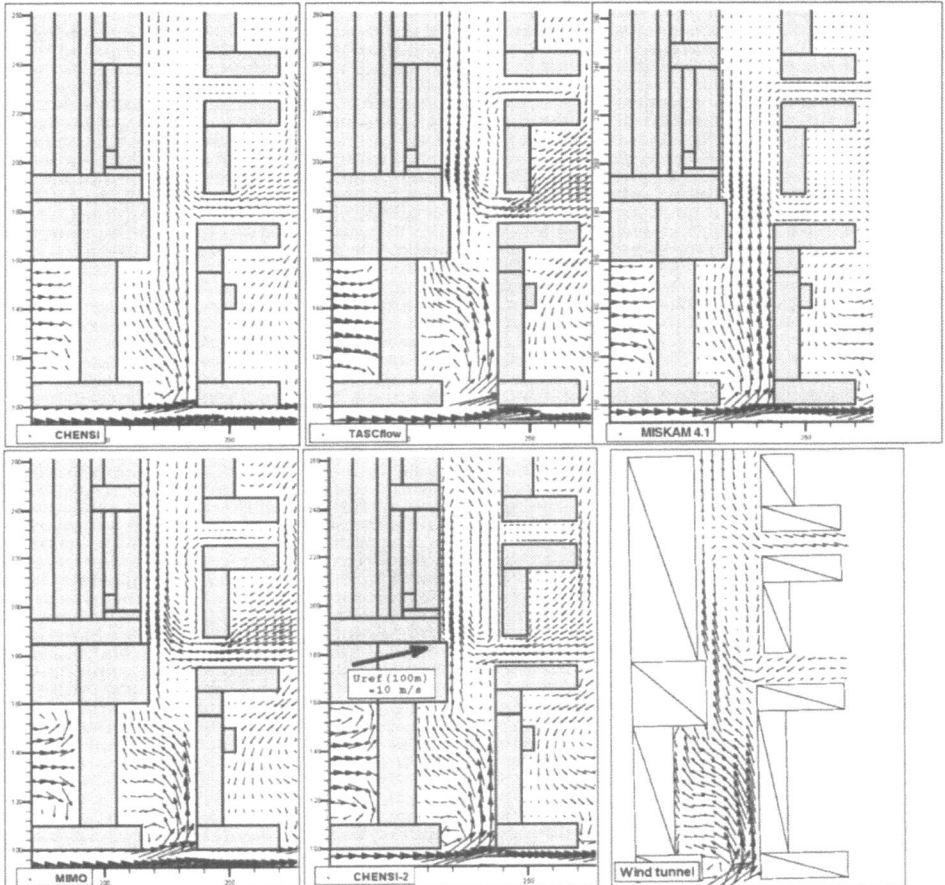

Figure 3. Flow fields calculated by five CFD codes and measured in the wind tunnel for approaching flow from 240° in a horizontal plane 10 m above ground.

south end of the street is not reproduced by all the codes and some codes show additional vortex structures in the central part of the street, which is not present in the wind tunnel. CHENSI calculates very small horizontal flow rates especially near the observation point, which will lead to high pollutant concentrations as shown later (see Figure 7, wind direction 280°). Due to the very time consuming measuring procedure only the three above documented wind directions could be investigated in the wind tunnel.

The differences in the flow field will be emphasised during the dispersion calculation and lead to more pronounced differences in the concentration fields, since relatively small changes in wind direction and wind speed cause quite different transport of the traffic pollution. For comparison we use the dimensionless concentration c* defined as $c^* = \frac{c \cdot u \cdot L}{q}$; where c is the modelled concentration including

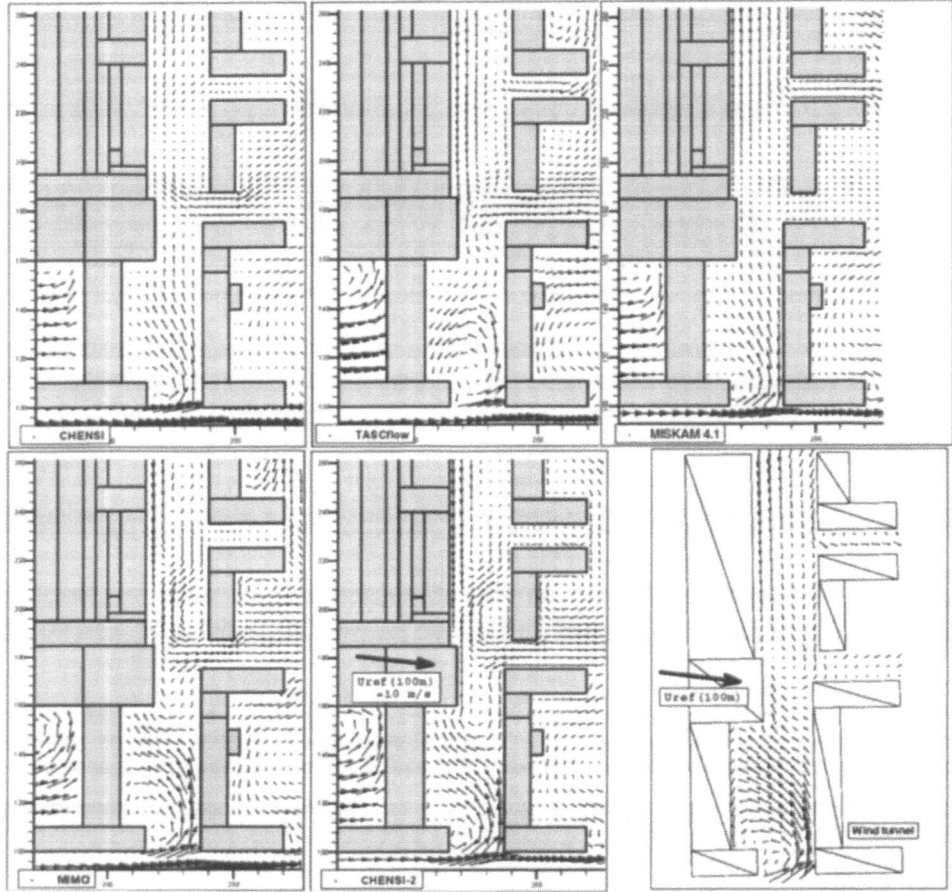

Figure 4. Flow fields calculated by five CFD codes and measured in the wind tunnel for approaching
flow from 260° in a horizontal plane 10 m above ground.

dimensions, u is the wind speed at a reference point (here 10 m s^{-1} at 100 m
height), L is a scaling length usually the width or the height of the street canyon
(here 20 m), q is the emission flux per unit length.

As an example Figure 6 shows the dimensionless concentration fields calculated
by four CFD codes for approaching flow from 260° based on the flow fields plotted
in Figure 4. Both the general structures of the concentration field – as position of
the maximum and ratio between concentration at opposite sides of the street – and
the concentration level are very different. Comparing the concentration predicted
by different codes at one particular point may lead to substantial discrepancies as
shown later. Concentration measurements from field and wind tunnel experiments
are usually only available for one or a few points, as it is the case for Göttinger

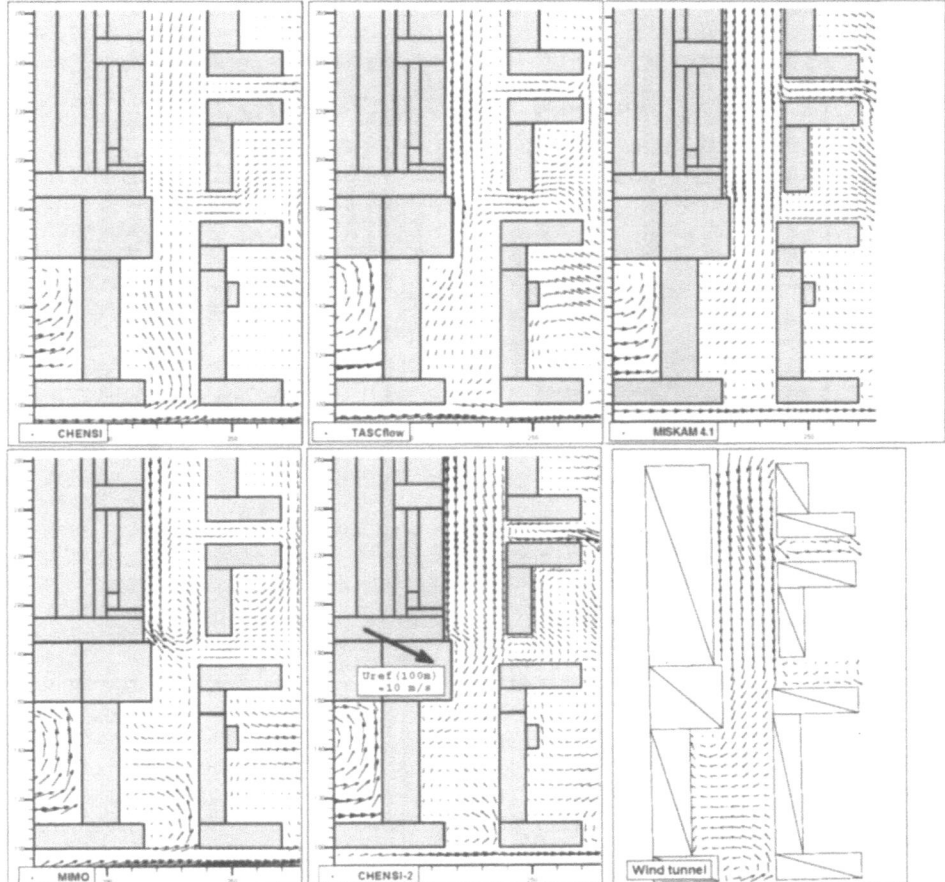

Figure 5. Flow fields calculated by five CFD codes and measured in the wind tunnel for approaching flow from 280° in a horizontal plane 10 m above ground.

Strasse. More comprehensive validation data for complex geometry will hopefully available in the near future. (e.g. CEDVAL database*).

The dimensionless concentrations observed or calculated at the observation point (see position in Figure 2) are illustrated in Figure 7 as a function of the wind direction. Agreement is found for the general shape of the dependence for most of the codes, i.e., low concentrations for wind direction from 60 to 90°, when the observation point lies in the so-called windward side and high concentrations for the leeward situation for wind direction from 230 to 290°. It is observed that the differences between the concentrations calculated by the different codes for a specific wind direction are within a range of factor 2 up to factor 7. The results from the field and wind tunnel measurements are typically in the middle of the range given

* http://www/mi.uni-hamburg.de/cedval/

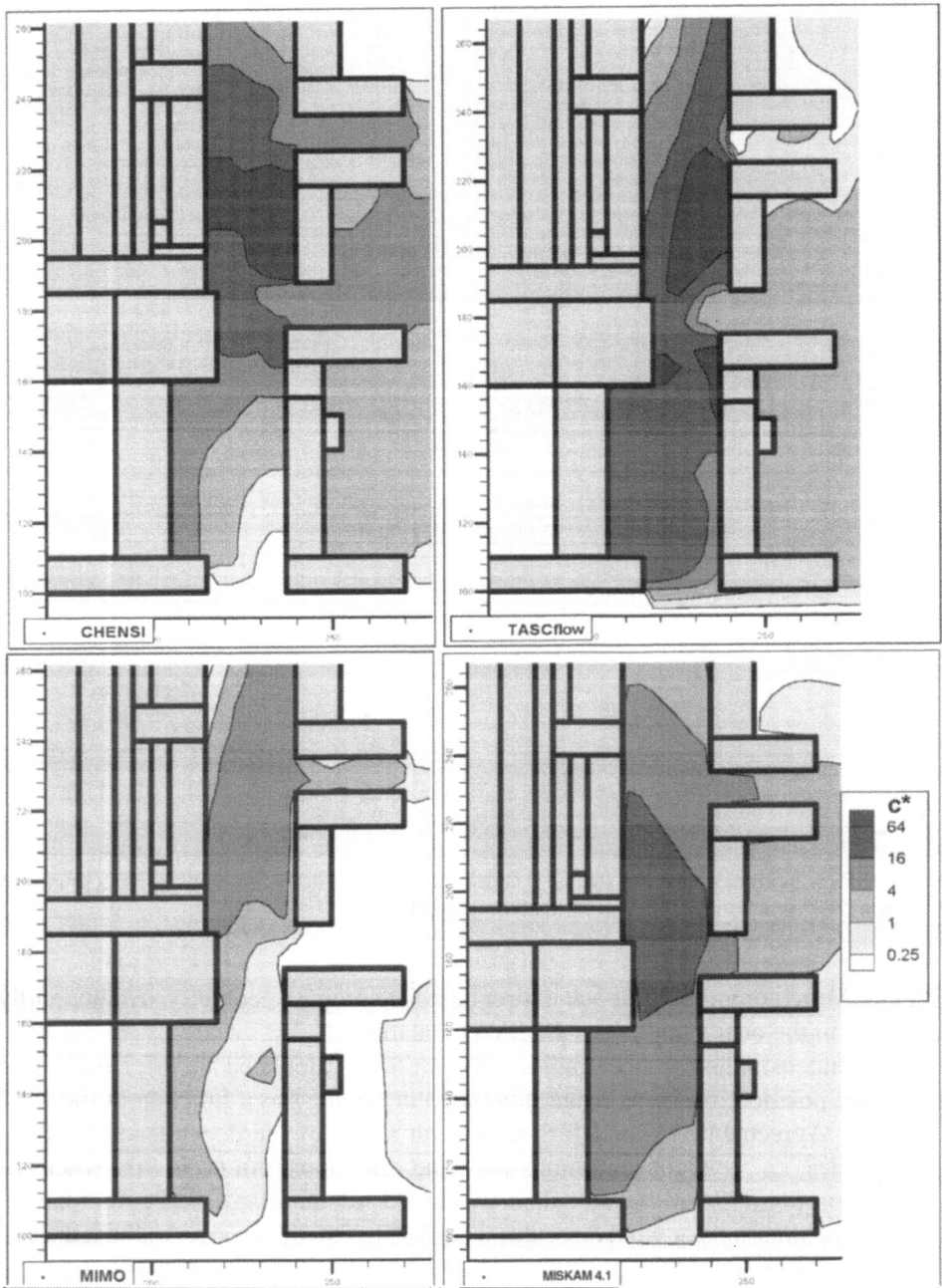

Figure 6. Contour plots for the dimensionless concentration calculated by four CFD for approaching flow from 260° in a horizontal plane 10 m above ground.

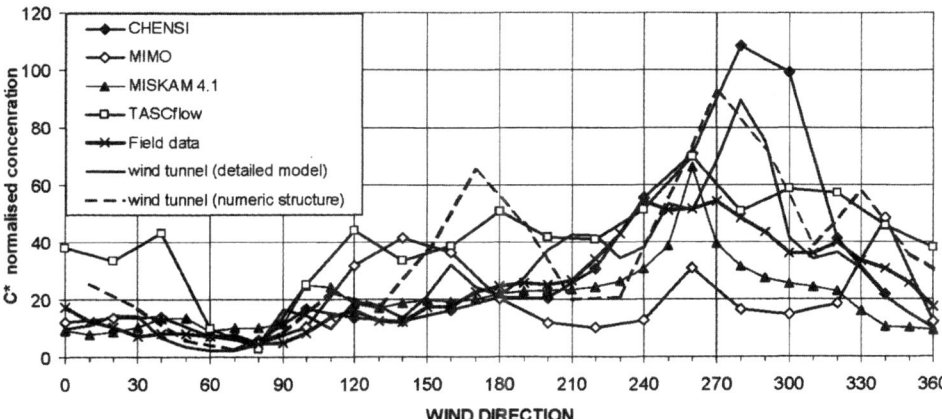

Figure 7. Dimensionless concentrations in the dependence on the wind direction. Calculated by four CFD codes, measured in the field and in the wind tunnel. The 'detailed model' contains all geometrical details normally modelled in the wind tunnel, while 'numeric structure' stands for a wind tunnel imitation of the coarser building configuration used by the CFD codes.

by the CFD codes. The building configuration used for the CFD codes is a coarse approximation of the reality in terms of fine structures like slanted roofs and in the modelled area. This could cause some systematic differences between the codes and field data and the wind tunnel measurements with the detailed model. The influence of the resolution of the geometrical details on the results was simulated in the wind tunnel by measurements with precisely the coarse building configuration used for the CFD codes (see results labeled with 'numeric structure' in Figure 7).

The reasons for the discrepancies between the codes should be found in order to use these codes in practical applications. This set-up was the most complex one in the series of test cases of the inter-comparison exercise performed and shows as expected the greatest quantitative differences among the model results. It is clear that the measuring point lies in a region with very complex flow patterns and strong gradients in flow and concentration fields. Thus any small difference in the predicted direction of the impinged flow may lead to a quite different flow pattern at this location resulting to large discrepancies among the predicted values. As suggested in Sahm *et al.* (2002) the different numerical implementation of the wall boundary conditions in the used CFD codes could explain part of the differences in the flow and concentration field that were observe in this complex 3-D case. The codes do as well differ in the employed numerical schemes, which will contribute to the observed differences. The size and resolution of the grid is constantly a sensitive point in CFD applications. The grid chosen in this complex case is a compromise between desired fine resolution and available computational power. A higher grid resolution would probably have influence on the results, some codes might be more sensitive to a coarse grid resolution, e.g. side streets with only 3–4 grid boxes in the cross section. We recommend a resolution of the smallest geometrical details (buildings or openings) with at least four cells. In the scope of this paper it is not possible to quantify the relative contributions of the mentioned error sources.

4. Conclusions

The presented results of a model inter-comparison exercise illustrate that the application of CFD codes to a well-defined, geometrically complex, test case can reproduce the general flow field measured in the wind tunnel. However, for details of the flow field – as the location of the centre of vortex structures and the flow velocity – differences were found. Consequently the calculated concentration of vehicular exhausts at a location close to building irregularities show large differences that may reach a factor of 7. As reasons for these discrepancies the influence of the boundary conditions, the numerical schemes and the grid resolution are discussed. It is suggested that the accuracy of the CFD modelling result for only one location affected by local gradients should be treated with special care. For practical purpose, it is possible that an estimation of averages in time (over different inflow situations) or averages in space (to avoid local gradients) is more appropriate.

This model exercise suggests that CFD modelling may be used for recommendations for detecting suitable monitoring sites in order to obtain a representative picture of the air quality in a street canyon. The strong local gradients observed close to irregularities in the building configuration (intersections, gateways, towers, corners etc.) should be avoided.

Moreover this comparison exercise indicated, that the CFD community working on the urban atmosphere and pollutant transport/dispersion problems still needs a standardised quality assurance procedure which would give clear guidance to developers and users as to how properly assure the quality of their models and results.

Acknowledgements

The authors wish to express their appreciation for support from the European Commission's Training and Mobility of Researchers Programme (TMR) within the frameworks of the European Research Network on 'Optimisation of Modelling Methods for Traffic Pollution in Streets' (TRAPOS). The authors wish to acknowledge the provision of input data for the model intercomparison by Drs Chauvet, Liedke and Schatzmann of the University of Hamburg and Dr. Müller of the State Environmental Agency of Lower Saxony (NLÖ Hannover). Complete computer support was given to the French team by the Scientific Council of Institut de Développement et de Recherche pour l'Informatique Scientifique – IDRIS – (CNRS), Orsay, France.

References

Chauvet, C., Leitl, B. and Schatzmann, M.: 2001, 'High resolution flow measurements in an idealised urban street canyon', *Water, Air and Soil Pollut.*, this issue.

Ketzel, M., Berkowicz, R. and Lohmeyer, A.: 2000, 'Dispersion of traffic emissions in street canyons – Comparison of European numerical models with each other as well as with results from wind tunnel and field measurements', *Environ. Monit. Assess.* **65**, 363–370.

Liedke, J., Leitl, B. and Schatzmann, M.: 1998, 'Car Exhaust Dispersion in a Street Canyon – Wind Tunnel Data for Validating Numerical Dispersion Models', *2nd East European Conference on Wind Engineering (EECWE)*, Prague, 7–11 September 1998.

NLÖ: 1993, *Lufthygienisches Überwachungssystem Niedersachsen – Luftschadstoffbelastungen in Straßenschluchten*, edited by Niedersächsisches Landesamt für Ökologie, D-30449 Hannover, ISSN 0945–4187.

Schädler, G., Bächlin, W., Lohmeyer, A. and van Wees, T.: 1996, 'Vergleich und Bewertung derzeit verfügbarer mikroskaliger Strömungs- und Ausbreitungsmodelle', in Berichte Umweltforschung Baden-Württemberg, PEF 2 93 001 (FZKA-PEF 138).

Sahm, P., Louka, P., Ketzel, M., Guilloteau, E. and Sini, J.-F.: 2002, 'Intercomparison of numerical urban dispersion models – Part I: Street canyon and single building configurations', *Water, Air and Soil Pollut.*, this issue.

SATELLITE AND GROUND BASED MONITORING OF AEROSOL PLUMES

MARTIN DOYLE* and STEPHEN DORLING

School of Environmental Sciences, University of East Anglia, Norwich, U.K. (author for correspondence, e-mail: m.doyle@uea.ac.uk, fax: +44 1603 507719)*

Abstract. Plumes of atmospheric aerosol have been studied using a range of satellite and ground-based techniques. The Sea-viewing Wide Field-of-view Sensor (SeaWiFS) has been used to observe plumes of sulphate aerosol and Saharan dust around the coast of the United Kingdom. Aerosol Optical Thickness (AOT) was retrieved from SeaWiFS for two events; a plume of Saharan dust transported over the United Kingdom from Western Africa and a period of elevated sulphate experienced over the Eastern region of the UK. Patterns of AOT are discussed and related to the synoptic and mesoscale weather conditions. Further observation of the sulphate aerosol event was undertaken using the Advanced Very High Resolution Radiometer instrument (AVHRR). Atmospheric back trajectories and weather conditions were studied in order to identify the meteorological conditions which led to this event. Co-located ground-based measurements of PM_{10} and $PM_{2.5}$ were obtained for 4 sites within the UK and $PM_{2.5/10}$ ratios were calculated in order to identify any unusually high or low ratios (indicating the dominant size fraction within the plume) during either of these events. Calculated percentiles of $PM_{2.5/10}$ ratios during the 2 events examined show that these events were notable within the record, but were in no way unique or unusual in the context of a 3 yr monitoring record. Visibility measurements for both episodes have been examined and show that visibility degradation occurred during both the sulphate aerosol and Saharan dust episodes.

Keywords: aerosol optical thickness, AVHRR, Saharan dust, SeaWiFS, horizontal visibility, sulphate aerosol

1. Introduction

It has become increasingly accepted over recent years that long-range transport of particulate matter plays an important role in the contribution to exceedences of both National and European particulate standards (King and Dorling, 1997; Beverland *et al.*, 2000). Secondary particulates are of concern as they have the potential to be transported for many hours and days depending on the particulate composition/size distribution and the dominant meteorological conditions. Primary particulates, and in particular Saharan dust, can also cause exceedences of particulate standards within the UK. Particulates of this nature are perhaps of more concern, as abatement strategies are harder to implement with regard to semi-natural events such as this. In this study, aerosol optical thickness was studied for two recent plumes of sulphate aerosol and Saharan dust, observed using the NASA Sea-viewing Wide Field of View Sensor (SeaWiFS). The sulphate aerosol

Water, Air, and Soil Pollution: Focus **2**: 615–629, 2002.

plume of May 2000 and the Saharan dust plume of March 2000 provide a unique opportunity to use satellite data to observe the behaviour and effects of such plumes on a synoptic scale. Corresponding ground based measurements of PM_{10} and $PM_{2.5}$ were also analysed and the $PM_{2.5/10}$ ratios observed during these events are placed in the context of measurements made over a 3 yr period. Visibility measurements from meteorological monitoring sites throughout the UK have been examined in order to identify the visibility degradational effects of the material within each plume.

2. Methodology

Two particulate events were examined using the techniques outlined below; a sulphate aerosol episode on the 14th May 2000 and a Saharan dust episode on the 3rd March 2000.

2.1. SEAWiFS

The SeaWiFS sensor was launched aboard the SeaStar spacecraft in August 1997 as part of NASA's Mission to Planet Earth programme. The purpose of SeaWiFS data is to examine oceanic factors that affect global change and to assess the oceans' role in the global carbon cycle, as well as other biogeochemical cycles, through a comprehensive research program. In order to retrieve this information, an atmospheric correction must be applied to remove the effects of Rayleigh scattering and the reflectance and absorption by aerosols and clouds, which account for up to 90% of the received signal (McClain et al., 2000a). The SeaWiFS atmospheric correction algorithm uses two near infrared (NIR) bands, 765 and 865 nm, to estimate the aerosol optical properties and then extrapolate these into the visible (Gordon and Wang, 1994) as the scatterings are wavelength dependent. Gordon and Wang 1994, describe in detail the basis of the correction algorithm, the technicalities of which shall not be entered into within this paper.

The SeaWiFS data was obtained from the Goddard Space Flight Centre and was visualised using the SeaWiFS Data Analysis System (SeaDAS). This Linux based program allows visualisation of all routinely collected parameters from SeaWiFS through the use of a user-friendly graphical user interface to Interactive Data Language (IDL) software.

2.1.1. *Difficulties in AOT Retrieval*

Aerosol optical thickness at 865 nm, $\tau a(865)$, is routinely retrieved from SeaWiFS measurements although it is reasonably straightforward to retrieve aerosol optical thickness for any of the SeaWiFS wavelengths (McClain et al., 2000b). At present, aerosol optical thickness can only be accurately retrieved from the SeaWiFS data over the sea surface. The relatively uniform surface reflectance of the oceans makes the atmospheric corrections less complex than those over spatially and temporally heterogeneous land surfaces. Dense cloud formations can also obscure the retrieval of the SeaWiFS data and atmospheric correction cannot be undertaken if a clear view of the ocean is unavailable.

2.2. ADVANCED VERY HIGH RESOLUTION RADIOMETRY

Another method which has also been employed as an additional means of tracking airborne particulate in this study, is the use of visible wavelength imagery (Advanced Very High Resolution Radiometry (AVHRR)). The AVHRR is a broadband, five channel scanner, sensing in the visible, near-infrared, and thermal infrared portions of the electromagnetic spectrum. This sensor is carried on NOAA's Polar Orbiting Environmental Satellite (POES), and imagery was supplied by the Dundee Satellite Receiving Station (http://www.sat.dundee.ac.uk/).

2.3. GROUND BASED MONITORING OF PARTICULATES

Sulphate aerosol and Saharan dust fall into very different size fractions of PM_{10}. Due to this difference in size, an analysis of $PM_{2.5}$ and PM_{10} data was undertaken for both the Saharan and sulphate aerosol events in order to identify how the relationship between fine and more coarse particulate concentrations changed during these periods. Ground-based particulate data from 3 South East of England sites (Marylebone Road (London), London Bloomsbury and Rochester (East of London)) were used in order to calculate ratios of $PM_{2.5}$ to PM_{10}. In this way, specific events can be characterised by the percentage of $PM_{2.5}$ which contributes to the total concentration of PM_{10}. It could be expected that events in which sulphate aerosols are more abundant would yield a relatively higher ratio of $PM_{2.5}/PM_{10}$ than an event in which the main component of the measured PM_{10} was coarser (e.g. Saharan dust). Routine measurement of $PM_{2.5}$ began in the UK in the summer of 1997, enabling an examination of $PM_{2.5}/PM_{10}$ ratios over a period of over 3 yr. The events, which we describe in this paper, are placed in the context of this 3 yr record.

Sulphate data for Stoke Ferry in Norfolk (eastern England) has also been obtained and is examined in reference to the sulphate aerosol episode.

2.4. VISIBILITY MEASUREMENTS

Changes in the atmospheric pollutant mix can result in changes to the optical properties of the atmosphere and hence changes in the visibility. Rayleigh scattering of solar radiation occurs when gas molecules are smaller than the radiation wavelength and scattering occurs in all directions. However, when aerosol particles with similar sizes to the radiation wavelength are present, most of the light is scattered forwards. This Mie scattering gives polluted atmospheres their greyish and hazy appearance.

TABLE I

UKMO synoptic code for reporting present weather conditions (codes 00 to 05)

Code	Present weather
00	Cloud development not observed or observable
01	Clouds dissolving or becoming less developed
02	State of sky on the whole unchanged
03	Clouds generally forming or developing
04	Visibility reduced by smoke haze
05	Haze

In the United States much attention has been paid to the visibility degradational effects of atmospheric aerosols. The Interagency Monitoring of Protected Visual Environments (IMPROVE) network in the US has been monitoring visibility trends in over 20 areas since 1987 in order to identify present visibility levels and any sources of anthropogenic impairment to the visibility resource. Studies into atmospheric visibility have been carried out within the UK (Lee, 1990; DTI, 1997) although a network such as IMPROVE is currently not in place within the UK and visibility trends are not routinely examined within studies of fine particulate matter. As part of this study, horizontal visibility measurements were examined from several UK Meteorological Office (UKMO) surface observation sites across the United Kingdom in order to identify any decrease in visibility due to an increased loading of aerosols within the atmosphere. Several of the sites chosen for analysis are located along the Eastern side of the UK in order to capture the effects of the sulphate aerosol episode, which was observed by SeaWiFS and AVHRR imagery to be located over the North Sea along the eastern Seaboard of the United Kingdom.

Decreased visibility measurements can be attributable not only to an increased loading of particulate matter within the atmosphere. Precipitation such as rain, mist and fog can also act to decrease the horizontal visibility distance. At many UK Meteorological Office (UKMO) surface observation sites information on multiple meteorological parameters is collected on an hourly basis. Simultaneous observations of horizontal visibility and present weather conditions, based on the SYNOP code (UKMO hourly surface observations), were obtained for this study with only codes of 00 to 05 being used in further analysis. Codes of 06 and above relate to localised dust suspension and precipitation effects which may distort the results if left in the analysis. Table I shows the codes used in the analysis.

Visibility observations can be taken at night with the use of lights at known distances from the observer. However, these may be more unreliable than daytime measurements due in part to the methods used.

2.5. ATMOSPHERIC BACK TRAJECTORIES

Atmospheric back trajectories were computed for both the Saharan dust episode and the sulphate aerosol episode. These were calculated using the US National Oceanic and Atmospheric Administration Air Resources Laboratory (NOAA – ARL) model (HYSPLIT 4,1997). Back trajectories were calculated for T-72 from the times that the episodes were observed. March 2nd/3rd 2000 (Saharan dust episode) and May 14th 2000 (sulphate aerosol episode) were chosen as these are the days in which i) maximum concentrations of PM10 were monitored across the UK (Saharan dust) and ii) the sulphate aerosol plume was observed by Sea-WiFS and AVHRR imagery. Back trajectories were calculated using gridded FNL meteorological data for the northern hemisphere. This data is based on National Centers for Environmental Prediction (NCEP) outputs. Within trajectory calculations, the u, v and vertical velocity components (w) of the wind are taken directly from the meteorological model. Other vertical motion calculation methods within the model include isobaric, isentropic, constant density and constant sigma above terrain options although the default (model output) was used in this analysis. The trajectory receptors chosen for the sulphate episode related to the rough outline of the sulphate plume along the Eastern UK as observed from the SeaWiFS imagery (Figure 1). These same receptor locations were then used in the Saharan dust study as no distinct plume was observable from SeaWiFS imagery during this episode.

3. Results and Discussion

All figures included within this paper can also be viewed in colour at http://www. uea.ac.uk/env/pubs/mdø1/.

3.1. SULPHATE AEROSOL EPISODE

Aerosol optical thickness (AOT) for the sulphate aerosol episode can be seen in Figure 1. It may be noted that the values of AOT are dimensionless and, in addition to the AOT being a total column amount, cannot be directly compared with ground-based measurements of aerosols. The plume, visible in the North Sea, has associated values of 0.1–0.3. The largest value for this particular satellite pass is 1.3 (a dust plume from the west African coast) and when no aerosols are detected, the value within the region of interest is zero.

Figure 2 shows the synoptic chart for the 14th May 2000. The low-pressure system centred to the west of Ireland and the high-pressure system centred to the west of Denmark, acted together to transport European derived aerosol into the North Sea region, as can be seen in Figure 1. Numerous meteorological observing stations along the east coast of the UK also reported haze during this episode.

Figure 3 shows AVHRR imagery for the sulphate aerosol episode and shows the plume over the North Sea. It is possible to see the plume over the land surfaces to

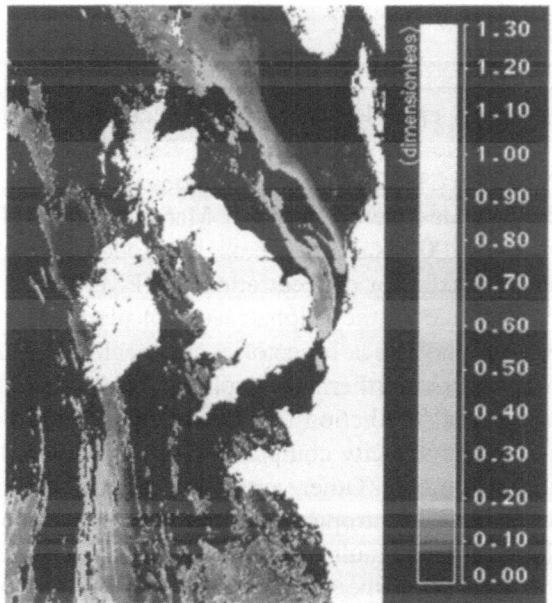

Figure 1. Aerosol optical thickness 14th May 2000 1300 h.

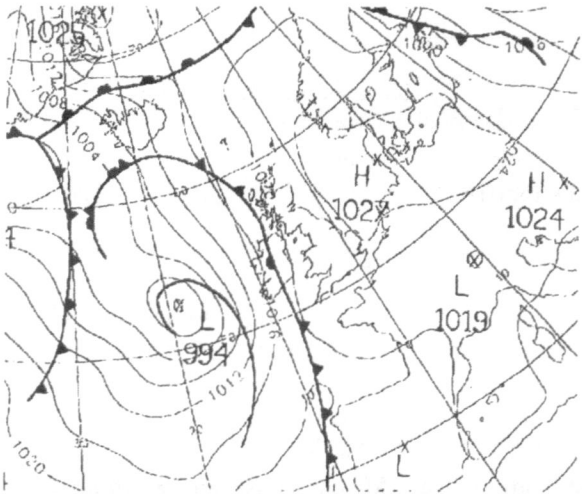

Figure 2. Synoptic chart for 14th May 2000 1200 h.

some extent, a feature which is not visible using the SeaWiFS imagery alone. The plume can be seen to stretch from Northern France, Belgium and the Netherlands up to the Shetland Islands. The extent to which the plume reaches over the land surfaces to the west is slightly more uncertain as definite boundaries cannot be observed against the land surfaces using this imagery.

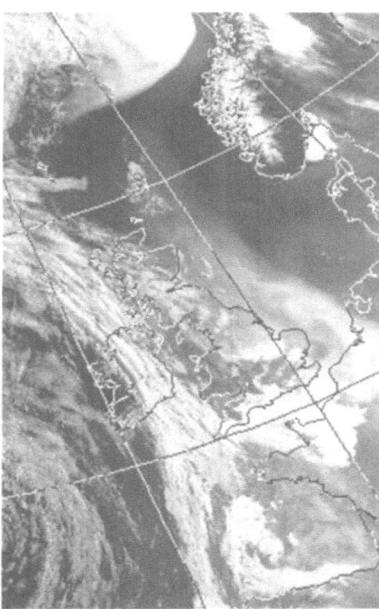

Figure 3. AVHRR Imagery (Channel 1) 14th May 2000.

Sulphate data from Stoke Ferry in Norfolk (eastern England) shows that on the 14th May 2000, the highest sulphate levels of the entire record for 2000 were monitored (Figure 4), indicating that this was an unusual event at this location when taken in the context of the year long record.

Visibility measurements during the sulphate aerosol episode can be seen in Table II. It can be observed that the visibility during the sulphate aerosol episode is considerably lower when compared with the average for the whole of May 2000, except at 3 sites: Fylingdales and Bridlington in northeast England, and Shaw-bury in mid-west England. Both Fylingdales and Bridlington did not report the occurrence of haze during the 14th May giving a very coarse indication of the 'edge' of the sulphate plume (The shortest distance between a station reporting haze (Bridlington) and one not reporting haze (Waddington) was approximately 80 km). From the SeaWiFS image in Figure 1 a more definite edge to the plume can be observed in comparison with the AVHRR imagery (Figure 3) where it is more difficult to distinguish the boundaries of the plume.

Back trajectories calculated for T-72 from 1200 h on the 14th May 2000 (Figure 5) show that potential precursor source regions for this plume cover much of central Europe. All trajectories were launched at 1000m above ground level.

3.2. SAHARAN DUST EPISODE

AOT for the Saharan episode, which shows a much different, patchier aerosol distribution, can be seen in Figure 6. Values of aerosol optical thickness around

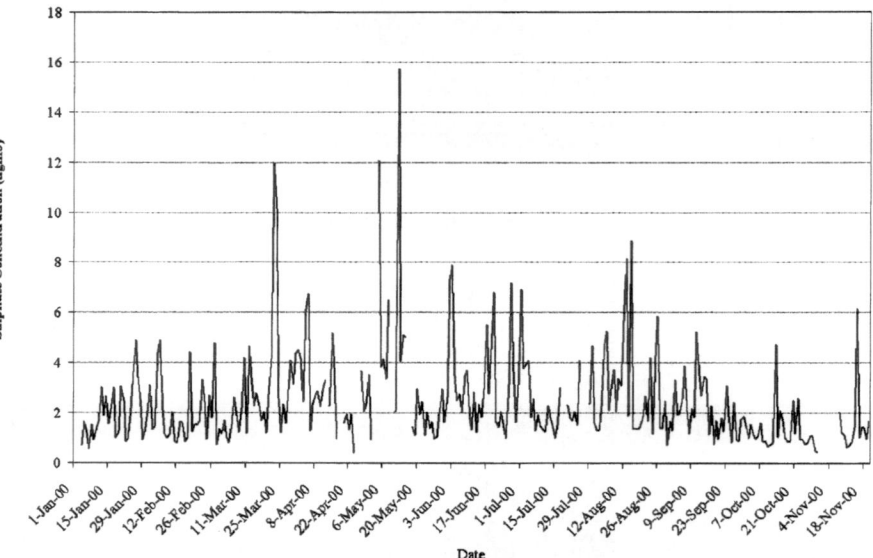

Figure 4. Sulphate data for Stoke Ferry 2000.

TABLE II

Visibility measurements at UKMO stations: Average for May 2000 and
during the sulphate aerosol episode (14th May 2000)

UKMO Station	Average horizontal visibility May 2000 (km)	Horizontal visibility 14th May 2000 (km)
Wattisham	19	14
Hemsby	15	8
Waddington	18	5
Bridlington	15	15
Fylingdales	21	25
Heathrow	21	15
Plymouth	14	10
Shawbury	21	30

the UK during this event are between 0.1 and 0.2. Automatic pollution monitors
across the UK measured high levels of particulates on this day and on the previous
evening, with some areas experiencing exceedences of the UK particulate standard,
although heavy rainfall was also experienced. This may have affected the aerosol
distributions, and the associated cumulus clouds would also have obscured the
SeaWiFS sensors view of the oceans around the UK. Figure 7 shows PM10 data

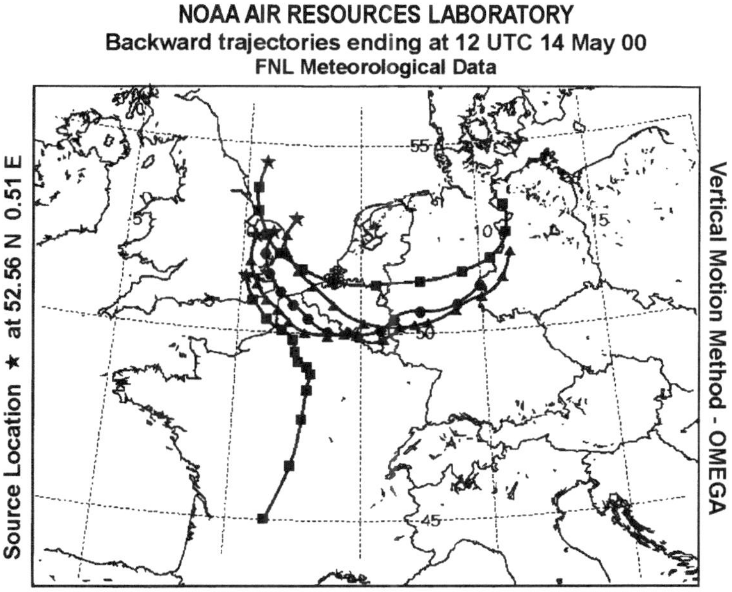
NOAA AIR RESOURCES LABORATORY
Backward trajectories ending at 12 UTC 14 May 00
FNL Meteorological Data
Source Location ★ at 52.56 N 0.51 E
Vertical Motion Method - OMEGA
55
50
45
5
10
15

Figure 5. Atmospheric back trajectories for 1200 h 14th May 2000.

for the 2nd and 3rd of March 2000 and highlights those periods when rainfall acted to decrease the ambient concentrations of PM10 within the atmosphere. In Figure 8, the massive dust cloud can be seen blowing off the African continent on the 1st of March, 2 days before the highest levels were observed in the UK.

Visibility measurements on this day (filtered for midday measurements and SYNOP codes of 05 or less) show that the horizontal visibility was not decreased on the 3rd March and was in fact, considerably larger when compared with the average for the entirety of March 2000. Table III shows data from UKMO stations around the UK. As noted previously, the maximum concentrations of PM10 were experienced during the evening of the 2nd 3rd March and by midday on the 3rd, concentrations of PM10 at all but one of the sites examined were below the Air Quality Strategy Objective of 50 μg m^{-3}. The uncertainties of using night time measurements of horizontal visibility has been raised previously, but in this case must be used in order to capture the effects of this material on the visibility distance. Table III contains data filtered for 2200 h on the 2nd March and also 0400 h on the 3rd March. Decreased visibility during this evening, in comparison to the average for March 2000, is notable at Plymouth (10 pm and 4 am), St. Bees (10 pm), Shawbury (10 pm), Wattisham (4 am) and Hemsby (4 am). Increased visibility distances at midday on the 3rd March may be due to the fact that little of this dust was held in suspension for any period of time within the atmosphere due to the effect of precipitation acting to wash this material from the plume.

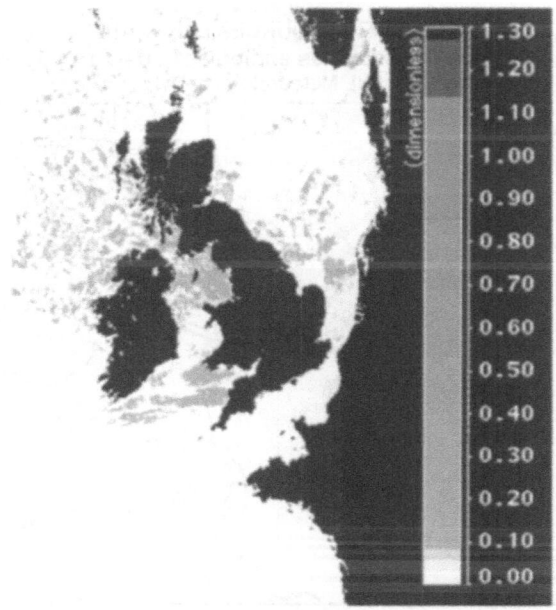

Figure 6. Aerosol optical thickness 3rd March 2000 1300 h.

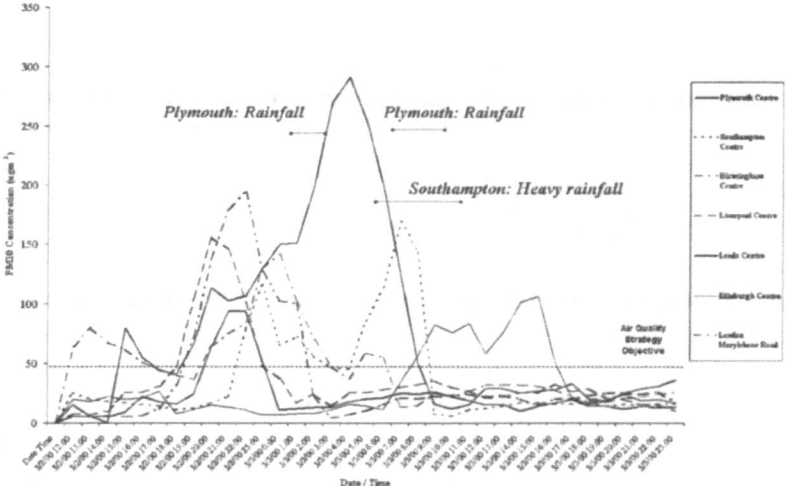

Figure 7. PM10 concentrations and rainfall events across the UK on 3rd March 2000.

Back trajectories were calculated for T-72 from 0400 h on the 3rd March 2000 (Figure 9). Trajectories were launched at this time after taking into account the period of maximum PM10 concentrations experienced throughout the UK. The source region of this dust appears to be the Atlantic ocean, where Saharan dust was being transported over the preceding days. Models such as the Navy Aerosol Analysis and Prediction System (NAAPS) run by the Naval Research Laboratory

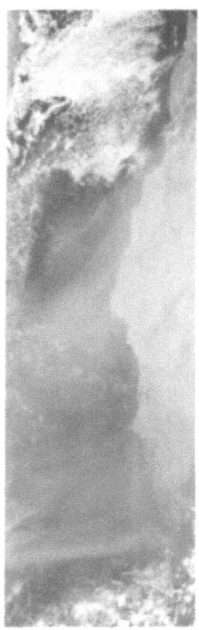

Figure 8. Saharan dust storm, SeaWiFS Image from 1st March 2000 1300 h (Level 1A Global Area Coverage image derived from bands 1, 5 and 6).

TABLE III

Visibility measurements at UKMO stations: Average for March 2000 and during the Saharan dust episode (2nd/3rd March 2000)

UKMO Station	Average horizontal visibility March 2000 (km) (12Z < 05)	Horizontal visibility 3rd March 2000 (km) (12Z < 05)	Horizontal visibility 2rd March 2000 (km) (10Z < 05)	Horizontal visibility 3rd March 2000 (km) (4Z < 05)
Manston	20	30	25	20
Walton-on-the-Naze	12	18	16	16
Wattisham	17	35	30	12
Hemsby	14	23	19	11
Bridlington	17	20	20	-
Fylingdales	24	28	-	-
Heathrow	21	30	-	-
Shawbury	22	25	10	-
Plymouth	14	17	8	5

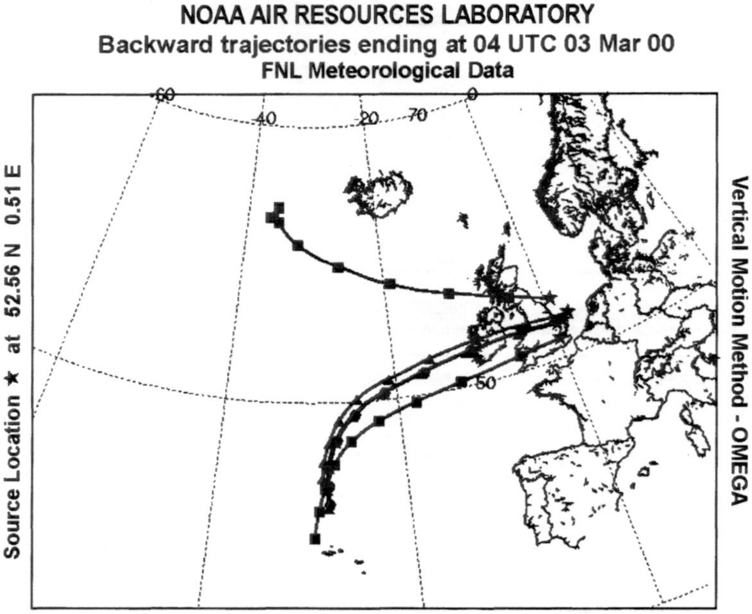
NOAA AIR RESOURCES LABORATORY
Backward trajectories ending at 04 UTC 03 Mar 00
FNL Meteorological Data
Source Location ★ at 52.56 N 0.51 E
Vertical Motion Method - OMEGA
60
40
20
70
0
50

Figure 9. Atmospheric back trajectories for 0400 h 3rd March 2000.

(NRL) in Monterey California, show good visual comparisons between modelled dust and dust monitored by the SeaWiFS instrument on the 28th February 2000. These images can be seen at http://www.nrlmry.navy.mil/aerosol/Case_studies/20000226_sahara/ (viewed Jan 2002). Visual comparison of these images, Figure 8 and the back trajectories in Figure 9 indicate that the material monitored in the UK on the 3rd March was in fact Saharan dust.

3.3. PM2.5/PM10 RATIOS

PM2.5/PM10 ratios for each site (Marylebone Road, London Bloomsbury and Rochester) were calculated for the entire 3 yr period for which this data is available and plotted as a cumulative frequency chart. Percentiles were calculated for the sulphate aerosol and Saharan episodes in order to identify whether or not these events were in any way extreme within the broader context of the 3 yr record. The $PM_{2.5/10}$ ratios during the sulphate aerosol episode fell between the 76th and 83rd percentiles of all measurements, indicating that although the majority of the material monitored during this event was fine particulate, the event itself was in no way unique when looking at the 3 yr record. Similarly, the ratios calculated for the Saharan episode fell between the 29th and 44th percentiles. This material was indeed coarser than that in the sulphate aerosol episode, but again, this event was not unusual in the broader context of the monitoring record. The cumulative frequency chart can be seen in Figure 10.

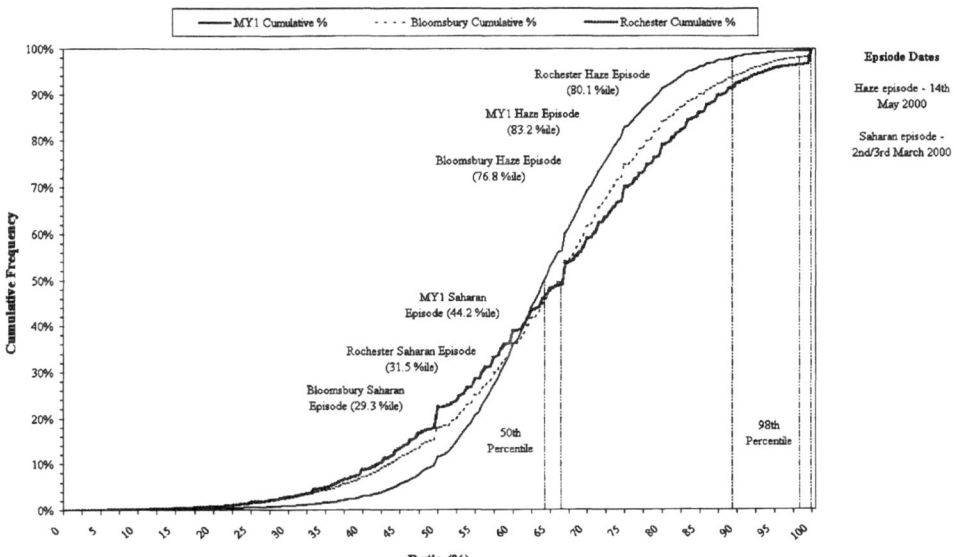

Figure 10. Cumulative frequency chart of $PM_{2.5/10}$ ratios and calculated percentiles for London Bloomsbury, Marylebone Road and Rochester.

4. Conclusions

- AVHRR imagery and aerosol optical thickness derived from the SeaWiFS sensor for the sulphate aerosol episode show a large, fine particulate plume emanating from continental Europe on the 14th May 2000. This seems to have been brought about by the synoptic situation on that day. Atmospheric back trajectories calculated for this day indicate source regions of this pollution in central Europe.

- A combination of SeaWiFS imagery and back trajectory analyses show that material monitored at many UK air pollution monitoring stations on the 2nd and 3rd of March 2000 was Saharan dust.

- The PM2.5/10 ratios during the sulphate aerosol episode were at the 80th percentile of all measurements taken in the 3 yr period. This shows that the material monitored was relatively fine but not unusually so.

- Particulates during the Saharan episode were coarser, the PM2.5/10 ratios accounting for around the 30th percentile of all measurements, although once more, the ratio was not extreme at the monitoring sites examined in the UK.

- Horizontal visibility measurements show the visibility degradational effects of both fine sulphate aerosol and Saharan dust.

- Aerosol optical thicknesses for the Saharan episode are similar to those in the sulphate aerosol episode, although some of the particulate was removed by precipitation associated with the depression.

The suite of tools used in this study gives an interesting view of two pollution events. However, both events give rise to the same problems in terms of meeting particulate objectives. Whether the pollution is anthropogenic or naturally produced, the results at monitoring sites tend to be the same, that is elevated levels of PM10, which no manner of local measures can reduce. It is therefore important to characterise these long-range particulates in order to identify where management measures do need to take place. Remote sensing and ground-based measurements of these events can go some way to characterising particulate pollution and perhaps even identifying the sources if it becomes feasible to use this promising technique over land surfaces. Remote sensing of air pollution over land surfaces using other satellite platforms has been undertaken (Sifakis *et al.*, 1998) and shows great promise in studies of this kind. The advent of new satellite sensors aboard the NASA TERRA (MODIS and MISR sensors) and European Space Agency (ESA) EnviSat (SCIAMACHY sensor) satellites promise to be able to vertically resolve much of the information obtained by sensors such as SeaWiFS into the troposphere. This will allow easier comparison between ground based monitoring of particulates and those derived by remote sensing techniques. A logical aim of this work is an operational air quality forecasting system informed by data assimilation of this remotely sensed imagery and corresponding ground based measurements.

Acknowledgements

The authors would like to thank the SeaWiFS Project (Code 970.2) and the Distributed Active Archive Center (Code 902) at the Goddard Space Flight Center, Greenbelt, MD 20771, for the production and distribution of these data, respectively. These activities are sponsored by NASA's Mission to Planet Earth Program.

AVHRR Imagery courtesy of the Dundee Satellite Receiving Station, Dundee University, Scotland (http://www.sat.dundee.ac.uk/)

References

Beverland, I. J., Tunes, T., Sozanska, M., Elton, R. A., Agius, R. M. and Heal, M. R.: 2000, 'Effect of long-range transport on local PM10 concentrations in the UK', *Int. J. Environ. Health Res.* **10**, 229–238

Department of Trade and Industry.: 1997, 'Trends in UK Visibility Between 1980 and 1996', ETSU Report Number- ETSU N/01/00047/REP

Gordon, H. R. and Wang, M.: 1994, 'Retrieval of water leaving radiance and aerosol optical thickness over the oceans with SeaWiFS: A preliminary algorithm', *Appl. Opt.* **33**, 443–452.

HYSPLIT4 (HYbrid Single-Particle Lagrangian Integrated Trajectory) Model, 1997. Web address: http://www.arl.noaa.gov/ready/hysplit4.html, NOAA Air Resources Laboratory, Silver Spring, MD.

King, A. M. and Dorling, S. R.: 1997, 'PM10 particulate matter-the significance of ambient levels', *Atmos. Environ.* **31**, 2379–2381.

Lee, D.: 1990, 'The influence of wind direction, circulation type and air pollution emissions on summer visibility trends in Southern England', *Atmos. Environ.* **24A**(1), 195–201

McClain, C. R., Ainsworth, E. J., Barnes, R. A., Eplee, Jr, R. E., Patt, F. S., Robinson, W. D., Wang, M., Bailey, S. W., Hooker, S. B. and Firestone, E. R. (eds): 2000a 'SeaWiFS Postlaunch Calibration and Validation Analyses, Part 1', NASA Tech. Memo. 2000–206892, Vol. 9.

McClain, C. R., Barnes, R. A., Eplee, Jr, R. E., Franz, B. A., Hsu, N. C., Patt, F. S., Pietras, C. M., Robinson, W. D., Schieber, B. D., Schmidt, G. M., Wang, M., Bailey, S. W., Werdell, P. J., Hooker, S. B. and Firestone, E. R. (eds): 2000b 'SeaWiFS Postlaunch Calibration and Validation Analyses, Part 2', NASA Tech. Memo. 2000–206892, Vol. 10.

Naval Aerosol Analysis and Prediction System (NAAPS) Naval Research Laboratory (NRL); Available: http://www.nrlmry.navy.mil/aerosol/ [viewed Jan 2002].

Sifakis, N. I., Soulakellis, N. A. and Paronis, D. K.: 1998, 'Quantitative mapping of air pollution density using Earth observations: a new processing methods and application to an urban area', *Int. J. Rem. Sens.* **19**(17), 3289–3300.

THE ASSESSMENT OF ATMOSPHERIC POLLUTION USING SATELLITE REMOTE SENSING TECHNOLOGY IN LARGE CITIES IN THE VICINITY OF AIRPORTS

D. G. HADJIMITSIS[1]*, A. RETALIS[2] and C. R. I. CLAYTON[1]

[1] *Department of Civil & Environmental Engineering, University of Southampton, Highfield, Southampton, SO17 1BJ, U.K.;* [2] *National Observatory of Athens, Institute for Space Applications and Remote Sensing, Metaxa & Vas. Pavlou, Palea Pendeli, GR 152-36, Athens, Greece*
(* author for correspondence, e-mail: dhadjimitsis@cytanet.com.cy, fax: + 44 2380 677519)

Abstract. This paper investigates the potential of using satellite remotely sensed imagery for assessing atmospheric pollution. A novel approach, which comprised radiative transfer calculations and pseudo-invariant targets for determining aerosol optical thickness has been developed. The key parameter for assessing atmospheric pollution in photochemical air pollution studies is the aerosol optical thickness. The need for identifying suitable pseudo-invariant objects in satellite images of urban areas is of great interest for increasing the potential of earth observation for monitoring air pollution in such areas. The identification of large water bodies and concrete aprons that can serve as suitable dark and bright targets respectively in different geographical areas was demonstrated in this study. This study added evidence on the correlation found between the visibility values measured at Heathrow Airport area during satellite overpass and aerosol optical thickness derived from Landsat-5 TM band 1 images.

Keywords: aerosol optical thickness, atmospheric pollution, bright targets, dark targets, eutrophic waters, pseudo-invariant targets, remote sensing

1. Introduction

Atmospheric pollution in large cities is one of the major issues to be tackled by local and global communities due to its widespread existence, which can have a deleterious impact on human life. Air quality monitoring stations have been established in major cities and provide means for alert. But measuring stations are widely distributed and do not provide sufficient tools for mapping atmospheric pollution since air quality is highly variable. However, earth observations made by satellite sensors are likely to be a valuable tool for assessing and mapping atmospheric pollution due to their major benefit of providing complete and synoptic views of large areas in one snap-shot. The use of earth observation to assess and map the atmospheric pollution in different geographical areas and especially in cities has received considerable attention from researchers who developed a variety of techniques (for example, Kaufman *et al.*, 1990; Sifakis and Deschamps, 1992; Retalis *et al.*, 1999; Wald and Balleynaud, 1999; Wald *et al.*, 1999). The key parameter for assessing atmospheric pollution in photochemical air pollution studies is the aerosol optical thickness (Kaufman *et al.*, 1990), which is also the most important unknown parameter in every atmospheric correction algorithm for solving the radiative transfer (RT) equation and removing atmospheric effects from satellite remotely sensed images.

This paper presents a new method of assessing atmospheric pollution in large cities with emphasis in areas in the vicinity of large international airports using satellite remote sensing technology. The method involves the use of suitable pseudo-invariant targets as a tool for determining the aerosol optical thickness from satellite remotely sensed data. The main sources of formation of aerosols (particulates) come from emissions of gasses mainly due to man-made and natural sources (Department of the Environment, 1996), therefore the use of aerosol optical thickness was found in the literature to be a valuable parameter to assess air pollution (for example, Sifakis and Deschamps, 1992; Retalis, 1998). The determined aerosol optical thickness is therefore used for assessing atmospheric pollution from satellite imagery. The proposed method was initially applied to three Landsat-5 TM images of London Heathrow Airport area (UK) and has now also been applied in the Hellinikon Airport area (Greece).

2. Methodology

The basic philosophy of the proposed method is to (a) identify suitable pseudo-invariant targets (PITs) in the satellite images which can serve as the suitable dark targets for the determination of the aerosol optical thickness; (b) set up the key assumptions for solving the RT equation as described by Forster (1984) and Hill (1993); and (c) determine the aerosol optical thickness from Landsat TM band 1 image.

2.1. SELECTION OF SUITABLE PSEUDO-INVARIANT TARGETS

Pseudo-invariant targets may be man-made or water features, such as concrete, asphalt, car parks and roads, buildings or lakes, whose reflectance do not change spectrally from image to image (Schott *et al.*, 1988; Berger, 1989; Caselles and Garcia, 1989; Hill and Sturm, 1993; Hadjimitsis *et al.*, 1999). Concrete airport runways/aprons (bright objects) and water storage reservoirs (dark objects) were found to be suitable targets because of their suitability in terms of size, uniformity and reflectance characteristics (Hadjimitsis *et al.*, 1999). Such features are common in many geographical areas. Spectral reflectance values can be found either from the literature or from ground measurements campaigns (Arenz *et al.*, 1996, Hadjimitsis, 1999; Hadjimitsis *et al.*, 2000).

The most suitable PIT in the scene should be a large water body, preferably inland waters because of their suitability in terms of size and reflectance characteristics (Hadjimitsis, 1999; Hadjimitsis *et al.*, 1999; Hadjimitsis *et al.*, 2000). Large eutrophic inland waters (such as reservoirs, lakes and dams) due to the high turbidity levels, significant populations of algae and of their high extinction coefficients makes them suitable for dark targets since bottom effects are eliminated (Arenz *et al.*, 1996; Hadjimitsis, 1999; Hadjimitsis *et al.*, 1999; Hadjimitsis *et al.*, 2000; Retalis *et al.*, 2000).

2.2. DETERMINATION OF AEROSOL OPTICAL THICKNESS

The proposed method is based on the following key assumptions: (a) the reflectance of the PIT does not vary between the images and the reflectance at ground level can be obtained either from ground measurements or from the literature (i.e. 'standard values'); (b) the effects of multiple scattering are neglected; (c) the images are acquired over a short interval time (Hadjimitsis, 1999; Retalis, 1998; Retalis *et al.*, 1999); and (d) the relative humidity does not have a significant effect on aerosols.

Based on the above assumptions, the satellite images of the investigated area should be geometric corrected and then converted to radiance values as a part of the pre-processing steps of image processing (Mather, 1999). In this case, Landsat-5 TM images were geometric corrected using ground control points (GCP's) as described by Mather (1999). Updated Landsat-5 TM calibration constants were used to convert the digital numbers to the at-satellite radiance (Hill, 1993). The Landsat TM band 1 (0.45–0.52 μm) images were used since the absorption was assumed to be minimum and therefore the effect of ozone and water vapour absorption was considered negligible (Forster, 1984).

Large PIT's identified in the scene, based on available records of their spectral signature, were used as controllable targets. The scene-measured (at-satellite) radiance of the PIT was determined. Ozone transmittance (t_{O3}) and water vapour transmittance (t_{H20}) are equal to 1 since their associated absorption is considered negligible in 0.45–0.52 μm region. The assumed value for the selected PIT (e.g. inland water bodies) was taken from in-situ measurements or from the literature. By applying the atmospheric correction mathematical approach and solving the radiative transfer equation as shown by Hadjimitsis (1999), the aerosol optical thickness could be found for every image date. Figure 1 summarises the proposed methodology in steps.

3. Results and Discussion

The proposed method was initially applied to three multi-temporal Landsat-5 TM images of London Heathrow Airport area (UK) and was then applied to two Landsat-5 TM images of Hellinikon Airport area (Athens, Greece).

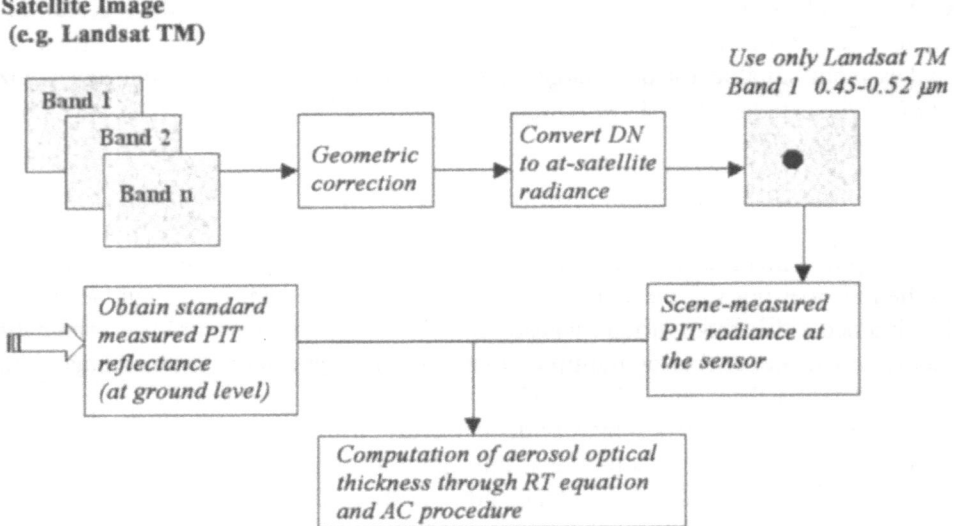

Figure 1. A brief description of the proposed methodology in steps (AC = atmospheric correction, RT = radiative transfer, PIT = pseudo-invariant target).

3.1. ASSESSMENT OF ATMOSPHERIC POLLUTION FOR THE HEATHROW AIRPORT AREA (UK)

The proposed approach was applied to Landsat-5 TM band 1 (0.45–0.52 μm) images of the London Heathrow area (see Figure 2) acquired on May 17th, June 2nd and July 4th, 1985. The images under investigation had approximately the same solar zenith angles and were acquired over a short interval time.

London Heathrow Airport is situated close to the London orbital motorway, the M25, which is one of the most heavily used motorways in Europe and which contributes significantly to the air pollution levels in the area (Stebbings *et al.*, 1999). Large airports such as the Heathrow airport provide a number of different emission sources. The principal source is aircraft and secondary sources are ground support equipment, aircraft servicing vehicles etc. From the emissions of NOx and PM_{10} for the range of sources (for example, airport road vehicles, auxiliary power units, engine testing, aircraft pollutants) found by Stebbings *et al.* (1999), aircraft emission was a significant polluting parameter in the area of Heathrow Airport. Large eutrophic reservoirs, the Lower Thames Valley reservoirs, which are located near the Heathrow Airport (see Figure 2), were used as suitable pseudo-invariant dark targets for determining the aerosol optical thickness.

The 'standard' value of the reservoir's reflectance at ground level was obtained from a series of ground measurements acquired in these reservoirs and from other inland waters such as dams in different geographical areas using a GER-1500 field spectro-radiometer (Hadjimitsis, 1999). Based on the fact that the 'darkest pixel'

Figure 2. Partial scene: Landsat-5 TM image of Heathrow Airport area (UK) acquired on June 2nd, 1985.

TABLE I

Calculated aerosol optical thickness for the three Landsat-5 TM band 1 images of Heathrow Airport area (Hadjimitsis, 1999)

Image date	Solar zenith angle	Visibility (km)	Calculated aerosol optical thickness (τ_α)
17/05/1985	37.02°	13.2	0.58
02/06/1985	34.60°	26.2	0.13
04/07/1985	34.92°	7.5	0.76

should be the most black target i.e. represent the lowest reflectance values, the minimum values from the range of ground values found by Hadjimitsis (1999) was selected. This gave a reflectance of 3% in TM band 1 (0.45–0.52 μm). The aerosol optical thickness was determined for each image such as shown in Table I.

There is evidence that visibility is related to the aerosol optical thickness, as shown from Forster (1984). Therefore, the available visibility values (see Table I) recorded at the Heathrow Meteorological station during the satellite overpass can be used as a mean of judge the determined aerosol optical thickness. By relating the determined aerosol optical thickness with the visibility values shown in Table

I, a linear regression was fitted with a correlation coefficient $r^2 = 1.00$ (Hadjimitsis, 1999). The observed significance for the regression model was $0.01 < 0.05$. From Table I, it is apparent that for the image acquired on 04/07/1985, the aerosol optical thickness was significantly increased.

By examining the available meteorological data from the Heathrow Meteorological Station for the months May, June and July, it was found that the highest visibility values were occurred during the first two weeks of June (Hadjimitsis, 1999) and the lowest visibility values were occurred between the following periods: 10/05/1985–25/05/1985 and 18/06/1985–10/07/1985. Therefore, the determined aerosol optical thickness data found at the satellite overpasses, support this finding as shown from the prevailing visibility values for every image data.

By applying the proposed method in two sets of image data acquired on 1985 and 1986 (March, June), it was found the determined aerosol optical thickness for the images acquired in 1986 were higher than those acquired on 1985 (Hadjimitsis, 1999). Indeed, Munday (1990) provide analytical tables, which explain the emissions of pollutants by source of pollution for the years 1978–1998 in the United Kingdom. From these data, it was found that during the above months and especially during the whole year in 1986 the atmosphere was more polluted than the atmospheres in 1985 and 1984 due to the occurrence of high values of CO_2 aircraft emissions.

3.2. ASSESSMENT OF ATMOSPHERIC POLLUTION FOR THE ATHENS AREA (GREECE)

Two Landsat-5 TM band 1 images of Athens area (Figure 3) acquired on August 15th and 31st 1988 were used (Retalis, 1998; Retalis *et al.*, 1999). Two cases have been investigated using two different PITs; firstly, using Marathon Lake (max. depth: 54 m, eutrophic) as a PIT in order to examine the atmospheric pollution near Pendeli; and secondly, using concrete aprons of Hellinikon Airport as PIT's in order to examine the atmospheric pollution in the surroundings of the airport.

In the first case, the 'standard' value of the ground reflectance for the Marathon Lake was taken as 3% in the Landsat TM band 1 (Hadjimitsis, 1999). Marathon Lake was characterised as eutrophic inland water based on available historical water quality databases (Hadjimitsis *et al.*, 1999). Therefore, the 3% reflectance value in Landsat TM band 1 was sufficient. For the second case, an average value of 14% in the Landsat TM band 1 was used as the 'standard' value of the Hellinikon Airport concrete aprons (Hadjimitsis *et al.*, 1999).

For both cases, calculation of the aerosol optical thickness (τ_α) for the Landsat-5 TM-1 images of Athens areas (Hadjimitsis, 1999) showed a heavily polluted atmosphere occurring on August 31st 1988. For example, it was found that for the images acquired on August 15th and 31st, in both cases, the maximum determined aerosol optical thickness was 0.40 and 0.80, respectively, in Athens region (including Hellinikon Airport and Pendeli areas). Using Tatoi Airport (see Figure

Figure 3. Landsat-5 TM image of Athens (Greece) acquired on August 15th, 1988.

3) as another example of a bright PIT, the same values of aerosol optical thickness were found for both images. The prevailing meteorological conditions, along with the analysis of the corresponding radio soundings (see Table II), indicated a strong inversion on August 31st, conditions supportive for the development of an atmospheric pollution episode.

Moreover using the technique developed by Retalis *et al.* (1999) for assessing the distribution of aerosols in the area of Athens with the use of Landsat TM band 1 images, it was found that the overall maximum difference in optical thickness between these two images was $\Delta \tau_\alpha = 0.35$ (see Figure 4). With the application of the proposed method a close optical thickness difference of $\Delta \tau_\alpha = 0.40$ agrees with the one found by Retalis *et al.* (1999). The proposed methodology was also applied

TABLE II

(a) Air pollution measurements and (b) prevailing meteorological condition on 15th and 31st
August 1988 in Athens (Retalis, 1998)

a. Maximum values of air pollution measurements					
Image date	CO (mg m^{-3})	NO$_2$ (μg m^{-3})	O$_3$ (μg m^{-3})	SO$_2$ (μg m^{-3})	Smoke (μg m^{-3})
15/08/1988	5.1	153	172.1	59	42
31/08/1988	15.6	561	276	183	241

b. Meteorological data				
Image date	T ($^{\circ}$C)	R. H. (%)	Wind velocity (m.s^{-1})	Wind direction
15/08/1988	29.7	38.6	5.3	N-NE
31/08/1988	28.0	50.6	1.7	W-SW

in another set of images for the same area acquired on 26/04/1994 and 13/06/1994.
By comparing the determined aerosol optical thickness with those obtained from
Retalis *et al.* (1999), again a close value was found. The results shown by Retalis
(1998) provide evidence to support our findings. Indeed, Retalis (1998) found that
the aerosol optical thickness distribution in the area of Athens was in close relation
to the ground air pollution measurements not only for the above scenes acquired
on August 1988, but also for the two additional scenes acquired on 26/04/1994 and
13/06/1994.

4. Conclusions

The proposed methodology allows the assessment of atmospheric pollution using Landsat TM band 1 satellite remotely sensed images acquired over a short
interval of time. It has been shown that determined aerosol optical thickness in
the Heathrow area is well correlated to the visibility data acquired at the time
of satellite overpass. Furthermore for the Athens region, the determined aerosol
optical thickness agrees with the prevailing air pollution data.

The proposed method has a number of advantages such as: ease of computation;
such very dark or/and very bright targets are common in many geographical areas;
the method is an image-based technique since the assumed dark or bright target
reflectance value can be found from the literature or other in-situ campaigns; the
methodology is a useful procedure for determining the aerosol optical thickness as
a part of any atmospheric correction algorithm.

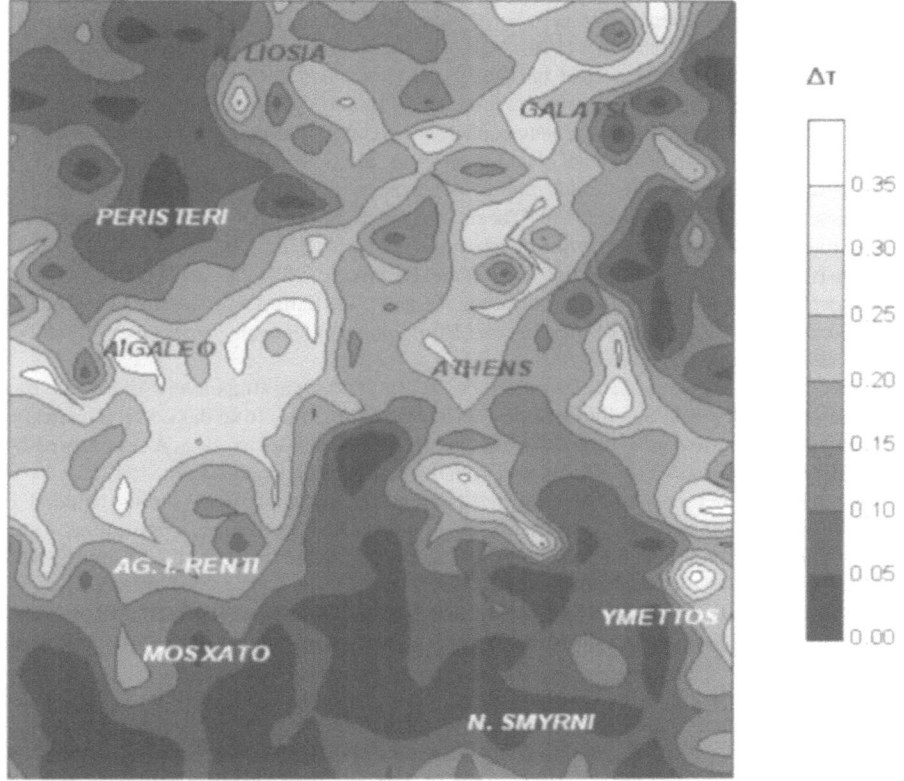

Figure 4. Isopleths indicating differences in optical thickness ($\Delta\tau$) for Athens on August 31st, 1988.

Acknowledgements

The authors acknowledge the support of the NERC Equipment Pool for Field Spectroscopy (EPFS) (University of Southampton) for the provision of the GER1500 field spectro-radiometer (EPFS loan reference: 267.1196). Thanks are also due to Thames Water Utilities Ltd (UK), which provides all the resources during the ground measurements campaign.

References

Arenz, R. F., Lewis, W. M. and Saunders, J. F.: 1996, 'Determination of chlorophyll and dissolved organic carbon from reflectance data for Colorado reservoirs', *Int. J. Rem. Sens.* **17**(8), 1547–1566.

Berger, H. F.: 1989, 'Multi-Temporal Analysis of Forest Areas in the Surroundings of Innsbruck, Austria', *Proceedings of a Workshop on Earthnet Pilot Project on Landsat Thematic Mapper Applications*, in T. D. Guyenne and G. Calabresi (eds), 'A Pilot Project Campaign on Landsat Thematic Mapper Applications', ESA SP-1102, pp. 293–298.

Caselles, V. and Garcia, M. J. L: 1989, 'An alternative simple approach to estimate atmospheric correction in multi-temporal studies', *Int. J. Rem. Sens.* **10**(6), 1127–1134.

Department of the Environment: 1996, 'Airborne Particulate Matter in the United Kingdom, December 1996, Third Report of the Quality of Urban Air Review Group, The University of Birmingham, Institute of Public and Environmental Health, School of Biological Sciences.

Forster, B. C.: 1984, 'Derivation of atmospheric correction procedures for Landsat MSS with particular reference to urban data', *Int. J. Rem. Sens.* **5**(5), 799–817.

Hadjimitsis, D. G.: 1999, 'The Application of Atmospheric Correction Algorithms in the Satellite Remote Sensing of Reservoirs', PhD Thesis, University of Surrey, School of Engineering in the Environment, Department of Civil Engineering, Guildford, UK.

Hadjimitsis, D. G., Hope, V. S., Clayton, C. R. I. and Retalis, A.: 1999, 'A New Method of Removing Atmospheric Effects using Pseudo-Invariant Targets', in 'Earth Observations from Data', *Proceedings of the 25th Annual Conference and Exhibition of the Remote Sensing Society*, University of Wales and Swansea, pp. 633–641.

Hadjimitsis, D. G., Clayton, C. R. I. and Hope, V. S.: 2000, 'The Importance of Accounting for Atmospheric Effects in Satellite Remote Sensing: A Case Study from the Lower Thames Valley Area-UK', *Space and Robotics 2000*, February 2000, New Mexico, U.S.A., sponsored by the ASCE and co-sponsored by NASA and SANDIA National Laboratories, pp. 194–201.

Hill, J.: 1993, 'High Precision Land Cover Mapping and Inventory with Multi-Temporal Earth Observation Satellite Data', PhD Thesis, Institute for Remote Sensing applications, Environmental mapping and modeling unit, Joint Research Centre, Published by the Commission of the European Communities, Luxembourg.

Hill, J. and Sturm, B.: 1993, 'Radiometric correction of multi-temporal Thematic Mapper data for use in agricultural land cover classification and vegetation monitoring', *Int. J. Rem. Sens.* **12**(7), 1471–1491.

Kaufman, Y. J., Fraser, R. S. and Ferrare, R. A.: 1990, 'Satellite measurements of large-scale air pollution: methods', *J. Geophys. Res.* **95**, 9895–9909.

Mather, P.: 1999, *Computer Processing of Remotely-Sensed Images*, John Wiley and Sons, Chichester, UK.

Munday, P. K: 1990, *UK Emissions of air pollutants 1970–1988*, Warren Spring Laboratory Report No. LR 764 (AP).

Retalis, A.: 1998, 'Study of Atmospheric Pollution in Large Cities with the Use of Satellite Observations: Development of an Atmospheric Correction Algorithm Applied to Polluted Urban Areas', PhD Thesis, Department of Applied Physics, University of Athens.

Retalis, A., Cartalis, C. and Athanassiou, E.: 1999, 'Assessment of the distribution of aerosols in the area of Athens with the use of Landsat TM', *Int. J. Rem. Sens.* **20**(5), 939–945.

Retalis, A., Hadjimitsis, D. G. and Clayton, C. R. I.: 2000, 'The Use of Large Eutrophic Inland Water Bodies as Suitable Dark Targets in the Assessment of Atmospheric Pollution Using Satellite Remotely Sensed Imagery', *5th Pan Hellenic Scientific Conference on Meteorology-Climatology and Atmospheric Physics*, 28–29 September 2000, Thessaloniki (proceedings in progress).

Schott, J. R., Salvaggio, C. and Wolchock, W. J.: 1988, 'Radiometric scene normalization using pseudo-invariant features', *Rem. Sens. Environ.* **26**, 1–16.

Sifakis, N. and Deschamps, P. Y.: 1992, 'Mapping of air pollution using SPOT satellite data', *Photogr. Engin. Rem. Sens.* **58**, 1433–1437.

Stebbings, T., Simms, K. L. and Crimes, S.: 1999, 'Air Quality Management for a Large International Airport the Experience at Heathrow', *Proceedings of 2nd International Conference*, Urban air quality, Computer Science School of the Technical University of Madrid, pp. 81–82.

Wald, L. and Balleynaud, J. M.: 1999, 'Observing air quality over the city of Nantes by means of Landsat thermal infrared data', *Int. J. Rem. Sens.* **20**(5), 947–959.

Wald, L., Basly, L. and Balleynaud, J. M.: 1999, 'Satellite Data for the Air Pollution Mapping', *Proceedings of the 18th EARseL Symposium on Operational Sensing for Sustainable Development* (Enschede, the Netherlands, 11–14 May 1998), in G. J. A. Nieeuwenhuis, R. A, Vaugham and M. Molenaar (eds), 'Operational Remote Sensing for Sustainable Development', pp. 133–139.

ICAROS: AN INTEGRATED COMPUTATIONAL ENVIRONMENT FOR THE ASSIMILATION OF ENVIRONMENTAL DATA AND MODELS FOR URBAN AND REGIONAL AIR QUALITY

D. A. SARIGIANNIS[1]*, N. SOULAKELLIS[2], K. SCHÄFER[3], M. TOMBROU[4], N. I. SIFAKIS[5], D. ASSIMAKOPOULOS[4], M. LOINTIER[6], A. DANTOU[4] and M. SAISANA[1]

[1] *Institute for Health and Consumer Protection, European Commission – DG JRC, B-1050 Brussels, Belgium;* [2] *Department of Geography, University of the Aegean, Mytilene, Greece;* [3] *Fraunhofer Institute for Atmospheric Environmental Research, Garmisch-Partenkirchen, Germany;* [4] *Department of Physics, University of Athens, Greece;* [5] *Institute for Space Application and Remote Sensing, National Observatory of Athens, Greece;* [6] *Institut de Recherche pour le Développement, Montpellier, France*
(* *author for correspondence, e-mail: dimosthenis.sarigiannis@cec.eu.int*)

Abstract. Integrated environmental management in urban areas is nowadays considered a sine qua non objective of Community and national environmental and development policies. A large amount of scientific information on the state of the environment is now available from a large pool of data sources. This work presents an innovative method for integration of these data sources and effective coupling of environmental information with appropriate models and decision-support tools. State-of-the-art Earth observation techniques, ground-based air quality measurements, atmospheric transport and chemical modelling, and multi-criteria decision-aid systems are used in an integrated information fusion environment in support of environmental and health impact assessment and decision-making at the urban and regional scales. Results of the pilot application of the method in the area of Lombardy in Northern Italy demonstrate the validity and usefulness of this novel approach.

Keywords: aerosol optical thickness, air quality, environmental management, information fusion

1. Introduction

Environmental systems and, consequently, environmental management problems have several attributes that render their formal representation particular and the proposed solutions multidisciplinary. Some of these distinctive features are highlighted below:

- Dynamics. Environmental systems evolve with time and many of these have memory (for instance, accumulation of toxic pollutants with biocumulative effects).
- Spatial coverage. Modelling environmental processes rigorously needs linking spatially referenced variables with time referenced ones. Data are stored in spatial databases often via a geographic information system (GIS) to assist in spatial analysis.

Water, Air, and Soil Pollution: Focus **2**: 631–640, 2002.
© 2002 *Kluwer Academic Publishers.*

- Complexity. Environmental systems are complex, involving interactions between physical-chemical and biological processes with social ones. Models of such systems require multi-disciplinary approaches, the success of which depends heavily on the establishment of common grounds for communication among different scientific disciplines and various classes of stakeholders.
- Randomness. Many environmental processes are stochastic. Parameter uncertainty characterises the models representing them.
- Periodicity. Many environmental processes are periodic in time, adding thus a further degree of complexity in parameter calibration and validation.
- Heterogeneity and scale. Processes in different media may have quite varying characteristic time and space scales.
- Paucity of information. Observational data on environmental systems, particularly on internal states, typically suffice only for the characterisation of simply parameterised models.

The users of environmental management decision-support systems vary, including scientists, managers (decision-makers) and concerned stakeholders. Each of the users categories have different needs, knowledge acquisition and communication standards, and final objectives.

Integrated environmental management requires simultaneously considering many information classes, including environmental quality data, impact pathway models, economic analyses, the respective regulatory framework, and the priorities of the stakeholders involved in the decision-making process or affected by it. Assimilating this information in a comprehensive, yet functional framework is needed for the development of user-friendly decision support tools, which foster public dialogue and socially compatible environmental problem solving (Sarigiannis, 1999).

The main feature of this work is the development of a state of the art methodology and of the respective tools for integrating the variety of information classes affecting decision-making on environmental problems and related health concerns.

The purpose of the work presented herein was the assimilation of various data sources (in situ measurements, Earth observation and simulation modelling) to develop links between these three basic data types through an open-ended integrated computational environment for improved air quality assessment and monitoring. In this way various environmental information sources are integrated into a unique environmental information processing tool (ICAROS), which could be used for pollution monitoring, extreme incident forecasting and strategic environmental assessment in the urban environment (Sarigiannis et al,. 1998).

The ICAROS computational environment is mainly addressed to environmental and transport policy- and decision-makers at the urban and regional levels. Decisions in the area of environmental protection in general, and in particular in air pollution control and abatement need to take into account information ranging from validated technical data to economic and social factors. Technical information on the status of the environment needs however to be processed before it can be used in order to integrate in the most efficient manner the array of different data types characterising environmental systems. In the case of air pollution control and abatement this information includes data on the causes of pollution, on atmospheric transport processes, and on the effect pollution has on the affected ecosystem and human health.

The user requirements for this information include:

a) dynamic character of the data used as input to the evaluation of the actual state of pollution;

b) validated detailed spatial distribution of pollution during its generation and transport;

c) timeliness in providing the necessary input to the decision-makers;

d) accuracy in the amplitude and spatial-temporal profile of pollution and its effects on humans and the ecosystem;

e) objectivity in the assessment of the sources, distribution and impacts of pollution (especially when environmental treaties and relevant legislation at the national or Community level are implemented)

Earth observation can help reduce the errors inherent to dynamic atmospheric transport models through the provision of high-resolution satellite-derived environmental and ancillary information (Ferrare *et al.*, 1990; Sifakis and Deschamps, 1992). Furthermore, EO-derived information can be used for the verification of the degree of implementation of international environmental treaties (e.g. LTRAP) and Community directives by signatory parties and member states. Air quality monitoring through ground-based networks can be optimised with regard to spatial configuration and typology through the assistance of EO-derived information on pollution loading. Through a knowledge base using EO-derived data and the appropriate inference engine the necessary input to the users community can be provided timely and economically with reduced uncertainty. In addition, EO-derived data can serve as an objective information source in order to streamline and optimise the spatial configuration of ground-based monitoring networks. The information derived from the satellite signal relates in particular to pollutants identified as key threats to human health, such as very fine particulate matter and nitrous oxides, or principal acidification factors, such as sulphates.

2. Methodology

EO data were used directly as input information into the ICAROS database, and indirectly by providing the inputs used to form ancillary information layers. Data that could be derived from EO images were distinguished into the four stages of the pollution cycle, namely, emission, transport, ambient air concentration and deposition.

a) First, concerning emissions, the main input of satellite data was in terms of land cover/use databases that provide information on the localisation of the potential sources. The European inventory CORINE Land Cover, that describes the spatial distribution of items relevant to different emission levels into the air, was used for this purpose. Additional land cover classes to describe the particularities of topical environment were devised as appropriate.

b) Second, with regard to tracking transported pollution, simulation models received a spatially resolved benchmarking by the results of this process. Satellites can provide snapshots of plumes and can depict the dispersion patterns and real pathway (versus the estimated by the simulation) of particulate tracers of pollution (Sifakis *et al.*, 1998). MSR sensors can occasionally allow a more õdynamicŕ follow up (2–4 images/day) of long-range transport of plumes but with much coarser spatial resolution.

c) Third, concerning air pollutant concentrations in the troposphere, information was retrieved on the atmospheric loading of pollution in terms of columnar particulate optical thickness, which corresponds to particles correlated, in many cases, with ammonium nitrates and sulphates. Integrating this information in the ICAROS platform in a geographical information system (GIS) allowed tracking particulate concentrations, with a resolution of 450–600 m (i.e. equivalent to circa 300 sampling points for a 100 km^2 area). It is noteworthy that the calculated total (capital and operational) cost of the method described here in an area of 50–100 km^2 ranges from three to ten times lower than the cost of a ground-based monitoring network with the same spatial resolution.

d) Fourth, concerning deposition, indicators of the degradation of the health state of natural ecosystems were evaluated by HSR by applying the Normalised Difference Vegetation Index. This information, coupled with aerosol optical thickness maps supported zoning for environmental protection and management in Lombardy (Sarigiannis and van Wijk, 2000).

Atmospheric turbidity is indicative of the particulate matter loading in the atmosphere. It does not provide, however, information on the chemical nature of the substances present in atmospheric aerosol. Still, particulate matter is a significant tracer for pollutant dispersion, and turbidity is an appropriate pollution indicator due to its strong correlation with the concentration of small particles (Sloane, 1983; Seinfeld and Pandis, 1998). The relevance of secondary aerosols to air quality is well established. Airborne particulate matter is known to have adverse impacts on human health, materials and visibility. Among atmospheric aerosols of all sizes, particles in the accumulation mode (i.e. particles with diameter in the range of 0.1 μm $\leq d_p \leq 2$ μm) are the most critical ones with respect to health (Lipfert, 1994; Pope *et al.*, 1995), visibility (Pilinis, 1989), and adverse effects caused by acid precipitation. Several surveys of the current state of knowledge of atmospheric aerosols indicate that most of the secondary aerosol mass, i.e. particles formed by gas-to-particle conversion processes in the atmosphere, falls within this range. Although numerous chemical species have been identified in secondary aerosols (Varela, 1994), the prevalent ones are sulphates, nitrates, ammonium and water (Seinfeld and Pandis, 1998). Therefore, the formation of secondary aerosol from gaseous H_2SO_4, HNO_3, NH_3 and H_2O has been the subject of much theoretical (Pham *et al.*, 1995) and experimental investigation. Mathematical models have been developed to study aerosol formation in plumes and in urban areas. These atmospheric Gas-Aerosol Equilibrium Models calculate the partitioning of the total sulfate, nitrate, ammonium and water in an air parcel assuming that the system is in thermodynamic equilibrium (minimum Gibbs free energy of the system) (Pilinis and Seinfeld, 1987). Knowledge of temperature, relative humidity and gaseous ammonia concentration is assumed. In most ambient situations, it is expected that assumptions required for equilibrium to hold are valid since the characteristic time for mass transfer to and from an aerosol particle is of the order of a fraction of a second (Pilinis *et al.*, 1995; Schwartz, 1996).

The flowchart of the ICAROS information fusion algorithm is given in Figure 1. Earth observation (EO) data provide direct information on atmospheric turbidity, as measured by the optical thickness of the aerosol layer in the atmosphere. Furthermore, EO would give information on land use/cover including seasonal changes (e.g. snow coverage and surface albedo) and the landscape of the area of interest through the creation of digital elevation models (DEM). The profiles of aerosol optical thickness derived from EO refer to the total atmospheric column. It is reasonable to assume, however, that the majority of the pollutants of interest for air quality assessment remain within an atmospheric layer that spans from the ground till the mixing height in the atmosphere. Mixing height is calculated from meteorological data, based either on observation data or on meteorological models such as MM5. This information is then used to correct the optical thickness values derived from EO image processing (Fraser *et al.*, 1984). This is based on the assumption that almost all the tropospheric aerosol affecting adversely human health is withheld within the well-mixed layer of the lower atmosphere. The optical thickness of tropospheric aerosol (AOT) is calculated from the signal captured by the satellite as an integral over the total height of the atmosphere (from the surface of the Earth to the satellite sensor). Hence, by dividing the value of AOT calculated by the satellite signal with the mixing layer height, AOT can be corrected to reflect the relative importance of lower atmospheric aerosol, i.e. the one of greatest relevance vis-à-vis pollution loading and adverse health effects.

Air quality measurements from fixed monitoring networks and/or from ground-based experimental campaigns are stored in an air quality database. The monitoring network data are crosschecked and validated with ad hoc field campaigns. These campaigns are designed so as to: (a) fill the information gaps in the existing monitoring networks, and (b) validate and improve the quality of the historical data, i.e. data retrieved through the existing monitoring stations over time. These automatically retrieved data serve as input to a chemical model (ARES) used for the transformation of primary pollutants such as NO_x and SO_2 into secondary aerosol. The sulphate/nitrate/ammonium system is, of course, an idealised one for representing actual ambient aerosols. The presence of other species such as those in the gaseous HCl-particulate chloride system can be expected to alter the sulph-

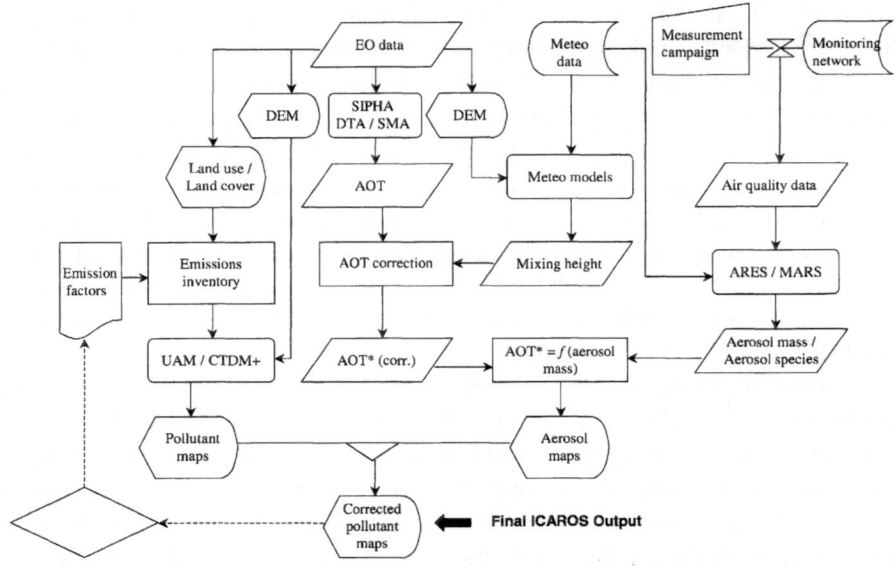

Figure 1. ICAROS information fusion algorithm.

ate/nitrate equilibriums. Also, the existence of a stable organic (surfactant) film on aerosol particles would reduce the rate of mass transfer between the gas and liquid phases and might even affect the position of equilibrium through solvation and solution activity effects. Similar comments apply in the case of the existence of a solid elemental carbon or mixed elemental-organic carbon core. Nevertheless, the sulphate/nitrate/ammonium system equilibrium is of key importance to analysing ambient data. The chemical equilibrium model predicts the quantity and composition of secondary atmospheric aerosols containing sulphate, nitrate and ammonium compounds. A thermodynamic approach is adopted to predict the chemical composition of multiphase aerosols containing water, $(NH_4)_2SO_4$, NH_4HSO_4, $(NH_4)_3H(SO_4)_2$, H_2SO_4, HNO_3, and NH_4NO_3. Table I (a and b) describes the way the various chemical species are treated in ARES in the gas, aqueous and solid phases. The presence and amounts of these species in a particular aerosol/gas system are dictated by the total concentrations of H_2SO_4, HNO_3, NH_3, and relative humidity.

In order to predict the quantity and composition of such an aerosol system, a gas-phase chemical model that estimates the concentrations of formation of nitric acid and sulphuric acid from SO_2 and NO_2 precursors must be coupled to the equilibrium description of the aerosol. The atmospheric reactions leading to the production of these precursors are delineated in Figure 2.

Through this process and measurements of PM_{10} and $PM_{2.5}$, the amount of atmospheric aerosol and the chemical species comprising it are calculated. Through a

TABLE I

Treatment of the aerosol chemical composition in the ARES model

(a) Ammonia-rich environment

Relative humidity (%)	Species
100 ↕ 80	$(NH_4)_2SO_{4(aq)}$ $NH_4NO_{3(aq)}$ $H_2O_{(l)}$
80 ↕ 62	$(NH_4)_2SO_{4(s)}$ $(NH_4)_2SO_{4(aq)}$ $NH_4NO_{3(aq)}$ $H_2O_{(l)}$
62 ↕ 0	$(NH_4)_2SO_{4(s)}$ $NH_4NO_{3(s)}$

(b) Ammonia-deficient environment

Relative humidity %	Species — Ratio of total ammonia to total sulphate		
	0 ←→ 1	←→ 1.5	←→ 2
100 ↕ 80	$NH_4HSO_{4(aq)}$ $H_2SO_{4(aq)}$ $NH_4NO_{3(aq)}$ $HNO_{3(aq)}$ $H_2O_{(l)}$	$(NH_4)_2SO_{4(aq)}$ $H_2SO_{4(aq)}$ $NH_4NO_{3(aq)}$ $HNO_{3(aq)}$ $H_2O_{(l)}$	
80 ↕ 69		$(NH_4)_2SO_{4(s)}$ $(NH_4)_3H(SO_4)_{2(aq)}$ $NH_4HSO_{4(aq)}$ $NH_4NO_{3(aq)}$ $HNO_{3(aq)}$ $H_2O_{(l)}$	$(NH_4)_2SO_{4(s)}$ $(NH_4)_3H(SO_4)_{2(aq)}$ $NH_4NO_{3(aq)}$ $HNO_{3(aq)}$ $H_2O_{(l)}$
69 ↕ 40		$(NH_4)_3H(SO_4)_{2(s)}$ $NH_4HSO_{4(aq)}$ $NH_4NO_{3(aq)}$ $HNO_{3(aq)}$ $H_2O_{(l)}$	$(NH_4)_2SO_{4(s)}$ $(NH_4)_3H(SO_4)_{2(s)}$
40 ↕ 0	$NH_4HSO_{4(s)}$ $H_2SO_{4(aq)}$ $HNO_{3(aq)}$ $H_2O_{(l)}$	$(NH_4)_3H(SO_4)_{2(s)}$ $NH_4HSO_{4(s)}$	

$SO_2 + OH \xrightarrow{M,k_1} HOSO_2$ $k_1 = 9 \cdot 10^{-13}$ cm^3/(mole·s) (at 298 K)
$HOSO_2 + O_2 \rightarrow HO_2 + SO_3$ (rapid)
$SO_3 + H_2O_g \rightarrow H_2SO_4$ (rapid)
$NO_2 + OH \xrightarrow{M,k_2} HNO_3$ $k_2 = 1.1 \cdot 10^{-11}$ cm^3/(mole·s) (at 300 K)

The production of $HOSO_2$ and HNO_3 in moles/cm^3 within an hour (for ex. from 7:00 to 8:00) will thus be given by:

$$[HOSO_2]_8 = \int_0^t k_1 \cdot [OH^-][SO_2]_t \, dt = k_1[OH^-]\left\{[SO_2]_8 t - \frac{1}{2}k_1[OH^-][SO_2]_8 t^2\right\}$$

$$[HNO_3]_8 = \int_0^t k_2 \cdot [OH^-][NO_2]_t \, dt = k_2[OH^-]\left\{[NO_2]_8 t - \frac{1}{2}k_2[OH^-][NO_2]_8 t^2\right\}$$

where t=3600 s

$[OH^-] \approx$ $\begin{cases} 5 \cdot 10^6 \text{ moles/cm}^3 \text{ for winter (daytime)} \\ 10 \cdot 10^6 \text{ moles/cm}^3 \text{ for summer (daytime)} \end{cases}$

$[SO_2]_8$: average ground-based concentration of SO_2 from 7:00 to 8:00 at 298 K in moles/cm^3
$[NO_2]_8$: average ground-based concentration of NO_2 from 7:00 to 8:00 at 298 K in moles/cm^3

From the above equations it can be deduced that $[HOSO_2]=[H_2SO_4]$; therefore

$$\mu g/m^3 \; SO_4^{2-} = (mole/cm^3 \; H_2SO_4) \cdot MW_{SO4} \cdot 10^{12}$$

Likewise:
$$\mu g/m^3 \; HNO_3 = (mole/cm^3 \; HNO_3) \cdot MW_{HNO3} \cdot 10^{12}$$

Figure 2. Gas-phase chemical model determining secondary aerosol speciation (formation of H_2SO_4 and HNO_3 from NO_2 and SO_2).

statistical model, the corrected aerosol optical thickness (EO-derived) is correlated to its mass (derived from ARES) and aerosol maps of the area of interest are drawn.

EO may support the application of atmospheric transport models for the evaluation and forecasting of air pollution incidents. EO-derived information on land use/cover can be the basis for the creation of emission inventories and for digital elevation maps. This is particularly important in areas where reliable emission inventories do not exist, or where they are out of date. Land use classes include, following the CORINAIR methodology of the European Environment Agency, industrial, commercial and residential (urban fabric) areas, agricultural and forest land, pastures, road and, more generally, transport networks, energy generation facilities. For each of these land use classes emission factors derived at the respective geographical scale are applied. The product of the intensity of the activity (unit of activity output/unit area) and the corresponding emission factor gives the emission intensity (emissions/unit area). High-resolution satellite information provides up-to-date estimates of the area covered by the main emitters. Multiplying emission intensity with the respective surface area it is possible to estimate the emissions of specific pollutants in each parcel of land.

A key issue in the calculation of the emissions inventory is the calculation of emission factors for the classified activities. In the case of Lombardy, the national emission factors estimated for Italy were adapted by the Regional Administration to reflect the technology and energy/fuel mix characterising the regional economy (Caserini, 2000). Applying the appropriate emission factors to the land use classification derived from the EO-based information, the emission inventory may be created. This serves as input to the appropriate atmospheric model. The output is time-dependent maps of the horizontal and vertical distribution of pollutants in the atmosphere. Additionally, secondary information such as pollutant deposition maps can be derived. Comparison of the air quality information stemming from the fusion of EO and ground data and the output of the atmospheric models gives corrected maps of surface concentration of pollutants over the whole domain of interest. Based on these corrections, the emission factors used to reckon the emission inventory could be appropriately calibrated. Once the emission inventory is calibrated, models of pollutant transport and transformation in the troposphere can be used to accurately predict extreme pollution incidents.

3.   Results and Discussion

The ICAROS model/data fusion methodology was applied in the extended area of the Po valley in northern Italy with a major focus on the urban area of Brescia. An indicative result of the EO-derived calculation of aerosol optical thickness at the urban region scale is given in Figure 3.

The significant advantage of the ICAROS method for Earth observation data assimilation to derive information on air quality is that it maintains a high level of spatial resolution (approximately 500 m) over large areas, while being able to zoom into the urban scale without loss of accuracy. Thus, the well-known adverse effects of scale-change in nested atmospheric models and the consequent diffusion of numerical error are avoided. Figure 4 shows the calculation of aerosol optical thickness over the city of Brescia. This representation highlights the level of detail that can be attained.

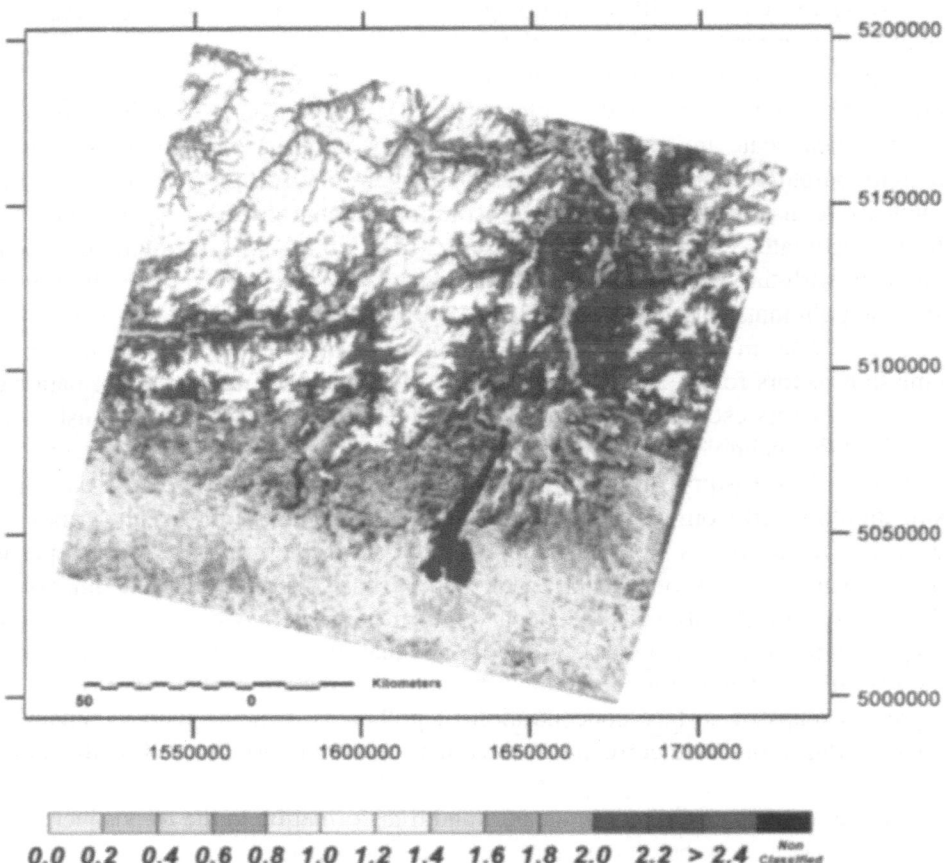

Figure 3. Aerosol optical thickness over the extended area of the Po valley in Lombardy. Landsat data recorded on 6/02/1998 at 9:40 U.T.

An array of different semi-empirical models were statistically investigated in order to best describe the physical-chemical links between aerosol optical thickness and the chemical composition of the aerosol mix in the lower atmosphere over the area of interest. The most significant ones and their correlation coefficients are shown herein. The prefix A in front of the various chemical species in the models below denotes the presence of these compounds in the aerosol phase. They are the main components of secondary atmospheric aerosol as calculated by the chemical model employed in this work.

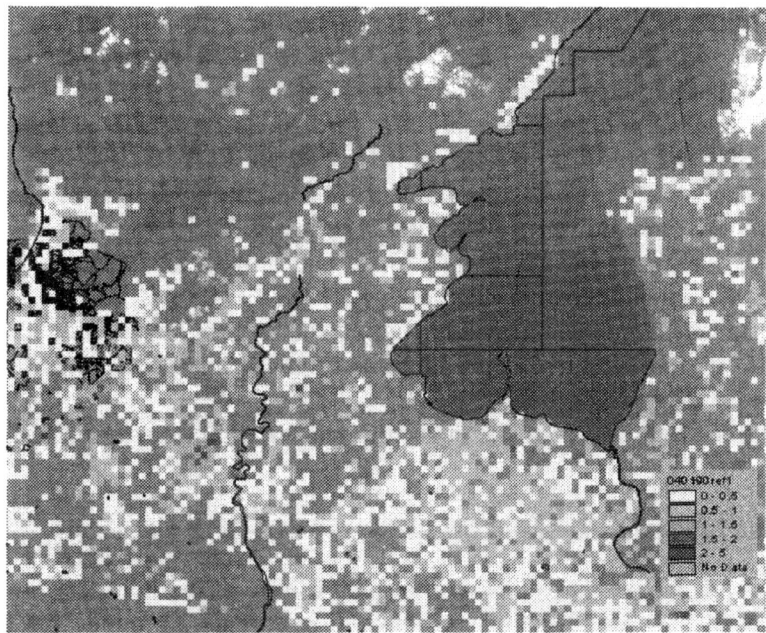

Figure 4. Aerosol optical thickness over the city of Brescia in Lombardy (data retrieved from SPOT on 4/1/1990).

Different Models investigated in the case of the central Po Plain

	R^2
1. $e^{\delta i} = a + b_1 ASO_4 + b_2 ANO_3 + b_3 ANH_4 + b_4 RH + b_5 NO_2$	0.76
2. $e^{\delta i} = a + b_1 ASO_4(1\text{-}RH)^{-1} + b_2 ANO_3(1\text{-}RH)^{-1}$	0.62
3. $e^{\delta i} = a + b_1 ASO_4(1\text{-}RH)^{-0.67} + b_2 ANO_3(1\text{-}RH)^{-0.67}$	0.65
4. $e^{\delta i} = a + b_1 ASO_4 + b_2 ANO_3 + b_3 ASO_4 RH^2 + b_4 ANO_3 RH^2$	0.65
5. $e^{\delta i} = a + b_1 ASO_4 + b_2 ANO_3 + b_3 ASO_4(1\text{-}RH)^{-1} + b_4 ANO_3(1\text{-}RH)^{-1} + b_5 NO_2$	0.79
6. $e^{\delta i} = a + b_1 ASO_4 + b_2 ANO_3 + b_3 ASO_4(1\text{-}RH)^{-1} + b_4 ANO_3(1\text{-}RH)^{-1} + b_5 NO_2, 0.4 < RH < 0.9$	0.88
7. $e^{\delta i} = a + b_1 ASO_4 + b_2 ANO_3 + b_3 ASO_4(1\text{-}RH)^{-1} + b_4 ANO_3(1\text{-}RH)^{-1} + b_5 NO_2, 0.6 < RH < 0.9$	0.94
8. $e^{\delta i} = a + b_1 ASO_4 + b_2 ANO_3 + b_3 ANH_4 + b_4 RH + b_5 NO_2, 0.6 < RH < 0.9$	0.86

The exponential form of all these models reflects well the expected theoretical relation between optical thickness (δ_i) and the concentration of the optically active compounds that constitute the secondary aerosol mass in the lower atmosphere. The limitations of model validity within a specific range of relative humidity values (see models 6 and 7 above) reflect the sub-set of air quality data used as input for the derivation of the respective model.

The most accurate model is number 7 above, with a correlation coefficient squared (R^2) of 0.94. This model encompasses both nitrates and sulphates present in the aerosol phase, as well as atmospheric vapour (as measured by relative ambient humidity).

An estimation of the total error of the approach followed here for the description of air pollution via use of optical thickness maps can be given from the following:

1. The algorithms calculating the aerosol optical thickness from the visible and near-infrared channels of the sensors on board SPOT and Landsat satellites have a standard error of the order of 5% (S1 = 0.05)
2. The chemical model for calculation of formation of sulphate and nitrate acids from SO_2 and NO_2 has an error of 2% (S2 = 0.02)
3. ARES model for the estimation of aerosol concentrations is a thermodynamic model; therefore no verification results are available.
4. Spatial interpolation via kriging estimates the temperature and absolute humidity with a standard error of the order of 5% (S3 = 0.05)
5. The statistical models correlating optical thickness values with the aerosol formed by the measured concentrations of gaseous pollutants has an error of the order of 4% (S4 = 0.04)

Thus, the overall error of the algorithm can be estimated from the following equation

$$Overall\ St.\ Error = 1 - \{(1 - (S1 + S2 + S3))^*(1 - S4)\} = 15.52\%$$

The results shown above demonstrate the validity of the ICAROS information fusion model. This conclusion is corroborated by extensive ground-truth data (obtained from circa 120 monitoring stations) in a large geographical area (156 × 100 km) over a ten-year period (the data covered both winter and summer seasons in the period from 1990 to 1999). The spatial resolution of the optical thickness data calculated by applying the ICAROS information fusion algorithm (the size of each pixel) is 600 × 600 m. This is higher than the best meso-scale air pollution transport models, which typically attain a spatial resolution of 1 × 1 km.

The results of this work have shown that fusion of satellite-derived data with output of atmospheric transport models and selected ground-based measurements provides a reliable, timely and cost-effective assessment of air quality at both the urban and regional scales.

4. Conclusions

The main conclusions of the work presented in this paper can be outlined as follows:

- Very encouraging statistical correlations between atmospheric model results, ground-based measurements and output of satellite image processing algorithms have been obtained. They indicate the good agreement between primary pollutant measurements on the ground and reckoned optical thickness of atmospheric aerosol, based on satellite-derived images. Further photochemical modelling of the secondary aerosol formation in the troposphere is required in order to best exploit the potential of Earth observation for air quality assessment.

- The integrated computational environment for air quality assessment envisaged in this project needs to follow an open-ended architecture. The system architecture is designed, it has been implemented in an object-oriented metalanguage (ILOG) and preliminary functionality tests have been done. Further refinements of the ICAROS software platform are currently pursued.
- The ICAROS computational environment is mainly addressed to environmental and transport policy- and decision-makers at the urban and regional levels. ICAROS has demonstrated that Earth observation can help reduce the errors inherent to dynamic atmospheric transport models by calibrating and validating them using high-resolution satellite information and the appropriate information processing algorithms. Furthermore, EO-derived information being a direct depiction of reality, it can be used for the verification of the degree of implementation of international environmental treaties (e.g. LTRAP) and Community directives by signatory parties and member states. Through a knowledge base using EO-derived data and the appropriate inference engine the necessary input to the users community can be provided timely and economically with reduced uncertainty.

Acknowledgements

This project was partially supported financially by the European Commission under grant ENV4-CT97-0417. The authors would like to thank Ing. Percesepe, Mr. Bissolati and Ing. Bonetti from ASM Brescia for the operation and data management of the ground-based air pollution monitoring network in Brescia, as well as Ing. Zabot, Giudici, Zanella and Volta of the Regional Plan of Air Quality in Lombardy for their kind collaboration and constructive comments to our work. They would also like to thank Drs. Alan Cross and Michel Schoupe for their support throughout the duration of the ICAROS project.

References

Caserini, S.: 2000, 'Emissions inventory in Lombardy', in G. Volta and G. Ballarin-Denti (eds), *Regional Plan for Air Quality*, Final Report, Fondazione Lombardia per l'Ambiente, Milan, Italy.
Ferrare, A., Fraser, R. S. and Kaufman, Y. J.: 1990, 'Satellite measurements of large scale air pollution: Measurements of forest fire smoke', *J. Geophys. Res.* **95**, 9911–9925.
Fraser, R. S., Kaufman, Y. I. and Mahoney, R. L.: 1984, 'Satellite measurements of aerosol mass and transport', *Atmos. Environ.* **18**, 2577–2584.
Lipfert, F. W.: 1994, *Air Pollution and Community Health: A Critical Review and Data Sourcebook*, Van Nostrand Reinhold, New York.
Pilinis, C. and Seinfeld, J. H.: 1987, 'Development and evaluation of a Eulerian photochemical gas-aerosol model', *Atmos. Environ.* **21**, 2453–2466.
Pilinis, C.: 1989, 'Numerical simulation of visibility degradation due to particulate matter: model development and evaluation', *J. Geophys. Res.* **94**, 9937–9946.

Pilinis, C., Pandis, S. N. and Seinfeld, J. H.: 1995, 'Sensitivity of direct climate forcing by atmospheric aerosols to aerosol size and composition', *J. Geophys. Res.* **100**, 18739–18754.

Pham, M., Müller, J.-F., Brasseur, G. P., Granier, C., Mégie, G.: 1995, 'A three-dimensional study of the tropospheric sulphur cycle', *J. Geophys. Res.* **100**, 26061–26092.

Pope, A. C., Thun, M. J., Namboodir, M. M., Dockery, D. W., Evans, J. S., Spezer, F. E. and Heath, C. W. Jr.: 1995, 'Particulate air pollution as a predictor of mortality in a prospective study for US adults', *Am. J. Respir. Crit. Cara Med.* **151**, 669–674.

Sarigiannis, D. A., Asimakopoulos, D. N., Bonetti, A., Huynh, F, Lointier, M., Schäfer, K., Sifakis, N. I., Soulakelis, N. and Tombrou, M.: 1998, 'ICAROS – Integrated computational assessment via remote observation system', *Proc. Ann. Conf. of Rem. Sens. Soc.*, Chatham Maritime, Kent, UK.

Sarigiannis, D. A., Saisana, M. and Triacchini, G.: 1999, 'Optimisation Criteria of the Air Quality Monitoring Network in the Region of Lombardy Using Satellite Derived Data', in *Final Report of the Regional Plan for Air Quality*, Fondazione Lombardia per l'Ambiente, Milan, Italy.

Sarigiannis, D. A.: 1999, 'Data Fusion and Inference Systems for Environmental Decision Support', *Proc. 11th AAAI Conference*, Orlando, Florida.

Sarigiannis, D. A. van Wijk, L.: 2000, 'Use of earth observation to assess ecosystem health', in G. Volta and G. Ballarin-Denti (eds), *Regional Plan for Air Quality*, Final Report, Fondazione Lombardia per l'Ambiente, Milan, Italy.

Schwartz, S. E.: 1996, 'Cloud droplet nucleation and its connection to aerosol properties', in M. Kulmala and P. E. Wagner (eds), *Nucleation and Atmospheric Aerosols*, Elsevier, Oxford, 770–779.

Seinfeld, J. H. and Pandis, S. N.: 1998, *Atmospheric Chemistry and Physics*, John Wiley, U.S.A.

Sifakis, N. I. and Deschamps, P. Y.: 1992, 'Mapping of air pollution using SPOT satellite data', *Phot. Eng. and Rem. Sens.* **LVIII**, 1433–1437.

Sloane, C. S.: 1983, 'Optical properties of aerosols: comparison of measurements with model calculation', *Atmos. Environ.* **17**, 409–419.

Varela, J. R.: 1994, 'Photochemical modelling of pollution scenarios in Mexico City', in H. Power, N. Moussiopoulos and C. A. Brebbia (eds), *Urban Air Pollution*, Computational Mechanics Publications, Boston, U.S.A.

COMPARISON OF OPEN-PATH AND POINT MEASUREMENTS OF A GASEOUS POLLUTANT IN THE VICINITY OF A MODEL BUILDING

I. MAVROIDIS[1]* and R. F. GRIFFITHS[2]

[1] *CERTH/CPERI, 6th km Harilaou-Thermi Rd, GR-57001, Thermi-Thessaloniki, Greece;*
[2] *Environmental Technology Centre, UMIST, Manchester, M60 1QD, U.K.*
(* author for correspondence, e-mail: imavr@tee.gr, tel: +30 - 93-7163 266. Current address:
20 Kalogera Street, GR 11361, Athens, Greece)

Abstract. Field experiments were carried out in the vicinity of an isolated building using open-path and point monitoring air pollution detectors to measure concentrations of a tracer gas released from a source located either on the building or at the downwind face of the building. Experiments were conducted under neutral or slightly unstable weather conditions. Either three or five fast-response point monitors were located along the beam path of an open-path instrument, in the wake of the building. Experiments were performed with different source locations, detector locations and building orientations to the wind. Visual comparison of data time series showed that data from the two types of instruments followed similar trends. The best degree of correlation occurred for an averaging time similar to the response time of the open-path monitor. Mean concentrations detected by the open-path instrument were generally in very good agreement with the path averaged mean concentrations from the point monitors. The point monitors were also capable of providing time-resolved data needed to describe detailed concentration fluctuation statistics.

Keywords: air pollution, building, long-path, open-path, point monitors, wake, DOAS

1. Introduction

The growing need for pollution monitoring and control has led to increased research on existing pollution monitors and to the introduction of new techniques. Gas monitoring has traditionally been conducted in the form of 'point monitoring', i.e. as a measurement at a single location in space. Point monitoring is suitable for very localised gas measurements or for measurements where the gas concentration is homogeneous over the region being monitored. However, point monitoring may not be adequate for measuring poorly-mixed gases in industrial or urban areas. Examples of this include fugitive emissions from industrial plants and leakages from chemical storage tanks, since in these cases it will be a matter of chance whether or not a particular plume of gas intersects a measuring point. The need to find new methods to complement and supplement the traditional 'point monitoring' methods has led to the development of new techniques of remote and open-path (or long-path) optical measurement. These techniques employ a beam of radiation, usually infrared or ultraviolet, which is emitted along a path that can be from a few metres to a few kilometres in length. Two of the most successfully used open-path

Water, Air, and Soil Pollution: Focus **2:** 655–667, 2002.
© 2002 *Kluwer Academic Publishers.*

techniques are: (a) Differential Optical Absorption Spectroscopy (DOAS), which is an active remote sensing technique – using primarily IR and near UV light – where absorption spectra are generated for a long-path and compared to library spectra for various known gases, and (b) Fourier Transform InfraRed (FTIR) spectroscopy, which is an analytical technique utilising an interferometer with moving and stationary mirrors to 'transform' the spectral distribution of wavelengths in the IR into a form that can be mathematically converted to a conventional infrared intensity spectrum.

Application areas of long-path monitoring systems include 'boundary fence' monitoring of gas concentrations beside industrial plants, roads or landfill sites, gas leakage detection around industrial plants, using single straight beams or a 'ring fence' beam configuration that surrounds a plant region or tank farm, gas leakage monitoring within industrial buildings, tunnels or mines, fire-warning within large buildings, through detection of combustion gases, monitoring at places of difficult access or containing severe flammable or toxic hazards and estimation of gas flux or total site emission rates, using concentration and meteorological data. Recent literature investigates a variety of open-path instrument applications, such as measurements of motor vehicle emissions (Jimenez et al., 2000), emission rate apportionment from fugitive sources (Hashmonay et al., 1999), area measurements of ammonia emissions from spreading of manure (Galle et al., 2000), measurements of greenhouse gas emissions from waste water treatment plants (Eklund, 1999) and measurements of gases emitted from swine confinement facilities (Childers et al., 2001; Todd et al., 2001). Brocco et al. (1997) and Kourtidis et al. (2000) present measurements of hydrocarbons and other pollutants of common interest, such as NO_2, which were carried out using open-path techniques in Rome, Italy, and in Thessaloniki, Greece, respectively. Although several advantages of the open-path monitoring techniques have already been recognized (e.g. Partridge, 1990–91; Grant et al., 1992), open-path monitors have certain limitations and, therefore, their performance needs to be investigated in the field. Comparisons of open-path monitors with established point-monitoring techniques, such as canisters, have mainly been used for this purpose, especially by the U.S. Environmental Protection Agency (e.g. Grant et al., 1992). A methodology for comparison between point monitors and open-path monitors was proposed by McClenny et al. (1974), according to which the path averaged concentration of the open-path instrument is compared to the path averaged concentration obtained by moving a point monitor along the measurement path with a constant velocity. Some discrepancies between the results from the two measurement techniques were attributed to the interference from water and CO_2 concentrations in the atmosphere (especially if a long pathlength was used), as well as to calibration discrepancies of the open-path instrument. Russwurm et al. (1991) performed a comparison of an FTIR open-path monitor, with point measurements obtained using a number of canisters located along the path. The results showed very good agreement between the two methods. Stevens et al. (1993) compared a DOAS open-path instrument with U.S. Federal

Reference Method (FRM) point monitors and the results showed in general very good agreement between the point measurements and the open-path system, with the correlation coefficient being sometimes greater than 0.9. However, during some short periods the DOAS detected significantly lower concentrations than the FRMs and this was attributed to atmospheric inhomogeneities and to the influence of meteorological conditions on the performance of the open-path instrument. The results presented in the literature suggest that the main limitations of the open-path monitors are that they can not recognize spatial inhomogenities of pollutant concentrations, that they are influenced by atmospheric conditions (e.g. humidity) and that they have higher detection limits than point monitors.

In the experiments described in this paper a fast-response point monitoring technique was used together with an open-path monitor. Both are designed for the measurement of gaseous contaminants rather than particulates. Experiments were conducted in the wake of a model building, where the concentration distribution is relatively homogeneous.

2. Methodology

The open-path monitor used for the experiments was the Hawk long-range gas monitor developed by the National Physical Laboratory (NPL). This is a portable, double-ended gas monitor that measures a wide range of gases, including pollutants like hydrocarbons, SO_2, NO_2, HCl, CO and CO_2. The sensitivity of the Hawk is typically better than 2% of full scale of gas detection range at 350 m. The minimum response time for gas detection is 6 sec (Siemens Plessey Controls Ltd, 1992). The calibration and the performance of the Hawk are checked by using an integrated zeroing reflector and gas cell. The Hawk system utilises the effect of the presence of abnormal gases in the atmosphere on the absorption of the near infrared (IR) portion of a light beam. Discrimination between different classes of gases relies on the different wavelengths at which IR absorption occurs. To reject the interfering components, a range of wavelengths is chosen so that a sloping side of the target gas absorption curve is scanned, to produce a differential measurement. The amount of the target gas in the beam determines the steepness of the absorption curve, and hence the size of the differential measurement. By the same token, if another gas absorbs infrared radiation between these wavelengths, but there is no net differential absorption signal, the interfering gas is rejected. This is particularly useful in eliminating atmospheric gases and vapours. The point monitor used in the experiments was the Ultra Violet Ion Collector (UVIC). The UVIC detector is a fast-response photo-ionization detector with a useful calibratable range from about 0.01 to 1000 ppm by volume and has a response time of about 0.02 sec. The development of the UVIC detector is described in detail in Griffiths et al. (1998).

The field experiments were conducted at Altcar Field Site, on the northwest coast of England. The surrounding terrain is generally flat. The model building used for the experiments is a rotatable building with a rectangular shape, and its dimensions are 10.96 m long × 4.24 m wide × 1.56 m high. Using the rotatable building it was possible to conduct experiments with the wind at any chosen orientation to the building. In these experiments two configurations were used, with the wind normal to either its long or its short face. The main parameters that varied between different experiments were building orientation, source location and downwind detection distance. The tracer gas used for the experiments was propylene, which is a typical hydrocarbon released from petrochemical plants and can be detected by both the UVIC and the Hawk instruments. The gas source was located either on top of the centre of the building roof or at the centre of the downwind building face. Nine experiments were conducted. The experimental set-up is shown in Figure 1 and the experimental conditions are described in Table I. The Hawk and retroreflector were positioned at an appropriate distance downwind of the building, with the detection path normal to the wind direction. The pathlength of the Hawk detector ranged between 23 and 43 m. An array of either three or five UVICs was positioned along the Hawk path, in such a way that the central UVIC and the centre of the building were on a straight line parallel to the mean wind direction. The next two UVICs were placed one on each side of the central UVIC, and at a distance of 1.5 m from it. When five UVICs were used the last two UVICs were placed one on each side of the central UVIC and at a distance of 3.0 m from it. All detectors were located at building height (H), except in experimental case H04, when they were located at a height of 2.58 m (1.65H). The Data Acquisition System recorded at a sampling frequency of 200Hz. Meteorological data were acquired using an Ultrasonic Anemometer. All the experiments were performed in slightly unstable or neutral atmospheric conditions, corresponding to Pasquill stability classes C and D respectively. Meteorological data are shown in Table I, together with the experimental conditions.

3. Results and Discussion

Raw data were acquired as time series in millivolts and were transformed to concentration data through appropriate calibration. Figure 2 shows 5 min of raw data from the Hawk and from the centrally positioned UVIC detector. These data show that a peak in the UVIC signal is usually followed by a peak in the Hawk signal. However, there seems to be some delay in the response of the Hawk to a peak recorded by the UVIC. This delay, of approximately 5 sec, is due to the Hawk response time, which is stated by the instrument manufacturer to be about 6 seconds (Siemens Plessey Controls Ltd, 1992). Although the two data sets follow similar trends, it is clear that the fast-response point monitor provides better temporal and spatial resolution of the fluctuating gas concentrations.

Calibrated data from the two instruments were averaged over 1 min periods. The simultaneously measured 1 min mean concentrations for the UVIC instruments

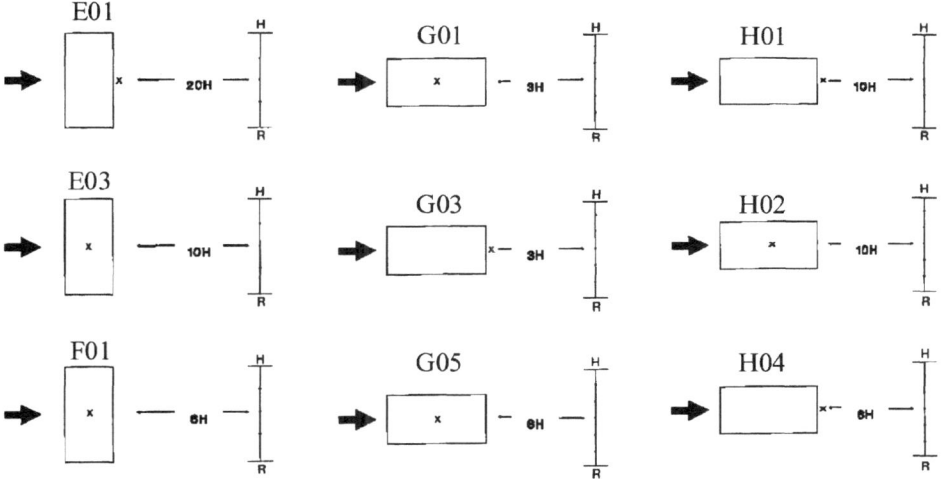

Figure 1. Experimental set-up for the nine experimental cases.

TABLE I

Experimental and meteorological conditions for the nine experimental cases (the front face of the building is at 0 degrees to the wind)

Experimental case	Detection distance (m)	Number of UVIC detectors	Hawk pathlength (m)	Mean wind speed (m.s^{-1})	Mean wind direction (degrees)
E01	20H	3	42.1	3.3	−5.6
E03	10H	3	41.0	N.A.	N.A.
F01	6H	3	38.0	3.9	−11.6
G01	3H	5	37.4	5.3	+ 4.6
G03	3H	5	37.4	5.3	+ 10.2
G05	6H	5	39.7	5.3	+ 10.9
H01	10H	5	23.7	4.2	+ 8.4
H02	10H	5	23.7	4.2	+ 8.6
H04	6H	5	25.4	4.6	+ 4.3

located along the path were summed and divided by the number of UVIC detectors, thus providing time series of the 1 min path averaged UVIC data. The path averaged UVIC data were compared to the 1 min average Hawk data. Plots of Hawk and path averaged UVIC data for two experimental cases are presented in Figure 3. Hawk data are presented in ppm.m while UVIC data are presented in ppm, thus allowing only a qualitative comparison of the results. In case G05 comparison of the two time series of data shows that concentrations measured by the two instruments

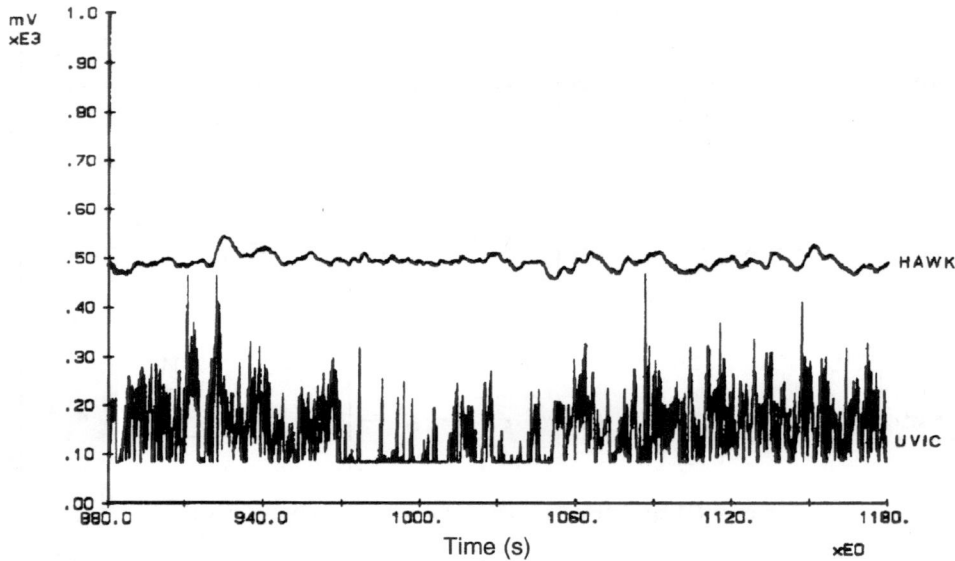

Figure 2. Data time series from a centrally positioned UVIC and the Hawk (experimental case H02).

follow a very similar trend. This indicates that the long-path instrument can, in general, provide very good qualitative information on path averaged concentration changes, which compares very well to the path averaged data from point monitors placed along its path. However, in case H01, which is also shown in Figure 3, data from the two instruments did not follow similar trends, at least for small periods during the experiment. This is attributed to the higher degree of variation of the lateral wind direction observed during this experiment, which was up to 35 degrees off the normal to the upwind obstacle face. It is then expected that some of the UVIC detectors would have been located outside the plume for parts of the experiment, therefore influencing the path averaged UVIC concentration data. In this respect, the use of an open-path instrument renders concentration measurements easier and more efficient, since it is possible to monitor adequately a larger area with just a single instrument. On the other hand, the fast-response point monitor provides further, localised, information on spatial concentration variations along the path and on concentration fluctuation statistics.

To further compare concentration measurements from the two types of detectors, the dependence of the correlation between UVIC and Hawk data on the concentration averaging time was investigated. Figure 4 shows the correlation coefficient between UVIC and Hawk data as a function of averaging time for experimental case G03 (both for central UVIC and path averaged UVIC concentrations). Both path averaged UVIC and central UVIC concentrations correlate quite well with Hawk data, with path averaged UVIC concentrations showing a slightly better correlation, as expected. In both cases the correlation increased with averaging

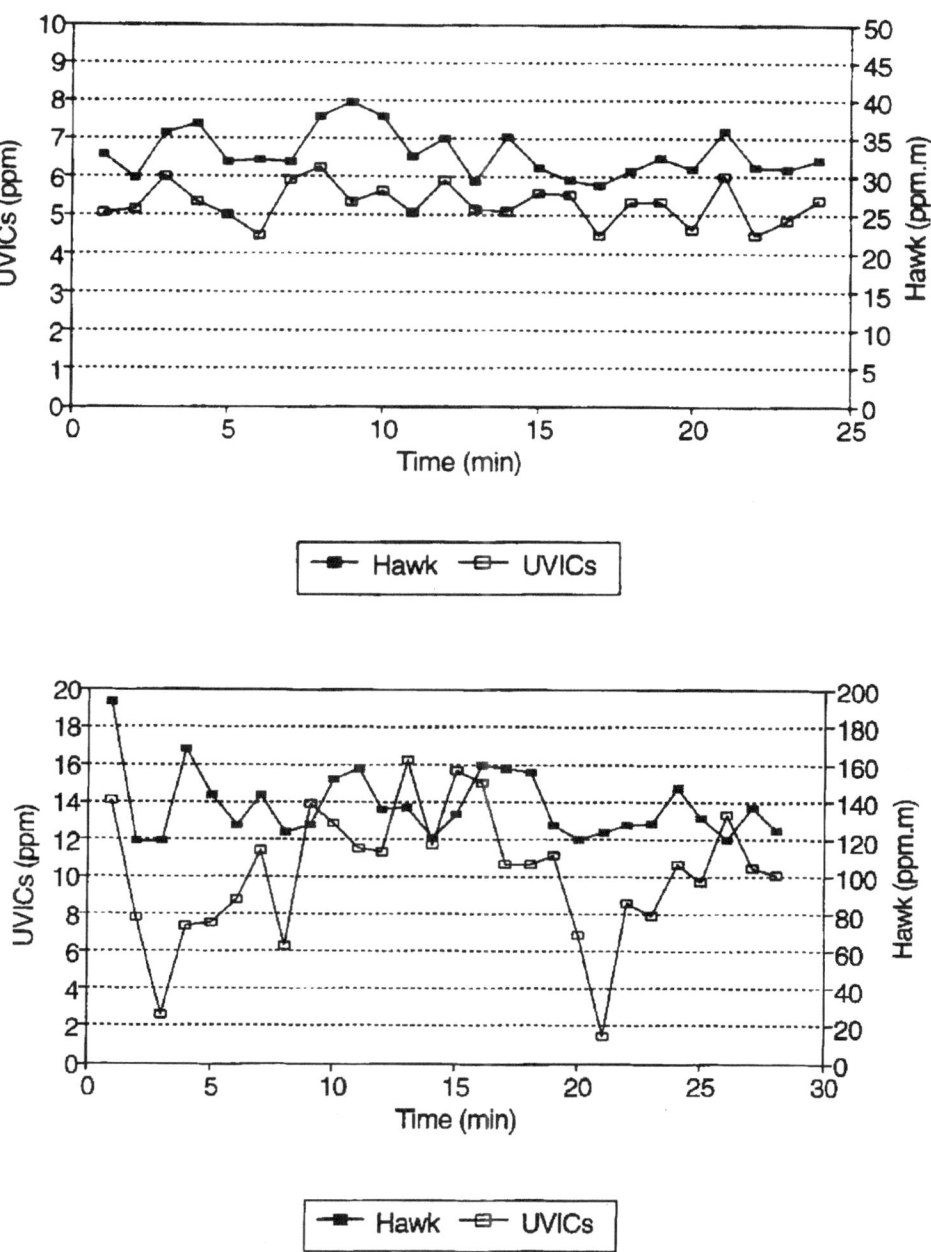

Figure 3. Comparison of path averaged UVIC and Hawk concentration data for experimental cases G05 (top) and H01 (bottom).

time, and reached a maximum at an averaging time of approximately 10 sec, which is comparable with the response time of the Hawk as stated by its manufacturer (6 sec). It seems reasonable that for averaging times shorter than the response time of the Hawk, the correlation between data from the UVICs and the Hawk will be low. Further increase of the averaging time did not result in an increase of the correlation between the two datasets. Actually, in some cases the correlation coefficient decreased for long averaging times, and this may be attributed to the fact that during longer periods the wind direction may have been more variable and thus not all the point monitors may have been located inside the plume during the entire period. Figure 5 shows the correlation of concentrations from two UVIC detectors (located at opposite lateral positions relative to the axis defined by the centre of the building and the mean wind direction) as a function of the averaging time. These results show clearly the effect of the meandering of the wind. The two data sets are negatively correlated for all averaging times, since, due to the wind meandering and the subsequent plume meandering, an increase of the concentrations measured at one side of the central UVIC would correspond to a decrease of the concentrations measured at the other side. Therefore, when one of these two detectors measured high concentrations it is likely that the other would not, since they were located at the two opposite edges of the mean plume. For longer averaging times the effect of the wind meandering was more pronounced and therefore the absolute value of the correlation coefficient increases. It should be stressed that the above examination of the effect of averaging time on the correlation of the two types of data does not address the question of what averaging time is appropriate for a given application. Clearly the averaging time applied in a particular application must be chosen to reflect the time-scale of the response that is of interest. More information on the effect of averaging time on mean concentrations and on the time scaling of small-scale field trials is provided in Macdonald et al. (1997) and in Mavroidis and Griffiths (2001).

Statistical parameters such as mean concentrations, intermittency, peak-to-mean concentration ratios and concentration fluctuation intensities were derived for all the experimental cases. Table II shows concentration statistics from a representative experimental case. Intermittency (I) is defined here as the proportion of time for which concentration is at or below a threshold value, which in this case is nominally zero concentration. The UVIC detectors located at the edge of the array of five UVICs measured the lowest mean concentrations. The mean concentrations measured from the three central instruments suggest that the plume in the wake of the building was to some extent bifurcated. This effect is attributed to the horseshoe vortex structure developed at the base of the front face of the rectangular building, and is in agreement with previous observations from field experiments (Macdonald et al., 1998). Intermittency values were in general low, since the plume was quite well mixed in the wake of the obstacle. Intermittency was greater at the two edges of the plume, where UVICs 1 and 5 were positioned. It is important to note that the Hawk measurements showed zero intermittency in all nine experiments, since there

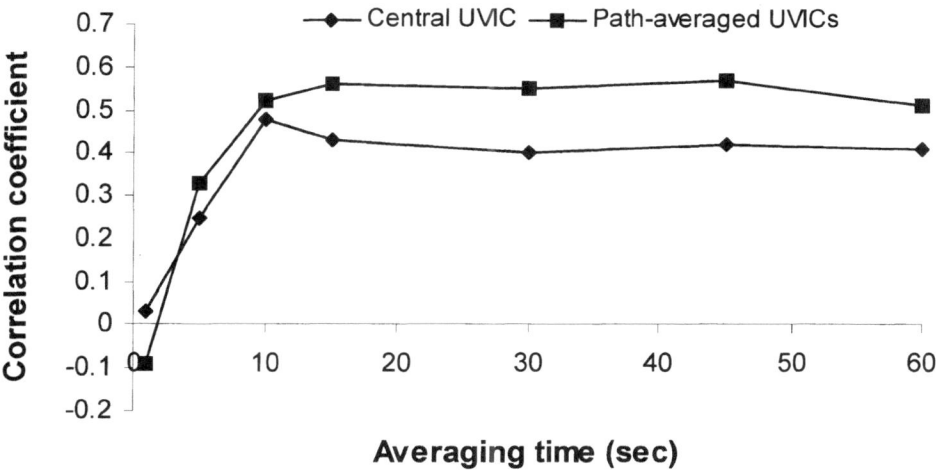

Figure 4. Correlation coefficient between UVIC and Hawk data as a function of averaging time (experimental case G03).

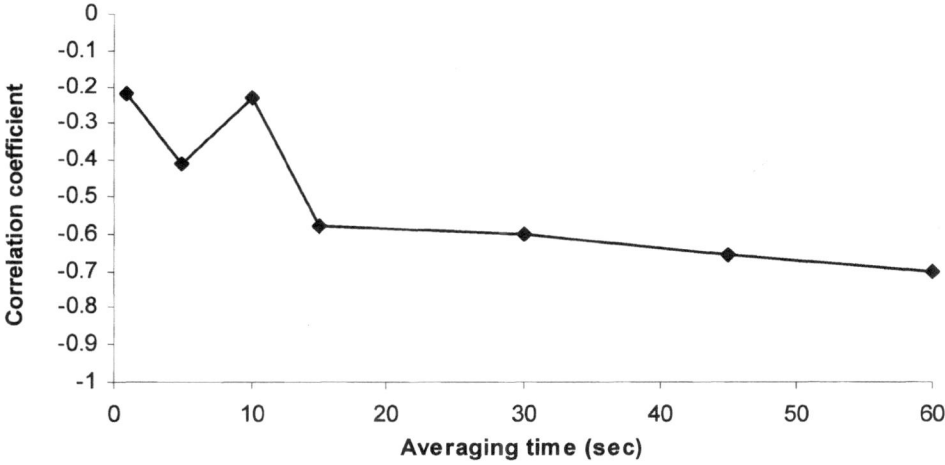

Figure 5. Correlation coefficient between two UVIC detectors (located at opposite lateral positions relative to the axis defined by the centre of the building and the mean wind direction) as a function of averaging time (experimental case G03).

was always some gas passing through the optical path, and also due to the longer response time of the Hawk instrument. Therefore, the long-path detector was not capable of providing any information on intermittency. For the same reason, the concentration fluctuation intensity and the peak-to-mean concentration ratio were much lower for the Hawk than for the UVICs. The peak-to-mean ratio (P/M) and the concentration fluctuation intensity measured by the UVIC detectors were greater towards the edge of the plume (UVICs 1 and 5), since the presence of gas is more intermittent there (Hinds, 1969). To examine the effect of intermittency

TABLE II

Concentration statistics for experimental case G01. The UVIC detectors are numbered consecutively, starting from one end of the Hawk beam

Detector	Mean concentration (ppm)[a]	Intermittency	Peak to mean ratio	Conditional peak to mean ratio	Concentration fluctuation intensity
UVIC 1	3.5	0.6	69.1	27.5	5.0
UVIC 2	16.5	0.4	36.4	20.5	2.0
UVIC 3	6.0	0.3	29.9	20.0	1.2
UVIC 4	15.1	0.6	43.6	18.8	2.3
UVIC 5	0.2	0.9	164.7	16.5	3.0
Average UVIC	8.3	-	-	-	-
Hawk[a]	55.7	0.0	3.2	3.2	0.2

[a] Mean concentration for the Hawk detector is in ppm.m.

on concentration fluctuation statistics, the conditional peak-to-mean ratio, which is equal to $P/M(1-I)$, is also shown in Table II. The variation of the conditional peak-to-mean ratio across the plume was much smaller, and this demonstrates the effect of intermittency on concentration fluctuation statistics. The results from the different experimental cases also suggest that as the downwind detection distance is increased, the peak-to-mean concentration ratio and the concentration fluctuation intensity decrease.

To compare quantitatively concentration data from the open-path instrument and the point detectors, the Hawk data was divided by an appropriate plume width. The plume width was calculated using an enhanced Gaussian plume model, evaluated by Huber (1984) and Fackrell (1984), which takes into account the effect of the building. The plume width ($2Y_o$) is equal to $4.3\sigma_y$, where σ_y is the lateral spread of the plume in the presence of the building (Pasquill and Smith, 1983). The wake region is divided into a near and a far region. If h is the building height, w the building width and x the downwind distance, then in the near wake region ($3 h < x < 10 h$), $\sigma_y = 0.35 w + (x - 3h)/15$, while for $x > 10 h$ a virtual source model is used, so that the spread matches the near wake model at $x = 10 h$. If the spread specified by the model is less than the undisturbed flow spread, then the latter is used. Table III presents the plume width calculated by the above model, together with the Hawk concentrations (both in ppm and ppm.m) and the UVIC path averaged concentrations, for all nine experimental cases. The last column shows the relative percent difference (RPD) between the Hawk and UVIC path averaged concentrations. It is clear that mean concentrations detected by the UVICs and the Hawk compare very well not only qualitatively but also quantitatively. The agreement of concentrations measured by the two types of instruments is very good, with the exception of

TABLE III

Comparison of Hawk and path averaged UVIC concentrations for all nine experimental cases

Experimental case	Hawk mean concentration (ppm.m)	Plume width (m)	Hawk mean concentration (ppm)	UVIC path averaged mean concentration (ppm)	R.P.D. (%)
E01	141.7	23.5	6.0	6.8	11.7
E03	95.1	19.0	5.0	5.3	5.1
F01	182.0	17.8	10.2	11.6	13.0
G01	55.7	6.4	8.7	8.3	5.2
G03	38.4	6.4	6.0	7.2	17.7
G05	33.0	7.7	4.3	5.2	9.4
H01	137.9	9.5	14.5	10.1	35.3
H02	162.3	9.5	17.1	14.5	16.5
H04	49.2	7.7	6.4	7.0	9.9

experimental case H01. As already discussed (Figure 3), during this experiment a very high degree of variation of the lateral wind direction was observed, which was up to 35 degrees off the normal to the upwind obstacle face. Therefore, some of the UVIC detectors would have been located outside the plume for a significant part of the experiment, thus influencing the path averaged UVIC concentration data.

4. Conclusions

In this paper a methodology for the intercomparison of concentrations simultaneously detected by an open-path detector and fast-response point monitors was established and tested. A number of tracer gas experiments were performed, with either three or five point monitors located along the beam path of an open-path instrument, in the wake of an isolated building. During the experiments different source locations, detector locations and building orientations to the wind were examined. Visual comparison of data time series collected by the open-path detector and point monitors located along its optical path showed that data from the two types of instruments follow similar trends. The main parameter influencing the correlation between the data of open-path and point monitors was found to be the meandering of wind direction. Quantitative comparison showed that the mean concentrations detected by the open-path instrument were very close to the path averaged mean concentrations from the point monitors. In general, the experiments have shown that if a short beam is used in a homogeneous environment, remote-sensing detection can produce excellent results. In certain cases, the use of open-path instruments may be the preferable method, given the convenience of being able to monitor a larger area with just a single instrument. However, such methods will not provide adequate resolution for impacts such as odour perception, since the measurement path is not appropriately matched to the very much smaller scale of the individual receptor, regardless of the instrument response time. For such applications fast-response point detectors are needed in order to match both the temporal and spatial scales of the receptor. Such instruments can provide the resolution needed for applications in which it is important to characterise features such as the fine-scale concentration fluctuation behaviour, and the intermittency. No single instrument type is adequate for all purposes, and the appropriate choice and combination of methods needs to be based on the particular requirements of the given application.

Acknowledgements

We would like to thank Shell Research Ltd, and especially Dr. I. Archibald, for providing the Hawk instrument and for their co-operation during the course of this work. The facilities at the Altcar site were provided by DERA Porton Down, under several research agreements.

References

Brocco, D., Fratarcangeli, R., Lepore, L., Petricca, M. and Ventrone T.: 1997, 'Determination of aromatic hydrocarbons in urban air of Rome', *Atmos. Environ.* **31A**, 557–566.
Childers, J. W., Thomson, E. L. Jr., Harris, D. B., Kirchgessner D. A., Clayton, M., Nataschke, F. and Phillips, W. J.: 2001, 'Multi-pollutant concentration measurements around a concentrated swine production facility using open-path FTIR spectroscopy', *Atmos. Environ.* **35**, 1923–1936.
Eklund, B.: 1999, 'Comparison of line- and point-source releases of tracer gases', *Atmos. Environ.* **33**, 1065–1071.
Fackrell, J. E.: 1984, 'An examination of simple models for building influenced dispersion', *Atmos. Environ.* **18**(1), 89–98.
Galle, B., Klemedtsson, L., Berqvist, B., Ferm, M., Tornqvist, K., Griffith, D. W. T., Jensen, N. O. and Hansen, F.: 2000, 'Measurements of ammonia emissions from spreading of manure using gradient FTIR techniques', *Atmos. Environ.* **34**, 4907–4915.
Grant, W. B., Kagann, R. H. and McClenny, W. A.: 1992, 'Optical remote measurement of toxic gases', *J. Air Waste Manage. Assoc.* **42**(1), 18–30.
Griffiths, R. F., Mavroidis, I. and Jones, C. D.: 1998, 'Development of a fast-response portable PID: model of the instrument response and validation tests in air', *Measure. Sci. Technol.* **9**, 1369–1379.
Hashmonay, R. A., Yost, M. G., Mamane, Y. and Benayahu, Y.: 1999, 'Emission rate apportionment from fugitive sources using open-path FTIR and mathematical inversion', *Atmos. Environ.* **33**, 735–743.
Hinds, W. T.: 1969, 'Peak-to-mean concentration ratios from ground-level sources in building wakes', *Atmos. Environ.* **3**, 145–156.

Huber, A. H.: 1984, 'Evaluation of a method for estimating pollution concentrations downwind of influencing buildings', *Atmos. Environ* **18**(11), 2313–2338.

Jimenez, J. L., McManus, J. B., Shorter, J. H., Nelson, D. D., Zahnizser, M. S., Koplow, M., McRae, G. J. and Kolb, C. E.: 2000, 'Cross road and mobile tunable infrared laser measurements of nitrous oxide emissions from motor vehicles', *Chemos.-Glob. Change Sci.* **2**, 397–412.

Kourtidis, K., Ziomas, I., Zerefos, C., Gousopoulos, A., Balis, D. and Tzoumaka, P.: 2000, 'Benzene and toluene levels measured with a commercial DOAS system in Thessaloniki, Greece', *Atmos. Environ.* **34**, 1471–1480.

Macdonald, R. W., Griffiths, R. F. and Cheah, S. C.: 1997, 'Field experiments of dispersion through regular arrays of cubic structures', *Atmos. Environ.* **31**(6), 783–795.

Macdonald, R. W., Griffiths, R. F. and Hall, D. J.: 1998. 'A comparison of results from scaled field and wind tunnel modelling of dispersion in arrays of obstacles', *Atmos. Environ.* **32**(22), 3845–3862.

Mavroidis, I. and Griffiths, R. F.: 2001, 'Local characteristics of atmospheric dispersion within building arrays', *Atmos. Environ.* **35**, 2941–2954.

McClenny, W. A., Baumgardner, R. E. Jr., Baity, F. W. and Gray, R. A.: 1974, 'Methodology for comparison of open-path monitors with point monitors', *JAPCA* **24**(11), 1044–1046.

Partridge, R. H.: 1990/91, 'Long-path monitoring of atmospheric pollution', *Measure. Cont.* **23**, 293–298.

Pasquill, F. and Smith, F. B.: 1983, *Atmospheric Diffusion*, Ellis Horwood Ltd, Chichester.

Russwurm, G. M., Kagann, R. H., Simpson, O. A., McClenny, W. A. and Herget, W. F.: 1991, 'Long-path FTIR measurements of volatile organic compounds in an industrial setting', *J. Air Waste Manage. Assoc.* **41**(8), 1062–1066.

Siemens Plessey Controls Ltd.: 1992, *Hawk. Operation and Maintenance Manual*, March 1992.

Stevens, R. K., Drago, R. J. and Mamane, Y.: 1993, 'A long-path differential optical absorption spectrometer end EPA-approved fixed-point methods intercomparison', *Atmos. Environ.* **27B**(2), 231–236.

Todd, L. A., Ramanathan, M., Mottus, K., Katz, R., Dodson, A. and Mihlan, G.: 2001, 'Measuring chemical emissions using open-path Fourier transform infrared (OP-FTIR) spectroscopy and computer assisted tomography', *Atmos. Environ.* **35**, 1937–1947.

THEORETICAL INVESTIGATION OF URBAN AIR QUALITY MANAGEMENT SYSTEMS PERFORMANCE TOWARDS SIMPLIFIED STRATEGIC ENVIRONMENTAL PLANNING

K. D. KARATZAS

Metron 19–21, 55132 Thessaloniki, Greece
(e-mail: kkara@auth.gr, fax: +30 310 996012)

Abstract. Urban Air Quality Management Systems are software tools that combine air quality models with various software modules like geographical information systems, databases, expert systems and statistical analysis tools. Such systems try to interpret as accurately as possible the complex interactions between various atmospheric, emission, land use and topographic parameters involved in the air pollution management problem, in order to provide support for environmental strategic planning and decision making. As this process involves a huge set of parameters, some of which may only be roughly estimated, air quality management systems tend to aggregate parameters in order to simplify their analysis and make it more effective and operational. Yet, this aggregation may lead to deviations in the analysis results, as proved in this theoretical article, and thus influence the decision making and strategic planning process. According to the findings of this article, such a policy making process should focus primarily on short term measures when dealing with air pollution episode management.

Keywords: decision making, strategic planning, urban air quality management system

1. Introduction

The air quality system of the urban and regional troposphere is a complex network of multiple competing interactions within and among chemical and physical atmospheric processes (Dennis *et al.*, 2000). Contemporary Urban Air Quality Management Systems (UAQMS) on the other hand are characterized from the combined use of state-of-the-art informatics and air quality simulation models: Monitoring networks provide on-line air quality information to a central system that archives them with the aid of databases. Processing tools like spreadsheets, dedicated statistical analysis modules and graphics modules are used to either analyse existing air quality or to forecast the future air quality. For the latter, sophisticated air quality modelling tools are extensively being used. In this way, UAQMS are continuously trying to establish a robust, integrated, environmental management life cycle for the urban area(s) of interest (Fedra and Haurie, 1999). In addition, expert systems are used to support decision-making, while geographical information systems are becoming a common platform for treating spatially distributed data.

Water, Air, and Soil Pollution: Focus **2**: 669–676, 2002.
© 2002 *Kluwer Academic Publishers.*

Although a great effort has been invested in the validation and the harmonisation of tools used, it is not common to treat the UAQMS as a black box (regardless of its various components), and to try to investigate its overall behaviour, and thus reveal its limitations. In the present article, a black-box approach is used in order to perform a mathematical, system-behaviour investigation, in order to achieve a quantitative approach for the systems' deviation from the 'real world', as a result of the (necessary) aggregation performed in systems' input and components. It is so proven that such UAQMS have an exponential deviation from the 'real world', as a function of time. As a consequence of this finding, some suggestions are introduced for the formulation and usage of such systems.

2. Aggregation in UAQMS

Ideally, an UAQMS should be capable of completely reproducing and handling all features of the 'real world', providing scientists and decision-makers with the ability to investigate all Air Quality Management (AQM) related parameters 'in vitro'. The complexity and the dynamics of the AQM problem, the number of variables involved and the requirements for massive information and computational power availability, make this approach impossible. Therefore, the construction and use of a generalized, 'simplified' system, that results from the aggregation of a detailed, 'realistic' UAQMS is essential.

The aggregation can be viewed as a simplification and summarisation of the parameters involved in the system (e.g., representation of actual emission inventory data from emission category data, classification of meteorological conditions, etc.). In addition, the aggregation can also be realized as a simplification of the equations describing the actual UAQMS (Axtell, 1992). This can mean, for example, the use of a Gaussian dispersion calculation methodology instead of an Eulerian, the use of one dimensional instead of multidimensional cost functions in order to evaluate the cost of emission abatement strategies, etc. In the frame of the approach used in this article, the UAQMS is considered to be black box. In this black box, all AQM related data are entered (meteorological conditions, emissions, etc.) and parameters describing the quality of air are 'produced', like concentration values of specific pollutants, population exposure parameters, etc.

2.1. VARIABLES AND FUNCTIONS IN AGGREGATED UAQMS

An UAQMS generally requires dynamic variables as an 'input'. Each one of these time dependent input variables is influencing the state (e.g. the 'output' parameters of the black box as previously defined) of the UAQMS, according to Equation (1):

$$\mathbf{x}(t) = [x_1(t), x_2(t), \ldots, x_n(t)] \in \Re^n, \tag{1}$$

$\Re^n$, = the space of AQM related parameters ,

where for instance x_1: meteorological values, x_2: emission values, etc. Therefore, UAQMS aggregation functions can be defined as follows:

$$\mathbf{a}(\mathbf{x}) = [a_1(\mathbf{x}), a_2(\mathbf{x}), \ldots, a_m(\mathbf{x})] \in \mathfrak{R}^m \ , \tag{2}$$

R^n, = the space of AQM aggregation functions .

These functions are linking the parameters of the actual UAQMS to the parameters of the aggregated UAQMS. Thus, the dynamic behaviour of the initial system, which is described by the following transformation vector

$$\mathbf{f}(\mathbf{x}) = \big[f_1(\mathbf{x}), f_2(\mathbf{x}), \ldots, f_n(\mathbf{x}) \big] \in \mathfrak{R}^n \ ,$$

$\mathfrak{R}^n$, = the space of the AQM status functions ,

is now 'replaced' by the behaviour of the aggregated synoptic system, which is described by the corresponding aggregated transformation vector:

$$\mathbf{g}(\mathbf{y}) = \big[\mathbf{g}_1(\mathbf{y}), \mathbf{g}_2(\mathbf{y}), \ldots, \mathbf{g}_m(\mathbf{y}) \big] \in \mathfrak{R}^m \ ,$$

R^n, = the space of the aggregated AQM status functions.

This transformation defines two sets of status functions: the set A of x over y which is symbolized by x_y and the set A* of $\dot{x}$ over $\dot{y}$, which is symbolized by $\dot{x}_y$.

By corresponding the element α of x_y to the element $\alpha^* = f^{-1} \mathrm{o} a \mathrm{o} g$ of $\dot{x}_y$, a new function results where:

$$\forall a \in x_y \ , \quad a^* = h(s) = f^{-1} \mathrm{o} a \mathrm{o} g \ \in \ \dot{x}_y \ ,$$

and:

$$h(a_1) = h(a_2) \Rightarrow f^{-1} \mathrm{o} a_1 \mathrm{o} g = f^{-1} \mathrm{o} a_2 \mathrm{o} g \Rightarrow$$

$$f \mathrm{o} f^{-1} \mathrm{o} a_1 \mathrm{o} g \mathrm{o} g^{-1} = f \mathrm{o} f^{-1} \mathrm{o} a_2 \mathrm{o} g \mathrm{o} g^{-1} \Rightarrow i_x \mathrm{o} a_1 \mathrm{o} i_y = i_x \mathrm{o} a_2 \mathrm{o} i_y \Rightarrow a_1 = a_2$$

In addition, $\forall \varphi \in \dot{x}_y \exists f^{-1} \mathrm{o} a \mathrm{o} g \in \dot{x}_y$ so that $h(f^{-1} \mathrm{o} a \mathrm{o} g) = \varphi$. Therefore, the function h is a one to one function of x_y over $\dot{x}_y$ and thus these two data sets have the same number of elements, meaning that the behaviour of the aggregated system has a one to one relationship with the behaviour of the initial system. Based on the above, the dynamics of the UAQMS can be described with the aid of a function of its aggregated variables, as follows:

$$\dot{\mathbf{y}}_i(t) = \mathbf{g}_i \left[a_1(\mathbf{x}(t)), a_2(\mathbf{x}(t)), \ldots, a_m(\mathbf{x}(t)) \right] \in \mathfrak{R}^m \ , \quad \forall i \in M \ . \tag{3}$$

Equation (3) can also be expressed in a vectorized form as follows:

$$\dot{\mathbf{y}}_i(t) = \mathbf{g}\,[\mathbf{a}(\mathbf{x}(t))] \in \Re^m\,, \qquad \forall i \in M\,. \tag{4}$$

By differentiating Equation (4), an expression for the dynamic behaviour of the input variables of the aggregated UAQMS way be introduced as follows:

$$\left[\dot{\mathbf{y}}_i(t)\right] = \sum_{j=1}^{n} \frac{\partial a_i(\mathbf{x})}{\partial \mathbf{x_j}} \frac{dx_j(t)}{dt} = \sum_{j=1}^{n} \frac{\partial a_i(\mathbf{x})}{\partial \mathbf{x_j}} f_j(\mathbf{x})\,, \qquad \forall i \in M\,, \tag{5}$$

or, written in a vectorized form:

$$\left[\dot{\mathbf{y}}_i(t)\right] = \mathbf{D_x}(\mathbf{a}(\mathbf{x}))\mathbf{f}(\mathbf{x}(t))\,, \qquad \forall i \in M\,, \tag{6}$$

Vector $\mathbf{D_x}(\mathbf{a}(\mathbf{x}))$ is a $m \times n$ matrix of partial differentials of the aggregated functions of the input values of the initial UAQMS, having as elements:

$$[\mathbf{D_x}(\mathbf{a}(\mathbf{x}))]_{i,j} \equiv \frac{\partial a_i(\mathbf{a})}{\partial \mathbf{x_j}}\,. \tag{7}$$

3. Behaviour of an Aggregated versus the Initial UAQMS

In the rest of this article the difference between the output of the aggregated system compared to the initial system is investigated. For this reason, the Lipschitz condition is being used, as introduced in the Appendix. By using the fact that for two functions the norm of their sum is less or equal to the sum of their norms, and also that the function $\mathbf{g}(\mathbf{y})$ satisfies the Lipschitz condition of order 1 (Axtell, 1992), the following inequality results from the above equations:

$$\|\mathbf{y}(t) - \mathbf{a}(\mathbf{x}(t))\| \leq$$

$$\int_0^t L\,\|\mathbf{y}(s) - \mathbf{a}(\mathbf{x}(s))\|\,ds + \int_0^t \|\mathbf{g}(\mathbf{a}(\mathbf{x}(s)) - [\mathbf{D_x}(\mathbf{a}(\mathbf{x}(s))]\,\mathbf{f}(\mathbf{x}(s))\|\,ds \tag{8}$$

According to the Gronwall inequality (Teman, 1988), if $\zeta(t)$, $\psi(t)$ and $k(t)$ are continuous non-negative functions for $t \in [\alpha, \beta]$ such that

$$\zeta(t) \leq \psi(t) + \int_\alpha^\beta k(\tau)\zeta(\tau)d\tau \quad \forall t \in [\alpha, \beta]\,,$$

then

$$\zeta(t) \leq \psi(\alpha)\exp\left[\int_\alpha^t k(\tau)d\tau d\tau\right] + \int_\alpha^t \psi'(\tau)\exp\left[\int_\alpha^t k(\vartheta)d\vartheta\right]d\tau \quad \forall t \in [\alpha, \beta]\,.$$

If $\psi(t)$ is the second integral on the right hand side of Equation (8) and $\zeta(t)$ is considered to be equal to the scalar quantity on the left hand side of Equation (8), and by substituting $k(t)$ with L, Equation (8) can be rewritten as follows:

$$\|\mathbf{y}(t) - \mathbf{a}(\mathbf{x}(t))\| \leq \int_0^t \exp\left[L(t-s)\right] \|\mathbf{g}(\mathbf{a}(\mathbf{x}(s))) - [\mathbf{D_x}(\mathbf{a}(\mathbf{x}(s)))]\mathbf{f}(\mathbf{x}(s))\| \, ds \, . \quad (9)$$

An upper bound of the norm on the right part of this expression is calculable through the sup norm. This permits the equation to be rewritten as

$$\|\mathbf{y}(t) - \mathbf{a}(\mathbf{x}(t))\| \leq \frac{\exp(Lt) - 1}{L} \sup_x \|\mathbf{g}(\mathbf{a}(\mathbf{x}(s))) - [\mathbf{D_x}(\mathbf{a}(\mathbf{x}(s)))]\mathbf{f}(\mathbf{x}(s))\| \, . \quad (10)$$

Equation (10) describes the fact that the difference between the behaviour of an aggregated and a non-aggregated UAQMS grows exponentially with time. Thus, the more we simplify-summarize the parameters included in an UAQMS, the less valid are the analysis results obtained.

4. Discussion

The fusion of software components towards the compilation of a contemporary UAQMS should take into account all major parameters involved in the AQM cycle in an urban area. Aggregation of parameters is among the most preferred mechanism that such systems adopt in operational basis. The simplification resulting from such an aggregation has already been proven to lead to false 'political' decisions based on initially 'rational' environmental measures, like the adoption of traffic restriction measures in the case of ozone-related air pollution episodes associated with NOx versus VOC sensitive areas in cities. This result is supported by both from theoretical studies (Friedrich and Reis, 2000) and case-study evidence, as in the city of Athens (Moussiopoulos et al., 1997). In addition, the findings of the work presented in this article provide independent support to already published research results stating that integrated ozone abatement strategies should take into account that an emission intervention which is effective on the regional scale may not necessarily be effective for a city and its surroundings (Moussiopoulos et al., 2000)

Another issue that should be addressed is the use of long-term spatial and temporal averages of air quality indicators, as a counter balancing measure in the adoption of 'false' decisions. Such averages can not be used alone, but should be supported by short term, hot-spot related air quality indicators, as suggested by the Air Quality Framework Directive (96/62 EC), and as a consequence, are also influenced by the findings demonstrated above. Suggestions are also made for the used of 'decentralized', decoupled modular elements of UAQMS, that are operated separately by experts of each specific scientific field, in order to avoid aggregation errors: this approach is also lacking the fact that the aggregation 'errors' do not result from the coupling and simplification in the combination of modules, but are implanted within the structure of each module, and thus are independent of their coupling and only depend on their combined usage.

Taking into account that UAQMS are used for the support of strategic planning and policy making, the findings of the theoretical investigation of the behaviour of such systems when aggregation-simplification is examined, support the idea of using such systems for short term detailed air pollution episode management. For long term planning, the use of systems that suggest an upper limit of overall emissions in the urban area of interest can be examined as an effective approach (Zerlauth and Schubert, 1999), in combination with efficient, detailed analysis of local situations with the aid of appropriate analytical tools.

5. Conclusions

Contemporary UAQMS address all information relevant with the urban AQM problem and provide access to appropriate tools while supporting effective decision making. The complexity and the dynamics of the AQM problem, the number of variables involved and the computational power, make the simplification-aggregation of such systems a necessity. Therefore, the use of meteorological and emission related scenarios in UAQMS is very common, and usually results in the simplification-minimization of the parameters involved. According to the previous analysis, the use of scenarios that classify and simplify the behaviour of an UAQMS has as a consequence deviations from the results obtained by the initial system. This deviation grows exponentially with time. Therefore, special care should be taken for the formulation and usage of scenarios within UAQMS, in relation to their desired usage: the scenario approach appears to be less valid for long term strategic planning and better for emergency response planning. This result contradicts current, commonly accepted strategies, according to which the scenario approach is widely used for long term strategic planning. For the latter, the use of 'detailed-non aggregated' methods like a target-based approach seem to be the most appropriate solution (Elsom, 1996).

Acknowledgements

The author would like to express his thanks to the anonymous reviewers for their valuable suggestions.

Appendix

Definition: Lipschitz condition of order a (Morgan, 1996):

Let $\mu: \Re^n \to \Re^n$ be a mapping with $z_1, z_2 \in \Re^n$. Then μ is said to satisfy a Lipschitz condition of order a if positive numbers L and α exist such that:

$$\|\mu(\mathbf{z}_1) - \mu(\mathbf{z}_2)\| \leq L \|\mathbf{z}_1 - \mathbf{z}_2\|^a , \quad \forall \mathbf{z}_1, \mathbf{z}_2 \in \Re^n .$$

According to Axtell (1992), it can be assumed that aggregated system functions like the function $\mathbf{g}(\mathbf{y})$ previously introduced satisfy the Lipschitz condition of order 1, e.g.

$$\|g(\mathbf{y}_1) - g(\mathbf{y}_2)\| \leq L \|\mathbf{y}_1 - \mathbf{y}_2\| , \quad \forall \mathbf{y}_1, \mathbf{y}_2 \in \Re^n .$$

By integrating the difference in the vector fields and by taking the norms of the result, we yield:

$$\|\mathbf{y}(t) - \mathbf{a}(\mathbf{x}(t))\| = \left\| \int_0^t \left[\mathbf{g}(\mathbf{y}(s) - \mathbf{D_x}\mathbf{a}(\mathbf{x}(s))\mathbf{f}(\mathbf{x}(s)) \right] ds \right\| .$$

Adding and subtracting $\mathbf{g}[\mathbf{a}(\mathbf{x}(s))]$ to each term in the right hand of the above equation:

$$\|\mathbf{y}(t) - \mathbf{a}(\mathbf{x}(t))\| =$$

$$\left\| \int_0^t \left[\mathbf{g}(\mathbf{y}(s) - \mathbf{g}(\mathbf{a}(\mathbf{x}(s)))) \right] ds + \int_0^t \left[\mathbf{g}(\mathbf{a}(\mathbf{x}(s))) - \mathbf{D_x}(\mathbf{a}(\mathbf{x}(s))\mathbf{f}(\mathbf{x}(s)) \right] ds \right\|$$

References

Axtell, R. L.: 1992, 'Theory of Model Aggregation with Dynamical Systems with Application to Problems of Global Change', *Ph.D. Thesis*, Carnegie Mellon University, UMI Dissertation Services, Order Number: 9419180.

Dennis, R., Arnold, J. and Tonnesen, G.: 2000, in A. Saltelli, A. Chan and E. M. Scott (eds), *Sensitivity Analysis*, Wiley Series in Probability and Statistics, pp. 329–353.

Elsom, D.: 1996, *Smog Alert – Managing Urban Air Quality*, Earthscan, London, 226 pp.

Fedra, K. and Haurie, A.: 1999, 'A decision support system for air quality management combining GIS and optimisation techniques', *Environ. Model. Assess.* **12**, 125–146.

Friedrich, R. and Reis, S. (eds): 2000, *Tropospheric Ozone Abatement*, Springer-Verlag, Berlin, Heidelberg, ISBN 3-540-66614-1.

Morgan, F.: 1996, 'What is a surface?', *Amer. Mathemat. Monthly* **103**, 369–376.

Moussiopoulos, N., Karatzas, K. and Theodoridis, G.: 1997, 'Air quality in Athens', in *Measurements of Air Pollutants at the Stack and in Ambient Air (Emission and Immision)*, Pollutec '97, Paris, October 1997, pp. 73–84.

Moussiopoulos, N., Sahm, P., Tourlou, P. M., Friedrich, R., Simpson, D. and Lutz, M.: 2000, 'Assessing ozone abatement strategies in terms of their effectiveness on the regional and urban scales, *Atmosph. Environ.* **34**, 4691–4699.

Teman, R.: 1988, *Infinite-dimensional Dynamic Systems in Mechanics and Physics. Applied Mathematical Sciences*, Vol. 68, Springer-Verlag, New York.

Zerlauth, A. and Schubert, U.: 1999, 'Air quality management systems in urban regions: An analysis of RECLAIM in Los Angeles and its transferability to Vienna', *Cities* **16**(4), pp. 269–283.

DESIGNATING AIR QUALITY MANAGEMENT AREAS (AQMAS) IN THE UK: IMPLICATIONS FOR SECURING UK AIR QUALITY OBJECTIVES

N. K. WOODFIELD[1], J. W. S. LONGHURST[1]*, C. I. BEATTIE[1] and D. P. H. LAXEN[2]

[1] *Air Quality Research Group, Faculty of Applied Science, University of the West of England, Frenchay Campus, Coldharbour Lane, Bristol BS16 1QY, U.K.;* [2] *Air Quality Consultants Ltd., 12 St. Oswalds Road, Bristol, BS6 7HT, U.K.*

(* *author for correspondence, e-mail: james.longhurst@uwe.ac.uk; Fax: 00 44 (0) 117 344 3668*)

Abstract. Across the United Kingdom, the majority of local authorities have now completed their first phase of local air quality review and assessment work, as required under the Air Quality Strategy for England, Scotland, Wales and Northern Ireland (DETR, 2000a). Emerging from this first phase work is an anticipated suite of over 110 *Air Quality Management Areas* (AQMAs). These areas are identified locations where one or more of the national air quality objectives are predicted to exceed by specific target dates, and their spatial extent and shape is emerging as highly variable. Local authorities are guided to use a variety of scientific tools to underpin the scientific assessments, and a consideration of uncertainty in both the tools used and subsequent delineation of AQMAs is likely to affect the emerging management areas significantly. With *subsidiarity* underpinning the process of local air quality management (LAQM), local decision-making is anticipated to influence the outcome of the LAQM process in its entirety, with the declaration of AQMAs necessitating the preparation and implementation of air quality action plans. UK experience of the effective management of local air quality, through the designation of AQMAs, demonstrates a valuable framework for other European countries developing mechanisms to manage air quality locally.

Keywords: Air Quality Management Areas (AQMAs), Air Quality Action Plans (AQAPs), decision-making, local air quality management, uncertainty

Acronyms used within the paper

AQAP   Air Quality Action Plan
AQMA   Air quality management areas
AQO   Air quality objective
DETR   Department of the Environment, Transport and the Regions
LAQM   Local Air Quality Management

Water, Air, and Soil Pollution: Focus **2**: 677–688, 2002.
© 2002 *Kluwer Academic Publishers.*

1. Introduction

Local government in the United Kingdom is responsible for the implementation of the local air quality management (LAQM) elements of the Air Quality Strategy for England, Scotland, Wales and Northern Ireland (DETR, 2000a). Following the first phase of a scientific review and assessment process, local authorities have a duty to declare *Air Quality Management Areas* (AQMAs) in locations where specific air quality objectives are predicted to exceed by future target dates. The Air Quality (England) Regulations 2000 (HM Government, 2000), and those of the other UK devolved administrations, specify air quality objectives for seven pollutants (nitrogen dioxide, carbon monoxide, lead, PM10, benzene, 1,3-butadiene and sulphur dioxide). Local authorities in England, Scotland and Wales have a duty to *work towards* achieving the national air quality objectives (AQOs) based upon the air quality standards. The AQOs are health-based, and allow for considerations of cost and benefit, feasibility and overall practicability of actually achieving the air quality standards.

The designation of an AQMA is a statutory requirement (DETR, 2000b) and local authorities have a period of four months from identifying locations where any AQOs are predicted to exceed, to officially declare an AQMA(s). The process of determining AQMA boundaries is emerging as highly variable. Some authorities anticipate declaring a much larger area than that defined by the scientific assessment process of identifying areas of AQO exceedences, and some authorities are anticipating designating their whole authority an AQMA. Methods for determining the exact boundary of an AQMA are also highly variable, and various local factors, including the local authority political framework and collaborative working may influence AQMA spatial and boundary decisions.

By early December 2000, approximately 44% of local authorities in Great Britain had officially completed their first phase of air quality review and assessment work (Woodfield, 2000a). Of these authorities, approximately 45 authorities (11% of all GB local authorities) had identified potential locations where AQO exceedences are predicted. By April 2001, approximately 86% of authorities had completed their assessment work. Even within the small subset of local authorities anticipating declaring AQMAs, various interpretations of the scientific assessment outcomes, and subsequent decision-making following consultation has resulted in a variety of AQMA outcomes.

With just over 20 officially declared AQMAs in the UK, and over ninety declarations expected in the spring and summer of 2001, this paper examines the anticipated AQMAs within local authorities across the UK, and explores the methodologies emerging in their physical designation and influences in the boundary decision-making underway. Evidence will be provided from an initial officer questionnaire survey and also appraisal work underway with respect to all local authorities in the UK for which AQO exceedences are predicted, and for whom AQMAs are a requirement. This work is part of a five-year programme of research investigating the evolution of the LAQM process (Woodfield *et al.*, 2000), which intends to examine the scientific assessment process and political decision-making processes involved in declaring AQMAs in the UK.

2. Methodology

A local authority officer survey and an appraisal of local authority air quality scientific assessments is underway, and is focused upon those local authorities who, on concluding their first phase of air quality assessment work, have predicted potential AQO exceedences. The UK national archive of air quality review and assessment literature is used to arrive at a subset of local authorities predicting exceedences of one or more of the AQOs. A questionnaire was devised in July 2000, to consider the anticipated AQMAs to emerge from the assessment work, prior to the official declaration of the AQMAs. Subsequently, a questionnaire was sent to every local authority that anticipated declaring following consultation.

Such authorities have a duty to designate AQMAs, where exposure to predicted AQO exceedences might be prejudicial to public health. Local authority officers are being surveyed with respect to AQMAs anticipated within their authority, and more specifically on the likely spatial extent of an AQMA(s) and specific boundaries to be used. Local authority officers are also asked for their perception of, amongst other criteria, the likely influences in the AQMA decision-making process, AQMA concerns amongst colleagues and the scientific assessment underpinning the designation of AQMAs.

In conjunction with this survey, an appraisal of the air quality assessments, of those local authorities anticipating declaring, is underway to consider, amongst other criteria, how the treatment of modelling uncertainty is taken into account in identifying areas of AQO exceedences, and also in delineating AQMAs. The scientific appraisal work also considers the mechanisms used to present the outcomes of the assessment process, and includes a judgement as to the thoroughness of validation and degree to which a precautionary approach is taken. With the first phase of scientific assessment process not yet concluded in the UK, the outcomes of the both the appraisal process and officer survey reported in this paper as preliminary findings only, from observations from a smaller sample of UK local authorities than all those anticipating declaring.

3. Appraisal and Survey Results and Discussion

From the 110 authorities so far surveyed, 74 authorities have responded, representing approximately 68% of authorities anticipating AQMAs, and the appraisal results provided in this paper account for approximately 50% of the 110 local authorities as part of the survey. Both subsets of local authorities includes district, unitary, and metropolitan authorities from London, England, Scotland and Wales.

TABLE I

Combinations of air quality objectives predicted to exceed for which AQMAs are required in the UK, within individual local authorities

Air quality objective(s)	% of local authorities
Nitrogen dioxide (NO_2) annual mean (ann.) only	34%
NO_2 (ann.) and PM10 24 hr mean (24 hr)	29%
NO_2 (ann.) and NO_2 hourly mean (hr)	11%
NO_2 (ann.) and PM10 annual mean (ann.)	6%
NO_2 (ann.), PM10 (24 hr) and PM10 (ann.)	4%
PM10 (24 hr) only	2%
NO_2 (ann.), NO_2 (hr), PM10 (24 hr) and PM10 (ann.)	2%
NO_2 (ann.) and sulphur dioxide (SO_2) 15 min mean (15 min)	2%
Other predicted objective exceedence combinations	10%

Table I illustrates the combination of pollutant objectives for which exceedances are predicted, as identified through the appraisals. A third of those authorities declaring are doing so for exceedences of the NO_2 long term objective alone, as a result of traffic emissions, and almost a third are declaring for the long term NO_2 average and the short term PM10 objective. The majority of authorities in this latter category anticipate declaring AQMAs encompassing both objectives. There are no combined pollutant objectives for UK local authorities to work towards, such as a combined SO_2 and PM10 objective. All objectives are for singular pollutants.

With respect to the tools used to predict future AQO exceedences, the vast majority of local authorities undertook modelling, and in the main complex dispersion modelling, alone. Very few used continuous or real time monitoring data alone and subsequent correction factors to identify possible future exceedences, as illustrated in Table II. Most of the models used for predicting future concentrations were advanced and complex dispersion models, with chemical, topographical and meteorological components. A variety of models are commercially available for local authorities to use (DETR, 2000c). Examples of intermediary modelling tools used include the Design Manual for Roads and Bridges (DMRB), and Aeolius, and examples of the more advanced modelling tools used include ADMS-Urban and Aermod. Very few local authorities used simple screening tools alone, and a number used a combination of either complex or simple model and monitoring data to identify potential localised exceedences.

Preliminary observations from the appraisal work indicate that most authorities have addressed modelling uncertainty, as illustrated in Table III, using a variety of methods. The treatment of uncertainty has come under increasing scrutiny as authorities explore the need to delineate AQMA boundaries from modelling outputs,

TABLE II

Use of specific tools for the review and assessment process

Tool criteria	% of local authorities
Modelling tools only	72%
Modelling and monitoring tools	22%
Monitoring tools only	6%

TABLE III

Consideration of uncertainty in modelling and monitoring work

Uncertainty criteria – modelling	% of local authorities
Not considered	7%
Considered insignificant	1%
(a)1 standard deviation (1SD), (b) 2 standard deviations (2SDs)	(a)15%, (b) 12%
Considered (a) ± 20%, (b) ± 30%	(a) 11%, (b) 3%
NSCA methodology for treatment of uncertainty[a]	34%

Uncertainty criteria – monitoring	% of local authorities
Not considered	1%
Assumed (a) 20%, (b) 30%	(a) 5%, (b) 1%
Accepted supplier quality assurance and control	51%
(a) Looked for bias, (b) accounted for bias	(a) 5%, (b) 8%

26% of authorities surveyed made no response to questions above.
[a] Methodology in National Society for Clean Air Guidance on Declaring AQMAs (NSCA, 2000).

and as a consequence of this, various approaches have been applied to the use of uncertainty for determining the extent and shape of AQMAs. Most have adopted a precautionary approach in identifying both AQO exceedence areas and delineating AQMAs, taking into account, for example, variable meteorological circumstances, assumptions of modelling input data and variable street topographies.

Many local authorities have used standard deviations to address modelling uncertainty, and have also used standard deviation from modelling specifically to inform an AQMA boundary. The methodology for determining AQMA boundaries as illustrated in the National Society for Clean Air's Guidance on Declaring AQMAs (NSCA, 2000) has been used to delineate proposed AQMAs by a third of local authorities surveyed. This involves the calculation of an uncertainty value, U, from modelling standard deviations and means of observed data, which is then applied to the concentration of the air quality objective under consideration.

TABLE IV

UK anticipated designation (spatial extent and boundaries) of AQMAs

AQMA spatial criteria	Yes	No
Designate exactly where AQO exceedence predicted?	13%	77%
Designate close as practically possible to AQOE area?	41%	48%
Designate a smaller area than AQO exceedence area?	4%	85%
Designate a larger area than AQO exceedence area?	30%	59%
Designate whole authority?	15%	74%
AQMA boundary criteria	Yes	No
Local Authority Ward boundaries?	4%	85%
Administrative boundaries?	4%	85%
Physical boundaries (roads, embankments, railway lines etc.)	26%	63%
Mixture of boundaries?	12%	77%
AQMA spatial extent/boundary influenced by other authorities?	24%	65%

11% of authorities surveyed made no response to questions above.

Having identified areas of predicted AQO exceedence, local authorities are asked to consider criteria for determining likely AQMA(s) with respect to the spatial extent and likely boundaries to be used, as part of the survey work. Table IV indicates that designated AQMAs are likely to be larger than the actual AQO exceedence areas identified, and in many cases are anticipated to encompass the whole local authority area. Some authorities are emerging as favouring discreet AQMAs, for which specific actions are required to address exceedences in each designation. The use of physical boundaries are anticipated more so than administrative boundaries (other than borough boundaries) in delineating AQMA(s), and both the AQMA spatial extent and boundary used by an individual authority is anticipated to influence decisions in neighbouring authorities in almost a third of authorities surveyed.

The need to address traffic congestion is clearly the greatest influence in decisions on the spatial extent of anticipated AQMAs, as illustrated in Table V. This is to be anticipated with traffic emissions accountable for over 95% of anticipated AQMAs, either in combination or their entirety, in the UK. Formal designations are also considered to improve the environmental profile of an authority and thus potentially influence the AQMA spatial extent. Pressures are, within some local authorities, being exerted by individual businesses and interested parties to have their property included within proposed AQMA(s) so as to ensure their involvement in subsequent local air quality action planning processes. Conversely, within other local authorities, pressures to not include specific locations, properties and recept-

TABLE V

Influences in AQMA decision-making in the UK

Decision-making criteria	Yes	No
Local need to address traffic congestion?	58%	28%
Political Party and/or new Cabinet style government?	30%	57%
Improve environmental profile of authority?	28%	58%
Local economic stability?	23%	64%
Potential for local regeneration?	22%	65%
Local electorate pressure to declare?	12%	74%
Employment prospects within the authority?	8%	78%

14% of authorities surveyed made no response to questions above.

ors are emerging as influences in the potential shape, extent and use of boundaries for delineating local AQMA(s).

When surveyed as to whether planning and other colleagues within the authority have specific concerns with respect to the local air quality management process, the majority of authorities preparing to designate AQMAs are concerned with the potential impact of any such designation on planning blight, as indicated in Table VI. Property blight was the main concern demonstrated during workshops provided in various regions of the UK over the course of the year 2000 (Woodfield, 2000b).

The scientific uncertainty of tools, particularly for advanced dispersion modelling tools, used for air quality assessment and prediction is also a concern of the colleagues of those undertaking the assessment process. However, the majority of officers specifically involved in using such tools consider the scientific assessment process to be sufficiently robust for the declaration of AQMAs, as illustrated in Table VII.

Whilst the scientific assessment process, now almost complete across the UK, is resulting in the identification of predicted exceedences of the air quality objectives across areas of the UK, the transposition of such exceedances into anticipated and eventual AQMAs is the result of further non-scientific processes. The first is an assessment of the exposure of members of the public to such predicted exceedences. Where no such exposure exists, relevant to the individual objective timescale, an authority is not required to declare an AQMA (DETR, 2000b). The appraisal of the assessment outcomes has so far revealed that only two thirds of local authorities refer to a risk of public exposure, explicitly or otherwise, in relation to their local air quality management work.

A second process influencing the ultimate AQMA designation is one of political decision-making. Democratically elected council members have the responsibility in most local authorities to consider and thus confirm the final AQMA outcomes within the authority. Local authority officers are asked to consider the level to

TABLE VI

Concerns with respect to potential AQMAs raised by internal colleagues

Potential concern criteria	Yes	No
Possible blight of properties?	81%	15%
Financial implications of declaring?	55%	41%
Scientific uncertainty of tools?	64%	32%
Process for EHOs only?	19%	77%
Complexity of process?	39%	57%
Apparent lack of robustness of process?	20%	76%
No concerns?	3%	93%

4% of authorities surveyed made no response to questions above.

TABLE VII

Consideration of the scientific assessment process

Scientific assessment criteria	Yes	No
Sufficiently robust for the authority to declare an AQMA(s)?	69%	26%
Sufficient guidance provided to undertake scientific assessment?	49%	46%
Sufficiently resourced to undertake scientific assessment?	38%	57%

5% of authorities surveyed made no response to questions above.

which they perceive the relevant council members to understand the various elements of the air quality review and assessment process and AQMA declaration process. Table VIII illustrates this perceived understanding. Collaboration across the authority and a need to engage the public was perceived as most understood by members, who were considered less able to understand the science underpinning AQMAs and the requirements of the areas.

The local authority officers are asked to consider the involvement of various professions with the authority, with respect to both the scientific assessment process and the anticipated AQMA designation process, and the involvement of specific agencies external to the authority. The Environment Agency (in England and Wales) or Scottish Environmental Protection Agency (SEPA), and the Highways Agency (in England and Wales), and the equivalent in London and Scotland, are the two important agencies with whom authorities are required to collaborate in their LAQM work. Tables IX and X illustrate the results from the survey.

TABLE VIII

Perceived understanding by council members of elements of the LAQM and AQMA process

Criteria	Average score
Need for effective consultation with the public?	2.34
Requirements of the scientific assessment process?	2.43
Need for collaboration across authority departments?	2.52
Need for collaboration with external agencies	2.73
Resource implications of the assessment process?	2.82
Requirements within an AQMA?	2.88
Science underpinning AQMA?	2.97

(Average scores on a scale of 1–5, with 1 = well informed, 2 = quite well informed, 3 = some understanding, 4 = little understanding, 5 = no awareness at all).
10% of authorities surveyed made no response to questions above.

TABLE IX

Involvement of local authority officers with the air quality scientific assessment process and anticipated AQMA designation process

Profession	Average score assessment process	Average score AQMA designation
Environmental Health Officers?	1.15	1.16
Transport Planning Officers?	2.97	2.50
Strategic Planning Officers?	3.49	2.95
Development Control Planners?	3.78	3.16
Local Agenda 21 Officers?	3.79	3.75
Economic Development Offices?	4.25	3.75

(Average scores on a scale of 1–5, with 1 = very involved, 2 = quite well involved, 3 = some involvement, 4 = little involvement, 5 = no involvement at all).
10% of authorities surveyed made no response to questions above.

4. Conclusions

The focus of the local air quality management process in the UK is the identification of local air pollution hot spots, through the scientific assessment process, and subsequent designation of AQMAs and preparation of air quality action plans (AQAPs) to articulate air quality improvements. This requires both sound science to underpin the process, and effective collaboration, consultation and wide engagement with respect to the management process, and more importantly action planning process, to secure local air quality improvements (Beattie *et al.*, 2000).

TABLE X

Involvement of external agencies with the air quality scientific assessment process and anticipated AQMA designation process

Environment agency involved	Percentage involved	Highways agency	Percentage
National	18%	National	23%
Regional	48%	Regional	23%
National and Regional	7%	National Regional	4%
No involvement	27%	No involvement	50%

10% of authorities surveyed made no response to questions above.

From the results presented, preliminary observations indicate that most AQMAs are to be based upon predictions derived from the use of advanced modelling techniques, with very few authorities basing such predictions on monitored data and the subsequent use of correction factors to predict future concentrations. Uncertainty in both modelling and monitoring tools has been addressed by the majority of authorities moving towards AQMA declaration. However, there is, as expected, a wide spectrum of treatment of uncertainty. Indeed, modelling techniques and the consideration of uncertainty have varied considerably between consultants instructed by local authorities to undertake the modelling component of the assessment work, and the treatment of uncertainty has come under increasing scrutiny as authorities explore the need to delineate AQMA boundaries from modelling. As a consequence of this, various approaches have been applied to the use of uncertainty for determining the extent and shape of AQMAs. The expectation of government was that local authorities would use recognised models (DETR, 2000c) validated against quality assured monitoring data and other model inter-comparison exercises. Most local authorities, in following the government guidance, have addressed model validation, using continuous monitoring data sets locally to do so.

The majority of local authorities have adopted a precautionary approach in identifying both AQO exceedence areas and delineating AQMAs, taking into account, for example, variable meteorological circumstances, assumptions of modelling input data and variable street topographies. Even within the small subset of officially designated AQMAs to date, a quarter of the local authorities have taken the precautionary approach of declaring the whole authority an AQMA, thereby reducing any potential risk of failing to identify any locations of public exposure to predicted AQO exceedences.

Methods for determining the exact boundary of an AQMA are also highly variable, with some authorities choosing to use administrative boundaries, and others choosing physical boundaries and features, such as roads, railway lines and rivers. Various local factors, including the local authority political regime, may influence

AQMA boundary decisions, and such factors may impact on the overall effectiveness and implementation of measures within an Action Plan, to subsequently deliver improved local air quality (Beattie *et al.*, 2000).

The Environmental Health and Protection profession within local government has facilitated and steered the air quality management work in the UK to date. With the need to address local and indeed regional traffic-related problems emerging as a real catalyst for seeking air quality improvements, the transport planning profession within local government is evidently more involved with the management process than other authority colleagues. However, with respect to larger trunk roads emerging as giving rise to local pollution hot spots, the need to involve the Highways Agency (or equivalent) will become particularly important as local authorities seek to prepare their action plans. Local authority involvement with the Highways Agency, at the regional or national level, has been slow to date, and low compared to involvement with the Environment Agency.

Collaboration with external agencies, neighbouring local authorities, stakeholders and the wider public is understood to be an important aspect of the local air quality management process, by officers and politicians alike. Local actions and initiatives will be required to assist with the improvement of local pollution hot spots, which cannot otherwise be articulated through national measures (Beattie *et al.*, 2001). Effective collaboration will in future, therefore, determine the extent to which such actions to mitigate local pollution hot spots will succeed.

A variety of approaches in both identifying locations of predicted AQO exceedences and the subsequent designation of AQMAs are emerging as the first phase of the assessment of local air quality draws to a conclusion in the UK. With over 110 AQMAs anticipated, of which only just over 20% are officially declared, the process of official designation is in its infancy. However, early indications suggest wide variation in the spatial extent of AQMAs, irrespective of the extent of public exposure to predicted AQO exceedences, and as a consequence of differing interpretation of largely modelling outcomes and use of uncertainty. This variability reflects the principal of subsidiarity and illustrates the importance of local decision-making, which is to become more so as UK local authorities move closer to preparing their local air quality action plans.

References

Beattie, C. I., Longhurst, J. W. S. and Woodfield, N. K.: 2000, 'Air quality management: Challenges and solutions in delivering air quality action plans', *Ener. Environ.* **11**(6), 729–747.

Beattie, C. I., Longhurst, J. W. S. and Woodfield, N. K.: 2001, 'Air quality management: Evolution of policy and practice in the UK as exemplified by the experience of English local government', *Atmos. Environ.* **35**, 1479–1490.

Department of the Environment, Transport and the Regions (DETR) in partnership with the Scottish Executive, the National Assembly for Wales and the Department of the Environment for Northern Ireland: 2000a, *The Air Quality Strategy for England, Scotland, Wales and Northern Ireland. Working Together for Clean Air*, CM 4548, The Stationary Office Ltd., London, 192 pp.

Department of the Environment, Transport and the Regions (DETR): 2000b, LAQM.G1(00) *Framework for Review and Assessment of Air Quality*, The Stationery Office Ltd., London.

Department of the Environment, Transport and the Regions (DETR): 2000c, LAQM.TG3(00) *Review and Assessment: Selection and Use of Dispersion Models*, The Stationery Office Ltd., London.

HM Government: 2000, *The Air Quality (England) Regulations 2000*. Statutory Instruments, 2000, No. 928, The Stationary Office Ltd., London.

National Society for Clean Air and Environmental Protection (NSCA): 2000, *Air Quality Management Areas: Turning Reviews into Action*, NSCA. Brighton, UK.

Woodfield, N. K., Beattie, C. I. and Longhurst, J. W. L.: 2000, 'The air quality strategy for England, Scotland, Wales and Northern Ireland – An evolving process', *J. Inst. Environ. Sci.* **9**(3), 5–8.

Woodfield, N. K.: 2000a, 'Air Quality Review and Assessment Report to Department of the Environment', Transport and the Regions (DETR) and UK Devolved Administrations. (December 2000).

Woodfield, N. K.: 2000b, 'Air Quality Management Areas in the UK: Workshop Outcomes', (www.uwe.ac.uk/aqm/centre/AQMAs/workshop.html).

INTEGRATED ASSESSMENT OF ABATEMENT STRATEGIES TO IMPROVE AIR QUALITY IN URBAN ENVIRONMENTS, THE USIAM MODEL

ANTONIO MEDIAVILLA-SAHAGÚN*, HELEN M. APSIMON and RACHEL F. WARREN

Integrated Assessment Unit of the Environmental Measurement, Modelling and Assessment Group (EMMA) – Air Pollution, Imperial College of Science, Technology and Medicine, T.H. Huxley School, Room 426, RSM Building, Prince Consort Road, London SW7 2BP, U.K.
(* author for correspondence, e-mail: amediavi@imp.mx, present address: Mexican Petroleum Institute, Eje Central Lazaro Cardenas 152, AP. Postal 14-805, CP 07730, Mexico DF, Mexico)

Abstract. Planning effective strategies to combat air pollution in a major city such as London requires integration of information on atmospheric concentrations and where they exceed prescribed air quality standards, detailed data on emissions and potential measures to reduce them including costs, and a good understanding of the relative contributions of different emission sources to pollutant concentrations plus the remaining background. The Urban Scale Integrated Assessment Model (USIAM) is designed as a tool to integrate such information, and to explore and assess a variety of potential strategies for improving air quality. It is based on the same principles as the Abatement Strategies Assessment Model (ASAM) that has been used in the UN Economic Commission for Europe. To start with the USIAM model is being developed with respect to the particulate PM_{10} only, and in particular the primary particulate contribution. The secondary particulate is treated as part of the background superimposed on the primary particulate concentrations; this may need to be treated more specifically at a later stage, particularly with respect to nitrate formation over the city. The USIAM model therefore sets out to examine a selection of severe episode conditions as well as long-term annual average concentrations, and aims to find strategies that are successful in eliminating exceedance of the prescribed target concentrations. By ranking different options for abatement of emissions, for example in terms of cost or ease of implementation, the USIAM model can also select and prioritise different potential strategies.

Keywords: emission reduction strategies, integrated assessment, PM_{10}, re-suspended dust, urban air quality

1. Introduction

Planning effective strategies to combat air pollution in a major city such as London requires integration of information on atmospheric concentrations and where they exceed prescribed air quality standards, detailed data on emissions and potential measures to reduce them including costs, and a good understanding of the relative contributions of different emission sources to pollutant concentrations plus the remaining background.

Water, Air, and Soil Pollution: Focus **2:** 689–701, 2002.
© 2002 *Kluwer Academic Publishers.*

An ideal methodology to integrate such disparate sources of information into a working system within a cost effective framework is the emerging and growing research area named Integrated Assessment. This methodology was first applied in the Regional Air Pollution Information and Simulation (RAINS) model developed at the International Institute for Applied Systems Analysis (Alcano, 1990) focused on establishing science-based abatement strategies for acid rain in Europe.

This same methodology was applied then to construct the Abatement Strategies Assessment Model (ASAM) (ApSimon et al., 1994; 1996; Warren et al., 1999; 2000) used in the development of international protocols under the Convention on Long-Range Transboundary Air Pollution of the United Nations Economic Commission for Europe.

This study describes the Urban Scale Integrated Assessment Model (USIAM) which, based on the same principles as the ASAM model, was designed as a tool to investigate and assess a variety of potential strategies to reduce air pollution in Major Urban Developments. In the ASAM model, the aim was to set emission ceilings for each of 40 European countries to reduce acid deposition cost-effectively, whereas USIAM is concerned with attainment of air quality standards through combined action spanning the 33 local authorities in Greater London and parts of 19 districts in the counties of Kent, Surrey, Berkshire, Buckinghamshire, Hertfordshire and Essex which lie between the M25 motorway and the Greater London boundary.

From the UK and EU air quality standards already set, the Nitrogen dioxide and PM_{10} targets are likely to be the most difficult to achieve. To start with, the USIAM model is being developed with respect to the particulate PM_{10} only, and in particular the primary particulate contribution. The secondary particulate is treated as part of the background superimposed on the primary particulate concentrations; this may need to be treated more specifically at a later stage, particularly with respect to nitrate formation over the city. Both the European and UK air quality standards for PM_{10} are 40 $\mu g \cdot m^{-3}$ for the annual mean and a limit of 35 exceedances per year of 50 $\mu g \cdot m^{-3}$ over 24 h periods. The later is far more restrictive and corresponds to an annual mean of about 29 $\mu g \cdot m^{-3}$ (DETR, 2000b).

The USIAM model therefore sets out to examine annual average concentrations, and aims to find strategies that are successful in eliminating exceedance of the prescribed target concentrations. By ranking different options for abatement of emissions in terms of cost of implementation and population exposure, the USIAM model can also select and prioritise different potential strategies. To do this it requires appropriate information on different options for reducing emissions.

Thus USIAM is intended to provide a tool for integrating information on sources and pollution imported into the city, atmospheric dispersion and resulting concentrations relative to air quality standards, and the costs and benefits of different options for emission reduction. The model is described in more detail below.

2. The USIAM Model Information Elements

The application of USIAM requires the integration of the following five elements:

2.1. AIR QUALITY STANDARDS (AQS) AND METEOROLOGICAL CONDITIONS

For PM_{10}, the specification of an AQS target is required. This can be either the Limit Concentration Mean over fixed 24 hr periods, which for both Europe and the UK corresponds to 50 $\mu g \cdot m^{-3}$ not to be exceeded more than 35 times a year by 2005; or an Annual Average Mean, which for the same cases as above is set to 40 $\mu g \cdot m^{-3}$ achievable by 2005. The later is far more restrictive and corresponds to an annual mean of about 29 $\mu g \cdot m^{-3}$ (DETR, 2000a), so that is the one selected as USIAM target concentration. This standard is derived from a statistical analysis that correlates the measured number of days with PM_{10} concentrations above 50 $\mu g \cdot m^{-3}$ for a given year, versus the PM_{10} annual mean for that same year. The analysis includes the years between 1992 and 1999, and produces an annual average concentration 'equivalent' to the 24 h mean standard equal to about 29 $\mu g \cdot m^{-3}$ (DETR, 2000b).

The model also requires a full year meteorology field divided in hourly measurements, and obtained from a suitable monitoring station (e.g. Heathrow airport).

2.2. EMISSION INVENTORIES

The model requires an accurate emissions inventory with good spatial resolution, and a break down into different source categories. The inventory provided by the Greater London Authority (ex London Research Centre) has been used so far, and corresponds to 1996 emissions (London Research Centre, 1997). It includes both stationary and mobile sources spanning the area of interest (i.e. London Metropolitan area up to the M25 motorway). For current development the emissions sources are attributed to the respective area in which they lie; i.e. Central, Inner, Outer, Out of London or M25 motorway; and are differentiated into different classes.

The stationary sources include both the Industrial and Area sources; the former are differentiated by source height, the latter are assigned to 1 km^2 grid squares and include the Railway, Aircraft, Shipping and Minor Road Traffic, and the Domestic Combustion sources. In the Inventory, the mobile sources on major roads are assigned to road links, and are treated as street canyon sources by USIAM. These are differentiated in vehicle types comprising Motorbikes, Petrol Cars, Diesel Cars, Petrol Light Goods Vehicles (LGVs), Diesel LGVs, Buses, Rigid Heavy Goods Vehicles (HGVs) and Articulated HGVs.

The emissions data for each vehicle (vehicle counts) are differentiated into 3 temporal variations: AM Peak, Inter peak and PM peak. However, since the model is calculating annual average concentrations to compare against the target AQS, also in annual average units, the hour-by-hour variability in emissions is not important.

2.3. PRE-CALCULATED SOURCE RECEPTOR MATRIXES

These are pre-calculated for Annual Average meteorology, and represent the dispersion of unitary sources using the ADMS-Urban 1.53 model (McHugh *et al.*, 1997), relating concentrations contributed at specified receptor points across London per unit emission from each source. Since the model calculates dispersion of primary PM_{10}, no chemistry is involved, thus there is a linear relationship between changes in emissions and changes in concentrations. This linearity allows source apportionment for the concentrations at each receptor point, and the effects of emission abatement. In principle, receptor points may be specified anywhere (for example at monitoring stations), and allowance may be made for local circumstances to include street canyon effects and contributions at road site sites from local traffic. At present receptor points are taken at the centre of 1 km^2 grid squares across the domain (3172 receptor points).

Source-Receptor matrices are calculated for 13 different generic source types with unitary emissions (e.g. 1 Ton y^{-1}), used to calculate dispersion from all sources contained in the inventory. The matrices comprise 2 types of area matrices to model all area sources of the inventory, 8 types of point matrices for all point sources, and 3 types of street canyon matrices for the eight road source types. The particular parameters specified for each matrix are presented in the appendix at the end of this document. Besides, provision has also been made to reflect the size spectrum of particulates, divided into PM_{10}, $PM_{2.5}$, and $PM_{0.1}$ for the ultra-fine fraction, although this requires corresponding (as yet non existent) detail in the emission data.

It has to be noted that from the 23,925 street canyon sources contained in the inventory, only the ones in each grid of the domain with the highest PM_{10} emissions – i.e. one per grid square – were selected to be explicitly modelled as a street canyon as a way of representing hot spots in each square. The amount of street canyon sources selected is 1,991. All remaining road sources in each grid are treated as Area sources and used to calculate an 'Urban Background' concentration together with all other sources present in each 1 km^2 grid. The resulting concentration from the busiest street canyon is then added on top of the calculated background for each grid square as a 'worst case'.

2.4. INFORMATION ON ABATEMENT OPTIONS FOR MOBILE SOURCES AND THEIR COSTS

The aim of this information is to relate different policies to their effect on base case emissions. A wide range of alternative strategies can be applied to mobile sources, comprising three different general types of measures as follows (Atkins, 1999; 1998; Auto Oil Program II, 2000a, b, 1999a, b, c; Elsom, 1996; 1999):

- 'Emission Control' measures, which include measures focused on emission reduction from vehicles by: adding end of pipe technology (e.g. fitting particle traps), switching to alternative fuels – or no fuels at all –, and the enforcement of emission standards in predetermined areas (e.g. Low emission zones, which effectively accelerate the upgrading of the vehicle fleet to higher efficiency and stricter emission standards).

- 'Traffic Management and Improved Public Transport' measures, including measures focused on decreasing traffic flow levels, such as parking permit schemes, and public transport priority schemes for travel reduction times.
- 'Combined' measures, which try to combine the aggregated effect of two or more of the measures mentioned in both points above.

Each alternative comprises both emission reduction effects and cost of implementation. These data are used to construct an 'Emission Reduction Strategies Database' which is then utilized in an optimisation procedure that identifies the strategy, or the application sequence of strategies, that should be implemented in order to reduce concentrations and comply with AQS entailing the least possible cost.

2.5. INFORMATION ON ABATEMENT OPTIONS FOR STATIONARY SOURCES AND THEIR COSTS

The model is able to consider measures for stationary sources as well as mobile, but due to the much smaller impact these emissions have on the total concentration when compared with the mobile ones (LRC, 1997), they were left out of the optimisation analysis. These sources will become increasingly more important as the mobile source emissions are progressively controlled. For further development of the model, a limited number of selected sources and their associated technological measures will be included in the form of a cost curve. The latter are usually based on an annualised cost over a specified period (e.g. up to 2005). At the moment, other work is in progress in the Integrated Assessment Unit at Imperial College, as well as in AEA Technology (DEFRA, 2001) focused on the construction of UK and London specific cost curves.

3. USIAM Model Structure

Given all the information mentioned above, the USIAM model provides a tool for rapid screening of effective strategies for eliminating exceedance of PM_{10} air quality standards. Based on the pre-calculated source-receptor matrices from atmospheric dispersion modelling, a large number of different options can be examined in a matter of minutes, rather than weeks when compared with other models that deal with primary PM_{10}, which need to model each source explicitly (Mediavilla-Sahagún, 2002). Estimates of the costs of implementation for different transport emission reduction strategies are used to develop a ranking scheme that reflects overall effectiveness taking account of environmental targets at national and international level, and of population exposure. Promising strategies can then be examined in more detail using a full and more detailed treatment of atmospheric dispersion over an extended period.

A flow diagram representing the USIAM Model Structure can be seen in Figure 1 below. In this figure, the grey boxes represent the information sources mentioned above. The concentration in each grid square is built up as a sum of contributions, including background from sources not included in the inventory and secondary particulates, according to the following equation:

$$C_1 = C_{Local} + C_{UrbanB} + C_{Coarse} + C_{Secondary} \tag{1}$$

Where:

C_1 = Total Concentration in grid square number 1 (repeated for all the other 3171 receptor points)

C_{Local} = Contribution to concentration from the local Street Canyon (this does not influence neighbouring squares)

C_{UrbanB} = Contribution from the Urban Background, i.e. the influence of emissions from the entire domain to the particular grid square being calculated, including all area and minor road sources.

C_{Coarse} = Contribution from the component not directly related with vehicle emissions, (i.e. the component between 2.5 and 10 μm), constant over the whole domain and time scale modelled (urban scale, annual average) and equal to 8 μg·m^{-3} according to measurements taken by the monitoring sites operated by the former UK Department of the Environment, Transport and the Regions (DETR), as discussed in APEG (1999) and Stedman et al. (2001).

$C_{Secondary}$ = Contribution imported from other UK locations and Long Range international sources, assumed to be constant over the whole domain and equal to 10 μg·m^{-3}. This contribution is produced when ammonia reacts with nitric and sulphuric acids in the atmosphere forming an aerosol of ammonium nitrate and sulphate crystals, and is estimated from measurements of sulphate aerosols by the acid deposition network. (APEG, 1999).

4. Results

Some of the USIAM model results are presented below. Figure 2 presents USIAM's footprint output modelling the primary PM$_{10}$ component and adding the Coarse (8 μg·m^{-3}) and Secondary (10 μg·m^{-3}) components as a uniform blanket for the 1996 LRC inventory, annual average case meteorology. This is the output from the initialisation module (see Figure 1). The black concentric rings represent the areas as defined by the Mayor of London (Greater London Authority, 2001), being the first the Central area; the second the Inner (i.e. excluding the central area), and the third the Outer London areas (excluding the previous two). The section between the border of the Outer area and the limits of the domain excluding the fourth ring is the 'Not in London' area, and finally, the fourth ring represents the M25 motorway. As expected, some of the worst locations in terms of concentration are the whole central London area and the A4- M25-Heathrow area. In this plot, some 11% of the

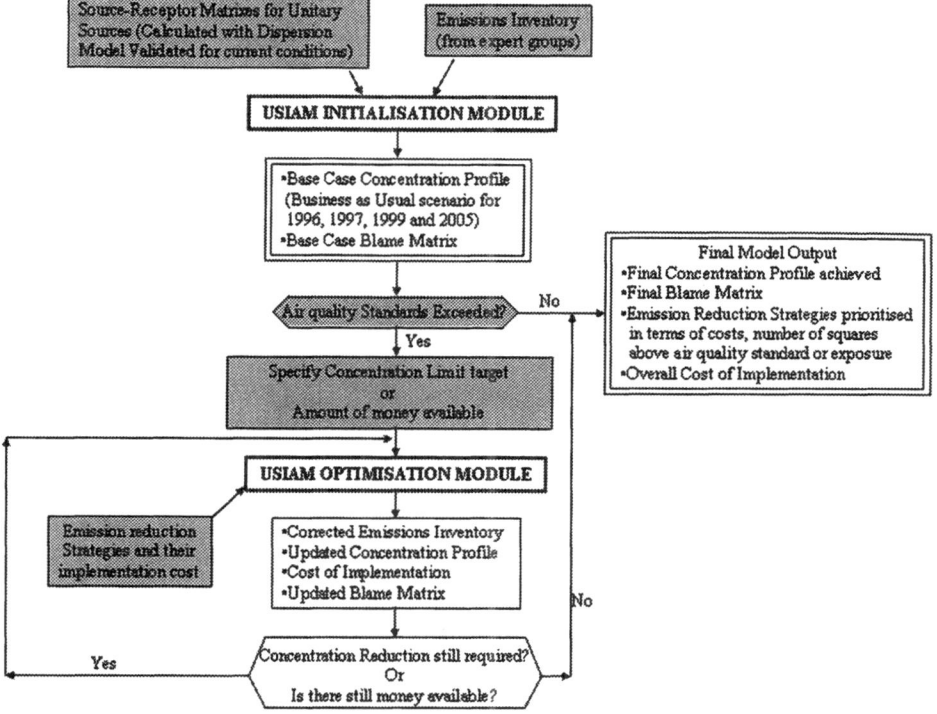

Figure 1. USIAM model structure.

grid squares considered by the model (i.e. the pale grey grids onwards according to the colour scale) are above the strictest standard of 29 μg·m^{-3}.

However, this 11% of the total grid squares represents 100% of the Central, 37% of the Inner and 28% of the M25 Motorway assigned grid squares.

A major concern when modelling primary PM$_{10}$ concentrations is the uncertainty in the emissions inventory. To exemplify this, if we analyse the emissions, the primary particulate is composed of two different fractions, a finer one below 2.5 μm originated mainly by vehicle exhausts, tyres and brakes, forming the majority of the London PM$_{10}$ Emissions Inventory; and a coarser one between 2.5 and 10 μm, not accounted for in the inventory (APEG, 1999; ApSimon *et al.*, 2001).

Observations of the coarser component indicated a value of 8 μg·m^{-3}, as mentioned above. Initially, this was assumed to be constant across the domain, as in Figure 2 and Equation 1.

It is still not clear what are the sources of this coarse contribution, which are not included in the emissions inventory. One possible source of this is re-suspended road dust. A first attempt at modelling this contribution has been made by taking approximate estimates of emission factors based on observations in other cities and discussions with other specialists. A revised version of Figure 2 in which this new

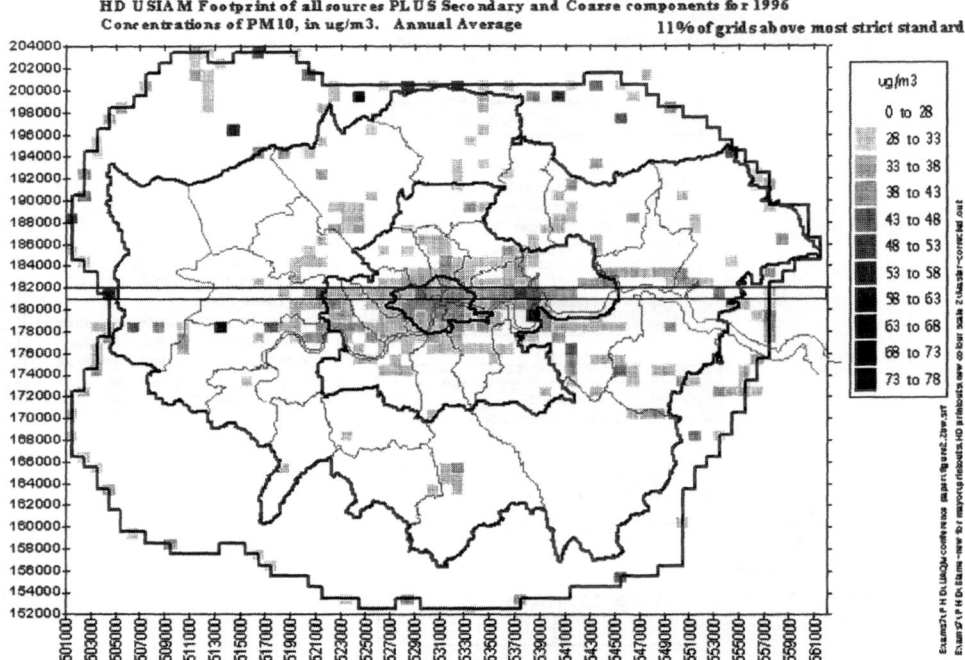

Figure 2. USIAM 1996 annual average concentration footprint for all sources plus Secondary and Coarse.

approach is taken is shown in Figure 3. The emission factors used were 0.6 g·km^{-1} for Heavy Goods Vehicles and 0.05 g·km^{-1} for cars, which in total is equivalent to about 60–65% of the vehicle exhaust PM$_{10}$ emissions for London. The resulting concentrations are similar to those presented in Figure 2 for the central area of London. This suggests that re-suspended dust could explain a high proportion of the coarse PM$_{10}$ matter in this area, although other sources such as Industrial activities and construction works also contribute.

It should be noted that if these assumptions are correct, it means that the concentration close to major roads and thus the street canyon contributions will increase, making them more difficult to attain air quality standards. Also, the application of measures to reduce emissions such as particulate traps will not have any effect on the coarse component contributions. However, it could also mean that some other measures, such as road cleaning might effectively help in reducing overall PM$_{10}$ concentrations at those locations, as discussed by Torp and Larsen (1996).

Another form of output from the USIAM model is the so-called 'Blame Matrices'. These matrices are the explicit result of the source apportionment concept presented earlier, and show the proportion of the total PM$_{10}$ concentration in each grid square that can be attributed to specific sources within a specific area. They

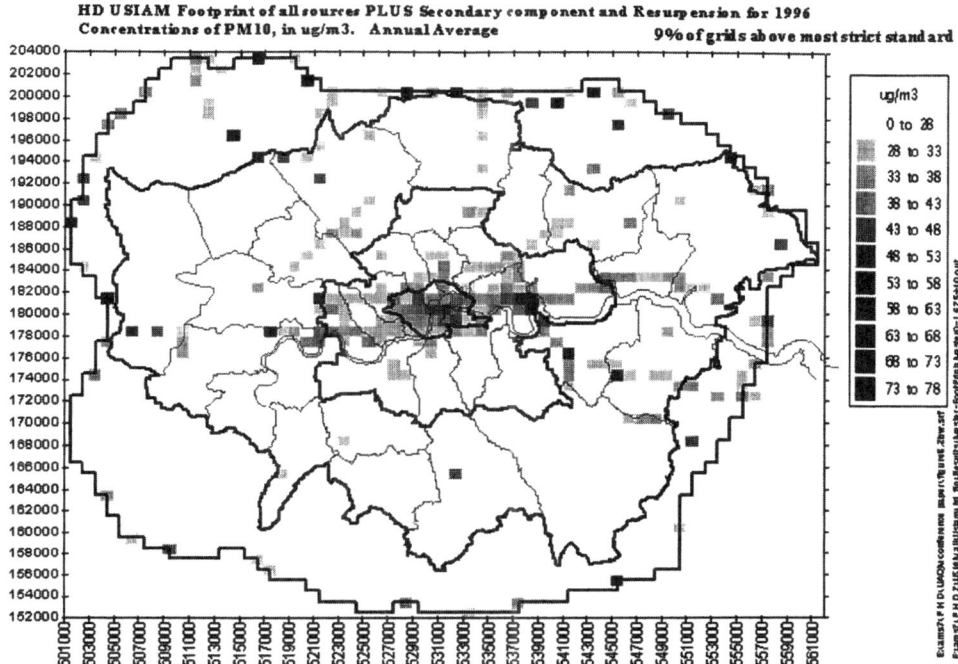

Figure 3. USIAM 1996 annual average concentration footprint for all sources plus Secondary and Resuspension.

also indicate the percentage contribution made by any particular source type within an area to the total concentration at any specific receptor point.

To illustrate this concept, a row of grid squares across central London was considered, represented by the horizontal band across Figure 2. The grid squares covered by this band have the designation numbers from 1770 to 1830 from left to right. At each of these grid squares, the contribution to the total grid concentration made by different source areas has been deduced and plotted against the grid square designation numbers mentioned above. This composes the 'Blame Matrix' for this particular cross section, and it is shown in Figure 4. In this figure, the bottom axis represents the grid square designations, and the left the PM_{10} concentration in $\mu g \cdot m^{-3}$. The top black segment of each column represents the contribution to the total grid square concentration from the Local Street Canyon, the next section down (dark grey) represents the contribution from the sources in the Outer London area, the pale grey the contribution from Inner London area and the white section the Central London area. The diagonal-bars segment represents the contribution from the M25 motorway, rather small compared with the others. All these are superimposed on a uniform blanket of Coarse and Secondary component contributions, represented by the bottom two segments.

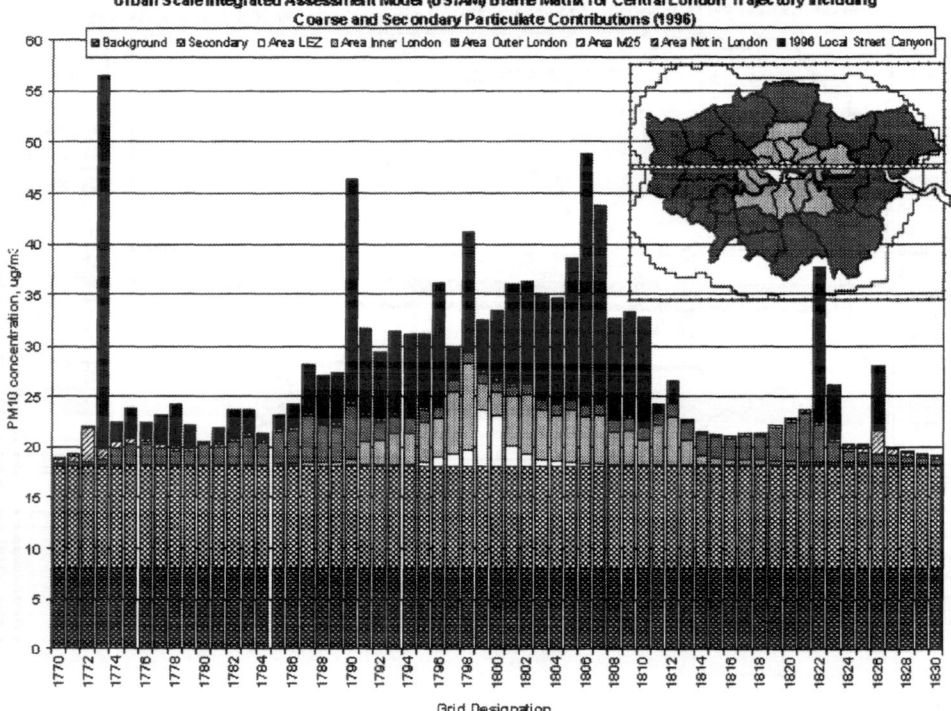

Figure 4. USIAM Blame Matrix for central London cross-section of Figure 2.

Finally, as an illustration of the kind of scenarios USIAM can analyse, Figure 5 shows the corresponding footprint equivalent to Figure 2, but considering the effect of fitting particle traps to all diesel vehicles, thus reducing PM_{10} emissions from these by 85%, together with the effect of fitting three way catalysts to all petrol vehicles and the better engine management systems that come with it, which represents a 40% reduction in PM_{10} emissions from these type of vehicles. Thus, this scenario represents the 'top limit' or best possible technological scenario for 1996. In this scenario, still some 2% of the grid squares considered are above the most strict standard, representing 20% of the total central area grid squares, and 7% of the M25 ones.

5. Conclusions

The USIAM model has been developed as an efficient tool for exploring abatement strategies since it takes a fraction of the time to run when compared with explicitly modelling all sources although produces similar results (Mediavilla-Sahagún, 2002; EMMA, 2001). Besides, it is an effective tool for source apportionment analysis and assessment, particularly useful when comparing the effects of different

Scenario 1. HD USIAM Footprint of all sources, Secondary and Coarse + Particle Traps and 3 way catalyst for 1996
Concentrations of PM 10, in ug/m3.   **2% of grids still above most strict standard**

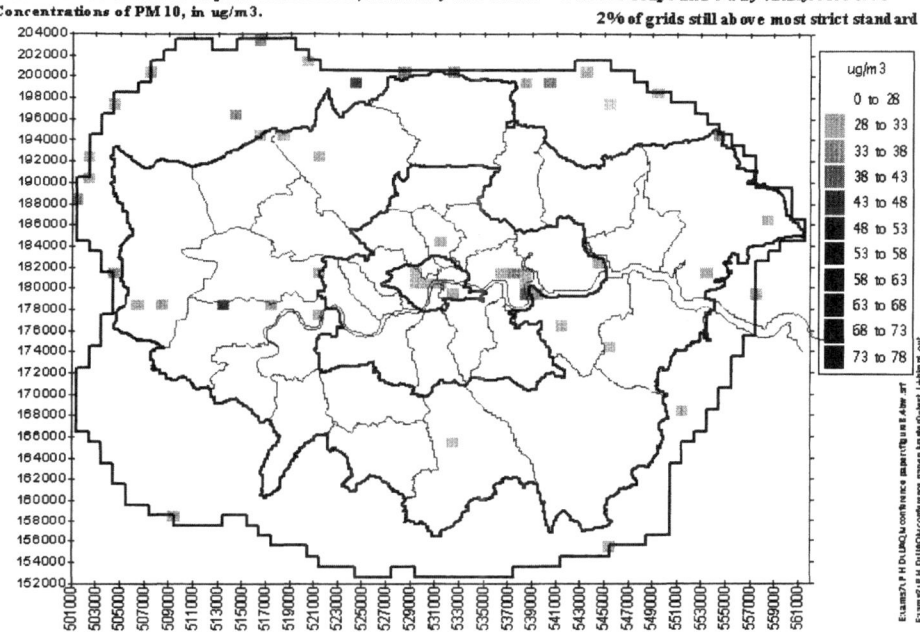

Figure 5. USIAM Scenario 1, Footprint concentrations for all sources as in Figure 2, PLUS particle traps fitted to all diesel vehicles and 3 way catalytic converters to all petrol vehicles.

strategies. The model will now be applied to a range of scenarios for London including both emission reduction and traffic control measures. These results are reported elsewhere (Mediavilla-Sahagún, 2002).

Appendix

Parameters specified in ADMS to construct the generic source-receptor matrices for:

TABLE I

Area sources

Type of source in the inventory	Source type	Height (m)	Exit velocity $(m.s^{-1})$	Temp. (°C)
Area	A	15	0	20
Minor road	A	5	0	20

TABLE II

Point sources

Stack height range in inventory (m)	Source type	Height (m)	Diameter (m)	Exit velocity $(m.s^{-1})$	Volume $(m^3 s^{-1})$	Temp. (°C)
0 < = Height < 9	P	8	0.35	8.25	0.75	45
9 < = Height < 12	P	10	0.3	7.5	0.53	200
12 < = Height < 26	P	15	0.5	12	2.34	300
26 < = Height < 46	P	36.6	0.5	15	2.95	250
46 < = Height < 63	P	55	5	17.1	336	127
63 < = Height < 85	P	70	1.35	2.84	4.01	157
85 < = Height < 150	P	100	2.74	17	100	230
150 < = Height < 220	P	199	6.5	19	630	130

TABLE III

Street canyon sources

Parameter	Location		
	Central and Inner London areas	Outer and not in London areas	M25 Motorway
Road Elevation (m)	0	0	0
Road Width (m)	24.5	15	30
Canyon Height (m)	18	12	3

Acknowledgements

This work is being sponsored by the Mexican Petroleum Institute (Instituto Mexicano del Petróleo).

References

Alcano, J., Shaw, R. and Hordijk, L. (eds): 1990, 'The RAINS Model of Acidification', *Institute of Applied Systems Analysis (IIASA)*, Kluwer Academic Publishers, Vienna, Austria.

APEG (Airborne Particles Expert Group): 1999, *Source Apportionment of Airborne Particulate Matter in the United Kingdom*, DETR, UK.

ApSimon, H. M., Warren, R. F. and Wilson, J. J. N.: 1994, 'The abatement strategies assessment model-ASAM: applications to reductions of sulphur dioxide emissions across Europe'. *Atmos. Environ.* **28**, 649–663.

ApSimon, H. M. and Warren, R. F.: 1996, 'Transboundary air pollution in Europe', *Ener. Pol.* **24**(7), 631–640.

ApSimon, H. M., Rejlova, K., Mediavilla-Sahagún, A., Gonzales, T., Warren, F. R. and Colvile, R.: 2001, 'Modelling Urban Background concentrations', Poster presented at the 3rd International Conference on Urban Air Quality, Loutraki, Greece.

(AOP II) Auto Oil Program II.: 1999a, 'Cost Effectiveness Study Part II: The TREMOVE model 1.3', European Commission, Standard and Poor's DRI and K.U. Leuven, Draft Final Report, August 1999.

Auto Oil Program II.: 1999b, 'Cost Effectiveness Study Part III: The Transport Base Case', European Commission, Standard & Poor's DRI and K.U. Leuven, Draft Final Report, August 1999.

Auto Oil Program II.: 1999c, 'Non Technical Measures', European Commission Final Report, November 1999.

Auto Oil Program II.: 2000a, 'Cost Effectiveness Study Part IV: Simulation and Integrated Assessment of Policy Measures', European Commission, Standard & Poor's DRI and K.U. Leuven, Draft Final Report, July 2000.

Auto Oil II Program II.: 2000b, 'A Report from the Services of the European Commission', European Commission Final Report, October 2000.

DEFRA (Department for Environment, Food and Rural Affairs): 2001, 'The Cost of Reducing PM_{10} and NO_2 Emissions and Concentrations in the UK: Part 1: PM_{10}', produced on behalf of the DEFRA by *AEA Technology*, October 2001, London, UK.

(DETR) Department of the Environment, Transport and the Regions UK.: 2000a, 'The Air Quality Strategy for England, Scotland, Wales and Northern Ireland', January 2000.

DETR.: 2000b, 'Review and Assessment: Pollutant Specific Guidance', Part IV The Environment Act 1995, Local Air Quality Management (LAQM.TG4(00), *DETR*, London, UK.

Elsom, D.: 1996, 'Smog alert: managing urban air quality', *Earthscan*, London, UK.

Elsom, D.: 1999, 'Development and implementation of strategic frameworks for air quality management in the UK and European Community', *J. Environ. Plann. Manage.* **42**(1), 103–121.

(EMMA) Environmental Measurement, Modelling and Assessment group-Integrated Assessment Unit: 2001, Personal communication and Meeting with Cambridge Environment Research Consultants (CERC) Representatives.

Greater London Authority: 2001, 'The Mayor's Draft Air Quality Strategy for public consultation', *Greater London Authority*, London, UK.

London Research Centre (LRC): 1997, 'London Atmospheric Emissions Inventory', (Release 2a 6th April 1998), London, UK.

McHugh, C. A., Carruthers, D. J. and Edmund, H. A.: 1997, 'ADMS-Urban: an air quality management system for traffic, domestic and industrial pollution', *Int. J. Environ. Poll.* **8**(3–6), 666–674.

Mediavilla-Sahagún, A.: 2002, 'Integrated Assessment Modelling Applied to Particulate Concentrations and Urban Air Quality Management', Ph.D. Thesis, University of London.

Stedman, J. R., Lineman, E. and Conlan, B.: 2001, 'Receptor modelling of PM10 concentrations at a UK national network monitoring site in central London', *Atmos. Environ.* **35**(2), 297–304.

Torp, C. and Larsen, S.: 1996, 'Modeling population exposure to air pollution near the road network of Norway, and the effects of measures to reduce the exposure', *Sci. Tot. Environ.* **189/190**, 35–40.

WS Atkins: 1998, 'Evaluation of transport measures to meet NAQS objectives – Stage 1 Final Report to the DETR', *WS Atkins Planning Consultants*, Epsom, Surrey, UK.

WS Atkins: 1999, 'Evaluation of transport measures to meet NAQS objectives – Stage 2 Final Report to the DETR', *WS Atkins Planning Consultants*, Epsom, Surrey, UK.

Warren, R. F. and ApSimon, H. M.: 1999, 'Uncertainties in integrated assessment modelling of abatement strategies: Illustrations with the ASAM model', *Environ. Sci. Pol.* **2**, 439–456.

Warren, R. F. and ApSimon, H. M.: 2000, 'The role of secondary particles in European emission abatement strategies', *Integr. Assess.* **1**, 63–86.

ATMOSPHERIC LEVELS OF NITROGEN OXIDES AT A GREEK OIL REFINERY COMPARED WITH THE URBAN MEASUREMENTS IN ATHENS

P. D. KALABOKAS[1]*, J. G. BARTZIS[2] and P. PAPAGIANNAKOPOULOS[3]

[1] *Academy of Athens, Research Center for Atmospheric Physics and Climatology, 131, Tritis Septemvriou, str., Athens, Greece;* [2] *National Center of Scientific Research 'Demokritos', INT-RP, Environmental Research Laboratory, 15310 Ag. Paraskevi Attikis, Greece;* [3] *University of Crete, Department of Chemistry, Heraklion, Greece*
(* author for correspondence, e-mail: phatmcli@otenet.gr, fax: +(301) 883 2048)

Abstract. Oil refining is among the industrial activities that emit considerable amounts of air pollutants into the atmosphere. Nitrogen oxides are important air pollutants that are emitted by oil refineries as products of combustion processes. The ambient air concentrations of nitrogen oxide (NO) and nitrogen dioxide (NO_2) were monitored continuously at a site close to an oil refinery, near the city of Corinth in Greece, during autumn 1997 together with the main meteorological parameters. The contribution of the oil refinery to the measured atmospheric levels of nitrogen oxides was estimated. The ambient air concentration of nitrogen oxides in the area surrounding the oil refinery were generally lower than the ambient air concentrations in the urban area of Athens in Greece, and the NO_2 levels were always below the existing air quality standards. The influence of the refinery emitted NO_x in the photochemical production of ozone seems to be more important in terms of human and vegetation exposure given the high ozone backgrounds measured in the area.

Keywords: industrial atmospheric pollution, nitrogen oxides, oil refinery, urban air pollution

1. Introduction

Nitrogen oxides (NO_x) play an important role in the physicochemical processes of the troposphere as in the presence of hydrocarbons and sunlight they contribute to the formation of ozone and other photochemical oxidants. They are also considered as hazardous substances (mainly NO_2) and for that reason they are monitored in the atmosphere of urban areas. NO_x are present in the urban and industrial atmosphere essentially as a result of human activities, associated with combustion processes utilizing fossil fuels (Finlayson-Pitts and Pitts, 1986; Warneck, 1988).

The operation of oil refineries, is associated with the emission of many pollutants into the atmosphere, originating from the different refining processes. The most important among them are hydrocarbons, NO_x, sulfur compounds and particulate matter (Westaway and Brockis, 1978). NO_x are emitted from oil refineries due primarily to the combustion of various hydrocarbon fuels. The formation of NO_x occurs either via the reaction of nitrogen and oxygen gases in the input air, or

Water, Air, and Soil Pollution: Focus **2**: 703–716, 2002.
© 2002 *Kluwer Academic Publishers.*

via the oxidation of nitrogen in the combustion fuel. The major sources of NO_x in an oil refinery are heaters and boilers, while NO_x are mostly in the form of NO at the time of release. The estimated amount of the refinery NO_x emissions is 2100 t yr^{-1} and the operated intensity of the refinery can be assumed as constant. The most important factors affecting the NO formation in a combustion process are the proportion of excess air, the flame temperature, and the residence time in the combustion zone. Although the NO_x emissions by a given heater can be reduced via adjusting the air supply, the actual levels of these emissions depend on the design of the burner and the furnace (Westaway and Brockis, 1978). In urban areas the main sources of NO_x are fuel combustion processes related with car traffic (Finlayson-Pitts and Pitts, 1986).

The air quality guidelines of the World Health Organization for NO_2 are 200 μg m^{-3} (104 ppb) for 1 hr values and 40 μg m^{-3} (21 ppb) for the annual mean (WHO, 1999). The European Union introduced the same values as standards. The hourly standard of are 200 μg m^{-3} (104 ppb) should not be exceeded more than 18 times per year (EU Directive 99/30).

The subject of this work is the determination of atmospheric NO_x levels in the area surrounding an oil refinery followed by a comparison with the NO_x levels at different urban sites in Athens, Greece. The refinery under study is among the largest oil refineries in Greece, located about 60 km west of the city of Athens and producing a wide spectrum of petroleum products with a nominal capacity of crude oil processing of 4.75 million tons per year. The location of the refinery is presented in Figure 1.

2. Methodology

The field measurements were carried out at a fixed site about 1 km west of the oil refinery boundaries during a two-week campaign in autumn 1997 (25 October–6 November). The ambient air concentrations of NO and NO_2 were recorded as 10 min averages together with the basic meteorological parameters (wind and temperature). A commercial chemiluminescence NO_x analyzer was used (Environment S.A. AC 31M) with a detection limit of about 1 ppb. The meteorological measurements were performed by a sonic anemometer, which has been crosschecked by conventional instrument during the whole period of the campaign. The instrument thresholds for wind direction and speed were 0.4°, and 0.05 m s^{-1}, respectively. The NO_x levels around the refinery were compared with the measurements collected during the same period at four air pollution monitoring sites in the urban area of Athens (Patission, Peireas, Peristeri, Smyrni) and the characteristic variations in the industrial and the urban sites were discussed.

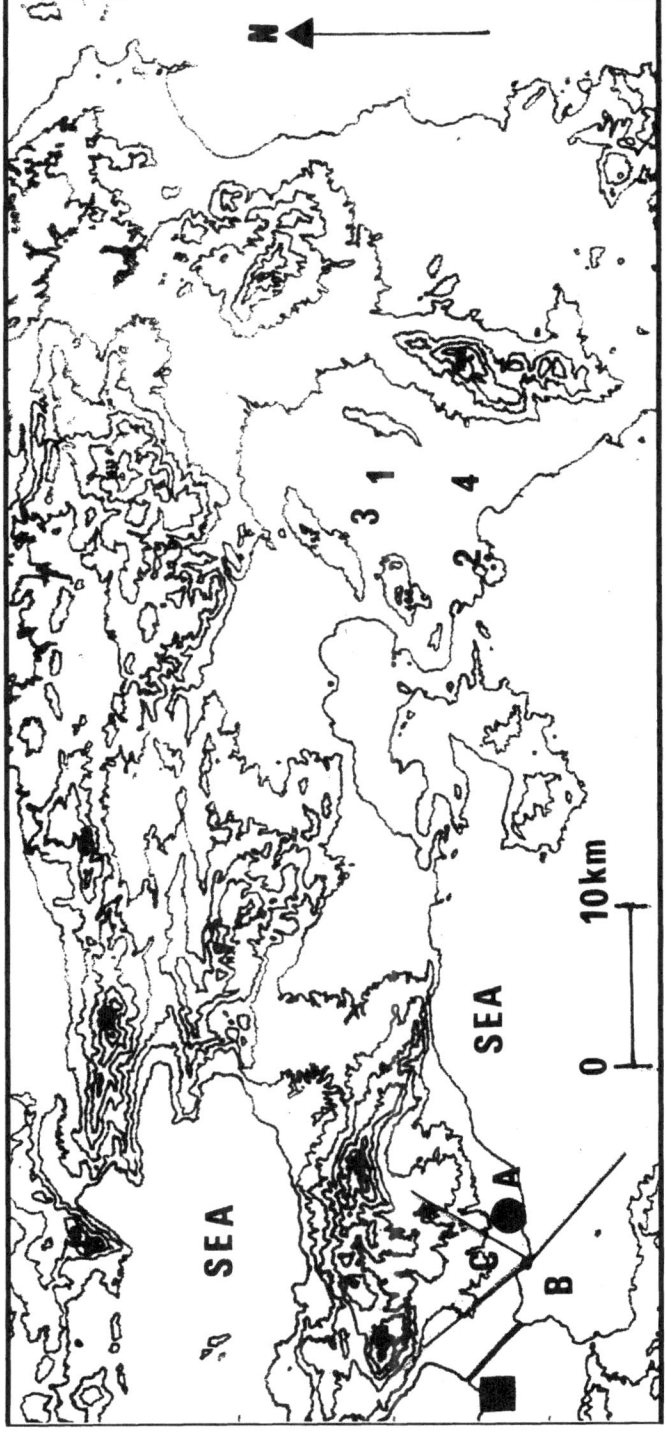

Figure 1. Map of the oil refinery area including height contours of 200 m. The oil refinery is marked with a black square. A, B, C, are the wind direction sectors as described in the text. The urban stations of Athens are: Patission (1), Peireas (2), Peristeri (3) and Smyrni (4).

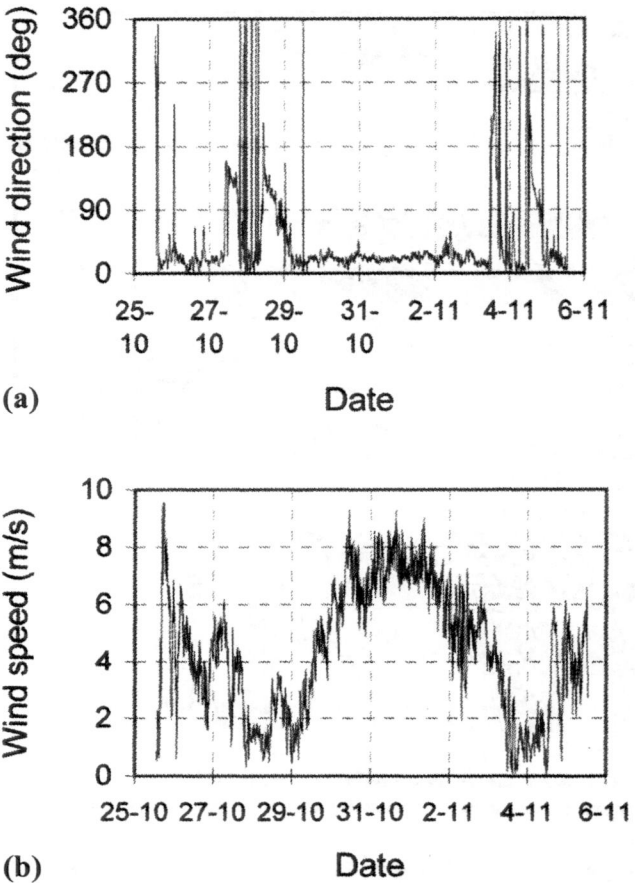

Figure 2. Variation of wind direction (a), and wind speed (b), during the campaign at the refinery.

3. Results and Discussion

The meteorological data are presented in Figures 2a (wind direction) and 2b (wind speed). The main directions of the winds during the campaign were NE and SE. The wind speed varied mostly from 2 to 7 m s^{-1}. The synoptic conditions during the examined period are close to the normal average mid-fall conditions consisting of upper-air trough occasional passages enhancing the surface winds. The observations of the site were consistent with the corresponding ones of the National Observatory of Athens, located close to the city-centre.

The NO and NO$_2$ measurements during the campaign are presented in Figures 3a and b, respectively. The NO values varied from 1 to 84 ppb with a mean value of 3.4±6.6 ppb and a median value of 1.5 ppb. The NO$_2$ concentrations varied from 1 to 48 ppb (mean value 9.9±8.7 ppb, median value 6.7 ppb). Since the air pollution and meteorological parameters were recorded every 10 min, a

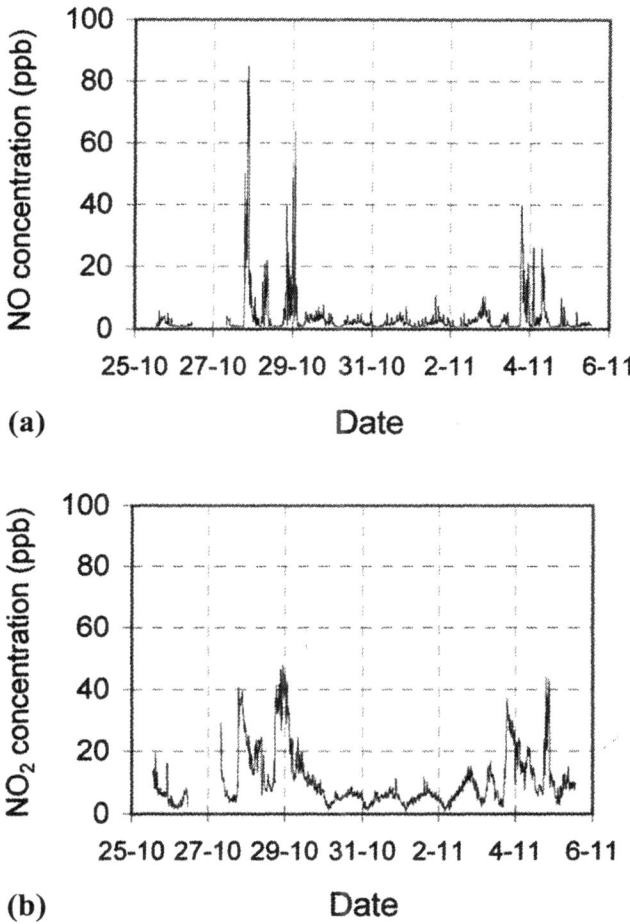

Figure 3. Variation of nitrogen monoxide (a), and nitrogen dioxide (b), during the campaign at the refinery.

sufficient number of data were collected at the end of two weeks campaign (about 1600 values). Therefore, considering the high variability of the wind direction during the campaign, it is possible to estimate the contribution of the oil refinery to the nitrogen oxides levels in the particular measuring site. From the above figures it can be seen that the highest nitrogen oxides concentrations were recorded at nighttime with low wind speeds and mostly with east winds blowing from the oil refinery direction. An additional remark is that due to the public holiday of 28 October (Tuesday) the site must have been subject to increased traffic due to the prolonged weekend.

As mentioned earlier emissions of NO_x originate also from car traffic. In fact, the measuring site was very close to the National highway connecting Athens to Peloponnisos and therefore influence from this type of emissions is expected. In

TABLE I

Average levels of nitrogen oxides per sector for low dispersion conditions (winds 0–4 m s^{-1}): Sector A, 30–130°, oil refinery and traffic pollution. Sector B, 130–320°, maritime air. Sector C, 320–30°, traffic pollution

	NO (ppb)	NO$_2$ (ppb)	NO$_x$ (ppb)	Number of measurements
Sector A	7.8±14.4	19.7±13.7	27.5	223
Sector B	1.0±0.4	6.9±3.9	7.9	95
Sector C	4.5±6.3	14.6±8.3	19.1	384

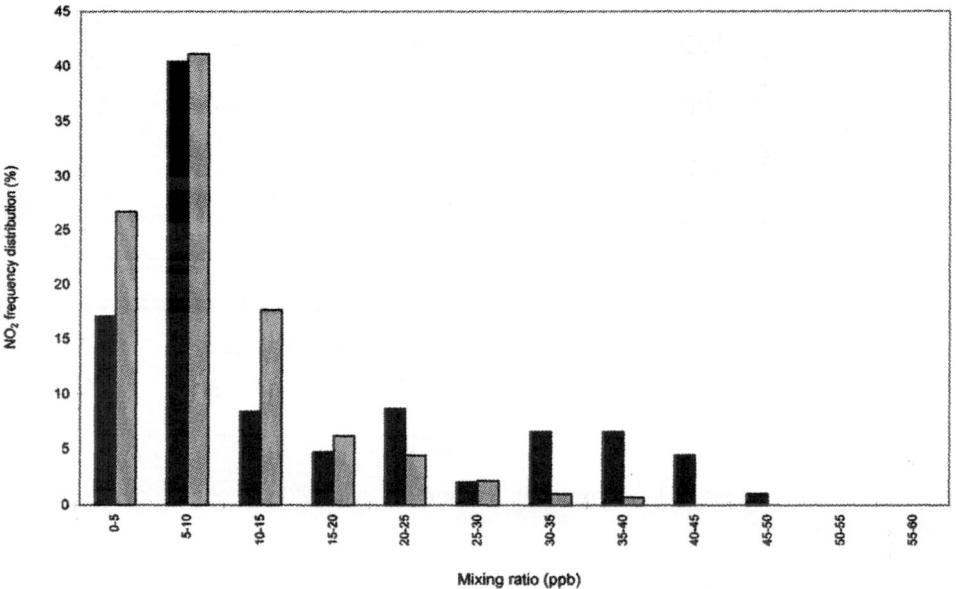

Figure 4. Frequency of occurrence of NO$_2$ values from sector A (oil refinery and traffic pollution, in black), and sector C (traffic pollution).

order to estimate the respective contributions of the refinery and the highway traffic to the measured atmospheric NO$_x$ levels, the data were divided according to the prevailing wind direction and then the average values influenced by each one of the above mentioned major emission sources were considered. Therefore, the prevailing wind directions were grouped to three wind sectors: (a) Sector A, 30–130°, which receives air masses both from the oil refinery and from the National highway. (b) Sector B, 130–320°, corresponding to air masses coming from the sea that are not directly polluted. (c) Sector C, 320–30°, which is influenced by the highway emissions only. The frequency distribution of values in the A, B and C sectors were 21, 7 and 72%, respectively. The average NO_x atmospheric concentrations in the above three sectors under low dispersion conditions (wind speeds 0-4 m s^{-1}) are presented in Table I. In this table, the difference in the NO_x levels of the air masses coming from the three wind sectors A, B and C is clearly observed. It should be first noticed that the NO_2 levels in sector B, influenced by the sea, are significantly lower than in sectors A and C, although they are relatively high for a rural area. This is probably due either to a local pollution recirculation caused by the sea-breeze, or to a long-range pollution transport from the urban areas of Athens (60 km to the west) and Corinth (15 km to the east) as shown in Figure 1. The NO_2 concentrations were higher than the NO ones in all sectors. The NO_x pollution that is attributed to the automobile emissions from the nearby National highway can be estimated by taking the difference between the average values of sectors C (highway) and B (sea). Similarly, the pollution originating from the oil refinery can be estimated by taking the difference between the average values of sectors A(highway + refinery) and C (highway). In general, the average atmospheric NO_x pollution caused by the highway traffic and the refinery emissions are of the same order, about 10 ppb for weak winds (0–4 m s^{-1}). The NO_x levels in air masses influenced by the oil refinery show high variability, as indicated by the higher standard deviations of the mean values of this sector (Table I). This pattern implies also the presence of very high NO_x values in this group, which is verified for the NO_2 values by plotting the concentration distribution separately for sectors A and C (Figure 4). Indeed, the highest NO_2values in the range of 30 to 50 ppb, which are about 10% of the total values, are coming exclusively from the oil refinery side. Nevertheless, the NO_x concentrations did not exceed the levels of the WHO guidelines or the EU standards, even if the 10 min average values are considered as hourly values.

In order to investigate in further detail the presence of high NO_x values, the average diurnal variations of NO and NO_2 at the measuring site were plotted in Figures 5a and b, respectively. In the same figures the diurnal profiles of the wind sectors A and C that are influenced by the oil refinery and the National highway respectively, were also plotted. In general, NO_2 values were high at nighttime, but a small peak was also observed in the morning hours. The nocturnal levels of the oil refinery influenced air are about 2–3 times higher than the concentrations due only to car traffic while the diurnal ones are almost equivalent for both wind sectors. The low dispersion conditions usually observed at night favor in general the increase of the ambient air pollutant concentrations but in urban profiles the nocturnal increase is not usually so high (Viras and Siskos, 1992).

For a better assessment of the NO_x air quality levels in the area surrounding the oil refinery the collected data of NO_x during the campaign period (25 October– 6 November, 1997), shown in Figure 6, in four air pollution monitoring sites in the urban area of Athens (Patission, Peireas, Peristeri, Smyrni) are comparatively examined. Patission is a central urban station located in a heavy traffic street. Peireas is also next to busy roads but on the same time is at the edge of the urban

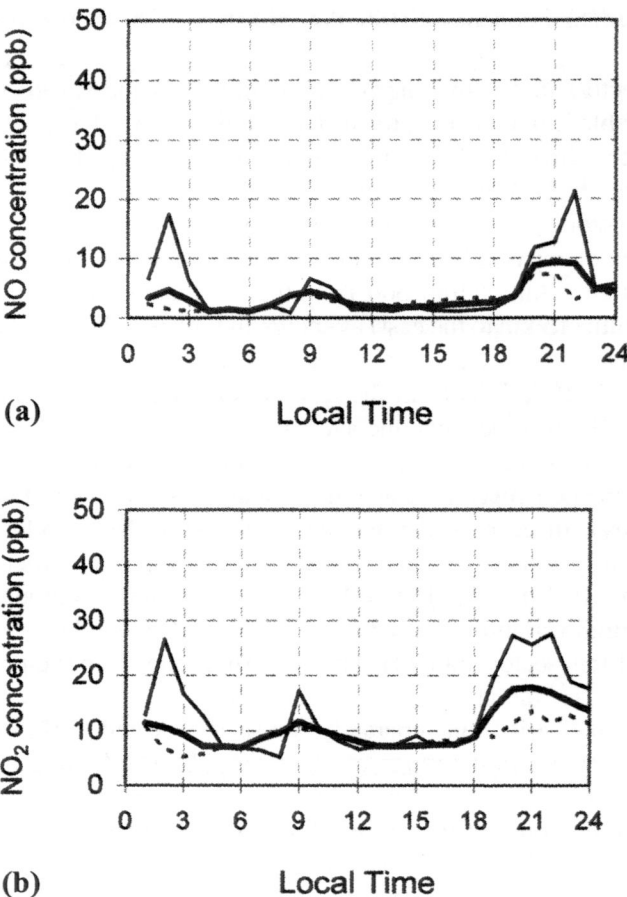

Figure 5. (a) Diurnal variation of the average NO concentration (bold solid line), the NO concentration in sector A (oil refinery and traffic pollution, solid line) and the NO concentration in sector C (traffic pollution, dashed line) during the campaign at the oil refinery. (b) Same as for (a) but for NO_2.

agglomeration close to the harbor. The stations of Peristeri and Smyrni are located in suburbs west and south of the center, respectively (Figure 1). They are also under the influence of traffic especially during the morning and the evening hours. The average atmospheric concentrations of NO_x in these sites are presented in Table II. The refinery NO_x levels are about one order of magnitude lower than the most polluted station (Patission) and about 3 times lower than the least polluted examined urban station (Peristeri), respectively. The NO average values at the oil refinery are 25 to 6 times lower in comparison to Patission and Peristeri, respectively. The NO_2 refinery values are about 8 times lower than the corresponding values of Patission and only about 40% lower than the corresponding urban values at Peristeri. For a more detailed comparison, especially of the high nocturnal values

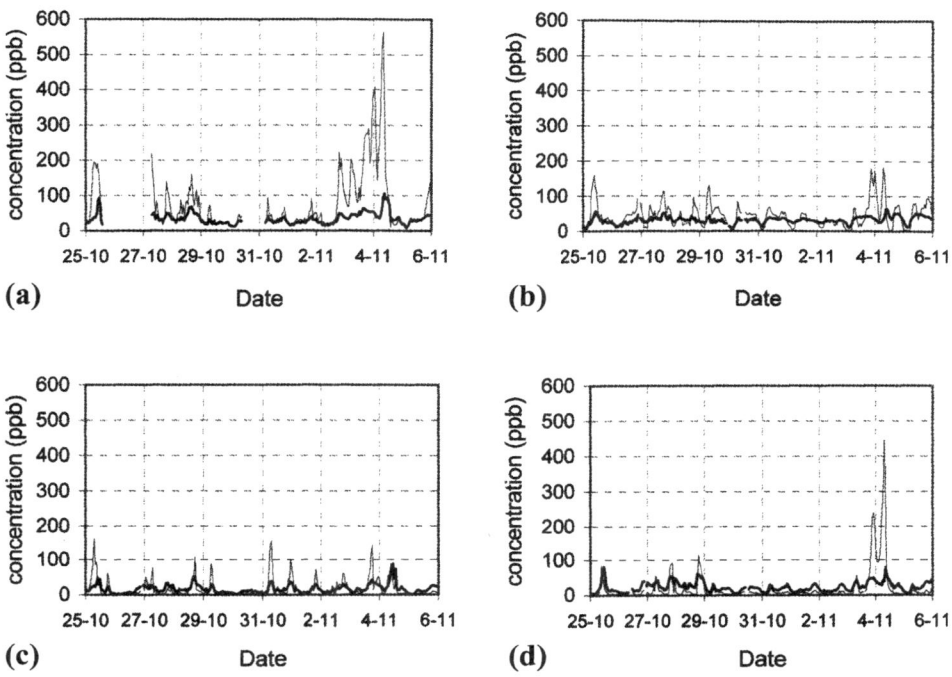

Figure 6. Variation of hourly values of NO (solid line) and NO$_2$ (bold solid line) in four air pollution sites Athens during the campaign period: Patission (a), Peireas (b), Peristeri (c) and Smyrni (d).

TABLE II

Average concentrations of nitrogen oxides at the oil refinery and at four sites in the urban area of Athens during the period 25 October–6 November, 1997

	Oil refinery	Patission	Peireas	Peristeri	Smyrni
		(ppb)			
NO	3.4	79	46	20	32
NO$_2$	9.9	35	33	17	24
NO$_x$	13.3	114	79	37	56

at the oil refinery, the diurnal variation of NO and NO$_2$ in the four urban sites in Athens is plotted in Figures 7a and b, respectively. It comes out that for both NO and NO$_2$ the nocturnal average value at the refinery becomes about the half of the night average at Peristeri with the lowest urban values but the NO average of sector A (oil refinery influenced) is at the same level with Peristeri. In order to have a complete picture of the concentration range found at the urban sites it has to be

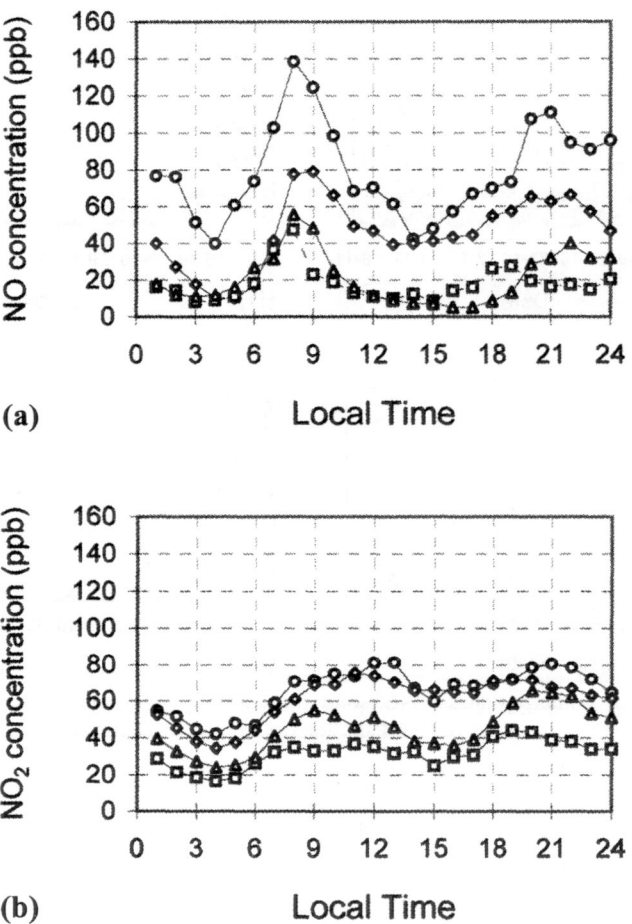

(a)   Local Time

(b)   Local Time

Figure 7. (a) Diurnal variation of the average NO concentration in four air pollution sites in Athens during the campaign period: Patission (cycles), Piraeus (diamonds), Smyrni (triangles) and Peristeri (squares). (b) Same as for (a) but for NO_2.

mentioned that the nocturnal NO levels at Patission were 5 times higher than in Peristeri while the nocturnal NO_2 levels were twice as high. For the urban sites the morning peak was equivalent or even less than the night one, which makes a difference with the NO_x profile obtained at the oil refinery showing much stronger peaks at night. The urban measurements in our campaign were in general below the air quality standards (WHO, 1999). However, there is a marked seasonal variation of NO_x in the urban area of Athens, with a maximum value in the December–January period, where the NO_x levels were on the average 20% higher than the mean concentrations during the autumn period (MinEnv Report, 1998; Kalabokas *et al.*, 1999), which suggests that the difference between the urban and the refinery

NO_x levels can be even higher in favor of the urban measurements during winter months assuming standard operating conditions at the refinery.

Considering in more detail the characteristics of the urban NO_x levels in Athens, an important fact is that their levels remained relatively unchanged during the first 11 yr period (1987–1997) of complete NO_x measurements. This happened despite the major pollution abatement measure consisting in the replacement of the approximately 30% of the old-technology passenger cars with new technology 3-way catalytic vehicles, during the period 1990–1992 (Kalabokas *et al.*, 1999). On the contrary the same measure proved to be very successful regarding the CO urban concentrations, which verifies that the 3-way catalyst converters are not as efficient for NO_x as for CO suggesting that the NO_x pollution problem tends to be more persistent. This picture is enhanced after examining the interannual variation of SO_2, the other major air pollutant. The SO_2 levels during the same period dropped by about 70% following reductions of the sulfur content in Diesel Oil (Kalabokas *et al.*, 1999).

A distinct difference between the oil refinery and the urban NO_x levels in Athens was found in the ratio NO/NO_2. On the average the NO_2 levels at the oil refinery were about 3 times higher than the NO mixing ratios. On the contrary, the NO mixing ratios in the four urban sites in Athens were higher by 20 to 120% than the NO_2 ones. This can be attributed to the fact that the NO in the urban areas was usually in excess relatively to ozone, thus the rapid conversion of NO to NO_2 through its reaction with ozone is delayed especially during the night due to the restricted availability of ozone. The relatively high rural ozone levels measured in the area, exceeding 60 ppb during summer (Kalabokas *et al.*, 1996, 2000; Kourtidis *et al.*, 1996; Kouvarakis *et al.*, 2000) should create favorable conditions for a rapid NO to NO_2 conversion around the refinery as ozone is always in excess. Therefore, the majority of NO_x species around the oil refinery are in NO_2 form, a more toxic atmospheric substance than NO with documented health effects, which has led to the implementation of air quality standards. Nevertheless, as mentioned the recorded NO_2 levels in the area were not high enough to create a pollution problem as far as human exposure is concerned.

The significantly lower levels of NO_x in the immediate vicinity of a large oil refinery in comparison to the urban concentrations is in agreement with the similar comparison made for hydrocarbon concentrations. In a 4-week measuring campaign taken place also in the same refinery the saturated and aromatic hydrocarbon levels measured always downwind of the plant are also by about an order of magnitude less than in the center of the urban areas of Athens and Thessaloniki (Kalabokas *et al.*, 2001). The average value of the sum of the three measured saturated hydrocarbons (hexane, heptane, octane) was at 2.5 ppbv while in the center of Athens and Thessaloniki the corresponding sum was at 7–8 ppbv. For the 5 measured aromatic compounds (benzene, toluene, ethyl-benzene, m-, p-, o-xylenes and 1,3,5-trimethyl-benzene) the average close to the refinery was at 4 ppbv while in the urban centers of the above cities the corresponding value was at 54–56 ppbv.

This large difference is somehow remarkable if it is kept in mind that that the raw material (crude oil) processed in large amounts (about 5 million tons per year), consists mainly of hydrocarbons. It comes out from all the above that the particular conditions concerning the emissions, (multiple sources) the atmospheric dispersion and the topography (street canyon effects) in urban areas lead to higher pollutant exposures of the population than in the vicinity of large industrial plants such as an oil refinery.

Another aspect of the impact of the NO_x emission from the oil refinery plant in combination with the hydrocarbon emissions and the high solar irradiation levels in the area is the expected surface ozone production. In this respect it has to be reminded that the rural ozone levels in the area are in general relatively high reaching the 60 ppb level in summer months during the afternoon hours (40–45 ppb in October) indicating that this is rather a regional pollution phenomenon (Kalabokas et al., 1996, 2000; Kourtidis et al., 1996; Kouvarakis et al., 2000). Critical ozone concentration levels are systematically reached in summer, if compared with the air quality standards for human health (8 hr average) or vegetation protection (WHO Report, 1987; EU Directive 99/30).

This fact leads to the following remarks: At first the low levels of the concentrations of the photochemical precursors NO_x and VOC originating from the operations of the oil refinery are not expected to produce significant amounts of ozone in comparison to the potential for photochemical ozone production of an urban area like Athens. But on the other hand any extra ozone amounts produced will be added to the already high background levels and might be very critical in terms of the violations of air quality standards and the subsequent human and vegetation exposure. An additional element is that on a regional scale the ozone production is in general NO_x-limited (Megie, 1996), which means that the NO_x emissions control the ozone production. The hydrocarbons are in general more abundant in rural areas influenced by the important biogenic emissions of hydrocarbons. In this respect it has to be mentioned that pine forests, which are potential emitters of biogenic hydrocarbons, cover the mountainous area north of the refinery. Therefore, it can be argued that the effects of the NO_x emissions from this particular oil refinery plant are more important, in terms of human exposure, for their contribution to the photochemical ozone production than the direct exposure in NO_x (with emphasis in NO_2), which seems to be quite low.

4. Conclusion

In this work the ambient air concentrations of NO_x (NO and NO_2) were monitored continuously at a site next to a large oil refinery in Greece, for two weeks during autumn 1997. The main meteorological parameters were also recorded in parallel. The results were compared with the measurements collected during the same period at four air pollution monitoring sites in the urban area of Athens and their common or different characteristics were discussed. The NO refinery values were in the range of 1 to 84 ppb with a mean value of 3.5 ppb and the NO_2 concentrations were in the range of 1 to 48 ppb (mean value at 10 ppb). The average atmospheric pollution in NO_x produced both by the oil refinery and the traffic emissions of the nearby National highway were in the order of 10 ppb under low dispersion conditions. However, the higher NO_2 values in the range of 30 to 50 ppb were coming exclusively from the direction of the oil refinery and they were recorded during nighttime. The average NO values at the oil refinery were 25 to 6 times lower than the levels at the four examined Athens urban stations while the average NO_2 values at the oil refinery were by 8 to 1 times lower. A major difference between the NO_x levels in the oil refinery and in Athens was the ratio NO/NO_2. This ratio was significantly smaller at the oil refinery than in Athens, which could be attributed to the high ozone abundance in the refinery area leading to a rapid conversion of NO to NO_2. Although the majority of NO_x species at the oil refinery were in the form of NO_2, in contrast to the situation in urban sites, this does not cause a major air pollution problem, since these values were always well below the existing air quality standards. The additional ozone expected to be produced from the NO_x emissions may be critical for aggravating problems in terms of human and vegetation exposure given the high ozone background levels in the area.

Acknowledgements

This study was supported by the Greek Ministry of the Environment (Directorate of Air Pollution Control), which is gratefully acknowledged, especially Dr. L. Viras. The authors would like also to thank Mr. Y. Xintavelonis, for technical assistance.

References

EU Directive 99/30. Council Directive 1999/30/EC. Official Journal, L163, pp. 41–60.

Finlayson-Pitts, B. and Pitts Jr., J.: 1986, *Atmospheric Chemistry: Fundamentals and Experimental Techniques*, A. Wiley Interscience Publication, J. Wiley & Sons, New York.

Kalabokas, P., Amanatidis, G. and Bartzis, J.: 1996, 'Rural ozone levels at an Eastern Mediterranean site (Attica, Greece)', in R. Bojkov and G. Visconti (eds), *Proceedings of the XVIII Quadrennial Ozone Symposium*, L'Aquila, Italy, September 1996, pp. 379–382.

Kalabokas, P. D., Viras, L. G. and Repapis, C. C.: 1999, 'Analysis of the 11-year record (1987–1997) of air pollution measurements in Athens, Greece. Part I: Primary air pollutants', *Global Nest: The International Journal* 1(3), 157–168.

Kalabokas, P. D, Viras, L. G., Bartzis, J. G. and Repapis, C. C.: 2000, 'Mediterranean rural ozone characteristics around the urban area of Athens', *Atmosp. Environ.* 34(29–30), 5199–5208.

Kalabokas, P. D., Hatzianestis, J., Bartzis, J. G. and Papagiannakopoulos, P.: 2001, 'Atmospheric concentrations of saturated and aromatic hydrocarbons around a Greek oil refinery', *Atmosph. Environ.* 35(14), 2545–2555.

Kourtidis, K., Ziomas, I., Zerefos, C., Balis, D., Suppan, P., Vassaras, A., Kosmidis, V. and Kentrarchos A.: 1996, 'On the Background Ozone Values in Greece', *EUR 17482 EN, 7th European*

Symposium on Physico-chemical Behaviour of Atmospheric Pollutants: The Oxidizing Capacity of the Troposphere, Venice, Italy, October 1996, pp. 456–461.

Kouvarakis, G., Tsigaridis, K., Kanakidou, M. and Mihalopoulos, N.: 2000, 'Temporal variations of surface regional background ozone over Crete Island in southeast Mediterranean', *J. Geophys. Res.* **105**, 4399–4407.

MinEnv Report: 1998, *Atmospheric Pollution in Athens – 1997. Directorate of Air Pollution and Noise Control*, Ministry of Environment (in Greek).

Viras, L. and Siskos, P.: 1992, 'Air pollution by gaseous pollutants in Athens, Greece', in J. O. Nriagu (ed.), *Gaseous Pollutants: Characterization and Cycling*, J. Wiley & Sons Inc., pp. 271–305.

Warneck, P.: 1988, *Chemistry of the Natural Atmosphere*, International Geophysical Series, Vol. 41, Academic Press Inc.

Westaway, M. T. and Brockis, G. J.: 1978, 'Petroleum refineries', in A. Parker (ed.), *Industrial Air Pollution Handbook*, McGraw-Hill, London, pp. 363–389.

WHO Report: 1999, *Air Quality Guidelines*, World Health Organization Publications, Protection of the Human Environment, Geneva.

COMPARISON OF FOUR MACHINE LEARNING METHODS FOR PREDICTING PM$_{10}$ CONCENTRATIONS IN HELSINKI, FINLAND

M. ZICKUS[1], A. J. GREIG[1]* and M. NIRANJAN[2]

[1] *Anglia Polytechnic University, Cambridge CBT 1PT, U.K.;* [2] *University of Sheffield, U.K.*
(* *author for correspondence, e-mail: e-mail: a.j.greig@apu.ac.uk, fax +44 01223 352973*)

Abstract. Machine learning methods can offer a practical alternative to deterministic and statistical methods for predicting air pollution concentrations. However, for a given data set, it is often not clear beforehand which machine learning method will yield the best prediction performance. This study compares the variable selection and prediction performance of four machine-learning methods of different complexity: logistic regression, decision tree, multivariate adaptive regression splines and neural network. The methods are applied to the task of predicting the exceedance of the European PM$_{10}$ daily average objective of 50 μg m^{-3} for a station in Helsinki, Finland. Our study shows that some predictors were selected by all models but that the different models also picked different variables. The performance of three of the four methods investigated was very similar, however, performance of the decision tree method was significantly inferior. Performance was sensitive to the learning sample size and time period used.

Keywords: air pollution prediction, empirical modelling, machine learning, PM$_{10}$

1. Introduction

Many large metropolitan areas experience elevated concentrations of PM$_{10}$ and the EC objective of 50 μg m^{-3} (running 24 h mean) is likely to be difficult to achieve by 2010 (De Leeuw *et al.*, 2000). Several previous studies on PM$_{10}$ in northern European cities have reported that re-suspended surface particles are an important local source of PM$_{10}$ (e.g. Pohjola *et al.*, 2000). In these northern cities PM$_{10}$ levels behave in a complex manner with, for example, street cleaning activities, the use of studded vehicle tyres and meteorological parameters all affecting both re-suspension and dispersion. For agencies charged with the responsibility of making daily air pollution forecasts the complex patterns of PM$_{10}$ concentrations are very difficult to predict accurately using available deterministic models (Berge *et al.*, 2001).

An alternative way to predict PM$_{10}$ concentrations is to use a data driven modelling approach. A review of the application of statistical models for predicting air quality can be found in Zickus (1999). These models are usually built in a trial and error process and the selection of the input variables is often based on correlation analysis and a prior knowledge of the atmospheric processes. However, if there are a large number of input variables available and non-linear interactions are present, the trial and error approach is tedious and does not guarantee the optimal selection of model structure and input variables.

Water, Air, and Soil Pollution: Focus **2:** 717–729, 2002.
© 2002 *Kluwer Academic Publishers.*

As data sets have grown in size and complexity a large number of automatic data analysis and knowledge discovery tools have been developed. These are generally described as machine learning (ML) methods. The ML methods allow automatic searches for relevant input variables and the development of an optimal model structure within the limits of the concept of the applied method. However, empirical comparisons of existing ML methods have shown that each method has a selective superiority; that is, it may be best for some but not all tasks. With a given data set, it is often not clear beforehand which method will yield the best performance for the task in hand. In such cases one must search through the available methods to find the one that produces the best prediction (Brodley, 1993).

This article compares the performance and discusses the application of four machine learning methods, of different complexity, for predicting exceedance of the PM_{10} daily average objective of 50 μg m^{-3} for a Helsinki dataset. There were three objectives to this research. The first objective was to determine which meteorological and temporal variables were the most useful for predicting the exceedance of the PM_{10} daily average. The second objective was to find the ML method that resulted in the most accurate prediction. The third objective was to investigate how the prediction capabilities of the various methods altered when different sized learning samples, drawn from different time periods, were used. In order to provide more insight into the strengths and weaknesses of each model's performance a case by case analysis of successes and failures to predict daily PM_{10} exceedances are analysed and discussed below.

2. Data for the Comparison

The study used daily averages of PM_{10} mass concentration measured at the urban traffic station Töölö located in Helsinki city centre over the four year period (1996–1999). Details of the location of these monitoring stations are described in Pohjola *et al.* (2000). To complement this dataset the study also used a concurrent meteorological dataset for Helsinki airport containing 12 surface, sounding and pre-processed meteorological variables. These were wind speed, wind direction, friction velocity, inverse of Monin-Obukhov length, pressure, relative humidity, amount of precipitation, temperature, dew point temperature, wet bulb temperature, total cloudiness and state of the ground. The task set for the 4 machine learning methods selected for this study was to predict the exceedances of the PM_{10} daily average objective of 50 μg m^{-3} in the year 1999. The ML models were developed from observations taken over the three previous years, that is 1996–1998.

2.1. VARIATION OF PM_{10} CONCENTRATION

Figure 1 represents the seasonal variation of PM10 concentrations (monthly averages) and the number of daily exceedances (i.e. > 50μg m^{-3}) in each month for

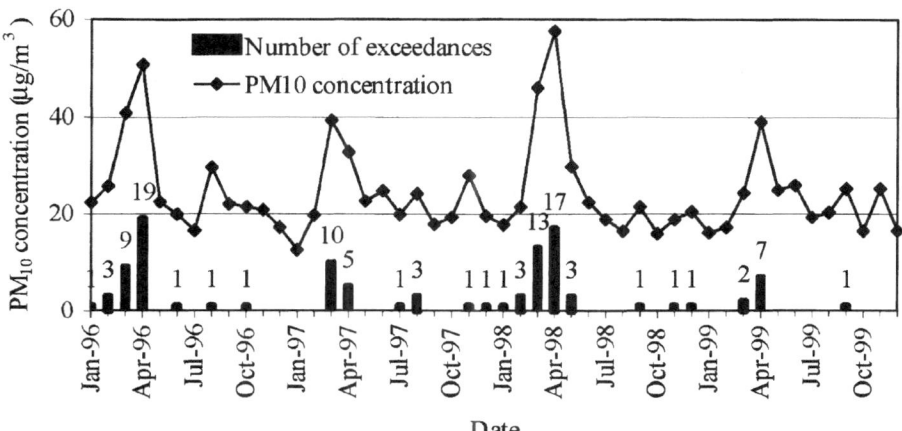

Figure 1. Seasonal variation of PM_{10} concentration (monthly data) and the number of daily exceedances in each month for the four year study period.

the four year study period. One can see that the largest number of exceedances always occur during March and April. Previous studies of PM_{10} concentrations in Helsinki have attributed this spring increase in PM_{10} concentrations to dust re-suspension, which takes place after the snow has melted and the streets dry out. The metal studs in vehicles winter tyres are also thought to contribute to the high PM_{10} concentrations recorded during this period. High springtime values tend to persist until the streets are swept clean of the sand which has accumulated during the winter and vehicles revert to non-studded summer tyres (Pohjola *et al.*, 2000).

2.2. DATA PRE-PROCESSING

To create a target variable the continuous output of PM_{10} concentration was cat-egorised as a dichotomous variable that indicated whether or not exceedance of the PM_{10} objective had taken place. It is not advisable to use all the available variables in the model unless the amount of data available is very large. ML al-gorithms are able to select the important input variables, however, since many of the meteorological variables are related to the same synoptic process they are strongly inter-related. With limited data ML algorithms cannot distinguish between random statistical correlations and salient correlations between inputs and outputs (Kennedy *et al.*, 1999). To avoid this problem, an initial screening of available input variables was carried out in order to reduce their number and select only those which were potentially relevant. In this way several derived meteorological vari-ables, including dew point temperature, wet bulb temperature and friction velocity were discarded.

The average PM_{10} concentration between hours 1 and 15 on the day before the day to be predicted was added to the subset of input variables. The reason for using a 1–15 h average instead of a 24 h average was to incorporate the possibility of using the models in an operational mode. Two temporal variables, such as month and weekday, were also introduced. Time variables were assumed to be able to explain the annual variation of PM_{10} concentrations and the daily variation of traffic intensity.

In total the whole potential subset of input variables contained 9 meteorological variables, 2 temporal variables and the average PM_{10} concentration between 1–15 h on the previous day. For each data set, the few missing feature values were replaced by the linear interpolation of nearby values.

2.3. LEARNING AND EVALUATION DATA SETS

For model development and evaluation the original data were divided into two disjointed sets – a learning set and an evaluation set. The learning set was further split into a training set and a test set. The training set was used to optimise the model's architecture and free parameter settings. By using test data the model parameters were adjusted to achieve the maximum performance on the test data set. When the best architecture and training parameters had been selected the final performance results were obtained by applying the model on the unseen evaluation set.

To compare the prediction capabilities of the four methods on different sized learning samples and different time periods three different learning data sets were created as shown in Table I. Dataset one and two allowed an investigation of whether learning from longer periods of observations significantly increased model accuracy. The goal of the third data set, which contained learning data from only March and April 1996–1998, was to indicate whether learning from data from the period of highest PM_{10} concentrations can improve the model's prediction accuracy. The training and testing data samples were created by randomly selecting observations from the learning data set. The training set contained 70% of records and the testing set the remaining 30% of records.

It was noted that the probability of exceedance, (P), in the evaluation sets was significantly lower than in the learning data sets. It is known that the unequal distribution of class probabilities in the learning and evaluation data sets may reduce prediction accuracy. However real data sets are likely to have different numbers of exceedances in different years, depending, for example, upon the persistence of particular meteorological conditions. It was therefore considered important to evaluate the 4 methods on data sets with an unknown number of exceedances as this would help highlight the robustness of a method and its practical applicability for air quality predictions.

To allow for fair comparison, each of the various classifier construction methods was applied using the same experimental conditions. Each method was allowed

TABLE I

Learning and evaluation data samples

Data set	Learning sample				Evaluation sample			
	Period	Obs.	N	P	Period	Obs.	N	P
1	1996–1998	1096	96	0.09	1999	365	10	0.03
2	1997–1998	730	61	0.08	1999	365	10	0.03
3	March and April of 1996–1998	183	73	0.40	March and April of 1999	61	9	0.15

(Obs. – number of observations; N – number of exceedances; P – probability of exceedance, Obs/N)

to select important predictors by method specific built in algorithms. The results below report the prediction performance of each method on the independent evaluation dataset.

3. Machine Learning Methods

This study investigated four machine learning methods: logistic regression (LR), multivariate adaptive regression splines (MARS), decision tree (DT) and neural net (NN). The selection was based on the different level of complexity they used in representing data structure and the differences in the way they selected input variables for the model. When selecting an appropriate ML method it is very important to find an optimum balance between robustness and flexibility. If the relationships between variables can be well described by bivariate interactions or the signal to noise ratio in the data is too small to allow detection of more complex structure, then a simple model, such as LR, usually describes data better than a complex one. On the other hand, more sophisticated and flexible ML methods such as MARS and NN can successfully discover nonlinear relationships, although they are very prone to overfitting. If the chosen model fits its learning data too well, then it will not merely be fitting the underlying process but will also be fitting the chance events which led to that the particular data set being generated. Since the future data will have different values for chance factors the prediction will be poor (Berthold and Hand, 1997). The following sections describe the ML methods applied to the Helsinki data stressing their conceptual differences, the way they select input variables and how they avoid overfitting.

3.1. LOGISTIC REGRESSION

Logistic regression is a form of parametric regression, it is used when the dependent variable is a dichotomy and the independents are continuous variables, categorical variables, or both. The expected probability of a dichotomous outcome is based on the logistical (sigmoid) cumulative function. To construct the LR model backward and forward stepwise variable selection algorithms were used. These are a built-in function of the SPSS program (SPSS, 1994). For the Helsinki dataset, the results obtained from both backward and forward selection were identical and therefore only the forward selection results are quoted.

In its standard form logistic regression is restricted to interactions between pairs of variables (bivariate interactions), however interactions involving more variables (higher-order interactions) are likely to be more realistic. The search and inclusion of variable interaction terms in LR is possible, however, it has to be done manually which is time consuming and usually suboptimal.

3.2. DECISION TREE

Decision trees are a way to represent rules underlying data with hierarchical sequential structures that recursively partition the data. The application of a regression tree method for predicting ozone concentrations has been previously reported by Gardner and Dorling (1999). In this study a CART algorithm was used to construct a binary decision tree with a Gini index based splitting rule (Breinman *et al.*, 1997).

The decision tree was generated via a recursive partitioning algorithm which determines the rules associated with each binary split in the tree. Initially all the data were used to determine the rule, which splits the data into the two most dissimilar sets. The best splitting criteria for each node was found by minimizing the Gini index, a measure of node impurity, which reaches a value of zero when only one class is present at a node. Following the initial rule, each subset was recursively split until a large tree was developed, with each terminal node containing only a few observations. The oversized tree was then pruned until the generalisation performance, described by either evaluating the error rate on a test set or by cross-validation, indicated no overfitting.

Decision trees readily lend themselves to being displayed graphically, helping to make them easier to interpret. However, due to discontinuous approximation at the partitions DT are notoriously weak at capturing strong linear structure and can produce a very large tree in an attempt to represent very simple relationships. Furthermore, a DT is often unable to identify the interactive effects of multiple inputs.

3.3. MULTIPLE ADAPTIVE REGRESSION SPLINES

The multiple adaptive regression splines method (MARS) is a multivariate nonparametric regression procedure first proposed by Friedman (1991). The MARS method approximates complex relationships by a series of linear regressions (splines) on different intervals of the independent variable. Knots mark the end of a region of data where a distinct linear regression is run, that is, where the behaviour of the modelled function changes. In order to model the concept of knots and piecewise linear regression splines, MARS uses the so-called basis functions, which re-express independent variables by mapping them to new variables. By mixing the different types of basis functions and providing adequate values for knots it is possible to approximate any functional shape. The correct values of knots and basis functions in the MARS model are found by searching in a stepwise manner. In each step the regression coefficients are found by the ordinary least squares method. In this study the MARS method is implemented via the POLYMARS algorithm (Kooperberg *et al.*, 1997).

MARS is a very flexible model building and data analysis tool as it can adapt any functional form. It is suitable both for predicting outcomes and for exploratory data analysis.

3.4. NEURAL NET

An artificial neural net (NN) is a network of many very processors or 'neurons', which, by learning from data, can detect relationships among variables and predict outcomes. It is a one of the most flexible non-linear regression methods and may capture the required complexity of a situation, however NN are prone to overfitting and sensitive to noise. Because neural nets typically have very many free parameters, and because so many combinations of parameters result in similar predictions, the parameters become uninterpretable and the network serves as a black box estimator and does not aid in understanding the underlying process generating the data. A detailed review of applications of NN to atmospheric predictions can be found in Gardner and Dorling (1998).

In this study a popular type of NN, a feedforward network with backpropagation, is used as the learning algorithm. A genetic algorithm (GA) is used to select a suitable subset of variables. GAs are stochastic general purpose search algorithms that have been applied to a wide range of machine learning problems. They work by evolving a population of chromosomes, each of which encodes a potential solution to the problem at hand. The task of a GA is to find a well fitting chromosome through the application of selection and perturbation operators. Goldberg (1989) provides a comprehensive description of GAs.

4. Results and Discussion

4.1. INPUT VARIABLE SELECTION

The input variables selected by each investigated ML method as important predictors of PM10 exceedance are presented in Table II. It shows the level of agreement between methods. Almost all methods selected month, PM_{10} concentration on the previous day, wind speed and humidity variables as important predictors. Several algorithms also selected cloud cover and state of the ground for different learning

samples. Interestingly, precipitation, which is likely to effectively washout particles
from the ambient air, was selected as an important predictor only by the neural
net. The other methods usually replaced precipitation by humidity, which is highly
correlated with precipitation. The LR, CART and MARS methods persistently
selected the same variables from all three learning data sets (except month, in
third data set), while the NN method selected an increasingly larger number of
variables from the second and third learning data sets, compared to the first data
set. It can be speculated that the NN method, being more flexible, was able to
distinguish the more complex effects that meteorological parameters were having
on PM_{10} concentrations. However the larger number of input variables used by the
NN method also made it more sensitive to errors in the input data.

4.2. PREDICTION PERFORMANCE

To compare the prediction performance of the 4 methods several performance evaluation indices were computed as recommended by the US EPA (US EPA, 1999).
A case by case analysis of the success and failure of each method to predict exceedances was also conducted. The results of the prediction performance analysis
are summarised in Table III. The index of success IS (IS = TP/(TP + FP + FN))
indicates how well the PM_{10} exceedances were predicted. It is not affected by a
large number of correctly forecasted non-exceedances and therefore it is useful for
evaluating rare events. The false alarms value, FAR (FAR = FP/(FP + TP)) indicates
the fraction of predictions of PM_{10} exceedances that did not actually occur. False
alarms are highly undesirable for practical reasons since the prediction of exc-
cedance usually instigates expensive public information or emission management
procedures. Overall accuracy A, indicates the fraction of predictions that correctly
predicted an event (exceedance) or a non-event (no exceedance) (A = (N − FP −
FN)/N).

The following observations may be drawn from the table. The CART method
produced the weakest results for all learning data sets. One reason for this could be
that the CART model had selected the lowest number of input variables. It can be
concluded that although the decision tree method can be more readily interpreted it
could not capture accurately the underlying relationship between both the meteorological and temporal predictor variables and PM_{10} exceedances. The predictions
of the CART method were worse then using a naïve persistency method. For this
reason the performance of only the three remaining methods (LR, NN and MARS)
are discussed below.

Using the first and largest learning data set and the success index as a measure of
performance the LR, NN and MARS based methods results were very similar. The
MARS method showed the best performance on the basis of predicting the lowest
number of false alarms and accuracy. The prediction performance of the NN and
LR methods was the same. Using the second, smaller learning data set the neural
net method out-performed both the LR and MARS method, as measured by the suc-

TABLE II

Input variables selected by the different methods

Method	Month	Week-day	Pm_{10} Day-1	W-Spd	W-Dir	1/Monin-Obuk	Pres-sure	Rel. humid.	PPT	Temp	Cloud	SOG at noon	Tot. number of selected variables
Data set 1													
LR	x		x	x				x			x		5
NN	x		x	x				x		x		x	6
CART	x		x	x				x					4
MARS	x		x	x				x			x	x	6
Data set 2													
LR	x		x	x				x			x		5
NN		x	x	x			x	x	x			x	7
MARS	x		x	x				x			x	x	6
CART	x		x	x				x					4
Data set 3													
LR	–		x	x				x			x		4
NN	–	x	x			x	x	x	x			x	7
MARS	–		x	x				x					3
CART	–		x	x				x					3

[a] Key to variables: Month; weekday; previous day PM_{10} concentration; wind speed; wind direction; inverse of Monin–Obukhov length at noon; pressure; relative humidity; amount of precipitation; temperature; cloudiness; state of the ground at noon.

TABLE III

Successful predictions of exceedances (x) and performance indices of the four methods tested

Exceedance on day number:	89	90	96	98	102	109	110	111	112	250	Performance indices					
Measured PM$_{10}$ concentration, μg m^{-3}	58	52	50	61	57	67	70	56	55	51	TP	FN	FP	IS	FAR	A
Correct prediction (x)																
Data set 1																
LR	x			x		x	x	x	x		6	4	4	0.43	0.40	0.978
NN	x			x		x	x	x	x		6	4	4	0.43	0.40	0.978
CART	x					x	x	x	x		5	5	9	0.26	0.64	0.962
MARS				x		x	x	x	x		5	5	2	0.42	0.29	0.981
Data set 2																
LR	x					x	x				3	7	2	0.25	0.40	0.975
NN	x		x			x	x	x	x		6	4	4	0.43	0.40	0.978
CART	x			x							2	8	9	0.11	0.82	0.953
MARS						x	x	x	x		4	6	0	0.40	0.00	0.984
Data set 3																
LR	x					x	x			na	3	6	2	0.27	0.40	0.869
NN	x			x		x	x	x	x	na	5	4	4	0.38	0.44	0.869
CART		x								na	1	8	4	0.08	0.80	0.803
MARS	x	x	x	x	x	x	x	x	x	na	9	0	10	0.47	0.53	0.836
Persistency model		x				x	x	x			4	6	6	0.25	0.60	0.967

TP – number of correct predictions of exceedance; FP – number of false alarms; FN – number of missed exceedances; IS – index of success; FAR – false alarms value; A – overall accuracy.

cess index. The performance of the MARS method, however, was only marginally worse and had a significantly lower fraction of false alarms. The performance of the NN and MARS methods is almost the same for both the first and second learning data sets, however, performance of the LR method deteriorates significantly in the second data set. Using the third learning data sets, which correspond to the months of the highest PM$_{10}$ concentrations, the MARS method had a significantly higher success index than the other methods. However, if performance is rated according to the accuracy and fraction of false alarms it does slightly worse than that both the LR and NN based methods.

The results show that the prediction performance of LR, NN and MARS methods are very similar. If small differences are not important in practical applications, one may wish to select a classifier based on other criteria such as the smallest number of input variables or ease of interpretation. According to these criteria the LR and MARS based methods selected the smallest number of input variables and are the easiest to physically interpret.

4.3. CASE BY CASE ANALYSIS OF THE PREDICTION PERFORMANCE

In terms of predicting exceedances on specific days the prediction performance of each method was surprisingly similar (see Table III). Almost all successfully predicted the exceedances which occurred on days 89 and 90 and the four day pollution episode which occurred between days 109–112. During these days the measured concentrations of PM_{10} was significantly above the EC objective of 50 $\mu g\ m^{-3}$. With the exception of MARS on the third dataset all methods failed to predict exceedances on days 90, 96, 102 and 250. During these days the PM_{10} 24h objective was only slightly exceeded.

In order to determine the reasons why such unanimously poor performance occurred when PM_{10} concentrations were only slightly above the objective the average meteorological situations during each of the exceedances has been analysed (Table IV). On day 96 the high PM_{10} concentrations corresponded with relatively high wind speeds which may have mislead the methods. On days 96, 102 and 250 a relatively low PM_{10} concentration on the previous day may have resulted in the failure to predict the exceedance.

It would be possible to capture the exceedances on days 90, 96, 102 and 250 (when the PM_{10} 24h objective was only slightly exceeded) by decreasing the cut-off value of the classification threshold. However, the trade off would be to increase the number of false exceedance predictions (false alarms). The selection of the optimal operating threshold, that is the threshold which produces an acceptable fraction of

TABLE IV

Meteorological conditions during the days exceedances were missed by the four methods tested

Day	Measured PM_{10} conc. $(\mu g\ m^{-3})$	Previous day PM_{10} conc. $(\mu g\ m^{03})$	Wind speed $(m\ s^{-1})$	Humidity $(\%)$	Precipitation $(mm\ h^{-1})$	Cloudiness $(fraction)$
90	52.8	49.7	2.6	80.4	0.05	0.8
96	50.3	22.4	1.4	95.8	0.26	1.0
102	57.2	22.0	2.7	74.5	0.21	0.5
250	50.6	31.0	1.6	62.6	0.00	0.3

predicted exceedances and an acceptable number of false alarms, is probably best made by the agencies charged with the responsibility of making daily air pollution forecasts and able to view the wider implications of false alarms to the community.

5. Conclusions

Machine learning methods can be used as a powerful data analysis tool where underlying deterministic equations are unknown or very complex. The results obtained from such objective process analysis may not only improve the performance of air quality forecasts but also help to understand which factors affect PM_{10} concentrations.

This paper has compared the prediction performance of four machine learning methods. All the investigated methods selected as important, almost the same input suite of pollution and meteorological variables, however the neural net method selected the largest number making it more sensitive to errors in the input data. It was established that in the analysed data (Helsinki 1996–1999) 3 of the methods investigated (NN, MARS and LR) performed similarly while performance of the CART method was significantly inferior. The neural net and multivariate adaptive regression splines models developed from data for two and three year periods performed similarly, however performance of the logistic regression based model was significantly reduced for the two years length learning data. None of the investigated methods had a markedly superior performance, as described by the computed performance measures, or for different learning samples.

If small differences are not important in practical applications, it may be wise to select a classifier based on other criteria such as smallest number of input variables or interpretability. The LR and MARS models are more readily interpreted and selected fewer input variables. Interpretation, however, may be the key, as the models may ultimately be used to support decision-making and, from a more general perspective, to gain a better understanding of the factors influencing PM_{10} levels.

Acknowledgements

This study has been part of the EU funded project APPETISE 'Air pollution episodes; modelling tools for improved smog management'. The data sets used was provided by the Finish Meteorological Institute under the terms of the APPETISE project.

References

Berge, E., Walker, S-E., Sorteberg, A., Lenkopane, M. L., Eastwood, S., Jablonska, H. J. and Ødegaard, M.: 2001, 'A Real Time Operational Forecast Model for Meteorology and Air Quality During Peak Air Pollution Episodes in Oslo, Norway', *Proceedings of 3th International Conference on Urban Air Quality*, Loutraki, Greece, March 2001.

Berthold, M. and Hand, D. (eds): 1999, *Intelligent Data Analysis*, Springer.

Breiman, L., Friedman, J., Olshen, R. and Stone, C.: 1984, *Classification and Regression Trees*, Wadsworth International Group.

Brodley, C. E.: 1993, 'Addressing the selective superiority problem: Automatic algorithms/model class selection', in P. Utgoff (ed.), *Proceedings of the Tenth International Conference on Machine Learning*, pp. 17–24.

De Leeuw, F., Moussiopoulos, N., Bartonova, A. and Sahm, P.: 2000, 'Air Quality in Larger Conurbations in the European Union', European Topic Centre on Air Quality.

Friedman, J. H.: 1991, 'Multivariate adaptive regression splines (with discussion)', *Ann. Statis.* **19**, 1–141.

Gardner, M. and Dorling, S., 1998: 'Artificial neural networks (the multi-layer perceptron) – a review of applications in the atmospheric sciences', *Atmos. Environ.* **32**, 2627–2636,

Gardner, M. and Dorling, S.: 1999, 'Statistical surface ozone models: an improved methodology to account for non-linear behaviour, *Atmos. Environ.* **34**, 21–34.

Goldberg, D. E.: 1989, *Genetic Algorithms*, Reading, MA: Addison Wesley.

Kennedy, R. L., Yuchun, L., van Roy, B., Reed, C. and Lippman, R.: 1997, 'Solving Data Mining Problems with Pattern Recognition', The Data Warehousing Institute Series.

Kooperberg, C., Smarajit, B. and Charles, J.: 1997, 'Polychotomous regression', *J. Amer. Stat. Assoc.* **92**, 117–127.

Pohjola, M., Kousa, A., P.Aarnio, P., Koskentalo, T., Kukkonen, J. and Karppinen, A.: 2000, 'Meteorological interpretation of measured urban $PM_{2.5}$ and PM_{10} concentrations in Helsinki Metropolitan Area', *Air Pollution* **VIII**, 679–698.

SPSS, *User Manual, Version 9.0*.

US EPA: 1999 'Guideline for Developing an Ozone Forecasting Program', EPA-454/R-99-009.

Zickus, M.: 1999, 'Influence of Meteorological Parameters on Urban Air Pollution and Its Forecast', PhD. Thesis, Department of Physics, Vilnius University, 105 pp. Available on Internet: http://195.194.93.120/thesis/.

INTEGRATED ASSESSMENT OF AIR POLLUTION ABATEMENT STRATEGIES IN URBAN AREAS: APPLICATION TO THE GREATER ATHENS AREA

P. M. TOURLOU*, P. SAHM and N. MOUSSIOPOULOS

Laboratory of Heat Transfer and Environmental Engineering, Aristotle University Thessaloniki, Box 483, 54 006 Thessaloniki, Greece
(* author for correspondence, e-mail: evelina@aix.meng.auth.gr, fax +30 31 996012)

Abstract. This paper presents an integrated system for the assessment of technical and non-technical measures that are put forward in order to reduce air pollution levels in urban areas. In contrast to the majority of the currently employed assessment tools, this system allows for the evaluation of any proposed air pollution control measure in terms of its combined impact on air quality and social welfare, by correlating the environmental and economic aspects of alternative air pollution abatement solutions. Based on the multi-pollutant, multi-effect concept, the system presented aims in providing policy-makers with a reliable tool for the objective assessment of the most cost-effective packages of measures, the latter being allocated according to the particular features and needs of the areas examined.

Keywords: air pollution abatement, air quality modelling, cost-benefit analysis, cost-effectiveness theory, economic benefit/damage, emission control, integrated assessment

1. Introduction

Over the last decades, human activities led to the continuous increase of air pollutant emissions into the atmosphere. As a result, the atmospheric content has changed. Nowadays, air pollution on the urban scale has become the source of a range of problems: Health risks mostly associated with inhalation of gases, accelerated deterioration of building materials, damage to historical monuments and to vegetation within and near the cities. In order to tackle these problems, efficient, long-term air pollution abatement strategies need to be assessed and implemented.

Undoubtedly, the adoption of an air pollution abatement strategy has the ultimate target to reduce sufficiently air pollution levels in the areas of implementation. Forming long-term, successful air pollution control strategies requires, however, knowledge of the costs associated with their implementation, the economic benefits that might result from the reduction of both the quantities of pollutants emitted and their concentrations in the atmosphere, as well as other possible benefits (or damages) arising from the adoption of the proposed strategies. In this sense, the selection of the most appropriate and most efficient bundles of emission control measures needs to satisfy a variety of criteria and, thus, should be conducted on the basis of multi-criteria analyses and decisionþmaking theories (Yu, 1990).

In contrast to the above remarks, until recently the impact of air pollution control strategies was primarily assessed in terms of their influence on the pollution levels of the area considered. The economic impact arising from the application of the proposed measure(s) was, in general, neglected. If we accept the environment as an economic commodity, then it becomes self-evident that any form of environmental degradation comprises an economic commodity itself. Nevertheless, since the value and the contribution of the environment are systematically underestimated, the monetary valuation of environmental degradation is, in many cases, still not feasible. At the same time, however, mankind incurs an important financial burden caused by the decrement of social prosperity when the latter is not compensated.

2. An Integrated Approach for Forming Efficient Air Pollution Abatement Strategies: Basic Structure and Components

The system developed and presented in this paper involves a three-step analysis to be followed for conducting an integrated assessment of various air pollution abatement strategies. Based on the principles of the cost-effectiveness theory, the first step has a twofold purpose: (a) to provide information on the total implementation costs and on the impact of the proposed strategy on the emission sources of the area examined, and (b) to form the basis for constructing cost-curves for selected pollutants, the latter allowing for a comparative assessment of alternative solutions proposed by correlating the relative abatement costs with the benefits in terms of the achieved emission reductions in the area of interest (Friedrich and Reis, 1999; cf. section 2.1). Accordingly, the proposed measures/strategies are evaluated in terms of their impact on air quality. The assessment is performed by constructing the emissions inventory for each emission situation examined (e.g. by quantifying the impact of the measure on the pollution sources and reproducing the spatial and temporal resolution of pollutants emissions) and, then, applying the European Zooming Model (EZM) system. The latter is a comprehensive group of models for simulating the transport and chemical transformation of air pollutants on the local-to-regional scale (Moussiopoulos, 1995). In the last step, the economic benefits on account of air quality improvement with regard to human health, human productivity and other ecosystems are evaluated. For the cost-benefit analysis the damage function approach is adopted (cf. section 2.3).

The interrelation between the basic components of the integrated assessment system is presented in Figure 1. As shown, the assessment of the proposed strategies is conducted (a) through the comparison of the current air pollution levels with the concentration levels resulting after the hypothetical implementation of the abate-

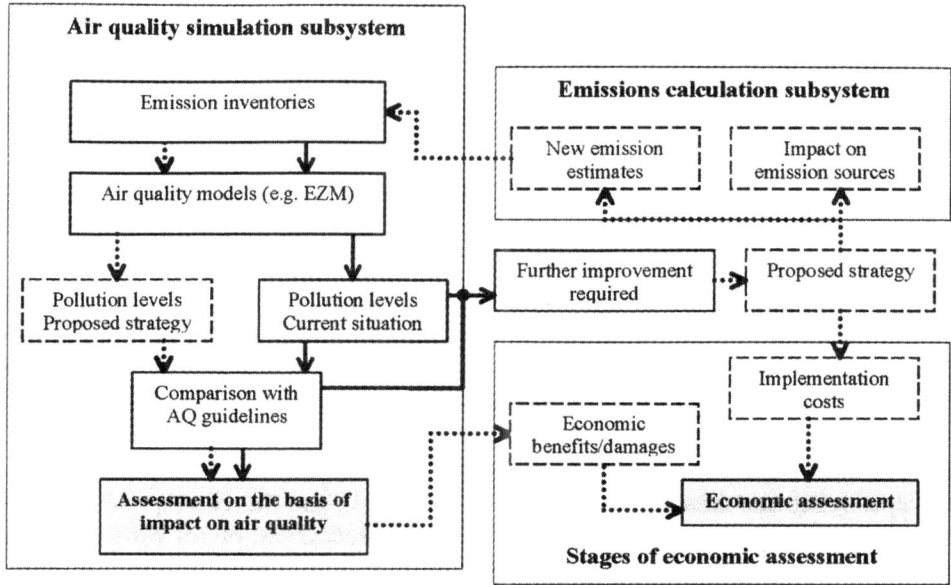

Figure 1. Flow diagram of the system for the assessment of pollution control strategies.

ment measure, and (b) having as a criterion the minimisation of the total costs required for, or arising from, the adoption of the measures.

2.1. ESTIMATION OF THE TOTAL COSTS AND BUILT-UP OF POLLUTANT SPECIFIC COST-CURVES

The total implementation cost of a pollution abatement strategy is estimated by summing up the fixed (C_{fixed}) and variable ($C_{variable}$) cost components. The former account for the investment and labour costs, whereas the latter, comprising of time-dependent quantities, include the annual amounts of operation and maintenance, energy and recovered costs.

If alternative measures are to be compared by the system, the optimum solution at this step is defined as the measure that may be enforced with the least cost while resulting to maximum reductions of the air pollutants emitted. In this case, the cost-curves for selected pollutants are constructed. A pollutant specific cost-curve correlates the total annual cost ($C_{total,an}$) with the corresponding emission reduction attained under each measure adopted. For estimating the total annual cost the uniform annual cost method is adopted, according to which the fixed cost components are distributed to the useful life years of the measure (Kolb and Scheraga, 1992; Friedrich and Reis, 1999).

$$C_{fixed,an} = C_{fixed} \cdot \frac{r}{1 - (1 + r)^{-n}}$$

and, accordingly, added to the variable cost components

$$C_{total,an} = C_{fixed,an} + C_{variable}$$

where:

$C_{fixed,an}$ annual fixed expenditure required by the end of each year of the useful life of the measure [monetary units/year]

r   annual discount rate [–]

n   expected useful life of the measure [years]

For the built-up of a cost-curve, the measures examined are classified in an increasing order based on the cost required for the unitary reduction of the emissions of each pollutant considered. Thus, the measures that appear first on the cost-curve represent the most 'effective' ones in terms of their combined economic and environmental (emission) impacts. A cost-curve may be additionally used, either for specifying the measure (or bundle of measures) and the corresponding costs required for achieving a predetermined emission reduction, or for estimating the potential emission reduction achieved through the adoption of specific measures under a given budget.

2.2. IMPACT ON AIR QUALITY

For assessing the impact of each measure on the air pollution levels of the area considered the corresponding emission inventory is constructed, providing temporally and spatially disaggregated information on the emission reduction attained with the adoption of the measure. The starting points for the built-up of the emission inventories are the total emission reductions depicted in the cost-curves. For measures addressed to the road traffic sector (i.e. introduction of advanced technology vehicles in the car fleet), an emission calculation module is applied. The latter estimates on a road-to-road basis the road traffic emissions based on the COPERT methodology (EMEP/CORINAIR, 1996; Eggleston et al., 1993). Accordingly, simulations of pollutant dispersion and chemical transformation are performed with the EZM system. Core models of the EZM system are the non-hydrostatic mesoscale model MEMO for the production of 3-D wind and meteorological fields (Kunz and Moussiopoulos, 1995), the photochemical dispersion models MARS and MUSE which produce 3-D pollutant concentration fields (Moussiopoulos, 1995) and the OFIS model for the long-term simulation of population exposure to ozone levels (Sahm and Moussiopoulos, 1999).

2.3. ECONOMIC BENEFITS ARISING FROM THE ADOPTION OF THE MEASURE: EXTERNAL COSTS

The estimation of the economic benefits (or damages) resulting from the implementation of the measure is based on the results attained with the application of the

EZM system. The assessment system allows for the monetary evaluation of specific health endpoints and human productivity caused by reductions (or increases) in air pollutants ambient concentrations. Impacts (and therefore benefits/damages) to other than human receptors (e.g. buildings, crops, etc.) are not accounted for. The assessment is based on the concept of the damage function approach (Cumberland and Kahn, 1982; Dixon *et al.*, 1994), the latter enabling the internalisation of the external costs caused by any form of environmental degradation (Dixon *et al.*, 1994; World Bank, 1998). The calculation is conducted with the aid of selected concentration- and exposure- response functions for specific pollutants and information on the monetary valuation of specific health endpoints. These are summarised by Tourlou (2000). It is pointed out that the concentration-response functions that are currently included in the system's database correlate ozone, nitrogen dioxide and carbon monoxide impacts with a variety of health endpoints ranging from hospital admissions due to respiratory illness or asthma symptoms to premature mortality. Thus, the total benefit ($C_{total,benefit}$) is expressed as a function of the number of the unfavorable implications over all health endpoints avoided (or not) after a measure's adoption ($\Delta_{cases,i}$), i = 1, ..., n) and the monetary value contributed to each of these implications ($C_{benefit,i}$). $\Delta_{cases,i}$ are calculated through the following expression:

$$\Delta_{cases,i} = R_{i,p} \times \Delta conc_p \times pop$$

where:

$R_{i,p}$ correlation coefficient between the pollutant's p concentration variation and the probability of experiencing or avoiding a specific health implication *i* (concentration – response function)

$\Delta conc_p$ change in pollutant's p concentration after the adoption of the measure

pop   population units exposed to pollutant p

Krupnick and Cropper (1992) define $C_{benefit,i}$ as the amount a person is willing to pay in order to avoid experiencing an incidence of a specific health endpoint (WTP value).

As regards human productivity, the suggestion of Crocker and Horst (1981) is adopted, which correlates the daily wage variations of employees working outdoors with changes in daily average ozone concentrations (cf. Tourlou, 2000).

3. Application of the system to the Greater Athens Area

In order to confirm the applicability and reliability of the system an application to the Greater Athens area (GAA) was performed. The topography of the region along with wind statistics for the period of simulation are shown in Figure 2.

The system was applied for analysing and evaluating the impact of various air quality regulations concerning technical measures for the reduction of NO_x and

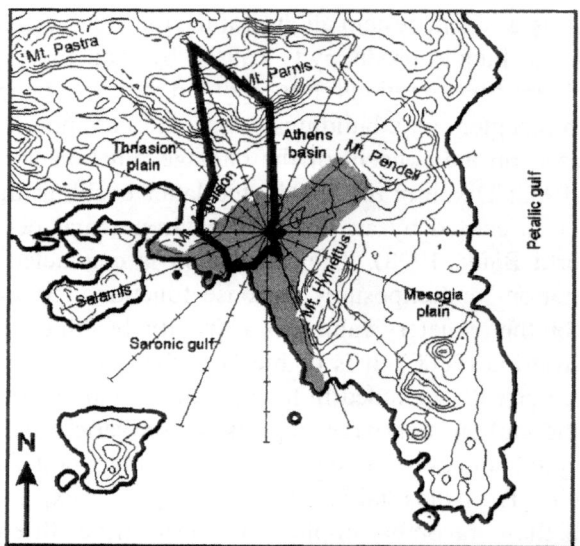

Figure 2. Topography of the Greater Athens area (GAA) and statistical data for the wind direction (the continuous line shows the frequency of the specific wind direction) during the period 1 April – 30 September 1990. Residential areas are grey. Altitude isopleths are contoured at 100 m.

NMVOC emissions from the most important emission sources of the area (e.g. road traffic, industrial and commercial units and use of solvents). These legislative modulations have already been, or will be implemented in Athens by the year 2010. Apart from the 'business as usual' scenario which assumes full compliance with the European legislation until the year 2010, two hypothetical situations involving 50% reductions of the total NOx and NMVOC emissions on top of the 'business as usual' scenario were simulated. The assessment was performed for the period 1990–2010, considering the year 1990 as the 'base case' scenario to serve as the reference for the evaluation of the proposed measures.

3.1. IMPLEMENTATION COSTS AND EXPECTED EMISSION REDUCTIONS IN THE PERIOD 1990–2010

In order to achieve full compliance with the European legislation, as recently amended through Directives 99/30, 99/13 and 99/32, various measures have already been, or need to be adopted, in Athens for the reduction of pollutant emissions from road traffic, energy and solvent-use sectors. In total 50 emission control measures were analysed by estimating their implementation costs and the corresponding emission reductions achieved with their adoption. Accordingly, the total costs for reaching compliance by the year 2010 along with the expected reductions in NO_x and NMVOC emissions in Athens were calculated on an annual basis for the period 1990–2010. It should be noted that the requirements of the 'business as usual' scenario (e.g. reaching compliance with the European legislation by the year 2010) claimed for complete implementation of the measures addressed to the transport sector and partial adoption of the measures proposed for the other two sectors. In order to obtain the corresponding values for the additional scenarios for the year 2010, the cost-curves for the period 1990–2010 were constructed and the potential of either imposing full adoption of the measures examined or introducing new measures was investigated (Tourlou, 2000).

The resulting cost-curves for NO_x and NMVOC for the road transport sector are shown in Figure 3. Figure 4 presents the NO_x and NMVOC cost-curves for the energy and the solvent use sectors, respectively. The application point of the EZM system under each scenario examined is marked on each cost-curve.

According to Figure 3, the implementation of the 'business as usual' scenario imposed the adoption of all the transport-related measures (cf. application point of the EZM system), with a corresponding cost of approx. 200 million Euro. In order to achieve full compliance with the European legislation however, additional measures addressed to the energy and solvent use sectors had to be adopted, thus increasing the cost of the specific scenario by almost 10 million Euro/year (cf. Figure 4).

The simulation of the 50% NO_x reduction on top of the 'business as usual' scenario imposed a 19500 Mg/year reduction on the total 2010 NO_x emitted quantities. Since the application of the 'business as usual' scenario claimed for full adoption of the road transport measures included in the cost-curves of Figure 3, this additional reduction in NO_x emissions could only be achieved by implementing the measures proposed for the energy sector. However, the measures in the NO_x cost-curve for the energy sector lead to an overall reduction of approximately 15000 Mg/year and therefore, failed to meet the scenario requirements (cf. Figure 4, upper part). This lead to the conclusion that additional measures were required for reaching the emission reduction target of this scenario. On the contrary, the reduction of NMVOC emissions by 50% was accomplished by the adoption of a relatively limited number of the measures introduced for the solvent use sector (cf. Figure 4, lower part). The overall results of the cost-effectiveness analysis are summarised in Figure 5.

As mentioned, the annual implementation costs of the 'business as usual' scenario are estimated to almost 210 million Euro. Further reductions of the NO_x and NMVOC emissions induce increased costs, which in the case of NO_x will exceed by far the amount of 270 million Euro per year (cf. previous paragraphs).

3.2. IMPACT ON AIR QUALITY

For assessing the impact of the 2010 emission reduction scenarios on the air pollution levels in the GAA the EZM system was applied for a 3-month summer period (between 1 April – 30 September), assuming meteorological conditions as in the year 1990 (cf. Figure 2). In order to form a basis for comparing the evolution of

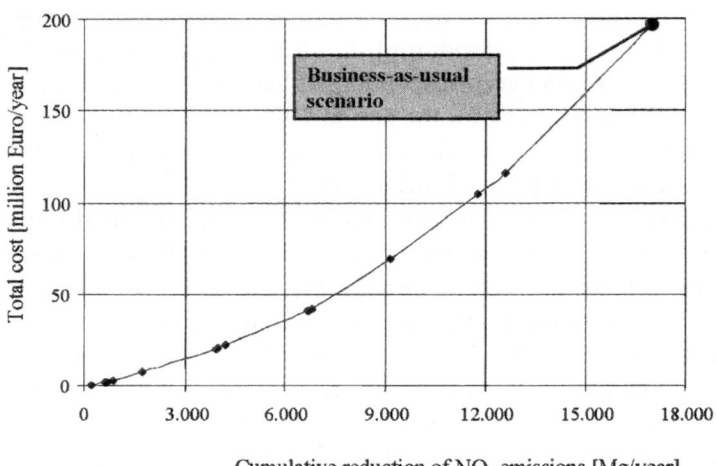

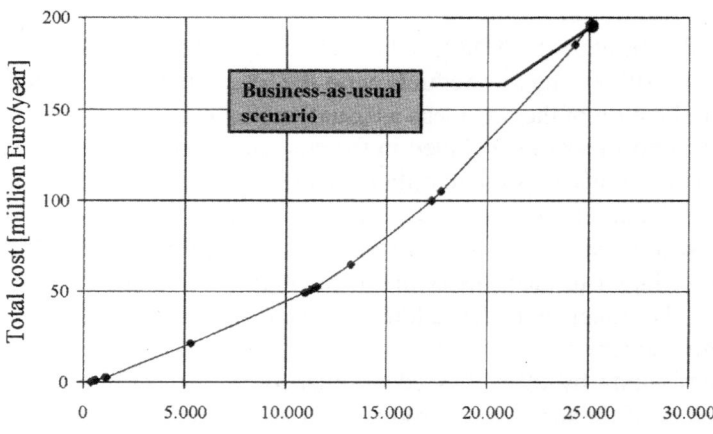

Figure 3. Cost-curves for NO$_x$ (upper part) and NMVOC (lower part) for the road transport sector in the GAA. Each point on the curves indicates an emission reduction measure. The beginning of the axes corresponds to the 'base case' scenario (year 1990).

pollutant concentrations in the area of interest, simulations were also performed considering the emission situation during 1990.

Selected model results with regard to the impact of the 2010 scenarios on ozone levels are presented in Figure 6. According to model results, in the time horizon of 2010, considering the 'business as usual' scenario the situation is expected to ameliorate. These benefits are not counterbalanced by the rather insignificant ozone increase in the urban area itself. A further reduction of 50% in NO$_x$ emissions has an adverse effect on the ozone threshold exceedances in the urban area, while in

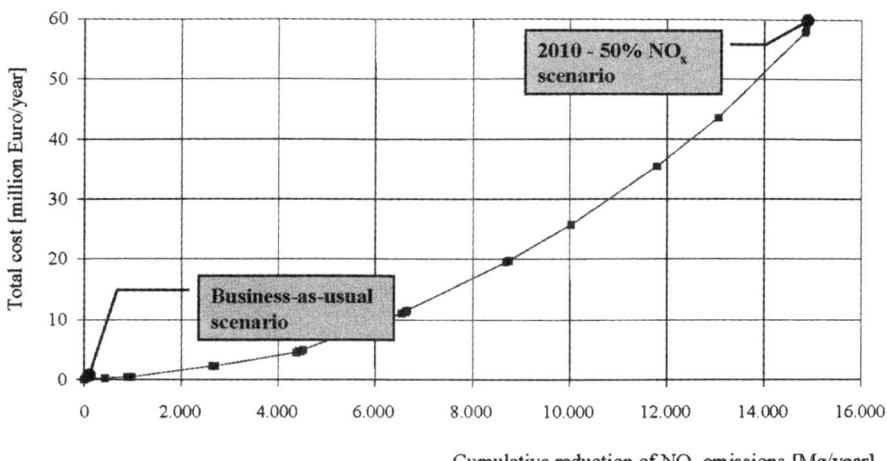

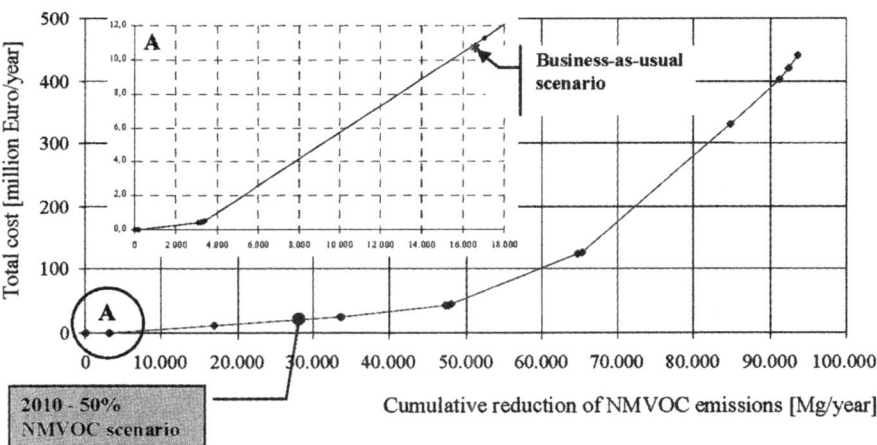

Figure 4. Cost-curves for NO$_x$ (energy sector) and NMVOC (solvent use sectors) in the GAA (upper and lower part, respectively). Each point on the curves corresponds to an emission reduction measure. The beginning of the axes corresponds to the year 1990.

the rest of the domain the situation remains more or less unchanged as compared to the 'business as usual' scenario. On the contrary, an additional 50% reduction in NMVOC emissions results in reductions of both the frequency and the degree of the exceedances in the urban area compared to the 'business as usual' scenario.

As regards the impact of the 2010 scenarios on NO$_2$ levels, it was concluded that both the maximum hourly and the daily average NO$_2$ concentrations were significantly reduced. The most significant impact was attained with the 50% reduction in NO$_x$ emissions in 2010.

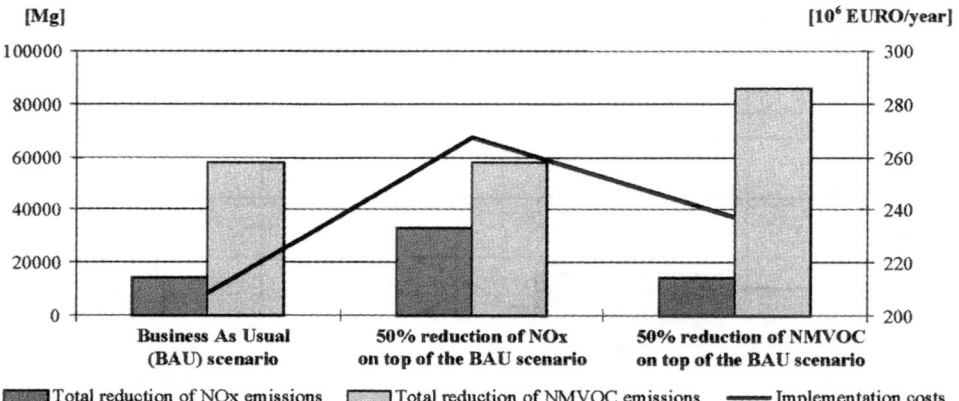

Figure 5. Impact of the emission reduction scenarios on the total NO$_x$ and VOC emissions in the GAA (bars, scale on left axis) and corresponding implementation costs (line, scale on right axis).

3.3. COST-BENEFIT ANALYSIS

Simulations with the EZM formed the basis for performing a cost-benefit analysis, the latter being oriented towards various effects of ozone, NO$_2$ and CO to specific health endpoints and workers productivity. The assumptions adopted and the input parameters used for performing this analysis are discussed in detail elsewhere (Tourlou, 2000). As shown in Table I, this analysis certified that the adoption of all three future scenarios results in quite important economic benefits.

The most significant impact was attained with the 50% reduction in NO$_x$ emissions on top of the 'business as usual' scenario. At the same time, however, the implementation of the specific scenario was found to be relatively expensive as compared to the other scenarios evaluated (cf. section 3.1).

All three scenarios resulted in economic losses with regard to the 'workers productivity' category. In contrast to the rest implications examined, the cost-benefit analysis here was based on the daily average ozone concentrations. These, however, were found to slightly increase for most of the days of the simulation period as compared to the base case situation (Friedrich and Reis, 1999; Tourlou, 2000).

4. Overall Assessment of Application Results

An application of the integrated assessment system was performed for the GAA with the aim to analyse the impact of various on going or proposed air quality regulations invoked through the implementation of technical emission reduction measures to the dominant emission sources of the area. The simulation of the air pollution levels in the year 2010 showed that, although the situation is expected to ameliorate with the hypothetical adoption of the measures proposed, air quality exceedances are still predicted in Athens. Among the emission situations examined,

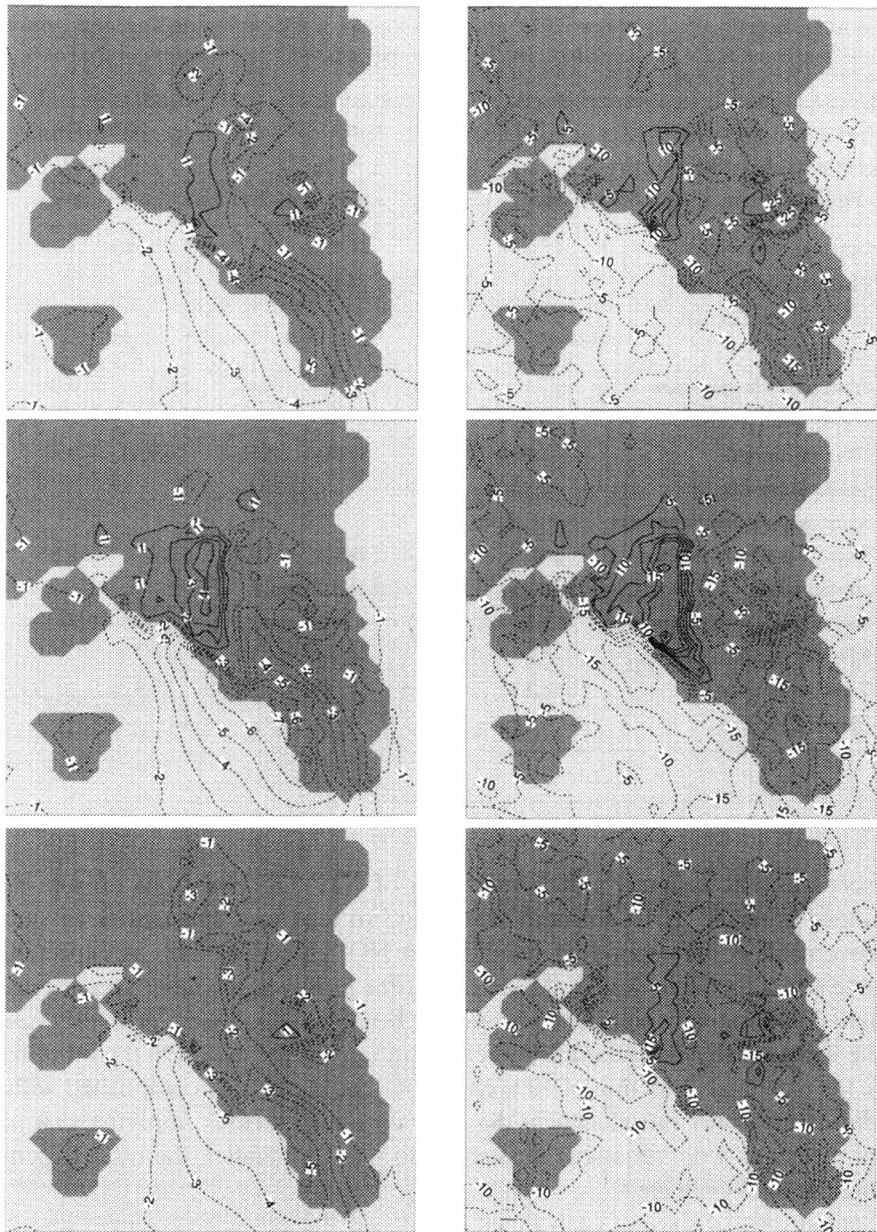

Figure 6. Upper part: Calculated differences between the 'business as usual' 2010 and the base case scenarios in terms of AOT60 values (left) and number of days of exceedance of the running 8h average of 120 μg m^{-3} (right) for the GAA. Middle part: Corresponding differences between the 2010–50% NO$_x$ and the base case scenarios. Lower part: Corresponding differences between the 2010–50% NMVOC and the base case scenarios. Dashed curves denote decrease, while continuous curves denote increase of the values.

TABLE I

Resulting benefits for the three 2010 emission scenarios. Maximum and average annual economic benefits for the inhabitants (including employees working outdoors) of the GAA

Categories of health endpoints and other implications	2010 – BAU		2010–50% NO$_x$		2010–50% NMVOC	
	Max.	Ave.	Max.	Ave.	Max.	Ave.
Premature mortality	32,970	9,060	47,750	11,780	41,940	9,720
Hospital admissions (all health endpoints included)	210	52	270	65	250	55
Emergency home visits due to asthma crises	7	0.8	5.2	0.6	6.5	0.9
Days of limited productivity due to acute respiratory symptoms	7,165	930	5,480	550	6,950	950
Days of partially limited productivity	244	31	188	18.8	237	32.5
Days with asthma symptoms	5,610	728	4,290	430	4,440	745
Workers productivity	*–5.0*	*–1.5*	*–6.9*	*–3.2*	*–4.7*	*–1.5*
Total annual benefit [thous. Euro]	46,200	10,800	57,970	12,840	53,820	11,500

the scenario that assumes a 50% reduction of NMVOC emissions on top of the 'business as usual' 2010 scenario was found to lead to a substantial air quality improvement. A corresponding reduction in NO$_x$ emissions had, as expected, an adverse effect on ozone levels. At the same time however, the latter scenario resulted into comparatively significant economic benefits, which were primarily caused by the substantially decreased NO$_2$ concentrations in the area. On the other hand, since the expected reduction of NO$_x$ emissions could not be accomplished with the bundle of measures introduced in the NO$_x$ cost-curves (thus implying that additional measures were required), this scenario was, in parallel, the most expensive among the ones evaluated.

Comparison of the net cash flows (results of the cost-effectiveness and cost-benefit analyses) indicated the 'business as usual' scenario as the most profitable solution, without however obtaining a significant eminence against the rest scenarios. The air quality criterion though, was not met; air quality limits were still exceeded in the area.

It should be stressed out that a straightforward comparison of the implementation costs with the corresponding benefits obtained for the scenarios examined

shows that the costs far outweigh the economic benefits. The reader should however, bear in mind that the cost-benefit analysis was based on the monetary evaluation of a restricted number of health endpoints, thus focusing on human receptors
only. The fact that neither the impacts (and therefore benefits or damages) to other-
than-human receptors, nor all pollutants have been taken into account for this
analysis, implies that the calculated benefits may be significantly underestimated.

Prior to concluding, there is an additional issue that needs to be briefly addressed: the accuracy of the results obtained through the economic analyses. The
determination of all possible errors and the range of uncertainty introduced to the
system in every stage of the evaluation procedure are out of the scope of this
paper. Having in mind though, that the results produced by the system are subjects to several uncertainties (e.g. inaccurate implementation cost values, empirical
exposure-response functions and WTP values, etc.), the costs and benefits calculated for each application must be considered as indicative rather than the actual
values.

5. Conclusions and General Remarks

An integrated assessment system has been built-up for the evaluation of air pollution abatement strategies taking into account their impact on air quality and the
costs arising from or due to their implementation. The basic innovative elements
of the system are:

- Quantification of the total cost required for the implementation of a measure.
- Comparative evaluation of alternative abatement solutions on the basis of area
  – and pollutant – specific cost-curves (e.g. comparison of (mutually divergent
  or not) investment options on the basis of the cost-effectiveness theory).
- Assessment of the measures through the application of a comprehensive air
  quality model system (e.g. evaluation of the impacts on ambient pollution
  concentrations).
- Having as starting point the fact that the environment and, thus, any form of
  environmental degradation comprise economic commodities, the concept of
  the external cost caused by any production procedure or anthropogenic activity
  is introduced and analysed, the aim being to quantify the economic benefit
  resulting from the application of any emission or pollution abatement option.

As regards the cost-benefit analysis, it is pointed out that it is currently conducted without accounting for the consequences of air pollution in other than human
receptors. The analysis is also oriented towards the implications of three pollutants
to specific health endpoints and workers productivity, whereas side effects are experienced due to the exposure to various air pollutants. One should bear in mind
that crucial parameters for performing sound cost-benefit analyses are the user's
ability and expertise, so as to be capable of evaluating the impact of the measure in

an integrated manner; that is, to determine the activity sectors that are affected by the measure and calculate the total benefit/damage arising from its implementation.

It can be generally concluded that the system forms a reliable tool for the objective assessment of the most cost-effective packages of measures, the latter being allocated according to the particular features and needs of the areas examined.

Acknowledgements

Part of this work was performed in the framework of the INFOS project (EC – DGXII, No. ENV4-CT96-0264, 1996–99). We would like to express sincere thanks to Drs. R. Friedrich, D. Papameletiou, S. Reis and D. Simpson for the valuable support and the provision of data for this project.

References

Crocker, T. D. and Horst, R. L.: 1981, 'Hours of work, labor productivity and environmental conditions: A case study', *Rev. Econ. Stat.* **63**, 361–368.

Cumberland, J. H. and Kahn, J. R.: 1982, 'The estimation of marginal damage functions: An assessment of the information requirements', *Northeast Reg. Sci. Rev.* **12**.

Dixon, J. A., Scura, L.F., Carpenter, R. A. and Sherman, P. B.: 1994, *Economic Analysis of Environmental Impacts*, Earthscan Publications, London.

Eggleston, S., Gaudioso, D., Gorissen, N., Joumard, R., Rijkeboer, R. C., Samaras, Z. and Zierock, K.-H.: 1993, 'CORINAIR Working Group on Emission Factors for Calculating 1990 Emissions from Road Traffic', Volume 1: *Methodology and Emission Factors*, Commission of European Communities Document, ISBN 92-826-5571-X, Luxembourg, 116 pp.

EMEP/CORINAIR: 1996, *Atmospheric Emission Inventory Guidebook*, G. McInnes (ed.), First Edition, European Environmental Agency, Copenhagen.

Friedrich, R. and Reis, S.: 1999, in R. Friedrich and S. Reis (eds), *Tropospheric Ozone Abatement – Developing Efficient Strategies for the Reduction of Ozone Precursor Emissions in Europe*, Springer Publications.

Kolb, J. A. and Scheraga, J. D.: 1992, 'Discounting the benefits and costs of environmental regulations', *J. Policy Anal. Manage.* **9**, 381–390.

Krupnick, A. J. and Cropper, M.: 1992, 'The effect of information on health risk valuation', *J. Risk Uncert.* **5**, 29–48.

Kunz, R. and Moussiopoulos, N.: 1995, 'Simulation of the wind field in Athens using refined boundary conditions', *Atmos. Environ.* **29**, 3575–3591.

Moussiopoulos, N.: 1995, 'The EUMAC Zooming Model, a tool for local-to-regional air quality studies', *Meteor. Atmos. Phys.* **57**, 115–133.

Sahm, P. and Moussiopoulos, N.: 1999, 'The OFIS Model: A New Approach in Urban Scale Photochemical Modelling', EUROTRAC Newsletter 21/99, EUROTRAC-ISS, Garmisch-Partenkirchen, 22–28.

Tourlou, P. M.: 2000, 'Integrated Assessment of Photochemical Pollution Abatement Strategies in Urban Areas', PhD Thesis, Department of Mechanical Engineering, Polytechnic School, Aristotle University Thessaloniki, 149 pp.

World Bank: 1998, *Pollution Prevention and Abatement Handbook*, Washington, D.C.

Yu Po, L: 1990, *Forming Winning Strategies – An integrated Theory of Habitual Domains*, Springer-Verlag, Berlin, Heidelberg.

A REAL-TIME OPERATIONAL FORECAST MODEL FOR METEOROLOGY AND AIR QUALITY DURING PEAK AIR POLLUTION EPISODES IN OSLO, NORWAY

ERIK BERGE[1]*, SAM-ERIK WALKER[2], ASGEIR SORTEBERG[1], MOTHEI LENKOPANE[2], STEINAR EASTWOOD[1], HILDEGUNN I. JABLONSKA[2] and MORTEN ØDEGAARD KØLTZOW[1]

[1] *The Norwegian Meteorological Institute, P.O.Box 43, Blindern, N-0313, Oslo, Norway;*
[2] *Norwegian Institute for Air Research. P.O.Box 100, N-2027, Kjeller, Norway*
(* author for correspondence, e-mail: erik.berge@met.no, fax: +47-22963050)

Abstract. A real-time operational forecast model for meteorology and air quality for Oslo, Norway is presented. The model system consists of an operational meteorological forecasts model and an air quality model. A non-hydrostatic model operated on two different domains with 1 and 3 km horizontal resolution is nested within the routine meteorological forecast model, which is run for North West Europe with 10 km horizontal resolution. The meteorological data are applied to an air quality model of Oslo with a 1 km grid and sub-grid treatment of line and point sources. Results from 22 days during the winter season 1999–2000 are presented and discussed. Prediction of wind speed and directions and relative humidity are clearly improved by increasing the horizontal resolution of the meteorological model. Temperature inversion strengths are however considerably overestimated. The predictions of PM_{10} corresponds best with measurements on winter days with wet or frozen surfaces in the city. On dry days, especially during spring time with a large deposit of accumulated dust on the roadside, the model under predicts the PM_{10} concentrations considerably. It is in partic- ular recommended to improve the description of the PM_{10} source strength in order to enhance the precision in the air quality forecasts.

Keywords: air quality and meteorological model, forecast models, urban air quality, wintertime

1. Introduction

High levels of PM_{10} and NO_2 are observed every winter in Norwegian cities during temperature inversions with weak winds and little vertical mixing. High levels of PM_{10} also occur during dry weather conditions in particular during spring, when large amounts of particulate matter have accumulated along the roads due to win- tertime road maintenance and usage of studded tires. Important measures to reduce the effects of wintertime air pollution are to issue forecasts and to employ traffic re- strictions on days when high concentrations are expected. In order to support the air quality forecasts, a real-time operational forecast model has been developed. The meteorological part of the model is based on the MM5-model (Grell *et al.*, 1994) operated with 1 km horizontal resolution and nested within the operational fore- cast model of the Norwegian Meteorological Institute. The meteorological forecast

Water, Air, and Soil Pollution: Focus **2**: 745–757, 2002.
© 2002 *Kluwer Academic Publishers.*

data are further employed by the air quality modeling system developed at the Norwegian Institute for Air Research (AirQUIS, see Slørdal and Tønnesen, 1999), which predicts the air pollution levels. During the winter 1999–2000 the forecast model was run operationally for 22 cases favorable to high levels of air pollution in Oslo. The results of the simulations are presented and discussed in this paper. The forecast results were distributed automatically at about 07:30 in the morning to the local authorities with a forecast valid for the next day. Traffic restrictions were applied during one of the cases. A more detailed discussion of the model results during the day with traffic restrictions is also given.

2. Meteorological Forecast for the City of Oslo

The operational runs with MM5 were based on initial and boundary fields from HIRLAM (High Resolution Limited Area Model, Källén, 1996) started at 00 UTC and run for a 48 h prognosis with 10 km horizontal resolution covering North West Europe and adjacent seas. The MM5 was nested within the HIRLAM 10 km model (henceforth denoted H10) employing firstly one domain with 3 km horizontal resolution and secondly 1 km horizontal resolution over the Oslo area (see Figure 1). MM5 was operated on the second day of the HIRLAM prognosis (+ 24 to + 48 prognosis). The reason for this was simply the need to save computer time and the fact that the local authorities mainly needed the quantitative air quality forecast for the second day in order to prepare for daily forecasts or possible traffic restrictions.

Table I presents a statistical overview of the results for 22 cases during the period November 1 1999 to April 30 2000 for the two stations Blindern and Valle Hovin close to the center of Oslo (see Fig. 1). The bias of the wind speed is 0.51 $m.s^{-1}$ at Valle Hovin while the bias is close to zero at Blindern. The BIAS, MAE and STDE are all reduced significantly compared with the H10. The scatter plot presented in Figure 2 also clearly shows an improvement in the wind speed forecast when shifting from H10 to MM5. Furthermore, the difference in predicted and observed wind direction is less than 20 degrees in about 30% of the cases for MM5 and 10% for H10. About 50% of the wind direction data from MM5 are within 40 degrees of the observed direction. The respective number for H10 is 25%. We can conclude that the forecasted wind speed and direction are improved by shifting from H10 to MM5 1 km. We expect a large part of the improvements to be due to increasing the horizontal resolution, since the topography around Oslo has a large impact on the local winds. Small scale circulation patterns with a horizontal scale of just a few kilometers develop in the Oslo region and therefore cannot be captured by H10. In addition, the two models employ different numerical solvers and physical packages (see Grell *et al.*, 1994 and Källén 1996 for details about the models). H10 employs a semi-lagrangian solver for the momentum equations, while MM5 uses an explicit method except for the acoustic waves which are solved by an implicit method. Both models have been run with 1. order PBL-schemes.

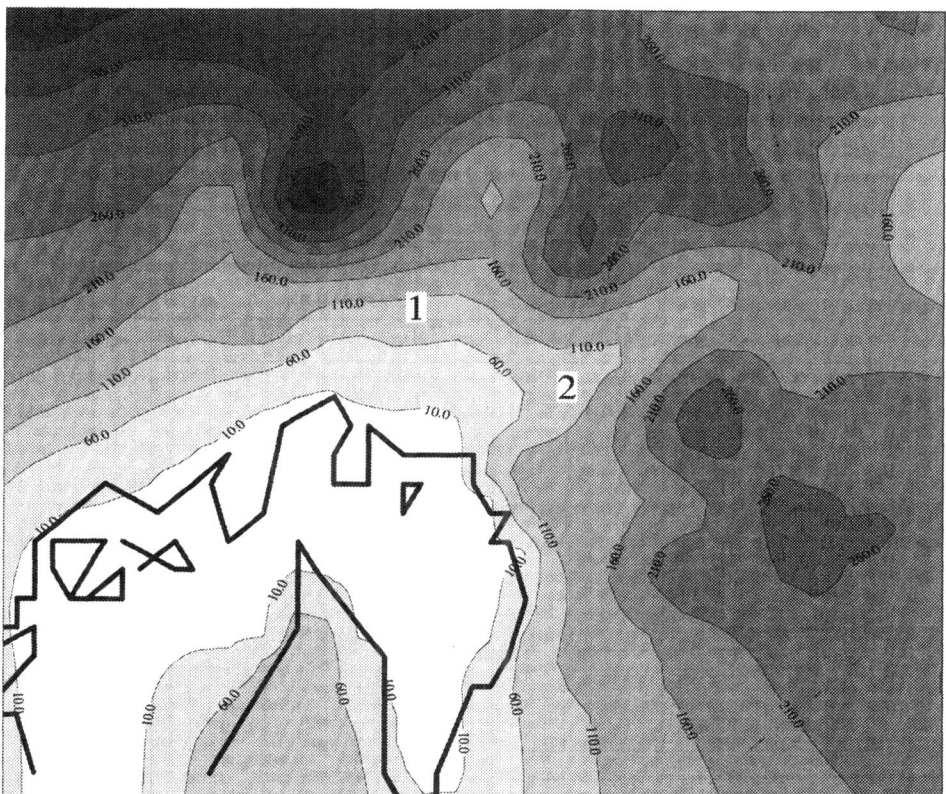

Figure 1. Inner modeling domain for MM5 (1 km horizontal resolution). Blindern is located at 1 and Valle Hovin is located at 2. Domain size is 31 * 31 km.

However, how important differences in formulation of the numerical solutions and the physical processes could be for the differences in the results has not been studied yet. Such a study would ideally require the set up of H10 on a 1 km horizontal grid, but since H10 is a hydrostatic model it is not recommended to apply this model at 1 km horizontal scale in complex terrain.

The temperature is however systematically underestimated in MM5 with a BIAS of –4.46 °C at the 2 m level at Valle Hovin and –4.39 °C at Blindern. We also note that a large part of this underestimation arises from too low temperatures in the HIRLAM model. However, at 25 m height at Valle Hovin the BIAS is reduced to –1.62 °C. Therefore, the near surface inversion strength is too strong in MM5. However, the daily cycles of near surface temperatures are qualitatively realistic (see also Figure 8 as an example). The standard deviation in the temperature is reduced for the 1 km compared to the 10 km horizontal resolution. Relative humidity is on average overestimated, but also for this component the predictions are improved by changing from 10 to 1 km resolution. In particular the daily cycle of relative humidity is considerably more realistic in MM5 compared to H10 (see

TABLE I

Results from 22 days of forecasted (+ 24 to + 48) wind speed (W), temperature (T) and relative humidity (RH) for HIRLAM with 10 km horizontal resolution (H10) and MM5 with 1 km horizontal resolution. The temperature and wind speed measured at 25 m are compared with the model output at 23 m (MM5) and 30 m (H10) for every 3 h

Station	Para-meters	Mean error (BIAS)		Mean absolute error (MAE)		Standard deviation (STDE)	
		H10	MM5	H10	MM5	H10	MM5
Valle Hovin	W25 m	1.71	0.51	2.11	1.17	1.92	1.32
Blindern	W10 m	0.73	−0.08	1.21	0.95	1.34	1.22
Valle Hovin	RH2 m	15.63	13.06	17.42	15.21	17.77	13.47
Blindern	RH2 m	11.36	5.67	15.32	12.17	17.42	15.15
Valle Hovin	T25 m		−1.62		3.04		3.20
Valle Hovin	T2 m	−3.34	−4.46	4.34	4.71	4.01	3.16
Blindern	T2 m	−3.67	−4.39	4.48	4.80	3.86	3.46

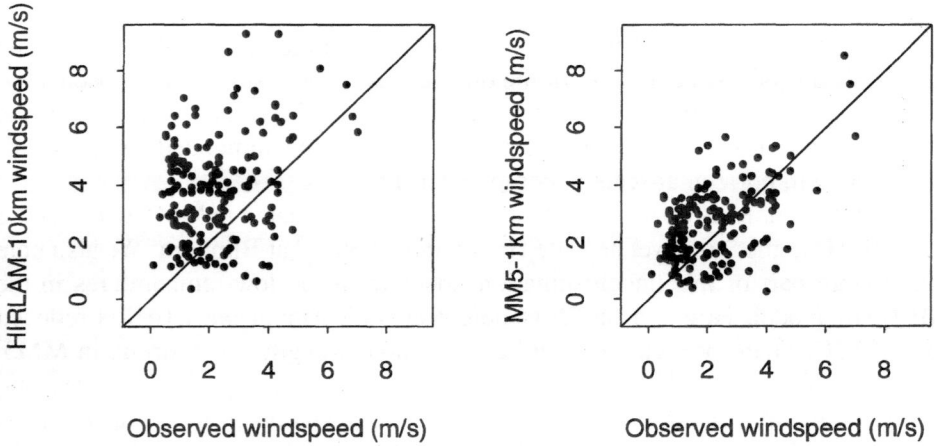

Figure 2. Observed and modeled wind speeds at Valle Hovin for 22 days during the winter 1999–2000.

Figure 9 as an example). Our analysis of the meteorological forecasts with 1 km horizontal resolution clearly indicates an improvement compared with the forecasts based on 10 km horizontal resolution.

3. Air Quality Forecast for the City of Oslo

The meteorological data are utilized to run an air quality modeling system called AirQUIS (Air QUality Information System) developed by the Norwegian Institute of Air Research (NILU) in cooperation with NORGIT Senteret A/S in Norway. The core features of the system are an air quality model and an emission inventory database and emission model. Additional features are linkage to a measurement database, statistical and graphical tools etc. The air quality model in AirQUIS consists of a combined eulerian/lagrangian dispersion model called EPISODE (Larssen *et al.*, 1994). Exposure calculations are also performed based on population distributions. The latter are improved by performing sub-grid scale calculations based on individual treatment of line-sources and building points within each Norwegian city.

The emission inventory database in AirQUIS typically consists of a set of area-, point- and line-sources. Area-sources are associated with emissions from domestic heating sources, or from traffic on smaller roads, distributed in a grid, point sources with emissions from industry or industrial activities, and line sources with emissions from larger individual roads. The exact position of the road, together with information about road width, steepness, speed limit, amount of daily traffic and vehicle composition, is stored in the database for each road. The AirQUIS database also contains daily and weekly traffic patterns for each road type, in addition to emission factors dependent on compound and vehicle type (gasoline-or diesel-driven cars, buses, trucks etc.). Emission of particles (PM_{10} and $PM_{2.5}$) is calculated dependent on the vehicle composition and traffic speed. The direct particle emissions from the exhaust gases, and small particles from road dust ($PM_{2.5}$) are calculated separately. The coarse fraction ($PM_{10} - PM_{2.5}$) is then calculated dependent on the fraction of vehicles using studded tires, taking into account the degree of dryness on the roads based on the meteorological information and the contribution from re-suspension of particles (Tønnesen, 2000).

The AirQUIS/EPISODE model uses a positive definite 4th order Bott scheme (Bott, 1993) for the horizontal transport of air pollutants. Vertically a simple upwind scheme is used (Grønskei *et al.*, 1993; Larssen *et al.*, 1994). Only the horizontal wind fields from Hirlam10 and MM5 are used directly. The vertical part of the wind field is generated internally in the AirQUIS/EPISODE model by integrating the horizontal divergence vertically. In this way the model operates locally using a 3D divergence free wind field for the transport.

The air quality forecasting system has first of all been set up to predict PM_{10} levels since the local authorities are more concerned about this component than

TABLE II

Statistical parameters for PM_{10} 24 h average values

Parameter	Unit	Kirkeveien	Furuset
N		21	20
Observed average	$\mu g\ m^{-3}$	57,2	56,9
Predicted average	$\mu g\ m^{-3}$	36,3	70,7
BIAS		0,37	−0,24
$\sigma_{observed}$	$\mu g\ m^{-3}$	26,8	23,2
$\sigma_{predicted}$	$\mu g\ m^{-3}$	11,3	33,4
Observed maximum	$\mu g\ m^{-3}$	126,8	127.5
Predicted maximum	$\mu g\ m^{-3}$	61,6	125,2
RMSE	$\mu g\ m^{-3}$	37,9	47,2

other chemical species. Additionally, any restriction on traffic would be based on 24 h average PM_{10} values rather than shorter term (hourly) values. Our evaluation of the air quality forecasts are therefore based on 24 h average PM_{10} values during the 22 episodes for which the whole model system was operated. Table II presents a few statistical parameters for the 24 h average PM_{10} values at the two stations Kirkeveien and Furuset. Kirkeveien is situated close to a main road in the outer part of the city center. Furuset is situated close to a highway east of the city.

The observed average concentrations are similar at the two stations, however the model underpredicts the PM_{10} values at Kirkeveien (37%) while an overprediction (24%) is seen at Furuset. The standard deviation in the model data (σ_p) is smaller than the observed value (σ_o) at Kirkeveien while the opposite is the case at Furuset. Similarly the root mean square error (RMSE) is larger at Furuset than at Kirkeveien. A further examination of the 22 cases is given in Figures 3 and 4. During the first four episodes (from 10 November 1999 to 2 December 1999) the conditions were rather dry in the city, i.e. there was little or no water, ice or snow on the roads. Similar dry conditions also prevailed from 23 to 31 March 2000. At Kirkeveien the model clearly underpredicts the 24 h PM_{10} concentrations during these dry periods. However, for the episodes from 16 December 1999 to 14 of March 2000 the modeled and the observed values compare quite well. In this period the surface was covered or partly covered by snow and ice and hence the source for road dust was suppressed. Apparently, the predictions at Kirkeveien coincide much better with the observations on days with wet surfaces. We believe this is linked to the description of the source strength of PM_{10} from the surface on dry days. In particular, the re-suspension of dust deposited at the roadside is difficult to describe accurately in the model. At Furuset the differences between the modeled and observed value are less systematic and local conditions close to the station are likely to be of more importance. Additionally, more detailed studies (not

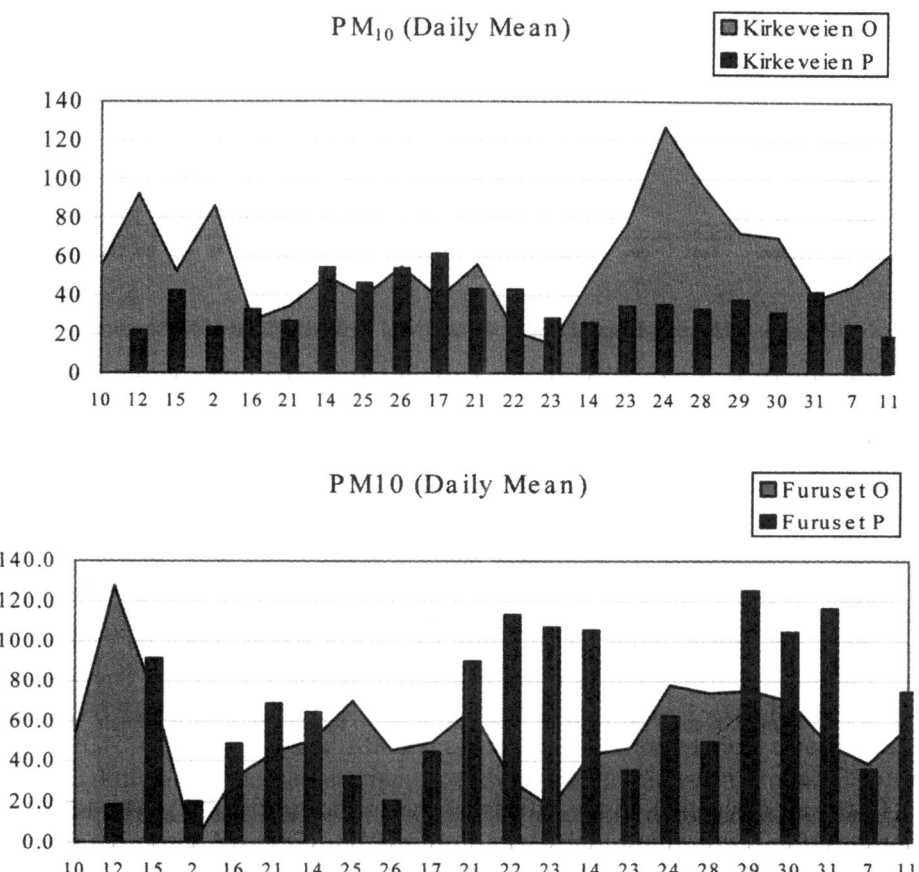

Figure 3. 24 h average PM_{10} values (in $\mu g\ m^{-3}$) for 22 cases at the stations Kirkeveien (upper panel) and Furuset (lower panel). The data starts 10 November 1999 and ends 11 April 2000. The continuous line represents the measurements (O). The model predictions (P) are given by the bars.

shown here) of the predicted wind direction indicate that the differences between observed and predicted PM_{10} concentrations can often be explained by the fact that the predicted wind is an average for a 1 km * 1 km area and therefore could deviate from the local wind direction close to the measuring stations (see also next section).

4. Results from a Day with High PM_{10} Concentrations and Traffic Restrictions

30 March 2000 the speed limits on the main roads in Oslo were reduced from 80 to 60 km h^{-1} aiming at reducing the PM_{10} levels during this day (see Bedre byluft,

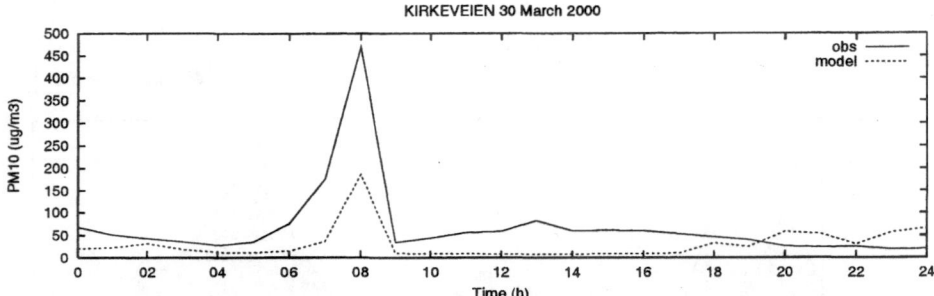

Figure 4. Hourly observed and modeled PM$_{10}$ 30 March 2000 at Kirkeveien.

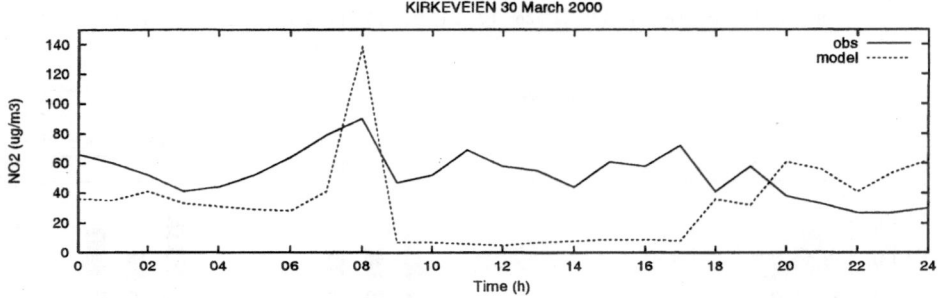

Figure 5. Hourly observed and modeled NO$_2$ 30 March 2000 at Kirkeveien.

2000, for an evaluation of the effects of the traffic restrictions). The restriction followed after several days with high levels of PM$_{10}$. Figure 4 shows the observed PM$_{10}$ levels for the station Kirkeveien during 30 March 2000. The station Kirkeveien is situated about 3 m on the southern side of a busy road with two lanes in each direction. There are buildings on each side of the road. The highest values were observed in the morning rush hours with a maximum value of nearly 500 μg m^{-3} at 0800 local time (LC). Note that the afternoon peak value was considerably lower. The corresponding NO$_2$ values from the same site are given in Figure 5. The NO$_2$ levels varied less than PM$_{10}$, but peak values linked to the morning and afternoon rush hours were observed. The weather conditions were dry and sunny as explained below, which would allow for some photochemical activity during daytime.

4.1. METEOROLOGICAL SIMULATIONS

A high pressure system was situated over southern Norway and the Oslo region on 30 March 2000. This gave rise to dry and sunny conditions together with weak winds. Inversions prevailed during nighttime while a well mixed PBL was established over land during daytime. Local circulation systems developed with kata-

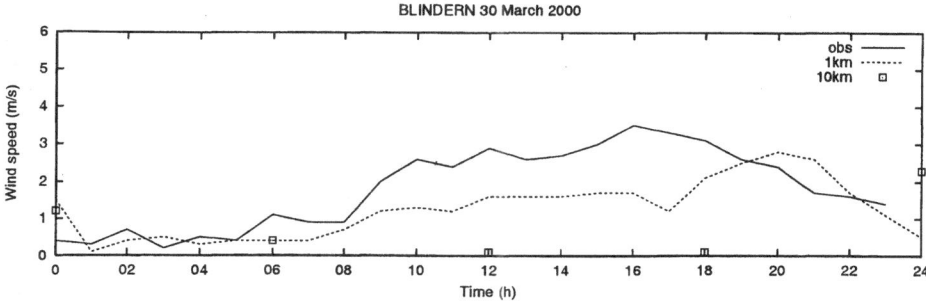

Figure 6. Observed (obs) and modeled (1 km = MM5 and 10 km = H10) wind speed at Blindern 30 March 2000.

batic winds from the hillsides toward the fjord during night (see Figure 1 for a description of the topographical features) and a sea breeze during daytime directed up the hillsides. Figures 6–9 show the meteorological observations and model simulations for Blindern, a meteorological station situated approximately 1 km north of the air quality station Kirkeveien. The model results are for the forecasting period + 24 to + 48 h. Wind speeds of 1 m.s^{-1} or less were recorded from midnight until 0800 LT (Figure 6). The wind direction was mostly from NE to NW veering toward E and SE (Figure 7). After 0900 LT winds were stronger (2–3 m.s^{-1}) and from S and SW. The wind did not veer back to north before 2300 LT. At that time wind speeds were lower again. MM5 predicted the wind speed and direction well during nighttime and in the morning. The shift in wind direction around 0800 LT was also well described, but the wind speeds were too low during daytime. The H10 predicted the same wind direction (W and SW) during the whole day, while wind speeds were very low (close to calm during daytime). MM5 was able to generate a local sea breeze (not shown) during daytime and thereby enhanced the wind speeds compared to the 10 km run. As we see, the shift from katabatic flow in the morning to a local sea breeze at 0800 to 0900 LT was well picked up by MM5. However, the afternoon katabatic flow was established around 1800 LT about 4–5 h too early. Sensitivity studies have shown that this was partly due to an overestimation of the snow cover in the surrounding hills. This forces the onset of the katabatic winds to be predicted too early. A model simulation including an improved snow cover (not shown) delayed the onset of the katabatic winds for about two hours.

The MM5 predicted temperature coincided very well with the observed temperature at Blindern. The larger scale H10 underestimates the temperature especially during daytime. Both models gave too high a figure for relative humidity in the night and the morning. At noon MM5 predicts a realistic relative humidity, but the onset of the increase came too early. At 1800 LT considerably higher humidity values were found in MM5 compared with the observations. This could be linked to the early onset of the katabatic winds in MM5. Vertical mixing is weak in the katabatic flow, hence the surface evaporation could moisten the air close to the

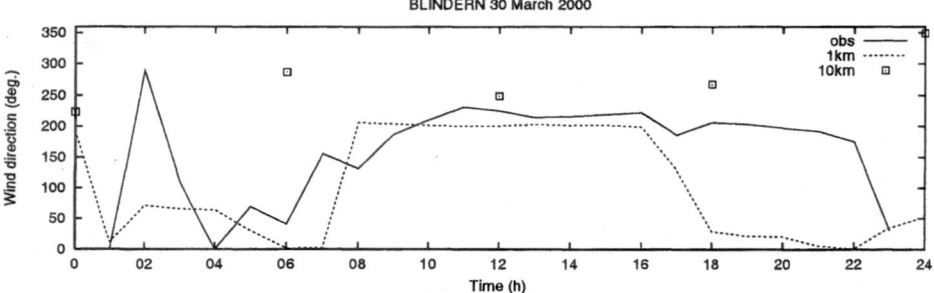

Figure 7. Observed (obs) and modeled (1 km = MM5 and 10 km = H10) wind direction at Blindern 30 March 2000.

surface more rapidly. Late in the day (at 2200 LT), when the local katabatic winds again were observed at Blindern the models corresponded much better with the observations.

4.2. PM$_{10}$ AND NO$_2$ SIMULATIONS

The high PM$_{10}$ values at 0700 and 0800 LT (see Figure 4) coincided with the peak emissions in the morning and with the shift from local drainage flow to the local sea breeze. At this time the winds were very weak and turned nearly 180 degrees. The nighttime inversion still persisted at this time. No PM$_{10}$ peak value was observed during the evening rush hours (1600 to 1700 LT) since the wind and thereby the turbulent mixing was stronger at this time. The model predicts the diurnal behavior in the PM$_{10}$ values qualitatively well, although the values were underestimated nearly for all hours. This was also the trend for the 22 daily averages presented in Figure 3 for Kirkeveien during days with dry and snow free conditions. It is quite likely that this is linked to the PM$_{10}$ emission strength which is difficult to estimate correctly as it depends on the magnitude of the accumulated storage of particles along the roads, the dryness of the surface, the particle size distribution of the storage, the efficiency of the wind in picking up dust, local road maintenance and cleaning routines etc. In addition, the local turbulence and wind fields at Kirkeveien were not measured and may have deviated somewhat from the 1 km spatial averages obtained from MM5.

From Figure 5 we observe that NO$_2$ varied much less during the day. The rush hour maximum was only twice as high as the nighttime minimum. As for PM$_{10}$ the peak was measured at 0800 LT. An afternoon peak at 1700 LT was also recorded. The model underestimated the NO$_2$ levels somewhat the first night, while a slight overestimation was seen the second night. The timing of the morning peak was correct although too large a value was modeled. During daytime the model strongly underestimated the NO$_2$ levels. Since NO$_2$ is coupled to the atmospheric chemical activity and the available ozone the behavior is more difficult to interpret since in particular no ozone measurements were available. Although the AirQUIS model employs a simplified chemical scheme for NO_2 the strong underestimation during daytime can probably not be explained by discrepancies in the chemistry only. We observe that both modeled PM_{10} and NO_2 were modeled at very low levels from 0900 to 1700 LT. Studies of the variability of the concentrations as a function of the MM5 wind directions (not shown) showed a large sensitivity to the wind direction. Since Kirkeveien is situated in a street canyon small shifts in the wind direction may give rise to large differences in the calculated concentrations at one side of the road.

5. Summary and Conclusions

A real-time operational forecast model for meteorology and air quality for Oslo, Norway, has been presented. The model was operated for the +24 to +48 forecasting period for 22 cases with high levels of air pollution in the Oslo region during the winter season 1999–2000. Results from the forecast model have been evaluated by use of two meteorological and air quality measuring sites. At one of the air quality sites the 24 h PM_{10} average was well simulated during days with snow cover and wet or icy roads. The highest PM_{10} values on dry days were however underestimated. This is probably linked to rather large uncertainties in the PM_{10} source strength during dry days. Also uncertainties in the predicted wind directions are important. At the second site an over prediction of PM_{10} was revealed. A larger random variability of the difference in observed and measured values was found which may indicate that local conditions are more important at this site. An improvement in the forecasted wind speed and direction and relative humidity was achieved by using the MM5 model with 1 km instead of the HIRLAM model with 10 km horizontal resolution. The strength of the temperature inversion was however overestimated near the ground by MM5 and a larger negative bias in the 2 m temperature was found compared with the 10 km model. However, the random error in the temperature predictions was reduced when the horizontal resolution was increased. An episode with high PM_{10} values during 30 March 2000 was further investigated. A realistic picture of the shift between nighttime katabatic winds and daytime see breeze was found. However, MM5 predicted the onset of evening katabatic flows too early. A clear linkage of the strength and timing of the cold drainage flows to the snow cover was found. The Air Quality Model gave a qualitatively realistic picture of the PM_{10} values, but the levels were generally underestimated. Less satisfactory correspondence was found for NO_2 on this particular day.

Further improvements of the temperature and inversion predictions of the HIRLAM and the MM5 model during wintertime are recommended. These studies should include improvements of the parameterization of the vertical eddy diffusion and improved descriptions of the surface characteristics such as snow cover, soil

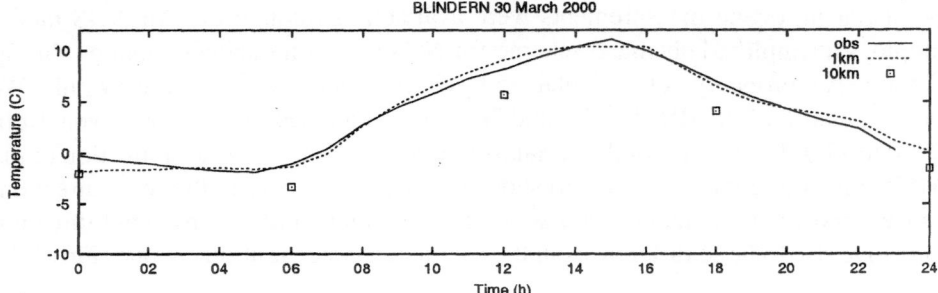

Figure 8. Observed (obs) and modeled (1 km = MM5 and 10 km = H10) temperature at Blindern 30 March 2000.

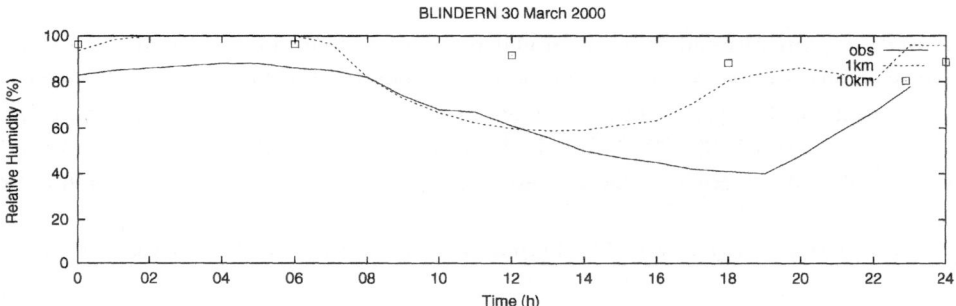

Figure 9. Observed (obs) and modeled (1 km = MM5 and 10 km = H10) relative humidity at Blindern 30 March 2000.

temperatures and soil humidity. There is also a clear need for an improved description of the PM_{10} sources especially on days when surfaces are dry. Attention should also be given to the linkage of the local dispersion modeling to the modeled wind speed and direction. More attention could also be given to the NO_2 chemistry and the selection of background ozone values.

References

Bedre byluft.: 2000, 'Evaluering av akutttiltaket nedsatt hastighet', Statens Vegvesen Oslo, Norway. 38 pp. (in Norwegian).

Bott, A.: 1993. 'The monotone area-preserving flux-form advection algorithm. Reducing the time-splitting error in two-dimensional flow fields', *Month. Weath. Rev.* **121**, 2637–2641.

Grell, G. A., Dudhia, J. and Stauffer, D. R.: 1994, 'A Description of the Fifth-Generation Penn State/NCAR Mesoscale Model (MM5)', NCAR Technical Note, NCAR/TN-397 + IA, 114 pp.

Grønskei, K. E, Walker, S. E. and Gram, F.: 1993, 'Evaluation of a model for hourly spatial concentrations distributions', *Atmos. Envir.* **27B**, 105–120.

Källén, E.: 1996, 'Hirlam Documentation Manual System 2.5', Available at the Norwegian Meteorological Institute, Oslo, Norway.

Slørdal, L. H. and Tønnesen, D.: 1999, 'Konsentrasjonsfordeling av NO_2, PM_{10} og $PM_{2.5}$ i sterke forurensningsepisoder i Oslo, Drammen, Bergen og Trondheim', Report OR 24/99, Norwegian Institute for Air Research, Lillestrøm, Norge (in Norwegian).

Larssen, S., Grønskei, K. E., Gram, F., Hagen, L. O. and Walker, S. E.: 1994, 'Verification of urban scale time dependent dispersion model with subgrid elements in Oslo, Norway', in S.-E. Gryning and M. M. Millan (eds), *I: Air Pollution Modelling and Its Application*, New York, Plenum Press, pp. 91–99.

Tønnesen, D. A.: 2000, 'Programdokumentasjon VLUFT, versjon 4.4'. TR 7/2000. (in Norwegian).

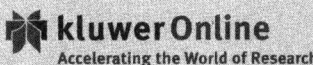